Primers in Biology

Immunity

The Immune Response in
Infectious and Inflammatory Disease

Anthony L DeFranco

Richard M Locksley

Miranda Robertson

NSP
New Science Press Ltd

OXFORD
UNIVERSITY PRESS

Managing Editor: Karen Freeland
Project Editors: Kerry Gardiner and Joanna Miles
Editorial Assistant: Mariam Orme
Design and Illustration: Matthew McClements, Blink Studio Ltd
Copy Editor: Bruce Goatly
Indexer: Liza Furnival
Production Director: Adrienne Hanratty

Distributors:

Inside North America:

Sinauer Associates, Inc., Publishers,
23 Plumtree Road, PO Box 407, Sunderland, MA 01375, USA
orders@sinauer.com
www.sinauer.com

Outside North America:

Oxford University Press
Saxon Way West
Corby, Northants
NN18 9ES
UK

Customers in the UK may use the OUP
freepost address:
Oxford University Press
FREEPOST NH 4051
Corby, Northants NN18 9BR
bookorders.uk@oup.com
www.oup.co.uk

ISBN-13: 978-0-9539181-0-2   (paperback) New Science Press Ltd
ISBN-10: 0-9539181-0-6

ISBN-13: 978-0-19-920614-8   (paperback) Oxford University Press
ISBN-10: 0-19-920614-7

ISBN-13: 978-0-87893-179-8   (paperback) Sinauer Associates, Inc.
ISBN-10: 0-87893-179-1

British Library Cataloguing-in-Publication Data

A catalogue record for this book is available from the British Library

Published by New Science Press Ltd
Middlesex House
34-42 Cleveland Street
London W1P 6LB
UK
www.new-science-press.com

in association with
Oxford University Press
and
Sinauer Associates, Inc., Publishers

Printed by Stamford Press PTE Singapore

15 14 13 12 11 10 9 8 7 6 5 4 3 2 1

iv

The Authors

Anthony L DeFranco graduated from Harvard College in biochemistry and molecular biology in 1975 and did his PhD on bacterial chemotaxis with Daniel E Koshland Jr at the University of California at Berkeley before turning to his present principal research interest, the activation of B lymphocytes, as a postdoctoral fellow in the laboratory of William E Paul at NIH in 1979. He is currently Chairman of the Department of Microbiology and Immunology at the University of California San Francisco Medical School where his research interests are the mechanisms of signaling by the B cell antigen receptor and Toll-like receptors, and B cell autoimmunity.

Richard M Locksley graduated from Harvard College in biochemistry in 1970 and in medicine from the University of Rochester in 1976. He was at the Moffitt Hospital in San Francisco for four years as a Medical Resident, trained in infectious diseases at the University of Washington for three years and then returned to the University of California in San Francisco where he served as Chief of the Division of Infectious Diseases from 1986–2004. He is currently the Sandler Distinguished Professor of Medicine and Microbiology and Immunology, Director of the Sandler Asthma Basic Research Center and an HHMI Investigator at UCSF. The principal focus of his research is on cellular immune responses in infectious and inflammatory disease.

Miranda Robertson studied psychology at Birkbeck College London and at the University of Chicago in the 1960s without graduating from either and then spent the greater part of a quarter of a century on the editorial staff of Nature, ultimately as its Biology Editor, since when she has had the privilege of working on behalf of Garland Publishing Inc and Current Biology Ltd with the authors of several outstanding textbooks. She is now the Managing Director of New Science Press Ltd.

Primers in Biology:
a note from the publisher

section heading   one-sentence subheading

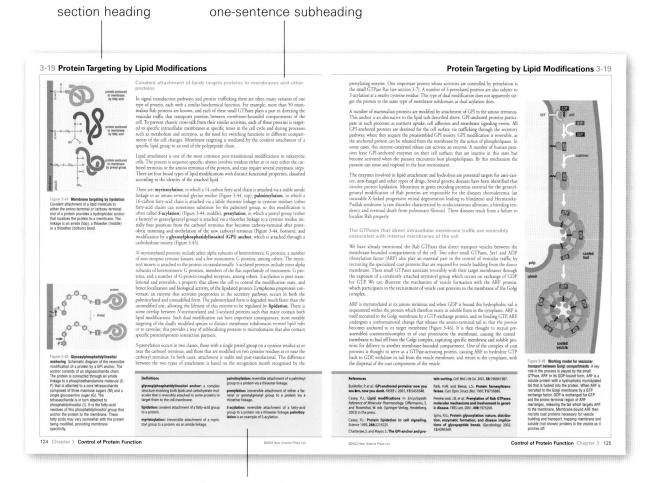

bottom margin:
definitions and references

Immunity: The Immune Response in Infectious and Inflammatory Disease is part of a series of books constructed on a modular principle that is intended to make them easy to teach from, to learn from, and to use for reference, without sacrificing the synthesis that is essential for any text that is to be truly instructive. The diagram above illustrates the modular structure and special features of these books. Each chapter is broken down into two-page sections each covering a defined topic and containing all the text, illustrations, definitions and references relevant to that topic. Within each section, the text is divided into subsections under one-sentence headings that reflect the sequence of ideas and the global logic of the chapter.

The modular structure of the text, and the transparency of its organization, make it easy for instructors to choose their own path through the material and for students to revise; or for working scientists using the book as an up-to-date reference to find the topics they want, and as much of the conceptual context of any individual topic as they may need.

All of the definitions and references are collected together at the end of the book, with the section or sections in which they occur indicated in each case. Glossary definitions may sometimes contain helpful elaboration of the definition in the text, and references contain a full list of authors instead of the abbreviated list in the text.

The picture above, which is reproduced here and on the cover with the kind permission of Kurt Dittmar and Manfred Rhode, German Research Centre of Biotechnology, Germany, shows a dendritic cell interacting with T lymphocytes and illustrates the single most important conceptual advance in immunology since the formulation of clonal selection theory by Frank MacFarlane Burnet in the mid-20th century and of MHC restriction by Rolf Zinkernagel and Peter Doherty in the 1970s.

Clonal selection theory states that the specificity of lymphocytes for foreign antigen is achieved by selection from a population of lymphocytes bearing receptors of indiscriminate specificity. MHC restriction is the term originally used by Doherty and Zinkernagel for the MHC-dependence of antigen recognition by T lymphocytes. These extraordinary insights focused the attention of immunologists on the adaptive immune system until in the late 20th century Charlie Janeway drew attention to what he called immunology's dirty little secret: an antigen injected experimentally into an animal to evoke an adaptive immune response will have no effect unless it contains or is accompanied by components that are recognized by the phagocytes of innate immunity. He thus refocused attention on the innate immune system and its central role in activating adaptive immune responses, subsequently shown to be largely the specialized territory of the dendritic cell, and established what is now a central tenet informing all of cellular immunology.

Preface

Life expectancy for a child born in the developed countries at the dawn of this century is 30 years greater than for forebears born 100 years ago. Most of this dramatic increase in human health is due to the decreasing toll of infectious disease, through public health measures to limit the spread of infections, through antibiotics to control them, and through vaccination to enlist immune defense in their prevention. Despite this progress, tremendous challenges remain, particularly in tropical areas where the scourges of infant diarrhea, malaria and tuberculosis, to name just a few, have been joined in recent decades by AIDS. Advances in understanding immune defenses against actual pathogens offer the real hope of progress against these hugely prevalent diseases, and of being prepared for the future, whether it bring the next influenza pandemic or another newly emerging pathogen.

The goal of this book is to describe how the immune system defends us against pathogens of all types. Immunological research in the last decade has seen the advent of sophisticated genetic manipulations in mice and techniques for imaging *in vivo*, which with the completion of the human and mouse genomic sequences, the elucidation of over 120 causes of genetic immune deficiency in humans, and the study of real infections in experimental models, to name just a few advances, have led to breathtakingly rapid progress. Our aim has been to draw on information from all these sources for what it tells us about how the immune system really works *in vivo*.

Pathogens are continually evolving to evade immune defenses. In response, the immune system has evolved many mechanisms for detecting infection, including mechanisms for generating diversity in recognition molecules during the lifetime of an individual that can match those of most infectious agents for mutating the molecules recognized; and has developed a remarkable diversity of responses tailored to distinct types of infections. At the same time, the power of this system lends it considerable potential for destruction and each mechanism for defense against infection is balanced by elaborate controls that prevent damage to the host. Immune pathology occurs where these systems fail. We strongly believe that this system cannot be understood without an appreciation of how all the elements work at the cellular and molecular levels. Therefore, we have kept in focus throughout this book the molecular interactions and the special properties of molecules and signaling pathways that underlie the essential properties of immune cells and their responses. These molecules are increasingly the targets of new therapeutic agents.

Of course, immunology remains a very active field and we have struggled to balance the need to acknowledge new discoveries with our responsibility, in an introductory text, to present a picture representing conclusions whose relevance *in vivo* is established. Some important questions in immunology remain unanswered but cannot be ignored, and we have taken care to indicate where we are venturing into supposition or even occasionally speculation.

We wish to acknowledge with much gratitude the help of many immunologists who have helped us make this book as accurate as we can: all those to whom we are indebted are listed in the Acknowledgements. We also want to offer special thanks to several teachers of immunology who helped us develop the book for the student coming to this topic for the first time by reading versions at varying stages of polish and telling us exactly what they thought; and to many medical and graduate students taught by two of us for invaluable feedback over many years.

We all wish to thank the staff at New Science Press in London—Karen Freeland, Kerry Gardiner and Joanna Miles who provided unflaggingly cheerful, energetic, and meticulous assistance throughout, and Mariam Orme, who helped with final preparation for the press. We especially thank our illustrator Matthew McClements, whose artistic talents combined with scientific understanding made him a true pleasure to work with. Finally, Tony DeFranco and Rich Locksley would like to thank family members who have had the patience to endure our long hours working on this book, and lab members who have also had to be patient.

Anthony DeFranco

Richard Locksley

Miranda Robertson

⬡ Online resources for Immunity

For everybody

All of the 367 colour illustrations in this book are freely available in the Oxford University Press Online Resource Centre and on the New Science Press website and can be downloaded for use in teaching.

Visit http://www.oxfordtextbooks.co.uk/orc/defranco/ or
http://www.new-science-press.com/browse/immunity/resources

For instructors

For instructors adopting the book for courses with enrolments of fifteen or more students:

Free access to

• the full text online for a year, for personal use only

• updates – revised, expanded, or new sections and updated references available online only

• PowerPoint functionality allowing instructors to compile any selection of illustrations into a slide show

Visit http://www.oxfordtextbooks.co.uk/orc/defranco/ to register for access to the instructor resources.

Access the resources, once registered, by visiting
http://www.new-science-press.com/browse/immunity/resources

Acknowledgements

The following individuals provided expert advice on entire chapters or parts of chapters:

Chapter 1 Leslie Berg, University of Massachusetts Medical School; Eric Brown, University of California, San Francisco; John Cohen, University of Colorado Health Sciences Center; Jason Cyster, University of California, San Francisco; Siamon Gordon, University of Oxford; Jonathan Howard, University of Cologne; Norman Iscove, The Ontario Cancer Institute, University of Toronto; Garnett Kelsoe, Duke University; Paul Kincade, Oklahoma Medical Research Foundation; Eric Pamer, Memorial Sloan-Kettering Cancer Center; Jennifer Punt, Haverford College; Caetano Reis e Sousa, Cancer Research UK London Research Institute; Ken Shortman, Walter and Eliza Hall Institute of Medical Research; Ralph Steinman, Rockefeller University; Steve Weinstein, San Francisco State University

Chapter 2 Neil Barclay, University of Oxford; Leslie Berg, University of Massachusetts Medical School; Eric Brown, University of California, San Francisco; John Cohen, University of Colorado Health Sciences Center; Anne Cooke, University of Cambridge; Richard Hynes, Massachusetts Institute of Technology; Pablo Irusta, Georgetown University; Lewis Lanier, University of California, San Francisco; Warren Leonard, National Heart, Lung, and Blood Institute, National Institutes of Health; Klaus Ley, University of Virginia; Grant McFadden, Robarts Research Institute; John O'Shea, National Institute of Arthritis and Musculoskeletal and Skin Diseases, National Institutes of Health; Jennifer Punt, Haverford College; Steven Rosen, University of California, San Francisco; Timothy Springer, Harvard Medical School; Andreas Strasser, Walter and Eliza Hall Institute of Medical Research; Carl Ware, La Jolla Institute for Allergy and Immunology, University of California, San Diego; Steve Weinstein, San Francisco State University; Albert Zlotnick, Neurocrine Biosciences

Chapter 3 Leslie Berg, University of Massachusetts Medical School; Bruce Beutler, Scripps Research Institute; Eric Brown, University of California, San Francisco; Michael Carroll, Harvard University; John Cohen, University of Colorado Health Sciences Center; Anne Cooke, University of Cambridge; Alan Ezekowitz, Harvard University; John Gallin, National Institute of Allergy and Infectious Diseases, National Institutes of Health; Steve Gerondakis, Walter and Eliza Hall Institute of Medical Research; Sankar Ghosh, Yale University School of Medicine; Doug Golenbock, University of Massachusetts Medical School; Marie Hardwick, Johns Hopkins Bloomberg School of Public Health; Pablo Irusta, Georgetown University; David Levy, New York University Medical Center; Michael Malim, King's College London; Grant McFadden, Robarts Research Institute; Ruslan Medzhitov, Yale University School of Medicine; Gabriel Nuñez, University of Michigan Medical School; Jennifer Punt, Haverford College; Ken Reid, University of Oxford; Fred Rosen, Harvard University; Philippe Sansonetti, Institut Pasteur; Michael Selsted, University of California, Irvine; Andreas Strasser, Walter and Eliza Hall Institute of Medical Research; Steve Weinstein, San Francisco State University; Bryan Williams, Lerner Research Institute

Chapter 4 James Allison, Memorial Sloan-Kettering Cancer Center; Leslie Berg, University of Massachusetts Medical School; John Cohen, University of Colorado Health Sciences Center; Anne Cooke, University of Cambridge; Peter Cresswell, Yale University School of Medicine; Ronald Germain, National Institute of Allergy and Infectious Diseases, National Institutes of Health; Mitchell Kronenberg, La Jolla Institute for Allergy and Immunology; Lewis Lainer, University of California, San Francisco; Andrew McMichael, University of Oxford; Ira Mellman, Yale University School of Medicine; Jennifer Punt, Haverford College; David Raulet, University of California, Berkeley; Caetano Reis e Sousa, Cancer Research UK London Research Institute; Kenneth Rock, University of Massachusetts Medical School; John Trowsdale, University of Cambridge; Colin Watts, University of Dundee; Jonathan Yewdell, National Institute of Allergy and Infectious Diseases, National Institutes of Health

Chapter 5 Carrie Arnold, Stanford University; Mike Bevan, University of Washington School of Medicine; John Cohen, University of Colorado Health Sciences Center; Anne Cooke, University of Cambridge; Jerry Crabtree, Stanford University; Daniel Cua, Schering-Plough Biopharma; Jason Cyster, University of California, San Francisco; Mark Davis, Stanford University; Richard Flavell, Yale University School of Medicine, Ron Germain, National Institute of Allergy and Infectious Diseases, National Institutes of Health; Laurie Glimcher, Harvard Medical School; Siamon Gordon, University of Oxford; Stephen Jameson, University of Minnesota; Stefan Kaufmann, Max Planck Institute for Infection Biology; Robert Kastelein, Schering-Plough Biopharma; Richard Kroczek, Robert Koch Institute; Klaus Ley, University of Virginia; Judy Lieberman, Harvard Medical School; Averil Ma, University of California, San Francisco; Charles Mackay, Garvan Institute of Medical Research; Phillipa Marrack, National Jewish Medical and Research Center and University of Colorado Health Sciences Center; Andrew McMichael, University of Oxford; Muriel Moser, Université Libre de Bruxelles; Tony Pawson, Mount Sinai Hospital Research Institute, Steven Reiner, University of Pennsylvania; Caetano Reis e Sousa, Cancer Research UK London Research Institute; Arlene Sharpe, Harvard Medical School; Jonathan Sprent, Scripps Research Institute; Kai-Michael Toellner, University of Birmingham; David Tough, GlaxoSmithKline Medicines Research Center; Arthur Weiss, University of California, San Francisco; Ian Wilson, Scripps Research Institute; Weiguo Zhang, Duke University

Chapter 6 Jay Berzofsky, National Cancer Institute, National Institutes of Health; Edward Clark, University of Washington School of Medicine; Marcus Clark, University of Chicago; John Cohen, University of Colorado Health Sciences Center; Anne Cooke, University of Cambridge; Mike Gold, University of British Columbia; John Kearney, University of Alabama at Birmingham; Tomohiro Kurosaki, RIKEN Research Center for Allergy and Immunology; Ian MacLennan, University of Birmingham; David Parker, Oregon Health and Science University; Jonathan Poe, Duke University; Jeff Ravetch, Rockefeller University; Clifford Snapper, Uniformed Services University of the Health Sciences; David Tarlington, Walter and Eliza Hall Institute of Medical Research; Tom Tedder, Duke University; Matthias Wabl, University of California, San Francisco; Jenny Woof, University of Dundee

Chapter 7 Leslie Berg, University of Massachusetts Medical School; Mike Cancro, University of Pennsylvania; B.J. Fowlkes, National Institute of Allergy and Infectious Diseases, National Institutes of Health; Richard Hardy, Fox Chase Cancer Center; Paul Kincade, Oklahoma Medical Research Foundation; Michel Nussenzweig,

Rockefeller University; Shiv Pillai, Harvard Medical School; Jennifer Punt, Haverford College; Terry Rabbitts, Medical Research Council Laboratory of Molecular Biology; Ellen Robey, University of California, Berkeley; Fred Rosen, Harvard Medical School; David Schatz, Yale University School of Medicine; Mark Schlissel, University of California, Berkeley; Edvard Smith, Karolinska Institutet

Chapter 8 Leslie Berg, University of Massachusetts Medical School; John Cohen, University of Colorado Health Sciences Center; Jason Cyster, University of California, San Francisco; John Kearney, University of Alabama at Birmingham; Mitchell Kronenberg, La Jolla Institute for Allergy and Immunology; Shiv Pillai, Harvard Medical School; David Raulet, University of California, Berkeley

Chapter 9 Bruce Beutler, Scripps Research Institute; Eric Brown, University of California, San Francisco; Pascale Cossart, Institut Pasteur; Stanley Falkow, Stanford University; Brett Finlay, University of British Columbia; Douglas Golenbock, University of Massachusetts; Stefan Kaufmann, Max Planck Institute for Infection Biology; Roy Mariuzza, University of Maryland Biotechnology Institute; Daniel Portnoy, University of California, Berkeley; Philippe Sansonetti, Institut Pasteur; Patrick Schlievert, University of Minnesota; Luc Teyton, Scripps Research Institute; Elaine Tuomanen, St. Jude Children's Research Hospital

Chapter 10 Rafi Ahmed, Emory University; Don Ganem, University of California, San Francisco; Grant McFadden, Robarts Research Institute; Andrew McMichael, University of Oxford; Bruce Walker, Harvard Medical School; Robin Weiss, University College London; David Woodland, Trudeau Institute

Chapter 11 Neil Gow, University of Aberdeen; Rick Maizels, University of Edinburgh; Ed Pearce, University of Pennsylvania; Luigina Romani, University of Perugia; James Stringer, University of Cincinnati

Chapter 12 Mark Anderson, University of California, San Francisco; Jeff Bluestone, University of California, San Francisco; Philippa Marrack, National Jewish Medical and Research Center and University of Colorado Health Sciences Center; David Nemazee, Scripps Research Institute; Shimon Sakaguchi, Kyoto University; Ron Schwartz, National Institute of Allergy and Infectious Diseases, National Institutes of Health; Ethan Shevach, National Institute of Allergy and Infectious Diseases, National Institutes of Health; David Wofsy, University of California, San Francisco

Chapter 13 Melissa Brown, Northwestern University; Jack Elias, Yale University School of Medicine; Matthias Goebeler, University of Heidelberg; Thomas MacDonald, Barts and The London, Queen Mary's School of Medicine and Dentistry; Thomas Platts-Mills, University of Virginia; Fiona Powrie, University of Oxford; Jeff Ravetch, Rockefeller University; Reuben Siraganian, National Institute of Dental and Craniofacial Research, National Institutes of Health; Ludvig Sollid, University of Oslo; Warren Strober, National Institute of Allergy and Infectious Diseases, National Institutes of Health; Craig Svensson, University of Iowa; David van Heel, Barts and The London, Queen Mary's School of Medicine and Dentistry

Chapter 14 Hugh Auchincloss, Harvard Medical School; Jean-Laurent Casanova, Necker University Hospital; Alain Fischer, Necker University Hospital; Philip Greenberg, University of Washington School of Medicine; Jan Holmgren, Göteborg University; Alan Krensky, Stanford University; Gus Nossal, University of Melbourne; Jennifer Puck, University of California, San Francisco; Robert Schreiber, Washington University School of Medicine; Edvard Smith, Karolinska Institutet; David Tough, GlaxoSmithKline Medicines Research Center

We are grateful to the following for providing or permitting the use of illustrations:

Front cover and frontispiece image Courtesy of Kurt Dittmar and Manfred Rhode.

Figure 2-10 Structural model of the CD2–CD58 complex. Kindly provided by Jia-huai Wang and Ellis Reinherz. Reprinted from *Cell*, volume 97, Wang, J.-h., Smolyar, A., Tan, K., Liu, J.-h., Kim, M., Sun, Z.-y.J., Wagner, G. and Reinherz, E.L.: **Structure of a heterophilic adhesion complex between the human CD2 and CD58 (LFA-3) counterreceptors**, pages 791–803, ©1999, with permission from Elsevier.

Figure 2-14b Conformational change in integrins between low-affinity and high-affinity states. Kindly provided by Jian-Ping Xiong. Reprinted with permission from Xiong,J.-P., Stehle, T., Diefenbach, B., Zhang, R., Dunker, R., Scott, D.L., Joachimiak, A., Goodman, S.L. and Arnaout, M.A.: **Crystal structure of the extracellular segment of integrin αVβ3**. *Science* 2001 294:339–345 ©2001 AAAS.

Figure 2-15a Binding of a sugar residue by the C-type lectin domain of E-selectin. Kindly provided by Heide Kogelberg and Ten Feizi. Reprinted from *Curr. Opin. Struct. Biol.*, volume 11, Kogelberg, H. and Feizi, T.: **New structural insights into lectin-type proteins of the immune system**, pages 635–643, ©2001, with permission from Elsevier.

Figure 4-12 Pockets in the peptide-binding groove of an MHC class II molecule. Courtesy of David Margulies.

Figure 5-6 Binding orientation of the TCR to MHC molecules. Kindly provided by Markus Rudolph and Ian Wilson. Reprinted from *Curr. Opin. Immunol.*, volume 14, Rudolph, M.G. and Wilson, I.A.: **The specificity of TCR/pMHC interaction**, pages 52–65, ©2002, with permission from Elsevier.

Figure 7-29 Autoimmunity to multiple organs in mice deficient in Aire. Courtesy of Mickie Chang, Jason DeVoss and Mark Anderson.

Figure 12-19 Characteristic butterfly skin rash of a female SLE patient. Kindly provided by David Wofsy. ©1972-2004 American College of Rheumatology Clinical Slide Collection. Used with permission.

Figure 13-9 Photograph of the back of an individual showing a hives reaction. Courtesy of Tim Berger and Kari Connolly, Department of Dermatology, University of California, San Francisco.

Figure 13-15a Celiac disease. Courtesy of Coeliac UK.

Figure 14-0 Key historical figures. Panels (a) and (b) courtesy of the Wellcome Library, London. Panels (c) and (d) ©The Nobel Foundation.

Contents summary

Contents in full

T lymphocytes activated by dendritic cells differentiate and migrate to B cell follicles and sites of infection

Activated B lymphocytes differentiate into plasma cells and secrete antibody

At the end of an immune response, the vast majority of the antigen-specific cells die and surviving cells become resting memory cells

Overview of the Immune Response

1

The immune system of vertebrates is an extraordinarily diverse inventory of specialized molecules and cells that continuously patrol the blood and tissues for invaders, and protect the epithelial surfaces which are under constant assault from infectious agents in the environment. In this chapter we introduce the specialized features of the molecules, cells and anatomical adaptations that are the basis of immunity, and the principles underlying their operation.

(a)

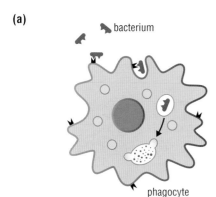

bacterium

phagocyte

(b)

B lymphocyte

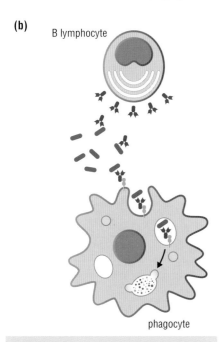

phagocyte

Figure 1-1 An example of an innate and an adaptive immune response (a) A phagocyte of the innate immune system recognizes a conserved surface component of a bacterium, and ingests and destroys it. **(b)** A lymphocyte of the adaptive immune system produces antibodies (purple) that recognize a variable component of the surface of a bacterium by means of a binding site that is itself highly variable. A non-variable portion of the antibody is then recognized by a receptor (green) on the phagocyte, which is thereby activated to ingest the bacterium and destroy it. In this way the adaptive immune system enables microorganisms that have masked the conserved components recognized by innate immune cells to be destroyed by these cells.

The immune system protects us against infectious organisms

The microorganisms that have evolved to exploit the advantages of growth in the vertebrate body are traditionally divided into four main groups: viruses, bacteria, fungi, and parasites, although parasites themselves comprise two phylogenetically quite distinct groups—protozoa and worms. Together, these microorganisms are capable of colonizing most of the tissues of the body, where they may proliferate in the cytoplasm or intracellular vesicles of cells, in the interstitial spaces between cells, in the blood or lymph, or on epithelial surfaces. Many of these microorganisms coexist harmlessly with their hosts, especially at epithelial surfaces. Those that cause disease usually have specialized mechanisms for breaching epithelial barriers and often for entering cells, and are collectively known as **pathogens**. The immune system of vertebrates embraces an extraordinarily wide range of defensive devices including intracellular sensor and signaling molecules that are common to all cells, and a specialized system of hematopoietic cells and serum proteins that together provide mechanisms for recognizing all of the diverse pathogenic organisms in any of the sites they colonize, and in most cases successfully eliminating them.

The first critical barrier to infectious disease is the skin. Cuts, abrasions, more serious wounds and burns expose the tissues to bacterial and fungal infection that can be lethal. Where the skin is intact, the main portals of entry of microorganisms into the body are the mucosal epithelia of the gastrointestinal, respiratory and urogenital tracts. These surfaces are particularly rich in immune defense mechanisms. The importance and efficacy of the large and diverse repertoire of immune mechanisms can be seen in the consequences of inherited immune deficiency and of acquired immune deficiency syndrome (AIDS) caused by the human immunodeficiency virus (HIV). Hereditary immune deficiencies are usually fatal in childhood; HIV ultimately kills by destroying one of the major cell classes of the immune system.

Immune mechanisms are divided into those of innate immunity and those of adaptive immunity

Immunity requires the recognition and elimination or containment of infectious organisms. This is achieved by two systems broadly classified as **innate immunity** and **adaptive immunity**. The innate immune system consists of molecules and cells that distinguish host cells from those of infectious agents, in part by recognizing conserved constituents of microorganisms; they are activated within hours of contact and their efficacy is not significantly increased by previous exposure. By contrast, the **lymphocytes** of the adaptive immune system, and the antibodies they produce, can recognize an essentially unlimited number of different targets but become effective only after a delay of two to four days on first encounter with a given microorganism. Lymphocytes and lymphocyte products specific for that microorganism then persist as **immune memory**, however, and are rapidly protective on reexposure to the same infectious agent. This property of the adaptive immune system is the basis for the protective effect of vaccination. Most of the mechanisms for eliminating infectious organisms are provided by the innate immune system: they may be invoked either directly through recognition of the microorganism by the innate immune system itself, or indirectly through recognition by cells of the adaptive immune system that then activate innate immune defenses (Figure 1-1).

Innate immune defense begins in the skin and the epithelia of the respiratory, intestinal, urinary and reproductive tracts with antimicrobial peptides that are thought to be important for protection against bacterial and fungal infection. Innate immunity in the blood and tissues is provided principally by *phagocytic cells* that recognize surface components of bacteria and engulf them; and

Definitions

adaptive immunity: immune responses mediated by **lymphocytes** and their products and requiring activation by **innate immune mechanisms** on first encounter with **antigen** but acting immediately on subsequent encounters.

antibodies: highly variable proteins produced by the **B lymphocytes** of the immune system and that recognize **antigen** and target it for destruction.

antigen: any molecule or part of a molecule recognized by the variable antigen receptors of **lymphocytes**.

B lymphocytes: lymphocytes that when activated differentiate into antibody-secreting cells.

chemokine: any of a family of closely related small, basic cytokines whose main function is as chemoattractants. The name is a contraction of chemotactic **cytokine**.

complement system: serum proteins activated directly or indirectly by conserved surface features of microorganisms.

cytokine: polypeptide signaling molecule that participates in immune responses.

dendritic cells: specialized cells that ingest debris and infectious agents in the peripheral tissues and migrate to lymphoid tissues where they present fragments of the ingested particles for recognition by T lymphocytes in the activation of adaptive immune responses.

effector T cells: T cells that secrete cytokines (in the case of helper T cells) or deliver cytotoxic signals or effector molecules (in the case of cytotoxic T cells) immediately on activation through recognition of peptide–MHC by

by the proteins of the **complement system**, which circulate in the blood and are activated by microbial membranes to destroy pathogens either directly, or indirectly by recruiting phagocytes.

The phagocytes of innate immunity have a pivotal role in immune responses, in two ways. First, on activation by microbial surfaces they release **cytokines** and **chemokines** that amplify the response to infection. Cytokines and chemokines are signaling molecules with a wide range of functions in immunity: the cytokines and chemokines produced by phagocytes act by increasing the permeability of blood vessels and changing their adhesive properties, as well as by direct signaling, to recruit additional cells and molecules of the immune system to sites of infection. These actions of cytokines and chemokines are collectively known as the **inflammatory response** and are responsible for the redness, swelling and pain that are associated with bacterial infection in tissues. Cytokines released by activated phagocytes are known as **inflammatory cytokines** and are the major cause of immune pathology such as septic shock and rheumatoid arthritis.

The other critical role of phagocytes is to activate the adaptive immune response. Adaptive immune responses are mediated by lymphocytes, which fall into two major classes: **T lymphocytes** and **B lymphocytes**. T lymphocytes kill virus-infected cells and activate other cells of the immune system including phagocytes, which they arouse to kill intracellular bacteria, and B lymphocytes, whose major function is to secrete **antibodies**, highly specialized proteins that recognize microorganisms and target them for destruction (see Figure 1-1). The highly variable surface receptors of lymphocytes do not, however, distinguish the components of infectious agents from any other molecules: it is dependence of the adaptive immune response on the phagocytes of innate immunity that focuses these responses on infectious microorganisms. This is the crucial role of highly specialized cells known as **dendritic cells**. These cells begin life as phagocytes, but when activated by conserved components of microorganisms, or by inflammatory cytokines released by macrophages, they become dedicated to displaying variable components of the ingested microorganisms for recognition by T lymphocytes, and of activating lymphocytes that specifically recognize them. This leads to the proliferation of the lymphocyte, and its differentiation into an **effector T cell**; that is, a cell capable of killing or activating other cells: this sequence of events is illustrated in Figure 1-2.

These critical interactions between activated dendritic cells and lymphocytes occur in a specialized system of widely disseminated **lymphoid tissues** adapted for this function. Foreign particles derived from microorganisms in body tissues can reach the lymphoid tissues from the bloodstream or through specialized *lymphatic vessels*, which drain tissue fluid (the *lymph*) into the lymphoid organs; or they can be carried in by dendritic cells migrating in with the lymph. Lymphocytes enter lymphoid tissues from the bloodstream, returning to the bloodstream through the lymph if they are not activated. The migration of cells of the immune system through the lymphoid tissues, and their interactions with one another, are guided and controlled by adhesive cell surface molecules whose expression is regulated by cytokines, and by chemokines.

An antigen is operationally defined

Two terms are widely used in immunology to refer to molecules or parts of molecules recognized by lymphocytes and should be clarified at the outset in this book. The first, and most widely used, is **antigen**, which was originally defined as anything recognized by an antibody and is now used more generally for any molecule recognized by the antigen receptors of lymphocytes, or by antibodies produced by lymphocytes. The second term is **epitope**, which means the specific part or feature of an antigen that is recognized by a lymphocyte or an antibody. An antigen may thus have several epitopes.

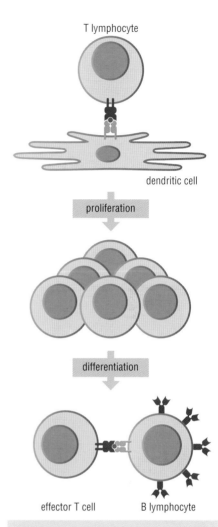

T lymphocyte

dendritic cell

proliferation

differentiation

effector T cell   B lymphocyte

Figure 1-2 Lymphocyte activation by a dendritic cell A dendritic cell displays a component of a microorganism (red dot) for recognition by a T lymphocyte bearing a receptor (purple) for that component, and stimulates it to proliferate and differentiate into an effector cell. When the effector cell recognizes the same antigen displayed on the surface of a B lymphocyte, the T cell activates the B lymphocyte, which in turn proliferates and secretes antibodies that recognize the microorganism.

---

their antigen receptors.

epitope: molecular feature of an antigen that is specifically recognized by a lymphocyte or an antibody.

immune memory: rapid response of the **adaptive immune system** to exposure to **antigens** previously encountered.

inflammatory cytokines: cytokines that are released by phagocytes of the **innate immune system** in the presence of microorganisms, or by activated lymphoid cells, and that act on blood vessels and cells of the immune system to induce or amplify immune responses.

inflammatory response: release of **cytokines** by leukocytes at a site of infection causing dilatation and increased permeability of blood vessels and the recruitment of immune cells.

innate immunity: immune responses mediated by cells and molecules recognizing conserved features of microorganisms and activated immediately on encounter with them.

lymphocytes: white blood cells bearing highly variable receptors for **antigen** that circulate in the blood and lymph and are the mediators of **adaptive immunity**.

lymphoid tissues: specialized tissues in which **lymphocytes** mature and immune responses are initiated.

pathogen: any microorganism that causes disease.

T lymphocytes: lymphocytes that mature in the thymus and different classes of which mediate cytotoxic responses against cells infected with viruses, activate **B lymphocytes** to produce antibodies, and activate phagocytes to ingest and destroy microorganisms.

The cells of the immune system originate in the bone marrow

All cells of the immune system derive from a common **hematopoietic stem cell (HSC)** in the bone marrow which also gives rise to red blood cells (**erythrocytes**) and the **platelets** that are responsible for blood clotting. A stem cell gives rise on each division to another stem cell and a **progenitor cell** committed to producing related cells of different types that comprise the different cell lineages that descend from the original stem cell (Figure 1-3).

There are two main lineages of immune cells: the **myeloid lineage**, which gives rise to the *inflammatory cells* of the innate immune system, as well as other innate immune system cells that are resident in tissues; and the **lymphoid lineage**, which gives rise to the lymphocytes of the adaptive immune response (Figure 1-4). The functional characteristics of the myeloid and lymphoid cells are described in the next sections; here we summarize the lineage relationships of these cells and briefly outline what is known about the growth factors that drive the differentiation of these cells from their precursors, and how their numbers can be adjusted to meet the requirements of physiological stress or infection.

Cytokines produced by bone marrow stromal cells drive the differentiation of immune cells from their common hematopoietic precursor

The differentiation of blood cells in the bone marrow is directed by **hematopoietins**, a large class of cytokines with important functions in the activation of mature immune cells as well as in their early differentiation. A second major class of cytokines, the *proinflammatory cytokines*, functions mainly, as its name suggests, in the activation of innate responses to infection (see section 1-0). The different structural and functional classes of cytokines and their receptors are discussed in detail in Chapter 2; here we are concerned with a limited number of hematopoietins that are produced by tissue cells and direct the differentiation of cells in the lymphoid and myeloid lineages. Many of these cytokines are also produced by immune cells, notably T cells and macrophages, once an immune response has been generated, and increase the production of both myeloid and lymphoid cells in the presence of infection.

Apart from erythropoietin and thrombopoietin, which direct the differentiation of erythrocytes and megakaryocytes respectively (see Figure 1-4), almost all the hematopoietins are termed either *interleukins (IL)* or *colony-stimulating factors (CSFs)*. Interleukins were originally identified as molecules mediating signaling between immune cells; colony-stimulating factors were discovered as growth factors that could induce the production of colonies of specific blood cell types in culture. Bone marrow *stromal cells*—that is, the cells that provide the tissue matrix of the bone marrow—also play a crucial part in supporting hematopoiesis through the production of hematopoietins. The main hematopoietins implicated in the differentiation of each hematopoietic cell type are indicated in Figure 1-4.

The production of cells of different lineages is tightly regulated by the needs of the organism, which are reflected in signals from the peripheral tissues. Hematopoietic growth factors are produced constitutively, often by a number of different cell types, but in the presence of physiological stress or infection this production can be greatly increased through inducible production of specific growth factors by other tissues. Thus, for example, infection induces the production of granulocyte colony-stimulating factor (G-CSF) by large numbers of immune and non-immune cells, and thus of neutrophils which, as we shall see in the next section, are essential in early responses to infection.

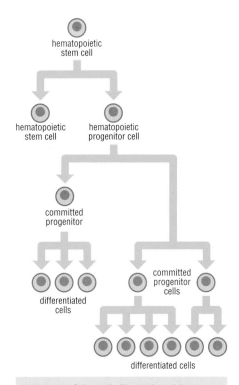

Figure 1-3 Schematic illustration of the pattern of production of progenitor cells and differentiated cells from stem cells The hematopoietic stem cell divides to replace itself and produce a progenitor cell with the potential to give rise to all of the hematopoietic lineages, by generating further progenitor cells committed to progressively narrower ranges of differentiated fates.

Definitions

colony-stimulating factor (CSF): growth factor or cytokine that induces the differentiation from a multipotent precursor of one or a few specific cell types.

CSF: see **colony-stimulating factor**.

erythrocytes: red blood cells containing the oxygen-carrying protein hemoglobin.

hematopoietic stem cell (HSC): self-renewing cell that gives rise to all the red and white blood cells.

hematopoietins: cytokines that promote proliferation and/or lineage-specific differentiation of hematopoietic cells.

HSC: see **hematopoietic stem cell**.

lymphoid lineage: hematopoietic cell lineage containing the lymphocytes of the immune system.

myeloid lineage: hematopoietic cell lineage containing the phagocytic and inflammatory cells of the immune system.

platelets: cell fragments that are shed from megakaryocytes in the erythroid lineage and induce blood clotting.

precursor cell: cell that is committed to a single specific differentiated state but has not yet terminally differentiated.

progenitor cell: committed cell giving rise to a lineage of related but distinct cells.

It is thought that hematopoietic progenitor cells express receptors for many cytokines that direct distinct pathways of differentiation, and can thus be induced to generate cells of different lineages depending upon which cytokines are present. **Colony-stimulating factors (CSFs)** are cytokines directing the differentiation of a single lineage or a limited number of lineages. Engagement of surface receptors by a specific colony-stimulating factor results in down-regulation of other receptors instructing different cell fates, thus ensuring the commitment of the cell to a specific differentiated fate. Cells that have become committed to a specific differentiated fate but do not yet express all the specialized characteristics of the terminally differentiated cell are known as **precursor cells**.

Reserves of hematopoietic cells are rapidly mobilized from the bone marrow during stress

The cells of many hematopoietic lineages, such as red blood cells and most myeloid cells, do not proliferate once they have disseminated to the blood and tissues, and conditions of increased demand, such as bleeding or infection, must be met by increased supply from the bone marrow. For red blood cells, many more precursor cells are generated than differentiate and leave the bone marrow, and the cells that remain normally undergo programmed death: under stress, however, these reserve cells are activated to leave and replace lost red blood cells, allowing the increased need to be met much more rapidly than would be possible by stimulation of proliferation from stem cells. Similarly, in the presence of infection, proinflammatory cytokines induce the production of myeloid cells and accelerate their differentiation.

Figure 1-4 Hematopoietic cell lineages
Both the red and the white blood cells are believed to be derived from a common hematopoietic stem cell (HSC) via committed progenitors (not shown) that give rise to the erythroid, myeloid and lymphoid lineages. The main hematopoietins necessary for growth and differentiation of different lineages and cell types are indicated. The erythroid lineage gives rise to the erythrocytes and to megakaryocytes which shed fragments that form the platelets that initiate blood clotting. The myeloid lineage gives rise to the phagocytic and inflammatory cells of innate immunity. The lymphoid lineage gives rise to the T and B cells of adaptive immunity and to natural killer (NK) cells which are specialized cytotoxic cells. Both lymphoid and myeloid cell lineages can give rise to dendritic cells. G-CSF is granulocyte colony-stimulating factor, so called because it induces the differentiation of neutrophils which are also known as granulocytes. M-CSF is macrophage colony-stimulating factor, so called because it induces the differentiation of macrophages; and GM-CSF is granulocyte-macrophage colony-stimulating factor, so called because it induces the differentiation of both granulocytes (which comprise neutrophils, basophils, eosinophils and mast cells) and macrophages. IL-3, IL-5, IL-7 and IL-15 are interleukins which we discuss in Chapter 2. Flt3L is Fms-like tyrosine kinase 3 ligand; TPO is thrombopoietin; EPO is erythropoietin; SCF is stem cell factor, which also plays a part in the differentiation of non-hematopoietic cells.

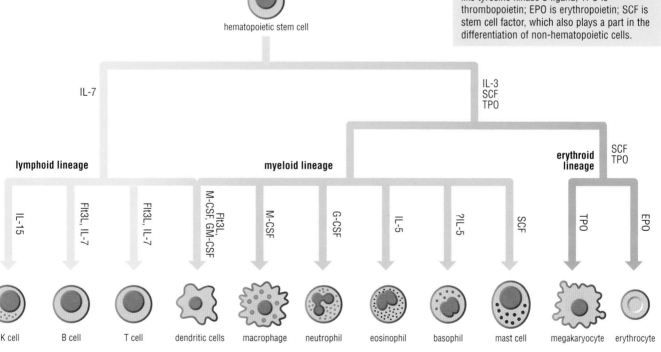

References

Alberts, B. *et al.*: **Histology: the lives and deaths of cells in tissues** in *Molecular Biology of the Cell* 4th ed. (Garland Science, New York, 2002), 1283–1296.

Kaushansky, K.: **Lineage-specific hematopoietic growth factors.** *N. Engl. J. Med.* 2006, **354**:2034–2045.

Kondo, M. *et al.*: **Biology of hematopoietic stem cells and progenitors: Implications for clinical application.** *Annu. Rev. Immunol.* 2003, **21**:759–806.

Leonard, W.J.: **Type I cytokines and their receptors** in *Fundamental Immunology* 4th ed. Paul, W.E. ed. (Lippincott-Raven, New York, 1999), 741–774.

Shortman, K. and Liu, Y.-J.: **Mouse and human dendritic cell subtypes.** *Nat. Rev. Immunol.* 2002, **2**:151–161.

Traver, D. *et al.*: **Development of CD8α-positive dendritic cells from a common myeloid progenitor.** *Science* 2000, **290**:2152–2154.

Immune cells may differentiate in several stages

Some cells of the immune system complete their differentiation in the bone marrow. Others, however, go through one or more stages of further maturation after they have left the bone marrow to enter the blood or tissues; and the lymphocytes of the adaptive immune system do not acquire their differentiated functions until they have encountered antigen. Figures 1-5 and 1-6 indicate the sites of intermediate and final differentiation of the cells of the myeloid and lymphoid lineages, respectively. On final differentiation they become **effector cells**, capable of activating or destroying other cells, or both.

In this section, we briefly introduce the effector functions of all of the major types of immune system cells. Neutrophils, eosinophils, basophils and mast cells are described in more detail in later chapters in the context of the immune responses to which they contribute; and the specialized classes and subsets of lymphocytes are discussed in the chapters on adaptive immunity. Macrophages and dendritic cells are versatile cells of the innate immune system that are found in several different forms with distinct functional properties, and we review these in the next section of this chapter.

Neutrophils, macrophages and dendritic cells are phagocytic cells

The **phagocytic cells** of the immune system recognize microorganisms through receptors that bind to conserved microbial components and include the neutrophils and macrophages of innate immunity, which are effector cells specialized to internalize and destroy microorganisms, and dendritic cells, which internalize microorganisms for presentation to T lymphocytes.

Neutrophils are the front-line effector cells of innate immunity: after differentiation they circulate for a few hours before entering the tissues where they ingest infectious microorganisms and kill them by means of a battery of microbicidal products stored in specialized vesicles. These vesicles appear as granules when stained with neutral dyes and for that reason neutrophils are also known as **granulocytes**. The other characteristic feature of neutrophils is a multilobed nucleus and they are therefore sometimes also called **polymorphonuclear leukocytes**. Neutrophils are short-lived and die after two or three days.

Macrophages are long-lived cells that provide immune surveillance in the tissues and play an important part in tissue maintenance. They derive from **monocytes** that circulate in the blood, differentiating as they leave the bloodstream. Like neutrophils, they ingest and destroy microorganisms, and both neutrophils and macrophages release inflammatory cytokines. Macrophages also participate as effector cells in adaptive immune responses when activated by T lymphocytes, or by antibodies secreted by B cells.

The principal function of dendritic cells is the induction of adaptive immunity. Like macrophages, they function in the immune surveillance of tissues, where as *immature dendritic cells* they operate as phagocytes, internalizing microorganisms and other particles. But while the function of macrophages is to destroy the microorganisms they ingest, the function of dendritic cells is to display the ingested particles on their surface for recognition by T lymphocytes, and after a period of days to weeks in the peripheral tissues dendritic cells undergo differentiation into *mature dendritic cells*, which are no longer phagocytic and leave the peripheral tissues to migrate into the lymphoid organs where they will encounter recirculating T cells. Mature dendritic cells are characterized by the long processes, or branches, for which they are named: these enable them to make contact with several T cells simultaneously.

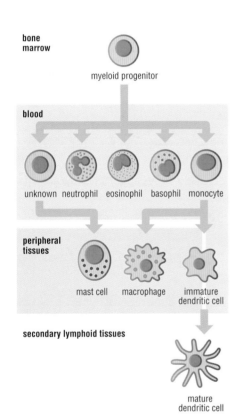

Figure 1-5 Cells of the myeloid lineage

Basophils, mast cells and eosinophils release inflammatory and cytotoxic mediators from intracellular stores

Basophils, **mast cells** and **eosinophils** all have a special role in the protection of epithelial surfaces, especially the mucosa of the gastrointestinal, respiratory and urogenital tracts: mast cells have a sentinel role; basophils and eosinophils are circulating cells recruited from the bloodstream. They can recognize microorganisms directly, or they can be activated by complement or by products of lymphocytes. Like neutrophils, they are characterized by densely staining cytoplasmic granules containing inflammatory and cytotoxic mediators, and they are also referred to as granulocytes. But whereas neutrophils destroy internalized microorganisms by delivering them to cytotoxic compartments inside the cell, eosinophils, basophils and mast cells function mainly in immune defense against parasites that are too large to be internalized by phagocytes, and they release the contents of their granules to the exterior on activation, thereby either creating an environment hostile to invading organisms or (in the case of eosinophils) directly killing the parasites. Basophils and mast cells release molecules including histamine that are clinically important as the mediators of allergic and pathological inflammatory responses: the coughing, sneezing, vomiting and other expulsive responses that accompany allergies may reflect mechanisms that evolved to expel parasites.

Lymphocytes detect antigen by means of variable receptors

Resting lymphocytes have no distinctive visual characteristics. The two major types of lymphocytes are B lymphocytes, which mature in the bone marrow, and T lymphocytes, which mature in the **thymus**, a specialized lymphoid organ in the chest. The principal function of B lymphocytes is to secrete antibodies. T lymphocytes fall into two major functional classes: **cytotoxic T cells**, which kill cells infected with viruses or with bacteria specialized to colonize the cytoplasm of cells; and **T helper cells**, which activate other cells of the immune system, including the B cells that secrete antibody.

Lymphocytes recognize antigen by means of receptors that are generated during their differentiation through a unique mechanism of genetic recombination that confers effectively unlimited diversity in antigen binding specificity. Each individual lymphocyte expresses antigen receptors of only one specificity, but the total population of lymphocytes in any given individual is collectively capable of recognizing virtually any antigen.

Mature lymphocytes that have not yet encountered antigen are known as **naïve lymphocytes**. Differentiation into effector cells is stimulated by encounter with antigen and is preceded by a phase of vigorous proliferation that selectively expands the numbers of those lymphocytes with receptors specific for the inducing antigen. This *clonal selection* of antigen-specific lymphocytes is discussed in a later section of this chapter.

Naïve B lymphocytes, whose antigen receptors are a membrane-bound form of antibody (more properly known as *immunoglobulin*), differentiate after activation into **plasma cells** that secrete immunoglobulin molecules of the same specificity as the membrane-bound immunoglobulin.

The lymphoid lineage contains a third major cell type that lacks antigen-specific receptors and is capable of killing virus-infected cells immediately without prior activation and clonal expansion. These cells are generally regarded as part of the innate immune system and are known as **natural killer (NK) cells**. They distinguish infected from uninfected cells by a mechanism that depends on inhibition of killing by surface molecules that are encoded in the *major histocompatibility complex* (see section 1-5) and are expressed on all normal healthy cells.

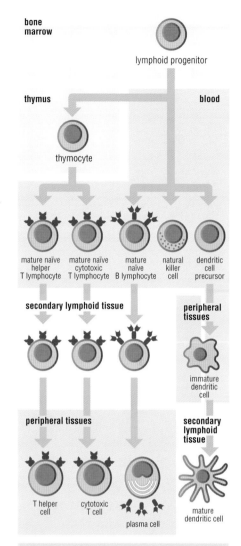

Figure 1-6 Cells of the lymphoid lineage

that have not yet encountered antigen.

natural killer (NK) cells: cytotoxic lymphocytes lacking antigen-specific receptors but with invariant receptors that detect infected cells and some tumor cells and activate their destruction.

neutrophils: phagocytic cells that circulate in the blood and detect microorganisms by means of receptors that recognize conserved components.

NK cells: see **natural killer cells**.

phagocytic cells: cells that recognize and ingest molecules and particles including microorganisms and destroy them.

plasma cells: terminally differentiated B lineage cells secreting large quantities of antibody.

polymorphonuclear leukocytes: another name for **neutrophils**.

T helper cells: T lymphocytes that activate other cells of the immune system, including phagocytes, **mast cells**, **basophils**, **eosinophils** and B cells, which when

activated differentiate into antibody-producing cells.

thymus: primary lymphoid organ in which T lymphocytes mature.

1-3 Macrophage and Dendritic Cell Subsets

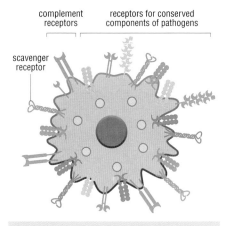

complement receptors

receptors for conserved components of pathogens

scavenger receptor

Figure 1-7 The array of receptors expressed by macrophages Macrophages display receptors that recognize conserved components of microorganisms as well as scavenger receptors that recognize particles released by dead or damaged tissues, and receptors for complement proteins and for antibodies (not shown). A similar array of receptors can be expressed on the surface of dendritic cells.

Macrophages and dendritic cells are versatile cells of innate immunity

All cells of the innate immune system have a range of receptors through which they can be signaled by cytokines, by conserved components of microorganisms, by complement components and by antibodies produced by B lymphocytes, and produce responses that may be distinct for signals through different receptors. Macrophages and dendritic cells are particularly versatile. Macrophages and most immature dendritic cells are sentinel cells that reside in tissues, and their receptors and functional properties can vary, sometimes considerably, depending on the tissue in which they are resident. Moreover, both of these cell types sample tissues not only for evidence of microbial invaders but also for normal tissue debris, which they recognize through a variety of *scavenger receptors* specific for molecules diagnostic of cells that have undergone programmed death in the normal course of tissue turnover. Figure 1-7 is a schematic representation of a macrophage showing a selection of the diverse receptors expressed by these cell types; some of these are phagocytic receptors specialized for internalization of particles and microorganisms; some modulate the behavior of the cell; and some do both. We shall describe all of these receptors in greater detail in later chapters. Both types of cells include resident cells that populate tissues in the absence of infection, and also cells with distinct properties that are recruited from circulating monocytes by inflammatory cytokines during immune responses.

The response of both macrophages and dendritic cells to harmless debris is distinct from their response to microbial components. Macrophages, as we shall see in Chapter 3, have important functions in scavenging debris and in tissue repair and maintenance as well as in defense against infection, and on recognition of normal tissue debris they do not generate toxic antimicrobial compounds or release inflammatory mediators, but instead produce antiinflammatory responses. The specialized function of dendritic cells is as dedicated **antigen-presenting cells** that display antigen for recognition by naïve T lymphocytes and activate their differentiation into effector cells. Activation of naïve T cells requires specialized activating signals that must be displayed by the dendritic cell along with the antigen, and these are elicited on the dendritic cell on recognition of conserved microbial components: in this way dendritic cells provide a crucial link between the recognition of microbial components and the activation of adaptive immune responses. In the absence of these signals, presentation of tissue components by dendritic cells is thought to help suppress immune attack on body tissues. We discuss the central role of dendritic cells in initiating and regulating adaptive immune responses in Chapters 4 and 5.

Macrophages have distinct specializations in different tissues

Macrophages are ubiquitous sentinel cells with representatives in tissues ranging from the brain, which is relatively protected and inaccessible, to the gut, which is continuously bathed in food particles and microbes. The predominant macrophages in the brain are known as **microglia**, and have long branching processes. They accumulate at sites of neurological damage where they are thought to participate, like other macrophages, in clearance of debris and tissue repair. In the lung, where they are resident in the alveolae and are therefore known as **alveolar macrophages**, they are constantly exposed to infectious and non-infectious particles in the inhaled air and one of their major functions is disposal of inhaled non-infectious particles. Because any swelling in the lung may compromise respiration, alveolar macrophages are subject to specialized suppressive control by local epithelium that helps to ensure that inflammatory responses are elicited only in the presence of infection.

The liver and the spleen have essential functions both in the routine clearance of dead and damaged erythrocytes and in the rapid destruction of microorganisms entering the blood, and

Definitions

alveolar macrophages: macrophages in the lung.

antigen-presenting cells: cells capable of displaying antigen for recognition by T cells and of activating naïve T cells.

dermal dendritic cells: immature dendritic cells resident in the dermal layer of the skin: they are the **interstitial dendritic cells** of skin.

interstitial dendritic cells: immature dendritic cells in

the peripheral tissues, mucosa or dermal layers of the skin. Also known as tissue-resident dendritic cells.

Kupffer cells: macrophages in the liver.

Langerhans cells: immature dendritic cells in the epidermal layers of the skin.

microglia: prevalent type of macrophages in the brain, characterized by long processes.

plasmacytoid dendritic cells: cell type of the dendritic cell family that produces very large amounts of type 1

interferons upon contact with viruses.

both of these organs contain specialized populations of macrophages adapted to these functions. In the liver, these are known as **Kupffer cells**, and play a crucial part in preventing the dissemination of bacteria through the bloodstream. The spleen, which we discuss in detail later in this chapter, is the lymphoid organ responsible for responses to infectious organisms in the blood and contains three specialized populations of macrophages, two of which have special functions in controlling the availability of iron, which is scavenged from hemoglobin from dead erythrocytes and which bacteria require for growth. The third specialized type of splenic macrophage resides in a region known as the *marginal zone*, strategically placed at the edge of the sinuses in which the blood flows through lymphoid areas: these *marginal zone macrophages*, like the Kupffer cells of the liver, are critical for destruction of bacteria entering the bloodstream.

We shall see in Chapter 5 that the effector responses of macrophages are also profoundly influenced by signals from effector T cells, which can induce greatly increased microbicidal activity or elicit non-inflammatory responses at mucosal barriers depending on the nature of the infection.

Dendritic cells can vary widely in form and function

In the absence of infection, immature dendritic cells populate peripheral tissues, including lymphoid tissues, where they have sentinel roles as immature cells as well as antigen-presenting functions on maturation. Dendritic cells are especially numerous in epithelia and at mucosal surfaces, and they were first recognized in the skin in the 19th century, long before anything was known about their function, and named **Langerhans cells**. It is now known that Langerhans cells are a specialized subset of dendritic cells in the keratinized epidermis (Figure 1-8), where, unlike other immature dendritic cells, which reside in tissues for days to weeks before moving to lymph nodes, they reside for months. A second population of skin-resident dendritic cells is found in the dermal layer: these are known as **dermal dendritic cells**, or **interstitial dendritic cells**, which is a more general term for tissue-resident antigen-sampling cells. Although all interstitial dendritic cells ingest particles for display to naïve T lymphocytes and thus have a common function, the specialized receptors they express on their surface vary considerably and this, like the variation in macrophage specialization we have just discussed, is thought to be a response to the different tissue environments they occupy.

In the mucosal epithelium of the gut, dendritic cells are concentrated at specialized sites of antigen collection that overlie lymphoid tissues (we describe these later in this chapter), as well as dispersed beneath the epithelial cells, while some have specialized surface properties that enable them to extend their long dendritic processes between the cells of the epithelium and into the lumen to sample antigens (Figure 1-9). These cells are directly exposed not only to ingested antigens but to the thousands of bacteria that normally and harmlessly inhabit the gut—these are known as *commensal bacteria*—as well as to any invading pathogens. It is thought that they may have a specialized role in preventing inflammatory responses to harmless gut residents.

Although we have defined dendritic cells as antigen-presenting cells, there is one important cell type that is classified as a dendritic cell but does not present antigen efficiently to naïve T cells and has a distinct specialized role in immune responses. These are **plasmacytoid dendritic cells**, which, as their name suggests, have some features in common with the plasma cells of the lymphoid lineage although, like most other dendritic cells, they can differentiate from either lymphoid or myeloid precursors. These cells, unlike other dendritic cells, do not reside in tissues but circulate in blood and on encounter with viruses produce large quantities of specialized antiviral cytokines known as *type 1 interferons*, which they secrete on entry into lymphoid tissues, where they are crucial in promoting adaptive responses to viral infection, as we shall see in Chapter 5.

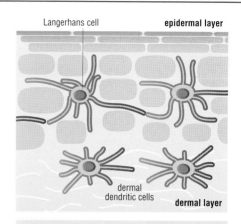

Figure 1-8 Dendritic cells in skin The two dendritic cell subsets that are resident in skin are the Langerhans cells, which populate the epidermal layer, and the dermal dendritic cells, which are found in the dermal layer.

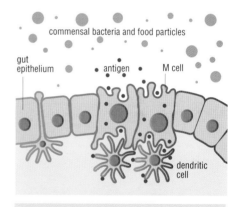

Figure 1-9 Dendritic cells in the epithelium of the small intestine *M cells* are specialized cells that deliver antigen from the lumen of the gut to the underlying tissue, where dendritic cells cluster. Some dendritic cells under the gut epithelium extend their branching processes between the epithelial cells, where they can sample the gut contents.

References

Ardavín, C.: **Origin, precursors and differentiation of mouse dendritic cells.** *Nat. Rev. Immunol.* 2003, **3**:582–590.

Asselin-Paturel, C. *et al.*: **Mouse type I IFN-producing cells are immature APCs with plasmacytoid morphology.** *Nat. Immunol.* 2001, **2**:1144–1150.

Brown, E.J. and Gresham, H.D.: **Phagocytosis** in *Fundamental Immunology* 5th ed. Paul, W.E. ed. (Lippincott Williams & Wilkins, Philadelphia, 2003), 1105–1126.

Gordon, S. and Taylor, P.R.: **Monocyte and macrophage heterogeneity.** *Nat. Rev. Immunol.* 2005, **5**:953–964.

Lambrecht, B.N.: **Alveolar macrophage in the driver's seat.** *Immunity* 2006, **24**:366–368.

Rescigno, M.: **CCR6+ dendritic cells: the gut tactical-response unit.** *Immunity* 2006, **24**:508–510.

Serbina, N.V. *et al.*: **TNF/iNOS-producing dendritic cells mediate innate immune defense against bacterial infection.** *Immunity* 2003, **19**:59–70.

Shortman, K. and Liu, Y.-J.: **Mouse and human dendritic cell subtypes.** *Nat. Rev. Immunol.* 2002, **2**:151–161.

Steinman, R.M. *et al.*: **Tolerogenic dendritic cells.** *Annu. Rev. Immunol.* 2003, **21**:685–711.

The versatility and selectivity of adaptive immunity are guaranteed by selective processes acting on a highly variable receptor repertoire

Two cardinal features of adaptive immunity made it for a long time one of the most mystifying systems in biology. The first is the ability of lymphocytes to recognize almost any antigen on first exposure and respond rapidly and specifically on a second or subsequent exposure. The second is **immune tolerance**, whereby despite the apparently unlimited receptor repertoire of lymphocytes, damaging responses to self molecules and cells are avoided. Tolerance can be explained in part by the requirement for signals from innate immune cells, which are activated only in the presence of infection; but this is not sufficient to prevent the activation of lymphocytes that bind strongly to self determinants.

The conflicting needs for specific recognition of microbes and tolerance to self are met by **clonal selection** acting on lymphocytes, each one of which has receptors of only one specificity, although collectively they represent a receptor repertoire of extraordinary diversity. The mechanisms by which receptor diversity is generated and the receptor repertoire is subsequently refined by clonal selection are outlined below and described in detail in Chapter 7.

Variable receptors for antigen are generated by rearrangement of lymphocyte DNA during ontogeny

The antigen receptors of lymphocytes belong to a large family of cell surface molecules known for historical reasons as the **immunoglobulin superfamily**. Because immunoglobulin is available in large quantities as a soluble protein, it was the first molecule of the superfamily whose structure and sequence were determined, and hence although it is probably one of the most recently evolved members of the family, the entire family bears its name. The antigen receptor of B cells, or the **B cell antigen receptor**, is membrane-bound immunoglobulin, often called **surface immunoglobulin**; the antigen receptor of T cells is usually called the **T cell receptor** (**TCR**). The structure of each of these receptors is schematically illustrated in Figure 1-10. The unique feature of the antigen receptors is that each chain contains a highly variable antigen-binding region (the **variable** or **V region**) and an invariant region known as the **constant region** (**C region**). These two regions are separately encoded in the germline DNA. During lymphocyte ontogeny, the DNA encoding the receptors undergoes programmed rearrangement whereby the sequences encoding the antigen-binding portions (V regions) of the receptor molecules are assembled from separate segments that recombine in different combinations from large pools containing up to 200 different segments, and are brought together with the sequences encoding the C regions. The DNA rearrangements required to generate a complete gene encoding one chain of an antigen receptor molecule are illustrated schematically in Figure 1-11.

(a) B cell antigen receptor (surface immunoglobulin)

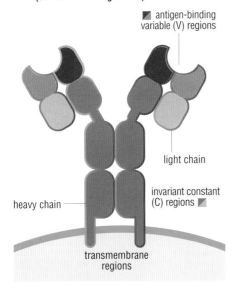

antigen-binding variable (V) regions

light chain

heavy chain

invariant constant (C) regions

transmembrane regions

(b) T cell receptor (TCR)

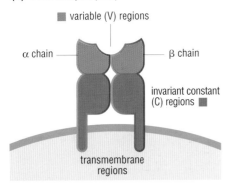

variable (V) regions

α chain

β chain

invariant constant (C) regions

transmembrane regions

Figure 1-10 Antigen recognition structures of the antigen receptors of lymphocytes Each antigen receptor contains two types of chains with different variable regions. The cell surface immunoglobulin **(a)** that serves as the antigen receptor of B cells has four chains, two identical **light chains** and two identical **heavy chains**, each with an invariant or constant region (grey) and a variable region (red) that participates in the antigen-binding site. The T cell receptor for antigen **(b)** is similarly constructed of constant (grey) and variable regions (red), but has only one of each of its two kinds of chains (α and β chains) and is a smaller molecule. Antigen receptors of both T and B cells include several additional chains that transmit the activation signal into the cell interior. These are omitted here for simplicity.

Definitions

B cell antigen receptor: the complex of membrane immunoglobulin that recognizes antigen and a heterodimer of Igα and Igβ that signals antigen recognition in B lymphocytes.

clonal deletion: elimination of potentially self-reactive lymphocytes. Immature lymphocytes undergo programmed cell death after binding to antigen; in this way, cells bearing receptors that recognize self are deleted before they are capable of participating in immune responses. This is a major mechanism of

immune tolerance.

clonal expansion: the selective proliferation of mature naïve lymphocytes that encounter antigen. Only those lymphocytes bearing receptors specifically recognizing antigen are activated to proliferate and differentiate into effector cells.

clonal selection: the process whereby potentially self-reactive lymphocytes are eliminated during ontogeny whereas mature lymphocytes recognizing non-self antigens are selectively expanded.

constant region (C region): region of a lymphocyte receptor for antigen that does not participate in antigen binding and does not vary between cells of different antigen specificities.

heavy chain: the immunoglobulin heavy chains are the larger of the two kinds of polypeptide chains in the immunoglobulin molecule. Each light chain has a **variable region** contributing to the antigen-binding site, and a **constant region** that mediates the effector function of the molecule.

immune tolerance: non-responsiveness of the adaptive

Figure 1-11 Schematic diagram of DNA rearrangements required to generate a complete sequence encoding a chain of an antigen receptor In germline DNA, the variable region of an antigen receptor chain is encoded in at least two segments: here, a V segment and a J segment (where J stands for joining). In some chains, there is a third segment, the D segment (for diversity), between the V and the J (not shown). The J segments are usually immediately upstream of the C segment. During lymphocyte ontogeny, one of the V segments at random undergoes a specialized recombination reaction and is joined to one of the J segments, also randomly selected. The assembled VJC sequence is then transcribed into RNA and the sequence between the J segment and the C segment, including any intervening additional J segments, is eliminated as an intron during RNA processing.

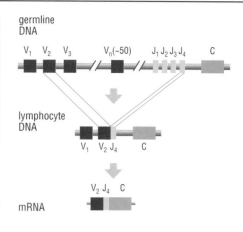

Because the different segments can recombine in many different ways, and each receptor contains two different chains that are independently generated by such recombination, this developmental process alone generates a receptor repertoire that can be in the region of 2,000,000 or more different receptors: specialized enzymes that edit the junctions between the recombining segments further increase diversity. However, in each cell the DNA rearrangements are controlled so that receptors of only one specificity are generated (Figure 1-12a). Subsequent selective processes therefore act on a population of cells with a highly diverse repertoire of receptors but each cell has receptors of only one specificity.

Selection by antigen leads to self tolerance and to the specific recognition of almost any foreign organism or particle

Early in lymphocyte ontogeny, after receptor gene rearrangement and expression but before final maturation, antigen binding initiates programmed cell death. This means that any lymphocyte whose receptors recognize self antigens is eliminated, a process known as **clonal deletion**. Although this is a critical step in ensuring self tolerance there are also mechanisms that operate later to inactivate mature cells recognizing self antigens not encountered by lymphocytes early in ontogeny.

Conversely, when a mature T lymphocyte binds antigen in the presence of an appropriate signal from a dendritic cell, or a mature B lymphocyte binds antigen in the presence of a signal from an antigen-specific T cell, it is activated to proliferate, giving rise to **clonal expansion**. This is the basis for antigen-specific adaptive immune responses and immune memory.

Receptor repertoire selection by these two processes, which together comprise clonal selection, is schematically illustrated in Figure 1-12b and c. Receptor repertoire selection during T cell maturation has an additional step that ensures that T cells can recognize antigen displayed on other body cells: this is discussed in Chapter 7.

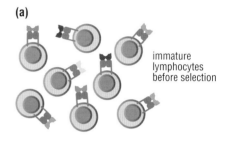

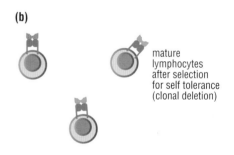

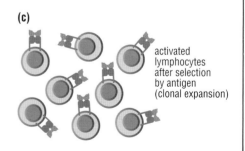

Figure 1-12 Clonal selection of antigen-specific lymphocytes (a) The lymphocytes of any given individual collectively express a large number of different receptors generated by the assembly of separate gene segments. All the receptors on any individual cell are however identical. Some cells bear receptors that recognize self antigens; lymphocytes that encounter antigen while still immature are eliminated **(b)**: this is a critical mechanism for ensuring self tolerance. Once mature, lymphocytes activated by recognition of antigen and appropriate signals from dendritic cells or T cells proliferate **(c)** and differentiate into effector or memory cells: this is the mechanism for antigen-specific immune responses and immune memory. Although only the lymphocytes that have undergone clonal expansion are shown in (c), the other lymphocytes that have escaped clonal deletion are still present as naïve cells that continue recirculating.

---

immune system to an antigen.

immunoglobulin superfamily: family of proteins containing at least one immunoglobulin-like domain and for which the Ig-like domain is a major structural element.

light chain: the immunoglobulin light chains are the smaller of the two kinds of chains in the immunoglobulin molecule. Each has a **variable region** contributing to the antigen-binding site, and a **constant region** containing a cysteine by which it makes a disulfide bond with the constant region of the **heavy chain**.

surface immunoglobulin: see **B cell antigen receptor**.

T cell receptor (TCR): The complex of variable chains whereby T cells recognize antigen and signaling chains whereby antigen recognition is signaled to the cell interior.

TCR: see **T cell receptor**.

variable region (V region): region of a lymphocyte receptor for antigen that participates in antigen binding and varies between cells of different antigen specificities.

References

Burnet, F.M.: *The Clonal Selection Theory of Acquired Immunity* (Cambridge University Press, London, 1959).

Frazer, J.K. and Capra, J.D.: **Structure and function of immunoglobulins** in *Fundamental Immunology* 4th ed. Paul, W.E. ed. (Lippincott-Raven, New York, 1999), 64–70.

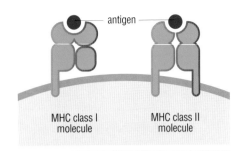

Figure 1-13 Schematic diagram of MHC class I and MHC class II molecules The two classes of MHC molecules are membrane proteins with similar structures, and each binds peptide fragments of antigen picked up inside cells in a cleft formed from the membrane-distal extracellular domains.

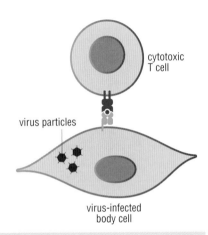

Figure 1-14 Recognition of antigen and MHC class I molecules by cytotoxic T cells Peptide fragments of pathogens, such as viruses, that replicate in the cytoplasmic compartments of cells are carried to the cell surface by MHC class I molecules and recognized by cytotoxic T cells.

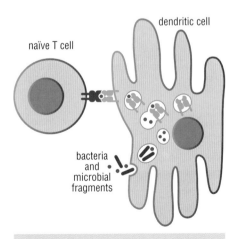

Figure 1-15 Recognition of antigen and MHC class II molecules on dendritic cells by naïve helper T cells Naïve helper T cells are first activated when they recognize antigen (here derived from internalized bacteria) bound to MHC class II molecules on the surface of dendritic cells.

T lymphocytes recognize fragments of antigen carried to the cell surface by MHC molecules

The surface immunoglobulin molecules that serve as the antigen receptors of B cells are adapted to recognize intact antigen in the extracellular spaces of the body, where the secreted form of the same immunoglobulin will later bind to it, thereby providing defense against extracellular pathogens. Effector T cells, by contrast, do not interact directly with intact antigen but must interact with other cells, in most cases by direct contact: thus cytotoxic T cells kill infected body cells, and two major functions of helper T cells are to activate B cells and macrophages. T cells are focused on their target cells by cell surface proteins, known as **MHC molecules** because they are encoded in the **major histocompatibility complex (MHC)**, that carry fragments of antigen, usually peptide fragments of proteins, from internal compartments of the target cell to the cell surface and display them for recognition by the T cell. All naïve T cells are activated by antigen fragments displayed on MHC molecules on the surface of dendritic cells, and after differentiation into effector cells they are triggered to kill or activate cells displaying the same complex of MHC and antigen (see Figure 1-2).

The major histocompatibility complex was originally discovered as the region of the genome determining tissue (histo-) compatibility between donor and recipient in tissue and organ grafts, and the MHC molecules were identified by genetic analysis as the principal markers that distinguished the tissues of different individuals. Graft rejection is caused by a massive immune attack on the cells of mismatched tissue grafts, and for a long time after this was discovered it was unclear why the immune response to mismatched tissue should be so aggressive. This mystery was solved with the realization that T cells detect antigen as cell-surface fragments bound to MHC molecules, and are therefore biased to recognize these molecules on the surface of cells. Because MHC molecules must be able to bind a very wide range of antigenic fragments, every individual has several distinct variants and thus it is very unlikely that any two individuals' MHC molecules will be an exact match. Immune responses are triggered by any departure from normal, either because of a foreign antigen bound to the MHC molecule or because the MHC molecule itself is foreign. It subsequently became clear that there are two classes of MHC molecules, MHC class I molecules and MHC class II molecules (Figure 1-13), that monitor different internal compartments of cells and are recognized by distinct classes of T cells.

MHC molecules are surface indicators of the proteins present in the interior compartments of cells

The two internal compartments of cells that need to be monitored for infection are the cytoplasm, where all viruses and some important bacterial pathogens replicate, and the vesicles of the endosomal/lysosomal pathway, which contain internalized antigens derived from extracellular pathogens, including most bacteria and extracellular virus particles. In the specialized case of macrophages, these compartments may harbor bacteria adapted to survive phagocytosis. The cytoplasmic and vesicular compartments are surveyed by MHC class I and MHC class II molecules respectively.

MHC class I molecules are expressed on virtually all body cells and bind peptides generated by cytoplasmic proteases from cytoplasmic proteins. Their critical role is to display antigens derived from pathogens that replicate in the cytoplasm, and antigen displayed on MHC class I molecules is recognized by cytotoxic T cells, which kill the infected cells (Figure 1-14). This system thus ensures the destruction of cells in which cytoplasmic pathogens are replicating and is believed to be essential to the elimination of some of these pathogens.

Definitions

CD: see **cluster of differentiation**.

cluster of differentiation (CD): the basis of a system for identifying cell surface molecules of immune cells by the use of antibodies and in which each molecule is given a specific number prefixed by CD to form the basis of a systematic nomenclature. The term cluster reflects the fact that each molecule is usually recognized by a group, or cluster of antibodies; and the appearance of the molecules usually reflects different differentiated states of the cell, hence differentiation.

Surface marker molecules of immune cells of different types and at different stages of differentiation or activation have been identified in this way and can be used to classify cells, or to follow their progress through development or their activation status.

coreceptor: (of T lymphocytes) receptor on a T cell that recognizes invariant parts of MHC molecules and forms a recognition complex with the antigen receptor and contributes to intracellular signaling. The coreceptor CD8 binds to MHC class I molecules and is generally expressed on cytotoxic cells; the coreceptor

MHC class II molecules are normally expressed only on cells of the immune system, and in particular on B cells, macrophages and dendritic cells, where they bind peptides generated from internalized antigen in the endosomal compartments of cells. Antigen derived from internalized microorganisms by lysosomal proteases and displayed by MHC class II molecules on dendritic cells is recognized by naïve T helper cells at the initiation of immune responses (Figure 1-15). B cells and macrophages are two of the major targets of effector T helper cell actions. Macrophages, as we have seen, are phagocytic cells that have many receptors that specifically recognize conserved components of microorganisms: these receptors can trigger phagocytosis, and thus many of the peptides returned to the cell surface of a macrophage on MHC class II molecules in the presence of infection are likely to derive from internalized microorganisms. T cells recognizing the complex of microbial peptide and MHC class II molecule activate the microbicidal machinery of the macrophage. This is particularly important for the destruction of bacteria that shelter and proliferate in the internal vesicles of these cells (Figure 1-16). B lymphocytes are not phagocytic, but they internalize, by receptor-mediated endocytosis, antigen bound to their surface immunoglobulin molecules and display fragments derived from this internalized antigen on MHC class II molecules. Helper cells that recognize the displayed antigen then activate the B cell to produce antibodies against it (Figure 1-17).

The differential recognition of MHC class I and class II molecules thus serves to focus the different types of T cells on antigens generated in different compartments of cells and thereby direct their different effector actions to the appropriate targets.

Cell surface markers of different T cell classes reflect the differential recognition of MHC class I and class II molecules

The major classes of T cells are often referred to as CD4 and CD8 T cells. CD4 and CD8 are cell surface molecules that were originally identified by the systematic analysis of lymphocyte surfaces with antibodies: **CD** stands for **cluster of differentiation**, and is a reference to the way in which antibodies have been used in a systematic way to characterize the molecules on the surface of immune cells. The use of antibodies for identifying immune cells of different types, and in different states of differentiation or activation, has been very important in the growth of understanding of the functions of lymphocytes, and we shall see in the course of this book that many important immune molecules were discovered in this way and are still known only by their CD numbers. CD4 and CD8 are among these. They were first recognized as markers of helper and cytotoxic cells, respectively, before any function was assigned to them, but are now known to be **coreceptors** for MHC class II and MHC class I molecules respectively: CD4 recognizes invariant features of MHC class II molecules and CD8 invariant features of class I molecules. T cells acquire these coreceptors as part of their maturation in the thymus, where their fate as cytotoxic or helper cells is also determined, as we shall see in Chapter 7; and they play an important part in signal transduction on antigen binding by T cells, which we describe in Chapter 5.

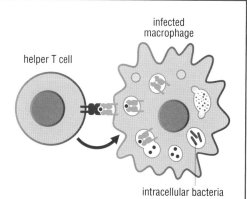

helper T cell

infected macrophage

intracellular bacteria

Figure 1-16 Recognition of antigen and MHC class II molecules on macrophages by helper T cells Helper T cells recognize fragments of bacteria that have been internalized by macrophages and have undergone proteolytic digestion to generate fragments that are carried to the cell surface by MHC class II molecules. The helper cell is then stimulated to activate the macrophage to destroy the internalized bacterium. This is particularly important for defense against bacteria that are able to grow in the internal vesicles of macrophages.

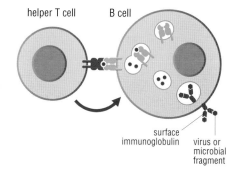

helper T cell   B cell

surface immunoglobulin   virus or microbial fragment

Figure 1-17 Recognition of antigen and MHC class II molecules on B cells by helper T cells B cells internalize antigens bound to their surface immunoglobulin and deliver them to the endosomal compartments of the cell where they are digested into fragments. The antigen fragments then bind to MHC class II molecules which carry them to the cell surface where they are recognized by helper T cells, which then activate the B cell to proliferate and differentiate into an antibody-secreting plasma cell.

CD4 binds to MHC class II molecules and is generally found on helper cells.

major histocompatibility complex (MHC): cluster of genes encoding the *classical* and many *non-classical* **MHC molecules** and other structurally unrelated molecules, many with important functions in immunity.

major histocompatibility complex (MHC) molecules: cell surface glycoproteins encoded in the **major histocompatibility complex** and which bind degraded fragments derived from intracellular proteins and display them on the cell surface.

MHC: see **major histocompatibility complex.**

References

Margulies, D.H.: **The major histocompatibility complex** in *Fundamental Immunology* 4th ed. Paul, W.E. ed. (Lippincott-Raven, New York, 1999), 263–285.

Zinkernagel, R.M. and Doherty, P.C.: **The discovery of MHC restriction.** *Immunol. Today* 1997, **18**:14–17.

1-6 The Lymphoid System and Lymphocyte Circulation

There are two types of lymphoid tissues

The lymphoid system is traditionally divided into primary and secondary lymphoid tissues. **Primary lymphoid tissues** are the tissues in which lymphocytes are generated and differentiate into mature naïve lymphocytes: these are the bone marrow for B cells, and the bone marrow and the thymus for T cells; they will be discussed in Chapter 7. **Secondary lymphoid tissues** are the tissues in which immune responses are initiated, and the **lymphatic vessels** that connect them to the tissues and the bloodstream and thus to sites of infection (Figure 1-18).

Secondary lymphoid tissue brings antigen together with lymphocytes

The adaptive immune system pays two penalties for the extraordinary diversity of the antigen receptors of lymphocytes. First, unlike those on cells of the innate immune system, the antigen receptors of lymphocytes do not distinguish microbial products from harmless ones; and second, only a very small number of lymphocytes express receptors of any given specificity. The first of these disadvantages is overcome by the requirement for cells of the innate immune system, and in particular dendritic cells, that have been activated by microbial products. The second disadvantage is overcome by the specialized architecture of the secondary lymphoid tissues, which are organized to collect antigen and antigen-presenting cells and ensure that each naïve lymphocyte is continually brought into contact with them, so that clonal selection can occur, with expansion in numbers of antigen-specific lymphocytes.

Secondary lymphoid tissues (Figure 1-19) consist of the **lymph nodes**, which are clustered at sites such as the groin, armpits and neck and along the small intestine, and collect antigen from the tissues; the **spleen**, which collects antigen from the bloodstream; and the **mucosa-associated lymphoid tissues (MALT)**, which collect antigen from the respiratory, gastrointestinal and urogenital tracts and are particularly well organized in the small intestine, in structures known as **Peyer's patches**. The lymph nodes are connected to the tissues and the bloodstream by a system of lymphatic vessels. The **afferent lymphatics** drain extracellular fluid (**lymph**) from the tissues, including mucosal tissues, into the lymph nodes; and the **efferent lymphatics** carry the lymph out of the secondary lymphoid tissues and ultimately into a collecting vessel known as the **thoracic duct** (or for lymph nodes in the neck, the *cervical duct*), and thence through the heart and into the bloodstream.

When mature naïve lymphocytes leave the bone marrow and thymus, they enter the bloodstream and then circulate continuously through the lymph nodes, which they enter by migrating through the walls of specialized blood vessels, leaving via the efferent lymphatic vessels, which ultimately return them to the bloodstream. Antigen is collected in secondary lymphoid tissues both directly through the blood, from which antigen reaches the spleen, and through the afferent lymphatic vessels, which drain infected tissues and deliver antigen to the lymph nodes; and more indirectly by three specialized cell types (including dendritic cells) that we discuss in the next section.

Chemokines direct the migration of lymphocytes and determine the specialized microenvironments in which they are activated

Secondary lymphoid tissues provide a highly structured microenvironment adapted to foster the cell–cell interactions required to initiate an adaptive immune response. All secondary lymphoid tissues have distinct regions specialized to support the activation of T cells and B cells, although

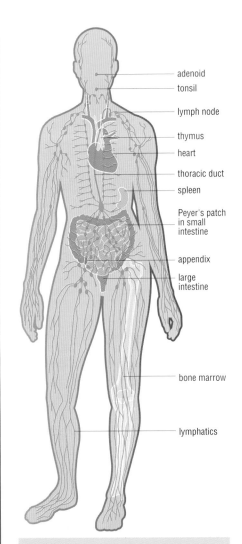

adenoid
tonsil
lymph node
thymus
heart
thoracic duct
spleen
Peyer's patch in small intestine
appendix
large intestine
bone marrow
lymphatics

Figure 1-18 The lymphoid organs The lymphatic vessels and secondary lymphoid tissues are shown in blue, with primary lymphoid organs in yellow.

Definitions

afferent lymphatics: lymphatic vessels entering **lymph nodes** from tissue spaces.

efferent lymphatics: lymphatic vessels leaving **lymph nodes** and returning **lymph** to the bloodstream.

follicle: (of lymphoid tissue) B cell area of **secondary lymphoid tissue**.

germinal center: site of vigorous proliferation of B cells

in the B cell follicles of secondary lymphoid organs.

lymph: fluid drained from the tissues and flowing through **lymphatic vessels**.

lymphatic vessels: system of vessels draining fluid (**lymph**) from the tissues and in which dendritic cells and antigens are delivered to **lymph nodes**.

lymph nodes: secondary lymphoid organs distributed widely in the body but especially in the groin, the axilla and the neck, and along the small intestine.

lymphoid chemokine: constitutively expressed chemoattractant molecule that directs the migration of lymphocytes and dendritic cells into specialized regions of the **secondary lymphoid tissues**. Also known as *homeostatic chemokine*.

MALT: see **mucosa-associated lymphoid tissues**.

mucosa-associated lymphoid tissues (MALT): secondary lymphoid tissue in the walls of the gastrointestinal, respiratory and urogenital tracts.

PALS: see **periarteriolar lymphoid sheath**.

this is most clearly seen in the lymph nodes (see Figure 1-19a). B cells are usually concentrated in relatively well defined areas known as **follicles**, adjacent to or surrounded by T cell areas that may be more diffuse: in the lymph nodes, the T cell area is sometimes called the **paracortical area**; in the spleen, the bulk of which is formed from highly vascularized tissue known as the **red pulp**, the lymphoid tissue is known as the **white pulp** and each T cell area takes the form of a cuff, known as the **periarteriolar lymphoid sheath (PALS)**, round an arteriole from which antigen is delivered (see Figure 1-19b). In lymphoid tissue actively responding to infection, the B cell follicles are enlarged and have a perceptible central region known as the **germinal center** composed of rapidly proliferating B cells.

The specialized architecture of the secondary lymphoid tissues is determined by **lymphoid chemokines**, chemoattractant molecules that are differentially expressed in the T and B cell zones and direct the migration of dendritic cells as well as T and B lymphocytes to these areas. Lymphoid chemokines and other specialized homing molecules also direct the migration of circulating lymphocytes through the specialized blood vessels through which they enter the lymphoid tissues, and regulate their exit, ensuring continued recirculation and surveillance of all lymphoid tissues in which antigen may be present. These processes are described in Chapter 2, and the control of lymphocyte homing is discussed in more detail in Chapters 5 and 6.

Secondary lymphoid tissue is sustained and can be induced by signals from immune cells

The chemokines that establish the architecture of the secondary lymphoid tissues, unlike the chemokines produced in the course of an immune response, are expressed constitutively. Once established however the lymphoid architecture may be maintained in part by normal lymphocyte traffic and interactions. This may explain why, in acquired immunodeficiency disease (AIDS), in which an entire class of T lymphocytes is destroyed, the organized architecture of the lymph nodes collapses. Conversely, at sites of chronic inflammation, organized lymphoid tissue may form: this is known as *tertiary lymphoid tissue*.

Figure 1-19 Secondary lymphoid tissues The specialized sites of T cell activation are colored purple; the sites of B cell activation are blue. In lymph nodes (**a**), spleen (**b**) and the Peyer's patches of the small intestine (**c**), the B cell zones are seen as defined follicles; functionally equivalent areas, not as well defined, are found in all mucosal lymphoid tissues. Proliferating B cells are concentrated in the germinal centers. The T cell regions (purple) are adjacent to the B cell follicles. In the spleen, these form the periarteriolar lymphatic sheaths (PALS) and are surrounded by a region known as the *marginal zone*, which we describe in the next section. In the lymph nodes, the T cell zones are separated from the medullary sinus, which forms the exit route into the efferent lymphatic vessels, by a region known as the *medullary cords*, which are rich in macrophages and plasma cells secreting large quantities of antibody.

(a) Lymph node

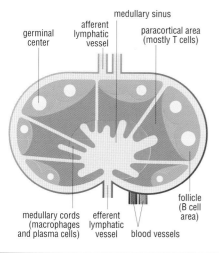

(b) Spleen

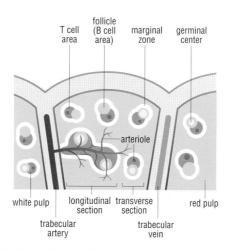

(c) Peyer's patch

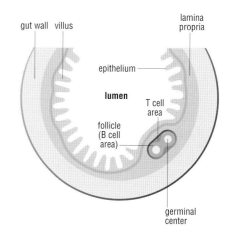

paracortical area: (of **lymph node**) area beneath the outer regions of the lymph node and where T lymphocytes accumulate.

periarteriolar lymphoid sheath (PALS): T cell area in the **spleen**, formed around arterioles.

Peyer's patches: organized regions of **secondary lymphoid tissue** in the wall of the small intestine.

primary lymphoid tissues: bone marrow and thymus, in which lymphocytes differentiate and mature.

red pulp: site of destruction of senescent red blood cells in the **spleen**.

secondary lymphoid tissues: tissues in which lymphocytes are brought together with antigen and adaptive immune responses are initiated.

spleen: secondary lymphoid organ in the abdomen collecting antigen from the bloodstream.

thoracic duct: main vessel collecting **lymph** for delivery to the heart.

white pulp: lymphoid area of the **spleen**.

References

Cyster, J.G.: **Chemokines, sphingosine-1-phosphate, and cell migration in secondary lymphoid organs.** *Annu. Rev. Immunol.* 2005, **23**:127–159.

Picker, L.J. and Siegelbaum, M.H.: **Lymphoid tissues and organs** in *Fundamental Immunology* 4th ed. Paul, W.E. ed. (Lippincott-Raven, New York, 1999), 449–531.

Dendritic cells and follicular dendritic cells collect antigen for recognition by T cells and B cells

The T cell and B cell zones of secondary lymphoid tissues are populated by networks of cells specialized to capture and present antigen for recognition by these two types of lymphocytes. In the T cell zones these cells are dendritic cells, whose origins and functions we have already described. Most of the dendritic cells in secondary lymphoid tissues are resident immature cells, which can capture antigen entering the lymphoid organ in blood or lymph, or directly from the lumen of mucosal tissues (see below), and then mature and display antigen fragments on surface MHC molecules for recognition by T cells. In peripheral lymph nodes, soluble antigen is thought to leak from conduits crossing the lymphoid tissue from sinuses into which lymph flows from the afferent vessels, and reach dendritic cells that cluster round the conduits. More important for T cell activation, however, are dendritic cells that have captured antigen and matured in the periphery at sites of infection and then migrate into the T cell zones of the lymphoid tissue in the afferent lymph. They converge with T cells arriving in the T cell zone through specialized blood vessels, the **high endothelial venules (HEV)**, to enter the lymphoid tissues.

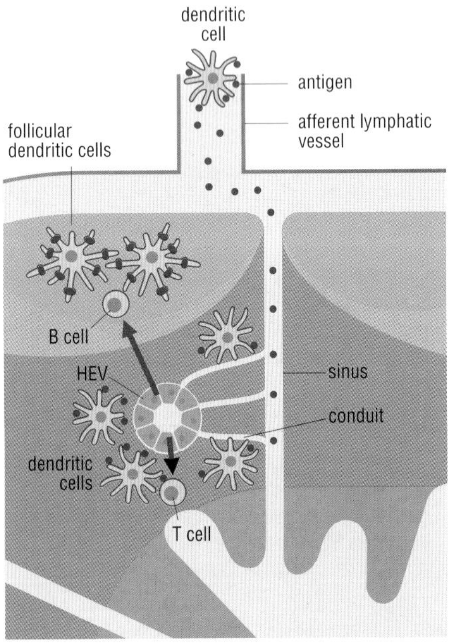

Dendritic cells are localized to the T cell areas and display antigen as peptide fragments bound to surface MHC molecules, for recognition by T lymphocytes. **Follicular dendritic cells**, which display antigen for recognition by B lymphocytes, are quite distinct both in lineage and in function from the dendritic cells that present antigen to T cells, although like them, they extend long processes and that is why they too are called dendritic cells. They are non-hematopoietic cells that, as their name suggests, are localized in the B cell follicles, where they display intact antigen, in complexes with complement components or antibodies that bind to receptors on the follicular dendritic cell surface. They play a central part in a process known as *affinity maturation* of antibody responses. Affinity maturation results in the production of antibodies with increased affinity for antigen and reflects a process in which the immunoglobulin genes undergo mutation and B cells with higher-affinity immunoglobulin resulting from these mutations are selectively induced to proliferate. This selective process occurs in the germinal centers of the B cell follicles, where B cells compete to bind to antigen trapped on follicular dendritic cells and are stimulated to proliferate, both by antigen binding and by signals from helper T cells that migrate into the follicle. We discuss this process in detail in Chapter 6.

The distribution of dendritic cells and follicular dendritic cells, and the distinct mechanisms of antigen capture in secondary lymphoid tissue, are illustrated for the lymph node in Figure 1-20.

The spleen is a critical filter for antigen in blood

The spleen is not fed by afferent lymphatic vessels and contains no high endothelial venules, and antigen enters in the blood through the splenic artery, along with the immune cells that populate it. The spleen is a highly vascularized organ one of whose major functions is the clearance of senescent or damaged red blood cells by specialized macrophages in the red pulp. The lymphoid tissue of the spleen, the white pulp, forms round arterioles that traverse the red pulp (Figure 1-21) and is surrounded by a specialized zone, the **marginal zone** (see Figure 1-21), that contains the **metallophilic** and **marginal zone macrophages** we have described in section 1-3, and specialized B cells that are adapted to produce rapid responses to bloodborne bacteria. These cells, and the immune cells of the white pulp, are fed by small branches from the arterioles that cross the white pulp to end in the **marginal sinus**, which runs between the marginal zone and the T cell area and delivers antigen through a leaky endothelium to the cells of the marginal zone.

Figure 1-20 Antigen capture in T and B cell areas of the lymph node Antigen enters lymph nodes through the afferent lymphatic vessels (of which there may be several for each lymph node), which empty into sinuses that are connected by specialized conduits to the high endothelial vessels (HEV), through which lymphocytes enter. Antigen leaking from the conduits may be captured by dendritic cells already present in the T cell areas of the lymph node, and often adhering to the conduits. Alternatively, dendritic cells may capture antigen in infected tissues in the periphery and migrate in the afferent lymphatic vessels into the draining lymph node, where they accumulate in the T cell areas. Follicular dendritic cells are localized to the B cell areas and capture antigen that is delivered to the lymph node in complexes with complement or antibodies, and display the antigen intact.

Definitions

follicular dendritic cell: specialized non-hematopoietic cell found in B cell areas of secondary lymphoid tissue and specialized for collecting antigen bound to antibody or complement components.

HEV: see **high endothelial venules**.

high endothelial venules (HEV): specialized blood vessels supplying lymph nodes and expressing adhesive molecules and chemokines specifically recognized by naïve circulating lymphocytes, which are thereby enabled to enter the secondary lymphoid tissue.

marginal sinus: (of spleen) blood vessels running between the **marginal zone** and the white pulp of the spleen.

marginal zone: narrow region at the outer boundary of the splenic lymphoid tissue, bounded by the **marginal sinus** and red pulp.

marginal zone macrophage: specialized macrophage in the **marginal zone** of the spleen thought to be important in resistance to bloodborne pathogens.

M cells: specialized epithelial cells in the small intestine that collect antigen at Peyer's patches.

metallophilic macrophage: specialized macrophage in the **marginal zone** of the spleen thought to be important in scavenging debris.

Figure 1-21 Specialized features of the spleen Antigen is delivered to the spleen directly from the bloodstream. Branches from the splenic artery, the trabecular arteries, run between the lobules of the spleen and branch in turn into arterioles ensheathed by lymphoid tissue that is surrounded by a specialized zone, the marginal zone. The marginal zone is separated from the periarteriolar lymphoid sheath by a vessel, the marginal sinus, into which smaller arterioles empty after traversing the lymphoid tissue and is the entry conduit for lymphocytes into the spleen. It contains specialized marginal zone macrophages, which are thought to help capture bloodborne pathogens and scavenge debris from blood, as well as marginal zone B cells, which are specialized for rapid antibody responses. The marginal sinus is lined on the other side by metallophilic macrophages, which are also thought to be important scavengers. The schematic drawing here represents the mouse marginal zone, which is the best studied. The marginal zone in humans is distinct in detail but is functionally equivalent.

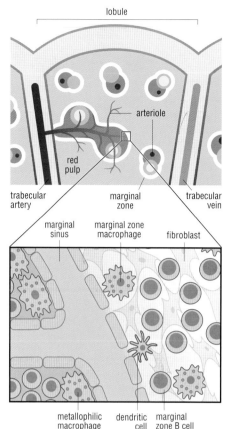

Whereas afferent lymphatic vessels deliver antigen only to local lymph nodes, the major function of the blood is to deliver oxygen and nutrients to tissues throughout the body, and pathogens entering the bloodstream may be rapidly and widely disseminated. Moreover immune responses occurring in the bloodstream are especially dangerous because of the effects of inflammatory cytokines on the permeability of blood vessels (see section 1-0), which if widely induced cause shock and circulatory collapse. We discuss the consequences of disseminated infection in Chapter 9.

Epithelial and tissue defenses ensure that antigen does not readily enter the bloodstream: the critical function of the spleen, which contains up to 25% of the body's lymphocytes, is to provide effective protection from pathogens breaching these defenses. In rare cases in which an individual is born without a spleen, or in more common cases in which the spleen is ruptured through injury and must be quickly removed to prevent bleeding into the abdominal cavity, individuals are susceptible to microorganisms to which they do not already have antibodies and require prompt antibiotic treatment to avoid fatal infection.

Antigen enters mucosal lymphoid tissues through specialized cells

Exposure to antigen in mucosal lymphoid tissues is directly from the lumen of the mucosal organ. The lumen of the small intestine in particular is constantly fed with antigenic material and is teeming with bacteria, and this is likely to be why the lymphoid tissues of this region are particularly highly organized. Antigen is delivered to mucosal lymphoid tissues through specialized epithelial cells, the **M cells**, which transport antigenic particles that can be as large as viruses and bacteria across the mucosal epithelium and deliver them to an adjacent layer of dendritic cells (Figure 1-22). The B cell follicles of gut lymphoid tissue are constitutively active, with germinal centers that reflect vigorous B cell proliferation, again probably because of the constant bombardment of the gut epithelium with ingested antigen and microorganisms. For the same reason, the small intestine is generously provided with lymph nodes, the *mesenteric lymph nodes*, that constitute the largest group of lymph nodes in the body. We shall see in Chapter 10 that the high level of activated immune cells in the lymphoid tissues of the gut is exploited by the human immunodeficiency virus (HIV) that causes AIDS and is specialized to proliferate in activated T cells.

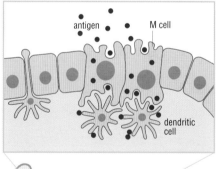

Figure 1-22 Antigen delivery into the mucosal lymphoid tissue of the gut Antigen is delivered to the lymphoid tissues of the gut by M cells, specialized epithelial cells that transport particles from the lumen into an underlying layer of dendritic cells. Note that B cell follicles in the gut are constitutively activated, as evidenced by germinal centers that reflect B cell proliferation. The small intestine (illustrated here) is liberally supplied with lymph nodes, the mesenteric lymph nodes.

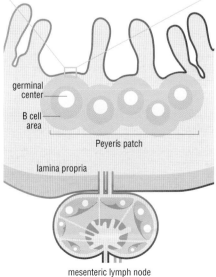

References

Chaplin, D.D.: **Lymphoid tissues and organs** in *Fundamental Immunology* 5th ed. Paul, W.E. ed. (Lippincott Williams & Wilkins, Philadelphia, 2003), 419–453.

Mebius, R.E. and Kraal, G.: **Structure and function of the spleen.** *Nature Rev. Immunol.* 2005, **5**:606–616.

Rosen, F. and Geha, R.: **Congenital asplenia** in *Case Studies in Immunology. A Clinical Companion* 4th ed. (Garland Science, New York, 2004), 1–6.

Steinman, R.: **Dendritic cells** in *Fundamental Immunology* 4th ed. Paul, W.E. ed. (Lippincott-Raven, New York, 1999), 547–573.

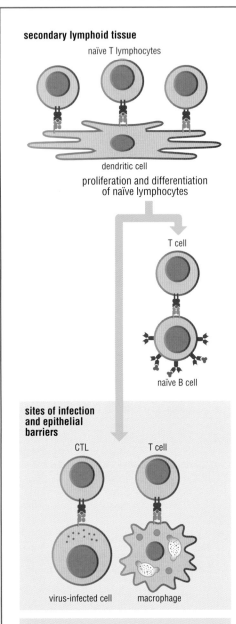

secondary lymphoid tissue

naïve T lymphocytes

dendritic cell

proliferation and differentiation
of naïve lymphocytes

T cell

naïve B cell

sites of infection
and epithelial
barriers

CTL   T cell

virus-infected cell   macrophage

**Figure 1-23 The initiation of an adaptive
immune response** Circulating naïve
T lymphocytes are arrested by binding to
antigen on dendritic cells in the T cell areas
of the secondary lymphoid organs and are
activated to proliferate. The progeny of each
rare antigen-specific lymphocyte then
differentiate into effector cells specific for the
same antigen. Effector T cells migrate from the
lymphoid organs to sites of infection in the
periphery and to epithelial barriers.

Innate immunity and specialized classes of lymphocytes provide front-line defense against invading microorganisms

We coexist peacefully with many species of microorganisms that harmlessly inhabit our skin and mucosal surfaces, especially those of the gut, and we shall see in Chapter 9 that the normal bacterial population at these surfaces is not merely harmless but important in controlling potentially harmful invaders. Pathogens, generally, do not remain at these barriers but invade the underlying tissues and cells, where some produce toxins that are directly harmful, or kill the cells they inhabit, while the disease caused by others is largely due to the damaging effects of the immune response itself. Microorganisms can gain entry to tissues through accidental damage to the barriers that normally restrain them, but most pathogens have specialized devices for entering the tissues or cells that they invade: these are discussed in Chapters 9 and 10 where the infectious strategies of bacteria and viruses are described.

Invading pathogens are met by a substantial armory of innate immune devices by which they can be repelled or destroyed. Phagocytes directly recognize and destroy many, while others are tagged by soluble molecules that coat them for recognition and destruction in the same way. Microorganisms entering cells are recognized by intracellular detectors that block their proliferation or trigger defensive or destructive reactions. All these devices are described in Chapter 3. Specialized classes of T lymphocytes that permanently populate epithelial barriers are equipped with invariant receptors and armed with effector molecules for immediate detection and destruction of infected cells; and specialized B lymphocytes in the spleen and body cavities are preprogrammed to produce antibodies of limited diversity in response to characteristic surface features of viruses and bacteria, and prevent these pathogens from establishing infection. These specialized classes of lymphocytes have properties that lie between those of the innate immune system and those of the adaptive immune system, and have only recently begun to be understood; they are discussed in Chapter 8. While these early responses provide immediate defense against infection, dendritic cells migrate to the secondary lymphoid organs where they will converge with recirculating lymphocytes to initiate the adaptive immune response.

T lymphocytes activated by dendritic cells differentiate and migrate to B cell follicles and sites of infection

The activation of T lymphocytes at the initiation of an adaptive immune response occurs at the surface of antigen-presenting dendritic cells in the T cell zones of the secondary lymphoid tissues. T cells circulating through the lymphoid tissues are arrested at the surface of the dendritic cell, and are activated through receptors for antigen and for specialized activating molecules on the dendritic cell surface to proliferate and differentiate into effector cells (Figure 1-23).

Among the important changes to the surface of effector T lymphocytes are changes in the expression of receptors for chemokines that direct their migration. The receptors for constitutively expressed chemokines that direct recirculation through the secondary lymphoid tissues are downregulated in favor of receptors for inflammatory chemokines that are induced by infection and direct the cells to sites of infection in the periphery, or receptors for chemokines expressed in the B cell follicles where T helper cells engage with activated B cells.

Most T cell effector actions require direct interactions with other cells. They are triggered when T cells recognize antigen displayed on the surface of the target cell, and involve the release of soluble mediators acting on the target cell. In the case of CD8 cytotoxic T cells, these are cytotoxic mediators that kill cells infected with microorganisms such as viruses and that betray their

Definitions

affinity maturation: the property of an immune response in which the average affinity of antibodies produced against an antigen increases as the response continues. This occurs over a period of several weeks.

memory cells: long-lived lymphocytes that differentiate during the clonal expansion of antigen-specific lymphocytes during a primary immune response and provide a rapidly activated effector pool on subsequent challenge with the same antigen.

presence through protein fragments that are carried to the surface of the infected cell and displayed there. Some helper T cells recognize protein fragments derived from microorganisms ingested by macrophages, and release cytokines that increase the bactericidal activity of these cells. Other helper T cells recognize protein fragments derived from antigen bound by surface immunoglobulin and internalized by B cells, and release cytokines that activate B cells to proliferate and differentiate into antibody-secreting plasma cells. As well as these focused effects on cells with which they directly interact, all types of T cells help to recruit innate effector cells appropriate for the destruction of the pathogen they specifically recognize: the differential tailoring of the adaptive immune response in this way is described in Chapter 5.

Activated B lymphocytes differentiate into plasma cells and secrete antibody

Naïve B lymphocytes, like naïve T lymphocytes, circulate continuously through the secondary lymphoid organs in the blood and lymph until they are arrested by binding to antigen. B cell responses to most antigens, including all protein antigens, require signals from T cells. These are initiated at the boundary of the B cell and T cell zones when B cells specifically bind antigen on their surface immunoglobulin molecules and internalize it, returning fragments to the cell surface where they are recognized by antigen-specific T cells which are thereby triggered to activate the B cell (Figure 1-24).

Some B cells proliferate in the region between the T cell zone and the B cell follicle and differentiate there into antibody-secreting plasma cells; others migrate into the lymphoid follicles and establish germinal centers—sites of rapid B cell proliferation within which changes occur in the immunoglobulin genes of the dividing cells. Some of these are mutations affecting the antigen-binding regions that can result in higher affinity for antigen, for which the B cells are then selected: this is known as the **affinity maturation** of antibodies. Others affect the effector actions of the antibodies secreted by the plasma cells that differentiate from germinal center B cells. These are specifically directed by the cytokines secreted by helper T cells and result in the production of antibodies of distinct *classes* determined by their constant regions (see Figure 1-10) which are of distinct types, or *isotypes,* with distinct effector properties adapted to provide defense of different types at different sites. The major classes of immunoglobulin, which will be described in detail in Chapter 6, are known as *immunoglobulin M (IgM)*, which is produced initially by all B cells and is especially important in the arrest of bloodborne infection; *immunoglobulin G (IgG)*, which is the predominant class in blood and tissues and comprises four distinct isotypes with different properties; *immunoglobulin A (IgA)*, which is specialized for protection of mucosal surfaces; and *immunoglobulin E (IgE)*, which is critical to defense against parasites. Like effector T cells, plasma cells migrate to sites appropriate to their functions, and depending on the effector properties of the antibodies they secrete may differentially home to mucosa or bone marrow.

At the end of an immune response, the vast majority of the antigen-specific cells die and surviving cells become resting memory cells

Immune responses are destructive and self-amplifying. In many diseases it is the immune response to infection and not the infectious organism itself that is the cause of the pathology. It is therefore not surprising that mechanisms have evolved for ensuring that they are also self-limiting. In the vast majority of cases, activation of an effector T lymphocyte by its target cell results in the death of the lymphocyte within a few days, usually through the induction of programmed cell death (*apoptosis*). Although some plasma cells migrate to the bone marrow and can persist there for years, many secrete antibody for only two or three days before undergoing apoptosis. The ability of the immune system to respond rapidly on reexposure to a given microorganism depends on long-lived plasma cells and the differentiation of a distinct subset of **memory cells** that join the recirculating lymphocyte population and assume effector functions immediately on reencounter with antigen. Memory responses are different in a number of important ways from the initial response to antigen. First, because the population of antigen-specific cells has already undergone expansion and most of the changes required for effector function, they are much quicker. Second, in the case of B cells, the antibodies have undergone affinity maturation and are more effective than those produced in a primary response. For these reasons, a second or subsequent infection with a given microorganism usually does not lead to disease. The development of vaccines that safely and efficiently induce immune memory is one of the most important goals of modern immunological research.

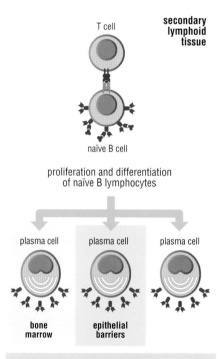

Figure 1-24 The induction of an antibody response by T cells Circulating naïve B lymphocytes are arrested by binding to antigen in the T cell areas of the secondary lymphoid organs. Antigen bound to surface immunoglobulin is internalized and fragments are returned to the surface of the B cell, where they are recognized by T helper cells which activate the B cell to proliferate. The progeny differentiate into plasma cells secreting antibody with the same antigen specificity as the surface immunoglobulin of their progenitor, and then disperse, some to the medulla of the lymph node, some to the mucosa, and some to the bone marrow from which antibodies can readily reach the bloodstream.

2

Signaling and Adhesive Molecules of the Immune System

Many central functions of the immune system depend on molecules belonging to structural families with related functions and properties. In this chapter, a number of these families are described along with the properties whereby they contribute to microbe recognition, intercellular communication, cell adhesion and cell migration. Individual members of these families reappear in many of the chapters of this book, and an understanding of the basic properties of a family provides the basis for an understanding of how individual members are adapted to their function in immune responses.

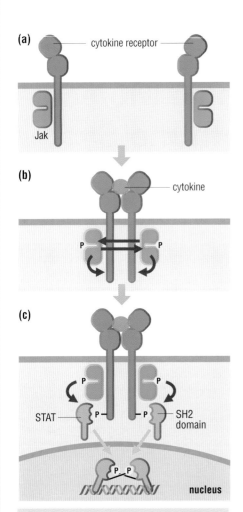

Figure 2-1 Signaling by phosphorylation
Some cytokine receptors have kinases associated with their cytoplasmic domains **(a)** and oligomerize on binding cytokine, bringing the two kinases into proximity and allowing them to phosphorylate and thereby activate one another **(b)**. They then phosphorylate the cytoplasmic domains of the receptor dimer, which acts as a scaffold recruiting downstream components in the pathway, including transcription factors called STATs. STATs bind to these phosphorylated sites through their SH2 domains and are then themselves phosphorylated by the kinases **(c)**. This leads to the dissociation of the STATs from the receptor, and dimerization through their SH2 domains: once dimerized they are able to bind DNA and activate the transcription of cytokine-regulated genes.

Immune cell migration and communication are controlled by specialized cell surface and signaling molecules

The hallmark of the immune system is its ability to recognize infectious agents and remove them, but this turns out to be an extraordinarily difficult charge. Infectious agents may enter the body in many places and replicate very rapidly, requiring rapid mobilization of immune cells to that location. Immune responses must be specialized to meet the challenge of the type of organism that is replicating, and the activity of the immune cells that are recruited to the site of infection must be exquisitely controlled to achieve optimum immune defense while minimizing damage to nearby healthy tissue. In the first chapter we have introduced the cells that are involved in immunity and their locations in the body. In this chapter we discuss key families of molecules that coordinate their function, concentrating on cell-surface molecules and intercellular communication molecules such as cytokines and chemokines. Strikingly, many of these molecules have been duplicated numerous times, leading to formation of extensive families of related molecules with highly related properties and functions. These molecules are widespread and pervasive and will be discussed in many places in this book. Understanding their general properties will make it easier to understand the functions of individual family members in particular contexts.

In the process of describing these important and pervasive families of molecules, we shall explain how their properties relate to their functions in the immune system, and therefore this chapter also describes some of the mechanisms of these functional properties, in particular, the regulation of cell–cell adhesion and the modulation of cell behavior by signaling pathways. Finally, both of these processes, as well as several families of molecules, combine in the mechanism by which immune cells leave the blood and enter tissues; how these various elements interrelate in this process is the subject of the final section of this chapter.

Adhesion molecules have important roles in promoting cell migration and cell–cell interaction

Many processes in immunity depend on adhesive interactions between cells. Migration of cells out of the blood and into lymph nodes or sites of infection depend on adhesive interactions with blood vessels; adhesive interactions between immune cells are essential during immune responses; and adhesion is important for internalization of infectious agents by phagocytes, which is one of the primary effector mechanisms by which microbes are killed.

In this chapter we shall describe four distinct types of cell–cell adhesion molecules and contrast their molecular properties. Of especial importance are the *integrins*, because their adhesive properties can rapidly be controlled to a very high degree by other receptors on the cell. For example, T cells can scan other cells through weak interactions until they recognize their antigen–MHC ligands, which leads to strengthened adhesiveness through integrins, promoting activating interactions between the two cells. Cell-surface protein ligands of integrins are generally members of the *immunoglobulin (Ig) superfamily*, which is a large and diverse group of proteins on the surface of many cells, but especially immune cells, named for their amino-acid and structural similarities to immunoglobulins. Members of a separate subfamily of the Ig superfamily, the CD2 subfamily, mediate adhesion between immune cells based on interactions between the same molecule on both cells or between different CD2 subfamily members on the two cells. These interactions are not as tightly regulated as integrin-mediated cell–cell adhesion, but they are also not as strong, so they are well suited for promoting the antigen scanning function of T cells mentioned above. Finally, a fourth very important group of cell adhesion molecules is composed of proteins with homology to the *C-type lectins*, which are carbohydrate-

Definitions

adaptor: component of a signaling pathway lacking enzymatic function but participating in signaling reactions by binding to other signaling molecules and thereby bringing them together.

GAP: see **GTPase-activating protein**.

GEF: see **guanine nucleotide exchange factor**.

G protein: any of a large class of GTPases that act as molecular switches that are active when bound to

GTP and inactive when bound to GDP. They may be heterotrimeric G proteins with α, β and γ subunits, which typically signal from seven-transmembrane receptors, or small G proteins of the Ras superfamily. G proteins are also known as **GTP-binding proteins** or guanine-nucleotide-binding proteins.

GTPase-activating protein (GAP): protein that accelerates the intrinsic GTPase activity of **G proteins**, thereby inactivating them.

GTP-binding protein: see **G protein**.

guanine nucleotide exchange factor (GEF): protein that facilitates the exchange of GDP for GTP in **G proteins**, thereby activating them.

protein tyrosine kinase: enzyme that phosphorylates tyrosine residues on other proteins and/or on itself.

protein tyrosine phosphatase: enzyme that removes phosphate groups from phosphorylated tyrosine residues on proteins. Protein tyrosine phosphatases oppose the action of **protein tyrosine kinases**.

scaffold: protein that binds two or more other proteins

recognizing proteins that, like the integrins, are important for directing migration through interactions with host cells, and for stimulating phagocytosis through recognition of microbes. Still other members of this family are primarily signaling receptors rather than adhesion receptors and some of these bind protein ligands rather than carbohydrate ligands.

Large families of signaling molecules control nearly all aspects of immune system development and function

Exquisite regulation is a hallmark of immune reactions, and cell-surface signaling receptors have a major role in this regulation. Among the main types of molecules that are central to immune system signaling are the Ig superfamily cell-surface receptors, the cytokines and their receptors, and the chemotactic cytokines or chemokines and their receptors.

Central to the function of signaling receptors is their mode of signal transduction. In general, related receptors signal by very similar mechanisms, as described in this chapter for the receptor families most commonly deployed in immune cells. Nonetheless, different receptors from the same family and expressed on the same cell can produce different responses. This occurs, for example, with cytokine receptors, because of distinct elements in the related signaling pathways they activate. In addition, it should be noted that a single receptor signaling mechanism can be translated into different biological responses by different cells, so that the ultimate outcome of signaling reactions is typically dependent on the differentiation state of the cell.

Signaling by most receptor types is mediated by the formation of multiprotein complexes that are induced on ligand binding by the receptor. These complexes contain signaling enzymes and their regulators or substrates, often brought together by molecules that have no signaling function themselves but serve as **adaptors** bringing together two or three other signaling molecules, or **scaffolds** that provide a platform on which several signaling molecules may aggregate. The resulting multiprotein complexes convey positional information to the cell that can direct exocytosis or migration. They also localize signaling enzymes near their substrates and in that way promote and may greatly amplify signaling events.

One of the most prominent intracellular signaling enzyme families, which we shall encounter in several signaling pathways in this book, is that of the **protein tyrosine kinases**, which attach phosphate groups to specific sites on other proteins and thereby either activate them or provide binding sites whereby further signaling components may be recruited, usually through phosphotyrosine-binding domains known as **SH2 (Src-homology 2) domains**. An example of such a pathway is shown for cytokine signaling in Figure 2-1. Tyrosine kinase-based signaling is opposed by **protein tyrosine phosphatases** that remove the phosphorylations. In other cases, signaling may be inactivated by the targeted degradation of specific components.

In many signaling pathways, and particularly those involving regulation of the cytoskeleton, small **G proteins**, or **GTP-binding proteins**, represent key molecular switches. We shall encounter one family of these, the Rho-family GTPases, when we discuss mechanisms of cell migration and phagocytosis. G proteins signal when they bind GTP and stop signaling when they hydrolyze the bound GTP to leave bound GDP. Receptors typically turn on signaling GTPases by localizing them together with **guanine nucleotide exchange factors (GEFs)** that induce a conformational change allowing exchange of bound GDP for GTP (Figure 2-2). Conversely, GTPases are often opposed by **GTPase-activating proteins (GAPs)**, which activate GTP hydrolysis, thereby inactivating the G protein. GTPases can regulate a wide variety of signaling events, including activation of protein kinases, as well as cytoskeletal remodeling.

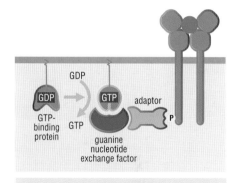

Figure 2-2 Recruitment of a GTP-binding protein by receptor signaling In this example, a phosphate group on the cytoplasmic domain of an activated receptor recruits an adaptor molecule that binds to a guanine nucleotide exchange factor, bringing it to the membrane and thereby into proximity with an inactive, GDP-bound GTP-binding protein that is held in the membrane by a lipid tail (GTP-binding proteins are commonly localized to the membrane in this way). The GTP-binding protein is then recruited to the guanine nucleotide exchange factor, which induces the release of bound GDP and binding of GTP, which activates the GTPase signaling molecule.

and increases the efficiency or specificity with which they act upon each other.

SH2 (Src-homology 2) domain: globular domains of slightly less than 100 amino-acid residues that have binding pockets for tyrosine-phosphorylated regions of proteins. Binding is generally dependent upon phosphorylation and the particular amino-acid residues found immediately downstream of the phosphorylation site. Specificity for the adjacent sequences can vary between different SH2 domains.

References

Bourne, H.R. *et al.*: **The GTPase superfamily: conserved structure and molecular mechanism.** *Nature* 1991, **349**:117–127.

Etienne-Manneville, S. and Hall, A.: **Rho GTPases in cell biology.** *Nature* 2002, **420**:629–635.

Mitin, N. *et al.*: **Signaling interplay in Ras superfamily function.** *Curr. Biol.* 2005, **15**:R563–R574.

Pawson, T.: **Specificity in signal transduction: from phosphotyrosine-SH2 domain interactions to complex cellular systems.** *Cell* 2004, **116**:191–203.

Pawson, T. and Scott, J.D.: **Protein phosphorylation in signaling – 50 years and counting.** *Trends Biochem. Sci.* 2005, **30**:286–290.

The Ig-like domain is an important element of structure for many proteins of the immune system

Many cell-surface and secreted recognition molecules of the immune system belong to a family of structurally related proteins called the **immunoglobulin (Ig) superfamily**, which contain a characteristic stably folded region of protein structure, the **Ig-like domain**. The domain was first recognized in early studies of the structure of antibody molecules in which it was observed that antibodies subjected to protease cleavage in appropriate conditions generate discrete protease-resistant cleavage fragments representing the different domains of the protein. Subsequently it was found that the domains were all made up of related sequences of about 110 amino-acid residues in length, now known as **Ig domains**.

Ig domains and their counterparts in other Ig superfamily proteins fold into related three-dimensional structures, composed of two beta sheets, each with three to five beta strands, and with the flat sides of the two sheets stacked against each other, referred to as **beta sandwich** structures (Figure 2-3a). Ig-like domains can be further subdivided on the basis of subtle details of the structure into C-type, V-type (based on the types found in antibody **constant (C)** and **variable (V) domains**; see Figure 1-10), and a third type, called the S (switched) type, in which one of the beta strands is located on the opposite side of the sandwich (Figure 2-3b).

Usually the Ig-like domain has a conserved disulfide bond linking the two beta sheets of the sandwich. This disulfide bond contributes to the very stable nature of Ig domains. Its stable nature makes the Ig domain useful for many functions in the extracellular environment, which is more destructive to proteins than that inside the cell; the most striking example is the antibody molecule, which can function in hostile environments such as the gut lumen. Perhaps for this reason, with rare exceptions, Ig superfamily members have their Ig-like domains located outside the cell.

The Ig superfamily contains a diverse group of molecular recognition elements of the immune system

In the immune system, Ig superfamily members have a disproportionately important role: the family includes the major antigen recognition molecules of adaptive immunity, the antibody and the T cell antigen receptor (TCR), as well as many other molecules with important immune system functions (Figure 2-4). The Ig superfamily is not restricted to immune function, however, and includes many molecules that function in other aspects of cell biology. It is currently estimated that more than 500 different proteins in the human genome contain Ig-like domains, although not all are properly called Ig superfamily members, since this family is limited to proteins that, like antibodies, have Ig-like domains as their primary structural motif and does not include proteins with an Ig-like domain and multiple other types of domains.

(a)

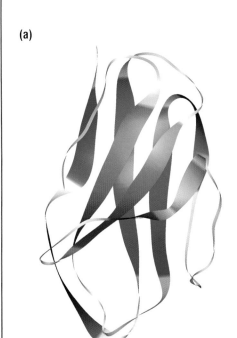

(b)

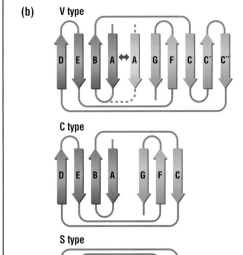

Figure 2-3 **The Ig-like domain (a)** Ribbon diagram of an Ig domain of the variable type. The beta strands of one sheet are shown in green and those of the second sheet shown in blue. **(b)** Two-dimensional topology tracings of the V domain, the C domain and the S domain variants of the Ig-like domain. In the C type there are two fewer beta strands than in the V type. The beta strands of the C-type domain are named A to G from amino terminus to carboxyl terminus. The two extra strands of the V-type domain are located between C and D and therefore are called C' and C''. The S or switched domain has the fourth strand switched from one beta sheet to the other. For this reason this domain has a C' strand and no D strand. The B and E beta strands in one sheet and the F and C strands in the other represent the core of the domain structure; other beta strands are on the edge of the molecule and are absent from some Ig superfamily members. The dimensions of the Ig domain are about 2 nm × 2.5 nm × 4 nm. (Panel a PDB 1b6d)

Definitions

beta sandwich: domain structure in which the flat sides of two beta sheets are stacked against one another, like two pieces of bread in a sandwich. **Ig domains** fold into this type of structure.

constant (C) domain: Ig-like domain of the type found in Ig constant regions.

framework residues: conserved amino-acid residues observed upon comparing Ig or TCR V regions with one another. These residues are involved in maintaining the structure of the **Ig domain**.

Ig domain: stably folded region of an immunoglobulin, as defined originally by protease resistance.

Ig-like domain: domain of protein structure with amino-acid sequence and structural homology to **Ig domains**. Ig-like domains can be identified by amino-acid sequence homology to those found in antibody molecules. Ig-like domains from diverse proteins seem to all fold into beta-sandwich structures, referred to as the immunoglobulin fold.

immunoglobulin (Ig) superfamily: family of proteins containing at least one immunoglobulin-like domain and for which the **Ig-like domain** is a major structural element. MHC proteins have an Ig-like domain, but their main structural element is a unique peptide-binding domain and therefore, it is best to consider MHC molecules as a separate family of proteins.

variable (V) domain: Ig-like domain of the type found in immunoglobulin variable regions. V domains have two more beta strands in the **beta sandwich** than do **C domains**.

A significant feature of Ig domains is that the amino terminus and the carboxyl terminus of the polypeptide chain are at opposite ends of the folded domain, so proteins composed of tandem Ig-like domains can evolve very easily by gene duplication, because attaching the polypeptides of the domains end to end does not disrupt the folded structure. As a consequence, Ig superfamily members range from those with a single Ig-like domain to those with many (see Figure 2-4).

Often Ig-like domains are found in homodimeric or heterodimeric proteins in which pairs of Ig-like domains are associated with one another. This is the case for the various domains of antibodies and T cell receptors, which are both heterodimers. Antigen binding by these molecules occurs in the paired amino-terminal domains, referred to as the variable domains because their amino-acid sequences vary considerably between different antibodies or between different TCRs. Binding of antigen occurs at the very end of these molecules, where loops connecting the beta strands of the two variable domains combine to interact with the antigen. Variations in the structures of the loops are tolerated within variable domains of antibodies and TCRs and in Ig-like domains generally because they do not affect the stability of the molecule which depends on the beta sandwich that forms the structural core. The amino-acid residues forming the core are highly conserved between Ig-like domains and allow their identification from amino-acid sequence data. In antibody molecule variable regions, such amino acid residues are often called **framework residues**. Similarly, in other Ig superfamily members, conserved residues form the basic beta-sandwich structure, but the surface residues can be varied to specify protein–protein and in some cases protein–carbohydrate interaction sites. Molecular recognition by Ig superfamily proteins can result from interactions on any of the six sides of the Ig domain sandwich; it is not always mediated by the loops at the amino-terminal end as it is in antibodies and TCRs.

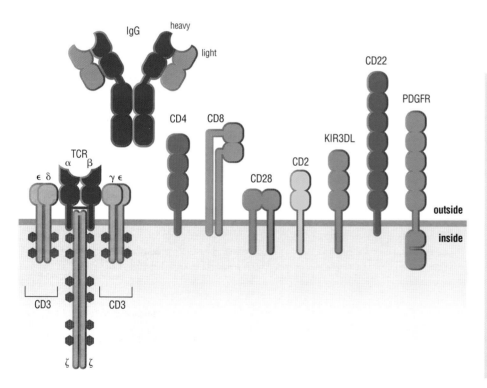

Figure 2-4 Representative examples of Ig superfamily members Ig domains are contained in at least 75 different combinations with other protein modules and over 500 distinct proteins in humans. Shown here are several different types, including an antibody; the T cell antigen receptor, which has transmembrane signaling chains some of which are also immunoglobulin related; the T cell coreceptors CD4 and CD8, which contribute to T cell signaling; a T cell costimulatory receptor (CD28), which receives critical activation signals from dendritic cells; an adhesion molecule (CD2); inhibitory receptors (KIR3DL and CD22); and a tyrosine kinase growth factor receptor (platelet-derived growth factor receptor, PDGFR). Ig domains are shown as squared off ovals here and throughout the book; the PDGF receptor also has a kinase domain in its cytoplasmic domain. The structure of the TCR and the function of T cell coreceptors and costimulatory receptors are described in Chapter 5.

References

Bork, P. *et al.*: **The immunoglobulin fold. Structural classification, sequence patterns and common core.** *J. Mol. Biol.* 1994, **242**:309–320.

Halaby, D.M. and Mornon, J.P.E.: **The immunoglobulin superfamily: an insight on its tissular, species, and functional diversity.** *J. Mol. Evol.* 1998, **46**:389–400.

Halaby, D.M. *et al.*: **The immunoglobulin fold family: sequence analysis and 3D structure comparisons.** *Prot. Eng.* 1999, **12**:563–571.

Williams, A.F. and Barclay, A.N.: **The immunoglobulin superfamily—domains for cell surface recognition.** *Annu. Rev. Immunol.* 1988, **6**:381–405.

Ig superfamily members mediate antigen recognition and many functional effects of antibodies

Lymphocyte antigen receptors are members of the Ig superfamily, and so are many other molecules involved in immune recognition events. For example, many of the effector functions of antibodies depend on their binding to cells and stimulating phagocytic or cytotoxic responses that destroy microbes. These responses are mediated by receptors on immune cells called **Fc receptors**, so named because they recognize the invariant, or constant, part of the antibody molecule, called the Fc region (described in more detail in Chapter 6), and almost all Fc receptors are Ig superfamily members. Antigen receptors and most Fc receptors are activating receptors in that they promote immune cell activity, but many Ig superfamily members have the opposite effect and block activation of immune cells: they are called **inhibitory receptors**.

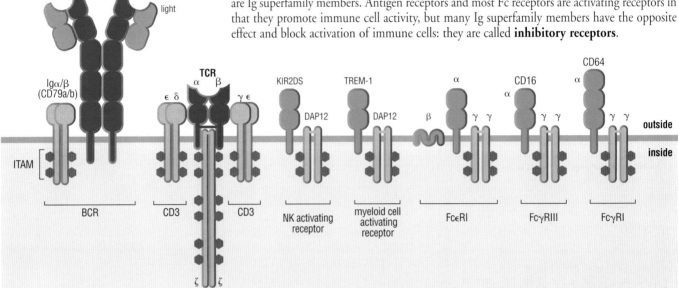

Figure 2-5 Multichain activating immune receptors of the Ig superfamily Antigen receptors of B cells (BCR) and T cells (TCR) are composed of ligand-binding units (membrane Ig and TCR α/β) and signaling subunits that contain ITAMs. These include the BCR signaling chains Igα and Igβ (also called CD79a and CD79b) and the three CD3 chains (called γ, δ and ε), each of which has a single extracellular Ig-like domain, and the ζ chain of the TCR, which does not. Similarly, activating Fc receptors and activating natural killer (NK) cell receptors, some of which are Ig superfamily members, often signal via associated chains with ITAMs.

Activating Ig superfamily receptors signal via intracellular tyrosine kinases

Antigen receptors, Fc receptors and some of the activating receptors of natural killer cells signal via a common mechanism that is dependent on a conserved amino-acid sequence motif, called the **immunoreceptor tyrosine-based activation motif (ITAM)**, which is typically present on a receptor-associated polypeptide (Figure 2-5). The ITAM contains two precisely spaced tyrosines within a consensus sequence (D/EXYXXL/IX$_7$YXXL/I, where Y is tyrosine, X is any amino acid, and L/I is leucine or isoleucine), which when phosphorylated provide a binding site for one of two closely related intracellular tyrosine kinases, *Syk* in most immune cells or *ZAP-70* in T cells (Figure 2-6, right-hand side), that have tandem SH2 domains spaced at exactly the right distance apart to dock onto the two phosphotyrosines, and that activate signaling events downstream of the receptor.

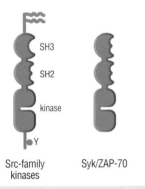

Figure 2-6 Structural elements of intracellular tyrosine kinases that participate in ITAM signaling The key intracellular tyrosine kinases that initiate immunoreceptor signaling are shown. Src-family kinases have an amino-terminal region containing fatty acid acylation sites that target the kinase to the plasma membrane, a kinase domain, and two protein interaction domains, an SH2 domain that binds to ITAM phosphotyrosines, and an SH3 domain, which recognizes proline-rich regions. The second important family of tyrosine kinases for ITAM signaling contains two members, Syk and ZAP-70; the latter is primarily expressed in T cells and the former is expressed in other hematopoietic cells. Syk and ZAP-70 lack a membrane targeting signal but have two SH2 domains that mediate recruitment of the kinase domain to signaling immunoreceptors.

Definitions

Fc receptors: immunoreceptors that mediate many of the effects of antibodies. These receptors are so named because the part of the antibody molecule they recognize is called the Fc region.

immunoreceptor tyrosine-based activation motif (ITAM): a sequence in the cytoplasmic domains of activating immunoreceptors that when phosphorylated becomes a binding site for the tandem SH2 domains of Syk and ZAP-70. The ITAM consensus sequence is D/EXYXXL/IX$_7$YXXL/I (one-letter amino

acid code; X = any amino acid).

immunoreceptor tyrosine-based inhibitory motif (ITIM): a sequence in the cytoplasmic domains of inhibitory immunoreceptors that, upon phosphorylation on tyrosine recruits signaling inhibitors (the consensus sequence is V/IXYXXL/V).

inhibitory receptors: receptors that block cell activation, generally when they bind their ligands. In immune cells the inhibitory function of these receptors is usually mediated by a consensus sequence in the cytoplasmic domain called the **ITIM**.

Figure 2-7 Signal transduction by ITAM-containing immunoreceptors Ligand-induced clustering of an immunoreceptor (in this example, the surface immunoglobulin (Ig) that serves as the B cell receptor for antigen) brings the receptor into proximity with a Src-family tyrosine kinase, which is held in the plasma membrane by a lipid tail and phosphorylates ITAM tyrosines in the receptor Igα and Igβ chains **(a)**. This creates a binding site for Syk, which binds to the doubly phosphorylated ITAM sequence **(b)**. Once bound, Syk becomes activated by tyrosine phosphorylation and it then phosphorylates other signaling proteins that ultimately bring about changes in the behavior of the cell.

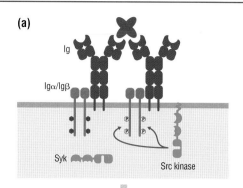

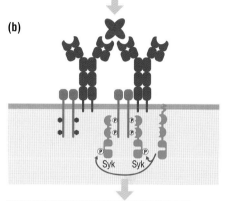

phosphorylation of scaffolds and adaptors, activation of signaling pathways

The ITAM phosphorylations leading to recruitment of the Syk and ZAP-70 kinases are mediated by another type of kinase, belonging to an important family of eight signaling kinases, the **Src-family tyrosine kinases**, which are localized to the plasma membrane by a covalently linked lipid group (Figure 2-6, left-hand side). The Src-family kinase is brought into proximity with the receptor on ligand binding, and phosphorylates the ITAMs, leading to the recruitment of Syk or ZAP-70, which is in turn phosphorylated and thereby activated by the Src kinase: this process is schematically illustrated for the B cell receptor for antigen in Figure 2-7.

We shall describe the signaling events downstream of the receptor in the context of individual activating receptors in later chapters. The ITAM signaling pathway is similar to other pathways that control growth, such as those activated by the epidermal growth factor (EGF) receptor or insulin receptor, the major difference being that individual elements of the signaling pathway (ligand binding, tyrosine kinase, and adaptor and scaffold functions) are dispersed onto separate polypeptides in the immunoreceptors. This separation of functions permits additional regulatory steps to control the process, particularly in the case of the T cell receptor.

Several viruses, including Epstein–Barr virus, which causes mononucleosis, and a herpes virus (HHV8) that causes Kaposi's sarcoma and B cell lymphoma in humans, have pirated ITAM sequences thereby activating these signaling events and promoting uncontrolled growth of the infected cell.

Immune cells express Ig superfamily inhibitory receptors that restrain immune responses

Immune responses must be regulated, to avoid damage to host cells or uncontrolled proliferation of activated immune cells. A large number of Ig superfamily members are specialized to inhibit activation of lymphocytes, phagocytosis by macrophages, or killing by natural killer cells for these reasons. These inhibitory receptors often contain a sequence motif, called the **immunoreceptor tyrosine-based inhibitory motif (ITIM)**, the consensus sequence for which is V/IXYXXL/V. This motif contains a single tyrosine that when phosphorylated serves as the docking site for an SH2 domain on either of two phosphatases (depending on the inhibitory receptor), SH2-containing tyrosine phosphatase 1 (SHP-1) or SH2-containing inositol phosphatase (SHIP). These phosphatases remove the activating phosphate groups added to components of the signaling pathway by the signaling kinases, thereby inhibiting activation. The inhibitory receptor ITIM is tyrosine phosphorylated by the Src-family kinases themselves, when it is brought into proximity with the ITAM-containing receptor: it then recruits and activates SHP-1 or SHIP, which in turn acts locally to inhibit the function of the activating receptor; one example of how this works in the regulation of B cell activation is shown in Figure 2-8.

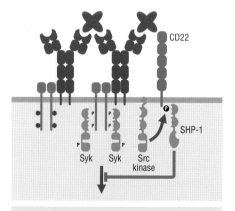

Figure 2-8 Signal inhibition by ITIM-containing immunoreceptors The inhibitory receptor CD22 binds to glycoproteins containing sialic acid, which are characteristic of host cells and largely absent from pathogens, and inhibits the response to these proteins. Here such a glycoprotein is shown bound both to the activating receptor and to CD22, bringing them into proximity. The ITIM tyrosine of the inhibitory receptor CD22 then becomes phosphorylated by Src-family kinases activated at the receptor, and this phosphorylated ITIM recruits the SH2-containing tyrosine phosphatase SHP-1, which counters the actions of Syk and/or Src kinases locally.

ITAM: see **immunoreceptor tyrosine-based activation motif**.

ITIM: see **immunoreceptor tyrosine-based inhibitory motif**.

Src-family tyrosine kinases: a family of eight membrane-linked intracellular protein tyrosine kinases first discovered as the product of a viral oncogene causing sarcoma in chickens. Src kinases participate in immune cell function primarily via **ITAM** and **ITIM** receptor signaling.

References

Ravetch, J.V. and Lanier, L.L.: **Immune inhibitory receptors.** *Science* 2000, **290**:84–89.

Sigalov, A.: **Multi-chain immune recognition receptors: spatial organization and signal transduction.** *Semin. Immunol.* 2005, **17**:51–64.

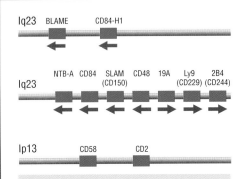

Figure 2-9 Organization of genes encoding CD2 family members Eleven CD2 family members are currently identified. They are encoded in two gene clusters, one on the long arm of chromosome 1 at 1q23 and one on the short arm of chromosome 1. Most of these family members are expressed on B cells, T cells, NK cells and/or myeloid cells.

Ig superfamily members have important roles in cell–cell adhesion of immune cells

Regulated cell–cell adhesion is a critical aspect of immune cell function. The Ig superfamily includes two groups of cell–cell adhesion molecules: those that are ligands for *integrins*, which we describe in the next section, and the **CD2 subfamily** of Ig superfamily members, which interact with each other.

The CD2 subfamily participates in the adhesion and activation of immune system cells

CD2 is the prototypical member of this 11-member subfamily, encoded in two clusters of highly related genes on human chromosome 1 (Figure 2-9). Structurally, the CD2 subfamily members are characterized by the presence of two Ig domains, the amino-terminal being a V-type domain and the carboxy-terminal being a C-type domain. The similarities in structure and in chromosome location both indicate that this subfamily arose by duplication of a single ancestral gene.

These molecules participate in cell–cell adhesion events via homophilic interactions (that is, interactions in which a CD2 family member on one cell interacts with the same CD2 family member on another cell) or heterophilic interactions between different family members (that is, interactions in which a CD2 family member interacts with another family member on the other cell). An example of a heterophilic interaction is the interaction between CD2, expressed on T cells, and CD58 which is expressed on antigen-presenting cells of various types. Recently a therapeutic drug that blocks the CD2–CD58 interaction has been developed to inhibit T cell activation in the chronic inflammatory skin disease psoriasis. This therapeutic agent, called Alefacept, consists of the extracellular portion of CD58 fused to the Fc region of one of the isotypes of IgG (IgG1). Ig fusion proteins of this type serve to give the recombinant molecule a long half-life in the blood, which is a characteristic of IgG antibodies, as described in Chapter 6. This therapeutic inhibits T cell activation by binding to CD2 on T cells and inhibiting cell–cell adhesion, and it can also induce immune effector cells with Fc receptors (see section 2-2) to kill some of the activated T cells, which would include those T cells driving the inflammation in psoriasis.

Structural analysis of the CD2–CD58 complex has shown that the binding involves interactions between the sides of the outer domains (Figure 2-10), mainly mediated by charge complementarity and exhibiting little surface shape complementarity. These features are thought to provide the observed rapid on-rate and rapid off-rate of binding of this interaction. These kinetic properties may be advantageous for T cells sampling antigen-presenting cells for the presence of the ligand for their antigen receptor, as described in Chapter 5.

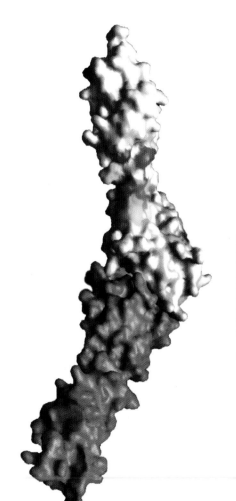

Figure 2-10 Structural model of the CD2–CD58 complex The model shown has human CD2 in light blue and human CD58 in orange. The interaction interface is hidden in this view, but it is characterized by charge–charge complementarity rather than extensive shape complementarity. The structure of the two interacting amino-terminal domains was solved by X-ray crystallography, as was the second domain of CD2. The second domain of CD58 is hypothetical, based on modeling and comparison with CD2. Molecular surface representation kindly provided by Jia-huai Wang and Ellis L. Reinherz. From Wang, J.-h. *et al.: Cell* 1999, **97**:791–803.

Definitions

CD2 subfamily: family of related Ig superfamily cell–cell adhesion molecules that bind either to themselves or to other family members. Most members contain two Ig domains and the outer domains are involved in binding.

SAP: see **SLAM-associated protein.**

SLAM-associated protein (SAP): a signaling adaptor molecule that associates with the intracellular domains of several **CD2 subfamily** members, including

SLAM (signaling lymphocytic activation molecule). SAP is also called SH2D1A. Defects in this molecule lead to increased susceptibility to Epstein–Barr virus infection.

X-linked lymphoproliferative disease: genetic disease caused by mutations in the gene encoding **SAP.**

CD2 family members have signaling properties as well as adhesion properties. Ligand engagement of CD2 and another subfamily member called signaling lymphocytic activation molecule (SLAM) can promote T cell activation via signaling events that they induce. SLAM (in common with six of the 11 CD2 family members) associates with an intracellular adaptor protein known as **SAP** (for **SLAM-associated protein**; also known as SH2D1A), the gene product defective in **X-linked lymphoproliferative disease**. In this disease, there is selective susceptibility to Epstein–Barr virus infections, which are normally controlled by cytotoxic T cells, suggesting an important role for CD2 family members in cytotoxic T cell function. Individuals with this disease also exhibit defects in antibody production, indicating a compromise of helper T cell function as well as of cytotoxic T cell function.

Several Ig superfamily members serve as cell-bound ligands for integrins

Several structurally distinctive Ig superfamily members serve as ligands for integrins in regulated cell–cell adhesion events important in lymphocyte activation and leukocyte migration out of the bloodstream, in which changes in the strength of adhesive interactions allow T cells to be activated by dendritic cells, and immune cells to migrate through the walls of blood vessels. The Ig superfamily members that engage in these interactions have between two and seven Ig domains, and the two most amino-terminal domains are responsible for integrin binding. Integrins, as is described more fully in the next section, are one of the major families of cell–cell and cell–extracellular matrix adhesion molecules of the body. The integrin–ligand pairs of particular importance for the immune system are listed in the next section. This specificity of molecular interaction is one part of a system that controls which immune cells are recruited to which sites of inflammation or peripheral lymphoid organs, a process we describe more fully in section 2-14.

One surprising feature of Ig-superfamily integrin ligands is the substantial variation in their molecular sizes (Figure 2-11). This suggests that they hold two cells apart by different distances. Although the significance of this is not established, it has been suggested that it may be important in the selective clustering of molecules in areas of close contact between T cells and the other immune cells with which they interact, as we shall see in Chapter 5.

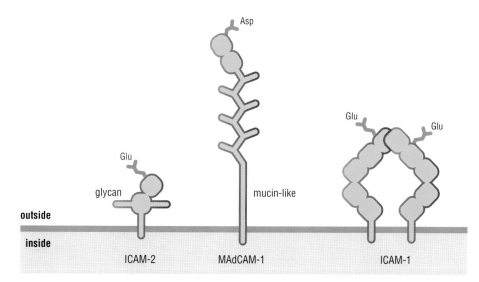

Figure 2-11 Molecular sizes of Ig subfamily members that bind to integrins Structural and biochemical studies have defined the sizes of ICAM-1, ICAM-2 and MAdCAM-1, and also key residues (glutamic acid, Glu, or aspartic acid, Asp) that interact with cognate integrins. Various features of these molecules indicate that these key residues are facing away from the cell such that the molecules interact with integrins on other cells rather than with integrins embedded in the same plasma membrane. Although ICAM-1 and ICAM-2 are made up of different numbers of Ig domains, for both it is the outermost Ig domain that interacts with the $\alpha_L\beta_2$ integrin and the atomic details of the interactions are very similar. ICAM: intercellular adhesion molecule; MAdCAM: mucosal addressin cell adhesion molecule. Adapted, with permission, from Wang, J.-h. and Springer, T.A.: *Immunol. Rev.* 1998, **163**:197–215.

References

Boles, K.S. *et al.*: **2B4 (CD244) and CS1: novel members of the CD2 subset of the immunoglobulin superfamily molecules expressed on natural killer cells and other leukocytes.** *Immunol. Rev.* 2001, **181**:234–249.

Engel, P. *et al.*: **The SAP and SLAM families in immune responses and X-linked lymphoproliferative disease.** *Nat. Rev. Immunol.* 2003, **3**:813–821.

Fraser, C.C. *et al.*: **Identification and characterization of SF2000 and SF2001, two new members of the** immune receptor SLAM/CD2 family. *Immunogenetics* 2002, **53**:843–850.

Song G. *et al.*: **An atomic resolution view of ICAM recognition in a complex between the binding domains of ICAM-3 and integrin $\alpha_L\beta_2$.** *Proc. Natl Acad. Sci. USA* 2005, **102**:3366–3371.

Springer, T.A. and Wang, J.-H.: **The three-dimensional structure of integrins and their ligands and conformational regulation of cell adhesion.** *Adv. Protein Chem.* 2004, **68**:29–63.

Veillette, A.: **SLAM family receptors regulate immunity with and without SAP-related adaptors.** *J. Exp. Med.* 2004, **199**:1175–1178.

Wang, J.-h. and Springer, T.A.: **Structural specializations of immunoglobulin superfamily members for adhesion to integrins and viruses.** *Immunol. Rev.* 1998, **163**:197–215.

Wang, J.-h. *et al.*: **Structure of a heterophilic adhesion complex between the CD2 and CD58 (LFA-3) counter-receptors.** *Cell* 1999, **97**:791–803.

2-4 Integrins in Immune Function

The Major Integrins of Lymphocytes and Phagocytes

Integrins	Cells expressing	Ligands
αLβ2 (LFA-1) (CD11a)	M, N, DC, T, B, NK	ICAM-1, 2, 3, 4, 5, JAM-1
αMβ2 (MAC-1, CR3) (CD11b)	M, N, NK	iC3b, ICAM-1, factor X, fibrinogen
αxβ2 (CD11c)	M, N, DC, NK	iC3b, fibrinogen
α1β1	M, eT	collagen, laminins
α2β1	B, M, eT, NK	collagen, laminins
α3β1	B	laminins
α4β1 (VLA-4)	B, eT, M, N, DC	VCAM-1, FN, MadCAM-1
α5β1	M, mT, DC	FN, fibrinogen
αvβ3	M, NK, T cell subsets	VN, PECAM, FN, fibrinogen, osteopontin, TSP, laminin
α4β7 (LPAM-1)	mT subset	MadCAM-1, VCAM-1
αEβ7	B, IEL	E-cadherin

Figure 2-12 Table of the major integrins of lymphocytes and phagocytes Shown is the integrin nomenclature based on identifying the α and β subunits of an integrin. Note that integrins expressed on lymphocytes often change on activation or entry into the memory phenotype to permit differential migration and/or binding to extracellular matrix components (collagens, laminins, fibronectin, and so on) at tissue sites of inflammation. B: B cell; DC: dendritic cell; FN: fibronectin; M: monocyte and/or macrophage; N: neutrophil; NK: natural killer cell; T: T cell (e: effector; m: memory; IEL: intraepithelial lymphocytes, which include several types of specialized T cells).
*The β2-chain integrins are also often referred to by their CD names: β2: CD18; αL: CD11a; αM: CD11b; αx: CD11c.

Integrins mediate cell–cell and cell–matrix contacts

The **integrins** comprise an evolutionarily conserved family of adhesion molecules characterized by a heterodimeric structure (with the two subunits being called α and β) and strong connections to the actin cytoskeleton. Mammals have 8 β chains and 18 α chains, and these chains can combine in various ways to form at least 24 integrins, a substantial number of which play important roles in immune function (Figure 2-12). They are typically named by listing the identities of their α and β chains, although many have alternative names including CD numbers. Many integrins bind **extracellular matrix (ECM)** proteins and their name thus derives from their function in integrating structures on the outside of the cell with the cytoskeleton inside the cell. Integrins are also involved in cell–cell adhesion, particularly in the immune system, and it is this latter type of function that we shall mostly encounter in this book, because it is important for the process by which immune cells leave the blood and enter tissues, as described in section 2-14; for cell–cell adhesion of immune cells with each other or with infected cells; and for phagocytosis of microbes leading to their destruction by phagocytic cells.

The adhesive properties of integrins are modulated by other signals coming into the cell

A key property of integrins is the regulation of their binding activity by signals coming into the cell via other receptors, such as antigen receptors and chemokine receptors (described in section 2-13). For example, to function properly, a cytotoxic T cell must adhere strongly to an infected cell until it has secreted the molecules that will kill the target, and then it must disengage and look for another infected cell. This regulated adhesion is due to integrins. Regulated adhesion through integrins is also important for blood cells, which must not adhere to one another or to vascular cells when they are circulating in the blood, but must bind avidly to endothelial cells to leave the bloodstream and enter a lymph node or a site of infection.

The regulation of ligand binding by integrins operates on two properties: conformational changes in integrin molecules that increase the strength of binding, or **affinity** for ligands; and changes in the lateral mobility of integrin molecules within the membrane that permit clustering of multiple integrins adjacent to one another (Figure 2-13). The latter mechanism is sometimes called avidity regulation, but more properly **avidity** refers to strength of binding due to the contribution of both the affinity of individual molecular interactions and from the number of discrete interactions; changes in lateral mobility of integrins can affect the latter component of avidity. When receptors on immune cells generate intracellular signals that activate integrin-based

Figure 2-13 Regulation of integrin-mediated cell–cell adhesion In an unstimulated T lymphocyte in the blood or in a lymph node (left panel), the integrins are almost all in the low-affinity form (bent molecules) and they are dispersed on the cell surface. Stimulation through the TCR or a chemokine receptor induces a conformational change in the integrins that increases their binding affinity and also frees them from barriers that limit their diffusion in the plasma membrane. These integrins now can diffuse until they bind their ligands on an opposing cell, leading to a bound state with multiple integrin molecules clustered near one another (right panel). Sufficient avidity to support strong cell–cell adhesion typically requires many molecules participating in the cell–cell adhesion, so lateral mobility of integrins is often required for adhesion to occur.

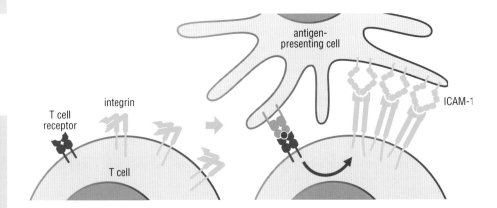

antigen-presenting cell

integrin

T cell receptor

T cell

ICAM-1

Definitions

affinity: strength of a nonconvalent binding interaction; the higher the affinity, the more likely two partners will exist in a complex.

avidity: increased apparent **affinity** of a molecule for its ligand due to the presence of multiple binding sites on both partners.

ECM: see **extracellular matrix**.

extracellular matrix (ECM): matrix of proteins that

forms in tissues between cells; it can include fibronectin, collagen, vitronectin, and so on.

inside-out signaling: process whereby stimulation of cells via receptors such as antigen receptors or chemokine receptors leads to intracellular signaling events that alter the **affinity** and/or lateral mobility of **integrin** molecules in the plasma membrane so as to increase adhesion of those integrins with ligands on other cells or of the **extracellular matrix**. The flow of information through the integrin is from its cytoplasmic tails to the extracellular portion.

adhesion by either of these mechanisms it is referred to as **inside-out signaling**, because information flow is from signaling events within the cell to changes in the binding properties of integrins to ligands outside the cell.

The affinity of integrins is thought to be determined by the conformation of the outermost domains of the α and β subunits, which in turn are strongly influenced by a marked conformational change between a sharply bent integrin structure, favoring the low-affinity state, and a straight conformation, favoring the high-affinity state (Figure 2-14). In the bent conformation, the end of the integrin may be as close as 5 nm from the cell surface, whereas in the straight conformation it extends about 20 nm out from the membrane. The conformational change between the bent and extended forms is thought to be controlled by the proximity of the cytoplasmic domains of the α and β subunits. When the cytoplasmic domains are adjacent to each other, the bent conformation is favored, whereas when they are separated, the straight conformation is favored. Consequently, binding of proteins to the cytoplasmic domains of integrins can regulate integrin affinity; one such protein is talin, which links integrins to the actin cytoskeleton and induces the high-affinity conformation. Also implicated as a mediator of inside-out signaling is the small GTP-binding protein Rap1 and its effector molecule RapL.

Changes in the lateral mobility of integrins allowing more of them to bind to the opposing cell or ECM also contribute to increased integrin adhesion induced by inside-out signaling. These changes probably involve remodeling of the actin cytoskeleton, although the molecular details are not well understood.

Integrin engagement regulates cell shape and also sends signals that support cell activation and survival

The initial effects of integrin engagement are the formation of clusters of integrins bound to ligands on the outside of the cell and the remodeling of actin cytoskeleton on the interior of the cells, which can alter the shape of the cell. Integrin β-chain cytoplasmic domains directly associate with a variety of proteins that mediate linkage to and/or induce assembly of actin filaments.

Integrin engagement is not only important for the adhesion it provides. Ligand-bound and clustered integrins generate signals within the cell that regulate a variety of cellular properties, including cellular activation, proliferation, and survival. This signaling property of integrins is called **outside-in signaling**, to distinguish it from the signaling by other receptors that regulates integrin adhesion by inside-out molecular events, as described above. This signaling is primarily mediated by two related protein tyrosine kinases, focal adhesion kinase (FAK) and Pyk2.

Integrins are differentially expressed to support specializations in immune cell function

Although integrins are important for processes involving rapid changes in cell adhesion, they are also versatile adhesion molecules that are differentially expressed in different cell types and in different stages of cell activation. For example, during the clonal expansion of the antigen-specific cells that occurs as part of the immune response, T cells change the integrins they express to ones that will support their ability to function at sites of infection. In addition, certain types of immune cells are tissue-resident cells, typically expressing integrins that help them adhere to the ECM in the corresponding location. Integrins are abundant proteins on the surface of most cells and these specializations of expression are often used in defining and isolating particular types of cells. For example, CD11c (αxβ2 integrin) is a commonly used marker of dendritic cells.

(a)

(b)

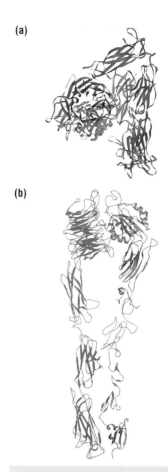

Figure 2-14 Conformational change in integrins between low-affinity and high-affinity states The low-affinity state of the αvβ3 integrin is favored by the bent configuration shown in **(a)** (PDB 1jv2), as revealed by X-ray crystallography, whereas the high-affinity configuration is favored by the extended conformation shown in **(b)** (model based on electron microscopy studies of a variety of integrins). Evidence also exists for an intermediate-affinity form in which the conformation has a smaller bend than the low-affinity state (not shown). Shown are ribbon diagrams of the polypeptide backbone of the α and β chain extracellular domains with the ends of the extracellular domains nearest the membrane at the bottom. The relationship between the outermost α and β domains is different in the two forms and this difference is what controls the affinity. Blue: α chain; orange: β chain. Part (b) kindly provided by Jian-Ping Xiong. From Xiong, J.-P. *et al.*: *Science* 2001, **294**:339–345.

integrins: large family of αβ heterodimeric molecules that participate in cell–cell and cell–matrix adhesion. Adhesion by integrins of cells of the immune system is generally regulated by activation signals or chemokine receptor signals.

outside-in signaling: intracellular signaling events that result from **integrin** binding to its extracellular ligands.

References

Carman, C.V. and Springer, T.A.: **Integrin avidity regulation: are changes in affinity and conformation underemphasized?** *Curr. Opin. Cell Biol.* 2003, **15**:547–556.

DeMali, K.A. *et al.*: **Integrin signaling to the actin cytoskeleton.** *Curr. Opin. Cell Biol.* 2003, **15**:572–582.

Grashoff, C. *et al.*: **Integrin-linked kinase: integrin's mysterious partner.** *Curr. Opin. Cell Biol.* 2004, **16**:565–571.

Hynes, R.O.: **Integrins: bidirectional allosteric signaling machines.** *Cell* 2002, **110**:7673–7687.

Hynes, R.O. and Zhao, Q.: **The evolution of cell adhesion.** *J. Cell Biol.* 2000, **150**:F89–F96.

Plow, E.F. *et al.*: **Ligand binding to integrins.** *J. Biol. Chem.* 2000, **275**:21785–21788.

Pribila, J.T. *et al.*: **Integrins and T cell-mediated immunity.** *Annu. Rev. Immunol.* 2004, **22**:157–180.

Schwartz, M.A.: **Integrin signaling revisited.** *Trends Cell Biol.* 2001, **11**:466–470.

Xiong, J.-P. *et al.*: **Crystal structure of the extracellular segment of integrin αVβ3.** *Science* 2001, **294**:339–345.

Carbohydrate-recognizing molecules have diverse roles in the immune system

Whereas Ig superfamily adhesion molecules and integrins mediate cell–cell adhesion via protein–protein interactions, many other molecules mediate protein–carbohydrate recognition. Proteins that recognize specific carbohydrate structures are referred to as **lectins**, and such proteins can have a wide variety of different structures. A particularly important family of lectins for many immune functions is the **C-type lectin** family, whose members contain a carbohydrate recognition domain with a characteristic three-dimensional fold and in which a calcium ion often contributes to saccharide binding (Figure 2-15). Simple carbohydrates can be combined to create a vast array of different oligosaccharide and polysaccharide structures, with a variety of functions including structural roles, participation in glycoprotein folding and targeting to specific locations within the cell, and serving as ligands for specific recognition events. In the immune system, carbohydrate recognition contributes both to immune cell migration between different places in the body and to discrimination between structures made by the host organism and those made by foreign organisms.

C-type lectins of the immune system can be soluble recognition molecules that bind to microorganisms and promote phagocytosis or complement activation, immune functions that are described in detail in the next chapter, or they can be surface molecules on immune cells that serve either as adhesion molecules or as receptors, the **C-type lectin receptors**. Examples of soluble C-type lectins are the *collectins*: these are highly multimerized structures that bind to foreign particles with polysaccharides and that serve as tags for uptake by phagocytes and/or for activation of the complement system. C-type lectin receptors of macrophages or dendritic cells are often specific for carbohydrates characteristic of microorganisms and not of host cells, or for carbohydrates indicative of host proteins or cells that have become senescent and need to be removed from the organism. On ligand binding, they mediate uptake into the cell for destruction and/or antigen presentation to T cells. A third important function of C-type lectins is in directing immune cell migration. This is the specialized function of *selectins*, which are transmembrane molecules containing C-type lectin domains. They bind to glycoproteins with highly regulated expression patterns and promote the migration of immune cells between the blood and either lymph nodes or sites of inflammation (see section 2-14). The main types of C-type lectins in the immune system are illustrated schematically in Figure 2-16.

(a)

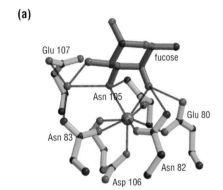

(b)

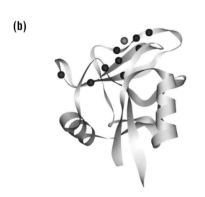

Figure 2-15 Binding of a sugar residue by the C-type lectin domain of E-selectin E-selectin is a molecule expressed on endothelial cells at sites of inflammation. Its expression is part of a coordinated process that brings leukocytes to sites of inflammation, as described in section 2-14. **(a)** Shown are the amino-acid side chains of the C-type lectin domain of E-selectin (green) that participate in binding to the fucose residue (red) within the oligosaccharide known as the sialyl Lewis^X blood group antigen (the rest of this oligosaccharide is not shown). A Ca^{2+} ion (green circle) is also involved in this interaction and it makes direct contacts both to the C-type lectin domain and to the fucose. Note that selectins also exhibit specificity for the peptide backbone to which the recognized oligosaccharide is attached (not shown). **(b)** A ribbon diagram representing the three-dimensional structure of the peptide backbone of the C-type lectin domain of E-selectin is shown. At the top of the structure is where ligand binds; at the bottom, the domain is connected to the rest of the protein. Red circles represent amino-acid residues implicated in ligand binding; the green circle is the Ca^{2+} ion that participates in ligand binding. Panel (a) kindly provided by Heide Kogelberg and Ten Feizi. From Kogelberg, H. and Feizi, T.: *Curr. Opin. Struct. Biol.* 2001, **11**:635–643. (Panel (b) PDB 1esl)

Definitions

C-type lectin: cell-surface or secreted protein that is characterized by a conserved three-dimensional fold and a somewhat conserved amino-acid sequence and that binds to carbohydrate ligands, typically in a calcium-dependent manner. Some immune receptors with this conserved structure have been shown to bind to protein ligands in a carbohydrate-independent manner and are called **C-type lectin-like receptors**.

C-type lectin-like receptors: cell-surface proteins containing **C-type lectin** domains that do not bind carbohydrate ligands and are thus distinct from **C-type lectin receptors**; C-type lectin-like receptors commonly mediate phagocytic or endocytic uptake and/or signaling into the cell.

C-type lectin receptors: cell-surface proteins that contain a **C-type lectin** domain that binds to carbohydrate ligands and that mediate phagocytic or endocytic uptake and/or signaling to the interior of the cell.

lectin: protein that binds specifically to particular polysaccharides or other carbohydrate structures. There are four major structural groups: **C-type lectins**, p-type lectins, I-type lectins and galectin-like lectins.

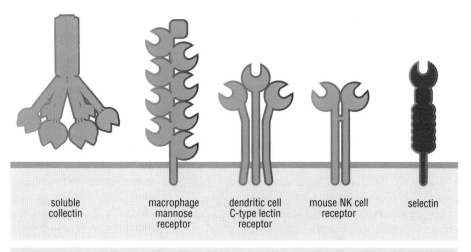

soluble
collectin

macrophage
mannose
receptor

dendritic cell
C-type lectin
receptor

mouse NK cell
receptor

selectin

Figure 2-16 C-type lectin molecules of the immune system Some examples of immune system molecules containing C-type lectin domains are shown.

Just as Ig superfamily members have evolved to perform a wide diversity of functions, not all molecules that contain domains with the C-type lectin fold are lectins; some mediate protein–protein interactions and do not require calcium for ligand binding. Because these receptors are not lectins, they are properly referred to as **C-type lectin-like receptors**, although we shall not generally make this distinction in this book, particularly as the ligands for molecules of this type are not always well enough defined for it to be known whether the C-type lectin domain-containing molecule recognizes carbohydrate or peptide.

Some examples of C-type lectins with important functions in the immune system are listed in Figure 2-17, along with reference to the section of the book where their function is described.

C-type Lectin Molecules of the Immune System

Category	Example	Ligands	Further description
soluble foreign recognition molecules	collectins	microbial cell walls	section 3-2
phagocytic receptors	mannose receptor, Dectin-1	microbes, virus particles	section 3-7
homing molecules	selectins	host cell-surface glycoproteins	section 2-14
adhesion molecules	DC-SIGN	ICAM-2, ICAM-3, HIV-1 gp120	section 10-4
NK cell-activating receptors	NKG2D	stress-induced proteins	section 8-2
NK cell-inhibiting receptors	Ly49 family	MHC class I molecules	section 8-1

Figure 2-17 Table of types of immune system molecules containing C-type lectin domains Specific examples and their ligands are shown, and the main sections in which they appear in the book are listed.

References

Cambi, A. and Figdor, C.G.: **Dual function of C-type lectin-like receptors in the immune system.** *Curr. Opin. Cell Biol.* 2003, **15**:539–546.

Geijtenbeek, T.B.H. *et al.*: **Self- and nonself-recognition by C-type lectins on dendritic cells.** *Annu. Rev. Immunol.* 2004, **22**:33–54.

Kogelberg, H. and Feizi, T.: **New structural insights into lectin-type proteins of the immune system.** *Curr. Opin. Struct. Biol.* 2001, **11**:635–643.

Lanier, L.L.: **NK cell recognition.** *Annu. Rev. Immunol.* 2005, **23**:225–274.

Polypeptide mediators are a principal means of regulating immune responses

Recognition of conserved elements of pathogens by innate immune cells or of antigen by lymphocytes leads to the production of polypeptide mediators called **cytokines** that regulate many aspects of immune responses. Many of these polypeptide mediators are secreted soluble molecules, but some are bound to cell membranes. Most cytokines, whether soluble or membrane-bound, act either on neighboring cells—this is known as **paracrine** signaling—or on the producing cell itself, an **autocrine** action. During a severe infection, soluble cytokines can also act systemically like endocrine hormones, for example to promote hematopoeisis in the bone marrow, increasing the production of effector leukocytes to fight the infection. Many soluble cytokines are referred to as **interleukins**, which is a standard nomenclature for these cytokines. The name reflects the discovery of the cytokines as signaling molecules made by leukocytes and acting on leukocytes; however, it is now known that this is too limited a view and cytokines can be secreted by non-immune cells or act on non-hematopoeitic cells. It is generally sufficient for a protein mediator to have a role in the immune response for it to be considered a cytokine, and the distinction between cytokine, growth or differentiation factor, and polypeptide hormone is not always a sharp one.

Cytokines act on many cell types in overlapping and coordinated ways

There are perhaps as many as 100 cytokines that act in the immune system, and many of them act on multiple target cells to induce multiple actions: this is known as **pleiotropy**. Conversely, a given cell may respond to several different cytokines in the same way: this is known as **redundancy**. The pleiotropy and redundancy of cytokine responses have made understanding the biological functions of cytokines a challenge for immunologists, but most cytokines are characterized by some functions in particular, and in this book we shall focus on these major functions, recognizing that a cytokine typically has additional roles. Often cytokines are produced in combinations that act in a coordinated way to promote particular immune responses, such as inflammation or activation and differentiation of B lymphocytes into antibody-secreting cells (Figure 2-18). A common response to a cytokine is induction or regulation of the production of other cytokines. For example, the proinflammatory cytokines TNF and interleukin-1 (IL-1) induce the production of proinflammatory cytokines from neighboring cells. This chain reaction may be needed to relay the signal until it reaches the nearest endothelial cells. Many cytokines promote immune responses; others negatively regulate immune responses.

Cytokines can be grouped into structurally related families that act via homologous receptors

Almost all cytokines can be grouped into six different structurally related families of molecules, each of which is recognized by a related family of receptors with shared signaling properties. These different groupings of cytokines and receptors are illustrated in Figure 2-19, and three

Cytokines and Immune Responses
Proinflammatory cytokines
TNF
IL-1
IL-6
chemokines (many)
Antiinflammatory cytokines
IL-10
IL-1ra
TGF-β
Inhibition of virus replication
IFN-α, -β
Macrophage-activating cytokines
IFN-γ
B cell-activating cytokines
IL-4
IL-5
IL-6
IL-21
T cell-activating cytokines
IL-2
IL-4
IL-12
IFN-γ
Eosinophil- and/or mast cell-activating cytokines
IL-3
IL-4
IL-13
IL-5

Figure 2-18 Table of involvement of cytokines in major immune responses Shown are some of the more important cytokines involved in promoting or inhibiting inflammation, activating macrophages, activating B cells, activating T cells, and activating eosinophils and mast cells. IFN: interferon; IL: interleukin; IL-1ra: IL-1 receptor antagonist; TGF-β: transforming growth factor β; TNF: tumor necrosis factor.

Definitions

autocrine: (of an extracellular signaling molecule) acting on the cell that secreted it.

cytokine: polypeptide signaling molecule that participates in immune responses. Cytokines often act locally (in an **autocrine** or a **paracrine** manner) but can act systemically. Most cytokines are secreted molecules, but membrane-bound versions also occur in most cytokine types.

interleukin: cytokine participating in immune respons-

es and originally thought always to be produced by leukocytes and to act on other leukocytes. Many cytokines have been given a systematic name of interleukin x, where x is a number from 1 to at least 29. It is now clear that often the range of action of interleukins extends to non-hematopoietic lineage cells.

paracrine: produced by one cell and acting on a nearby cell. This is thought to be the main mode of action of **cytokines**, although they can act in an **autocrine** fashion or systemically in an endocrine fashion.

pleiotropy: property whereby one agent may have

diverse effects on many different cell types.

redundancy: (of properties of molecules) property whereby more than one molecule seems to have the same function.

SMAD proteins: any of a related family of signaling proteins activated by TGF-β family receptors. Phosphorylation of SMAD proteins induces them to enter the nucleus and activate or inhibit the transcription of target genes.

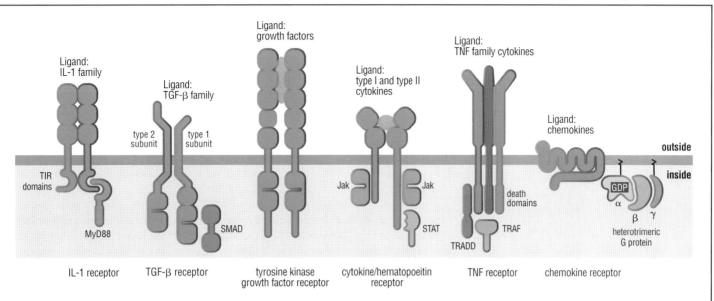

IL-1 receptor TGF-β receptor tyrosine kinase cytokine/hematopoeitin TNF receptor chemokine receptor
growth factor receptor receptor

major groups—the type I and II cytokine/hematopoietin superfamily, the chemokines, and the TNF superfamily—are further described in this chapter in separate sections. The remaining three types of cytokines are briefly described here.

The IL-1 family consists of IL-1α, IL-1β, IL-1ra (a natural IL-1 antagonist), IL-18, and six new family members found in the human genome (IL-1F5–10). The known receptors for the IL-1 family cytokines are heterodimers in which each chain has an amino-terminal Ig super-family extracellular domain. The intracellular domains of IL-1 family receptors are related to the cytoplasmic domains of another important family of receptors that we shall meet in Chapter 3. These are called *Toll-like receptors* and are not cytokine receptors but are specialized for recognition of characteristic features of microorganisms and are important in activating innate immune responses. Their cytoplasmic domains are called *TIR (Toll/interleukin 1 receptor)* domains, and their signaling mechanism is described in Chapter 3.

The TGF-β family of cytokines and differentiation factors plays a prominent part in the regulation of development and wound healing. TGF-β itself is also an immunosuppressive cytokine. This family of cytokines is recognized by heteromeric receptors containing two types of subunits known as type I and type II subunits, both of which have cytoplasmic domains containing serine/threonine protein kinase domains. They signal primarily by ligand-induced multimerization followed by phosphorylation and activation of the type I kinase domains by the type II kinase domains. The type I receptors then phosphorylate a family of transcriptional regulators called **SMAD proteins**, which then dimerize and regulate transcription of target genes.

The third type of cytokine discussed in this section acts via tyrosine kinase growth factor receptors. In these receptors, the ligand induces homo- or heterodimerization and the kinase domains cross-activate one another and also phosphorylate tyrosines in their cytoplasmic domains that serve as binding sites for SH2-containing signaling proteins. These receptors signal by a mechanism that is similar to that of ITAM-bearing Ig superfamily receptors (section 2-2). M-CSF, c-Kit, and Flt3, all of which are involved in hematopoiesis, bind to receptors of this type. The M-CSF receptor also regulates monocyte and macrophage survival and function.

Figure 2-19 Six major classes of cytokines and their receptors IL-1 family receptors and Toll-like receptors (TLRs) have homologous domains, called TIR domains, in their cytoplasmic tails and they signal by interacting with adaptor proteins, such as MyD88, that also contain the TIR domain. TGF-β receptors have serine/threonine kinases in their cytoplasmic domain and signal by phosphorylating SMAD transcriptional regulators. Several tyrosine kinase growth-factor-receptor family members have important roles in the immune system, particularly in hematopoiesis. Lymphocyte antigen receptors share many signaling features with these receptors. The family of cytokine/hematopoietin receptors that bind type 1 or type 2 cytokines signal primarily by the Jak–STAT signaling pathway and are discussed in the next two sections. TNF receptor superfamily members are discussed later in this chapter and can signal programmed cell death via death domains to adaptor proteins called TRADD and FADD that link to enzymes of the programmed cell death pathway, and/or via TNF-receptor-associated factors (TRAFs) to signal other cellular responses. Chemokine receptors, which we discuss later in this chapter, pass seven times through the membrane and are known as serpentine seven-transmembrane receptors. They couple to heterotrimeric G proteins. Note that with the exception of the serpentine chemokine receptors, signaling by cytokine receptors is triggered by ligand-induced dimerization or multimerization of the receptor.

References

Dunn, E. *et al.*: **Annotating genes with potential roles in the immune system: six new members of the IL-1 family.** *Trends Immunol.* 2001, **22**:533–536.

Paul, W.E.: **Pleiotropy and redundancy: T cell-derived lymphokines in the immune response.** *Cell* 1989, **57**:521–524.

Schlessinger, J. and Ullrich, A.: **Growth factor signaling by receptor tyrosine kinases.** *Neuron* 1992, **9**:383–391.

Wakefield, L.M. and Roberts, A.B.: **TGF-β signaling: positive and negative effects on tumorigenesis.** *Curr. Opin. Genet. Devel.* 2002, **12**:22–29.

Weiss, A. and Schlessinger, J.: **Switching signals on or off by receptor dimerization.** *Cell* 1998, **94**:277–280.

The family of type I cytokines is recognized by a conserved family of receptors

Many cytokines, belonging to a group often called **type I cytokines**, fold into a conserved four-helix-bundle globular structure and are recognized by a family of receptors, the **type I cytokine receptors**, that are characterized by beta-sandwich domains structurally similar to Ig-like domains and containing a conserved sequence motif, WSXWS. These receptors signal primarily through a pathway commonly known as the *Jak–STAT pathway* because it involves a family of intracellular kinases that are known as the **Jak (Janus kinase) tyrosine kinases**—so called because they have two kinase domains, Janus being a two-headed Roman god—and that activate members of a family of transcriptional regulators, the **signal transducer and activator of transcription (STAT)** transcriptional regulators (discussed further below). In addition to cytokines of immune function, signaling molecules acting through this pathway include hormones and growth and differentiation factors regulating hematopoiesis, also called **hematopoietins**. Many hematopoietins are produced during immune responses, and increase hematopoiesis.

(a)

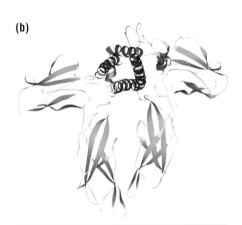

(b)

Interferons and IL-10 are related to type I cytokines and are also recognized by receptors signaling via the Jak–STAT pathway

Interferons and the closely related IL-10 family of cytokines are structurally similar and are more distantly related to the four-helix-bundle cytokines: they are known as **type II cytokines**. They act through receptors that also bind to Jaks and activate STAT transcriptional regulators, and are called **type II cytokine receptors**. These receptors are structurally related to the type I cytokine receptors but fall into a distinct group on the basis of their sequence and lack the WSXWS sequence motif. Both type I cytokine and type II cytokine receptors signal as homodimers or as heterodimers (Figure 2-20), and both, as we shall see in the next section, often have a signaling chain that is common to many receptors and a unique chain that confers specificity for different cytokines.

Figure 2-20 Dimerization of growth hormone receptor by growth hormone Ribbon diagram representations of growth hormone alone **(a)** and growth hormone bound to its receptor **(b)**. The lower structure shows growth hormone (orange), seen from a different orientation, bound to two molecules of growth hormone receptor (green). Note that a single molecule of growth hormone binds to the two growth hormone receptor molecules asymmetrically to dimerize them. The ribbon diagram representations of the structures illustrate the four-alpha-helix structure of type I cytokines and two-beta-sandwich domain structures common to the type I and type II cytokine receptors. The ligand binds near the junction of the two domains, a structural feature that is conserved in other family members. The plasma membrane would be at the bottom of the structure shown. (PDB 3hhr)

Type I and type II cytokine receptors signal principally via the Jak–STAT signaling pathway

The basic signaling mechanism of the type I and type II cytokine receptors is illustrated in Figure 2-21. The cytoplasmic domains of signaling receptors of this class bind stably to one of the Jak kinases, of which there are four: Jak1, Jak2, Jak3 and Tyk2. Ligand-induced dimerization (or in some cases higher-order multimerization) of the receptor brings the weakly active Jaks into proximity with one another, whereupon they transphosphorylate and activate one another. The activated Jaks then phosphorylate tyrosine residues in the cytoplasmic domains of the receptors; these sites recruit one or more of the seven members of the STAT family of transcriptional regulators, which bind to the phosphotyrosine residues through SH2 domains and are in turn activated by the activated Jaks. Some cytokine receptors can also activate other downstream signaling events characteristic of tyrosine kinase signaling pathways.

STATs are responsible for much of the specificity of cytokine receptor signaling

The tyrosine phosphorylation sites of the cytoplasmic domains of different cytokine receptor chains selectively recruit particular STAT family members by means of the distinct SH2 domains of the STATs. The recruited STATs then become tyrosine phosphorylated by Jaks

---

Definitions

hematopoietins: cytokines that promote proliferation and/or lineage-specific differentiation of hematopoietic cells.

Jak (Janus kinase) tyrosine kinase: an intracellular tyrosine kinase belonging to a small family with two kinase domains, one of which is the active tyrosine kinase and the second of which does not have catalytic activity. For this reason they are named for the two-headed Roman god of gates and doorways, Janus.

signal transducer and activator of transcription (STAT): any of a family of rapidly activated transcriptional regulators that are directly activated by cytokine and growth factor receptors.

SOCS protein: see **suppressor of cytokine signaling protein**.

STAT: see **signal transducer and activator of transcription**.

suppressor of cytokine signaling (SOCS) protein: any of a family of inhibitors of **Jak–STAT** signaling, also

called CIS (cytokine-inducible SH2-containing) proteins, that have a central SH2 domain and a conserved carboxy-terminal motif called the SOCS box that mediates ubiquitination of bound proteins.

type I cytokines: cytokines with substantial amino-acid homology that form related four-helix bundles.

type I cytokine receptors: receptors for **type I cytokines** that contain two or more beta-sandwich domains, one of which has a WSXWS motif, and that signal via **Jak tyrosine kinases** and **STAT** transcriptional regulators.

(see Figure 2-21). Which STAT is recruited depends on the receptor sequences surrounding the phosphorylation sites, and the specificity of the SH2 domain of the STAT molecule. Once phosphorylated, STATs dissociate from the receptor and then form STAT dimers by reciprocal interactions between the STAT SH2 domains and the phosphorylated sites on the partner STAT. Heterodimers can form as well as homodimers, if more than one type of STAT is activated by a particular receptor. For example, STAT1 and STAT2 are activated by the interferon-α receptor (which we discuss in Chapter 3), leading to the formation of heterodimers of STAT1 and STAT2. The range of different dimers that can be formed is limited, however; for example, STAT2 cannot form homodimers.

One cell can respond in different ways to multiple type I cytokines at once. For example, B cells respond in distinct ways to IL-4, IL-5 and IL-6, and the functions of these cytokines are complementary for promoting B cell activation. The specificity of the cellular responses to type I cytokines and interferons is determined primarily by cytokine receptor specificity for particular STATs, and the specificity of different STAT dimers for the upstream control regions of different target genes.

Jak–STAT signaling is negatively regulated by SOCS proteins

An important regulatory mechanism in cytokine receptor signaling is the induction by many cytokines of inhibitors of Jak–STAT signaling called **suppressor of cytokine signaling (SOCS) proteins**. SOCS proteins comprise a family of eight members, which contain SH2 domains and a conserved 50 amino-acid residue motif also found in other proteins, called the SOCS box. SOCS proteins can directly bind to and inhibit Jaks or STATs, and the SOCS box targets their binding partner for rapid degradation. SOCS proteins represent a mechanism of feedback inhibition of cytokine receptor signaling: they are not expressed in cells before cytokine stimulation, which induces their expression, which in turn inhibits that signaling.

Two other important negative regulators of cytokine/hematopoietin receptor signaling are the SH2-domain-containing tyrosine phosphatase 1 (Shp-1) and a family of proteins called protein inhibitors of activated STATs (PIAS) that bind activated STATs and hold them in the cytoplasm. The activity of some STAT proteins is enhanced by serine phosphorylation.

Figure 2-21 Basic Jak–STAT signaling pathway of type I cytokine receptors Ligand binding induces receptor chain oligomerization. Bound weakly active Jak kinases cross-activate one another, probably by phosphorylation of the activating site in their kinase domains. The active Jaks then phosphorylate key tyrosine residues in the cytoplasmic tail of the cytokine receptor. These sites attract signaling molecules, including STATs and other signaling molecules containing SH2 domains (not shown). Which signaling molecules bind depends on the amino-acid sequence surrounding the phosphorylation site and may vary between different cytokine receptors, permitting selective responses to different type I cytokines by the same cell. Bound STATs become tyrosine phosphorylated, after which they dissociate from the receptor and dimerize via their phosphorylation sites and the SH2 domains of the STATs. The STAT dimers rapidly enter the nucleus and activate the transcription of specific target genes.

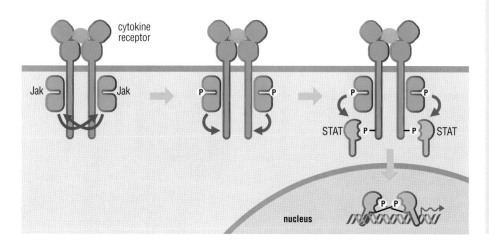

type II cytokines: cytokines including interferons, IL-10 and IL-10-like cytokines, that form an evolutionarily conserved group whose members are structurally similar to those of the **type I cytokines**.

type II cytokine receptors: receptors for **type II cytokines** that signal via **Jaks** and **STATs** and have structural similarity to **type I cytokine receptors**, but lack the WSXWS motif characteristic of the latter receptors.

References

Alexander, W.S. and Hilton, D.J.: **The role of suppressors of cytokine signaling (SOCS) proteins in regulation of the immune response.** *Annu. Rev. Immunol.* 2004, **22**:503–529.

Gadina, M. *et al.*: **Signaling by type I and II cytokine receptors: ten years after.** *Curr. Opin. Immunol.* 2001, **13**:363–373.

Leonard, W.J.: **Type 1 cytokines and interferons and their receptors** in *Fundamental Immunology* 5th ed.

Paul, W.E. ed. (Lippincott Williams & Wilkins, Philadelphia, 2003), 701–747.

Pestka, S. *et al.*: **Interleukin-10 and related cytokines and receptors.** *Annu. Rev. Immunol.* 2004, **22**:929–979.

Shuai, K.: **Modulation of STAT signaling by STAT-interacting proteins.** *Oncogene* 2000, **19**:2638–2644.

Shared Cytokine-Receptor Subunits

Shared receptor chain	Cytokines recognized
γ_c	IL-2, -4, -7, -9, -15, -21
IL-2Rβ	IL-2, IL-15
IL-4Rα	IL-4, -13
IL-13Rα1	IL-4, -13
β_c	IL-3, -5, GM-CSF
gp130	IL-6, -11, -27, -31, LIF, OSM, CNTF, CT-1, CLC
IL-12Rβ1	IL-12, -23
IL-10R2	IL-10, -22
IL-20R2	IL-20, -19, -24
IL-22R	IL-22, -24, -20

Figure 2-22 Table of use of shared receptor chains in the type I cytokine-receptor family Cytokine receptors that signal via the Jak–STAT pathway often share one or more components of their receptors, as indicated in this table. Often receptors sharing chains have overlapping or related functional properties. CLC: cardiotrophin-like cytokine; CNTF: ciliary neurotophic factor; CT-1: cardiotrophin-1; GM-CSF: granulocyte–monocyte colony-stimulating factor; LIF: leukemia inhibitory factor; OSM: oncostatin M.

Many cytokines and hematopoietins are recognized by structurally related receptors characterized by shared receptor subunits

In a few cases, the receptor for a type I cytokine is composed of a single type of chain that dimerizes on ligand binding and thereby activates a Jak kinase and downstream signaling events. More commonly, these cytokine receptors are composed of two different chains, one of which uniquely binds a single cytokine while the other is common to multiple receptors, often with functionally related properties (Figure 2-22). One such grouping contains the receptors that include the **common** cytokine-receptor **γ chain** (γ_c **chain**) and one or more cytokine-specific chains. These receptors, including the receptors for IL-2, IL-4, IL-7, IL-9, IL-15 and IL-21, are among the most important for immune function and are encountered in numerous places in this book (Figure 2-23). The γ_c chain and Jak3, which associates exclusively with it and is essential for its signaling function, are expressed only in hematopoietic cells and their functions are primarily related to lymphocyte development, growth and survival. Individuals with a defect in the γ_c component have no T cells or NK cells and have defective B cells, and suffer from a severe immunodeficiency that is known as **X-linked severe combined immunodeficiency** (X-SCID) because the γ_c gene is on the X chromosome. A very similar disease is seen in rarer individuals with deficiency in Jak3, which is encoded on an autosomal chromosome. The defect in T cell development seen in these individuals is due to an inability to respond to IL-7, which is required for T cell development, as described in Chapter 7. NK cells require IL-15 for their survival, and the inability to respond to this cytokine explains their absence. The B cells present in these individuals can make some antibodies but are defective in T cell-dependent antibody responses even if tested in the presence of normal T cells, probably reflecting significant roles for IL-2 and IL-21 in these responses.

Lymphocytes often have multiple γ_c-containing cytokine receptors but respond in different ways to the different members. This is because the γ_c chain promotes signaling by bringing Jak3 to the receptor complex, but the specificity for downstream signaling events is provided primarily by the cytokine-specific receptor subunit, which specifically recruits distinct STATs. The use of shared chains is extended even further by the IL-2 and IL-15 receptors, as they both use the IL-2Rβ chain and the γ_c chain, but they are differentiated by the use of unique third subunits, the IL-2Rα chain and the closely related IL-15Rα chain, which are not structurally related to other cytokine-receptor chains. These subunits are responsible for high-affinity binding of the respective cytokine but apparently do not contribute to signaling function; the IL-2Rβ chain controls the specificity of the signaling reactions and hence the ultimate biological responses.

Figure 2-23 Cytokine receptors containing the common γ chain (γ_c) Cytokine receptors that mediate the responses to IL-2, IL-4, IL-7, IL-9, IL-15 and IL-21 all contain unique chains and the γ_c chain. The γ_c chain binds Jak3 and contributes to signaling by these receptors primarily by bringing in Jak3 to collaborate with Jak1 bound to the other receptor chain. Which STAT is activated is determined by the cytokine-specific receptor chain, so that a single cell can respond in distinct ways to two or more cytokines via γ_c-containing cytokine receptors. The IL-2 and IL-15 receptors are unique in that they also share the IL-2Rβ chain and specificity for IL-2 or IL-15 is provided by the IL-2Rα chain or IL-15Rα chain, neither of which is structurally related to the Jak–STAT signaling cytokine-receptor family members. Note that IL-15Rα is made by the cell that makes IL-15, not by the responding cell, and therefore is more properly considered a part of the cytokine than a component of the cytokine receptor. The type I cytokine receptors shown here all have a conserved structure composed of two beta-sandwich domains. The carboxyl terminus of these domains has a characteristic WSXWS sequence motif that is part of the structure that does not contribute to cytokine binding.

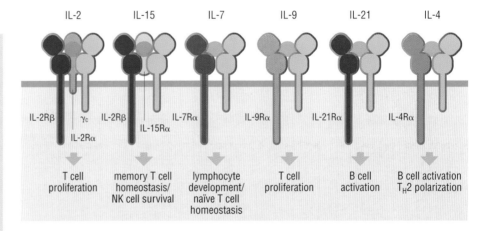

IL-2   IL-15   IL-7   IL-9   IL-21   IL-4

IL-2Rβ γ_c IL-2Rβ IL-7Rα IL-9Rα IL-21Rα IL-4Rα

IL-15Rα

IL-2Rα

| T cell proliferation | memory T cell homeostasis/ NK cell survival | lymphocyte development/ naïve T cell homeostasis | T cell proliferation | B cell activation | B cell activation T_H2 polarization |

Definitions

common γ (γ_c) chain: common component of the receptors for IL-2, IL-4, IL-7, IL-9, IL-15 and IL-21 and the component that is defective in patients with **X-linked severe combined immunodeficiency** (X-SCID).

γ_c chain: see **common γ chain**.

gp130: type I cytokine-receptor family member that mediates signaling by the IL-6 receptor and several other growth and differentiation factors of diverse function.

type II IL-4 receptor: cytokine receptor composed of the IL-4Rα chain which recognizes IL-4 and the IL-13Rα1 chain which recognizes IL-13 and exhibits high-affinity for both IL-13 and IL-4. The type II IL-4 receptor is expressed widely on non-hematopoietic cells.

X-linked severe combined immunodeficiency: severe immunodeficiency disease caused by defects in the gene encoding the cytokine-receptor γ_c **chain**. Affected individuals lack T cells and NK cells and have

Although IL-4 is among the cytokines recognized by γ_c-containing receptors, not all responses to IL-4 are defective in X-SCID patients' cells. This is because there are two types of IL-4 receptors: one that is composed of the IL-4Rα chain and the γ_c chain, and another that combines the IL-4Rα chain with another chain, called the IL-13Rα1 chain. This latter IL-4 receptor, which is known as the **type II IL-4 receptor**, also mediates responses to IL-13, which has IL-4-like function in cells expressing the receptor.

A second subclass of type I cytokine receptors includes the receptors for three hematopoietins that share the common cytokine-receptor β chain (β_c). These include the receptors for GM-CSF, IL-3 and IL-5, each of which is composed of a unique receptor chain and the β_c chain (Figure 2-24). All of these cytokines are active in hematopoiesis: IL-5 has a role in B cell activation and eosinophil development and activation; GM-CSF promotes early steps in hematopoeisis and has a variety of activating effects on myeloid cells, including dendritic cells; and IL-3 also promotes early hematopoiesis. In this subfamily, the β_c chain controls the specificity of signaling; differential biological roles for these hematopoietins are determined primarily by which cells express which unique receptor chains.

A third subclass of cytokine/hematopoietin receptors contains those related to the IL-6 receptor. This subfamily includes cytokines and growth and differentiation factors with functions in development, and the important inflammatory cytokine IL-6. In these receptors the cytokine-specific receptor chain is typically not involved in signal transduction and may not even span the membrane: these chains are shown in light orange in Figure 2-25. Indeed, in the case of the IL-6 receptor, which is composed of two signaling chains belonging to a cytokine receptor called **gp130** and two (the IL-6R chains) that bind IL-6, the IL-6R chains do span the membrane but their extracellular domains can also be released as soluble fragments. In this form IL-6R complexed with IL-6 can bind to gp130 molecules on other cells and thereby confer IL-6 responsiveness on them. We shall see in Chapter 3 that the release of soluble IL-6R in this way is important in the control of inflammation.

Other receptors closely related to gp130 have ligands that include IL-12 and IL-23. These cytokines are composed of a soluble cytokine-receptor-like chain related to the IL-6 receptor, bound to a four-helix-bundle cytokine (Figure 2-25). In this case, one of the type I cytokine-receptor subunits (called p40 and present in both IL-12 and IL-23) is made by the cytokine-secreting cell and combines with the cytokine-like subunit (called p35 in the case of IL-12). These then bind to gp130-like signaling receptors on the responding cell.

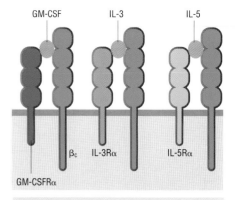

Figure 2-24 Hematopoietic receptors including the β_c chain The β_c chain is essentially a duplicated version of the two-domain type I cytokine receptors, such as the growth hormone receptor (Figure 2-20), and contains the WSXWS motif in its second and fourth beta-sandwich domains. The other chains have three domains, with the third having the WSXWS motif. For simplicity, bends between domains are not represented.

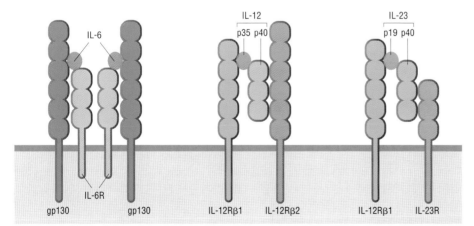

Figure 2-25 IL-6, IL-12 and IL-23 receptors The IL-6 receptor is composed of IL-6R and gp130. IL-6R has three beta-sandwich domains: an amino-terminal Ig-like domain followed by two domains related to other type I cytokine receptors. gp130 has six beta-sandwich domains: an amino-terminal Ig-like domain followed by two domains with strong homology to type I cytokine receptors (which are responsible for binding IL-6 together with IL-6R), followed by three fibronectin type III repeats. IL-12 and IL-23 are heterodimers of a four-alpha-helix-bundle type I cytokine molecule (p35 and p19 respectively) and a soluble type I cytokine-receptor-like molecule (p40). They are recognized by IL-12Rβ1 and IL-12Rβ2, which are both gp130-like, and by IL-12Rβ1 and IL-23R, which is like IL-6R.

defective B cells. An indistinguishable autosomal recessive form of SCID is caused by genetic defects in the gene encoding Jak3, a downstream signaling component of the pathway activated by γ_c receptors.

References

Boulay, J.-L. *et al.*: **Molecular phylogeny within type I cytokines and their cognate receptors.** *Immunity* 2003, **19**:159–163.

Leonard, W.J.: **Cytokines and immunodeficiency diseases.** *Nat. Rev. Immunol.* 2001, **1**:200–208.

Ozaki, K. and Leonard, W.J.: **Cytokine and cytokine receptor pleiotropy and redundancy.** *J. Biol. Chem.* 2002, **277**:29355–29358.

Trinchieri, G. *et al.*: **The IL-12 family of heterodimeric cytokines: new players in the regulation of T cell responses.** *Immunity* 2003, **19**:641–644.

Members of the TNF superfamily regulate a variety of immune and developmental events

Originally discovered because it can cause some tumors to regress, **tumor necrosis factor**, or **TNF**, is a key inflammatory cytokine and the prototype of a large family of secreted and membrane-bound cytokines that regulate both innate and adaptive immune responses and developmental events. Currently at least 18 individual TNF family members are known, with 25–30% amino acid similarity to one another. TNF superfamily members are trimeric molecules, usually homotrimers, and many remain membrane-bound and serve as cell-contact-mediated regulators. Other family members, including TNF itself, are cleaved from the membrane to release a soluble regulator that acts locally but can have systemic effects (as does TNF). The soluble and membrane-bound modes of signaling are illustrated schematically in Figure 2-26. Typically, membrane-bound forms of TNF family cytokines are stronger stimulants than are the released forms and in some cases the soluble form can antagonize responses to the membrane form.

TNF superfamily cytokines participate in a wide variety of immunological processes. TNF itself is a soluble mediator that acts principally to promote inflammation, both early in infection, when it is produced by innate immune cells, and later, when it is produced by T lymphocytes. It may also have a role in the killing of virus-infected cells. Many membrane-bound TNF family molecules participate in the interactions between immune cells that are necessary for initiating and regulating adaptive immune responses, and in activating effector actions such as antibody production by B cells or the destruction of microbes by phagocytic cells. Others such as *Fas ligand* and TRAIL have important functions in inducing a regulated form of cell death, *apoptosis*, which is a crucial mechanism both in the regulation of immune responses (through the death of activated cells) and in the destruction of infected cells by cytotoxic T cells and NK cells: the best-understood member of this subgroup is Fas ligand, which is produced by cytotoxic cells and acts through its receptor **Fas** to induce apoptosis.

The induction of inflammation by TNF, with the recruitment of immune cells to sites of infection, can be seen as the transient and reversible organization of tissue to foster interactions between immune cells; several other TNF family members also function in this way. **Lymphotoxin (LT)**, for example, is a membrane-bound TNF family signaling molecule that is critical to the development of the major secondary lymphoid organs: mice with mutations in either membrane-bound LTβ or its receptor, the LTβ receptor, have no lymph nodes or Peyer's patches and have a disorganized spleen. Similarly, the follicular dendritic cell network of B cell follicles in secondary lymphoid tissue requires TNF and LT for its development.

TNF-receptor superfamily members signal on oligomerization

The receptors for TNF family members, like the TNF family of signaling molecules, are a group of structurally related molecules, including at least 32 receptors and receptor decoy molecules. The latter are secreted versions of TNF receptors that inhibit responses by binding to the cytokine and preventing it from binding to cell-surface receptors.

Because the TNF family ligands are trimers, it was originally thought that ligand binding induced oligomerization of its receptor and this initiated signaling. Recently, however, it has been found that some TNF family receptors are preformed trimers on the cell surface, so it may be that ligand binding leads to a rearrangement of a preexisting oligomer and this structural change triggers interaction with intracellular signaling molecules.

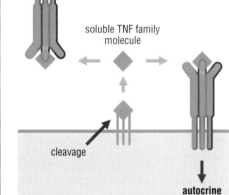

(a)
paracrine signaling

soluble TNF family molecule

cleavage

autocrine signaling

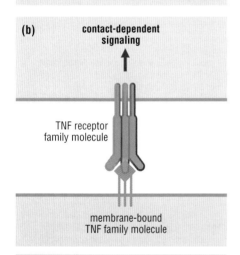

(b)
contact-dependent signaling

TNF receptor family molecule

membrane-bound TNF family molecule

Figure 2-26 Signaling by soluble and membrane-bound forms of TNF family molecules TNF family ligands are synthesized as type II membrane proteins (amino terminus inside cell, carboxyl terminus outside). In some cases **(a)** they are proteolytically cleaved from the membrane by proteases and can act distantly or locally either on a neighboring cell in a paracrine fashion or on the cell that produced it in an autocrine fashion. Frequently, TNF family members are retained on the plasma membrane and act on receptors via cell–cell contact **(b)**. In some cases the membrane-bound forms are biologically active and the secreted forms are not, in which case the soluble form can act as an antagonist (not shown).

Definitions

death domain: protein domain that is found in the cytoplasmic regions of members of a subfamily of the TNF receptor family that activate apoptosis, and in adaptor molecules, and that mediates the interactions between these receptors and their adaptors.

FADD: see **Fas-associated death domain**.

Fas: TNF receptor superfamily member that induces apoptosis in response to Fas ligand (FasL) made by T cells. Fas is one of the mechanisms by which virus-infected cells are killed and is also important for immune tolerance to self.

Fas-associated death domain (FADD): adaptor molecule that links death-domain-containing receptors of the TNF receptor superfamily to apoptosis.

LT: see **lymphotoxin**.

lymphotoxin (LT): TNF family member that can have two forms, a trimer of the secreted LTα subunit, which binds to the TNF receptors, and a heterotrimer of one α subunit and two membrane-bound β subunits

The extracellular domains of members of this family are all similar, but the cytoplasmic domains of the receptors fall into two classes, activating distinct pathways—one leading to apoptosis, and the other to the activation of transcriptional regulators controlling inflammatory and other immune responses (Figure 2-27). Thus, seven members of the family contain the **death domain**, a globular domain that was given this name because it is associated with signaling pathways leading to the death of the cell. On ligand binding by the receptor, the death domain recruits one or more intracellular adaptor molecules that also contain death domains, for example **Fas-associated death domain (FADD)** or **TNF receptor-associated death domain (TRADD)**. These adaptors recruit in turn members of a family of enzymes, the *caspases*, that activate the death of the cell by a mechanism that we will describe later in this chapter.

The other class of TNF receptors lacks cytoplasmic death domains; instead, it signals by means of interactions with a family of signaling adaptor molecules called **TRAFs**, for **TNF receptor-associated factors**, leading to the activation of transcriptional regulators (Figure 2-27c), especially NF-κB, which is discussed in the next section. There are six TRAFs and all except TRAF4 associate with at least one receptor of this family (the function of TRAF4 is unknown, however).

Variations in downstream signaling events from these receptors are depicted in Figure 2-27. Most receptors of this family seem to signal either to induce apoptosis through FADD and caspases, or to signal to transcriptional events through TRAFs, the major exception being TNF receptor 1 (TNFR1), which can signal through both types of pathways. In most cells, activation of the transcriptional pathway by TNFR1 inhibits the apoptotic pathway, so that the proapoptotic pathway induced by TNF is only operational if the transcriptional response is blocked, as can occur on infection of a cell with a virus that shuts off protein synthesis in the host. Thus, the dual mode of signaling by TNFR1 may reflect a role in antiviral immune defense.

We discuss one of the most important transcriptional pathways activated by TNF receptors in the next section, and the apoptotic pathway in section 2-11.

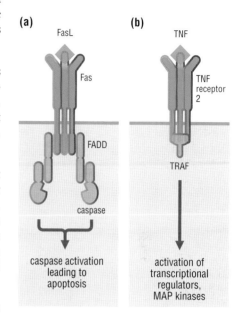

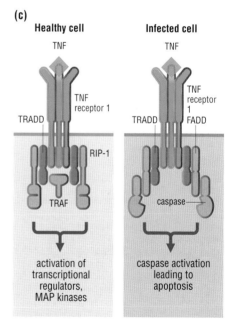

Figure 2-27 Signaling by TNF receptor superfamily members A subgroup of the TNF receptor family have a conserved 70-amino-acid globular domain in their cytoplasmic tail called the death domain. **(a)** Ligand binding by death-domain-containing receptors induces the formation of an oligomeric complex that in Fas and in most death-domain-containing receptors is composed of FADD (Fas-associated death domain protein) and caspase 8 (in mouse or human) or caspase 10 (in human only). This complex rapidly induces cell death if activated in sufficient quantities to overcome the levels of apoptosis inhibitors present in the cell (see sections 2-11 and 2-12). **(b)** In contrast, many TNF receptor superfamily members, including TNF receptor 2 (which is expressed mostly in lymphocytes), lack a death domain and instead have one or more short sequence motifs in their cytoplasmic domains that mediate the binding of signaling components called TRAFs (TNF receptor-associated factors). TRAFs in turn mediate the activation of NF-κB (see next section) and mitogen-activated protein (MAP) kinases, which are a family of serine/threonine protein kinases with important roles in intracellular signaling, often by activating transcriptional regulators. **(c)** TNF receptor 1, which is very widely expressed, is unusual in that it signals by both types of pathways. It first associates through its death domain with the adaptor TRADD (TNF receptor-associated death domain) which couples to TRAF2 and triggers transcriptional responses. Later, TRADD associates with FADD and possibly with caspase 8. In most cells, TNF does not induce apoptosis via the FADD/caspase 8 pathway unless transcription, protein synthesis or NF-κB is inhibited, any of which may happen in virus-infected cells. This is because several NF-κB-induced proteins prevent the death pathway from operating.

(LTα₁β₂), which binds to the LTβR and is important for the development of lymphoid structures.

TNF: see **tumor necrosis factor**.

TNF receptor-associated death domain (TRADD): adaptor molecule that links **TNF** receptor 1 to transcriptional activators or to caspases.

TNF receptor-associated factor (TRAF): one of a family of signaling components most of which associate with **TNF** superfamily receptors.

TRADD: see **TNF receptor-associated death domain**.

TRAF: see **TNF receptor-associated factor**.

tumor necrosis factor (TNF): prototype of a family of signaling molecules and a key initiator of inflammatory reactions.

References

Aggarwal, B.B.: **Signaling pathways of the TNF Superfamily: a double-edged sword.** *Nat. Rev. Immunol.* 2003, **3**:745–756.

Benedict, C.A. *et al.*: **Death and survival: viral regulation of TNF signaling pathways.** *Curr. Opin. Immunol.* 2003, **15**:59–65.

Bradley, J.R. and Pober, J.S.: **Tumor necrosis factor receptor-associated factors (TRAFs).** *Oncogene* 2001, **20**:6482–6491.

Locksley, R.M. *et al.*: **The TNF and TNF receptor superfamilies: integrating mammalian biology.** *Cell* 2001, **104**:487–501.

NF-κB activates inflammatory gene expression

A prominent feature of signaling by TNF-receptor family members is the activation of the transcriptional activator **nuclear factor κB (NF-κB)**. NF-κB activation is also a hallmark signaling event in the inflammatory response downstream of IL-1 receptors and the closely related Toll-like receptors, which are discussed in the next chapter. In addition, NF-κB is activated by a number of cellular stress response pathways and, as we shall see in Chapters 5 and 6, by antigen-receptor signaling in both T cells and B cells. Thus, NF-κB is central to the inflammatory response initiated by recognition of infection by innate immune mechanisms and in the adaptive immune response.

Among the molecules induced by NF-κB during an inflammatory response are cytokines, chemokines, effector molecules of immunity and prosurvival factors (Figure 2-28). In addition to countering the apoptotic signal coming from TNFR1, the prosurvival effects of NF-κB protect stimulated lymphocytes and other stressed cells. Mutations that fully inactivate NF-κB in mice are lethal during early development because of the essential role of this protein in cell survival after TNF-receptor stimulation. People with partial loss-of-function mutations in several components required for NF-κB activation have been found to be immunodeficient. In addition, these individuals have various ectodermal manifestations due to the role of the TNF-receptor family member EDAR in development of skin, teeth and hair, hence the name of the syndrome, *anhidrotic ectodermal dysplasia with immunodeficiency*.

NF-κB is regulated by inhibitory subunits

NF-κB comprises a small family of dimeric proteins constructed from five different gene products, all of which have a conserved region called the **Rel homology domain** which binds DNA. Three of these subunits, called c-Rel, p65 (or RelA) and RelB, also have an activation domain to stimulate transcription; the other two, p50 or NF-κB1 and p52 or NF-κB2, are initially synthesized as much longer precursors and lack a transactivation domain. NF-κB family members can form homodimers or heterodimers with each other, to produce gene regulatory complexes with different properties. For example, the p50–p50 dimer is a transcriptional repressor rather than an activator. Different NF-κB complexes have the same general DNA binding specificity, but may bind more strongly to at least partly different target genes.

The classical pathway of NF-κB activation is illustrated in the left panel of Figure 2-29, which shows an inflammatory response operating through a heterodimer of p50 and p65. NF-κB dimers are held in the inactive state by a family of inhibitors called **inhibitor of NF-κB (I-κB)**. Receptor signaling leads to activation of a multisubunit **I-κB kinase (IKK)** complex, which phosphorylates I-κB on two key serines. Phosphorylation of I-κB marks it for degradation by the *ubiquitin pathway* (Figure 2-29), so that the NF-κB dimer is liberated to translocate to the nucleus, bind DNA and activate transcription.

It is essential that the inflammatory actions of NF-κB are switched off once the inflammatory signal has ceased; because the inhibitor I-κB is degraded on NF-κB activation, new I-κB must be synthesized. There are three main members of the I-κB family, two of which, I-κBβ and I-κBε are synthesized constitutively and reestablish NF-κB inhibition with a relatively slow time course on cessation of signaling. Synthesis of the third, I-κBα, is under the control of NF-κB itself, and it is therefore produced in response to signaling: it enters the nucleus on synthesis, binds to NF-κB and shuttles it back to the cytoplasm via a nuclear export signal, switching off NF-κB action with a very short delay, thus making NF-κB activity self-limiting.

Gene Products Under the Control of NF-κB

Inflammatory cytokines

TNF

IL-1

IL-6

IL-12

Lymphotoxin α/β

GM-CSF

IFN-β

Chemokines

IL-8

MIP-1α

MCP

Eotaxin

Adhesion molecules

ICAM-1

VCAM-1

E-selectin

Immune effector molecules

FasL

iNOS

COX-2

β-defensins

Prosurvival molecules

Bcl-X$_L$

A1

c-IAP1, 2

Figure 2-28 Gene products under the control of NF-κB Examples of genes for which NF-κB is believed to be an important regulator of transcription. Note that different NF-κB dimers probably participate in the induction of at least partly distinct sets of target genes. The functions of these molecules are described elsewhere in this book, the prosurvival ones in the next two sections.

Definitions

I-κB: see **inhibitor of NF-κB.**

I-κB kinase (IKK): complex consisting of two related kinase subunits, called IKKα and IKKβ, and a scaffolding subunit, IKKγ or NEMO, that activates the transcriptional activator **NF-κB** by marking its inhibitor, **I-κB,** for degradation. There are two pathways of activation, the classical and the alternative pathway. IKKβ and IKKγ are necessary for responses via the classical pathway but not via the alternative pathway, which only requires IKKα. Conversely, IKKα is dispensable for the classical pathway. IKKγ (NEMO) is encoded on the X-chromosome and is partly defective in the disease X-linked anhidrotic ectodermal dysplasia with immunodeficiency.

IKK: see **I-κB kinase.**

inhibitor of NF-κB (I-κB): any of a family of proteins that inhibit the transcriptional activator **NF-κB** by binding to it and preventing it from translocating to the nucleus. It exists in multiple isoforms, all of which contain ankyrin repeat structures that mediate the interaction with NF-κB.

NF-κB: see **nuclear factor κB.**

nuclear factor κB (NF-κB): any of a small family of dimeric DNA-binding proteins that mostly function as transcriptional activators and have a central role in both innate and adaptive immune responses; originally described for their binding to the B site in the Igκ intronic enhancer.

Rel homology domain: 300-amino-acid-long homology region in all five subunits of the transcriptional activator **NF-κB.** This homology forms two immunoglobulin-like beta-sheet sandwich structures.

Some NF-κB heterodimers are self-inactivated

Although the mechanism illustrated on the left in Figure 2-29 is the usual, or *classical pathway* of NF-κB activation, there is an *alternative pathway* in which there is activation of an NF-κB dimer made up of RelB and p100, the unprocessed precursor of p52. Both p50 and p52, as we have mentioned, are synthesized as longer precursors (p105 and p100 respectively). The extended carboxy-terminal domains of these precursors are structurally homologous to I-κB and have the same function. Whereas the p105 precursor is believed to be constitutively processed to give p50, p100 is not; it combines with RelB and is inactive until the cell is stimulated via a subset of TNF receptor family members, including the receptor for the TNF family molecule BAFF (B cell-activating factor), which promotes the survival of mature B cells. As shown in the right-hand panel of Figure 2-29, receptor activation leads to proteolytic processing and degradation of the inhibitory carboxyl terminus of p100 and liberation of the remainder as the transcriptionally active p52–Rel-B complex.

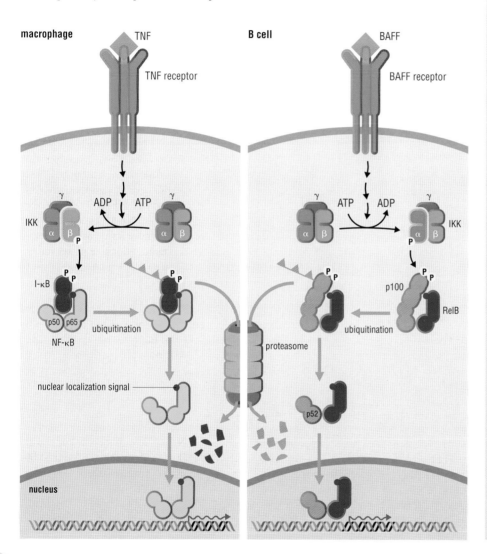

Figure 2-29 The classical and alternative pathways of NF-κB activation (a) Activation of NF-κB by TNF receptor 1 in an inflammatory pathway. TNF binding to the receptor recruits TRAF signaling components (shown in Figure 2-27), which in turn activate the I-κB kinase complex (IKK). IKK is composed of two protein kinase subunits (IKKα and IKKβ; also called IKK1 and IKK2) and a subunit called IKKγ, or NEMO, which acts as an adaptor for the kinase subunits to upstream activators (indicated by black arrows), in most cases. In this pathway IKKβ and its kinase activity are essential but IKKα is dispensable. IKKβ phosphorylates I-κB on two neighboring serine residues, which target I-κB for ubiquitination followed by degradation by the *proteasome*, a large cytosolic proteolytic enzyme complex. Removal of I-κB reveals a nuclear localization signal in the NF-κB subunits, allowing translocation to the nucleus, binding to DNA and activation of target gene transcription. NF-κB subunits also become phosphorylated (not shown), which modulates their transcriptional activity. (b) A survival pathway operating through NF-κB activation by the BAFF receptor of B cells, a member of the TNF receptor family. Here IKKα is essential but IKKβ and IKKβ are not. IKKα phosphorylates the carboxy-terminal I-κB-like region of p100, leading to ubiquitination and removal of the I-κB-like region, leaving a p52–RelB NF-κB complex that goes to the nucleus and activates gene transcription. The partial degradation of p100 by the proteasome is atypical; more frequently, the proteasome fully degrades a polyubiquitinated protein. Note that BAFF also activates the classical NF-κB pathway (not shown). The mechanisms by which receptors activate IKK are incompletely understood, but they differ between different receptor families and are described where the signaling mechanisms of particular receptor types are discussed.

References

Bonizzi, G. and Karin, M.: **The two NF-κB activation pathways and their role in innate and adaptive immunity.** *Trends Immunol.* 2004, **25**:280–288.

Ghosh, S. and Karin, M.: **Missing pieces in the NF-κB puzzle.** *Cell* 2002, **109**:S81–S96.

Hayden, M.S. and Ghosh, S.: **Signaling to NF-κB.** *Genes Dev.* 2004, **18**:2195–2224.

Hoffmann, A. *et al.*: **Genetic analysis of NF-κB/Rel tran-** scription factors defines functional specificities. *EMBO J.* 2003, **22**:5530–5539.

Hoffmann, A. *et al.*: **The IκB–NF-κB signaling module: temporal control and selective gene activation.** *Science* 2002, **298**:1241–1245.

Jacobs, M.D. and Harrison, S.D.: **Structure of an IκBα/NF-κB complex.** *Cell* 1998, **95**:749–758.

Puel, A. *et al.*: **Inherited disorders of NF-κB-mediated immunity in man.** *Curr. Opin. Immunol.* 2004, **16**:34–41.

2-11 Molecular Control of Apoptosis

Apoptosis has many important roles in immune function

Cell death generally takes one of two forms, **necrosis** or **apoptosis**. Necrosis occurs for example in cells subjected to an acute injury leading to cell rupture. Necrotic cells release cytoplasmic contents into the extracellular space and trigger inflammation by mechanisms we shall discuss in Chapter 3. In contrast, apoptosis is a regulated physiological process that is triggered by molecular events within the cell, Normally, the apoptotic cell is internalized and digested by phagocytes without release of cytoplasmic contents and triggering of the consequent inflammatory response (although we shall see in Chapter 3 that infected apoptotic cells do trigger inflammation because of the microbial components they contain).

Apoptosis occurs during normal cell turnover, but also in cells infected by viruses or invasive microbes. In infected cells, this can be mediated by stresses put on the cell by the infection or it can be induced by immune-system cells to fight the infection. Also, infections by extracellular microbes attract inflammatory leukocytes, whose killing mechanisms can spill over and damage bystander cells, inducing their apoptosis.

Apoptosis also has a crucial role in controlling the lifespan of immune-system cells. Immune cell numbers increase during infection and then are brought back to normal levels by increased apoptosis after the infection is cleared. Immune stimulation leads to the production of cytokines that promote the proliferation and/or longevity of particular immune cells, expanding their numbers. When the cytokines are used up and production stops, the cells responding to those cytokines may undergo apoptosis. Apoptosis is also critical in eliminating those lymphocytes with self-reactivity during their development (see section 1-4). Clearly, apoptosis is critical in many aspects of immune-system function, and its activation and regulation will be important issues in many of the processes discussed in this book.

Figure 2-30 Extrinsic and intrinsic pathways for inducing apoptosis In the extrinsic pathway, death receptors such as Fas and TNFR1 are oligomerized by interaction with extracellular ligands, which induce binding of adaptor molecules and procaspase 8 and/or procaspase 10 to the cytoplasmic death domains of these receptors. These procaspases have a low level of protease activity and cleave one another to generate the active initiator caspases. These caspases induce cell death by proteolytically activating effector caspases, such as caspase 3 and caspase 7, which then are primarily responsible for acting on the molecules that induce the cellular events associated with apoptosis. In the intrinsic pathway, perturbations of the mitochondria, as occur on withdrawal of survival factors such as cytokines, lead to the release of cytochrome c, which binds to Apaf-1 and leads to assembly of the oligomeric apoptosome including procaspase 9, which is allosterically activated and can now cleave itself to form caspase 9, as well as activating the effector caspases. The intrinsic pathway can also be initiated from the endoplasmic reticulum (not shown), and also involves additional events promoting cell death, as described in subsequent figures. Not all apoptosis via the intrinsic pathway is dependent on apoptosome components, however, indicating that there are alternative mechanisms by which effector caspases can be activated.

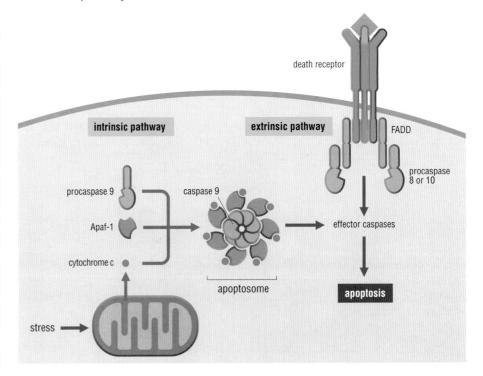

Definitions

Apaf-1: see **apoptotic protease activating factor 1**.

apoptosis: regulated cell death in which activation of specific proteases and nucleases leads to death characterized by chromatin condensation, protein and DNA degradation, loss of plasma membrane lipid asymmetry and disintegration of the cell into membrane-bounded fragments.

apoptosome: molecular complex of procaspase 9, **Apaf-1** and cytochrome c that assembles in response to

mitochondrial release of cytochrome c and results in the activation of caspase 9 and triggering of the **intrinsic pathway** of **apoptosis**.

apoptotic protease activating factor 1 (Apaf-1): an adaptor molecule that oligomerizes on interaction with cytochrome c, leading to the binding and activation of caspase 9.

caspase: any of a family of cysteine proteases that cleave after aspartic acid residues in proteins and have an essential role in either **apoptosis** or processing of inflammatory cytokines to active forms.

caspase activation and recruitment domain (CARD): domain found in several proteins that activate **apoptosis** or regulate NF-κB, with strong structural similarity to death domains and death effector domains. CARD domains generally associate with CARD domains of other proteins.

effector caspases: caspases that are activated by **initiator caspases**, amplifying the processes that result in the stereotypical features of **apoptosis**, including loss of plasma membrane asymmetry, nuclear condensation and endonuclease attack on nuclear DNA.

Apoptosis is triggered by a proteolytic cascade involving caspases

Apoptosis proceeds via a stereotypical sequence of events, including: loss of mitochondrial function; nuclear condensation; nuclease attack on genomic DNA between nucleosomes, generating a characteristic size ladder of DNA fragments; and loss of plasma membrane asymmetry resulting in exposure of phosphatidylserine on the outside of the plasma membrane. The changes in the plasma membrane of apoptotic cells are recognized early in the process by phagocytic cells, leading to their rapid clearance.

Apoptosis can be triggered by two pathways, called the **extrinsic pathway** and the **intrinsic pathway** (Figure 2-30), which differ in their upstream events but converge on the activation of a small family of proteolytic enzymes called **caspases**. The name derives from the fact that these enzymes are cysteine proteases and that they cleave after aspartic acid residues in their target proteins. Each pathway initiates apoptosis by triggering the activation of an **initiator caspase** that cleaves and activates the same downstream **effector caspases** (caspase 3 and caspase 7), which then perform cleavages that result in the manifestations of apoptosis (Figure 2-31).

The extrinsic pathway is one of the mechanisms whereby virus-infected cells are killed by cytotoxic T lymphocytes and their innate immune counterpart, natural killer cells (see section 1-2). This pathway is initiated by TNF-receptor family members that contain death domains in their intracellular domain (see section 2-9). Ligand binding induces assembly at the receptor cytoplasmic domain of a complex including adaptor molecules and **procaspase** 8 or procaspase 10 (see Figure 2-30). These procaspases have a low level of activity, but this activity is sufficient to process adjacent molecules efficiently into their highly active caspase form. The resulting caspase 8 and/or caspase 10 molecules then proteolytically process the effector procaspases, generating an amplifying cascade leading to apoptosis. In addition to the extrinsic pathway, cytotoxic T cells and natural killer cells induce apoptosis of virus-infected cells by another mechanism in which they introduce into the cytoplasm of the target cells a series of proteases, called granzymes. One of these, granzyme B, is an efficient activator of effector caspases, as an alternative to the extrinsic pathway. This killing mechanism is described more fully in Chapter 5.

The intrinsic pathway of apoptosis can be triggered by a variety of cell stresses or external signals. For example, internal disturbances in mitochondrial or endoplasmic reticulum function, or in DNA synthesis, can lead to the induction of apoptosis. These pathways may protect against neoplasia and virus infection by inducing death of cells with regulatory defects or of infected cells (for the latter see Chapter 3). The intrinsic pathway can also be triggered by loss of cytokine-receptor- or integrin-generated survival signals.

The intrinsic pathway is controlled in a more complex way than the extrinsic pathway, as described in the next section, but one of its important features is the assembly of the initiator procaspase 9 with a heptameric complex of the adaptor molecule **apoptotic protease activating factor 1 (Apaf-1)**. Assembly of this complex, called the **apoptosome**, is triggered by binding to Apaf-1 of cytochrome c, which is released from mitochondria as an early event in the induction of apoptosis. Binding of caspase 9 to the assembled Apaf-1–cytochrome c complex greatly increases its activity, allowing sufficient activation of the effector caspases to trigger apoptosis. Although the apoptosome strongly activates effector caspases, it is not absolutely essential to apoptosis via the intrinsic pathway, which can still occur in mice deficient in Apaf-1 or caspase 9, or through stresses to the endoplasmic reticulum independently of mitochondrial involvement. Thus, multiple biochemical reactions can trigger the intrinsic pathway of apoptosis, only some of which are understood.

Function of human caspases

Caspase	Function
1	cytokine maturation
2	initiator caspase (intrinsic pathway)
3	effector caspase
4	cytokine maturation
5	cytokine maturation
6	effector caspase
7	effector caspase
8	initiator caspase (extrinsic pathway)
9	initiator caspase (intrinsic pathway)
10	initiator caspase (extrinsic pathway)
12	cytokine maturation?
14	effector caspase?

Figure 2-31 Table of functions of human caspases Humans have at least 12 caspases involved in the control of apoptosis and/or the maturation of IL-1 family cytokines. These fall into three general groups: effector caspases, which generally have very short amino-terminal pro-domains; initiator caspases, which have long pro-domains containing either a **caspase activation and recruitment domain (CARD)** or a highly related domain called the death effector domain to interact with adaptor proteins involved in their activation, and the proinflammatory caspases (caspase 1, 4, 5). Caspase 2 seems to be an initiator caspase that can induce the release from mitochondria of apoptotic activators. Caspase 6 is an effector caspase that is thought to be downstream of the main effector caspases, caspase 3 and caspase 7, and targets mainly nuclear lamin, whereas caspases 3 and 7 have many targets that participate in apoptosis.

---

extrinsic pathway: (of **apoptosis**) pathway inducing apoptosis downstream of TNF-receptor family members with death domains in their cytoplasmic tails.

initiator caspases: caspases that can be activated by death-domain-containing receptors or by perturbations to internal compartments such as the mitochondria or the endoplasmic reticulum.

intrinsic pathway: (of **apoptosis**) pathway of inducing apoptosis resulting from cell stress or loss of survival signals from outside the cell (typically adhesion and/or cytokines).

necrosis: mode of cell-injury-induced cell death that often involves rupture of the plasma membrane and release of cytoplasmic contents into the extracellular space.

procaspase: precursor form of **caspase** with low or no activity. As with many other proteases, proteolytic cleavage removes an amino-terminal pro-domain. In the **initiator caspases**, this pro-domain includes the **CARD**.

References

Adams, J.M.: **Ways of dying: multiple pathways to apoptosis.** *Genes Dev.* 2003, **17**:2481–2495.

Danial, N.N. and Korsmeyer, S.J.: **Cell death: critical control points.** *Cell* 2004, **116**:205–219.

Fesik, S.W.: **Insights into programmed cell death through structural biology.** *Cell* 2000, **103**:273–282.

Martinon, F. and Tschopp, J.: **Inflammatory caspases: linking an intracellular innate immune system to autoinflammatory diseases.** *Cell* 2004, **117**:561–574.

Shi, Y.: **Mechanisms of caspase activation and inhibition during apoptosis.** *Mol. Cell* 2002, **9**:459–470.

Increased mitochondrial outer-membrane permeability is the major trigger of the intrinsic apoptosis pathway

The caspases are the major effectors of apoptosis, as described in the previous section, and are thus the target of regulatory mechanisms that initiate or block apoptosis. Key intracellular regulators of the intrinsic pathway include three groups: (1) a family of inhibitors of caspases, called the **inhibitors of apoptosis (IAPs)**; (2) several unrelated molecules released from the space between the two mitochondrial membranes into the cytoplasm to promote cell death by activating procaspase 9, by blocking IAP function and by directly acting in the nucleus (Figure 2-32); and (3) a family of proteins, the **Bcl-2 family** proteins, that regulate apoptosis upstream of the release of mitochondrial proteins and the consequent activation of caspases.

Figure 2-32 Release of mitochondrial matrix proteins into the cytoplasm induces apoptosis In most cases, the intrinsic pathway of apoptosis is initiated by the release of a set of mitochondrial proteins into the cytoplasm. Cytochrome c binds to Apaf-1, inducing assembly of the apoptosome and activation of caspase 9, which in turn activates the effector caspases, caspase 3 and caspase 7. Among the targets of the effector caspases is caspase-activated DNase (CAD), which is important for cleaving chromosomal DNA. Other mitochondrial proteins participating in apoptosis are two (Smac/Diablo and Omi/HtrA2) that inhibit the IAPs, which in turn inhibit active caspases 3, 7 and 9. Also released from mitochondria during the intrinsic pathway are AIF (apoptosis-inducing factor) and endonuclease G (endo G). These two proteins go to the nucleus, where they induce chromatin condensation and breaks in the DNA respectively. Upstream of mitochondrial release of these proteins are the Bcl-2 family members, which can be proapoptotic or antiapoptotic, and, in some cases, caspase 2. How caspase 2 is activated is not understood.

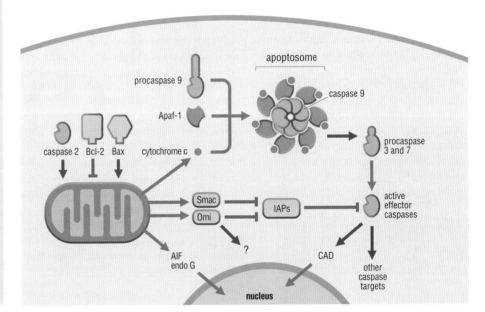

Bcl-2 is an antiapoptotic protein that was originally discovered as the product of a gene commonly found translocated to the Ig locus in follicular B cell lymphomas (Bcl stands for B cell lymphoma). Because the Ig locus is highly active in B cells, the translocation results in overexpression of the Bcl-2 protein, and it is now known that Bcl-2 is one of a family of anti-apoptotic proteins and protects the lymphoma cells from apoptosis, thereby promoting tumor growth. The Bcl-2 family has since been discovered to contain at least 17 proteins in mammals, some of which, like Bcl-2, are antiapoptotic, whereas the others promote apoptosis. They fall broadly into three structural and functional groups (Figure 2-33), consisting of a group of three directly death-inducing proteins; a group of six antiapoptotic proteins, including Bcl-2 itself,

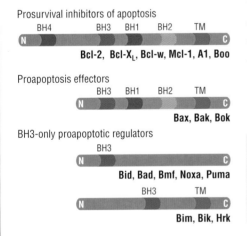

Prosurvival inhibitors of apoptosis

Bcl-2, Bcl-X$_L$, Bcl-w, Mcl-1, A1, Boo

Proapoptosis effectors

Bax, Bak, Bok

BH3-only proapoptotic regulators

Bid, Bad, Bmf, Noxa, Puma

Bim, Bik, Hrk

Figure 2-33 The Bcl-2 family of apoptotic regulators Bcl-2 family members fall into three groups, whose basic structural features are illustrated. One group, exemplified by Bcl-2 itself, is antiapoptotic, acting to inhibit the next group, exemplified by Bax, which has homology to Bcl-2 in three regions (called Bcl-2 homology or BH1, 2 and 3) and is proapoptotic, directly initiating the apoptotic program. The third group has homology only in the roughly 16-amino-acid BH3 region, and is proapoptotic, acting to sense cellular or environmental signals and regulate the antiapoptotic or apoptotic family members.

Definitions

Bcl-2: antiapoptotic protein of the **Bcl-2 family** of regulators of the intrinsic pathway of apoptosis, of which it was the first member to be discovered.

Bcl-2 family: a family of regulators of apoptosis with homology to **Bcl-2**, falling into three main structural and functional groups: a directly death inducing group of three members, an antiapoptotic group of six members including Bcl-2 itself, and the **BH3-only family**, which regulate members of the other groups to induce apoptosis in various conditions.

BH3-only family: family of stress-sensing proteins that induce apoptosis by inhibiting antiapoptotic **Bcl-2 family** members or, in one case, by activating apoptotic Bcl-2 family members.

IAP: see **inhibitor of apoptosis**.

inhibitor of apoptosis (IAP): any of a family of structurally related molecules that bind to and inhibit effector caspases and active caspase 9 as well as promoting their degradation by the proteasome. These proteins are thought to set a threshold for the activation of caspases to induce apoptosis.

References

Cory, S. and Adams, J.M.: **The Bcl2 family: regulators of the cellular life-or-death switch.** *Nat. Rev. Cancer* 2002, **2**: 647–656.

Green, D.R. and Kroemer, G.: **The pathophysiology of mitochondrial cell death.** *Science* 2004, **305**: 626–629.

Liston, P. *et al.*: **The inhibitors of apoptosis: there is more to life than Bcl2.** *Oncogene* 2003, **22**: 8568–8580.

Strasser, A.: **The role of BH3-only proteins in the immune system.** *Nat. Rev. Immunol.* 2005, **5**: 189–200.

that act to block the first group; and a group of eight sensors that respond to environmental or cellular signals and regulate members of the other groups. The sensor group of Bcl-2-related proteins shows sequence similarity to other family members only in an approximately 16-amino-acid-residue-long alpha-helical region called the Bcl-2 homology 3 region or BH3 (see Figure 2-33), and hence this group is referred to as the **BH3-only family**.

The six antiapoptotic Bcl-2 family members are expressed in unique patterns, but it seems that every mammalian nucleated cell must express at least one of these molecules to be viable. In contrast, Bax, Bak and Bok induce apoptosis. Bok is expressed only in reproductive cells, but Bax and Bak are widely expressed. Mice lacking either Bax or Bak are essentially normal, but mice lacking both are severely defective in the induction of apoptosis. For this reason it is thought that the intrinsic pathway is triggered mainly by Bax and Bak, and their functions are inhibited by the antiapoptotic Bcl-2 family members.

Some Bcl-2 family members bind constitutively to internal membranes, especially mitochondria and endoplasmic reticulum, whereas other family members are cytosolic in healthy cells and become attached to intracellular membranes in response to apoptosis-inducing stresses. One hypothesis is that Bcl-2 and the antiapoptotic family members stabilize the mitochondrial outer membrane, whereas Bax and Bak directly destabilize the mitochondrial membrane, promoting its rupture and the release of cytochrome c and other apoptosis-inducing proteins. It should be noted that Bcl-2 and other antiapoptotic family members also seem to block apoptosis initiating at the endoplasmic reticulum, which can be independent of mitochondrial rupture. The mechanism of this endoplasmic reticulum stress-induced intrinsic pathway is not well understood.

BH3-only family members are sensors of cellular stresses and promote apoptosis in response to a loss of survival signals, DNA damage, or other cellular stress. They act on other Bcl-2 family members by binding to them through their BH3 regions. With one exception they seem to act by inhibiting the prosurvival Bcl-2-like molecules. The exception, Bid, is thought to bind to Bax and promote apoptosis. The cellular stresses that activate BH3-only proteins, and some of the mechanisms by which they are thought to act, are summarized in Figure 2-34. Bim in particular is important for lymphocyte apoptosis.

Figure 2-34 The BH3-only family of apoptosis inducers and their regulation Illustrated are four means by which BH3-only protein function is regulated: protein cleavage, phosphorylation, sequestration, and transcriptional induction. **(a)** Bid is activated by cleavage by caspases or by granzyme B (introduced into target cells by cytotoxic T cells or natural killer cells). The cleaved form activates apoptosis, apparently by directly activating Bax. **(b)** Bad is regulated by phosphorylation, particularly by the protein kinase Akt downstream of receptor signaling. The phosphorylated form is bound to the chaperone-like protein 14-3-3. Dephosphorylation by unknown phosphatases leads to active Bad, which seems to inhibit the function of Bcl-2. **(c)** Bim is found in healthy cells bound to microtubules via dynein light chains. Cellular stress is postulated to release Bim from microtubules to induce apoptosis. Similarly, Bmf is associated with microfilaments (not shown) and may be released by certain cellular stresses. Increased active Bim also promotes apoptosis in self-reactive lymphocytes binding to self-antigen and in lymphocytes undergoing cytokine withdrawal. It is not known how Bim activity is increased in these situations, but there is evidence for the regulation of Bim by transcriptional induction and by phosphorylation, in addition to sequestration. Finally, several BH3 family members are controlled at the level of gene transcription. For example, **(d)** p53 senses unscheduled DNA synthesis, as occurs during infection with DNA viruses, and induces the synthesis of BH3-only family members Puma and Noxa. These BH3-only proteins can induce apoptosis, probably by inhibiting Bcl-2.

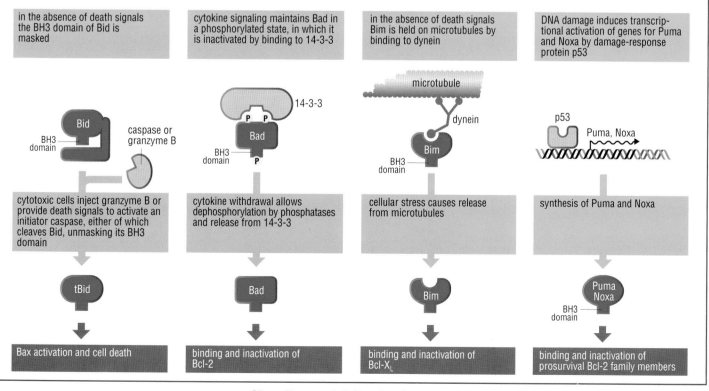

| in the absence of death signals the BH3 domain of Bid is masked | cytokine signaling maintains Bad in a phosphorylated state, in which it is inactivated by binding to 14-3-3 | in the absence of death signals Bim is held on microtubules by binding to dynein | DNA damage induces transcriptional activation of genes for Puma and Noxa by damage-response protein p53 |

| cytotoxic cells inject granzyme B or provide death signals to activate an initiator caspase, either of which cleaves Bid, unmasking its BH3 domain | cytokine withdrawal allows dephosphorylation by phosphatases and release from 14-3-3 | cellular stress causes release from microtubules | synthesis of Puma and Noxa |

| Bax activation and cell death | binding and inactivation of Bcl-2 | binding and inactivation of Bcl-X$_L$ | binding and inactivation of prosurvival Bcl-2 family members |

Figure 2-35 Structure of a CXC chemokine
Shown is a ribbon diagram of IL-8 (CXCL8).
The amino terminus is at the lower left of the
structure. The two cysteine residues with a
single intervening amino-acid residue are near
the amino terminus and the disulfide bonds in
which they participate are shown in yellow.
The main structural features of the chemokine
are a beta sheet (blue) and an alpha helix
(red). (PDB 1il8)

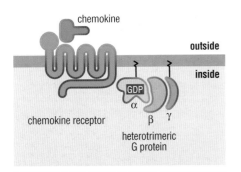

**Figure 2-36 Schematic representation of
chemokine receptor–chemokine interaction**
Chemokine receptors are seven-transmembrane
receptors that couple to heterotrimeric G
proteins, especially Gi. Chemokines interact
primarily with the globular amino-terminal
domain of chemokine receptors (on the left
side of the molecule, as shown) but can also
interact with the first extracellular loop of the
receptor (loop between the second and third
transmembrane segments). The amino-terminal
region of the chemokine often has an important
role in the interaction with the receptor.

Chemokines direct the migration of immune-system cells

Chemokines are a related family of at least 40 small, basic proteins that function in directing the migration of immune-system cells to sites of inflammation (**inflammatory chemokines**) or into and through lymphoid organs (**homeostatic** or **lymphoid chemokines**). Although a few chemokines are transmembrane proteins, the vast majority are secreted proteins that bind to the surface of endothelial cells or to the extracellular matrix at decreasing concentration with distance from the source and thus generate gradients up which cells can be directed to their source. Whereas inflammatory chemokines are produced only in response to infection, homeostatic chemokines are constitutively produced in lymphoid organs and serve to maintain the normal populations of migratory and resident cells there.

Structurally, there are two major and two minor families of chemokines, based on the relative positions of two highly conserved cysteine residues in the amino-terminal region of the protein. These two cysteines form disulfide bonds with cysteines located toward the carboxyl terminus (Figure 2-35). In one major family (the CC chemokines), the two cysteines are adjacent to one another, whereas in the other major family (the CXC chemokines) they are separated from one another by an intervening amino acid. In more minor variants, one chemokine has three residues between the two cysteines and is referred to as a CX_3C chemokine, and two chemokines lack the first conserved cysteine and therefore are referred to as either C chemokines or XC chemokines. Chemokines often have multiple names, including a standardized nomenclature based on their amino-terminal cysteines (CCL1 is CC chemokine 1; L is for ligand, as the chemokines are ligands for chemokine receptors). The amino-acid sequence amino-terminal to the two cysteines in chemokines is often critical for receptor binding and biological activity. In some cases, it is also a site of biological regulation. For example, CXCL4 is released by platelets, but the active form is generated by further proteolytic cleavage of the amino terminus by monocyte proteases made in response to tissue damage.

Chemokines act on cells through a family of at least 18 receptors, all of which are **seven-transmembrane G-protein-coupled receptors** (Figure 2-36). These receptors are so called because the polypeptide chain of the receptor crosses the plasma membrane seven times, and they signal through a family of G proteins, the **heterotrimeric G proteins** having three subunits, one of which is homologous to the small G proteins we encountered in section 2-0. Especially important is Gi, which controls cell migration. The bacterial pathogen *Bordetella pertussis* makes a toxin, pertussis toxin, that can enter cells and inactivate Gi, possibly as a way of decreasing leukocyte migration to sites of bacterial growth. The chemokine receptors are named on the basis of the family of chemokines they recognize (CCR1 is receptor 1 for CC chemokines, and so on). A list of known chemokine receptors, their ligands and their principal functions is shown in Figure 2-37. The large numbers of chemokines and chemokine receptors available to the mammalian immune system permit exquisite control over inflammatory processes because different chemokines attract different subsets of immune cells.

Dynamic regulation of chemokine receptor expression is another key element of control. For example, activation of lymphocytes leads to a change of expression from the chemokine receptors that promote migration through the spleen and lymph nodes, to the chemokine receptors that promote migration to sites of inflammation. Moreover, different functional classes of lymphocytes express distinct chemokine receptors directing them to inflammatory sites that are distinguished by different inflammatory chemokines, and in this way lymphocytes are differentially attracted to sites appropriate to their function.

Definitions

chemokine: any of a family of closely related small, basic cytokines whose main function is as chemoattractants. The name is a contraction of chemotactic cytokine.

heterotrimeric G protein: signaling G protein composed of three different subunits, an α subunit with GTPase activity, and associated regulatory β and γ subunits. These GTPases act to relay signals from seven-transmembrane receptors to downstream targets.

homeostatic chemokine: constitutively expressed chemoattractant molecule that directs the migration of lymphocytes and dendritic cells into specialized regions of the secondary lymphoid tissues. Also known as **lymphoid chemokine**.

inflammatory chemokine: chemokine that is produced at an inflammatory site and mediates the attraction of immune cells from the blood to that location. Different chemokines attract different types of cells and thereby dictate the nature of the inflammation.

lymphoid chemokine: see **homeostatic chemokine**.

seven-transmembrane G-protein-coupled receptor: receptor protein that crosses the cell membrane seven times and relays signals to the interior of the cell through **heterotrimeric G proteins**.

Human Chemokines and Chemokine Receptors

Chemokine	Chemokine receptor	Cells attracted
(a) Inflammatory		
CXCL1 (GRO α)	CXCR2	N
CXCL2 (GRO β)	CXCR2	N
CXCL3 (GRO γ)	CXCR2	N
CXCL5 (ENA-78)	CXCR2	N
CXCL7 (NAP-2)	CXCR2	N
CXCL6 (GCP-1)	CXCR1, CXCR2	N
CXCL8 (IL-8)	CXCR1, CXCR2	N
CXCL4	CXCR3B	eT, NK, DC
CXCL9 (Mig)	CXCR3, CXCR3B	eT, NK, DC
CXCL10 (IP-10)	CXCR3, CXCR3B	eT, NK, DC
CXCL11 (I-TAC)	CXCR3, CXCR3B	eT, NK, DC
CCL23 (MPIF-1)	CCR1	T_H1, M, DC
CCL15 (HCC-2)	CCR1, CCR3	eT, M, DC, Eos, Baso
CCL3 (MIP-1α)	CCR1, CCR5	T_H1, NK, M, DC, Eos
CCL14 (HCC-1)	CCR1, CCR5	T_H1, NK, M, DC, Eos
CCL7 (MCP-3)	CCR1, CCR2, CCR3	eT, NK, M, DC, Eos, Baso
CCL16 (HCC-4)	CCR1, CCR2, CCR5	T_H1, NK, DC, Eos, Baso
CCL5 (RANTES)	CCR1, CCR3, CCR5	eT, NK, M, DC, Eos, Baso
CCL8 (MCP-2)	CCR1, CCR2, CCR3, CCR5	eT, DC, Eos, Baso
CCL2 (MCP-1)	CCR2	eT, NK, M, DC, Baso
CCL13 (MCP-4)	CCR2, CCR3	eT, NK, M, DC, Eos, Baso
CCL24 (Eotaxin-2)	CCR3	T_H2, DC, Eos, Baso
CCL26 (Eotaxin-3)	CCR3	T_H2, DC, Eos, Baso
CCL18 (DC-CK-1, PARC)	CCR3 antagonist	?
CCL11 (Eotaxin)	CCR3, CCR5	eT, DC, Eos, Baso
CCL28	CCR3, CCR10	eT, LC
CCL4 (MIP-1β)	CCR5	eT, NK, M, DC
CCL20 (MIP-3α)	CCR6	memory T, B, DC
CCL1 (I-309)	CCR8	T_H2, M
CX3CL1 (fractalkine)	CX3CR1	eT, NK, M
XCL1 (lymphotactin, SCM1-α)	XCR1	eT, B, NK, N
XCL2 (SCM1-β)	XCR1	eT, B, NK, N
(b) Homeostatic		
CXCL12 (SDF1)	CXCR4	pre-B, HSC, thymocytes
CXCL13 (BLC, BCA-1)	CXCR5	B, some eT
CXCL16	CXCR6	CD8 eT, IEL, NKT, DC
CXCL14 (BRAK)	?	M
CCL17 (TARC)	CCR4	eT, DC, thymocytes, Baso
CCL22 (MDC)	CCR4	eT, DC, thymocytes, Baso
CCL19 (ELC, MIP-3β)	CCR7	T, B, DC
CCL21 (SLC)	CCR7	T, B, DC
CCL25 (TECK)	CCR9	CD8 eT, M, thymocytes
CCL27 (CTACK)	CCR10	eT, LC

Figure 2-37 Table of human chemokines and chemokine receptors Chemokines are listed with their receptors and the cells that they attract. The chemokines are divided roughly into inflammatory chemokines, which attract cells to sites of inflammation, and lymphoid chemokines, which are involved in migration to and within secondary lymphoid organs. Also included in the latter category are chemokines involved in hematopoiesis and in constitutive placement of tissue macrophages (CXCL14). Some chemokines (for example CCL18 and CCL25) are thought to have both inflammatory and homeostatic properties. B: B cell; Baso: basophil; DC: dendritic cell; Eos: eosinophil; HSC: hematopoietic stem cell; IEL: intestinal epithelial lymphocyte; LC: Langerhans cell; M: monocyte; N: neutrophil; NK: natural killer cell; NKT: NK T cell; T: T cell; eT: effector T cell; T_H1, T_H2: subsets of helper T cells. For the most part, non-immune cells that may respond (for example endothelial cells, fibroblasts and various cancerous cell types) are not listed. One or two frequently used common names for most chemokines are shown. These names are for the most part acronyms. For a most complete listing of these names and their derivations, see Zlotnik, A. and Yoshie, O.: *Immunity* 2000, **12**:121–127. All names are for human chemokines.

References

Cyster, J.G.: **Chemokines and cell migration in secondary lymphoid organs.** *Science* 1999, **286**:2098–2102.

Luster, A.D.: **Chemokines—chemotactic cytokines that mediate inflammation.** *N. Engl. J. Med.* 1998, **338**:436–445.

Mellado, M. *et al.*: **Chemokine signaling and functional responses: the role of receptor dimerization and TK pathway activation.** *Annu. Rev. Immunol.* 2001,

19:397–421.

Rot, A. and von Andrian, U.H.: **Chemokines in innate and adaptive host defense: basic chemokinese grammar for immune cells.** *Annu. Rev. Immunol.* 2004, **22**:891–928.

Sallusto, F. *et al.*: **The role of chemokine receptors in primary, effector, and memory immune responses.** *Annu. Rev. Immunol.* 2000, **18**:593–620.

Zlotnik, A. and Yoshie, O.: **Chemokines: a new classification system and their role in immunity.** *Immunity*

2000, **12**:121–127.

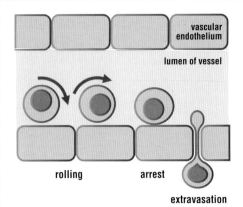

vascular endothelium

lumen of vessel

rolling   arrest

extravasation

Figure 2-38 Schematic diagram of the three steps whereby immune cells enter tissues Circulating cells driven by blood flow first begin rolling along the surface of the endothelium, then arrest, bind tightly and squeeze through the endothelial cells into the tissues.

Chemokines collaborate with other specialized molecules in directing the homing of lymphocytes and leukocytes to tissues

Chemokines are only one of three classes of molecules that direct the homing of immune cells to secondary lymphoid tissue or to infected tissues in the periphery. The other two are integrins, which we have already described (see section 2-4), and the class of C-type lectins known as *selectins* (see section 2-5), which recognize carbohydrate structures, sometimes called *addressins*, on cell surface proteins. Migration of immune cells from the bloodstream into tissues requires that the immune cells squeeze between the endothelial cells that make up the walls of the blood vessels supplying the target tissues. This occurs in a three-step process that is directed by these three molecules and is illustrated schematically in Figure 2-38. The first step is rolling, in which adhesive interactions between selectins and their ligands with a fast off-rate slow the progress of the circulating cell. The second is arrest, which occurs when chemokine receptors on the immune cell recognize chemokines bound to proteoglycan molecules on the endothelial cells and induce conversion of the immune-cell integrins to their high-affinity state, causing tight binding to adhesion molecules on the cells of the blood vessel. This allows the third step, which is transmigration of the immune cell between endothelial cells of the vessel wall. Different combinations of chemokines, integrins and selectins determine the homing of immune cells to different tissues and sites of infection.

We have already described the different chemokines and chemokine receptors, and the integrins and the adhesion molecules to which they bind, and touched on the way in which these are regulated during immune responses. In this section we describe the selectins and their ligands, whose regulated expression is equally important in orchestrating immune responses, and the molecules that direct the reverse process, by which lymphocytes leave lymphoid organs and return to the bloodstream.

Selectins are C-type lectins that bind to carbohydrate structures usually on cell surfaces

Selectins comprise a small group of C-type lectins containing three members: L-selectin, which is expressed on lymphocytes, especially on recirculating lymphocytes before they have been activated by antigen, and on monocytes and neutrophils; P-selectin, whose expression is induced by inflammation on endothelial cells and by activation on platelets; and E-selectin, which is also induced by inflammation but only on endothelium. L-selectin is essential for the migration of circulating lymphocytes into secondary lymphoid organs. P-selectin and E-selectin, as we shall see in the next chapter, are essential for leukocyte recruitment to infected tissues.

Selectins recognize carbohydrate groups on specific protein scaffolds, most commonly membrane-bound **mucins**, long thin proteins bearing many O-linked carbohydrate chains on serine or threonine residues: a selectin and its ligand are schematically represented in Figure 2-39. Selectin binding is determined by expression not only of the protein itself but also of the glycosyltransferase enzymes that add the carbohydrate groups recognized by the corresponding selectin. For example, the mucin **P-selectin glycoprotein ligand 1 (PSGL1)** is expressed on phagocytic leukocytes and lymphocytes. As its name suggests, it can bind P-selectin, and it is

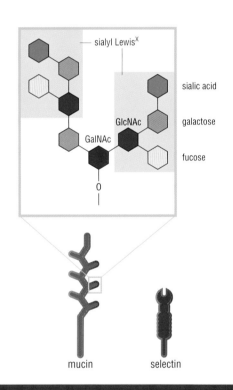

sialyl Lewis[X]

sialic acid

galactose

GlcNAc

GalNAc

fucose

0

mucin   selectin

Figure 2-39 Schematic diagram of a selectin and a mucin bearing selectin-binding carbohydrate groups Selectins (right) are C-type lectins with an amino-terminal lectin domain that binds to oligosaccharide structures on protein scaffolds (left) that are often membrane-bound mucins. Shown is an oligosaccharide containing two copies of a tetrasaccharide unit called sialyl Lewis[X] (in yellow squares).

Definitions

mucin: glycoprotein with many O-linked glycans on serine or threonine residues.

peripheral node addressin (PNAd): oligosaccharide ligand for L-selectin attached to any of several different protein backbones.

PNAd: see **peripheral node addressin.**

P-selectin glycoprotein ligand 1 (PSGL1): mucin of leukocytes that is the principal scaffold for the oligosaccharide ligand for P-selectin but can also serve as the

scaffold for the oligosaccharides recognized by E-selectin and L-selectin.

PSGL1: see **P-selectin glycoprotein ligand 1.**

selectin: any of a family of three structurally related lectins that mediate interactions between leukocytes or lymphocytes and endothelial cells in the tissues into which they migrate.

References

Hla, T.: **Dietary factors and immunological consequences.** *Science* 2005, **309**:1682–1683.

Ley, K. and Kansas, G.S.: **Selectins in T-cell recruitment to non-lymphoid tissues and sites of inflammation.** *Nat. Rev. Immunol.* 2004, **4**:325–335.

Rosen, S.D.: **Ligands for L-selectin: homing, inflammation, and beyond.** *Annu. Rev. Immunol.* 2004, **22**:129–156.

Schwab, S.R. *et al.*: **Lymphocyte sequestration through S1P lyase inhibition and disruption of S1P gradients.** *Science* 2005, **309**:1735–1739.

Figure 2-40 Molecular interactions that direct immune cell homing (**a**) A neutrophil binding to an endothelial cell before entry into a site of infection; (**b**) homing interactions between a lymphocyte and an endothelial cell in a lymph node. Note that in the case of homing to infected tissues the selectin is expressed on the endothelial cell, whereas in the case of the lymphocyte entering a lymph node the selectin is expressed on the lymphocyte.

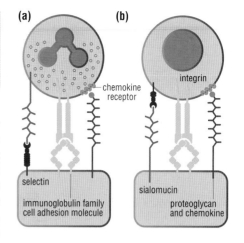

important in the entry of leukocytes into infected tissue sites where P-selectin is induced by inflammatory signals. It is, however, also expressed by recirculating naïve lymphocytes, which do not bind P-selectin and do not enter infected tissues. But whereas phagocytes, which must always be ready to respond to infection, constitutively express the glycosyltransferases required to add the groups that P-selectin recognizes, circulating lymphocytes do not express these enzymes until they have been activated against an infectious microorganism, when they cease to circulate between secondary lymphoid organs and instead migrate to sites of infection. Moreover despite its name, PSGL1 can also serve as a scaffold for carbohydrate groups recognized by E- and L-selectins, in endothelial cells which express the appropriate glycosyltransferases.

Conversely, different protein scaffolds may carry the same carbohydrate groups and bind the same selectin. An example of this is found in the glycoprotein ligands for L-selectin, which direct circulating lymphocytes into secondary lymphoid tissues. L-selectin recognizes both a sulfated oligosaccharide and elements of the protein backbone to which it is attached. This combination ligand is also known as **peripheral node addressin (PNAd)**; it is carried by several different protein scaffolds expressed on the endothelial cells of the blood vessels supplying lymph nodes. These include a mucin called CD34, and MAdCAM, which we have already encountered as an adhesion molecule (see Figure 2-11). PNAd binding to L-selectin directs circulating lymphocytes to peripheral lymph nodes; MAdCAM is expressed specifically in lymph nodes associated with mucosal epithelia and directs homing lymphocytes to those lymph nodes.

Whereas chemokine receptors and integrins are always expressed on the immune cell, with the chemokine and the integrin ligand on the endothelial cell, selectins and their ligands can be expressed on either partner in the interaction. This is because the rolling of the circulating cell can be supported by either arrangement, whereas arrest and entry require signaling to the interior of the migrating cell. Figure 2-40 shows schematically the three interacting pairs for a phagocyte expressing a selectin ligand and a recirculating lymphocyte expressing a selectin. We shall see in more detail in the course of this book how the expression of these homing molecules is regulated to direct the activities of immune cells. The main cell types expressing the different selectins, with their expression patterns, and their carbohydrate ligands, are summarized in Figure 2-41.

Circulating lymphocytes express a regulatable receptor that directs their exit from lymphoid tissues

The effectiveness of lymphocyte surveillance for antigen requires not only that circulating lymphocytes are efficiently directed into the secondary lymphoid tissues, but also that if they do not encounter antigen, they leave to explore other lymphoid tissues where it may be present. Exit from secondary lymphoid tissues is guided not by chemokines but by a gradient of the lipid sphingosine 1-phosphate (S1P), which lymphocytes recognize by means of a G-protein-coupled receptor like the chemokine receptors (see Figure 2-36). S1P is a secreted product of intracellular sphingosine metabolism that exists in higher concentrations in blood and lymph than tissues where it is very low. Recognition by the sphingosine 1-phosphate receptor-1 (S1P$_1$) causes lymphocytes to migrate up concentration gradients of the lipid, and thus out of the lymphoid tissues where it is low and into the blood (via the lymph, in the case of lymph nodes) where it is high. Once in the presence of high concentrations of S1P, however, the receptor is down-regulated, so that circulating lymphocytes can leave the bloodstream again to enter another lymph node. Inside the lymphoid tissue, S1P concentration is again low, receptor expression recovers, and the lymphocytes escape once more into the bloodstream. Similarly, lymphocytes that have been activated in the lymphoid tissues to differentiate into effector cells (see section 1-8) can escape into the bloodstream and thence into the infected tissues where they are required.

The concentration of S1P in lymphoid and peripheral tissues is kept low by an enzyme, S1P lyase, whose activity requires vitamin B$_6$. Deficiency in this vitamin, which can occur in rare cases of malabsorption from the gut, or after drug treatment, leads to immune deficiency that is thought to result from the retention of lymphocytes in the secondary lymphoid tissues.

Selectin Expression and Ligands

L-selectin

Expression	Ligand
myeloid cells (constitutive)	sulfated glycosaminoglycans on various proteins including CD34, PSG-1 and MAdCAM
naïve T and B cells (constitutive)	
central memory T cells (constitutive)	

E-selectin

Expression	Ligand
skin endothelium (constitutive)	sialylated Lewis^X glycans
inflamed endothelium (inducible)	

P-selectin

Expression	Ligand
choroid plexus (constitutive)	sialylated Lewis^X glycans on PSGL-1 and other proteins
lung endothelium (constitutive)	
platelets (inducible)	
inflamed endothelium (inducible)	

Figure 2-41 Table of selectins and their expression patterns and ligands In the case of L-selectin and E-selectin, constitutive means constitutively synthesized and expressed on the cell surface, and inducible refers to synthesis. In the case of P-selectin, synthesis in platelets and endothelial cells is constitutive but translocation from intracellular storage vesicles to the cell surface is inducible. L-selectin, P-selectin and E-selectin are often called CD62L, CD62P and CD62E respectively. The choroid plexus is a highly vascularized tissue in the ventricles of the brain and that produces the cerobrospinal fluid in which the brain and spinal cord are bathed.

3

Innate Immunity

Innate immune recognition of molecules characteristic of broad classes of microorganisms or viruses is critical to the early defense against infection and also promotes adaptive immune responses. In this chapter, some of the better-understood innate immune recognition mechanisms are described along with the microbial molecules they recognize and the major innate effector actions of immunity, including inflammation, microbial killing by phagocytes, and innate interference with viral growth.

3-0 Overview: Evolution and Function of Innate Immunity

Types of Pathogens and the Diseases They Cause

Type of pathogen	Disease
DNA viruses	
herpesviruses	mononucleosis, venereal disease
poxviruses	smallpox
adenovirus	respiratory disease
RNA viruses	
hepatitis C virus	hepatitis
poliovirus	polio
rhinovirus	common cold
measles virus	measles
Retroviruses	
human immunodeficiency virus (HIV)	acquired immunodeficiency syndrome (AIDS)
Gram-positive bacteria	
streptococci	pneumonia, scarlet fever, pharyngitis
staphylococci	severe skin and wound infection
Listeria	food poisoning
clostridia	gastro-intestinal disease
Gram-negative bacteria	
salmonellae	typhoid and food poisoning
neisseriae	gonorrhea and meningitis
Haemophilus	pneumonia
Yersinia	plague
Other bacteria	
mycobacteria	tuberculosis, leprosy
Mycoplasma	pneumonia
Chlamydia	blindness, sterility
Fungi	
Candida	mucosal or systemic infections
Aspergillus	respiratory infections

Figure 3-1 Table of examples of microorganisms belonging to the major groups of pathogens Microorganisms are listed with the main type of disease they cause.

Molecules Recognized by Innate Immunity

Molecular component	Type of organism
Nucleic acids	
double-stranded RNA (dsRNA)	viruses
CpG-containing DNA (CpG DNA)	bacteria, fungi, viruses
Cell wall components	
lipopolysaccharide	Gram-negative bacteria
lipoteichoic acids	Gram-positive bacteria
peptidoglycan	bacteria
flagellin	Gram-negative bacteria
lipoproteins	bacteria
mannose-, fucose-rich polysaccharides	bacteria, fungi
β-glucans	bacteria, fungi
Membrane properties	
exposed acidic phospholipids	bacteria, fungi (apoptotic cells)
Biosynthetic components	
N-formylated methionyl peptides	bacteria

Figure 3-2 Table of conserved components of microorganisms that are recognized by innate immune cells and molecules

Multicellular organisms have conserved mechanisms of innate immune recognition

Although multicellular organisms generally have an outer layer that provides a barrier to infection, defenses against microorganisms that enter through wounds or other breaches in the outer layer are essential to multicellular life and are found in plants and invertebrate animals as well as in vertebrates. The immune defenses of invertebrates and plants depend on mechanisms that have evolved to recognize molecular features typical of the pathogens that infect them, and effector mechanisms that are triggered by recognition. An analogous system exists in vertebrates, in which it is known as *innate immunity* (see section 1-0). Vertebrates later evolved a second system, providing a more sophisticated and flexible mechanism of recognition through the variable antigen receptors of the T and B lymphocytes of *adaptive immunity* (see section 1-0). The remarkable properties of this system are the subject of the next three chapters. In this chapter we describe the innate immune system, which provides the first line of defense against infection and on which, as we discuss later, the adaptive immune system depends for activation and for most of its effector functions.

Innate immunity remains effective despite rapid evolution of microbes and viruses

Innate immunity has a critical role in providing protection early in infection. Immunodeficient individuals who lack particular components of innate immunity, or mice in which the genes encoding them have been ablated, are particularly sensitive to characteristic types of pathogens, or in extreme cases become susceptible to infection with microorganisms that are not normally pathogens.

Microbes and viruses, with their short generation times and mechanisms for horizontal gene transfer and gene capture, evolve much more rapidly than vertebrate animals, and thus one might expect that innate immune mechanisms would rapidly become obsolete for protection against microorganisms that have specialized to infect multicellular organisms. Indeed, pathogens that cause disease in humans have in general developed mechanisms for evading at least some elements of innate immunity, and this is what has driven the evolution of adaptive immunity.

Nonetheless, immune defenses remain largely effective even in the face of concerted evolutionary selection for evasion by pathogens, as described in Chapters 9–11. In part this is because if a pathogen kills all of the individuals that it infects it may become an evolutionary dead-end. An example of this selective pressure was seen when the poxvirus myxoma virus was introduced into the feral European rabbit populations of Australia in 1950 in a failed attempt to control these populations. Myxoma virus kills more than 99% of the European rabbits it infects, in contrast to its minimal effects on its natural host, the South American rabbits, in which it causes a benign cutaneous fibroma. The introduction of myxoma virus killed large numbers of feral rabbits, but not all. Surviving rabbits were found to harbor an attenuated myxoma virus that not only failed to kill infected rabbits but induced an adaptive immune response that was protective against the original myxoma virus.

The second reason for the continued effectiveness of innate immune mechanisms is that these mechanisms have evolved to focus on components of microorganisms or viruses that cannot easily be changed. The major groups of common human pathogens are listed in Figure 3-1, with some important examples of each: molecules characteristic of each group that are recognized by innate immune mechanisms are listed in Figure 3-2. Double-stranded RNA, for example, is an obligatory replication intermediate for RNA viruses and its recognition is an important trigger for antiviral innate mechanisms. Many bacteria are classified by their cell walls as **Gram-positive** or **Gram-negative**, terms that refer to the staining properties of the cell walls when the bacteria are prepared for microscopy. These distinct staining properties correspond to the distinct cell wall structure and composition of the two groups of bacteria (Figure 3-3). Recognition of Gram-negative bacteria is accomplished through recognition of the **lipopolysaccharide (LPS)** component of their outer membrane, which is required for the viability of most Gram-negative bacterial species. Similarly, recognition of Gram-positive bacteria occurs through conserved molecules of their cell wall, such as **lipoteichoic acid (LTA)** and **peptidoglycan**, which are required for their integrity. The highly conserved structures recognized by innate immunity have been referred to as *pathogen-associated molecular patterns*

(PAMPs) and the innate immune receptors that recognize them as *pattern recognition receptors*. In this book we shall not use these terms, because vertebrate innate immune recognition depends on conserved molecules of microbes or viruses, not of pathogens specifically, and in most cases these recognized entities are specific types of molecules, rather than molecular patterns (the exception being carbohydrate recognition, in which the pattern concept is in some cases a useful one).

Innate immune recognition mechanisms predate adaptive immunity and instruct it

We begin this chapter with the defense mechanisms that operate at the epithelium—the first and most critical barrier against infection—before describing the soluble molecules that recognize and bind to microorganisms that penetrate this barrier. These molecules promote the recognition and ingestion of bound microorganisms by the phagocytic cells of innate immunity, the *neutrophils* and *macrophages*. Neutrophils and macrophages destroy the ingested particles and at the same time, through receptors that specifically recognize microbial signature molecules, are activated to produce cytokines that mediate *inflammation*, a process in which changes to blood vessel walls and the production of chemokines act to recruit more immune cells to sites of infection. Finally, we describe the innate immune response to viruses, which includes inflammation but ultimately may require induction of apoptosis of the infected cell, to limit viral replication.

Many of these innate immune recognition and effector mechanisms evolved before the evolution of vertebrates: related molecules are found in contemporary invertebrates and in our own innate immune systems. Thus, adaptive immunity arose in an environment in which innate immunity already existed, and indeed innate immune recognition has a key role in instructing adaptive immunity, both in signaling when an immune response should be made and in directing the nature of that immune response: these functions of innate immune recognition will be described in Chapters 4 and 5.

Figure 3-3 Features of bacterial cell walls that are recognized by components of innate immunity Innate immune recognition mechanisms focus on highly conserved features of groups of pathogens that are very hard for these organisms to change. Examples of different targets of innate immunity are shown, including lipopolysaccharide (LPS), peptidoglycan, lipoproteins, lipoteichoic acid (LTA) and porin.

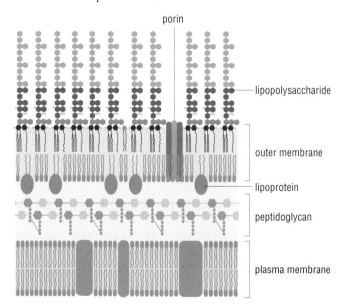

porin

lipopolysaccharide

outer membrane

lipoprotein

peptidoglycan

plasma membrane

teichoic acid

peptidoglycan

lipoteichoic acid

plasma membrane

Gram-negative bacterial cell wall

Gram-positive bacterial cell wall

Definitions

Gram-negative bacteria: major subgroup of bacteria characterized by the presence of an outer membrane bilayer made up of phospholipids and **LPS**.

Gram-positive bacteria: major subgroup of bacteria lacking an outer membrane but typically containing a thick **peptidoglycan** layer.

lipopolysaccharide (LPS): major component of the outer membrane of **Gram-negative bacteria** and an important recognition element for innate immunity.

lipoteichoic acid (LTA): cell wall constituent of **Gram-positive bacteria**.

LPS: see **lipopolysaccharide**.

LTA: see **lipoteichoic acid**.

peptidoglycan: rigid polymer of repeating disaccharides cross-linked by short peptides that is a major structural element of most types of bacterial cell walls.

References

Fearon, D.T. and Locksley, R.M.: **The instructive role of**

innate immunity in the acquired immune response. *Science* 1996, **272**:50–53.

Hoffmann, J.A. *et al.*: **Phylogenetic perspectives in innate immunity.** *Science* 1999, **284**:1313–1318.

Janeway, C.A. Jr: **The immune system evolved to discriminate infectious nonself from noninfectious self.** *Immunol. Today* 1992, **13**:11–16.

Ross, J. and Tittensor, A.M.: **The establishment and spread of myxomatosis and its effect on rabbit populations.** *Philos. Trans. R. Soc. Lond., B, Biol. Sci.* 1986, **314**:599–606.

Human Antimicrobial Peptides

Family/member	Distribution
α-defensins	
HNP1-4	all present in neutrophils
HD5	Paneth cells of small intestine, airway epithelium, urogenital epithelium
HD6	Paneth cells of small intestine, airway epithelium
β-defensins	
HBD1	lung, other epithelia
HBD2	lung, other epithelia
HBD3	skin keratinocytes, airway epithelium
HBD4	neutrophils, epithelium of testes, stomach, uterus, lung, kidney
HBD5	male reproductive tract
HBD6	male reproductive tract
Cathelicidins	
LL37/hCAP-18	neutrophils, lung epithelium, mast cells, monocytes/ macrophages

Figure 3-4 Table of human antimicrobial peptides and their primary locations Note that additional human β-defensins are found in the genome but are not yet characterized.

Figure 3-5 Structural features of defensins Defensins have characteristic structures containing three beta-strands, whereas cathelicidins have a different structure containing alpha-helical regions (not shown). (PDB 1fd4)

The epithelial cell layer provides the first barrier to most infections

Epithelial layers of the skin, gastrointestinal tract, genitourinary tract, and respiratory tract provide a physical barrier to infection. In many sites, microorganisms are tolerated at the outside surface of these epithelial layers, but in other locations, such as in the inner airways and the small intestines, the function of the interface with the environment (gas exchange, uptake of nutrients) would be compromised by microbes. For that reason the epithelial cells in these locations produce peptide antibiotics called **antimicrobial peptides** that inhibit bacterial and fungal cell growth. The low pH of the stomach is also a barrier to microbial colonization, as are hydrophobic agents such as fatty acids and bile salts, which coat certain mucosal surfaces. The mucosal layer of the gastrointestinal tract is rich with intraepithelial lymphocytes, mostly T cells, which are thought to have a role in host defense at this barrier and are discussed in Chapter 8. The mechanisms by which pathogens get past the epithelial barrier are discussed in Chapters 9–11.

Antimicrobial peptides are a widespread defense mechanism

Plants, insects and vertebrates all employ peptide antimicrobial agents to help protect themselves from infection by bacteria, fungi, yeasts, protozoans and enveloped viruses. These molecules fall into a variety of different family groupings, but they are generally characterized by being moderately sized peptides (fewer than 100 amino acid residues in length) that are often highly cationic and amphipathic in nature.

In mammals, antimicrobial peptides are expressed both by epithelial cells in select locations to prevent bacterial colonization and by neutrophils to contribute to killing of microbes after phagocytosis (see section 3-9). Three major groups of these peptide antimicrobial agents are currently recognized: the **defensins**, of which there are two classes, α-defensins and β-defensins, and the **cathelicidins** (Figure 3-4). The defensins are 29–35 amino acids long and have a three-stranded beta-sheet structure including three disulfide bonds (Figure 3-5). The structures of α-defensins and β-defensins are similar, but the disulfide bond connections differ slightly. The cathelicidins are of similar size, but have a distinct structure, including alpha-helical regions. Neutrophils contain members of each class of antimicrobial peptide. Epithelial cells in various locations express different subsets of antimicrobial peptides. Specialized cells called **Paneth cells** at the base of the crypts of the small intestines (Figure 3-6) express two α-defensins. In Paneth cells, the synthesis and secretion of defensins are induced by innate immune recognition of bacterial components. The importance of antimicrobial peptides for immune defense at this location is illustrated by a mouse strain lacking the enzyme that cleaves defensin precursors in Paneth cells and that therefore cannot make active defensins in the small intestine. These mice are susceptible to oral doses of *Salmonella typhimurium* 10-fold lower than those required to cause disease in normal mice.

Each of these antimicrobial peptides has widespread activity against different microorganisms, but conversely, different peptides have distinct abilities to kill various infectious agents, often based on subtle structural differences. For example, two murine gut α-defensins, cryptdin-2 and cryptdin-3, differ at only a single amino-acid position, and both are effective at killing the parasite *Giardia lamblia* at the trophozoite stage of its life cycle, but only cryptdin-3 can kill the gut bacterium *Escherichia coli*.

The cationic and amphipathic character of defensins is believed to enable them to bind to acidic phospholipids in the membranes of microorganisms and then to enter membranes and generate pores in a process that requires the presence of a membrane potential (Figure 3-7).

Definitions

antimicrobial peptides: peptide antibiotics that provide defense against microbes and viruses by interacting with membranes of infectious agents and increasing their permeability. Human antimicrobial peptides are members of either the α-**defensin**, β-defensin or **cathelicidin** families.

cathelicidins: family of cationic **antimicrobial peptides**.

defensins: any of a family of cationic **antimicrobial peptides** of vertebrates with three disulfide bonds and a largely beta-sheet structure. Humans have α- and β-defensins; some other mammals also have circular θ defensins. Cationic, cysteine-rich antimicrobial peptide molecules also called defensins are found in insects and plants, although it is not known whether they are derived from a common ancestral gene.

Paneth cell: specialized cell type of the epithelial layer of the small intestines. Paneth cells are localized to the base of the crypts of the intestines and secrete **antimicrobial peptides**.

References

Ayabe, T. *et al.*: **Secretion of microbicidal α-defensins by intestinal Paneth cells in response to bacteria.** *Nat. Immunol.* 2000, **1**:113–118.

Lehrer, R.I.: **Primate defensins.** *Nat. Rev. Microbiol.* 2004, **2**:727–738.

Ouellette, A.J. and Selsted, M.E.: **Paneth cell defensins: endogenous peptide components of intestinal host defense.** *FASEB J.* 1996, **10**:1280–1289.

Figure 3-6 Paneth cells in the small intestine Paneth cells are specialized cells that secrete α-defensins at the bottom of the crypts formed by the infolding of the epithelium of the small intestine. Paneth cells synthesize and secrete defensins in response to the presence of bacterial components.

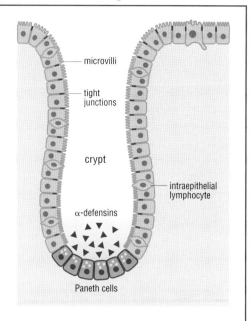

Most microbes have membranes rich in acidic lipids, whereas the lipids on the outsides of metazoan membranes are less negatively charged, being composed primarily of zwitterionic phospholipids such as phosphatidylcholine and the neutral lipid cholesterol. The outsides of higher eukaryotic cells are highly negatively charged, but much of this charge comes from carbohydrates such as the sialic acid on glycoproteins and glycolipids. Moving the charge away from the membrane may afford some protection to these cells from the toxic actions of the antimicrobial peptides, and indeed, most antimicrobial peptides more readily kill microbial cells than host cells.

Quite high concentrations of antimicrobial peptides are needed to kill infectious agents, generally around 1 μM. These peptides are expressed at high levels in neutrophils, where they are stored in specialized organelles, the *primary* and *secondary* granules. Defensins account for about 5% of the protein of human neutrophils, and are estimated to be present at about 10 mg/ml in their primary granules, which would put them in the mM range. Infectious agents are exposed to these defensins when the vesicles in which they are internalized fuse with the granules, but little dilution would be expected to result from this; and in addition to killing microbes directly, the pores formed by antimicrobial peptides probably promote the killing action of other neutrophil effector molecules (see section 3-9).

Not all microbes are equally easily killed by antimicrobial peptides. In part this may reflect alterations to microbial cell membranes that confer resistance. For example, *Staphylococcus aureus* modifies the membrane phospholipid phosphatidylglycerol with L-lysine to reduce the negative charge of its plasma membrane and this makes it insensitive to defensins. Mutation of the gene encoding this enzyme attenuates virulence in mice and makes these bacteria more sensitive to killing *in vitro* by human neutrophils. Analogously, the Gram-negative organism *Pseudomonas aeruginosa* can modify its LPS by acylation of the lipid A moiety and this is associated with resistance to cationic antimicrobial peptides. This modification is an induced response to an environment with low extracellular Mg^{2+} ion and Ca^{2+} ion concentrations, corresponding to the conditions in the lung. Other mechanisms of bacterial resistance to antimicrobial peptides include efflux pumps and the formation of biofilms, a colonial state in which the organisms embed themselves in a protective polysaccharide matrix.

In addition to their direct role in attacking microbes, antimicrobial peptides also have a large variety of effects on cells of the immune system. For example, some are chemoattractants for neutrophils. Presumably, this is a mechanism to attract neutrophils to a site where these molecules are being released from cells actively fighting an infection. They also attract dendritic cells, which are thereby able to take up antigen at the site of infection and initiate adaptive immune responses, as we shall see in Chapter 4.

Figure 3-7 Mechanism of defensin killing of microbes Defensins are highly amphipathic and positively charged molecules. They seem to make pores in membranes approximately 20 Å in diameter. It is thought that their positively charged regions interact with the negatively charged phospholipids of microbial membranes, after which they insert their hydrophobic regions into the membrane and oligomerize to form pores, disrupting the normal function of the membrane. The pores also allow entry of other effector molecules of immune defense that can contribute to killing microbes.

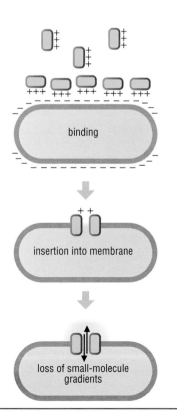

Peschel, A. *et al.*: **Staphylococcus aureus resistance to human defensins and evasion of neutrophil killing via the novel virulence factor MprF is based on modification of membrane lipids with L-lysine.** *J. Exp. Med.* 2001, **193**:1067–1076.

Sansonetti, P.J.: **War and peace at mucosal surfaces.** *Nat. Rev. Immunol.* 2004, **4**:953–964.

Travis, S.M. *et al.*: **Antimicrobial peptides and proteins in the innate defense of the airway surface.** *Curr. Opin. Immunol.* 2001, **13**:89–95.

Wilson, C.L. *et al.*: **Regulation of intestinal α-defensin activation by the metalloproteinase matrilysin in innate host defense.** *Science* 1999, **286**:113–117.

Yang, D. *et al.*: **Multiple roles of antimicrobial defensins, cathelicidins, and eosinophil-derived neurotoxin in host defense.** *Annu. Rev. Immunol.* 2004, **22**:181–215.

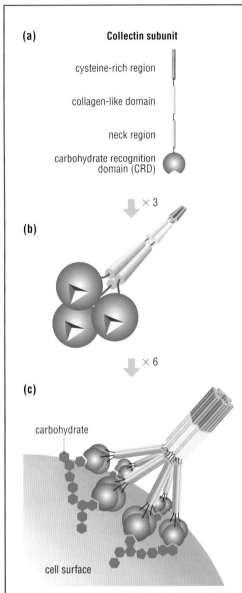

(a) Collectin subunit

cysteine-rich region

collagen-like domain

neck region

carbohydrate recognition
domain (CRD)

× 3

(b)

× 6

(c)

carbohydrate

cell surface

Figure 3-8 Structures of the collectins
(a) Each polypeptide chain of the collectin family of proteins is composed of an amino-terminal cysteine-rich region followed by a collagen-like region, an alpha-helical neck region and a carboxy-terminal globular domain that is a C-type lectin and is referred to as the carbohydrate recognition domain (CRD). **(b)** The polypeptide chains trimerize by coiled-coil interactions of the alpha-helical neck regions to form the basic subunit of structure. **(c)** Both mannose-binding lectin (MBL) and surfactant protein A (SP-A) are hexameric structures in which six of the basic subunits are linked together via the amino-terminal cysteine-rich regions and collagen-like regions to give hexameric units with 18 polypeptide chains. The resulting multiple arrays of CRDs give high-avidity binding to the repetitive macropattern of polysaccharide ligands in microbial cell walls. Surfactant protein D (SP-D) forms a tetramer of the trimeric basic units (12 polypeptide chains). Ficolins are also hexamers or tetramers of a similar structure, the main difference being the type of domain that comprises the carbohydrate recognition domain.

Collectins and ficolins activate innate immune protective mechanisms at mucosal surfaces and in the bloodstream

Two of the most important defenses against bacteria, as we shall see later in this chapter, are phagocytosis by macrophages and neutrophils, and the activation of a system of serum proteins called *complement* that destroy some bacteria directly and are important in increasing the effectiveness of phagocytic uptake of many others. The **collectins** and **ficolins** are two families of soluble proteins present at mucosal surfaces and/or in the bloodstream that recognize distinctive carbohydrate configurations that occur on the surfaces of microbes but not on the cells of the multicellular organism, and couple this recognition to uptake by phagocytes or activation of the complement system. Collectins and ficolins, as well as being major components of the complement system, are **opsonins**, a general term for components of the immune system that coat microorganisms and stimulate uptake by phagocytes. In humans, three collectins and two ficolins are known to participate in immunity. Several additional members of these families have been discovered recently, but their functions are not yet known.

Collectins and ficolins have a common architecture but distinct recognition domains

Both collectins and ficolins have an amino-terminal cysteine-rich region that seems to form inter-chain disulfide bonds, followed by a collagen-like region, an alpha-helical coiled-coil neck region and finally a carbohydrate-recognizing region that in collectins is a C-type lectin domain (see section 2-5) referred to as the carbohydrate recognition domain (CRD) and in ficolins is a fibrinogen domain. The structural and functional properties of these molecules are illustrated in Figure 3-8 for a collectin. The alpha-helical coiled-coil neck region mediates trimerization of the polypeptide chains, whereas the collagen-like region mediates the formation of higher-order structures. Different collectins or ficolins exhibit distinctive higher-order structures, typically either tetramers of trimers or hexamers of trimers. The grouping of large numbers of binding domains allows these molecules to bind effectively to microbial cell walls despite a relatively low intrinsic affinity of each individual subunit for carbohydrates.

Each globular head of a collectin or a ficolin binds to a single sugar residue, typically at the end of the polysaccharide chain. Ficolins typically bind *N*-acetyl glucosamine (GlcNAc), whereas the collectin *mannose-binding lectin* prefers D-mannose, but neither binds sialic acid nor D-galactose, sugars that commonly terminate mammalian cell-surface glycoproteins and glycolipids. In this way, collectins and ficolins recognize a general pattern of cell-surface carbohydrates that distinguishes many microbes from host cells (see Figure 3-8). Particle recognition is translated into immune action either by phagocytes, which are thereby stimulated to internalize bound microorganisms, or by conformational changes in the collectin or ficolin that activate prebound protease subunits (Figure 3-9) and initiate the complement cascade, as we shall see in the next sections.

Two lung surfactants are mucosal collectins

The epithelium of the respiratory tract is lubricated and protected by a layer of phospholipids and proteins known as **surfactants**. Two of these, surfactant protein A (SP-A) and surfactant protein D (SP-D), are collectins and participate in immune defense. SP-A has been found in all the major vertebrate groups, whereas SP-D has not yet been found in fish. Although they are expressed at their highest level in the lung, these collectins are also found at lower levels at other mucosal surfaces.

Definitions

capsular polysaccharide: cell-surface polymers of repeating oligosaccharide units, usually linked through phosphodiester bonds, that form a capsule on the surface of many pathogenic bacteria and protect bacterial cells from recognition by phagocytes. For this reason the presence of a capsule is often associated with virulence.

collectin: any of a family of structurally related, carbohydrate-recognizing proteins of innate immunity, including **mannose-binding lectin** and **surfactant**

proteins A and D. The name refers to the presence of a collagen-like domain and a C-type lectin domain.

encapsulated bacteria: bacteria with a thick polysaccharide capsule covering up other cell wall structures, such as LPS or peptidoglycan.

ficolin: any of a family of a structurally related, carbohydrate-recognizing proteins of innate immunity, containing a collagen-like domain and a fibrinogen domain. The name is derived from these domains.

mannose-binding lectin (MBL): a **collectin** family

SP-A is a hexamer of trimers (18 polypeptide chains; see Figure 3-8), whereas SP-D is a tetramer of trimers (12 polypeptide chains), with very long collagen arms connecting the four sets of trimers. Both SP-A and SP-D bind a variety of microbial carbohydrate and phospholipid structures, but their preferences are different and this seems to have a major impact on their physiological binding properties. For example, *Klebsiella pneumoniae*, a pathogen that infects individuals with weakened immune systems, can mask its surface lipopolysaccharide (LPS) with a thick polysaccharide capsule: bacteria that do this are said to be **encapsulated bacteria**, and the polysaccharide is referred to as the **capsular polysaccharide**. *Klebsiella* rapidly switch between encapsulated forms and non-encapsulated forms, in which LPS is exposed; apparently this switching is essential to allow infection of the lung. SP-A binds to approximately one-third of encapsulated *Klebsiella* serotypes, whereas SP-D binds to the non-encapsulated form by binding to the core region of its LPS. Some other microbes bind to both SP-A and SP-D.

Mice in which the gene for SP-A has been deleted exhibit increased susceptibility to a variety of infectious agents, including *Staphylococcus aureus*, group B streptococci, *Pseudomonas aeruginosa*, and respiratory syncytial virus. Evidence for the importance of SP-D in immunity has also been reported, although SP-D-deficient mice also develop a lung disease related to lipid problems. SP-A and SP-D function primarily by promoting phagocytosis.

Mannose-binding lectin provides innate immune protection against some bacterial infections

The two ficolins L-ficolin and H-ficolin and a third human collectin, **mannose-binding lectin (MBL)**, are serum proteins. All three activate complement. MBL is the best studied. Whereas the mucosal collectins and the two ficolins are present in normal individuals and are part of the immediate defense against infection, MBL is produced in greater amounts in response to infection, when, as we shall see in section 3-15, it is induced in the liver by cytokines released by macrophages, along with other proteins together known as *acute phase proteins*. MBL can bind a subset of Gram-negative and Gram-positive bacteria, fungi, and even some viruses such as influenza virus. MBL couples to phagocytosis via receptors that also bind SP-A, and has an important role in activating the complement system, as we shall see shortly (section 3-4).

About 4% of the human population are deficient in MBL, as a result either of transcription defects or of point mutations in the collagenous region of the molecule that seem to interfere with multimerization and consequently the ability to activate complement. These mutant forms of MBL are also lost from the blood more rapidly. People deficient in MBL exhibit a greater risk of infection, particularly early in life when antibody-mediated protection is at its lowest. For example, the odds of contracting meningococcal disease are sixfold greater in MBL-deficient children than in the general population, and it is estimated that approximately one-third of the total cases of meningococcal disease occur in the 4% of the population that are MBL-deficient. A similar odds ratio has been found for increased susceptibility to pulmonary tuberculosis in India.

(a) **mannose-binding lectin**

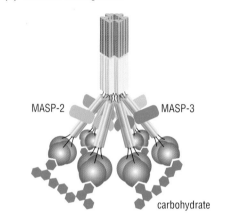

MASP-2   MASP-3

carbohydrate

(b) **H-ficolin**

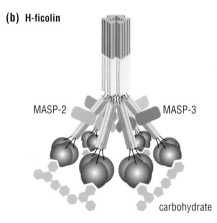

MASP-2   MASP-3

carbohydrate

Figure 3-9 MASPs couple MBL and ficolins to the complement pathway Mannose-binding lectin-associated serine proteases (MASPs) couple some collectins and some ficolins to the complement pathway. MBL, H-ficolin and L-ficolin bind MASPs. Before ligand binding, these MASPs lack significant protease activity. Binding of multiple carbohydrate recognition domains of the collectin or ficolin to a microbial cell surface leads to conformational changes in the molecule, perhaps due to the geometry of the ligand, which promotes proteolytic activation of the MASPs, which then can initiate the complement cascade, as described in later sections.

member that recognizes terminal sugars with equatorial hydroxyls in the C3 and C4 positions, such as mannose and fucose. Also called mannan-binding protein (MBP).

MBL: see **mannose-binding lectin.**

opsonin: any soluble molecule that recognizes and coats microorganisms and thereby stimulates internalization of the microorganism by phagocytes. A particle coated with opsonins is said to be opsonized.

surfactant: any of a number of lung proteins of diverse functions. The **collectins** surfactant protein A (SP-A) and surfactant protein D (SP-D) are important molecules of innate immunity.

References

Fujita, T. *et al.*: **The lectin-complement pathway – its role in innate immunity and evolution.** *Immunol. Rev.* 2004, **198**:185–202.

Kawasaki, T.: **Structure and biology of mannan-binding protein, MBP, an important component of innate immunity.** *Biochim. Biophys. Acta* 1999, **1473**:186–195.

Korfhagen, T.R. *et al.*: **Surfactant protein A (SP-A) gene targeted mice.** *Biochim. Biophys. Acta* 1998, **1408**:296–302.

Shepherd, V.L.: **Distinct roles for lung collectins in pulmonary host defense.** *Am. J. Respir. Cell Mol. Biol.* 2002, **26**:257–260.

Wright, J.R.: **Immunoregulatory functions of surfactant proteins.** *Nat. Rev. Immunol.* 2005, **5**:58–68.

3-3 Overview of the Complement System

The complement cascade links soluble factor recognition of microbes to mechanisms for their destruction

The **complement system**, also known as **complement**, consists of about 30 serum and membrane proteins that can mediate a variety of immune reactions, including triggering an inflammatory response, attracting phagocytes, promoting phagocytosis by opsonization, directly attacking the membrane of the cell or enveloped virus to which it is targeted, and stimulation of antibody production. Complement is so named because it was originally discovered, over 100 years ago, as an unknown component of fresh serum that enhances (or complements) the killing of bacteria by stored immune serum containing specific antibody, but—as we now know—not functional complement proteins. Although complement was discovered because of its actions in association with antibody, it has since become clear that complement is also activated by several innate triggers.

The active components of complement are generated from inactive precursors by a cascade of proteolytic reactions. These can be triggered in any of three ways: by carbohydrate-recognizing soluble factors including mannose-binding lectin (MBL) and ficolins bound to their ligands (the **lectin pathway**); by antibodies bound to antigen in immune complexes (the **classical pathway**); or directly at microbial cell surfaces (the **alternative pathway**). The classical pathway is so called because it was discovered first, and the alternative pathway because it was discovered as an antibody-independent way of activating complement. In evolution, however, the complement pathway clearly predated the appearance of antibodies, and it is now known that even the classical pathway can be activated independently of antibodies by **pentraxins**, a family of several pentameric proteins that bind to various lipids (the properties of individual pentraxins are discussed later in section 3-15).

Activation of complement by any of these pathways initiates a succession of cleavages in which each component in the pathway is cleaved into two fragments: a big fragment (the b fragment), which typically forms a subunit of the protease complex mediating the next cleavage in the pathway, and a small fragment (the a fragment), which in some cases is an inflammatory mediator. These reactions can conveniently be divided into early events, in which the complement components are not the same for the three pathways, and late events, in which they are identical. The function of the early events is to generate two functionally equivalent forms of a protease, known as *C3 convertase*, which initiates the late events to produce the effector components of complement. This sequence, known as the *complement cascade*, is outlined in Figure 3-10. The C3 convertases covalently attach to the cell surface at which complement activation was initiated, and in this way the effects of complement activation are confined to the infectious organism that triggered it.

Figure 3-10 Outline of the complement system Recognition of a foreign particle through mannose-binding lectin, ficolins, antibodies or pentraxins, or the alternative pathway, leads to activation of the complement cascade and generation of an enzyme, C3 convertase, which proteolytically cleaves complement component C3. C3 cleavage leads to four major consequences: (1) activation of endothelial cells and generation of chemo-attractants for leukocytes, leading to an inflammatory response; (2) tagging of the particle for uptake by phagocytes; (3) generation of the membrane-attack complex, which forms 7–10 nm diameter pores in membranes, breaking down their permeability barrier; and (4) stimulation of an antibody response. The specific components of the complement system that mediate these biological effects are listed. Their generation and properties are described in subsequent sections.

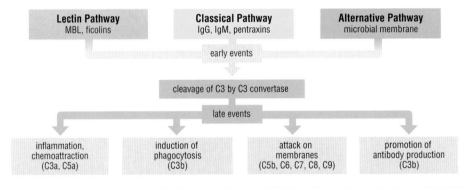

Components of the Early and Late Events in the Complement Cascade

Lectin Pathway	Classical Pathway	Alternative Pathway
Early events		
MASP-1, -2, -3	C1	factor D
C4	C4	C3b
C2	C2	factor B
Late events		
C5	C5	C5
C6, C7, C8, C9	C6, C7, C8, C9	C6, C7, C8, C9

Figure 3-11 Complement components activated in early and late events in the complement cascade Note that the active cleavage products generated in the complement cascade are not shown.

We shall discuss the functions of the complement components in detail in the sections that follow. Here we briefly introduce them, to indicate how they contribute to each pathway. The components of the classical pathway, which was the first to be described, are named C1 to C9, reflecting the order of their action except in the case of C4, which, it is now known, is activated before C2. All three pathways share the complement components mediating the late events. For the early events, the classical and lectin pathways are identical except for the first step: C1 binds antibodies or pentraxins and participates only in the classical pathway, whereas in the lectin pathway, the lectin innate immune components, and several bound protease subunits, are the equivalent of C1. In the alternative pathway, the functions of the early components C2 and C4 are performed instead by factor B, which is homologous to C2, and by C3, which is closely related to C4. The early events in each of the three complement pathways culminate in the generation of a protease called **C3 convertase** (described in detail in the next section), which cleaves C3. This is the pivotal event in the cascade, leading to the generation of the principal effector molecules (see Figure 3-10). The early and late components of each pathway are listed (without their active cleavage products) in Figure 3-11.

Inherited deficiencies in complement components lead to severe bacterial infections or to susceptibility to immune complex disease

Deficiencies in the alternative pathway of complement activation, in C3 or in the membrane-attack complex components often lead to severe bacterial infections. Deficiencies of C3 are the most severe, because of the central position of C3 in the complement cascade; they are also relatively rare. In deficiencies affecting the later components that form the membrane-attack complex, severe infections are mostly confined to various species of *Neisseria*, Gram-negative bacteria that cause gonorrhea and meningitis.

The complement system, as well as contributing to immune defense, is also important in the clearance of debris from cells that have died by apoptosis, and of complexes of antibody with soluble antigen whose accumulation would be destructive. This is one of the functions of the early components of the classical pathway, and may explain why individuals with deficiencies in C1 or C4 very often (in 75–90% of cases) develop *systemic lupus erythematosus*, an autoimmune disease in which antibodies are produced against components of the cell nucleus; complexes of these autoantibodies with the nuclear components are deposited in capillaries and joints, causing tissue destruction. Some important human diseases caused by complement deficiencies are listed in Figure 3-12. We discuss lupus erythematosus and its causes in more detail in Chapter 12.

Diseases Resulting from Complement Deficiencies

Deficiency	Disease susceptibility
Classical + lectin pathways	
C1, C4, C2	immune complex disease (SLE like)
MBL	increased infections, particularly early in life
Alternative pathway	
factors B, D, properdin	severe infections with *Neisseria*
Classical + lectin + alternative pathways	
C3 (factors H, I)	severe pyogenic infections, particularly with *Streptococcus pneumoniae*, *Neisseria meningitidis*, *Hemophilus influenzae*
Membrane-attack complex	
C5, C6, C7, C8, C9	severe infections with *Neisseria*

Figure 3-12 Table of diseases common in people deficient in complement components Individuals with defects in the early classical pathway components often develop systemic lupus erythematosus (SLE), an immune complex disease. Deficiencies of C3 are rare but exhibit the greatest susceptibility to bacterial infection, reflecting the central role of C3 in the complement cascade and the importance of phagocytosis. Note that deficiencies in factor H or factor I result in depletion of C3 from serum due to the loss of inhibition of the alternative pathway by these factors (as explained in section 3-4); thus, individuals lacking factor H or factor I are essentially C3-deficient and exhibit similarly severe bacterial infections. Relatively common complement deficiencies include C6 deficiency, which occurs in about 1/60,000 Caucasians, C7 deficiency, which occurs in about 1/25,000 Japanese, and C9 deficiency, which occurs in 1/1,000 Japanese. The latter was originally thought to be asymptomatic, but it is now recognized to increase the risk of Neisserial meningitis 700-fold in Japan, where this is a rare disease. C7 deficiency increases this risk 5,000-fold.

References

Fujita, T. *et al.*: **The lectin-complement pathway – its role in innate immunity and evolution.** *Immunol. Rev.* 2004, **198**:185–202.

Mold, C.: **Role of complement in host defense against bacterial infection.** *Microbes Infect.* 1999, **1**:633–638.

Walport, M.J.: **Complement.** *N. Engl. J. Med.* 2001, **344**:1058–1066 and 1140–1144.

3-4 Activation of the Complement Cascade

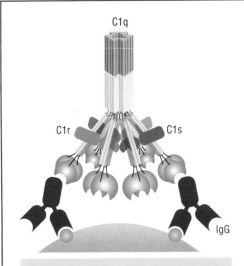

Figure 3-13 C1q/r/s complex that couples antibody or pentraxin binding to complement activation via the classical pathway

The C1q/r/s complex assumes a quaternary structure that is closely analogous to that of the mannose-binding lectin and H-ficolin complexes with MASPs, reflecting their analogous functional properties.

Initiation of complement activation by the lectin and classical pathways has substantially common features

The classical and lectin pathways of complement represent two closely analogous and evolutionarily related variations that couple different means of recognizing microbes or virus particles to the same complement effector functions. In the more ancient lectin pathway, soluble carbohydrate-binding innate immune components, namely collectins such as mannose-binding lectin or ficolins, recognize foreign polysaccharides based on specificity for their terminal sugar residues. These innate recognition elements are pre-bound to proteases, mannose-binding lectin-associated serine proteases (MASP) -1, -2 and/or -3 (see Figure 3-9), which become activated via conformational changes induced by particle binding and then cleave C4 and C2 to generate C4bC2b, which is one form of C3 convertase. In the classical pathway, recognition is via either antibody or the innate recognition molecules of the pentraxin family and **C1q** connects these components to the complement pathway. C1q is structurally similar to collectins and ficolins in its amino-terminal cysteine-rich domain and its central collagen-like domain, which together assemble it into a hexameric structure very similar to that seen in MBL and

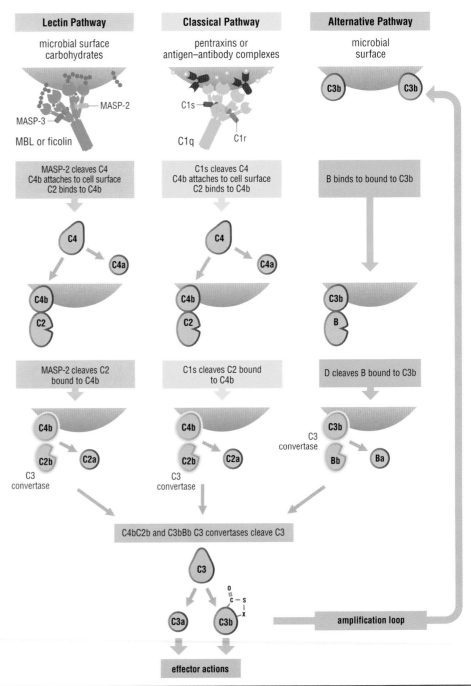

Figure 3-14 Activation of the complement cascade

The lectin pathway, the classical pathway, and the alternative pathway all lead to the production of C3 convertases, proteolytic enzymes that cleave C3. In the lectin pathway and the classical pathway, the convertase is generated by cleavage of C4, binding of C2 to the newly generated C4b and then cleavage of C2, resulting in the C4bC2b complex, which is the C3 convertase of these two pathways. These cleavages are conducted primarily by MBL-associated serine protease 2 (MASP-2) in the lectin pathway and the C1q-associated protease C1s in the classical pathway. C1q and its associated proteases are closely related to MBL and the ficolins and their associated proteases. In the alternative pathway, C3b or the water-reacted form of C3 recruits factor B, which then becomes a substrate for the protease factor D. This generates the alternative pathway C3 convertase, which is a complex between C3b and the large fragment of factor B (Bb). C3 is cleaved by the C3 convertases into its two major fragments, C3a, which promotes inflammation by triggering mast cells, and C3b, which is central to the other effector functions of complement. Note that because the alternative pathway is initiated by C3b, it provides an amplification loop for the other two pathways.

H-ficolin (Figure 3-13). Where these molecules differ is in their carboxy-terminal domains that bind to carbohydrates or to antibodies and pentraxins. The similarity of these proteins extends to the means by which they connect downstream to the complement pathway: C1q binds to two proteases, C1r and C1s, which are evolutionary variants of the MASPs and perform the same function. C1r is activated by C1q, and in turn cleaves and activates C1s, which then cleaves C4 and C2 (Figure 3-14).

Cleavage of C4 by MASP-2 or C1s produces C4b, which differs from C4 in two important respects. C4b contains a now reactive thioester group that attaches it covalently to the cell surface where it was generated. The second effect of the cleavage of C4 is to reveal an affinity for soluble C2, which is then cleaved by C1s in the classical pathway or MASP-2 in the lectin pathway. The resulting C4bC2b complex is the classical pathway form of C3 convertase, which cleaves C3. This sequence of events is illustrated in the left-hand and middle columns of Figure 3-14.

Complement activation is largely confined to cell surfaces

The effector actions of complement in promoting phagocytosis and attacking cell membranes require that the relevant complement components adhere tightly to the surface of the target cell. This is guaranteed for the large cleavage products of C4 and C3, which are structurally related. These complement components both contain a thioester bond that becomes highly reactive in the cleavage products C4b and C3b. In the fluid phase, this bond will react with H_2O ($t_{1/2} \approx 100~\mu s$), but if the cleavage occurs near a cell surface it can react with hydroxyl groups on cell surface carbohydrates, with consequent covalent attachment to the surface of the cell. In this way, the cleavage of C4 that is triggered by MBL or antibody bound to a cell surface leads to the covalent attachment of C4b to the same cell surface within microseconds; similarly the subsequent cleavage of C3 leads to rapid attachment of C3b to the same particle.

The alternative pathway of complement activation provides an amplification loop that is activated at cell surfaces

Although the alternative pathway of complement activation can be initiated independently of the other pathways, its principal significance is probably in the amplification of the complement cascade once C3b has been deposited through the lectin or classical pathways. Alternatively, a C3b-like molecule can be produced through a low level of spontaneous reaction of C3 with water that occurs continuously in plasma, followed by rapid binding of the molecule to a membrane (host cell membranes are protected from complement activation by specialized inhibitors that we discuss later). The next step is binding of factor B to C3b, which enables factor B to be cleaved by the serum protease factor D to generate Bb. C3bBb is the alternative pathway C3 convertase, so this leads to the cleavage of more C3 and hence a positive feedback loop and the amplification of the complement cascade (see Figure 3-14, right-hand column).

Activation of the alternative pathway can occur efficiently only at the surface of a cell. Most of the water-reacted C3 or C3b generated in the fluid phase is inactivated by the actions of two inhibitors, factors H and I. Factor H binds C3b and recruits factor I, which is a protease that cleaves C3b, inactivating its convertase function (Figure 3-15). However, when C3b is bound to a cell surface, factors H and I are less effective, possibly due to the action of a molecule called **properdin** (also called factor P), which stabilizes the C3bBb complex. To survive for long enough to bind factor B, C3b must attach to a cell surface with its reactive thioester. This is more likely to occur if it is generated at a cell surface through the activation of the lectin or classical pathways than if C3 has become activated by directly reacting with water in solution.

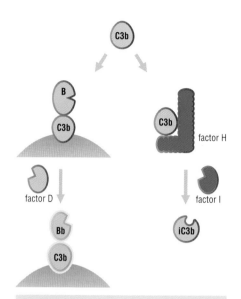

Figure 3-15 Amplification and inactivation of C3b C3b is produced by C3 convertase and can combine with factors B and D to create the alternative pathway C3 convertase. Thus, C3b is part of a positive amplification loop in the complement cascade. This positive amplification loop is restricted to non-self cell surfaces by the relative preferences of the positive acting factors B and D, on the one hand, and of the negative regulatory components factors H and I that proteolytically cleave and inactive C3b, on the other. Factor B and factor D actions are favored if C3b is tethered to a foreign cell surface, whereas factor H and factor I actions are favored for C3b that fails to attach to a cell surface because its thioester group has reacted with water. Note that factor H has an additional binding site specific for sialic acid-containing oligosaccharides. This promotes factor H binding to C3b that is attached to mammalian cells, which have sialic acids. This is one mechanism that helps protect host cells from complement attack. Additional regulatory steps that inhibit the complement cascade in various ways are described in section 3-6.

Definitions

C1q: complement component that binds to antibodies in immune complexes and activates the classical pathway of complement activation.

properdin: complement pathway component that binds to and stabilizes the alternative pathway C3 convertase, C3bBb.

References

Gadjeva, M. *et al.*: **The mannan-binding-lectin pathway of the innate immune response.** *Curr. Opin. Immunol.* 2001, **13**:74–78.

Matsushita, M. *et al.*: **Activation of the lectin complement pathway by ficolins.** *Int. Immunopharmacol.* 2001, **1**:359–363.

Nonaka, M.: **Evolution of the complement system.** *Curr. Opin. Immunol.* 2001, **13**:69–73.

The complement cascade promotes inflammation, phagocytosis, membrane attack and antibody production

C3b is not only a central component in the alternative pathway amplification loop; it is also critical to the effector functions of the complement cascade: these are summarized in Figure 3-16. C3b covalently attached to the surface of a microorganism serves as a tag for internalization by phagocytes. In addition, if it attaches close enough to either of the two types of C3 convertases it can bind C5 and promote proteolytic cleavage of this C5 by the C3 convertase to generate C5a and C5b. Thus, if positioned optimally, C3b alters the specificity of C3 convertases, making them also **C5 convertases**.

C5a is the major mediator of the inflammatory effects of complement. It acts on endothelial cells and mast cells to promote inflammation and it serves as a chemoattractant for neutrophils. These proinflammatory functions are shared by C3a, the small fragment released on cleavage of C3, but C3a is less potent. The larger fragment of C5, C5b, has structural similarity to C3b and C4b but lacks the reactive thioester group. C5b nucleates the formation of the **membrane-attack complex**: this complex contains C6, C7, C8 and multiple copies of C9, which assemble to form a transmembrane pore with an internal diameter of approximately 7–10 nm. These holes can promote leakage of intracellular components out of the cell or entry of various immune mediators through the outer membrane of bacteria. The membrane-attack complex is effective primarily against Gram-negative bacteria and enveloped viruses; the thick peptidoglycan layer of Gram-positive bacteria may prevent the membrane-attack complex from affecting these organisms.

Complement receptors mediate many of the actions of complement

Destruction by the membrane-attack complex is the only direct action of complement: all of the other functions of complement are mediated by the interaction of complement components with receptors on a variety of cell types. The specific receptors and some of their important properties are summarized in Figure 3-17. Complement receptors fall broadly into four structural and functional classes (Figure 3-18). CR1 and CR2 belong to a family called **SCR proteins** because they contain short consensus repeat sequences (SCRs) that are characteristic of many complement control proteins (we discuss these in the next section). Indeed, CR1 partly serves as a control protein. It binds C3b and recruits the C3b protease factor I, which shuts off the complement cascade by cleaving and inactivating C3b, and in so doing generates the biologically active cleavage fragments iC3b, which is an opsonin, and C3d, which contributes to adaptive immunity by promoting antibody production. CR1 on erythrocytes also has an important role in the clearance of small immune complexes from the blood by binding them and transporting

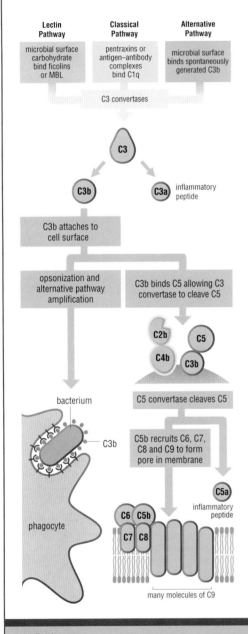

Figure 3-16 Late events and effector actions of the complement cascade C3 convertases of any of the three upstream complement pathways (see Figure 3-14) cleave C3 to generate C3a, an inflammatory peptide, plus C3b, which provides a tag for phagocytosis and can combine with factor B to amplify complement activation via the alternative pathway (not shown). C3b can also direct a C3 convertase complex to cleave C5, releasing C5a, a very potent proinflammatory peptide and neutrophil attractant, and C5b. C5b nucleates assembly of the membrane-attack complex, resulting in the formation of a pore in a membrane. The membrane-attack complex is effective against enveloped viruses and cells without thick peptidoglycan layers or polysaccharide capsules. Stimulation of phagocytosis is a major effector mechanism for many infectious agents.

Definitions

C5 convertase: any of several proteolytic enzymes that cleave complement component C5 to generate C5a and C5b. The enzyme is formed by the membrane attachment of a C3b molecule adjacent to either of the two types of C3 convertases. The C3b binds C5 which is then efficiently cleaved by the C3 convertase.

membrane-attack complex: membrane pore structure of 7–10 nm in internal diameter formed from complement proteins C6, C7, C8 and C9 recruited to a membrane by C5b, and that is important in immune defense against some classes of bacteria.

SCR protein: protein with short consensus repeats, which are seen on many complement regulatory proteins.

References

Gasque, P.: **Complement: a unique innate immune sensor for danger signals.** *Mol. Immunol.* 2004, **41**:1089–1098.

Guo, R.-F. and Ward, P.A.: **Role of C5a in inflammatory responses.** *Annu. Rev. Immunol.* 2005, **23**:821–852.

Helmy, K.Y. *et al.*: **CRIg: a macrophage complement receptor required for phagocytosis of circulating pathogens.** *Cell* 2006, **124**:915–927.

Major Complement Components and Receptors

Component	Role in Complement Cascade and Actions
C1qrs	binding to IgG or IgM or pentraxins and activation of classical pathway
MASP-1, -2, -3	proteases activated by MBL or ficolins to initiate lectin pathway
C2, C4	cleavage by C1s or MASP-2 generates C4b, which attaches to membranes, and C2b, which, associated with C4b forms the classical C3 convertase, the protease domain of which is contained in C2b
C3	cleaved by C3 convertases. C3b attaches to particles and promotes phagocytic uptake; C3b combines with factor B to form alternative pathway C3 convertase; C3b acts together with C3 convertases to form C5 convertases; C3a induces inflammation
factors B, D	B combines with C3b and is cleaved by D to yield the alternative pathway C3 convertase (C3bBb). The protease domain of this convertase is contained within the Bb subunit
properdin	binds to C3bBb (alternative pathway C3 convertase), stabilizing it
C5	cleaved by C5 convertase; C5b directs assembly of membrane-attack complex; C5a is proinflammatory peptide, neutrophil chemoattractant
C6, C7, C8, C9	associate with C5b and are assembled into the membrane attack complex, which forms pores in membranes. Multiple copies of C9 are incorporated into the pore structure
iC3b	cleavage fragment of C3b that opsonizes microbes, inducing phagocytosis
C3d	cleavage fragment of C3b that binds to CR2 and promotes antibody production by B cells

Receptor	Type	Ligand	Cells Expressing	Main Functions
CR1 (CD35)	SCR	C3b	B, E, M, N, Eos, FDC	promote cleavage of C3b and C4b, inactivate C3 convertases, limit alternative pathway, promote clearance of immune complexes
CR2 (CD21)	SCR	iC3b, C3d	B, FDC	promote antibody production
CR3 (CD11b/CD18)	integrin	iC3b, ICAM-1, β-glucans	NK, M, N, FDC, Ma	phagocytosis
CR4 (CD11c/CD18)	integrin	iC3b, fibronectin	NK, M, N, DC, Ma	phagocytosis
CRIg	Ig	C3b, iC3b	Kupffer	phagocytosis of particles in blood
C3aR	GPCR	C3a	Ba, Ma, SM	induce inflammation, smooth muscle contraction
C5aR (CD88)	GPCR	C5a	Ba, EC, Ma, M, N, SM	induce inflammation, chemotaxis, increase vascular permeability
C1qRp (CD93)	C-type lectin	C1q, MBL, SP-A	EC, M, N	phagocytosis
CD91, calreticulin	–	C1q, MBL, SP-A	M, NHC	phagocytosis of apoptotic cells

Figure 3-17 Table of complement components and receptors Listed in the top part of the table are the major components of the complement cascade and their main roles in the cascade and in the effector actions of complement. Not shown are the inhibitory regulators (see section 3-6). Note that the large fragment of C2 was originally named C2a, in contrast to every other large complement fragment, which are all b fragments. In recent years there have been efforts to regularize this aspect of complement nomenclature and call the large fragment C2b. Both nomenclatures can be found in the literature at this time. The regularized nomenclature is used throughout this book. The bottom part of the table shows the major complement receptors (CRs). Type refers to the type of molecule or domains present. B: B cell; Ba: basophil; DC: dendritic cell; E: erythrocyte; EC: endothelial cell; Eos: eosinophil; FDC: follicular dendritic cell; GPCR: G-protein-coupled receptor of the seven-transmembrane type; M: monocyte and/or tissue macrophage; Ma: mast cell; MBL: mannose-binding lectin; N: neutrophil; NK: natural killer cell; NHC: non-hematopoietic cells; SCR: short consensus repeat, a repeating motif in many complement control proteins; SM: smooth muscle cell; SP-A: surfactant protein A.

them to macrophages in liver and spleen. CR2 is the complement receptor that promotes antibody production by B cells on binding iC3b or the further breakdown product C3d, and we discuss its function in Chapter 6. CR3 and CR4 are integrins and mediate phagocytosis, as does CRIg, a member of the immunoglobulin superfamily, which is expressed selectively in Kupffer cells, the macrophages of the liver. C3aR and C5aR are G-protein-coupled receptors (GPCR) of the seven-transmembrane type (like chemokine receptors, see section 2-13) and, as their names suggest, they are receptors for C3a and C5a, the small soluble fragments released by the cleavage of C3 and C5, and induce the release of inflammatory mediators from mast cells or act on endothelial cells to induce inflammation directly. C5a is also a powerful chemoattractant for neutrophils and monocytes, acting via the C5a receptor. Finally, there are several receptors for C1q and the structurally related collectins mannose-binding lectin and surfactant protein A, generally known as C1qR. C1qRp is the best characterized: it is itself a C-type lectin (see section 2-5) and mediates phagocytosis.

The major complement components are also summarized in Figure 3-17. Several pivotal components are the targets of complement regulatory proteins, to which we now turn.

Figure 3-18 Structure and function of complement receptors Four structural classes of complement receptors are shown. CR1 and CR2 both regulate immune responses: CR1 binds to C3b and shuts off the alternative pathway complement cascade by promoting cleavage of C3b to iC3b and C3d; and CR2 binds these cleavage fragments and is thereby activated to promote antibody production by B cells. CR3 and CR4 also bind iC3b and induce phagocytosis, as does C1qR, which binds to C1q, MBL or SP-A. The seven-transmembrane receptors C3aR and C5aR bind inflammatory peptides and induce inflammation and neutrophil chemotaxis. SCR: small consensus repeat; GPCR: G-protein-coupled receptor; CLR: C-type lectin receptor.

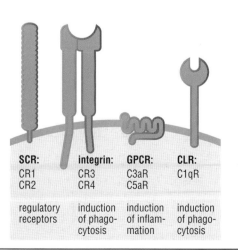

SCR:	integrin:	GPCR:	CLR:
CR1	CR3	C3aR	C1qR
CR2	CR4	C5aR	
regulatory receptors	induction of phago- cytosis	induction of inflam- mation	induction of phago- cytosis

3-6 Regulation of Complement and Evasion by Microbes

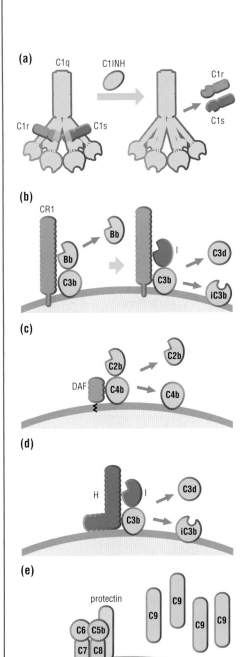

(a)

C1q   C1INH   C1r

C1r   C1s   C1s

(b)

CR1

Bb

Bb   I   C3d

C3b   C3b   iC3b

(c)

C2b   C2b

DAF   C4b   C4b

(d)

H   I   C3d

C3b   iC3b

(e)

protectin

C9   C9   C9   C9

C6 C5b

C7 C8

Complement activation is regulated by soluble factors and membrane-bound inhibitors

Because of the destructive nature of the complement system and its automatic amplification by the alternative pathway, stringent controls are necessary to prevent excessive damage to host tissues. Some of this control is exerted by soluble factors that prevent excessive activation of complement by circulating immune complexes that could trigger the classical pathway, or spontaneous activation of the alternative pathway in solution. A soluble negative regulator called **C1 inhibitor** inactivates C1 by inducing dissociation of the active C1r and C1s proteases from C1q (Figure 3-19a). This protease inhibitor also inhibits the kinin and coagulation cascades (see section 3-13). Genetic deficiency in one copy of the gene encoding this protein causes a sporadic inflammatory disease called hereditary angioneurotic edema. Another soluble inhibitor of the complement pathway is factor I, which we have already encountered, a protease that inactivates the C3 convertases by cleaving either C3b to shut off the alternative pathway, or C4b to shut off the classical pathway. Additional soluble inhibitors of complement are factor H, which targets factor I to C3b, as we have already seen; and a factor called **C4-binding protein (C4bp)**, which targets factor I to C4b.

Soluble regulators alone cannot prevent complement activation when C3b becomes attached to cell surfaces, however. We have already seen that activation and amplification by the alternative pathway is favored over inactivation by factors H and I when C3b is deposited on a microbial cell surface. The reactive thioesters of C4b and C3b form covalent bonds just as readily with host cell surfaces as they do with those of microorganisms. Host cells must therefore be protected from the destructive effects of complement, and this protection is provided by cell-surface molecules that rapidly and efficiently inactivate C3b and thus prevent the activation of the alternative pathway and the late events of the complement cascade. Protection of host cells is largely ensured by membrane-bound negative regulators, most of which are members of the SCR family of proteins, and which include **complement receptor 1 (CR1)(CD35)**. CR1 induces the dissociation of both classical and alternative pathway C3 convertases and recruits factor I, which cleaves C3b and C4b (Figure 3-19b). The membrane-bound SCR proteins **decay-accelerating factor (DAF)(CD55)** and **membrane cofactor of proteolysis (MCP)(CD46)** between them have similar functions to CR1: DAF induces dissociation of the convertases and MCP recruits factor I and promotes the proteolysis of C3b (Figure 3-19c). The soluble regulator factor H, which is also a member of the SCR family, can also act at mammalian cell surfaces by binding to sialic acid, which is common at the end of mammalian glycoproteins but generally absent from microbial surfaces (Figure 3-19d). **Protectin (CD59)** acts later in the pathway, preventing the assembly of the membrane-attack complex (Figure 3-19e).

These membrane-bound negative regulators can protect from alternative pathway activation and low levels of antibody-directed classical pathway activation, but not from high levels of autoantibody binding, as occurs in some autoimmune diseases (for example autoimmune

---

Figure 3-19 Mechanisms of action of some complement regulatory proteins (a) C1 inhibitor binds to activated C1r and C1s on C1q and causes their dissociation. (b) CR1 binds to and causes dissociation of either form of C3 convertase and recruits factor I, which cleaves and thereby inactivates C3b (or C4b, not shown). (c) DAF induces dissociation and thus inactivation of C3 convertase. (d) The soluble regulator factor H binds to C3b on sialic acid-containing cell surfaces (for example, host cells), and targets factor I to cleave C3b. The transmembrane protein MCP similarly promotes proteolysis of C3b and can also promote proteolysis of C4b (not shown). (e) Protectin binds to C7 and C8, preventing them from binding to C9 and nucleating assembly of the membrane-attack complex.

Definitions

C1 inhibitor: a soluble protein that dissociates active complement components C1r–C1s from C1q and thereby limits the classical pathway of complement.

C4-binding protein (C4bp): a soluble protein that binds complement component C4b and recruits factor I, which targets C4b for degradation, thereby limiting complement activation.

C4bp: see **C4-binding protein**.

CD35: see **complement receptor 1**.

CD46: see **membrane cofactor of proteolysis**.

CD55: see **decay-accelerating factor**.

CD59: see **protectin**.

complement receptor 1 (CR1); CD35: a membrane protein of phagocytes that targets complement component C3b for proteolytic inactivation, to limit activation of the alternative pathway.

CR1: see **complement receptor 1**.

DAF: see **decay-accelerating factor**.

decay-accelerating factor (DAF); CD55: a membrane protein that protects cells against attack by the alternative pathway of complement by binding C3b and causing dissociation of factor B.

MCP: see **membrane cofactor of proteolysis**.

membrane cofactor of proteolysis (MCP); CD46: a molecule that protects cells against attack by the alternative pathway of complement activation by binding to C3b and C4b and recruiting factor I to inactivate them by proteolysis.

Figure 3-20 Table of complement regulatory proteins The major soluble and membrane-bound negative regulators of the complement system are listed and their functions described.

hemolytic anemia). The importance of these proteins is illustrated by mice made genetically deficient for a protein called Crry, which has a role analogous to those of human DAF and MCP. Such mice die *in utero* as a result of complement-dependent attack at the feto-maternal interface. Thus, the alternative pathway of the complement system discriminates between self and pathogen in a negative way: all cells are attacked, but self cells have a protective mechanism. The complement regulatory proteins and their actions are summarized in Figure 3-20. Not surprisingly, professional pathogens often have developed molecules of their own to protect them from the alternative pathway of complement.

Many bacteria and viruses have evolved mechanisms for evading complement action

The significance of complement as a mechanism of immune defense is indicated by the severe and recurrent bacterial infections that are seen in individuals with genetic deficiencies of complement components, particularly C3. Conversely, professional pathogens have evolved a multitude of mechanisms for evading complement action. Some of the more common evasion strategies are illustrated in Figure 3-21. Both microbes and viruses have a variety of evasion mechanisms that for the most part parallel the intrinsic host mechanisms for limiting complement activation. Especially prevalent are mechanisms for interfering with the C3 convertases. These involve viral expression of genes that are similar to vertebrate regulators (CR1, DAF, MCP) and hence probably reflect gene capture. Herpes virus saimiri (HVS) encodes a homolog of CD59, which blocks formation of the membrane-attack complex. A number of bacteria also express surface proteins that can block this complex formation, including TraT of *Salmonella*, Ail of *Yersinia enterocolitica*, and SIC of *Streptococcus pyogenes*. Opportunism is common in the bacterial world: many bacteria have evolved means of capturing soluble host regulators of complement and concentrating them on their surface to limit complement action there. Viruses can be opportunistic as well: HIV-1 captures factor H via its gp41 glycoprotein. Some enveloped viruses also incorporate host membrane regulators of complement into their membranes when they bud from cells. The widespread nature of these mechanisms and their correlation with virulence in many cases argues for the importance of complement in immune defense, at least against those pathogens that have not yet developed strong means of preventing complement activation.

Inhibitors of Complement

Component	Mechanism of action
Soluble components	
C1 INH	dissociates active C1r–C1s complex from C1q
C4bp	binds to C4b and targets factor I to cleave it
factor H	binds to C3b and targets factor I to cleave it
factor I	protease that can inactivate C3b, C4b
Membrane-bound components	
CR1 (CD35)	complement receptor; induces dissociation of C3 convertases; targets factor I for cleavage of C3b, C4b
DAF (CD55)	induces the dissociation of the components of C3 convertases
MCP (CD46)	cofactor for factor I for cleavage of C3b, C4b
protectin (CD59)	interferes with assembly of membrane-attack complex by binding to C7, C8
Crry	similar to DAF and MCP (found in mouse not human)

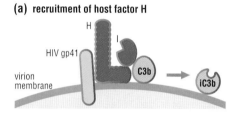

(a) recruitment of host factor H

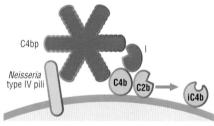

(b) recruitment of host C4bp

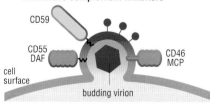

(c) incorporation of host membrane complement inhibitors

Figure 3-21 Microbes and viruses evade complement attack by multiple mechanisms
(a) HIV-1 and many bacteria make cell surface proteins that recruit factor H, which in turn recruits factor I, which cleaves and inactivates C3b. HIV-1 gp41 binds factor H, as do many *Streptococcus pyogenes* M protein family members, *Streptococcus pneumoniae* Hic protein, the *Neisseria* porin, Por1A, and *Borrelia burgdorferi* OspE. Some pathogenic bacteria such as *Neisseria gonorrheae* also incorporate sialic acid into their cell walls and recruit factor H in that way. **(b)** Similarly, some microbes make proteins that recruit C4b-binding protein (C4bp), which recruits factor I to degrade C4b and shut off the classical pathway C3 convertase. Examples include type IV pili of *N. gonorrheae* and filamentous hemagglutinin of *Bordetella pertussis*. **(c)** Enveloped viruses may acquire host membrane complement regulatory proteins upon budding. For example, HIV-1 particles include DAF (CD55). Shown is vaccinia, which buds in two forms, one from internal membranes (called intracellular mature virus), which lacks these components and is highly sensitive to complement lysis, and another from the plasma membrane (called extracellular enveloped virus), which is less stable but incorporates CD46, CD55 and CD59 and is resistant to complement lysis.

protectin (CD59): a membrane-bound complement regulatory protein that prevents formation of the membrane-attack complex.

References

Kotwal, G.J.: **Poxviral mimicry of complement and chemokine system components: what's the end game?** *Immunol. Today* 2000, **21**: 242–248.

Miwa, T. and Song, W.-C.: **Membrane complement regulatory proteins: insight from animal studies and relevance to human diseases.** *Int. Immunopharmacol.* 2001, **1**:445–459.

Rautemaa, R. and Meri, S.: **Complement-resistance mechanisms of bacteria.** *Microbes Infect.* 1999, 1:785–794.

Phagocytosis is a major mechanism for the destruction of microbes

The function of the collectins and the complement opsonins that we have described in the preceding five sections is to tag microbes that have penetrated the epithelial barrier, as well as apoptotic cells and other tissue debris, for destruction by *phagocytosis*. **Phagocytosis** is the process by which a single cell engulfs a large particle, generally defined as 1 μm in diameter or larger, and is important both for immune defense and for tissue repair and homeostasis. Many types of cells are capable of phagocytosis, but neutrophils, macrophages and dendritic cells are particularly active in this process and are known as professional phagocytes. Neutrophils are circulating cells that are recruited into the tissues at sites of infection and are highly specialized for the internalization and destruction of microorganisms; macrophages and dendritic cells are sentinel cells resident in the tissues: macrophages have an important role in tissue maintenance as well as in immune defense; the main function of dendritic cells is to initiate adaptive immune responses, and we shall defer detailed discussion of these cells until Chapter 4.

Phagocytosis is a receptor-mediated process in which specific recognition of a particle by receptors on the phagocyte triggers actin polymerization, engulfment of the particle and fusion of a vesicle containing the internalized particle with specialized intracellular organelles, *granules* and *lysosomes*, containing destructive enzymes and microbicidal products (Figure 3-22a). A related process is **macropinocytosis**, in which phagocytic cells take up large amounts of fluid in their extracellular environment, apparently without receptor recognition but also through a process that is dependent upon actin polymerization (Figure 3-22b).

Phagocytes have receptors for recognizing and internalizing microbes

Phagocytes, and particularly macrophages and dendritic cells, carry an array of receptors that recognize many microbial and other targets. Not all of these induce phagocytosis: the *Toll-like receptors (TLRs)*, for example, which we describe in subsequent sections, are an important family of signaling molecules that activate the production of inflammatory and antiviral cytokines and the maturation of dendritic cells but do not directly stimulate particle internalization. Receptors that directly induce internalization of bound particles are known as **phagocytic receptors** and fall into two categories: those that can directly recognize microbes, and those that bind to **opsonized** particles after soluble molecules have tagged them (Figure 3-23). Receptors that directly recognize microbes include the **mannose receptor** of macrophages (Figure 3-24), which recognizes branched α-linked mannose oligosaccharide chains found on the surfaces of many bacteria, fungi and protozoa but absent from most mammalian proteins, and **dectin-1**, a C-type lectin-like receptor that recognizes β-glucan polysaccharides found in various microbial cell walls. Interestingly, the mannose receptor also binds to some viruses, including influenza virus and HIV, presumably reflecting aberrant glycosylation of viral proteins for unknown reasons. A group of structurally diverse phagocytic receptors collectively known as **scavenger receptors**, by contrast, can often recognize components of apoptotic cells as well as components of pathogens. Two, both belonging to a structural family known as the *scavenger receptor A (SRA)* family, contribute in particular to defense against bacterial infection. The prototype of the SRA family, scavenger receptor AI, originally defined by the ability to bind oxidized lipoprotein components of atherosclerotic plaques, binds LPS, lipoteichoic acid and intact Gram-negative and Gram-positive bacteria. Mice lacking this molecule are more susceptible to infections with Gram-positive bacteria, including *Staphylococcus aureus* and the intracellular pathogen *Listeria monocytogenes*, which is also less readily killed by the mutant macrophages *in vitro*. A highly related receptor called MARCO is expressed on macrophages in the marginal zone of the spleen, which filters the blood, and also mediates binding of intact bacteria and may help remove them from the bloodstream.

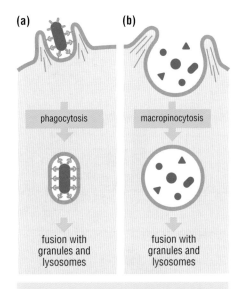

(a) phagocytosis

fusion with granules and lysosomes

(b) macropinocytosis

fusion with granules and lysosomes

Figure 3-22 Mechanisms of particle internalization by phagocytic cells
(a) In phagocytosis, binding of a phagocyte to a particle leads to engulfment in which the phagocyte extends its membrane around the particle, supported by actin polymerization and by insertion of new membrane coming from fusion with cytoplasmic vesicles (not shown).
(b) In macropinocytosis, a related process, large quantities of extracellular fluid and the particles it contains are engulfed in an actin-dependent process. In both cases, membrane vesicles containing the internalized particles fuse with intracellular organelles, granules and lysosomes, containing destructive enzymes and microbicidal products.

Definitions

dectin-1: a phagocytic C-type lectin-like receptor recognizing microbial β-glucan polysaccharides.

iC3b: a large proteolytic fragment of complement component C3b that has lost enzymatic activity in the complement pathway but is recognized by complement receptors 2, 3 and 4 (CR2, CR3, CR4) and activates **phagocytosis**.

macropinocytosis: process whereby cells take up large amounts of extracellular fluid independently of receptor

binding to ligands.

mannose receptor: endocytic receptor of macrophages and dendritic cells that binds to branched α-linked oligomannose-containing molecules through its C-type lectin domains. Also called CD206.

opsonized: bound by soluble recognition elements of the innate or adaptive immune system, such as antibody, **iC3b** and collectins, that are recognized by **phagocytic receptors**.

phagocytic receptor: cell surface molecule of phago-

cytes that binds microbes, viruses or apoptotic cells, either directly or through opsonins, and induces **phagocytosis**.

phagocytosis: receptor-mediated internalization of cells or other particles larger than 1 μm in diameter.

scavenger receptor: any of a structurally diverse group of receptors defined by their ability to bind polyanionic ligands such as oxidized low-density lipoprotein (LDL), as occurs in atherosclerotic plaques. Most scavenger receptors additionally bind either microbial ligands or apoptotic cells.

Phagocytic Receptors

Receptor	Type	Expression	Ligands
Innate immune receptors			
mannose receptor	C-type lectin	macrophages, DC	mannans
DEC 205	C-type lectin	DC	mannans
dectin-1	C-type lectin	macrophages	glucose-rich polysaccharides
CD14	leucine-rich repeats	macrophages, neutrophils	apoptotic cells, LPS
MARCO	SR-A	MZ macrophages	bacterial cell walls
scavenger receptor A I	SR-A	macrophages	apoptotic cells; LPS, LTA
CD36	SR-B	macrophages	apoptotic cells; parasitized RBCs
Opsonin receptors			
C1qRp (CD93)	C-type lectin	macrophages	C1q; collectins
CR3 ($\alpha M\beta 2$, CD11c/CD18)	integrin	macrophages, neutrophils	iC3b, β-glucans, ICAM1/2
CR4 ($\alpha x\beta 2$, CD11d/CD18)	integrin	macrophages, neutrophils, DC	iC3b, fibrinogen
FcγRI (CD64)	Ig, ITAM	macrophages, neutrophils	IgG, CRP, SAP
FcγRII (CD32)	Ig, ITAM	macrophages, neutrophils	IgG
FcγRIII (CD16)	Ig, ITAM	macrophages, neutrophils, NK	IgG, SAP
FcγRIV	Ig, ITAM	macrophages, neutrophils, DC	IgG2a, IgG2b
FcαR (CD89)	Ig, ITAM	macrophages, neutrophils, Eos	IgA
integrin $\alpha v\beta 3$	integrin	macrophages, platelets	thrombospondin-opsonized cells
Mer	RTK	macrophages	apoptotic cells

Figure 3-23 Table of phagocytic receptors Receptors are grouped into those that mediate phagocytosis by directly binding to the particle and those that mediate phagocytosis of opsonized particles, although CR3 fits in both categories. CRP: C-reactive protein; DC: dendritic cell; Eos: eosinophil; Ig: immunoglobulin superfamily domain; ITAM: immunoreceptor tyrosine-based activation motif; LPS: lipopolysaccharide; LTA: lipoteichoic acids; MZ: marginal zone of spleen; NK: natural killer cell; RBCs: red blood cells; RTK: receptor tyrosine kinase; SAP: serum amyloid protein; SR-A: scavenger receptor type A; SR-B: scavenger receptor type B. FcγRIV has been identified in mouse but not in humans.

Opsonin receptors, which recognize particles indirectly, fall into two categories: the complement receptors C1qRp, CR3 and CR4, and CRIg which bind to particles opsonized by complement components; and the *Fc receptors*, which bind to particles opsonized by antibodies, the versatile opsonins of adaptive immunity.

Although both microbes and apoptotic debris are targeted for destruction by phagocytosis, the mechanism and strength of the phagocytic response can vary, and this may depend upon whether the phagocytic receptor itself is specific for microbial components or whether the phagocytic stimulus coincides with signals from other receptors binding to microbial molecules. This can be illustrated by the case of CR3, which has two binding sites, one for **iC3b** and a distinct one for microbial β-glucan polysaccharides. Binding to these two sites has distinct functional consequences. Binding iC3b alone leads to a phagocytosis characterized by slow fusion to lysosomes and a relatively weak microbicidal response, whereas if both the iC3b and the β-glucan binding sites of CR3 are engaged, the microbicidal response is much more robust. This is a general feature of phagocytosis mediated by iC3b tagging of particles: it is greatly enhanced by simultaneous recognition of other immune ligands, either by CR3 itself or by another phagocytic receptor, such as the mannose receptor, which specifically recognizes microbial components. Alternatively, other signals such as proinflammatory cytokines can boost the phagocytic response to iC3b-opsonized particles. The relative mildness of the phagocytic response to iC3b alone presumably reflects the fact that the alternative pathway of complement can tag inert particles such as pollen and apoptotic cells as well as infectious agents.

We shall see in the next section that Fc receptors, which bind opsonins almost always specifically targeted at microorganisms, enable the especially rapid and destructive phagocytosis of particles.

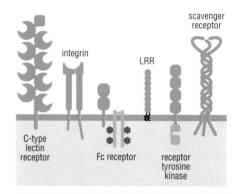

Figure 3-24 Different types of phagocytic receptors Shown are examples of phagocytic receptors of six different structural classes, including C-type lectin receptors (mannose receptor), integrins, immunoglobulin superfamily members (FcγRIII), leucine-rich repeat (LRR) receptors (CD14), receptor tyrosine kinases (Mer) and scavenger receptors.

References

Ehlers, M.R.: **CR3: a general purpose adhesion-recognition receptor essential for innate immunity.** *Microbes Infect.* 2000, **2**:289–294.

Linehan, S.A. *et al.*: **Macrophage lectins in host disease.** *Microbes Infect.* 2000, **2**:279–288.

Peiser, L. and Gordon, S.: **The function of scavenger receptors expressed by macrophages and their role in the regulation of inflammation.** *Microbes Infect.* 2001, **3**:149–159.

3-8 Mechanisms of Phagocytosis

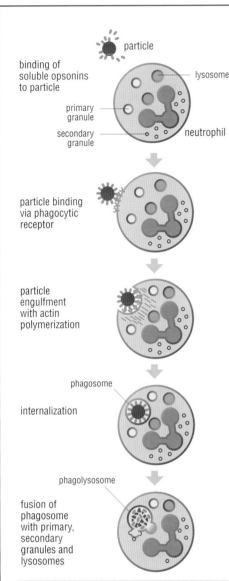

binding of soluble opsonins to particle

particle
lysosome
primary granule
secondary granule
neutrophil

particle binding via phagocytic receptor

particle engulfment with actin polymerization

phagosome

internalization

phagolysosome

fusion of phagosome with primary, secondary granules and lysosomes

Figure 3-25 Mechanism of phagocytosis
Binding of receptors on a phagocyte (in this case a neutrophil) to a particle leads to engulfment in which the phagocyte extends its membrane around the particle, supported by actin polymerization and by the insertion of new membrane coming from fusion with cytoplasmic vesicles (not shown). Membrane fusion then sequesters the particle in a phagosome inside the phagocyte. Subsequently, fusion of primary and secondary granules and lysosomes with the phagosome to generate a phagolysosome introduces toxic molecules promoting destruction of the microbe.

Phagocytosis proceeds by three distinct pathways

Phagocytosis proceeds by a series of steps illustrated in Figure 3-25 for a neutrophil. Receptor-mediated internalization of a particle (in this case opsonized) results in the formation of a vesicle, the **phagosome**, containing the internalized microbe. The phagosome then fuses with *primary* and *secondary granules*, and finally with lysosomes to form the fully destructive **phagolysosome**, which now contains antimicrobial peptides, lysosomal digestive enzymes, such as the proteases neutrophil elastase and cathepsin G, and highly toxic reactive oxygen and nitrogen compounds. The process is similar in macrophages and dendritic cells except that primary and secondary granules do not seem to be present in these cells.

The exact nature of the phagocytic process is dependent upon the receptor or receptors engaged by the particles. All phagocytic receptors activate actin polymerization through small GTPases belonging to a family known as the **Rho-family GTPases**, but this process is distinct for different receptors. When phagocytes engage particles primarily via complement receptor 3 (CR3), they internalize these particles via a relatively gentle process in which the particle seems to sink into the phagocytic cell, rather than being engulfed. This process is dependent on the small GTPase Rho; in this case, fusion of the early phagosome to lysosomes is quite slow, although it can be accelerated by recognition through an additional receptor such as a Toll-like receptor or C-type lectin, just as microbicidal activity of the phagocyte can be increased by simultaneous recognition of iC3b and microbial β-glucans, as we saw in the previous section.

By contrast, phagocytosis involving recognition by Fcγ receptors of particles opsonized by antibodies is a very vigorous process in which membrane is actively extended up along the particle in a process called zippering. These membrane extensions are driven not by Rho but by two other small GTPases of the same family, Rac and Cdc42, which induce local actin polymerization that pushes the membrane forward. The relative vigor of phagocytosis induced by Fc receptors probably reflects the fact that whereas complement opsonizes both pathogens and apoptotic cells, antibodies are generally specific for pathogens.

A third mechanism of phagocytosis resembles macropinocytosis and is seen when phagocytes encounter certain intracellular pathogens which may be stimulating the process as a means of entering the macrophage in a way that favors their survival.

Phagocytosis is a common target of immune evasion by pathogens, including microbes that establish intracellular infections and those that grow extracellularly. For example, *Yersinia* YopE, *Pseudomonas* ExoS and ExoT and *Salmonella* SptP are all toxins that block Rho-family GTPases. These toxins are introduced into host cells by a specialized needle-like secretion system (the *type III secretion system*) of these bacteria and hence act only on cells in physical contact with the bacterium. The *Clostridium* organisms that cause tetanus and botulism also make exotoxins that inactivate Rho-family GTPases, although these toxins enter cells by a different route.

Microbes that have evolved to be intracellular parasites require the early steps of phagocytosis to enter the cell but have generally developed mechanisms for disrupting or avoiding the later steps, either by rupturing the early phagosome and escaping into the cytoplasm (*Listeria monocytogenes*) or by blocking fusion to lysosomes (*Mycobacterium tuberculosis* and *Salmonella*). In *Mycobacterium*, signaling by the important T cell cytokine interferon-γ (IFN-γ) overcomes this block and allows killing of these microbes. Humans with genetic defects in the IFN-γ pathway typically have recurrent, severe infections with *Mycobacterium* species that are ordinarily less pathogenic for humans than *Mycobacterium tuberculosis* is.

Definitions

autophagosome: double-membrane-enclosed compartment generated in the cytoplasm as a response to either nutrient deprivation or a bacterium in the cytoplasm, a process known as **autophagy**. Autophagosomes fuse with endosomes and then lysosomes and so mature to a structure similar to the **phagolysosome**.

autophagy: a process activated during cellular starvation or in response to a damaged organelle that involves the sequestration and degradation of cellular organelles and cytoplasm for recycling of macromolecules.

phagolysosome: phagosome that has fused with lysosomes.

phagosome: vesicle generated by invagination and fusion of the plasma membrane of a phagocyte around a particle bound to phagocytic receptors.

phosphatidylserine: a phospholipid that is a major component of the internal half of the plasma membrane. In apoptotic cells, the asymmetry of phospholipids is lost as an early event, with phosphatidylserine appearing on the outside of the plasma membrane, where it is recognized by opsonins and phagocytic

receptors of macrophages or dendritic cells, leading to phagocytosis of the apoptotic cell.

Rho-family GTPases: family of small G proteins that includes Rac, Rho and Cdc42, all of which have primary functions involving regulation of actin polymerization or myosin-based contraction of actin filaments.

There is an alternative to phagocytic destruction for bacteria reaching the cytoplasm: these are targeted for encapsulation within a double membrane structure called the **autophagosome**, which is ordinarily involved in the removal of damaged organelles and in the catabolism of cell components as a response to starvation. This process, known as **autophagy**, may have a role in immune defense and is enhanced in response to IFN-γ and tumor necrosis factor (TNF).

Phagocytosis of apoptotic cells is usually antiinflammatory

Phagocytosis is also important for the rapid removal of apoptotic cells, and macrophages and dendritic cells have many receptors for recognizing them. Apoptosis of cells occurs in the course of embryonic development and during homeostatic renewal of many tissues, and thus can be a normal process, but it can also be induced after infection, for example by killing of virus-infected cells. Thus, immune system cells need mechanisms both for efficiently taking up apoptotic cells and for avoiding inflammatory responses to the apoptotic products of normal processes.

Cells undergoing apoptosis first of all signal their presence by releasing the lipid mediator lysophosphatidic acid (LPA), which attracts tissue phagocytes through a seven-transmembrane G-protein-coupled receptor. Once adjacent to the apoptotic cell, the phagocyte binds to it by a surprising multiplicity of mechanisms that result from the loss of phospholipid asymmetry of the apoptotic cell, in particular the externalization of **phosphatidylserine**, as well as the presence of surface LPA. Receptors thought to participate in phagocytosis of apoptotic cells include receptors for externalized phosphatidylserine, various scavenger receptors, and receptors for opsonins that bind to apoptotic cells, including C1q, iC3b, collectins, and the pentraxins CRP and SAP (Figure 3-26). In mammals, multiple parallel pathways seem to be operating. Internalization is generally mediated by the macropinocytosis-like pathway.

Whereas phagocytosis of microbes is typically coupled both to destructive events such as production of highly toxic oxidative compounds and to production of inflammatory cytokines and chemokines, phagocytosis of uninfected apoptotic cells is generally antiinflammatory, accompanied by the production of a number of immunosuppressive mediators including *TGF-β*, *IL-10* and prostaglandin E_2 (we shall meet these later when we discuss the regulation of inflammation). The intracellular parasite *Trypanosoma cruzii* seems to exploit the non-destructive nature of phagocytosis of apoptotic cells to establish itself inside macrophages during infection.

The choice between an antiinflammatory and a proinflammatory response depends on the nature of signals transmitted internally by the receptors that mediate the internalization of the particle, and on the presence or absence of ligands on the particle for signaling receptors that are specialized for recognition of signature molecules of pathogens; the most important of these, as we shall see later in this chapter, are the Toll-like receptors.

Inhibitory receptors can regulate phagocytosis

Phagocytosis is also negatively regulated by some receptors on phagocytes. For example, senescent erythrocytes are removed from the blood via phagocytes by the macrophage-lineage cells of the liver, the Kupffer cells. Healthy erythrocytes avoid this fate by expressing CD47, which binds to an immunoglobulin superfamily receptor on macrophages called SIRP-1α. Recognition of CD47 by SIRP-1α inhibits phagocytosis of erythrocytes by macrophages. This interaction increases the threshold for phagocytosis so that more positive signals are required to overcome this inhibition, ensuring that erythrocytes are not removed from the circulation prematurely.

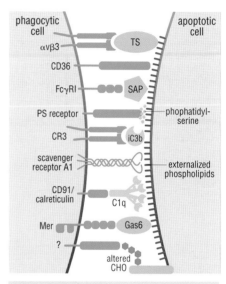

Figure 3-26 Phagocytosis of apoptotic cells
Apoptotic cells provide a number of tags for non-inflammatory phagocytosis, including externalized phosphatidylserine (PS), lysophosphatidylcholine and altered carbohydrates. Soluble factors that bind to apoptotic cells and promote their phagocytosis include serum amyloid protein (SAP), thrombospondin (TS) (a soluble protein, also found in extracellular matrix), collectins, annexin I (released from the apoptotic cell), Gas6, C1q and iC3b deposited by the action of complement. Receptors implicated in recognizing these opsonized apoptotic cells include scavenger receptors SR-A1 and CD36, integrins $\alpha v\beta 3$ or $\alpha v\beta 5$ and CR3, FcγRI and III, the collectin/C1q receptor CD91/calreticulin, the PS receptor and the receptor tyrosine kinase Mer.

References

Aderem, A. and Underhill, D.M.: **Mechanisms of phagocytosis in macrophages.** *Annu. Rev. Immunol.* 1999, **17**:593–623.

Coombes, B.K. *et al.*: **Evasive maneuvers by secreted bacterial proteins to avoid innate immune responses.** *Curr. Biol.* 2004, **14**:R856–R867.

Duclos, S. and Desjardins, M.: **Subversion of a young phagosome: the survival strategies of intracellular pathogens.** *Cell. Microbiol.* 2000, **2**:365–377.

Ernst, J.D.: **Bacterial inhibition of phagocytosis.** *Cell. Microbiol.* 2000, **2**:379–386.

Kirkegaard, K. *et al.*: **Cellular autophagy: surrender, avoidance and subversion of microorganisms.** *Nat. Rev. Microbiol.* 2004, **2**:301–314.

Lauber, K. *et al.*: **Clearance of apoptotic cells: getting rid of the corpses.** *Mol. Cell* 2004, **14**:277–287.

Levine, B.: **Eating oneself and uninvited guests: autophagy-related pathways in cellular defense.** *Cell* 2005, **120**:159–162.

Oldenborg, P.-A. *et al.*: **Role of CD47 as a marker of self on red blood cells.** *Science* 2000, **288**:2051–2054.

Scott, R.S. *et al.*: **Phagocytosis and clearance of apoptotic cells is mediated by MER.** *Nature* 2001, **411**:207–211.

Stuart, L.M. and Ezekowitz, R.A.: **Phagocytosis: elegant complexity.** *Immunity* 2005, **22**:539–550.

Williamson, P. and Schlegel, R.A.: **Hide and seek: the secret identity of the phosphatidylserine receptor.** *J. Biol.* 2004, **3**:14.

3-9 Destructive Mechanisms of Phagocytes

Contents of Neutrophil Granules
Primary (azurophilic) granules
defensins
lysozyme
bacterial-permeability-inducing protein
neutrophil elastase
cathepsin G
proteinase 3
myeloperoxidase
proteoglycans
Secondary (specific) granules
lactoferrin
cathelicidins
transcobalamin II
lysozyme
NADPH oxidase components

Figure 3-27 Table of contents of the primary and secondary granules of neutrophils Although primary and secondary granules have distinct properties, many of the contents of each granule are present at lower levels in the other type of granule. In addition, neutrophils have a third type of granule, called the tertiary granule.

Phagocytes have many mechanisms for killing internalized microbes

Whether it is a microbe or debris from an apoptotic cell, a particle that has been internalized by a phagocyte is destined for delivery to the phagolysosome, which, as we have seen (Figure 3-25), is derived in neutrophils from the fusion of the phagosome containing the particle with granules and lysosomes. The **primary granules** of neutrophils are a specialized type of internal vesicle somewhat related to lysosomes. They contain many microbicidal components including the antimicrobial peptides described in section 3-1 (which make up 5–10% of total neutrophil protein), several proteases (neutrophil elastase, cathepsin G and proteinase 3), **lysozyme**, which can degrade peptidoglycan, and bacterial-permeability-inducing protein (BPI), which can bind to LPS and make the outer membrane of Gram-negative bacteria permeable, which promotes killing in combination with other microbicidal agents (Figure 3-27). **Secondary granules** are a distinct type of internal vesicle of neutrophils that contain components of the enzyme **NADPH oxidase**, which is activated by signaling from phagocytic receptors to produce toxic effectors, as described below. As in other cells, neutrophil lysosomes contain a variety of digestive enzymes. The combined action of the components of the granules and lysosomes is highly effective in killing internalized bacteria or single-celled fungi or parasites. Macrophages do not have preformed microbicidal granules and kill internalized microbes less effectively than neutrophils unless they are stimulated by interferon-γ, which induces some of the mechanisms that preexist in neutrophils.

Reactive oxygen and nitrogen species are used to kill internalized microorganisms

A particular specialization of professional phagocytes is the generation of highly corrosive reactive oxygen and nitrogen compounds that are extremely toxic. The former are produced in the phagolysosome by NADPH oxidase (also called phagocyte oxidase) and are highly concentrated there to kill whatever has been internalized. In contrast, NO is produced in the cytoplasm by **inducible nitric oxide synthase (iNOS)** and then diffuses into the phagosome to react with internalized microbes. These reactions produce **reactive oxygen intermediates** and **reactive nitrogen intermediates**: the major such intermediates and their effects are summarized in Figure 3-28. Although the reactive nitrogen intermediates produced by iNOS are directly toxic, the reactive oxygen intermediates produced by NADPH oxidase also have indirect effects that may be equally important (see below). NADPH oxidase is a multicomponent enzyme (Figure 3-29) that preexists in neutrophils and is activated by signals from phagocytic receptors. As a result, this is a rapid response, particularly in neutrophils. Macrophages need prestimulation with interferon-γ for full induction of the enzyme, and this is believed to be a major reason for the importance of interferon-γ in promoting the effector function of macrophages. The NADPH oxidase complex consists of membrane-bound and cytosolic components that are assembled only on receipt of a phagocytic signal. The membrane-bound component,

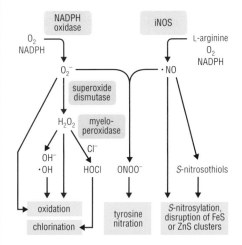

Figure 3-28 Generation of toxic oxygen and nitrogen compounds by phagocytes Phagocytosis and activation of macrophages and neutrophils are coupled to activation of the NADPH oxidase (also called phagocyte oxidase) and inducible nitric oxide synthetase (iNOS, also called NOS2). The resulting compounds are able to kill microbes by inducing oxidation, chlorination and nitration of molecules of microbes. Recently there has been evidence that the highly toxic molecule O_3 (ozone) is also produced by the combined action of NADPH oxidase and antibodies, but the exact reaction is not defined. O_2^-: superoxide anion; ·OH: hydroxyl radical; HOCl: hypochlorous acid; ·NO: nitric oxide; $ONOO^-$: peroxynitrite.

Definitions

CGD: see **chronic granulomatous disease**.

chronic granulomatous disease (CGD): genetic immunodeficiency disease caused by inactivating mutations of one of the subunits of **NADPH oxidase**. In this disease, certain types of bacteria cannot be killed adequately.

granuloma: tightly structured spherical inflammatory lesion characterized by central necrosis and containing microorganisms and infected macrophages in various stages of maturation and activation. Interspersed with these are activated CD4 T cells, and surrounding them are epithelioid fibroblasts and activated CD8 T cells. Granulomas ultimately undergo fibrosis and calcification of the outer mantle.

inducible nitric oxide synthase (iNOS): one of three isoforms of NOS, the enzyme that makes nitric oxide, and that is synthesized by phagocytic cells as part of the microbicidal response. The other forms are found in neurons and endothelial cells and are constitutively expressed and regulated by intracellular calcium levels.

iNOS: see **inducible nitric oxide synthase**.

lysozyme: enzyme that degrades the bacterial cell wall component peptidoglycan.

NADPH oxidase: phagocyte enzyme that upon induced assembly creates **reactive oxygen intermediates**, also called phagocyte oxidase.

primary granule: specialized intracellular vesicle in neutrophils derived from lysosomes and which contains antimicrobial peptides, proteases and other components; also called azurophilic granule because of its

cytochrome b_{558}, resides in secretory vesicles and secondary granules in resting phagocytes and is composed of $p22^{PHOX}$ and $gp91^{PHOX}$, of which $gp91^{PHOX}$ contains the FAD cofactor required for electron transfer to molecular oxygen; the cytosolic subcomplex consists of $p47^{PHOX}$, $p67^{PHOX}$ and $p40^{PHOX}$. Signaling by phagocytic receptors as they internalize particles induces the translocation of the cytosolic complex, along with the small GTPase Rac, to the membrane, where it associates with the cytochrome b_{558} subcomponent of the oxidase. This requires the phosphorylation on serine residues of the p47 subunit, which is responsible for directing the complex to the membrane.

As the enzyme is being assembled, the secondary granules containing it fuse with the phagosome, thereby targeting NADPH oxidase to the relevant cellular compartment; there it activates the rapid translocation of electrons into the phagosome, where they are transferred to O_2, giving rise to superoxide anion (O_2^-). The transfer of charge across the phagosome membrane in large amounts (estimated to be 1–4 moles/liter within the phagosome) must be compensated for by ion movements. These ionic changes, in turn, are important for releasing the primary granule proteases in an active form. These proteases are stored in the primary granules in a complex with granule proteoglycans, which inhibit their activity until changes in ionic conditions dissociate this complex. Thus, the phagocyte oxidase performs two functions: it generates toxic molecules and it promotes the activation of primary granule proteases that also promote microbe killing.

Genetic defects in any of the NADPH oxidase subunits with the exception of p40 give rise to a disease known as **chronic granulomatous disease (CGD)**. Particularly common are defects in gp91, because this protein is encoded by an X-linked gene. Patients with CGD have trouble clearing a relatively limited number of microbial infections, including *Staphylococcus aureus*, *Burkholderia cepacia* and the fungus *Aspergillus*, resulting in chronic infection with these organisms, which persist in **granulomas**: spherical aggregates of macrophages surrounding a central area containing live microbes (these are described more fully in Chapter 9, where we discuss tuberculosis). Infections seen in these patients are generally with organisms that express catalase, an enzyme that removes hydrogen peroxide. It is thought that although microbial catalase is not sufficient to protect against the high levels of reactive oxygen intermediates produced by a normal phagocyte oxidase system, it can protect against small amounts of H_2O_2 produced in the absence of functional enzyme.

Whereas the production of reactive oxygen intermediates can be quite rapid, the production of reactive nitrogen intermediates requires the transcriptional induction of iNOS gene expression, for example in response to LPS and interferon-γ. Thus, at least in some circumstances, the NADPH oxidase represents the initial killing mechanism, which is then followed by the slower and longer-lasting iNOS mechanism. Once synthesized, iNOS constitutively produces high levels of reactive nitrogen intermediates.

Genetic studies suggest that there is considerable redundancy between these two killing mechanisms: mice genetically deficient either in the NADPH oxidase or in iNOS have problems with limited types of pathogenic microbes. Mice lacking iNOS do poorly with a number of intracellular pathogens, perhaps because these pathogens can keep themselves out of phagosomes or alter normal phagosome fusion with lysosomes. As mentioned above, people with CGD have problems with some microbes and this pattern is generally reflected in mice deficient in gp91 or $p47^{PHOX}$. In striking contrast, mice doubly deficient for iNOS and gp91 often have severe abscesses with relatively common bacteria, such as *Escherichia coli*, even if they are maintained in a specific pathogen-free facility with antibiotics. In the absence of antibiotics, these animals quickly die from infections of commensal microbes.

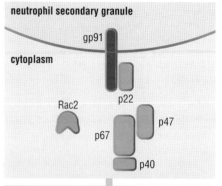

neutrophil secondary granule

cytoplasm

gp91

Rac2

p22

p67

p47

p40

fusion of secondary granule to phagosome, activation of Rac2, phosphorylation of p47

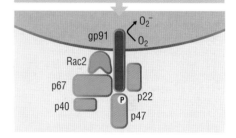

gp91

O_2^-

O_2

Rac2

p67

p40

P

p22

p47

Figure 3-29 Activation-dependent assembly of the NADPH oxidase In the quiescent state (upper), NADPH oxidase (also called phagocyte oxidase) exists as two components, a cytosolic complex of p67, p47 and p40 and a membrane-bound complex of gp91 and p22, also called cytochrome b_{558}. Activation of Rac2 and serine phosphorylation of p47 leads to the recruitment of Rac2 and of the cytosolic complex to the membrane (lower). The assembled complex is enzymatically active in producing superoxide anion, O_2^-.

dye-staining properties.

reactive nitrogen intermediates: highly reactive and toxic compounds generated from L-arginine by **inducible nitric oxide synthase (iNOS)**.

reactive oxygen intermediates: highly reactive and toxic compounds generated from molecular oxygen by **NADPH oxidase**, superoxide dismutase and/or myeloperoxidase.

secondary granule: specialized intracellular vesicle in neutrophils that contains the membrane-bound com-

ponents of the **NADPH oxidase**; also called specific granule.

References

Babior, B.M.: **NADPH oxidase.** *Curr. Opin. Immunol.* 2004, **16**:42–47.

Bogdan, C. *et al.*: **Reactive oxygen and reactive nitrogen intermediates in innate and specific immunity.** *Curr. Opin. Immunol.* 2000, **12**:64–76.

Faurschou, M. and Borregaard, N.: **Neutrophil granules and secretory vesicles in inflammation.** *Microbes Infect.* 2003, **5**:1317–1327.

Segal, A.W.: **How neutrophils kill microbes.** *Annu. Rev. Immunol.* 2005, **23**:197–223.

Tkalcevic, J. *et al.*: **Impaired immunity and enhanced resistance to endotoxin in the absence of neutrophil elastase and cathepsin G.** *Immunity* 2000, **12**:201–210.

Mammalian Toll-Like Receptors and their Ligands

TLR2+TLR1	bacterial lipoproteins
TLR2+TLR6	bacterial lipoproteins, lipoteichoic acid, yeast cell wall mannans
TLR2+?	GPI anchors (parasites), bacterial porins, HMGB1
TLR3	dsRNA
TLR4	LPS, HSPs, HMGB1, some viral proteins
TLR5	bacterial flagellin
TLR7	ssRNA (viral)
TLR8	ssRNA (viral)
TLR9	CpG-containing DNA (viral and bacterial)
TLR10	unknown
TLR11	*Toxoplasma* profilin
TLR12	unknown
TLR13	unknown

Figure 3-30 Table showing the ligand specificity of mammalian TLRs TLR1–9 seem to be closely homologous between mice and humans. The human genome has TLR10 but not TLR11, 12 and 13, whereas the mouse genome has TLR11, 12 and 13 but not TLR10. TLR2 functions in complementary pairs with TLR1 or TLR6. The response to diacylated bacterial lipoproteins from *Mycoplasma* requires TLR2 and TLR6, whereas the response to triacylated bacterial lipoproteins requires TLR2 and TLR1. In addition to recognizing certain RNA species, TLR7 and TLR8 also recognize synthetic imidazoquinolines, compounds that have some utility in topical treatment of viral infections. dsRNA: double-stranded RNA, a replication intermediate of RNA viruses; HMGB1: high mobility group box 1 protein, a chromatin-binding protein released by necrosis; HSP: heat shock proteins, molecular chaperones; LPS: lipopolysaccharide; ssRNA: single-stranded RNA.

Recognition of conserved microbial components by Toll-like receptors leads to inflammation and activation of sentinel immune cells

Microbes that penetrate an epithelial barrier and enter a tissue site are encountered by the three types of sentinel immune cells in the tissues: tissue macrophages, mast cells and immature dendritic cells. These sentinels must be able to distinguish between apoptotic particles generated by normal tissue turnover and particles that are indicative of infection. The molecules mainly responsible for making this pivotal distinction are those of the family of **Toll-like receptors (TLRs)**. Stimulation of macrophages or mast cells through their Toll-like receptors leads to the synthesis and secretion of proinflammatory cytokines and lipid mediators, thereby initiating the *inflammatory response* that recruits both soluble immune components and immune cells from the blood, and which we describe in later sections. TLR stimulation of dendritic cells induces the initiation of an adaptive immune response, as we shall see in the next chapter. In this section and the next we shall focus on the structural and functional features of this family of receptors that enable it to detect the presence of infection and to signal an appropriate response.

The Toll-like receptors were named after the fruit-fly receptor Toll, which was first discovered because it has an important role in early fly development and was later recognized as contributing to innate immunity in adult flies. TLRs are characterized by an amino-terminal extracellular domain composed of repeated motifs high in leucine and known as **leucine-rich repeats (LRRs)**, followed by a single transmembrane domain and a globular cytoplasmic domain called the **Toll/interleukin 1 receptor (TIR) domain**, or **TIR domain** that is also found in IL-1 receptors as well as in adaptors of the TLR signaling pathway, as we shall see in the next section. Thirteen TLRs have been identified in mammals so far. TLRs recognize constituents of microbial cell walls or pathogen-specific nucleic acids (Figure 3-30). Most of the determinants recognized by these receptors are molecules essential to the integrity, function or replication of microbes or viruses, and therefore the infectious agent cannot readily escape detection by changing them. For example, lipopolysaccharide (LPS), the major ligand for TLR4, is a central component of the outer membrane of Gram-negative bacteria (see Figure 3-3), and mutations that ablate the enzymes required for synthesis of LPS are lethal to most species of Gram-negative bacteria. Similarly, double-stranded RNA, the ligand for TLR3, is a central replication intermediate for all RNA viruses, so evasion of TLR3 recognition by these viruses is not easily achieved.

Accessory molecules aid Toll-like receptor recognition of some ligands

A striking feature of TLR ligands is their molecular diversity: they include nucleic acids, proteins, lipids and polysaccharides. How one family of receptors can recognize all of these different types of molecules is not well understood. Some mammalian TLRs are thought to bind directly to their innate immune ligands, but in other cases recognition is greatly facilitated by accessory proteins: one example of this is LPS binding by TLR4, to which two accessory proteins and a soluble lipid-transfer protein contribute.

Figure 3-31 Recognition of bacterial lipopolysaccharide by innate immune cells Lipopolysaccharide (LPS) monomers are extracted from bacterial membranes by the serum protein LPS-binding protein (LBP) which transfers the LPS monomer to a lipid-binding site on CD14 in the membrane of phagocytes. CD14 promotes the binding of LPS to the TLR4–MD-2 complex, which signals to the cell interior. In the absence of CD14, TLR4–MD-2 can still function with some forms of LPS or with much higher LPS concentrations.

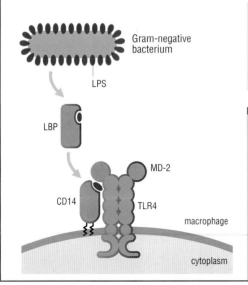

Definitions

CpG-containing DNA: DNA containing unmethylated C followed by G. Note that the sequence of bases adjacent to the CG motif also affects the stimulatory activity.

LBP: see **LPS-binding protein**.

leucine-rich repeat (LRR): unit of protein structure in which there are many repeats of a basic unit of approximately 25 amino acids.

lipoarabinomannan: a major immunostimulatory component of the lipid-rich mycobacterial cell wall,

containing phosphatidylinositol linked to the carbohydrates mannose and arabinose.

LPS-binding protein (LBP): a lipid transfer protein of serum that can extract monomers of LPS from bacterial membranes and deliver them to the innate immune receptor CD14.

LRR: see **leucine-rich repeat**.

MD-2: a polypeptide that associates with the extracellular domain of the **Toll-like receptor** TLR4 and is required for LPS responsiveness.

The requirements for LPS recognition are illustrated in experiments in which the gene encoding TLR4 is introduced into cells that do not normally express it. TLR4 confers responsiveness to LPS on such cells only in the presence of a second polypeptide, called **MD-2**, which binds to the extracellular domain of TLR4 and enables it to bind to the relatively conserved inner lipid-containing region of the lipopolysaccharide (the outer polysaccharide region differs between different bacterial species).

Responses to forms of LPS from many but not all bacteria are also substantially increased by a second accessory molecule, CD14, a membrane protein expressed by monocytes, macrophages and neutrophils, that accepts LPS from a serum lipid transfer protein called **LPS-binding protein (LBP)**. Like most lipids, LPS is released from cells only in very small amounts. Efficient recognition of LPS by innate immune system cells therefore requires a mechanism for extracting LPS monomers and making them available to cells. This is the function of LBP, which exchanges monomers of LPS for other lipids present in its lipid-binding site and then carries the LPS monomer to cells where it is transferred to CD14, which presents it to the TLR4–MD-2 complex (Figure 3-31). Genetic removal of CD14 or blocking its function with monoclonal antibodies greatly reduces the sensitivity of macrophages or neutrophils to LPS, although some responses still occur with LPS from some bacteria. CD14 also promotes responses to other bacterial cell wall components, including lipoteichoic acid of Gram-positive bacteria, and **lipoarabinomannan** of *Mycobacterium tuberculosis*.

In other cases, phagocytic receptors contribute to the recognition for TLR signaling. For example, dimers of TLR2 and TLR6 respond to lipid-containing ligands such as lipoproteins and lipoteichoic acid, which they recognize with the help of a scavenger receptor called CD36, whereas the recognition of fungal cell wall polysaccharides by these TLRs is facilitated by the transmembrane C-type lectin molecule dectin-1.

Mammalian Toll-like receptors recognize their ligands on the cell surface or intracellularly

Whereas TLRs that recognize bacterial and fungal cell wall components are localized to the cell surface, TLRs that recognize viral or microbial nucleic acids are localized to intracellular membranes and are thought to encounter their ligands in phagosomes or endosomes (Figure 3-32). This localization is thought to be an adaptation ensuring that these receptors detect nucleic acids released from apoptotic host cells, microbial cells or virions only after phagocytosis and partial digestion of the ingested particles has released the nucleic acids. Cell-surface TLRs are also targeted to nascent phagosomes and upon ligand binding can signal there.

TLRs participating in innate immunity to viruses recognize a variety of nucleic-acid ligands produced by viruses, although the mechanism by which they distinguish cellular nucleic acids from viral nucleic acids is not always evident. The basis for discrimination is clearest for TLR3. This TLR recognizes double-stranded RNA (dsRNA), which is an obligatory replication intermediate for viruses with RNA genomes, whereas vertebrate cells express very little dsRNA. TLR7 and TLR8 recognize single-stranded RNA that is rich in guanosine, and it has not yet been established how viral RNAs are distinguished from cellular RNAs. TLR9 recognizes DNA motifs containing the dinucleotide CG where the C is unmethylated (**CpG-containing DNA**). This dinucleotide is about 10-fold under-represented in vertebrate DNA, and most such dinucleotides are methylated in vertebrate genomes where CpG methylation is a gene-silencing mechanism.

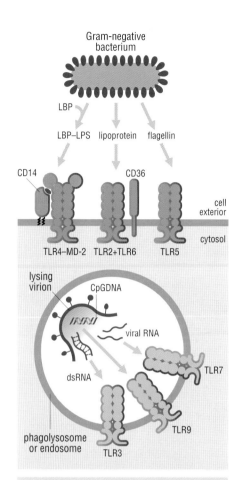

Figure 3-32 Cellular localization of TLRs
TLRs that recognize bacterial and fungal cell wall components, such as TLR4–MD-2, TLR5 and the heterodimers TLR2+TLR1 and TLR2+TLR6, are localized to the plasma membrane and can recognize ligands there. They are also delivered to nascent phagosomes, apparently by a mechanism that is independent of ligand binding. In contrast, TLRs recognizing nucleic acids (TLR3, TLR7, TLR8 and TLR9) are primarily or completely contained in intracellular membranes and unavailable for interaction with extracellular ligands. It is believed that these TLRs are delivered to phagolyosomes or late endosomes. Ligand recognition by these TLRs can often be blocked by agents that block the acidification of endosomes or phagosomes, such as chloroquine.

TIR domain: see **Toll/interleukin 1 receptor (TIR) domain**.

TLR: see **Toll-like receptor**.

Toll/interleukin 1 receptor (TIR) domain: domain responsible for transmitting signals downstream of IL-1 receptors and **Toll-like receptors**.

Toll-like receptors (TLRs): family of receptors that have leucine-rich repeats in their extracellular domains and the **TIR domain** in their cytoplasmic domains.

References

Cook, D.N. *et al.*: **Toll-like receptors in pathogenesis of human disease**. *Nat. Immunol.* 2004, **5**:975–979.

Poltorak, A. *et al.*: **Defective LPS signaling in C3H/HeJ and C57BL/10ScCr mice: mutations in *Tlr4* gene**. *Science* 1998, **282**:2085–2088.

Takeda, K. *et al.*: **Toll-like receptors**. *Annu. Rev. Immunol.* 2003, **21**:335–376.

Takeuchi, O. *et al.*: **Differential roles of TLR2 and TLR4 in recognition of Gram-negative and Gram-positive bacterial cell wall components.** *Immunity* 1999, **11**:443–451.

Ulevitch, R.J. and Tobias, P.S.: **Receptor-dependent mechanisms of cell stimulation by bacterial endotoxin.** *Annu. Rev. Immunol.* 1995, **13**:437–457.

Mammalian Toll-like receptors activate the production of proinflammatory cytokines and interferon

Toll-like receptors, as we have seen, recognize both bacterial and viral components. Correspondingly, they induce both antibacterial and antiviral responses. These include the production of proinflammatory cytokines, which are essential for coordinating responses to both bacterial and viral infections, and the *type 1 interferons*, which act against viruses. The type 1 interferons, *interferon-α* and *interferon-β*, are so called to distinguish them from the *type 2 interferon*, interferon-γ. The last of these is important for defense against intracellular pathogenic microbes and viruses; it is produced by T cells and NK cells, which we discuss later in this book. We shall describe the actions of the type 1 interferons later in this chapter, but because they are part of the response elicited by the TLRs, we describe here the signaling pathways leading to their production. The production of inflammatory cytokines and interferons depends upon the action of multiple types of transcription factors, including NF-κB (see section 2-10), and a family of transcription factors known as *interferon regulatory factors (IRFs)*. IRFs are so called

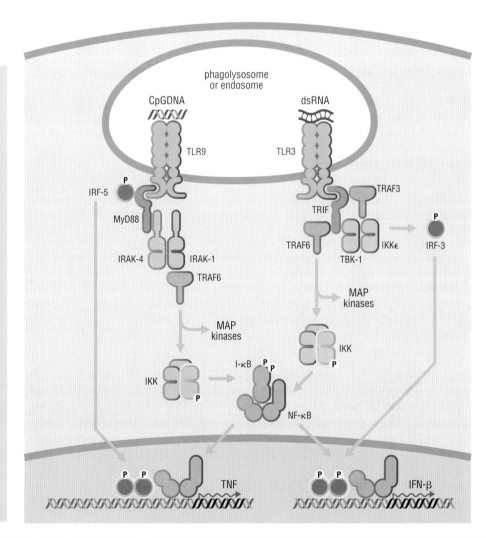

Figure 3-33 Two signaling pathways of mammalian TLRs TLR9 (left) signals primarily via the MyD88-dependent signaling pathway. In this pathway, ligand binding recruits the MyD88 adaptor to the TLR. MyD88 then recruits the related protein kinases IRAK-4 and IRAK-1 and the adaptor TRAF6. The IRAK1–TRAF6 complex dissociates from the receptor (not shown), perhaps allowing amplification of the signal, and then associates with additional signaling intermediates (not shown). This leads to activation of MAP kinase pathways (Erk, JNK and p38 MAP kinases) and the activation of the I-κB kinase (IKK) complex, which activates the transcription factor NF-κB (see section 2-10). MyD88 also interacts directly with the IRF-5 transcription factor, which is thought to be activated by phosphorylation, and then goes to the nucleus. In contrast, TLR3 (right) signals via the adaptor TRIF. TRIF primarily signals via TRAF3 and the kinases TBK-1 and IKKε, which phosphorylate IRF-3 on sites that mediate its translocation to the nucleus and transcriptional activity. The TRIF pathway also stimulates the TRAF6 pathway to NF-κB and MAP kinases, although to a lesser degree than the MyD88 pathway. In contrast to MyD88, TRIF associates directly with TRAF6. The MAP kinases signal to transcription factors and to cytoplasmic regulatory events but these are not shown, for simplicity. IFN-β: interferon-β; IKK: I-κB kinase complex (IKKε is closely related to the IKKα and IKKβ subunits of the IKK complex); TBK-1: TANK-binding kinase 1; TRAF6: TNF receptor associated factor 6.

Definitions

IL-1 receptor associated kinase (IRAK): any of four protein kinases, IRAK-1, -2, -M and -4. IRAK-1 and IRAK-4 contribute to TLR and IL-1 receptor signaling.

IRAK: see **IL-1 receptor associated kinase.**

mitogen-activated protein (MAP) kinase: any of a small family of closely related protein kinases activated in receptor signaling pathways downstream of Ras or Rho-family GTPases through related upstream kinases (MAP kinase kinases) that phosphorylate a characteristic

site in the kinase domain, resulting in the activation of the enzymatic activity of the MAP kinase. MAP kinases include Erk (extracellular signal regulated kinase), JNK (c-Jun N-terminal kinase) and p38 MAP kinase subfamilies. They regulate AP-1 type transcription factors and other events in the cell, including stability of the mRNA encoding TNF.

MyD88: see **myeloid differentiation factor 88.**

myeloid differentiation factor 88 (MyD88): an adaptor molecule that contains a TIR domain and mediates the major proinflammatory signaling pathway of TLRs.

TIR domain-containing adaptor inducing interferon-β (TRIF): an adaptor that mediates signaling by some TLRs to IRF-3 and transcription of type 1 interferons.

TRIF: see **TIR domain-containing adaptor inducing interferon-β.**

---

because they were first discovered as transcriptional activators of interferon genes, as we shall see in section 3-16, although they can also participate in the activation of other cytokine genes. In macrophages, TLR signaling activates *IRF-3*, which, with NF-κB, induces transcription of the interferon-β gene; in *plasmacytoid dendritic cells*, which have a special role in antiviral defense, signaling through TLR7 or TLR9 to IRF7 activates transcription of the interferon-α genes, as we shall see later. A third member of the IRF family, IRF-5, participates in TLR signaling to produce other cytokines.

Mammalian Toll-like receptors signal via two major pathways

TLRs induce proinflammatory cytokines and type 1 interferons through different signaling pathways, which are controlled by two different TIR domain-containing adaptor molecules, **MyD88** and **TRIF** (Figure 3-33). Both of these pathways culminate in the transcriptional activation of cytokine genes by NF-κB in cooperation with transcription factors of the IRF family (which we discuss in section 3-16), but whereas the MyD88 pathway activates IRF-5 or IRF-7 and culminates in the transcription of inflammatory cytokine genes or interferon-α, the TRIF pathway activates IRF-3 and culminates in the transcription of genes for interferon-β.

In the MyD88 pathway, which is activated by most TLRs and by IL-1 and IL-18 receptors, NF-κB activation begins with the assembly at the receptor of a complex including MyD88, the adaptor molecule TRAF6 (related to the TRAFs that participate in TNF receptor signaling; see section 2-9), and two closely related protein kinases belonging to a small family known as the **IL-1 receptor associated kinases (IRAKs)**. These kinases, IRAK-1 and IRAK-4, cooperate with TRAF6 to activate the I-κB kinase (IKK) complex, which in turn activates NF-κB (see section 2-10). TRAF6 also activates a set of related protein kinases, the **mitogen-activated protein (MAP) kinases**, which are involved in signaling by many types of receptors and promote activation of transcription factors as well as other events in the cell. Activation of IRF-5 is promoted by direct interaction with MyD88. IRF-5, NF-κB and other transcription factors then activate the promoters of the TNF and IL-1 genes. The biological importance of the MyD88 pathway is illustrated by the severe recurrent bacterial infections seen in rare individuals with defects in IRAK-4.

The TRIF pathway is more specialized and is activated only by TLR3 and TLR4. TRIF associates directly with TRAF6, which activates IKK and thus NF-κB, and also with TRAF3 and two protein kinases that are close relatives of the IKK kinase subunits, which in turn phosphorylate and activate IRF-3. IRF-3 and NF-κB then activate transcription of the interferon-β gene. The TRIF pathway is also important for dendritic cell maturation, which is important for activation of T cells, as described in Chapter 4.

Most known TLRs signal through MyD88; thus both bacterial and viral components induce proinflammatory cytokines. The exceptions are TLR3, which signals through TRIF (see Figure 3-33), and TLR4, which signals through both. Unlike the other TLRs, TLR4 does not bind directly to MyD88 and TRIF but is connected through two additional adaptors, TRAM and TIRAP, which also have TIR domains (Figure 3-34). TLR3 is specialized for recognition of double-stranded RNA, which is characteristic of viruses, so it is not surprising that it induces antiviral interferon-β. TLR4 recognizes both bacterial and viral molecules, and the versatility of its intracellular signaling pathway may reflect the need to respond to both; it may also be relevant that interferon-β, as well as combating viral replication, can (like interferon-γ) induce production of the microbicidal gas nitric oxide by macrophages (see section 3-9).

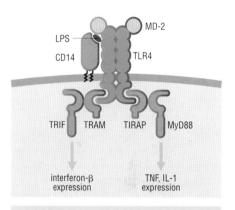

Figure 3-34 TLR4 signaling to both the MyD88 pathway and the TRIF pathway TLR4 associates directly with two TIR domain-containing adaptor molecules, TIRAP, which associates with MyD88, and TRAM, which associates with TRIF. In all cases, the TIR domains of these proteins mediate the associations. Thus, TLR4 signaling induces the transcription of both proinflammatory cytokines and interferon-β. TIRAP and TRAM are not known to connect to any additional signaling events beyond the MyD88 and TRIF signaling pathways.

References

Akira, S. and Takeda, K.: **Toll-like receptor signalling.** *Nat. Rev. Immunol.* 2004, **4**:499–511.

Beutler, B.: **Inferences, questions and possibilities in Toll-like receptor signalling.** *Nature* 2004, **430**:257–263.

Fitzgerald, K.A. *et al.*: **IKKε and TBK1 are essential components of the IRF3 signaling pathway.** *Nat. Immunol.* 2003, **4**:491–496.

Hoebe, K. *et al.*: **Identification of *Lps2* as a key trans-**ducer of MyD88-independent TIR signalling. *Nature* 2003, **424**:743–748.

Kawai, T. and Akira, S.: **Innate immune recognition of viral infection.** *Nat. Immunol.* 2006, **7**:131–137.

Picard, C. *et al.*: **Pyogenic bacterial infections in humans with IRAK-4 deficiency.** *Science* 2003, **299**:2076–2079.

Tabeta, K. *et al.*: **Toll-like receptors 9 and 3 as essential components of innate immune defense against mouse cytomegalovirus infection.** *Proc. Natl Acad. Sci.* USA 2004, **101**:3516–3521.

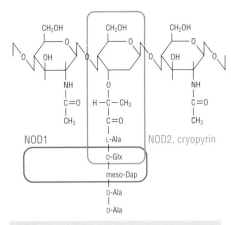

Figure 3-35 Ligands for NOD1, NOD2 and cryopyrin The repeating structure of peptidoglycan is shown. The peptide chains can vary somewhat between bacteria and they are linked to one another via peptide bonds (not shown) to create a two-dimensional polymer that provides structural rigidity to bacterial cell walls. Minimal substructures of peptidoglycan that can activate NOD1, NOD2 and cryopyrin are shown by the boxes. Note that the natural ligands are probably somewhat larger substructures. Dap: diaminopimelic acid; Glx: glutamic acid or glutamine.

NOD1, NOD2 and cryopyrin are intracellular sensors of peptidoglycan substructures

Most of the innate antibacterial defense mechanisms that we discuss in this chapter are adapted to recognize bacteria outside cells. This reflects the fact that most bacteria proliferate in extracellular spaces: even the intracellular Toll-like receptors largely recognize bacteria that originate outside the cell. This is by contrast with viruses, which replicate inside cells and, as we shall see later in the chapter, are detected by innate sensors in the cytoplasm. Intracellular innate immune sensors of bacterial infection do also exist, however: three related molecules, **NOD1**, **NOD2** and **cryopyrin** (also called **NALP3**), are specialized to recognize components of bacterial peptidoglycan, a widespread constituent of bacterial cells walls (see Figure 3-3).

The minimal structures recognized by NOD1, NOD2 and cryopyrin are illustrated in Figure 3-35. NOD1 recognizes dipeptides containing diaminopimelic acid, which are characteristic of peptidoglycan in many Gram-negative bacteria, whereas the minimal ligand for NOD2 and cryopyrin is **muramyl dipeptide (MDP)**, which is found in the peptidoglycan of virtually all Gram-negative and Gram-positive bacteria. There are various ways in which bacterial peptidoglycan fragments might enter the cytoplasm, where they can be detected by NOD1, NOD2 and cryopyrin. Some pathogenic bacteria are specialized to enter the cytoplasm of epithelial cells (*Listeria* is an example that we describe in Chapter 9). More commonly, phagocytic cells seem to have a mechanism to transport MDP ligands into their cytosol, and hence respond to the presence of extracellular bacteria or bacteria in phagosomes. Additionally, in some gut bacteria, pathogenicity is associated with specialized mechanisms for invading epithelia, for example by injecting toxins, and ligands seem to reach NOD1 in epithelial cells in this way.

Both NOD1 and NOD2 activate the key inflammatory transcription factor NF-κB (see section 2-10) and induce the production of inflammatory cytokines. NOD2 is expressed primarily by cells of the macrophage lineage and dendritic cells, and by the Paneth cells in the crypts of the small intestine, where it has a special role in inducing the secretion of α-defensins (see Figure 3-6); NOD1 has a broader expression pattern that includes epithelial cells and may be generally important in the defense of epithelial barriers. Increased susceptibility to *Crohn's disease*, one of the two main types of inflammatory bowel disease (see Chapter 13), is strongly associated with some moderately common point mutations in NOD2 that seem to abolish the recognition of MDP, suggesting that the protective role of NOD2 in the crypts is important for the health of the small intestine.

Whereas NOD1 and NOD2 are primarily involved in inducing cytokine gene transcription, cryopyrin is involved in a separate regulatory step specifically involved in the production of the inflammatory cytokine IL-1β and the interferon-γ-inducing cytokine IL-18, namely the

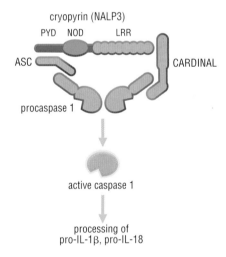

Figure 3-36 Inflammasome assembled in response to muramyl dipeptide The presence of fragments of peptidoglycan leads to the assembly of the active inflammasome in the cytoplasm. The scaffold cryopyrin (also called NALP3), assembles with adaptors ASC and CARDINAL, both of which have caspase recruitment domains (CARDs) that bind to the amino-terminal prodomain of procaspase 1. Clustering of procaspase 1 molecules leads to their cross-activation by proteolysis, generating active caspase 1; this processes precursor forms of IL-1β, which is proinflammatory, and IL-18, which promotes interferon-γ production in conjunction with IL-12. The inflammasome may be a larger structure with multiple copies of this complex held together, analogous to the apoptosome (see section 2-11). Note that cryopyrin belongs to an extensive family of NALPs, and at least one other, NALP1, also can form an inflammasome complex, probably in response to other upstream signals.

Definitions

cryopyrin: also known as **NALP3**, an intracellular sensor of peptidoglycan that acts as a scaffold to promote the activation of caspase 1, which is required for the processing of precursor forms of IL-1β and IL-18.

inflammasome: multiprotein complex that promotes inflammatory responses by processing precursor forms of inflammatory cytokines.

MDP: see **muramyl dipeptide**.

muramyl dipeptide (MDP): a substructure of bacterial peptidoglycan containing the sugar muramic acid linked to L-alanine and D-glutamic acid or D-glutamine.

NALP3: see **cryopyrin**.

NOD: see **nucleotide oligomerization domain**.

NOD1 (nucleotide oligomerization domain 1): an intracellular innate immune recognition molecule for dipeptides, which contain diaminopimelic acid, a modified amino acid found in peptidoglycan. Also called CARD4.

NOD2 (nucleotide oligomerization domain 2): an intracellular innate immune recognition molecule for muramyl dipeptide, a substructure of peptidoglycan. Also called CARD15.

nucleotide oligomerization domain (NOD): nucleotide-binding domain that regulates protein–protein interactions in a variety of proteins that participate in immunity and apoptosis.

proteolytic processing of precursor forms of these proteins by inflammatory caspases such as caspase 1. Sensing of MDP by cryopyrin causes it to assemble a multiprotein complex that promotes the activation of procaspase 1, as illustrated in Figure 3-36. This complex is called the **inflammasome**. Cryopyrin is only one of an extensive family of related proteins, called NALPs, and at least one other, NALP1, also can assemble into an inflammasome to generate active inflammatory caspases. It is hypothesized that all or many of the different NALPs are responsive to different upstream activators.

The NOD domain mediates ligand-induced signal generation

NOD1, NOD2 and crypyrin are structurally similar to one another and to a variety of other molecules present in the mammalian genome, some of which function in immunity and apoptosis, and also to a large number of molecules encoded in plant genomes that mediate resistance to plant pathogens (Figure 3-37). These proteins have a leucine-rich repeat (LRR) region that is thought to be responsible for specific ligand recognition, although it is also possible that recognition of peptidoglycan fragments is indirect, via another protein. Leucine-rich repeats are also important structural elements of the Toll-like receptors, as described earlier in this chapter (section 3-10). The designation NOD derives from a central nucleotide-binding domain that mediates oligomerization, hence **nucleotide oligomerization domain (NOD)**, and is the common structural motif for which the NOD proteins illustrated in Figure 3-36 are named. The NOD domain is thought to control signal transduction by inducing the oligomerization of the NOD protein, which promotes association of the amino-terminal domain, the caspase recruitment domain (CARD) in the case of Apaf-1 (see section 2-11) or the pyrin domain in the case of cryopyrin, with caspases (hence the name CARD).

In the case of NOD1 and NOD2, the CARD associates not with a caspase but with a protein kinase called RIP-2 (because of its sequence homology to RIP-1, which participates in TNFR1 signaling; see section 2-9), which connects NOD1 and NOD2 to NF-κB activation. The importance of the NOD domain in controlling signaling is illustrated by a rare genetic disease, *Blau syndrome*, which results from any of several point mutations in the NOD domain of NOD2 that cause it to be constitutively active and leads to multi-organ inflammatory disease. Analogous mutations in cryopyrin lead to three closely related inflammatory diseases in which the inflammasome is constitutively activated: familial cold autoinflammatory syndrome, Muckle–Wells syndrome, and chronic infantile neurological articular syndrome. Skin and joints are common sites of inflammation in all of these diseases.

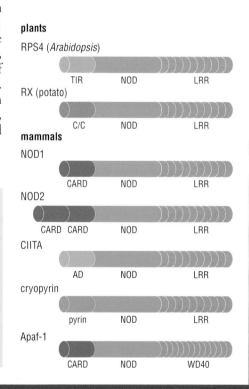

Figure 3-37 NOD-containing proteins involved in immunity and apoptosis NOD1, NOD2 and cryopyrin are intracellular sensors of bacterial infections that recognize substructures of peptidoglycan. NOD1 and NOD2 have amino-terminal CARD domains, a central nucleotide oligomerization domain (NOD), and a carboxy-terminal leucine-rich repeat region (LRR), which is thought to bind the ligand. The amino-terminal CARD couples to downstream signaling proteins. Cryopyrin (also called NALP3) has a similar structure but with a pyrin domain at its amino terminus. Many plant disease resistance proteins have a very similar structure, with NOD and LRR domains following either a TIR domain (see section 3-10) or a coiled-coil domain (C/C). Shown are two examples, but *Arabidopsis* has approximately 150 genes in its genome predicted to encode proteins of these two types. Molecules with related structures include CIITA, a transcriptional coactivator protein involved in class II MHC gene expression, and Apaf-1, which is a key regulator of apoptosis, as discussed in section 2-11. AD: transcriptional activation domain; pyrin: pyrin homology domain; WD40: WD40 repeats. Note that CIITA exists in different isoforms, one of which has a CARD at the amino terminus (not shown).

References

Eckmann, L. and Karin, M.: **NOD2 and Crohn's disease: loss or gain of function?** *Immunity* 2005, **22**:661–667.

Girardin, S.E. *et al.*: **Lessons from Nod2 studies: towards a link between Crohn's disease and bacterial sensing.** *Trends Immunol.* 2003, **24**:652–658.

Holt, B.F. 3rd *et al.*: **Resistance gene signaling in plants—complex similarities to animal innate immunity.** *Curr. Opin. Immunol.* 2003, **15**:20–25.

Inohara, N. and Nunez, G.: **NODs: intracellular proteins involved in inflammation and apoptosis.** *Nat. Rev. Immunol.* 2003, **3**:371–382.

Inohara, N. *et al.*: **Host recognition of bacterial muramyl dipeptide mediated through NOD2. Implications for Crohn's disease.** *J. Biol. Chem.* 2003, **278**:5509–5512.

Kobayashi, K.S. *et al.*: **Nod2-dependent regulation of innate and adaptive immunity in the intestinal tract.** *Science* 2005, **307**:731–734.

Martinon, F. and Tschopp, J.: **NLRs join TLRs as innate sensors of pathogens.** *Trends Immunol.* 2005, **26**:447–454.

Viala, J. *et al.*: **Nod1 responds to peptidoglycan delivered by the *Helicobacter pylori cag* pathogenicity island.** *Nat. Immunol.* 2004; **5**:1166–1174.

Innate immune recognition triggers an inflammatory response that focuses the immune system on the site of infection

When microorganisms first enter a tissue site, their presence is sensed by Toll-like receptors and the intracellular peptidoglycan sensors of the three types of immune cells embedded in tissues: the tissue macrophage, the mast cell and the immature dendritic cell. The latter cell type is involved in initiating adaptive immune responses, and its actions are described in Chapter 4. TLRs and NODs induce the production by macrophages and mast cells of lipid mediators, cytokines and chemokines that lead to a coordinated response on the part of blood vessels and leukocytes that is known collectively as **inflammation** (Figure 3-38). The inflammatory mediators have four main actions. First, they induce vascular permeability to allow the influx of soluble immune components from the blood. Second, they change the adhesive properties of the endothelium to attract more phagocytes to the site of infection. Third, they activate the incoming phagocytes to promote their microbicidal action. Fourth, they activate natural killer cells, enhancing their cytotoxicity and inducing the production of complementary cytokines.

The activation of endothelial cells, as well as allowing the influx of cells and soluble proteins from blood, triggers two proteolytic cascades, the **kinin cascade** and the **coagulation cascade**, that are part of an ancient response to tissue injury with counterparts in species as phylogenetically remote as *Drosophila* and the horseshoe crab. The kinin cascade, through the production of bradykinin, further increases vascular permeability, whereas the coagulation cascade leads to a clotting reaction that decreases the spread of infectious microorganisms away from the infection site. The influx of plasma complement components to the infected site further contributes to inflammation through complement activation by the microbe and the release of C5a (see section 3-5), which has proinflammatory effects on the endothelium and activates phagocytes and mast cells.

This coordinated inflammatory response is the hallmark of the innate immune response to microbial infection and is essential to its resolution. Inflammation is also damaging and painful, and many chronic disease states are due to inflammatory pathology. Some important examples of inflammatory diseases are described in Chapter 12. Control of inflammation by drugs is therefore an important clinical issue. Inflammation is also triggered by tissue injury directly, presumably reflecting a common association between breaches in the epithelial barrier and infections. In this case, inflammation may result from the coagulation cascade, recognition of necrotic cells, and/or the action of pain neurons, which release neuropeptides that activate mast cells. Indeed, mast cells often are closely associated with both blood vessels and neurons.

Cytokines and lipid mediators are important inducers of inflammation

The main mediators of inflammation are summarized in Figure 3-39. Particularly important for induction of inflammation by macrophages and mast cells are the **proinflammatory cytokines** TNF and IL-1, and a number of lipid mediators, including the prostaglandins, leukotrienes and platelet-activating factor. The prostaglandins and leukotrienes are produced from arachidonic acid,

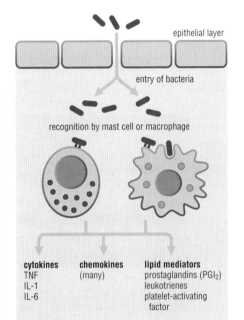

epithelial layer

entry of bacteria

recognition by mast cell or macrophage

cytokines	chemokines	lipid mediators
TNF	(many)	prostaglandins (PGI₂)
IL-1		leukotrienes
IL-6		platelet-activating factor

Figure 3-38 Innate immune initiation of inflammation Bacteria entering through the epithelial barrier are encountered by tissue macrophages or mast cells. Tissue macrophages recognize bacteria via Toll-like receptors and produce proinflammatory mediators including cytokines, neutrophil-attracting chemokines (IL-8 and other CXC chemokines) and lipid mediators of inflammation. Mast cells have a similar response, which includes also vasoactive amines such as histamine (not shown).

Definitions

coagulation cascade: proteolytic cascade triggered by damage to endothelium or by plasma entering a tissue site due to increased vascular permeability and that results in the generation of plasmin, which induces blood clotting.

edema: swelling of tissues due to increased vascular permeability.

inflammation: coordinated response to infection or tissue injury recognized since ancient Roman times as characterized by heat, pain, redness and swelling (*calor, dolor, rubor, tumor* in Latin).

kinin cascade: proteolytic cascade that is initiated by injury to tissue and ends with the production of bradykinin, which increases the permeability of blood vessels.

non-steroidal antiinflammatory drugs (NSAIDs): any of a class of drugs that inhibit inflammatory responses by inhibiting cyclooxygenases 1 and/or 2. Aspirin and most other NSAIDs block COX-1 and COX-2. COX-2-specific NSAIDs have been developed more recently;

they have fewer gastrointestinal side effects but may carry an increased risk of cardiovascular side effects.

NSAIDs: see **non-steroidal antiinflammatory drugs**.

proinflammatory cytokines: cytokines that act on endothelial cells to induce **inflammation**. Also known as inflammatory cytokines.

The Inflammatory Mediators of Innate Immunity

Mediator	Source	Functions
Proinflammatory cytokines		
TNF	macrophage, mast	↑vascular permeability, ↑endothelial cell adhesiveness, activation of phagocytes, ↑cytotoxicity of NK
IL-1	macrophage, mast, KC, EC	↑endothelial cell adhesiveness, chemokine production
IL-6	macrophage, mast, FB	promotes monocyte recruitment, systemic effects
IL-12	macrophage	↑IFN-γ production by NK, ↑NK cytotoxicity
IFN-γ	natural killer cell	enhanced phagocytosis and killing by phagocytes
chemokines	macrophage, mast, EC	attraction of neutrophils, monocytes, effector T cells
Lipid mediators		
PGI$_2$	macrophage	↑vascular permeability
PGE$_2$	macrophage, mast	↑vascular dilation, inhibits lipid mediator production
leukotrienes	macrophage, mast	↑smooth muscle contraction, vascular permeability
PAF	macrophage, mast	platelet aggregation, ↑endothelial cell adhesiveness
LysoPC	macrophage, mast	↑endothelial adherence, ↑macrophage activation chemotactic for macrophages and T cells
Vasoactive amines		
histamine	mast	↑vascular permeability, ↑vasodilation
Complement-derived mediators		
C3a	complement activation	↑vascular permeability, ↑vasodilation
C5a	complement activation	↑vascular permeability, ↑vasodilation, chemoattractant
Antiinflammatory cytokines		
IL-10	macrophages, T cells	↓production of TNF, IL-12, ↓MHC and B7 expression on macrophages
TGF-β	macrophages, T cells	antiinflammatory effects on endothelial cells, PMN, lymphocytes

Figure 3-39 Table of innate immune mediators of inflammation Listed are the principle inflammatory mediators produced after innate immune recognition of infection, including cytokines, lipid mediators and complement-derived mediators. Lipid mediators are produced instantaneously. Among the cytokines, TNF and IL-1 are produced most rapidly, whereas the antiinflammatory cytokines IL-10 and TGF-β are ordinarily produced later, with IL-6 and IL-12 appearing at intermediate times. The principal chemokines produced vary somewhat with the location. EC: endothelial cell; FB: fibroblast; KC: keratinocyte; LysoPC: lysophosphatidylcholine; MHC: major histocompatibility complex molecule; NK: natural killer cell; PAF: platelet-activating factor; PG: prostaglandin; PMN: polymorphonuclear cell.

which is released from cell membrane phospholipids by phospholipase A$_2$ (PLA$_2$). Arachidonic acid is then converted to prostaglandins by cyclooxygenase (of which there are two main isoforms, COX-1 and COX-2), or to leukotrienes by 5-lipoxygenase. Inflammatory stimuli both activate PLA$_2$ and induce the synthesis of COX-2. Many other cell types also produce prostaglandins with a variety of biological functions, often through COX-1. The biosynthetic pathways of these inflammatory mediators are summarized in Figure 3-40. Aspirin and related compounds, the **non-steroidal antiinflammatory drugs (NSAIDs)**, are inhibitors of the cyclooxygenases; they decrease inflammation but do not completely block it. The cyclooxygenases produce prostaglandin H$_2$, which is then converted to a handful of other prostaglandins by enzymes that are differentially expressed in different cell types. Each of the resulting mediators has specific receptors that mediate its action on target cells. For example, prostacyclin (PGI$_2$) mediates the increased vascular permeability response and resulting swelling referred to clinically as **edema**. Whereas prostaglandins in general are fast-acting inflammatory mediators, leukotrienes are slow-acting and are thought to be particularly important for chronic inflammation, as occurs in asthma.

Recently, some success has been obtained with treating the inflammatory diseases rheumatoid arthritis and inflammatory bowel disease with agents that block TNF (we discuss these in Chapter 12). These clinical observations demonstrate the importance of TNF as a mediator of inflammation.

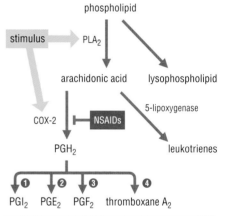

Figure 3-40 Arachidonic acid-derived lipid mediators A stimulus activates phospholipase A$_2$ (PLA$_2$), which releases arachidonic acid from phospholipids, some of which have this fatty acid esterified to the central carbon of the glycerol component. Arachidonic acid is then converted by cyclooxygenase (primarily the COX-2 isoform in immune cells) to prostaglandins, the exact nature of which is determined by cell type-specific expression of various prostaglandin synthetases (schematized as enzymes 1, 2, 3, 4) acting on the product of COX-2. The resulting prostaglandins diffuse to neighboring cells and are recognized by different specific receptors on those target cells. Arachidonic acid is also converted by 5-lipoxygenase to leukotrienes, which are slower-acting mediators of smooth muscle contraction and vascular permeability. Also released by PLA$_2$ are lysophospholipids, some of which are also inflammatory mediators. NSAIDs: non-steroidal antiinflammatory drugs.

References

Bley, K.R. *et al.*: **The role of IP prostanoid receptors in inflammatory pain.** *Trends Pharmacol. Sci.* 1998, **19**:141–147.

Dubois, R.N. *et al.*: **Cyclooxygenases in biology and disease.** *FASEB J.* 1998, **12**:1063–1073.

Feldmann, M. and Maini, R.N.: **Anti-TNFα therapy of rheumatoid arthritis: what have we learned?** *Annu. Rev. Immunol.* 2001, **19**:163–196.

Marshall, J.S.: **Mast-cell responses to pathogens.** *Nat. Rev. Immunol.* 2004, **4**:787–799.

Nathan, C.: **Points of control in inflammation.** *Nature* 2002, **420**:846–852.

Tilley, S.L. *et al.*: **Mixed messages: modulation of inflammation and immune responses by prostaglandins and thromboxanes.** *J. Clin. Invest.* 2001, **108**:15–23.

Leukocyte extravasation is a multi-step process

Controlling an infection in a tissue site requires an influx of neutrophils and other leukocytes that is orchestrated by inflammatory mediators, particularly the proinflammatory cytokines TNF and IL-1. These cytokines cause nearby endothelial cells to produce chemokines, which are tethered to proteoglycans attached to the luminal side of the endothelial cell, and to express adhesive molecules. The exit of leukocytes or lymphocytes from the blood into a tissue occurs via a multi-step process, which was outlined in section 2-14. The specific case of a neutrophil entering a site of inflammation via the multi-step process known as **extravasation** is illustrated in Figure 3-41.

In the absence of infection, neutrophils, monocytes and other white blood cells are swept through the circulation by fluid flow. TNF and IL-1 induce endothelial cells to translocate P-selectin (CD62P) rapidly from intracellular organelles to their cell surface and more slowly to express E-selectin (CD62E) after *de novo* synthesis. When leukocytes in the blood encounter P-selectin and E-selectin induced on endothelial cells by inflammatory cytokines, they bind through sulfated glycoproteins that are expressed constitutively, and their progress is thereby slowed. Selectins characteristically bind to their ligands with a moderate affinity and a rapid off-rate. In the blood, selectin binding is not strong enough to resist the fluid flow fully, and the consequence is a rolling behavior in which individual selectin molecules bind and let go as the cell moves along the surface of the endothelium. This is the first step in leukocyte recruitment. In the second step, sensing of chemokines tethered to the endothelial cells causes the rolling cells to flatten and at the same time induces a change in the integrins, which shift from a resting low-affinity state to an activated high-affinity state (inside-out signaling; see section 2-4). The high-affinity integrins of the leukocytes then recognize ICAM-1 and VCAM, which are also induced on the endothelial cells by the proinflammatory cytokines, resulting in firm adhesion of the leukocyte to the endothelium. The rare human immunodeficiency disease **leukocyte adhesion deficiency** is caused by genetic defects in this step, either in the β2-integrins that bind to ICAM-1 or in the fucosylation enzymes needed to synthesize the selectin ligands on the leukocytes. Individuals with this deficiency have an increased number of neutrophils in the blood but are highly susceptible to bacterial infections because these neutrophils are unable to enter sites of inflammation. The absence of Rac2, a Ras-like GTPase that functions in cell migration, leads to a similar phenotype in mice and humans. In the final step of leukocyte recruitment, the leukocyte finds a cell–cell junction in the endothelium and migrates between the endothelial cells and out of the blood vessel. Transmigration is mediated in part by homotypic interactions of the CD31 molecule (also called platelet/endothelial cell adhesion molecule-1, PECAM-1). CD31 is present on the surface of leukocytes and also on the endothelial cells at their sites of cell–cell contact. Blocking CD31 or deleting it in mice inhibits emigration of neutrophils into inflammatory sites substantially in some but not all tissues. Presumably there is a second mechanism for mediating this step of cell movement into inflammatory sites.

Chemokines and small-molecule chemoattractants direct phagocytes to the site of infection

At the initiation of an inflammatory response, chemokines bound to the endothelium have a critical role in directing leukocytes out of the blood and into the site of infection. The types of leukocytes and lymphocytes that are attracted to a particular site are determined largely by which chemokines are being made and displayed at that location. Gradients of chemokines and other chemoattractants then direct the migration of leukocytes to the main focus of infection

Definitions

diapedesis: process by which cells migrate between endothelial cells out of the blood into tissues.

extravasation: process of moving out of the blood through an intact endothelial layer and into the tissue, as occurs during inflammation or lymphocyte recirculation.

formyl peptide receptor (FPR): seven-transmembrane chemotactic receptor expressed on neutrophils and monocytes and mediating recruitment of these cells to sites of infection.

FPR: see **formyl peptide receptor**.

leukocyte adhesion deficiency: inherited immunodeficiency disease characterized by a defect in neutrophil adhesion and **extravasation**. Can result from genetic deficiency in the β2 integrin chain, in ICAM-1 or in fucosylation enzymes that contribute to the synthesis of ligands for selectins.

where their concentrations are highest. The initial infiltration into an acute inflammatory site is primarily composed of neutrophils, reflecting the induction of proinflammatory CXC chemokines, notably IL-8 (see section 2-13), by TNF and IL-1. Once in the site of inflammation, neutrophils respond preferentially to chemoattractants generated at the focus of infection. These include C5a, which is generated at the pathogen surface by the complement cascade, antimicrobial peptides released by neutrophils upon contacting bacteria, and f-Met-Leu-Phe, which is a conserved component of bacteria. Synthesis of all bacterial proteins is initiated with formyl methionine, and the amino-terminal-few amino-acid residues of many bacterial proteins are cleaved off to generate functional bacterial proteins, with *N*-formylated peptides as a necessary byproduct. In contrast, all proteins encoded in the eukaryotic cell nucleus begin with methionine. The **formyl peptide receptor (FPR)** is therefore another example of an innate immune recognition element. It is possible that necrotic cell lysis and release of formyl peptides derived from mitochondrial proteins could also attract neutrophils via this receptor.

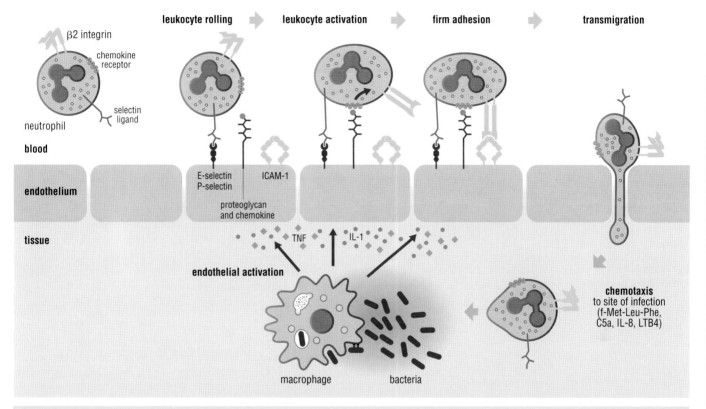

Figure 3-41 Recruitment of leukocytes to sites of infection Macrophage- or mast-cell recognition of a bacterial infection induces the production of TNF and IL-1, which act on neighboring cells to produce more of these cytokines until the response is propagated to the endothelial cells, which rapidly translocate P-selectin from intracellular storage vesicles to the plasma membrane and begin to synthesize E-selectin and integrin ligands such as ICAM-1 and VCAM. Also induced is the synthesis of chemokines, which bind to cell-surface proteoglycans that extend into the vessel. Leukocytes such as neutrophils flow through the blood vessel until they contact endothelial cells expressing selectins, which slows their movement with the fluid flow to a rolling motion. This allows the recognition of chemokines. The types of chemokines displayed determine the type of cell that will enter that tissue site. Typically early in a bacterial infection, neutrophil-attractive chemokines are expressed. Chemokine signaling activates integrins (β2 integrins in the case of neutrophils or monocytes), which mediate firm adhesion to the endothelium. This is followed by **diapedesis**, the process of moving between two endothelial cells, and chemotaxis within the tissue to locate the site of infection. Once in the tissue, the neutrophil migrates to the site of the microbes and attacks them.

References

Bunting, M. *et al.*: **Leukocyte adhesion deficiency syndromes: adhesion and tethering defects involving β₂ integrins and selectin ligands.** *Curr. Opin. Hematol.* 2002, **9**:30–35.

Foxman, E.F. *et al.*: **Integrating conflicting chemotactic signals. The role of memory in leukocyte navigation.** *J. Cell Biol.* 1999, **147**:577–588.

Mempel, T.R. *et al.*: ***In vivo* imaging of leukocyte trafficking in blood vessels and tissues.** *Curr. Opin.*

Immunol. 2004, **16**:406–417.

Rosen, S.D.: **Ligands for L-selectin: homing, inflammation, and beyond.** *Annu. Rev. Immunol.* 2004, **22**:129–156.

Rot, A. and von Andrian, U.H.: **Chemokines in innate and adaptive host defense: basic chemokinese grammar for immune cells.** *Annu. Rev. Immunol.* 2004, **22**:891–928.

Springer, T.A.: **Traffic signals for lymphocyte recirculation and leukocyte emigration: the multistep paradigm.** *Cell* 1994, **76**:301–314.

Inflammatory responses can be divided into two phases: the acute phase and the chronic phase

Inflammation triggered by bacterial infection is initially dominated by neutrophils. Later, the balance of cells changes, with monocytes and then lymphocytes coming into the inflammatory site. If the inflammatory insult is not resolved, the inflammation may become chronic owing to continued innate immune stimulation or T cell release of cytokines and chemokines.

The early shift from a neutrophil-dominated infiltrate to a monocyte-dominated one is in part mediated by IL-6. IL-6 is induced on a slower timescale than are TNF and IL-1 and, although it is generally thought of as proinflammatory, its actions are more complex and, as we see below, it contributes to the negative feedback that limits inflammatory responses. Neutrophils themselves have an important role in inducing the shift to monocyte recruitment, by shedding their IL-6 receptors as they enter a site of infection. Endothelial cells do not have IL-6 receptors and therefore cannot respond directly to IL-6. They do express the cytokine-receptor signaling protein gp130, however, and this protein, as described in section 2-8, enables them to respond to soluble IL-6 receptors complexed with IL-6. IL-6 receptors shed from neutrophils combine with the IL-6 released by macrophages, and the resulting complexes act on nearby endothelial cells to decrease the production of CXC chemokines and increase the production of the CC chemokines MCP-1 and MCP-3, which attract monocytes. This shift in cell types from neutrophils to mostly monocytes may serve to decrease the tissue injury consequent on inflammation, as monocytes participate in wound healing as well as immune defense and are less destructive than neutrophils.

Viruses have evolved devices for modulating inflammatory responses

Although we have primarily been considering the role of inflammation in immune defense against bacterial infections, inflammation is also likely to be important in defense against many viral infections. This is evident from the fact that several classes of viruses interfere with the induction of inflammation at multiple points. Several poxviruses make inhibitors of caspase 1, and thereby prevent the release of IL-1 from infected cells: in the case of myxoma virus, deletion of the gene encoding the inhibitor increases the inflammatory response to viral infection. Poxviruses also make soluble TNF receptors and soluble IL-1 receptors that block the actions of these cytokines. Especially common are viral chemokine receptors, which bind chemokines, blocking the normal mechanism for attracting inflammatory cells. Some viruses do the opposite, and encode activated chemokine receptors. It is thought that this facilitates dissemination of the virus out of the initial site of infection.

Proinflammatory cytokines act systemically to promote immunity

In a vigorous immune response, levels of proinflammatory cytokines rise in the blood to the point where they act systemically to promote immunity. TNF acts on fat and muscle cells to induce catabolism and mobilize energy stores. IL-1 and IL-6 act on the liver to induce the production of blood proteins that help fight infections: this is known as the **acute-phase response**. Acute-phase proteins include molecules that have a role in wound healing as well as in immunity to infection; they are listed in Figure 3-42. Although the immune components that are induced vary somewhat from species to species, these typically include mannose-binding lectin and some of the complement components, as well as members of the pentraxin family of proteins. Pentraxins are pentameric lipid-binding molecules. Three of these, namely C-reactive protein (CRP), serum amyloid protein (SAP) and PTX-3, bind to phospholipids on microbes and on apoptotic cells and stimulate phagocytosis and complement activation.

Acute-Phase Proteins
Opsonins
C-reactive protein
serum amyloid P component
mannose-binding lectin
Complement proteins
C2
C3
C4
C5
C9
factor B
C1 inhibitor
C4-binding protein
Coagulation proteins
fibrinogen α, β, γ
von Willebrand factor
Protease inhibitors
α_1-antitrypsin
α_1-antichymotrypsin
α_2-antiplasmin
plasminogen activator inhibitor I
heparin cofactor II
Scavenging proteins
haptoglobin
ceruloplasmin
hemopexin
Lipid-binding proteins
serum amyloid A
LPS-binding protein (LBP)

Figure 3-42 **Table of acute-phase proteins**

Definitions

acute-phase response: systemic response to infection mediated primarily by IL-1 and IL-6 acting on hepatocytes to increase the synthesis of blood proteins that include soluble recognition molecules for microbes or apoptotic cells, complement components, coagulation proteins, protease inhibitors, lipid-binding proteins and metal-binding proteins. These proteins are all likely to have value in helping fight infections.

References

Cartmell, T. *et al.*: **Circulating interleukin-6 mediates the febrile response to localised inflammation in rats.** *J. Physiol.* 2000, **526**:653–661.

Gabay, C. and Kushner, I.: **Acute-phase proteins and other systemic responses to inflammation.** *N. Engl. J. Med.* 1999, **340**:448–454.

Hurst, S.M. *et al.*: **IL-6 and its soluble receptor orchestrate a temporal switch in the pattern of leukocyte recruitment seen during acute inflammation.** *Immunity* 2001, **14**:705–714.

Kontoyiannis, D. *et al.*: **Impaired on/off regulation of TNF biosynthesis in mice lacking TNF AU-rich elements: implications for joint and gut-associated immunopathologies.** *Immunity* 1999, **10**:387–398.

Steel, D.M. and Whitehead, A.S.: **The major acute phase reactants: C-reactive protein, serum amyloid P component and serum amyloid A protein.** *Immunol. Today* 1994, **15**:81–88.

Tracey, K.J.: **The inflammatory reflex.** *Nature* 2002,

IL-1 and IL-6 (and to a smaller extent type 1 interferons and TNF) also act on the hypothalamus to induce fever, which is thought to have beneficial effects in fighting infections, at least at moderate levels. These cytokines induce the local production of PGE_2 as an intermediary signal, and for this reason the non-steroidal antiinflammatory drugs, which block prostaglandin synthesis (see section 3-13), inhibit fever as well as inflammation. The hypothalamic response to inflammatory cytokines is also important for inducing the release of glucocorticoids by the adrenal glands and thereby providing a feedback inhibitory loop, as we discuss below.

Inflammation is normally downregulated by a series of feedback inhibitory mechanisms

The proinflammatory reactions induced by TNF and IL-1 exhibit a positive feedback loop, in that both TNF and IL-1 induce the production of TNF and IL-1 in neighboring cells (Figure 3-43). This positive loop is a mechanism to propagate the response until it is large enough to attract sufficient leukocytes to control the infection. However, inflammatory reactions can be quite damaging to the surrounding tissue, and so mechanisms for limiting and downregulating inflammation are likely to be of great importance. For example, tissues such as lung, eye and brain, whose function is likely to be critically compromised by the swelling that accompanies inflammation, have special mechanisms for reducing the inflammatory response, although they cannot avoid it altogether in the presence of infection.

A number of inhibitory processes, operating both locally and systemically, are now recognized; these are illustrated in Figure 3-43. For example, TNF induces the shedding of TNF receptors, which decreases the sensitivity of the cell to TNF and at the same time reduces the stimulation of neighboring cells because the released receptors bind free TNF so that it cannot reach their surface TNF receptors. Also, there is delayed production by macrophages of IL-1 receptor antagonist (IL-1ra), a protein that competes with IL-1 for binding to IL-1 receptors but does not trigger their signaling. In addition, systemic levels of inflammatory cytokines trigger an inhibitory feedback loop from the hypothalamus, in which the adrenal medulla is induced to produce glucocorticoids; these inhibit inflammation by multiple mechanisms, including inhibition of the production of inflammatory cytokines. Synthetic glucocorticoids are frequently used as antiinflammatory drugs (described more fully in Chapter 14), a usage that is limited by their considerable side effects.

Another type of antiinflammatory mechanism is the production of antiinflammatory cytokines, of which the most important are IL-10 and TGF-β. Mice lacking IL-10 exhibit excessive inflammation of the gastrointestinal tract, whereas those lacking TGF-β exhibit inflammatory disease in many tissues. As mentioned above, IL-10 is made by macrophages that have been stimulated by TLRs, but more slowly than TNF or IL-1, and it promotes the wound-healing functions of macrophages at the expense of the microbe-killing functions, which are described in section 3-9. TGF-β is also produced in its active form by macrophages, particularly those ingesting apoptotic cells. Again, this can be seen as part of the wound-healing phase of the response.

Inflammation is also regulated by the nervous system. Pain-sensing neurons release neuropeptides such as substance P that induce release of inflammatory mediators by mast cells. Conversely, acetylcholine released by neurons of the vagus nerve act via nicotinic acetylcholine receptors on tissue macrophages (but not on monocytes) to inhibit the production of TNF and IL-1. This action may explain some of the effects of acupuncture.

420:853–859.

Zheng, H. *et al.*: **Resistance to fever induction and impaired acute-phase response in interleukin-1β-deficient mice.** *Immunity* 1995, **3**:9–19.

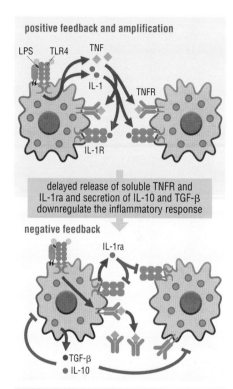

positive feedback and amplification

delayed release of soluble TNFR and IL-1ra and secretion of IL-10 and TGF-β downregulate the inflammatory response

negative feedback

Figure 3-43 Positive and negative feedback loops in the production of inflammatory cytokines Top panel: stimulation of the macrophage on the left by LPS induces the production of TNF and IL-1, which then act on neighboring cells to induce them also to produce TNF and IL-1, a process that propagates and expands the response. These cytokines also act back on the cell producing them in a positive feedback loop. Bottom panel: a delayed response to IL-1 and TNF produces negative feedback and shuts off the inflammatory response. TNF induces the proteolytic release of the extracellular domain of TNF receptors, which then act as inhibitors by binding TNF and preventing it from binding surface receptors on surrounding cells. In addition, cells that have shed their receptors are less responsive as a result of lower receptor expression levels. TNF also induces an intracellular inhibitor of TNF production, tristetraprolin, a protein that binds to an AU-rich motif in the 3′ untranslated region of the TNF mRNA and destabilizes it (not shown). IL-1 induces the production of IL-1ra, a homolog of IL-1 that is a receptor antagonist. Macrophages can also produce the cytokines IL-10 and TGF-β, which have a variety of anti-inflammatory actions including effects on cell types not shown (such as T cells and dendritic cells). Their overall effect is to suppress inflammation.

Interferon provides a critical defense against virus infection

The innate immune mechanisms we have considered so far are especially important in defense against bacterial and fungal infections. They also participate in defense against virus infections, for example through opsonization of virus particles by collectins or complement followed by phagocytosis and destruction of the virions. Although these mechanisms can decrease the numbers of viruses that infect cells, they do little to limit production of virus once it starts replicating in a cell. In this and the next sections we consider several innate immune mechanisms that are principally directed at this aspect of immune defense. We begin with the interferons.

There are two functionally overlapping but distinct types of interferons: the **type 1 interferons**, which include multiple variants of **interferon-α** and a single **interferon-β**, and **type 2 interferon** or **interferon-γ**. Each type of interferon is recognized by a selective heterodimeric receptor of the cytokine/hematopoietin receptor superfamily (see section 2-7), which signals via the Jak kinases and the STAT transcription factors (Figure 3-44). Interferon-γ functions in defense against intracellular infections by microbes and parasites as well as against viruses, whereas the function of the type 1 interferons is primarily in defense against viruses; we shall concentrate on the latter in this section. Mice genetically deficient in both receptor types or in STAT1 exhibit increased susceptibility to infections by a wide range of viruses, including vaccinia virus, herpes viruses, influenza virus, measles virus and vesicular stomatitis virus.

Interferons are produced by innate immune cells and by infected cells

Type 1 interferons are produced in two distinct ways: by infected cells detecting components of virus replication within them, and by innate immune cells detecting the presence of viruses through TLRs. As we saw in section 3-11, a subset of TLRs recognize pathogen-related nucleic acids and reside in the internal membrane-bound compartments of cells, where they can detect such molecules after virus particles or apoptotic virus-infected cells are internalized by

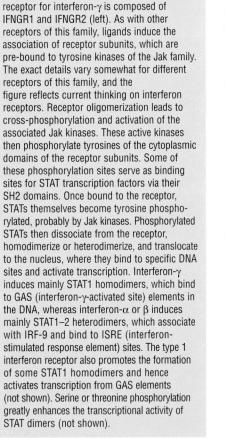

Figure 3-44 Signaling by receptors for interferons Type 1 and 2 interferons are recognized by heterodimeric receptors of the hematopoietin/cytokine receptor family. The receptor for interferons-α and β is composed of IFNAR1 and IFNAR2 (right), whereas the receptor for interferon-γ is composed of IFNGR1 and IFNGR2 (left). As with other receptors of this family, ligands induce the association of receptor subunits, which are pre-bound to tyrosine kinases of the Jak family. The exact details vary somewhat for different receptors of this family, and the figure reflects current thinking on interferon receptors. Receptor oligomerization leads to cross-phosphorylation and activation of the associated Jak kinases. These active kinases then phosphorylate tyrosines of the cytoplasmic domains of the receptor subunits. Some of these phosphorylation sites serve as binding sites for STAT transcription factors via their SH2 domains. Once bound to the receptor, STATs themselves become tyrosine phosphorylated, probably by Jak kinases. Phosphorylated STATs then dissociate from the receptor, homodimerize or heterodimerize, and translocate to the nucleus, where they bind to specific DNA sites and activate transcription. Interferon-γ induces mainly STAT1 homodimers, which bind to GAS (interferon-γ-activated site) elements in the DNA, whereas interferon-α or β induces mainly STAT1–2 heterodimers, which associate with IRF-9 and bind to ISRE (interferon-stimulated response element) sites. The type 1 interferon receptor also promotes the formation of some STAT1 homodimers and hence activates transcription from GAS elements (not shown). Serine or threonine phosphorylation greatly enhances the transcriptional activity of STAT dimers (not shown).

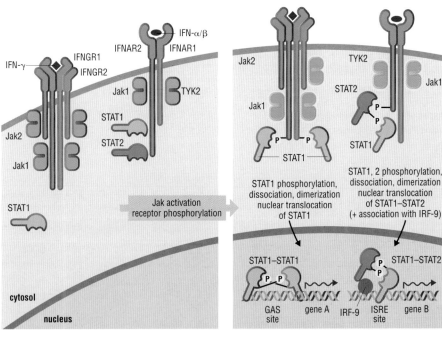

Definitions

interferon: any of several cytokines that act in immunity to inhibit viral replication or promote killing activity of macrophages, natural killer cells and cytotoxic T cells. Interferons are divided into two types: **type 1 interferons** (**interferon-α** and **interferon-β**) made by any virus-infected cell and by macrophages, dendritic cells and **plasmacytoid dendritic cells**, and acting to inhibit viral replication; and **type 2 interferon** (**interferon-γ**), which is made by natural killer cells and some T cells and whose main function is the activation of macrophages.

interferon-α: any of a large number of closely related **type 1 interferons**.

interferon-β: **type 1 interferon** that is made before **interferons-α** by virus-infected cells.

interferon-γ: cytokine made by T cells and natural killer cells that promotes killing of internalized microbes by macrophages and also has important antiviral defense roles (also called **type 2 interferon**).

interferon regulatory factor (IRF): any of a family of nine structurally related transcription factors, many of which play key roles in the induction of **interferon** and also in responses to interferon. IRFs also participate in TLR signaling.

Mda5: a cytoplasmic protein composed of a CARD and an RNA helicase domain that recognizes dsRNA in virus-infected cells and induces **type 1 interferon** production.

plasmacytoid dendritic cell: cell type of the dendritic cell family that produces very large amounts of **type 1 interferons** upon contact with viruses. These cells are found principally in the blood and home to secondary lymphoid organs after detecting viral nucleic acids

phagocytosis or macropinocytosis. Tissue macrophages, immature dendritic cells and **plasmacytoid dendritic cells** express various combinations of TLR3, TLR7, TLR8 and TLR9, and produce type 1 interferons in response to signaling by these TLRs. As we have seen, TLR3 recognizes double-stranded RNA (dsRNA), an obligatory replication intermediate of viruses with RNA genomes, and is dedicated to the production of interferon-β through the adaptor molecule TRIF, whereas other intracellular TLRs (TLR7, 8 and 9) signal the production of interferon-α through MyD88. In both cases, interferon gene expression is activated by **interferon regulatory factors (IRFs)**, a family of nine structurally related transcription factors that have an integral role in the production of type 1 interferons and in the response to them. Whereas signaling through TRIF activates IRF3, leading to the expression of interferon-β (see Figure 3-33), signaling through MyD88 activates IRF7, leading to the expression of interferon-α.

Plasmacytoid dendritic cells, which we have already briefly mentioned in section 3-11, produce by far the largest amounts of interferon in response to viral infection. These cells are not highly phagocytic and do not take up apoptotic cells, but rather are believed to take up virions by macropinocytosis or receptor-mediated endocytosis. They express TLR7 and TLR9 and make large amounts of interferon-α in response to internalized virus particles. Unlike other cells, they express IRF-7 constitutively and use it to couple TLR7 and TLR9 ligands (single-stranded RNA and CpG-containing DNA) to production of multiple interferons-α via the signaling component MyD88. We shall see later that interferon-α secreted by plasmacytoid dendritic cells is thought to be important in the regulation of adaptive immunity, reflecting the fact that type 1 interferons have immunoregulatory effects in addition to their ability to make cells refractory to virus replication.

The second way in which type 1 interferons are produced is by tissue cells when they are infected by a virus. Many cells sense virus replication in the cytoplasm and then make and secrete type 1 interferons, which then act on themselves and on neighboring cells to limit virus replication. Indeed, it is these two properties that led to the discovery of interferons about 50 years ago. This mode of interferon production also occurs through activation of the transcriptional regulators IRF-3 and IRF-7, in a two-step cascade (Figure 3-45). Double-stranded RNA produced during viral replication is recognized by the cytoplasmic RNA helicases **RIG-I** and **Mda5**, which initiate the activation of protein kinases related to the I-κB kinase (IKK) subunits. These kinases phosphorylate IRF-3, thereby inducing it to dimerize and translocate to the nucleus, where with other transcription factors it activates transcription of the interferon-β gene. Interferon-β then acts back on the same cell (as well as neighboring cells) via type 1 interferon receptors and a transcription factor complex consisting of STAT1, STAT2 and the constitutively expressed IRF-9 to induce production of transcription factor IRF-7, which can be activated by phosphorylation to promote synthesis of interferon-α genes, amplifying the response.

Figure 3-45 IRFs induce the production of type 1 interferons Virus infection is sensed by an RNA helicase RIG-I and the related molecule Mda5, which signal via an adaptor molecule bound to the outside of the mitochondria, mitochondrial antiviral signaling protein (MAVS), and two protein kinases, which phosphorylate the constitutively expressed IRF-3. This phosphorylation inactivates the nuclear export signal of IRF-3, allowing it to enter and remain in the nucleus. It binds to the virus-inducible enhancer of the interferon-β gene and drives transcription. Initial production of interferon-β then acts back on the cell via type 1 interferon receptors to induce the synthesis of IRF-7 via the STAT1–STAT2–IRF-9 complex. IRF-7 then forms homodimers or heterodimers with IRF-3, the activities of which are controlled by phosphorylation as for IRF-3 homodimers. These dimers bind to enhancers in the interferon-α genes, inducing their synthesis. Note that additional transcription factors participate in these gene inductions and are not shown. For example, also bound to the interferon-β promoter after virus infection are NF-κB and a heterodimer of c-Jun and ATF-2. How infection activates these transcription factors is less well established.

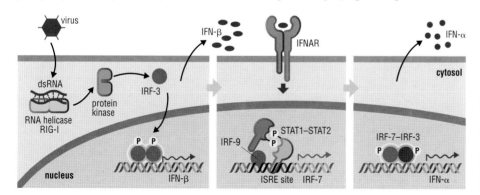

through TLR7 or TLR9. Also called interferon-producing cells.

RIG-I: a cytoplasmic protein composed of a CARD and an RNA helicase domain that recognizes dsRNA in virus-infected cells and induces **type 1 interferon** production.

type 1 interferon: any of **interferons-α** or **interferon-β**, which are important antiviral cytokines.

type 2 interferon: interferon-γ, a cytokine important in adaptive immunity and especially in macrophage activation.

References

Bogdan, C.: **The function of type I interferons in antimicrobial immunity.** *Curr. Opin. Immunol.* 2000, **12**:419–424.

Liu, Y.-J.: **IPC: professional type 1 interferon-producing cells and plasmacytoid dendritic cell precursors.** *Annu. Rev. Immunol.* 2005, **23**:275–306.

Plantanias, L.C.: **Mechanisms of type-I- and type-II-interferon-mediated signalling.** *Nat. Rev. Immunol.* 2005, **5**:375–386.

Stark, G.R. *et al.*: **How cells respond to interferons.**

Annu. Rev. Biochem. 1998, **67**:227–264.

Taniguchi, T. *et al.*: **IRF family of transcription factors as regulators of host defense.** *Annu. Rev. Immunol.* 2001, **19**:623–655.

van den Broek, M.F. *et al.*: **Immune defence in mice lacking type I and/or type II interferon receptors.** *Immunol. Rev.* 1995, **148**:5–18.

Yoneyama, M. *et al.*: **The RNA helicase RIG-I has an essential function in double-stranded RNA-induced antiviral responses.** *Nat. Immunol.* 2004, **5**:730–737.

Interferon primes cells to allow them to block virus replication

Interferons-α and β produced either by infected cells or by sentinel immune cells in the infected tissue act on nearby cells to inhibit virus replication by a multiplicity of mechanisms acting at virtually any step in the viral life cycle, from uncoating, transcription and translation to assembly and packaging. Three particularly well understood mechanisms involve a dsRNA-dependent protein kinase called **PKR**, 2′,5′-oligoadenylate synthetase (2′,5′-oligo A synthetase) together with **RNase L** and the **Mx proteins** (Figure 3-46). Whereas type 1 interferons induce all three of these responses, interferon-γ induces only PKR.

PKR is a protein kinase that uses a duplicated amino-terminal RNA-binding domain to distinguish between many viral dsRNAs and cellular dsRNAs such as tRNA. It is expressed at low levels in many cell types, where it may have additional functions in responding to cell stress, but only after induction by interferon-receptor signaling are levels high enough to be able to interfere with virus replication. Once activated, its main target is a key component of the machinery that translates mRNAs into proteins, the translation initiation factor eIF2α. PKR inhibits all protein synthesis, cellular and viral, so this is likely to result in cell death in addition to blocking virus replication. Note, however, that cells responding to interferon express elevated PKR, but it still needs to bind dsRNA to be active, ensuring that only infected cells stop their protein synthesis.

Similarly, interferon receptor signaling induces the expression of a small family of 2′,5′-oligoadenylate synthetases, but these enzymes synthesize their signaling molecule only after recognition of dsRNA. The synthesized oligoadenylate then activates a cellular RNase, RNase L,

Figure 3-46 Major antiviral actions of interferons-α and β Interferons-α and β bind to the interferon-α/β receptor and induce a transcriptional response that elevates the expression of PKR, bringing it to levels sufficient to inhibit protein synthesis after viral infection. Also transcriptionally induced are the Mx proteins and 2′,5′-oligoadenylate synthetase. The latter enzyme synthesizes the second messenger 2′,5′-oligoadenylate (oligo A) in response to dsRNA. This small molecule binds to and activates RNase L, which degrades mRNAs. These two mechanisms block virus replication at the cost of blocking cellular protein synthesis as well. The Mx proteins are related to the GTPase dynamin, which polymerizes to form a ring-like structure that is required for the budding off of endosomal vesicles as part of endocytosis. The Mx proteins also self-assemble into horseshoe or ring-like structures, which interfere with viral replication, apparently at multiple steps.

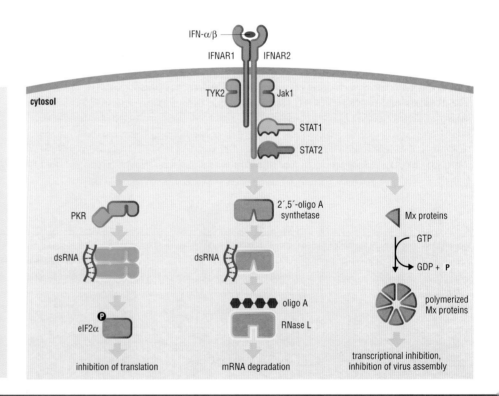

Definitions

Mx proteins: any of a series of antiviral proteins synthesized in response to interferon and structurally related to the GTPase dynamin.

PKR: double-stranded RNA-dependent protein kinase important in intracellular defense against viruses. PKR is also activated by other stimuli, including Toll-like receptors and stress, but its main function is blocking virus replication.

RNase L: RNase that is activated by 2′,5′-oligoadenylate,

which is synthesized in response to the presence in a cell of viral dsRNA. RNase L degrades both mRNA and ribosomal RNA and thus interferes with virus and host protein synthesis.

References

Levy, D.E. and García-Sastre, A.: **The virus battles: IFN induction of the antiviral state and mechanisms of viral evasion.** *Cytokine Growth Factor Rev.* 2001, **12**:143–156.

Weber, F. *et al.*: **Inverse interference: how viruses fight the interferon system.** *Viral Immunol.* 2004, **17**:498–515.

Williams, B.R.G.: **PKR; a sentinel kinase for cellular stress.** *Oncogene* 1999, **18**:6112–6120.

which degrades mRNAs and also ribosomal RNAs, which again will act to block all protein synthesis in the cell. Finally, a third antiviral mechanism induced by type 1 interferons is a family of proteins called Mx proteins. These proteins are structurally related to the protein dynamin, which has a structural role in pinching off membrane in endocytosis. Like dynamin, these are GTPases that assemble into multimeric structures in a regulated fashion. Exactly how they function is not well understood, but some are found in the nucleus and may block transcription, whereas others are in the cytoplasm and have other effects countering virus replication, such as interfering with virus particle assembly.

Viruses can evade the effects of interferon

The potency of the interferon system for limiting virus replication can be seen by the great susceptibility to virus infection in mice lacking either STAT1 or both types of interferon receptors. Despite this continued efficacy of the interferon system, viruses have developed many ways to block the interferon pathway at one step or another. Correspondingly, interferon is even more effective in limiting the replication of mutant viruses in which these blocking mechanisms are removed experimentally. Examples of viral evasion of interferon include influenza virus NS1 protein, which binds dsRNA and makes it less stimulatory, vaccinia virus E3L protein, which blocks the activating phosphorylation of IRF-3 and IRF-7, human herpesvirus 8, which encodes four IRF homologues that inhibit cellular IRFs, adenovirus E1A protein, which binds to STAT1 and blocks its function, and vaccinia virus B18R protein, which is a secreted molecule that binds to type 1 interferons and prevents their binding to INFAR1, to name just a few. A particularly common target of viral evasion is PKR. The known means by which viruses interfere with this antiviral mechanism are illustrated in Figure 3-47.

Interferons are used to treat some diseases

Given the importance of interferon for antiviral defense, it is not surprising that several chronic viral diseases can be treated by interferon administration. The most important of these are the chronic diseases caused by hepatitis C virus and hepatitis B virus. Some patients with the disease multiple sclerosis benefit from treatment with interferon-β. Although this disease is often considered to be an autoimmune disease, the lesions have a focal nature, perhaps indicative of a viral cause. In addition, a number of malignancies can be treated with interferons, including hairy cell leukemia, cutaneous T cell lymphoma, renal cell carcinoma and AIDS-related Kaposi's sarcoma. The last of these is a virally caused cancer, but for the others the reasons for the efficacy of interferons are less clear.

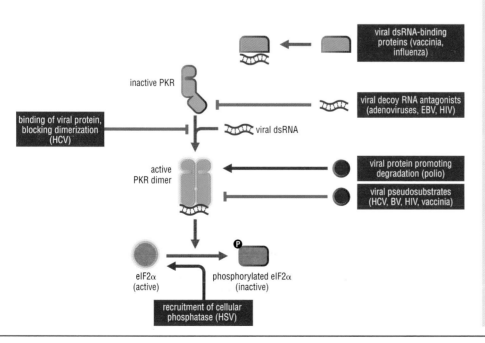

Figure 3-47 Evasion of PKR by viruses Many viruses have evolved mechanisms of blocking the ability of PKR to inhibit the translation of viral proteins in infected cells. In this pathway, dsRNA at least 25 nucleotides in length is recognized by the dsRNA-dependent protein kinase (PKR), which dimerizes to form the active kinase and then phosphorylates eIF2α, inactivating this critical translation initiation factor. Some viruses make proteins that bind to dsRNA and prevent it from being recognized by PKR. Other viruses make decoy RNA molecules that compete for binding to PKR but do not activate it. Another common strategy for viruses is to make a protein that interacts with PKR and inhibits it in one of several ways, including targeting PKR for proteasomal degradation, binding and preventing dimerization, or serving as a pseudosubstrate and preventing the phosphorylation of eIF2α. Additionally, a herpes simplex virus protein recruits a cellular phosphatase to eIF2α and promotes its dephosphorylation. BV: baculovirus; EBV: Epstein–Barr virus; HCV: hepatitis C virus; HIV: human immunodeficiency virus; HSV: herpes simplex virus.

Viral replication triggers apoptotic mechanisms in the infected cell

One of the most important ways in which the immune system deals with viruses is by killing the infected cell, and thereby limiting virus production. Cells killed in this way undergo apoptosis. Many of the mechanisms that trigger apoptosis probably either evolved as defenses against viral infection or were co-opted for that purpose soon after they were developed. For example, the **p53**-dependent pathway of apoptosis is typically triggered by unscheduled DNA replication, a hallmark of a cell infected by a DNA virus. Many DNA viruses express proteins that block p53 function. Another example is the murine caspase 12 pathway of apoptosis, which is initiated at the endoplasmic reticulum by irregularities there, possibly including high levels of viral glycoprotein synthesis or virus assembly. A general overview of apoptotic mechanisms active in killing virus-infected cells is depicted in Figure 3-48 (see also sections 2-11 and 2-12).

Thus, the cells of multicellular organisms have a number of intrinsic apoptotic pathways that function to sense viral infection and trigger apoptosis. Supplementing these intrinsic triggers are two additional strategies for killing virus-infected cells, one by means of signals from the outside of the cell acting via death-domain-containing receptors (Fas, TNF receptor 1 and the TRAIL receptors DR3 and DR4; see section 2-9) and the other involving a more direct mechanism of cell killing by the professional executioners of the immune system, natural killer cells and cytotoxic T lymphocytes.

Figure 3-48 Mechanisms of apoptosis induction in virus-infected cells Virus infection can trigger apoptosis by intrinsic mechanisms, such as perturbations of the endoplasmic reticulum during virus assembly, sensed by caspase 12, or recognition of unscheduled DNA synthesis by DNA viruses, which activates p53 to induce synthesis of the BH3-only proteins Puma and Noxa. Other BH3-only Bcl-2 family members are also thought to recognize various signs of intracellular stresses caused by high-rate virus replication. Many viruses have mechanisms for inhibiting these intrinsic mechanisms for induction of apoptosis, and some of these are described in Chapter 10. In addition, apoptosis of virus-infected cells can be triggered by the extrinsic pathway in which ligands for death receptors of the TNF receptor family are produced by innate or adaptive immune cells and stimulate death via the activation of caspase 8 or caspase 10. Finally, cytotoxic T lymphocytes and natural killer cells can introduce into virus-infected cells the proteolytic enzyme granzyme B, which can cleave Bid, inducing apoptosis, or can activate caspases directly. For details of the apoptotic pathways see sections 2-11 and 2-12. The mechanism of cytotoxicity by cytotoxic T cells is described in detail in Chapter 5.

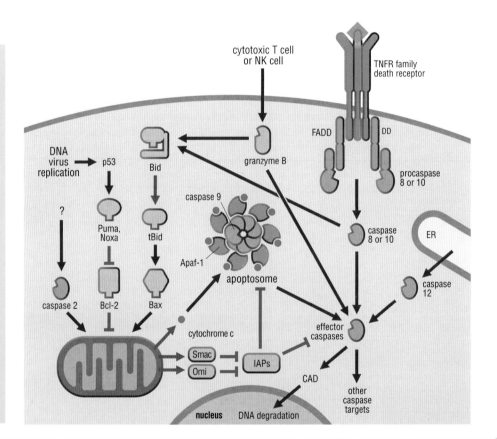

Definitions

p53: gene regulatory protein that orchestrates the long-term DNA damage response in multicellular organisms by activating a cell-cycle checkpoint and/or apoptosis in response to DNA damage or to DNA replication in an incorrect phase of the cell cycle.

Death domain receptors can induce death of virus-infected cells

Viral infection often leads to sensitivity of the infected cell to killing by ligands that interact with Fas or TNF receptor 1 (TNFR1). For example, in many cells, inhibition of protein synthesis sensitizes them to apoptosis induced by TNF. This results from a bifurcated TNFR1 signaling pathway, involving a delayed branch that directly activates apoptosis via caspase 8 and/or 10 and a more rapid branch that activates the transcription factor NF-κB, leading to induced synthesis of proteins that protect against apoptosis by the death pathway (Figure 3-49). In addition to its role in inducing inflammation in response to bacterial infections TNF is rapidly produced in response to virus infection, especially by macrophages and plasmacytoid dendritic cells, and hence is generally present at a site of virus replication. Synthesis of host cell proteins would be inhibited only in virus-infected cells; this could occur upon infection with a virus that blocks host protein synthesis, such as vaccinia virus, or by the inactivation of translation factor eIF2α by elevated and activated PKR. The significance of this antiviral mechanism is supported by the plethora of viral mechanisms for interfering with TNF in particular or with apoptosis induced by death-domain-containing receptors in general (see Figure 10-6). Fas also seems to be an important mechanism for killing virus-infected cells, at least in the case of adenovirus. Adenovirus infection induces Fas expression on hepatocytes, in part via the transcription factor NF-κB, and makes them sensitive to Fas-induced death. This is in contrast to the antiapoptotic role of NF-κB that is typical in TNF receptor responses. Interferon induces FasL expression by NK cells, providing the ligand to induce cell death. Adenoviruses express a protein complex, called RID, that causes the endocytosis and lysosomal degradation of Fas to counter this apoptotic mechanism. Fas is also important for restraining lymphocyte activation, as described in Chapter 12. As mentioned above, many of the viral antiapoptotic mechanisms block Fas killing as well as TNFR1 killing. Recently TRAIL has been implicated in apoptosis induced by reovirus infection. Thus, the death domain receptors as a class seem to have significant roles in antiviral innate immunity.

Viruses have evolved many mechanisms to block apoptosis

Although some viruses seem to require cellular apoptosis for virus release, most viruses block apoptosis, often by multiple mechanisms. Some representative examples are shown in Figure 10-6. Perhaps the most powerful immune mechanism for killing virus-infected cells is killing by natural killer cells and cytotoxic T cells via the perforin/granzyme mechanism. This effector mechanism is described in more detail in Chapter 5. TNF and interferons-α and β produced by macrophages or plasmacytoid dendritic cells can activate natural killer cells to provide an innate immune version of this important antiviral defense mechanism. The mechanisms by which natural killer cells recognize virus-infected cells are discussed in Chapter 8.

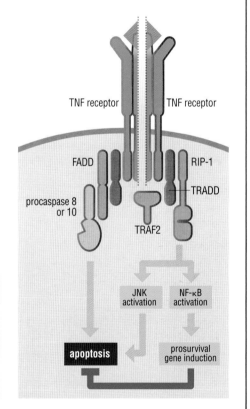

Figure 3-49 Signaling by death domain receptors TNF is a trimer that induces the oligomerization of TNFR1 into homotrimers or higher-order oligomers, leading to the formation of a larger protein assembly with the cytoplasmic tail. The association of TNFR1 with TRADD occurs between the latter's death domain and the death domain of the receptor. The TRADD death domain also associates with another adaptor molecule called FADD, which in turn associates with procaspase 8 or 10. A separate assembly of TNFR1 and TRADD with TRAF2 or related TRAFs and the serine/threonine protein kinase RIP1 leads to activation of NF-κB and JNK (one type of MAP kinase). NF-κB leads to prosurvival gene induction in most cell types, provided that protein synthesis is not blocked by PKR or by viral means. NF-κB can be proapoptotic via the induction of Fas in some cell types. JNK activation may promote apoptosis.

References

Benedict, C.A. *et al.*: **Death and survival: viral regulation of TNF signaling pathways.** *Curr. Opin. Immunol.* 2003, **15**:59–65.

Krajcsi, P. and Wold, W.S.M.: **Inhibition of tumor necrosis factor and interferon triggered responses by DNA viruses.** *Semin. Cell Dev. Biol.* 1998, **9**:351–358.

Kühnel, F. *et al.*: **NFκB mediates apoptosis through transcriptional activation of Fas (CD95) in adenoviral hepatitis.** *J. Biol. Chem.* 2000, **275**:6421–6427.

Lohrum, M.A.E. and Vousden, K.H.: **Regulation and function of the p53-related proteins: same family, different rules.** *Trends Cell Biol.* 2000, **10**:197–202.

Nakagawa, T. *et al.*: **Caspase-12 mediates endoplasmic-reticulum-specific apoptosis and cytotoxicity by amyloid-β.** *Nature* 2000, **403**:98–103.

Roulston, A. *et al.*: **Viruses and apoptosis.** *Annu. Rev. Microbiol.* 1999, **53**:577–628.

Tortorella, D. *et al.*: **Viral subversion of the immune system.** *Annu. Rev. Immunol.* 2000, **18**:861–926.

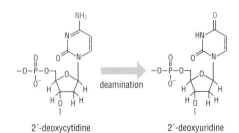

Figure 3-50 The conversion catalyzed by cytidine deaminase The enzyme deaminates cytidine residues in DNA or RNA (not shown) to generate uridine residues.

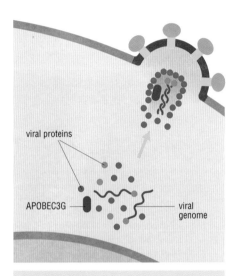

Figure 3-51 Incorporation of APOBEC3G into HIV virions The viral proteins are shown in the cytoplasm of an infected cell. APOBEC3G (purple) attaches itself to the capsid protein (blue) and is incorporated in the virion, where it introduces mutations into the viral genome. This process is blocked by the Vif protein of HIV-1.

Antiviral innate immune defenses include mechanisms directed at viral nucleic acid

Viral RNA, as we saw in sections 3-16 and 3-17, is vulnerable to recognition by the interferon pathway, which activates defense systems that interfere with protein synthesis and degrade mRNAs. This decreases virus production, but it also interferes with the synthesis of cellular proteins. Viral nucleic acid can also be directly targeted by several mechanisms in mammalian cells that either destroy it or introduce a catastrophic level of mutation. These powerful systems have driven the evolution of mechanisms whereby pathogenic viruses counter them.

APOBEC3G is a cytidine deaminase that mutates retroviral genomes

Mammalian cells express a small family (11 in humans) of **cytidine deaminases** with the function of mutating RNA or DNA. These enzymes convert cytidine residues to uridine residues in DNA or RNA (Figure 3-50). In RNA this represents a mutation; in DNA, which normally contains thymidine, not uridine, it causes mispairing that introduces mutations on DNA replication or through the action of DNA repair enzymes.

The first member of this family to be described was APOBEC1, a cytidine deaminase responsible for editing the mRNA encoding apolipoprotein B, which is important in cholesterol transport to the liver. This editing produces an altered transcript encoding a truncated form of the protein in the small intestine, where instead it functions in lipid absorption.

The second known family member is the activation-induced cytidine deaminase (AID), which we shall encounter in Chapter 6 when we discuss the mechanism whereby mutations are introduced into the immunoglobulin genes of B lymphocytes followed by selection for variant immunoglobulins with increased affinity for antigen (see section 6-11).

Additional cytidine deaminases, including **APOBEC3G** and at least three related APOBECs (3B, 3C and 3F), function in defense against retroviruses.

Individual cytidine deaminases restrict their action to particular targets by localization mechanisms that direct them to the relevant substrate: for example APOBEC1 binds to a targeting polypeptide that targets it specifically to the ApoB mRNA. Analogously, APOBEC3G is targeted to retrovirus nucleocapsids—the protein coats that form around the viral RNA—thus enclosing the editing enzyme with the viral genome (Figure 3-51). The result of this targeting is the release from the infected cell of virions that are normal in number but have extremely low infectivity due to the many mutations within their genome, the vast majority of which correspond to C to U changes that are introduced into the viral DNA during replication of the viral genome, when APOBEC3G can deaminate up to 10% of cytidines to uridines

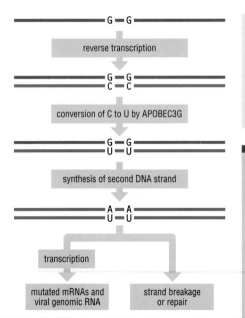

Figure 3-52 Mutation of retroviral DNA by APOBEC3G The first step in viral replication is reverse transcription of the viral RNA genome to generate a DNA copy. It is thought that APOBEC3G acts on the DNA copy at this stage to deaminate the cytidines. The next step is the synthesis of a second DNA strand to replace the RNA. This results in the replacement of the guanidines with adenines since uridine pairs with adenine. The resulting retrovirus DNA is believed to be a target for DNA repair enzymes, which remove the uridine residues. The resulting gap can be repaired, so that the cytidines are replaced, or it can lead to strand breakage, damaging the retrovirus DNA. Transcription of the mutated retroviral DNA leads to mutated mRNAs, mutated viral genomic RNA, and mutated viral proteins.

Definitions

APOBEC3G: one of a family of **cytidine deaminases** that includes APOBEC1, an mRNA editing enzyme, and AID, a DNA-editing enzyme that is responsible for immunoglobulin gene somatic hypermutation and class switch recombination. APOBEC3G targets retrovirus nucleocapsids for catastrophic mutation of retroviral DNA.

cytidine deaminase: enzyme that converts cytidine to uridine. The APOBEC1 family of cytidine deaminases act only on cytidines in nucleic acids, not on free cytosine or CTP.

Dicer: endonuclease that recognizes double-stranded RNA and cleaves it into short fragments 21–25 nucleotides in length, called **siRNAs.**

RISC: see **RNA-induced silencing complex.**

RNAi: see **RNA interference.**

RNA-induced silencing complex (RISC): multisubunit complex that binds **siRNA** and targets homologous mRNAs for nuclease digestion.

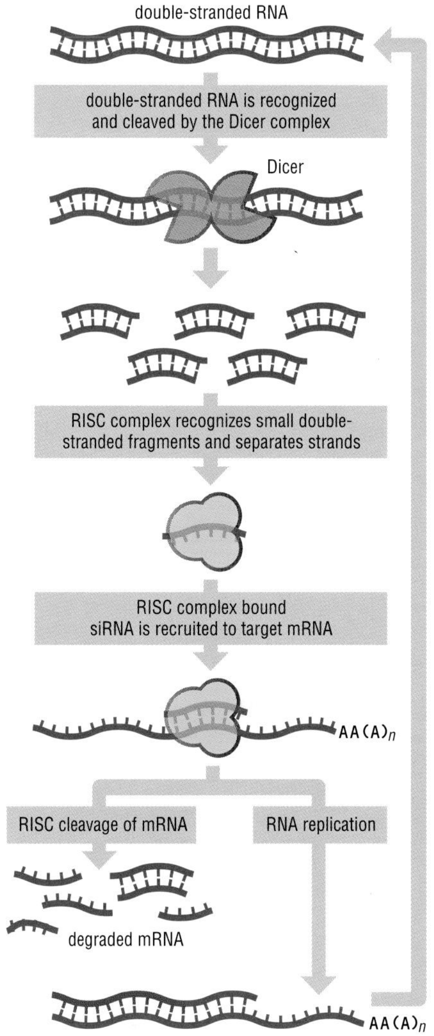

Figure 3-53 Biochemical mechanism of RNA interference Double-stranded RNA is recognized by Dicer and cleaved into short RNA duplexes (21–25 nucleotides in length with staggered ends), called siRNA. siRNA molecules bind to the RISC complex, which then identifies and binds to mRNAs with sequences homologous to one of the two siRNA strands. In mammals, the RISC complex then degrades the mRNA recognized. In some species, the small RNA then serves as a primer to allow RNA copying of the recognized mRNA, a process that amplifies the amount of dsRNA and, after Dicer action, the amount of siRNA. In plants, the amplified siRNAs are transported to neighboring cells and contribute to virus resistance.

(Figure 3-52). HIV-1, the virus that causes AIDS, encodes a protein *Vif (viral infectivity factor)* that protects the virus from this antiviral defense mechanism by binding to APOBEC3G and directing it to the proteasome, where it is degraded.

RNA interference is an antiviral defense mechanism in plants and probably in animals

RNA interference (RNAi) or RNA silencing is a phenomenon observed in plants, animals and some fungi in which the expression of double-stranded RNA (dsRNA) leads to inhibition of expression of mRNA molecules containing homologous nucleic acid sequences. RNAi has become extremely popular in recent years as an experimental means of reducing gene expression to test gene function. RNAi probably serves multiple functions, including gene regulation in development, immune antiviral defense and genome defense by suppressing transposon movement.

The basic mechanism of RNAi is understood in part and is illustrated in Figure 3-53. dsRNA, which is typically generated during virus replication, is first recognized by an endonuclease called **Dicer**, which cleaves it into short duplexes, generally 21–25 nucleotides in length, often just below the minimum size recognized by PKR of the interferon response pathway (see section 3-17). This reaction can thus destroy viral replication intermediates. The resulting short dsRNA duplexes are referred to as **small interfering RNAs (siRNA)** and have a further part in disabling the virus. They are recognized by a large multisubunit complex called the **RNA-induced silencing complex (RISC)**, reflecting the fact that triggers of the RNA silencing pathway induce increased expression of this complex in plants.

The RISC complex is targeted to mRNAs (or viral single-stranded RNA) with homology to one of the strands of the siRNA molecule bound to it, and it cleaves the corresponding RNA, thereby inactivating it. In some species of plants and invertebrates, the 21–25-nucleotide fragment paired to the mRNA is extended by RNA-dependent RNA synthesis, generating a longer dsRNA, which can then be cleaved to generate more siRNA. This amplification mechanism is generally not seen in mammalian cells, and presumably for this reason RNAi is less effective in decreasing gene expression in mammalian cells. In plants, these amplified siRNAs are transported to neighboring cells, increasing their resistance to virus replication.

The role of RNAi in antiviral defense is well established in plants and in some insects. For example, the flock house virus (FHV), which infects insects, has a gene called B2 that can block RNA silencing. Insect cells infected with FHV lacking B2 accumulate much higher levels of siRNAs corresponding to FHV sequences than do cells expressing wild-type virus, and much lower levels of genome-length RNA. Experimental disabling of the RNAi system restored levels of genome-length RNA to those seen with wild-type virus infection. Thus, FHV replication requires B2 only if the infected cell has an active RNAi system, indicating that the function of B2 is to block the RNAi system.

RNA interference (RNAi): mechanism by which short fragments of double-stranded RNA lead to the degradation of homologous mRNAs. Also called RNA silencing. RNAi is widely used experimentally to decrease the expression of a gene of interest to study its function. Many plant and some insect viruses have been shown to encode proteins that block RNAi.

siRNA: see **small interfering RNA.**

small interfering RNA (siRNA): double-stranded RNAs 21–25 nucleotides in length that participate in **RNA interference**.

References

Ahlquist, P.: **RNA-dependent RNA polymerases, viruses, and RNA silencing.** *Science* 2002, **296**:1270–1273.

Bieniasz, P.D.: **Intrinsic immunity: a front line defense against viral attack.** *Nat. Immunol.* 2004, **5**:1109–1115.

Harris, R.S. *et al.*: **DNA deamination mediates innate immunity to retroviral infection.** *Cell* 2003, **113**:803–809.

Mangeat, B. *et al.*: **Broad antiretroviral defence by human APOBEC3G through lethal editing of nascent reverse transcripts.** *Nature* 2003, **424**:99–103.

Plasterk, R.H.A.: **RNA silencing: the genome's immune system.** *Science* 2002, **296**:1263–1265.

Sheehy, A.M. *et al.*: **Isolation of a human gene that inhibits HIV-1 infection and is suppressed by the viral Vif protein.** *Nature* 2002, **418**:646–650.

Voinnet, O.: **RNA silencing as a plant immune system against viruses.** *Trends Genet.* 2001, **17**:449–459.

4

Adaptive Immunity and the Detection of Infection by T Lymphocytes

The T lymphocytes of adaptive immunity recognize infection indirectly, through highly diverse surface receptors that bind degradation fragments of microorganisms generated inside cells and complexed with specialized cell-surface molecules encoded in the major histocompatibility complex. The mechanisms for generating these fragments and loading them for surface display are adapted in different cells for the distinct functions of activating the distinct classes of effector T cells and of initiating adaptive immune responses by activating naïve T cells, which is the specialized function of dendritic cells.

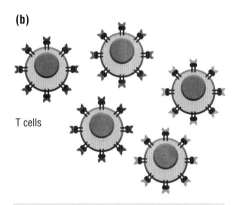

(a)

scavenger receptor

CR3 CR4

C-type lectin receptor (CLR)

Toll-like receptor (TLR)

C-type lectin receptor (CLR)

macrophage

(b)

T cells

Figure 4-1 Innate and adaptive immune recognition The innate immune recognition strategy is schematically represented here by the macrophage **(a)**, which has a variety of receptors of at least three types that are expressed on all macrophages and recognize conserved distinctive features of micro-organisms. The lymphocytes of adaptive immunity, schematically represented here as T cells **(b)**, have only one type of receptor but with a vast array of variants, each variant expressed on a different cell, and that are not pre-adapted to recognize any specific feature of a microorganism.

Adaptive immunity is distinct from innate immunity in three main ways

The innate immune response probably clears most infections before they cause perceptible disease. But many pathogens, as we have seen in Chapter 3, have evolved specializations for obstructing detection by innate immunity, or blocking its destructive machinery. Adaptive immunity has evolved to circumvent these evasive devices with a recognition system so versatile as to make it very hard for microorganisms to avoid detection, coupled to mechanisms for delivering them to the innate immune system for destruction, or for overcoming microbial blockade. The protection provided by the adaptive immune system is essential for survival: children born with defects that prevent its development must be given bone-marrow transplants or they die in infancy.

Adaptive immunity depends upon the specialized properties of lymphocytes and their products and is distinct in three ways from innate immunity. First, instead of having a limited repertoire of receptors that recognize conserved molecules of microorganisms (Figure 4-1a), lymphocytes have a single type of receptor but with an essentially unlimited repertoire of variants (Figure 4-1b) that can recognize virtually any molecule. Molecular determinants recognized by lymphocytes, as noted in Chapter 1, are known as *antigens*, an operationally defined term that reflects the versatility of lymphocyte receptors. The peak of this versatility is reached in the antibodies produced by B lymphocytes, which we discuss in Chapter 6. The second distinguishing feature of adaptive immunity is its delayed onset: adaptive immune responses do not take effect until a few days after infection, because of the need for expansion in numbers of the rare cells recognizing the antigens that invoke them. The third specialization of adaptive immunity is immune memory, whereby the adaptive response that takes days to develop on first encounter with a microorganism is effective almost immediately on subsequent infection with the same organism. This property underlies lasting immunity and the effectiveness of vaccination, which we discuss in Chapter 14.

In this chapter we are concerned with the detection of infection by T lymphocytes, which are responsible for adaptive immune responses to intracellular pathogens and are essential for activating antibody production in most B cell responses against extracellular pathogens. Antigen recognition by T cells is not as versatile as that of B cells. Whereas immunoglobulins can bind to antigen in any form, whether soluble or on the surface of a microorganism, and recognize carbohydrates and lipids as well as proteins, T cells recognize fragments of antigens, usually peptides derived from proteins, presented on the surface of cells by the specialized antigen-presenting molecules encoded by the major histocompatibility complex (see section 1-5). This reflects the requirements of the three major functions of T lymphocytes.

T lymphocytes kill cells infected with viruses and activate B cells and phagocytes

The major effector functions of T cells generally require that they recognize antigen in association with other cells. *Cytotoxic T cells (T$_C$)* directly kill cells infected with agents such as viruses whose proteins are replicated in the cytosol, and recognize MHC molecules complexed with degradation fragments of proteins derived from these organisms (Figure 4-2a). The other two major functions of effector T cells are performed by *helper T cells (T$_H$)*, which activate other cells of the immune system. B cells are induced to proliferate and differentiate into antibody-producing cells after antigen bound by their surface immunoglobulin is internalized and returned as degradation fragments on MHC molecules for presentation to T cells, which then activate the presenting B cell (Figure 4-2b). Similarly, phagocytes that have ingested bacteria signal the presence of the internalized microbe through fragments presented on MHC molecules at the cell surface, where they are recognized by helper T cells that can activate the phagocyte and thereby ensure the destruction of the microbe (Figure 4-2b). The activation of B cells and phagocytes by helper T cells depends on secretion of cytokines by the T cells that not only activate the presenting cell but can bias the response of that cell so as to direct the type of antibody secreted (in the case of a B cell) or the nature of the inflammatory response and the other immune cells recruited (in the case of a macrophage), depending upon the particular cytokines produced. We discuss in Chapter 5 how distinct types of immune response are coordinated by subsets of helper T cells producing distinct characteristic cytokines. The focus of this chapter is on the detection of antigen by T cells, and in this respect the key distinction is not between different subsets of helper T cells but between helper and cytotoxic cells.

The major classes of T lymphocytes are defined by the class of MHC molecule they recognize

The effector functions of cytotoxic T cells require that they detect antigen derived from viruses, which may proliferate in virtually any type of cell and from which peptide fragments are generated in the cytosol of infected cells. Helper cells, by contrast, are required to detect antigenic peptides principally on other cells of the immune system (phagocytes or B cells) where they are generated in the endosomal compartment of the cell, not in the cytosol. These distinct requirements of cytotoxic and helper cells are met by the existence of two types of MHC molecules with distinct functional properties. Peptide fragments generated in the cytosol are carried to the cell surface by *MHC class I molecules*, which are expressed on almost all cells, whereas peptides generated in the endosomal compartment are carried to the cell surface by the structurally homologous but distinct *MHC class II molecules*, which are normally expressed only on cells of the immune system. Correspondingly, each of the two major classes of T lymphocytes is adapted to recognize the appropriate peptide–MHC complex: cytotoxic T cells recognize peptides bound to MHC class I molecules (see Figure 4-2a) and helper T cells recognize peptides bound to MHC class II molecules (see Figure 4-2b). The differential recognition of MHC class I and MHC class II by the two T cell subclasses is ensured in part by the *coreceptors* CD4 and CD8. CD4 is expressed on helper cells, which are also known as CD4 T cells, and binds to invariant parts of the MHC class II molecule; and CD8 is expressed on cytotoxic T cells, which are also known as CD8 T cells, and binds to invariant parts of the MHC class I molecule. CD4 and CD8, as well as focusing cytotoxic and helper cells on the MHC molecules appropriate to their effector actions, collaborate with the T cell antigen receptor in signaling the T cell: we describe the specialized part played by these molecules in detail in Chapter 5.

Dendritic cells provide a bridge between innate and adaptive immunity

The chief advantage of the adaptive immune system is the flexibility of antigen recognition conferred by its variable receptors. However unlike the receptors of the innate immune system, the antigen receptors of lymphocytes, as a direct consequence of their versatility, do not distinguish molecules associated with microbial invaders from any other molecules. Constraining the effector actions of lymphocytes to invading microorganisms is the critical function of dendritic cells, which are required to activate naïve T cells and launch their differentiation into effector cells at the initiation of an adaptive immune response. Dendritic cells are phagocytic cells localized in body tissues and especially abundant in the mucosa, where, like macrophages and neutrophils, they internalize microorganisms through phagocytic receptors. They are also stimulated through TLRs recognizing conserved components of the microorganisms both to upregulate expression of MHC molecules of both classes and to express specialized molecules called *costimulatory molecules* that are necessary to activate the proliferation of naïve T cells. In this way, the presentation of antigen to the variable receptors of T cells is coupled to activation by molecules signaling infection. We begin this chapter by describing how dendritic cells perform this crucial bridging function.

(a) Cytotoxicity

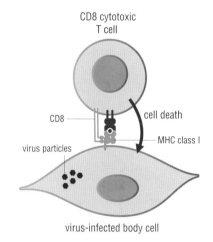

(b) Help

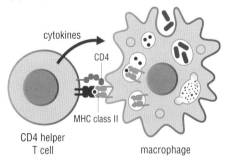

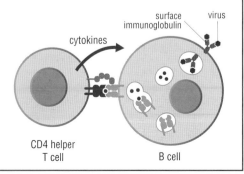

Figure 4-2 The main functions of T lymphocytes (a) Cytotoxic T cells recognize peptide antigens generated in the cytosol (not shown) from viruses, as illustrated here, or cytosolic bacteria, and presented by MHC class I molecules on the surface of infected cells: this activates the cytotoxic T cell to kill the infected cell, limiting the replication of the infected agent. **(b)** Helper T cells recognize peptide fragments generated from internalized microorganisms in phagocytes, or internalized antigen in B cells, and are thereby activated to secrete cytokines and activate these cells in turn to kill the microorganisms or to secrete antibodies, respectively.

4-1 Dendritic Cells in the Induction of Adaptive Immunity

The function of dendritic cells is to present antigen to naïve T cells

Dendritic cells are specialized to capture antigen in the peripheral tissues, process it for display on MHC molecules, and then migrate to secondary lymphoid organs where the displayed antigen fragments can be recognized by circulating naïve T cells. During their residence in the peripheral tissues, dendritic cells express a variety of phagocytic receptors that allow them to internalize both infectious and harmless particles from their environment, but they display only modest amounts of MHC class I molecules on their surface and almost no MHC class II molecules. At this stage they are known as **immature dendritic cells**. After a few days in the tissues, or when activated by signals from local infection, they undergo transformation into **mature dendritic cells**: their phagocytic activities cease or are greatly reduced, MHC class II molecules appear on their surface, MHC class I expression increases, and they adopt the characteristic branched morphology of the mature cell (see section 1-2). They are now specialized to present the antigen they have internalized in the periphery, and this transformation is accompanied by migration of the dendritic cells to the lymph nodes where they can engage with recirculating naïve T cells. Dendritic cells are thought to be the only cells that can effectively activate naïve T cells, and special adaptations of the properties they share with other phagocytes equip them for this role.

Immature dendritic cells comprehensively sample their tissue environment

Immature dendritic cells are extremely efficient phagocytic cells. Like neutrophils and macrophages, they can internalize microorganisms and cellular debris by receptor-mediated phagocytosis and endocytosis, as well by macropinocytosis, which is more indiscriminate (see section 3-7), thus ensuring capture of particles and soluble debris that may escape receptor-mediated internalization (Figure 4-3, top panel). A fraction of these particles will eventually be loaded onto MHC class II molecules in the endosomal compartments of the cell for transport to the plasma membrane; but in immature dendritic cells—which are also often known as **tissue-resident** or **interstitial dendritic cells**—the MHC class II molecules are held for days in these internal compartments and few reach the cell surface, while those that do are rapidly recycled. Tissue-resident dendritic cells express the chemokine receptors CCR1, CCR2 and CCR5, which recognize inflammatory chemokines released in infected tissues (see section 2-13), to which they are thereby attracted and where they internalize infectious particles or fragments derived from them.

Immature dendritic cells may remain in the tissues for some days (in some cases weeks) before undergoing maturation. With maturation, innate receptors and phagocytic activities are downregulated; the trapped MHC class II molecules are released to transport their cargo of peptides to the cell surface; and receptors for inflammatory chemokines are replaced by CCR7, which directs the migration of the dendritic cell to the T cell zones of the lymph nodes (Figure 4-3, bottom panel). Maturation of immature dendritic cells and migration to the lymph nodes occurs continuously, irrespective of the presence of infection. Dendritic cells that have matured in the absence of infection display fragments derived from cellular debris that are recognized by circulating T cells and are thought to play a part in the maintenance of tolerance, as we shall see later in the book. Such cells are said to have undergone *spontaneous maturation*, and do not activate effector differentiation of naïve T cells. This requires more dramatic changes in the dendritic cell that occur in an inflammatory environment on recognition of microbial components.

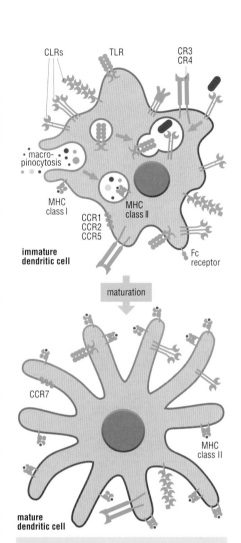

Figure 4-3 Distinctive features of immature and mature dendritic cells Immature dendritic cells (top) express a broad range of Toll-like receptors (TLRs) and C-type lectin receptors (CLRs) recognizing conserved microbial components and other debris, as well as complement and Fc receptors, and are highly active phagocytes, sampling their tissue environment through receptor-mediated endocytosis and phagocytosis, and through macropinocytosis, which enables them to engulf debris indiscriminately. They express receptors for inflammatory chemokines (CCR1, CCR2 and CCR5) which guide them to sites of infection. MHC class II molecules are synthesized but largely retained inside the cell or rapidly recycled from the cell surface. Maturation (bottom) is accompanied by the downregulation of receptors for microbial constituents and inflammatory chemokines, and of macropinocytosis; increased expression of peptide-loaded MHC class II molecules to the cell surface; increased expression of MHC class I molecules; and upregulation of CCR7, which directs the cells to secondary lymphoid tissues where they display peptide–MHC complexes to T cells.

©2007 New Science Press Ltd

Recognition of infection licenses dendritic cells to activate the effector differentiation of naïve T cells

Like macrophages, dendritic cells display many C-type lectin receptors (CLRs), as well as Toll-like receptors (TLRs) and receptors for inflammatory cytokines. The CLRs are thought to function chiefly in adhesive interactions and the internalization of microbial components and debris, although some also have signaling functions. The primary function of TLRs by contrast is to signal recognition of microorganisms, and they are critical in activating dendritic cells to drive the effector differentiation of T cells.

At a site of infection, TLRs of the dendritic cell directly signal the presence of a microorganism, while local epithelial cells and macrophages, which also express TLRs, are induced to produce cytokines, particularly IL-1 and TNF, which act on the dendritic cell and augment the signal from the TLRs. In viral infections, type 1 interferons produced by infected cells and plasmacytoid dendritic cells are an important signal; and for some microorganisms, signals from CLRs also play a part.

These signals from microbes and the cytokines they induce first of all bring about cytoskeletal changes in the dendritic cell resulting in a sharp but transient upregulation of phagocytic activity that ensures the internalization of the infectious agent for subsequent presentation to T lymphocytes, before phagocytic activity is shut down. The dendritic cell then enters a greatly accelerated and amplified maturation program, with the appearance of large numbers of MHC molecules of both classes on the cell surface and migration to the lymph nodes. Microbial signals also upregulate the surface expression of the **costimulatory molecules** B7-1 (CD80) and B7-2 (CD86), two closely related molecules that bind to a receptor called CD28 which collaborates with the T cell antigen receptor in the activation of naïve T cells. Naïve T cells cannot be activated by recognition of peptide–MHC complexes alone. The critical signals leading to T cell activation are illustrated in Figure 4-4.

When the critical role of dendritic cells in T cell activation was first recognized, maturation was thought to occur only in the presence of infection, and the term mature dendritic cell thus meant a dendritic cell capable of activating a naïve T cell, a sense in which it is still widely used. In this book however we shall use the term mature dendritic cell to include cells that have matured spontaneously, and we shall designate those that have matured in the presence of infection *mature activated dendritic cells*. It is thought that several factors conspire to equip dendritic cells uniquely to activate naïve T cells. Both macrophages and B cells also express high levels of MHC molecules of both classes as well as the B7 costimulatory molecules; but dendritic cells that have matured in the presence of infection have four or five times as much B7 on their surface and because they home to the T cell areas of secondary lymphoid tissues they are strategically placed to interact with T cells, several of which can engage at the same time with their long processes. They also deliver signals that are important in inducing effector differentiation of T cells.

Mature activated dendritic cells are distinguished not only by high surface levels of activating B7 and peptide-loaded MHC molecules, but by profound changes in the pathways for processing and presenting antigen. In order to understand the specialized properties and importance of these pathways, we must first examine the fundamental processes by which peptides are generated and loaded onto MHC molecules, beginning with the structural and functional properties of the MHC molecules themselves.

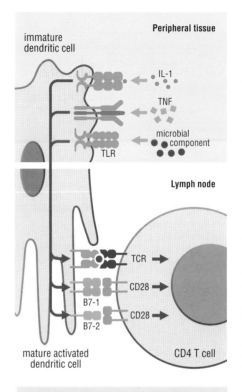

Figure 4-4 Critical signals leading to the activation of naïve T cells by dendritic cells Inflammatory signals from IL-1 and TNF, and microbial signals delivered through Toll-like receptors in the periphery (upper half of figure) induce the cell-surface expression of peptide-loaded MHC class II molecules and the immunoglobulin superfamily costimulatory molecules B7-1 and B7-2 on dendritic cells. Recognition of B7 molecules by the costimulatory receptor CD28 on the T cell is essential for activation of naïve T cells recognizing peptide–MHC complexes on the same dendritic cell in the lymph node (lower half of figure). MHC class I molecules on the dendritic cell are also upregulated (not shown).

enable them to internalize microbial and cellular components in tissues, and are positioned in T cell zones in secondary lymphoid organs where antigen internalized in the tissues is displayed on their surface for recognition by T cells. Dendritic cells that mature in the presence of infection also express high levels of **costimulatory molecules** and secrete inflammatory cytokines, and are the only cell type to be capable of efficiently activating naïve T cells.

tissue-resident dendritic cells: see **interstitial dendritic cells**.

References

Banchereau, J. and Steinman, R.M.: **Dendritic cells and the control of immunity.** *Nature* 1998, **392**:245–252.

Honda, K. *et al.*: **Selective contributions of IFN-α/β signaling to the maturation of dendritic cells induced by double-stranded RNA or viral infection.** *Proc. Natl Acad. Sci. USA* 2003, **100**:10872–10877.

Iwasaki, A. and Medzhitov, R.: **Toll-like receptor control of the adaptive immune responses.** *Nat. Immunol.* 2004, **5**:987–995.

Mellman, I. and Steinman, R.M.: **Dendritic cells: specialized and regulated antigen processing machines.** *Cell* 2001, **106**:255–258.

Reis e Sousa, C. *et al.*: **In vivo microbial stimulation induces rapid CD40 ligand-independent production of interleukin 12 by dendritic cells and their redistribution to T cell areas.** *J. Exp. Med.* 1997, **186**:1819–1829.

von Andrian, U.H. and Mempel, T.R.: **Homing and cellular traffic in lymph nodes.** *Nat. Rev. Immunol.* 2003, **3**:867–878.

4-2 The Structure and Function of MHC Molecules

Figure 4-5 Structure of MHC class I and class II molecules MHC class I and class II molecules are effectively heterotrimers, each comprising one alpha and one beta chain and a bound peptide. The alpha chain of the MHC class I molecule (dark red) contains both membrane-distal domains (α1 and α2) and one membrane-proximal domain (α3); the beta chain, β2-microglobulin (pale red), is encoded outside the MHC and has no transmembrane section. In MHC class II molecules, the alpha (dark blue) and beta (pale blue) chains have one membrane-distal and one membrane-proximal domain each and both have trans-membrane sections. In both classes of MHC molecules, the membrane-distal domains form the cleft in which peptides bind. **(a)** and **(d)** Structure graphics of MHC class I and class II molecules; the crystal structures from which the transmembrane domains were removed for purposes of crystallization, and these regions are therefore absent from these structures. **(b)** and **(e)** Schematic diagrams of MHC class I and class II molecules in the cell membrane. **(c)** and **(f)** The peptide-binding grooves of the MHC class I and class II molecules viewed from above. Note that in the class II molecule, both chains contribute to the peptide-binding groove but in the class I molecule only the alpha chain contributes. (Panel (a) PDB 1a1m, panel (d) PDB 1bx2)

The major histocompatibility molecules monitor the internal compartments of cells

Recognition of the immunological importance of the **major histocompatibility complex (MHC) molecules** began with research in the aftermath of the Second World War on the rejection of skin grafts by badly burned airmen. Investigations on the basis for the rejection of foreign tissue grafts led to the discovery of a family of molecules so variable that no two unrelated people are likely to have the same combination of variants and that can be distinguished by the immune system, which responds with great ferocity to those it recognizes as non-self. The complex of genes encoding these molecules is known as the *major histocompatibility complex* (histocompatibility means tissue compatibility). It was not until the early 1970s that the physiological role of the MHC molecules became clear, and they were recognized as molecules that display antigen to T lymphocytes, a function now known as **antigen presentation**. Grafts are rejected because (as we shall see in Chapter 14) a non-self peptide–MHC complex can look like a complex of self MHC plus foreign antigen. The variability of the MHC molecules, as we shall see shortly, is essential to their role in defense against infection.

We now know that the MHC molecules are a family of molecules that have evolved to bind to peptide (and occasionally lipid or carbohydrate) fragments of degraded molecules in the internal compartments of the cell and carry them to the cell surface for recognition by T cells. In a normal healthy cell, such fragments are derived from self components and are generally ignored by the immune system. In an infected cell, or a phagocytic cell in the presence of infection, some of the fragments will be derived from the infecting microorganism and recognized as foreign by T cells, which can thereby be activated to combat the infection.

Two major intracellular compartments may contain microorganisms, or fragments generated by their internalization: the cytosol, where viral proteins are synthesized and some important pathogenic bacteria replicate; and the endosomal–lysosomal system, which extracellular

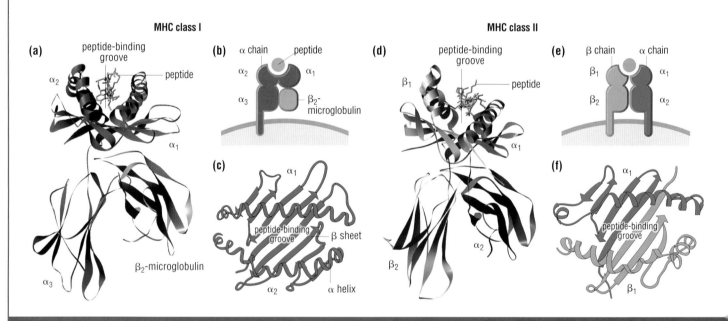

Definitions

antigen presentation: the binding of fragments of intracellular molecules, usually peptides derived from proteins of pathogens, by **major histocompatability complex (MHC) molecules** and their presentation on the cell surface for recognition by T cells.

β2-microglobulin: the smaller of the two chains of the **MHC class I molecule** heterodimer.

major histocompatibility complex (MHC) molecules: cell-surface glycoproteins encoded in the **major histo-** compatibility complex and which bind degraded fragments derived from intracellular proteins and display them on the cell surface.

MHC class I molecules: cell surface glycoproteins most of which are encoded in the **major histocompatibility complex** and most of which bind peptide fragments of cytoplasmic and secreted proteins and display them on the cell surface. An important function of these molecules in immunity is to signal the presence of viral infection to CD8 T cells.

MHC class II molecules: cell surface glycoproteins encoded in the **major histocompatibility complex** and most of which bind peptide fragments of proteins derived from internalized molecules, including those of extracellular pathogens, and display them for recognition by CD4 T cells.

peptide-binding groove: cleft on the membrane-distal surface of **major histocompatibility complex (MHC) molecules** that binds to fragments of intracellular or internalized extracellular molecules, usually proteins, and in which these fragments are displayed on the cell surface.

microorganisms enter through phagocytosis or endocytosis, and which, as we saw in Chapter 3, some bacteria colonize. These two compartments are monitored by two structural classes of MHC molecules: **MHC class I molecules** are specialized to sample the cytosolic compartment, and **MHC class II molecules** are specialized to sample the endosomal–lysosomal system. The distinct pathways by which antigenic fragments are generated and loaded in these compartments are described in later sections of this chapter. Here we describe the main structural features of the antigen-presenting MHC molecules and the implications for their function.

The two classes of MHC molecules have a common architecture but distinct functions in immunity

Each of the two classes of MHC molecules is a dimer composed of four extracellular domains: two membrane-proximal immunoglobulin domains and two membrane-distal domains. The distinctive feature of both molecules is the structure of the two membrane-distal domains of the molecule, which together form a groove, known as the **peptide-binding groove**, where peptides bind. Despite this general structural similarity, the distribution of the four domains between the two chains of each molecule is distinct.

The main structural features of each of the MHC molecules are illustrated in Figure 4-5. Each comprises an alpha chain and a beta chain, but whereas in MHC class II molecules the alpha and beta chains have one membrane-proximal, one membrane-distal, and one transmembrane domain each (Figure 4-5e), in MHC class I molecules the alpha chain is anchored in the membrane and contains a membrane-proximal and both membrane-distal domains, and the beta chain, which is called β_2-**microglobulin**, is a single domain that is encoded outside the MHC and has no transmembrane region (Figure 4-5b). As schematically indicated in Figure 4-5c and f, this means that whereas both chains of the MHC class II molecule contribute to the peptide-binding groove, the peptide-binding groove of MHC class I molecules is formed from the alpha chain only.

The unique structural feature common to both antigen-presenting MHC molecules is that the peptides they bind are integral to the structure of the molecules, which are unstable in their absence. Each molecule is therefore effectively a trimer of alpha chain, beta chain and peptide, and we shall see later that specialized chaperones are required to supervise their assembly.

Although they are structurally analogous, there are important functional distinctions between the two classes of MHC molecules. The MHC class II molecules monitor the endosomal compartments of cells and are normally expressed only on phagocytes, dendritic cells and B cells, all of which are specialized to internalize extracellular antigen and display it for recognition by CD4 T cells. The MHC class I molecules by contrast are expressed on all body cells except red blood cells, which cannot support the growth of viruses. These MHC molecules monitor the cytosol and secretory compartments of the cell, displaying peptides for recognition by CD8 T cells which destroy infected cells betrayed by the presence of foreign peptides.

To achieve effective surveillance of the internal compartments of cells, MHC molecules on any given cell must be able to bind a wide range of different peptides. The factors contributing to this versatility are discussed in the next section.

References

Garcia, K.C. *et al.*: **Structural basis of T cell recognition.** *Annu. Rev. Immunol.* 1999, **17**:369–397.

Mazumdar, P.M.H.: **History of immunology** in *Fundamental Immunology* 5th ed. Paul, W.E. ed. (Lippincott Williams & Wilkins, Philadelphia, 2003), 23–46.

4-3 The MHC and Polymorphism of MHC Molecules

Figure 4-6 The human and mouse MHC

A simplified diagram of the human MHC on chromosome 6 is shown in the top part of the figure, with the mouse MHC on chromosome 17 below it. Only those class III genes with known immune functions are shown. Note that the organization of the mouse and human MHCs is very similar, except that mouse class I genes have become separated at either end of the MHC. This is believed to reflect a chromosomal rearrangement that occurred after the divergence of mouse and human (rats are like mice). The gene for β2-microglobulin, which is encoded outside the MHC on human chromosome 15 and mouse chromosome 2, and five additional non-classical MHC class I genes—the CD1 genes, which are encoded outside the MHC on human chromosome 1—are not shown here. C4A, C2, Bf: genes for complement components; DN, DM, DO, M, O: non-classical class II genes; DP, DQ, DR, A, E: classical MHC class II genes; HLA-A, HLA-B, HLA-C, K, D, L: classical MHC class I genes; LMP: genes for components of the proteasome, the enzyme complex that degrades proteins into peptides in the cytoplasm; LT: genes for lymphotoxin; MICA, MICB, HLA-E, HLA-G, HLA-F, HFE, Q, T, M: genes for non-classical MHC class I molecules; TAP: genes for the transporter through which peptides from the cytoplasm enter the endoplasmic reticulum; TAPBP: gene for tapasin, a chaperone molecule involved in MHC class I peptide loading; TNF: genes for tumor necrosis factor, an important cytokine.

The major histocompatibility complex contains genes encoding many molecules with different functions in immunity

The **major histocompatibility complex (MHC)** contains at least 128 functional genes, more than 20% of which have functions in immunity, and is the most gene-dense region of the human genome. It is also the region with the most disease associations: most, if not all, autoimmune diseases are associated with genes in the MHC, reflecting the central role of MHC molecules in focusing immune responses. Two of the most important of these diseases, diabetes and rheumatoid arthritis, will be discussed in Chapter 12.

The major histocompatibility complex of human and mouse are schematically illustrated in Figure 4-6. For historical reasons, human MHC molecules are known as *HLA*, for human leukocyte antigen, and mouse as *H-2*, for histocompatibility 2, with the different genes indicated by letters as shown in the figure.

The MHC genes are traditionally divided into three classes: the MHC class I and class II genes, which encode the antigen-presenting MHC molecules; and the class III genes, a miscellany of genes encoding some molecules with important immune functions and others with no known immune function (see Figure 4-6). The gene encoding β2-microglobulin lies outside the MHC, on a different chromosome. All the other genes encoding chains of the class I and class II MHC molecules are present in several different copies within the MHC, and each cell expressing them displays several different MHC molecules. This is important for ensuring sufficient diversity to bind peptides from a wide range of microorganisms, an issue we discuss in more detail below.

The peptide antigen-presenting MHC molecules are known as **classical MHC molecules**. There are also structurally related molecules of both classes that do not function in the presentation of peptide antigens to T cells: these are known as **non-classical MHC molecules**. The non-classical MHC class II molecules (*DM* and *DO* in human) regulate peptide loading onto classical MHC class II molecules and they are discussed in section 4-4. The non-classical MHC class I molecules are more numerous and more diverse, and some are important in activating specialized classes of T cells; they are described in section 4-5, and their functions are discussed in Chapter 8. A notable distinction between classical and non-classical MHC molecules that bears on their different functions is the *polymorphism* of the classical MHC molecules.

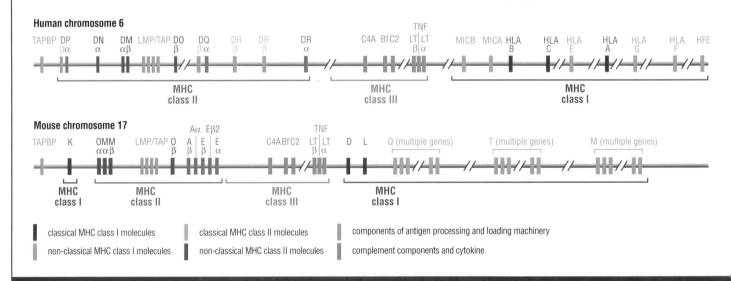

Definitions

allele: genetic variant (of gene).

classical MHC molecules: highly polymorphic cell-surface glycoproteins encoded in the **major histocompatibility complex** and whose function is to present peptide antigens to T cells.

heterozygosity: the possession by an individual of two different **alleles** of a gene.

major histocompatibility complex (MHC): cluster of genes encoding the **classical** and many **non-classical MHC molecules** and other structurally unrelated molecules, many with important functions in immunity.

MHC: see **major histocompatibility complex**.

non-classical MHC molecules: MHC molecules that are relatively nonpolymorphic and may have functions other than the presentation of antigen to T cells.

polygenic: encoded by more than one gene.

polymorphism: difference between individuals in a DNA or protein sequence in which the different sequences are present at a frequency greater than 1% in a population.

The classical MHC molecules are highly polymorphic

Many proteins are present in more than one genetic variant in human populations: the differences between such proteins are known as **polymorphisms**, and most polymorphic proteins are found in two or three variant forms. The classical MHC molecules are present in more than 500 different variants. Polymorphism on this scale is unknown in any other protein, and is believed to reflect selection for resistance to disease under pressure from highly mutable pathogens. Thus the evolution of diversity in peptide binding is driven by the diversity and mutability of the infectious agents that threaten the survival of animals.

Two lines of evidence support the view that MHC polymorphism is driven by the need to maximize peptide binding diversity. First, the polymorphisms are concentrated in the amino acids forming the peptide-binding groove (Figure 4-7). Second, when bound peptides are extracted from MHC molecules it can be seen that although each variant of a given MHC molecule binds a wide range of different peptides, those bound by different variants form distinct sets. It is thought that the clustering of the genes encoding the MHC molecules may contribute to their polymorphism by promoting the possibility of recombination that generates new variants.

We can now see why it is so difficult to find tissue matches for transplant patients. For simplicity, let us take the case of MHC class I molecules, in which only the alpha chain is polymorphic. There are three genes encoding MHC class I alpha chains: that is, the MHC class I alpha chain is **polygenic**, so each individual inherits six alpha chains—three from the mother and three from the father (Figure 4-8a). Because each alpha-chain gene can be any one of 90–500 different variants, or **alleles**, the likelihood of an individual's inheriting identical copies of any of the three MHC class I genes from both parents is very low, and the cells of most individuals are therefore likely to express six different MHC class I molecules (Figure 4-8b). For the same reason, no two individuals are likely to have the same six MHC molecules, and hence the problem of tissue matching. From the point of view of disease resistance, the important effect of MHC molecule polymorphism is to maintain a high probability that each individual will inherit a different allele from each parent: that is to say, the individual will be *heterozygous* for each MHC molecule. A striking example of the advantage of this **heterozygosity** is resistance to HIV, the virus that causes AIDS, which is discussed in Chapter 10.

Although MHC class II polymorphism is more complex than that of MHC class I, the effect is the same. There are three MHC class II molecules (so class II MHC molecules are also polygenic): in humans, these are known as DP, DQ and DR. Both chains contribute to the peptide-binding groove and both can be polymorphic, although the beta chains are for unknown reasons much more polymorphic than the alpha chains. DR is particularly complex. The DR alpha chain is not polymorphic at all but can be coupled with a DR beta chain encoded by any of four different genes, DRβ1, DRβ3, DRβ4 and DRβ5. Normal copies of chromosome 6 contain a DRβ1 gene, of which there are more than 300 alleles, and any one of the three other DRβ genes. Variation of this sort, in which particular genes may be present in one individual and absent from another, is known as *presence–absence polymorphism*.

Despite their general structural similarities (see Figure 4-5), the two classes of MHC molecules bind peptides in distinct ways. This reflects the distinctive processes by which peptides are generated for binding to class I and class II MHC molecules. Before examining these distinctions, however, we digress briefly to describe the non-classical MHC class I molecules.

References

Haldane, J.B.S.: **Disease and evolution.** *Ric. Sci.* 1949, **19 suppl.**:68–76.

Horton, R. *et al.*: **Gene map of the extended MHC.** *Nat. Rev. Genet.* 2004, **5**:889–899.

Margulies, D.H. and McCluskey, J.: **The major histocompatibility complex and its encoded proteins** in *Fundamental Immunology* 5th ed. Paul, W.E. ed. (Lippincott Williams & Wilkins, Philadelphia, 2003), 571–612.

Shiina, T. *et al.*: **An update of the HLA genomic region, locus information and disease associations: 2004.** *Tissue Antigens* 2004, **64**:631–649.

The MHC Sequencing Consortium: **Complete sequence and gene map of a human major histocompatibility complex.** *Nature* 1999, **401**:921–923.

Trowsdale, J. and Parham, P.: **Defense strategies and immunity-related genes.** *Eur. J. Immunol.* 2004, **34**:7–17.

(a) MHC class I

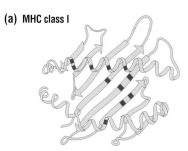

(b) MHC class II

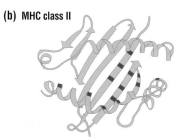

Figure 4-7 Distribution of polymorphic amino acids in the peptide-binding grooves of classical MHC molecules Polymorphic residues are indicated as red bands on the peptide-binding groove of **(a)** an MHC class I molecule and **(b)** an MHC class II molecule.

(a)

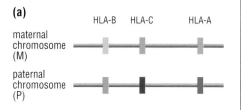

(b)

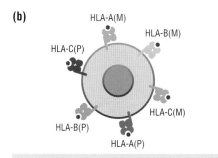

Figure 4-8 Effect of the polygenic and polymorphic character of MHC molecules (a) Three different genes encode the alpha chain of the MHC class I molecule (the beta chain, β2 microglobulin, is not polymorphic and does not contribute to the peptide-binding groove). Each individual inherits one copy of each of these genes from the father (P: dark blue, dark green and dark red) and one copy of each from the mother (M: light blue, light green and light red). Because there are many alleles of each gene at relatively high frequency in the human population, the likelihood of inheriting the same allele of any one of them from both parents is low. Therefore **(b)** most individuals have six different MHC class I molecules on their cells.

4-4 Non-Classical MHC Class I Molecules

The non-classical MHC class I molecules are a diverse group of structural variants on the classical MHC class I molecules

Figure 4-9 Table of non-classical MHC class I molecules CD1: cluster of differentiation 1 (see Chapter 1); DN (double-negative): T cells expressing neither CD4 nor CD8; EPCR: endothelial cell protein C receptor; FcRn: Fc immunoglobulin receptor; H2-M3: product of a gene within the mouse MHC; H60: a protein discovered as a minor histocompatibility antigen; HFE (formerly designated HLA-H): the product of a gene associated with hereditary hemochromatosis; HLA-E: product of one of the class I genes within the human MHC; HLA-G: product of a gene within the human MHC; MAIT: mucosal-associated invariant T cells; MICA and B: products of MHC class I chain-related (MIC) genes in the human MHC; MR1: products of MHC class I-related genes; Qa-1: functional homolog of HLA-E: with T10 and T22, product of a group of genes of largely unknown function designated T in the mouse MHC; RAE-1 (retinoic-acid early inducible 1): the products of genes induced by retinoic acid; T3/T18(TL): non-classical mouse MHC class I molecule; ULBP (UL16-binding protein): proteins with homology to RAE-1 that bind to products of cytomegalovirus; ZAG (Zn-α2-glycoprotein): a soluble protein implicated in lipid metabolism.

The functions and biological significance of most of the non-classical MHC class I molecules are, by contrast with the clearly delineated and central functions of the classical molecules, still relatively ill defined. Unlike the classical MHC class I molecules, the non-classical molecules do not belong to a single group of structurally and functionally homologous proteins, and the term loosely embraces several families of proteins that were discovered by various routes as the cell-surface products of genes with varying degrees of sequence and structural homology to the classical molecules. Some of them do not seem to be conserved between mouse and human: some mouse molecules seem to have no human homolog, and vice versa; and in some cases the same function seems to be performed by different non-classical MHC class I molecules in the two species. All of the non-classical MHC class I molecules have structural modifications, in some cases profound, that affect the cargo (if any) that they carry to the cell surface: the peptide-binding groove may be modified or effectively absent, some are not stabilized by β_2-microglobulin, and some are anchored to the membrane by glycosylphosphatidylinositol tails instead of transmembrane protein segments. Most have a specialized tissue distribution. They were originally distinguished from the classical MHC class I molecules by their lack of polymorphism, and although this distinction has been eroded by the discovery that some of them are quite polymorphic, the functional significance of this polymorphism is unclear since the receptors that are thought to recognize them are invariant. The known non-classical MHC class I molecules of mouse and human are listed, with a summary of what is known about them, in Figure 4-9. A few have established functions, not necessarily immune. Most of the rest are implicated in presenting non-peptide antigens to specialized classes of T cells, or in signaling tissue damage or abnormality, chiefly to natural killer cells or specialized classes of T cells. Because it is thought that they may stimulate immune responses at stages intermediate between innate and adaptive immunity, and in some cases to tumor cells, there is considerable interest in establishing the functions of these molecules. The best-studied of the non-classical MHC class I molecules are briefly described below. The specialized cells that recognize them will be discussed in Chapter 8.

Non-Classical MHC Class I Molecules

Non-classical MHC class I molecule	Functional binding groove	Polymorphism	Cargo	Tissue distribution	Interacting cell	Receptor/ligand	Function
human: CD1a-c	yes	no	lipid	dendritic cells B cells (CD1c)	γδ, DN and CD4 and CD8 αβ T cells	γδ and αβ TCR	presenting mycobacterial and self lipids
human: CD1d mouse: CD1D1 and 2	yes	no	lipid	B cells, monocytes, macrophages, dendritic cells double-positive thymocytes	NK T cells, αβ T cells	αβ TCR	?presenting mycobacterial and self lipids
mouse: H2-M3	yes	low	formylated peptides	ubiquitous	cytotoxic T cells	αβ TCR	presenting bacterial peptides
human: HLA-E mouse: Qa-1	yes	low	MHC class I leader peptide	ubiquitous	NK cells	CD94/NKG2	signaling normal MHC molecule expression
mouse: T10 mouse: T22	no	yes	none	immune cells (inducible)	γδ T cells	γδ TCR	unknown
human: HLA-G	yes	low	peptides	placenta macrophages	NK cells and T cells	ILT2, ILT4	?maternofetal tolerance
human: MICA and B	no	high	none known	epithelial tumor cells gut epithelium (inducible only)	NK cells, γδ T cells CD8 αβ T cells	γδ TCR? NKG2D	stress responses
human: ULBP mouse: RAE-1, H60, MULT1	no	low	none known	general, low-level except in embryonic and tumor cells	NK cells, CD8 αβ T cells, NKT cells, γδ T cells	NKG2D	stress responses NK activation CD8 T cell costimulation
human: EPCR mouse: EPCR	?no	no		endothelium	?neutrophils	activated protein C	anti-coagulant
human: MR1 mouse: MR1	yes	no	unknown	intestinal B cells	MAIT	some αβ TCR	unknown
human: FcRn mouse: FcRn	no	no	IgG	endothelium			IgG scavenging
human: HFE mouse: HFE	no	no	transferrin receptor	gut and liver			regulation of iron uptake
human: ZAG	not known	no	lipid	serum			?lipid metabolism
mouse: M1, M10	yes	no	unknown	vomeronasal organ	vomeronasal tissue		pheromone sensing
mouse: T3/T18(TL)	no	no	none	intestinal epithelium thymocytes, activated T cells, dendritic cells	IEL activated CD8 αβ T cells	CD8 αα	?survival of activated T cells

The peptide-binding grooves of non-classical MHC class I molecules are adapted to different specialized functions

With one exception, all of the non-classical MHC class I molecules described below are thought to engage invariant or relatively invariant receptors of cytotoxic immune cells. The exception is FcRn, also known as the neonatal Fc immunoglobulin receptor, which is expressed on vascular endothelium and maintains immunoglobulin G levels in the bloodstream, as we shall see in Chapter 6. In some species it also transports maternal immunoglobulin across the placenta, or from milk across the gut wall: hence its description as neonatal. Its peptide-binding groove is non-functional: it binds its cargo through contacts with its $\alpha 1$ and β_2-microglobulin domains.

HLA-E of humans and Qa-1 of mice are an example of functionally equivalent non-classical MHC class I molecules that have little sequence homology. Both are non-polymorphic and ubiquitous, and bind the leader peptides that target newly synthesized classical MHC class I molecules to the membrane of the endoplasmic reticulum. They are thought to contribute to surveillance for normal expression of MHC class I molecules on the surface of cells, and thereby help allow the detection of viral infection by natural killer cells, a role discussed in Chapter 8.

H2-M3 of mouse has no known equivalent in humans. It behaves like a classical MHC class I molecule except that its peptide-binding groove is specifically adapted to accommodate N-formylated (f-) peptides. These are characteristic of the amino termini of bacterial and mitochondrial proteins, and the molecule was originally discovered for its ability to present a mitochondrial peptide to cytotoxic T cells. It has since been shown also to present peptides from the cytoplasmic bacterium *Listeria monocytogenes*. Although this suggests a role for H2-M3 in resistance to bacterial infection, its importance is not yet established and mice lacking H2-M3 can survive *Listeria* infections.

The CD1 molecules are encoded outside the MHC; there are four on human chromosome 1 and two on mouse chromosome 3. They fall into two groups, represented by human CD1a–c (group 1), which are absent from mouse; and human CD1d (group 2), which has mouse homologs in CD1D1 and CD1D2. They are expressed on a range of hematopoietic cells (see Figure 4-9) and present mycobacterial glycolipids whose lipid tails bind in their specially adapted peptide-binding grooves (Figure 4-10). Although they are implicated in the response to mycobacteria by this and by other circumstantial evidence, it is not clear how important they are in the response to these infections. The specialized cells that recognize the CD1 molecules, and their significance in immunity, are discussed further in Chapter 8.

The MICA and MICB molecules in humans have no mouse homolog. They have an alpha chain only, with a narrowed groove that is thought to be nonfunctional. These molecules can play a part in activating T cells and natural killer cells against virus-infected or tumor cells through invariant receptors, and a search for a mouse equivalent led to the discovery of three mouse molecules that bind the same invariant receptors. These molecules, RAE-1, H60 and MULT1 (see Figure 4-9), are not encoded in the MHC and have relatively little sequence similarity either to the MIC molecules or to one another, but like the MIC molecules they do not associate with β_2-microglobulin and do not seem to have a functional peptide-binding groove. RAE-1 molecules consist of the $\alpha 1$ and $\alpha 2$ domains only, anchored in the membrane by a glycosylphosphatidylinositol tail. ULBP molecules (see Figure 4-9) seem to be human structural homologs of RAE-1, H60 and MULT1. All of these molecules are induced by injury and stress and are thought to activate specialized T cell subsets discussed in Chapter 8.

(a) Classical MHC Class I

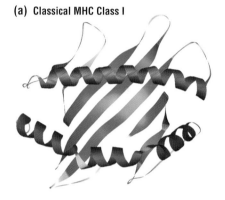

(b) CD1

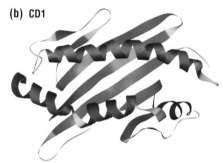

Figure 4-10 Comparison of a peptide-binding groove and a lipid-binding groove The peptide-binding groove of the classical MHC class I molecule **(a)** (PDB 1vac) is broader than the lipid-binding groove of the non-classical CD1 molecule **(b)** (PDB 1cd1). The lipid-binding groove is also deeper and is lined with hydrophobic side chains.

References

Braud, V.M. *et al.*: **Functions of nonclassical MHC and non-MHC-encoded class I molecules.** *Curr. Opin. Immunol.* 1999, **11**:100–108.

Cosman, D. *et al.*: **ULBPs, novel MHC class I-related molecules, bind to CMV glycoprotein UL16 and stimulate NK cytotoxicity through the NKG2D receptor.** *Immunity* 2001, **14**:123–133.

Diefenbach, A. and Raulet, D.H.: **Strategies for target cell recognition by natural killer cells.** *Immunol. Rev.*

2001, **181**:170–184.

Gumperz, J.E. and Brenner, M.B.: **CD1-specific T cells in microbial immunity.** *Curr. Opin. Immunol.* 2001, **13**:471–478.

Liaw, P.C.Y. *et al.*: **Identification of the protein C/activated protein C binding sites on the endothelial cell protein C receptor.** *J. Biol. Chem.* 2001, **276**:8364–8370.

Madakamutil, L.T. *et al.*: **CD8αα-mediated survival and differentiation of CD8 memory T cell precursors.** *Science* 2004, **304**:590–593.

Seaman, M.S. *et al.*: **MHC class Ib-restricted CTL provide protection against primary and secondary** *Listeria monocytogenes* **infection.** *J. Immunol.* 2000, **165**:5192–5201.

Treiner, E. *et al.*: **Selection of evolutionarily conserved mucosal-associated invariant T cells by MR1.** *Nature* 2003, **422**:164–169.

Ugolini, S. and Vivier, E.: **Multifaceted roles of MHC class I and MHC class I-like molecules in T cell activation.** *Nat. Immunol.* 2001, **2**:198–200.

Peptides bind in distinct ways to MHC class I and MHC class II molecules

The peptides that bind to the two classes of MHC molecules are generated in different compartments of the cell by different enzyme systems, as we shall see shortly, and the peptide-binding grooves of the two classes of molecules have distinct properties that are adapted to capture peptides from different sources in very different environments.

Figure 4-11, showing the peptide-binding groove viewed looking down at the top of the molecule, illustrates the most striking difference between the two classes. The peptide-binding groove of the MHC class I molecule is closed by salt bridges at both ends and accommodates only peptides 8–10 amino-acid residues in length, which must fit within the closed groove; the MHC class II peptide-binding groove by contrast is open at the ends, and peptides bound to MHC class II molecules are longer—usually 10–20 residues—and may extend beyond the ends of the groove. These differences in the characteristics of the peptide-binding grooves correspond to differences in the character of the peptides that bind in them and reflect the way in which peptides binding to the two classes of MHC molecules are generated. Peptides that bind to MHC class I molecules are the products of proteases in the cytosol and the endoplasmic reticulum that specifically tailor their degradation products to a uniform length, whereas the peptides that bind MHC class II molecules are the products of a battery of different proteases in the endosomal–lysosomal compartment, where proteins are ultimately degraded to their component amino acids, and the open peptide-binding groove of the MHC class II is adapted to capture intermediate degradation products before they are destroyed.

These differences are also reflected in distinctions in the way in which peptides are anchored in the grooves.

Variable pockets in the peptide-binding groove anchor specific side chains of the peptide in the groove

In both MHC class I and MHC class II molecules, peptides are bound in the grooves by both specific and invariant contacts. The specific contacts are determined by pockets in the floor of the groove into which side chains of the amino acids of the peptide can fit (Figure 4-12). The amino acids whose side chains fit in the pockets are known as **anchor residues**. MHC class I molecules also make invariant contacts with the amino and carboxyl groups at the ends of the

(a) MHC class I

(b) MHC class II

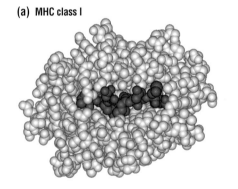
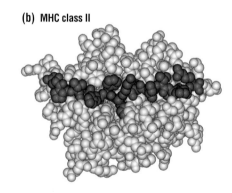

Figure 4-11 Peptide-binding groove with bound peptide Space-filling diagrams showing how a peptide fits in the peptide-binding groove of **(a)** (PDB 1a1m) an MHC class I molecule and **(b)** (PDB 1bx2) an MHC class II molecule. The view shown in both cases is of the surface seen when looking down at the cell surface. Note that the peptide fits inside the groove of the MHC class I molecule, whereas it extends beyond the ends of the groove of the MHC class II molecule.

Definitions

anchor residue: an amino acid in an antigenic peptide whose side chain binds in a pocket formed by specific amino-acid side chains in the peptide-binding groove of an MHC molecule. The composition of the pocket can vary in different polymorphic variants of the MHC molecule.

short peptides that they bind, and these are particularly important in anchoring the peptide, which may sometimes be accommodated in the closed groove by bulging up from the groove in the middle in the case of longer peptides. Peptides bound to MHC class II molecules by contrast lie flat in the groove and the invariant contacts in this case are between the peptide backbone and the floor of the peptide-binding groove.

The differences in binding properties between the different polymorphic variants of MHC molecules are conferred by differences in the binding pockets in the groove. MHC class I molecules have up to six of these; MHC class II molecules may have more. It is thought that polymorphic variants are generated in some cases by recombination events that result in the exchange of some pockets between different MHC molecules: recombination is more efficient than random mutation because it generates new combinations of structural features already selected to bind anchor residues and is less likely to result in destabilization of the molecule.

The anchor residues that fit into the pockets are often identifiable as *motifs*, defined in this context as preferred amino acids at specific positions, in peptides eluted from solubilized MHC molecules (Figure 4-13). Anchor residues are less easily identified in peptides bound to MHC class II molecules than in those bound to class I molecules, because any nine of the amino-acid residues of the peptide—which may be up to 30 residues long—may be the ones that bind in the groove; but motifs can be discerned and often include residues in the middle of the sequence. The identification of motifs that bind MHC molecules is of considerable interest in the design of vaccines based on components of pathogens, which we discuss in Chapter 14.

We now turn to the machinery whereby antigenic peptides are generated and loaded into the peptide-binding grooves of MHC molecules.

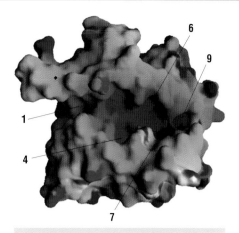

Figure 4-12 Pockets in the peptide-binding groove of an MHC class II molecule A structure graphic showing the major peptide-binding pockets of the human MHC class I molecule HLD-DR1. The pockets are numbered according to Stern, L.J. *et al.*: *Nature* 1994, **368**:215–221 and correspond to the binding sites for specific amino-acid side chains of the peptide. GRASP structure graphic courtesy of David Margulies. (PDB 1dlh)

N-terminus	S Y I G S I N N I	C-terminus
N-terminus	T Y Q R R A L V	C-terminus
N-terminus	G Y K D G N E Y I	C-terminus
N-terminus	R G Y V Y Q G L	C-terminus
N-terminus	H I Y E F P Q L	C-terminus
N-terminus	S I I N F E K L	C-terminus
N-terminus	S G P R K A I A L	C-terminus
N-terminus	S G P E R I L S L	C-terminus
N-terminus	V G P S G K Y F I	C-terminus

Figure 4-13 Peptides eluted from three mouse MHC class I molecules Each group of three represents peptides eluted from a different MHC molecule. The top two peptides are from different alleles of the class I D molecule and the bottom one is from the class I K molecule (see Figure 4-6). The single-letter code for amino acids is used, with the anchor residues in each peptide colored red. Where the anchor residues are not identical, they are chemically similar: for example, L (leucine), I (isoleucine) and V (valine) all have similar hydrophobic side chains, and phenylalanine (F) and tyrosine (Y) both have aromatic rings. The data in the figure are taken from Engelhard, V.H.: *Curr. Opin. Immunol.* 1994, **6**:13–23.

References

Bouvier, M. and Wiley, D.C.: **Importance of peptide amino and carboxyl termini to the stability of MHC class I molecules.** *Science* 1994, **265**:398–402.

Engelhard, V.H.: **Structure of peptides associated with MHC class I molecules.** *Curr. Opin. Immunol.* 1994, **6**:13–23.

Rudensky, A.Y. *et al.*: **Sequence analysis of peptides bound to MHC class II molecules.** *Nature* 1991, **353**:622–627.

Stern, L.J. *et al.*: **Crystal structure of the human class II MHC protein HLA-DR1 complexed with an influenza virus peptide.** *Nature* 1994, **368**:215–221.

Young, A.C.M. *et al.*: **The three-dimensional structure of H-2D^b at 2.4 Å resolution: implications for antigen-determinant selection.** *Cell* 1994, **76**:39–50.

Classical MHC class I molecules display peptides generated in the cytosol

Viruses can infect many cell types in virtually any tissue of the body. They are prevented from spreading by type 1 interferons produced by infected cells, by high-affinity antibodies against the extracellular viral particles, and by CD8 T cells which destroy the cells in which the virus replicates. The complete elimination of viral infections generally requires CD8 T cells, which are also essential for the control of viruses that adopt a latent state and cannot be eliminated: these functions of CD8 T cells are discussed in more detail later in this book. The presence of an intracellular infection is signaled to CD8 T cells by fragments of the infectious microorganisms carried to the cell surface by classical MHC class I molecules, which are present on virtually all body cells.

Although viruses are the largest class of intracellular microorganisms whose presence is signaled by MHC class I molecules, some intracellular bacteria are also flagged in this way. These belong to the relatively small class of bacteria that replicate in the cytosolic compartment of cells; one of the most important of these, *Listeria monocytogenes*, is discussed in Chapter 9. Bacteria that grow in the endosomal vesicles of phagocytic cells, which constitute a larger class, are detected by means of peptides displayed on MHC class II molecules, by a pathway that is described in the next section. The MHC class I pathway is adapted to the cell-surface presentation of peptide fragments generated in the cytosol or the lumen of the endoplasmic reticulum.

Enzyme complexes in the cytosol process proteins into peptides

Most of the peptides that are loaded onto MHC class I molecules are derived from proteins that are degraded in the cytosol by a giant cytosolic enzyme complex, the **proteasome**, which functions in normal protein turnover in all cells. A significant proportion of these peptides are thought to be generated from **defective ribosomal products (DRiPs)**, which are newly synthesized proteins that fail to fold properly, usually because they are defective in some way, and are therefore quickly tagged for degradation. This is believed to enable proteins from cytosolic pathogens to be detected immediately after their synthesis, and not only at the end of their useful lifetime when they have already performed their pathogenic function.

Figure 4-14 Peptide loading and antigen presentation by MHC class I molecules The alpha chain of the MHC class I molecule is synthesized on the endoplasmic reticulum (ER) and translocated through it so that the molecule is anchored in the ER with the α1, α2 and α3 domains extending into the lumen. Before binding to β2-microglobulin and peptide to form a stable complex, the alpha chain associates first with the chaperone protein Bip (not shown) and then with the membrane-bound chaperone calnexin. Peptides are generated from cytosolic proteins by degradation in the proteasome and in some cases also cytosolic aminopeptidases, and transported by the TAP transporter into the lumen of the ER where they may be trimmed by ERAP1. The initial complex of MHC class I alpha chain and calnexin is replaced by a larger complex containing the chaperones calreticulin and ERp57 and a specialized chaperone, tapasin, which associates with the MHC class I molecule and with the TAP transporter. This complex dissociates when a stable complex of MHC class I alpha chain, β2-microglobulin and peptide has formed and is released to migrate through the Golgi complex (omitted for simplicity) to the cell surface. The sequence schematically illustrated represents the events of the human pathway: in mouse, calnexin remains bound to the MHC class I alpha chain when it binds calreticulin. Peptides generated by proteases in the ER lumen can also be loaded onto MHC class I molecules (not shown).

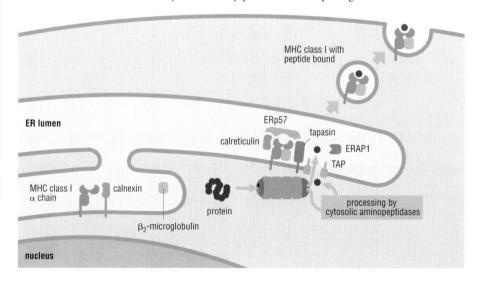

Definitions

calnexin: a membrane-bound general cellular chaperone that associates with MHC class I alpha chains immediately on synthesis.

calreticulin: a soluble general cellular chaperone homologous to **calnexin** that associates with newly synthesized MHC class I alpha chains when calnexin dissociates.

defective ribosomal products (DRiPs): newly synthesized proteins that fail to fold properly, usually because

they are defective, and are quickly degraded by the **proteasome**.

DRiPs: see **defective ribosomal products**.

endoplasmic reticulum aminopeptidase1 (ERAP1): an aminopeptidase that is resident in the endoplasmic reticulum and can trim peptides to 8–10 amino-acid residues long, to fit in the peptide-binding groove of MHC class I molecules.

ERAP1: see **endoplasmic reticulum aminopeptidase1**.

ERp57: a cellular thiol oxidoreductase that is thought to chaperone the formation of the intrachain disulfide bonds in newly synthesized MHC class I alpha chains.

proteasome: large, multisubunit enzyme complex that degrades cytosolic proteins into short peptides.

TAP: see **transporter associated with antigen processing**.

tapasin: a membrane-bound chaperone encoded in the MHC and that binds both to the MHC class I alpha chain and to the **TAP** transporter and facilitates peptide

The proteasome produces peptides many of which are more than 10 amino acids in length and must undergo further trimming before they can fit into the closed groove of the MHC class I molecule. Post-proteasomal trimming can occur in the cytosol, where aminopeptidases can further degrade the peptides, or in the lumen of the endoplasmic reticulum (ER), to which peptides must be transported in order to bind to MHC class I molecules.

A machinery encoded mainly in the MHC exists in the endoplasmic reticulum to load peptides onto MHC class I molecules

The alpha chain of the MHC class I molecule is extruded immediately on synthesis across the membrane of the endoplasmic reticulum, in which it remains anchored with its extracellular domain in the lumen (Figure 4-14) until it is pinched off into vesicles that migrate through the Golgi apparatus to the cell surface. The peptide-binding groove of the molecule is therefore never exposed to the cytosol, where proteins are synthesized and degraded by cytosolic enzyme complexes, and instead the products of these complexes are transported into the lumen of the endoplasmic reticulum by a specialized transporter, the **transporter associated with antigen processing (TAP)**, in the endoplasmic reticulum membrane. Both of the two subunits of TAP are encoded in the MHC (see Figure 4-6). Once in the lumen of the endoplasmic reticulum, peptides can be subjected to final trimming by an aminopeptidase called **ER aminopeptidase1 (ERAP1)**.

The assembly of a stable complex of MHC class I alpha chain, β_2-microglobulin and peptide is critical to the effective detection of infected cells and is an extensively chaperoned process (summarized in Figure 4-14) that begins when the newly synthesized alpha chain, bound to the membrane-bound chaperone **calnexin**, associates with a homologous soluble chaperone known as **calreticulin**, which forms part of a loading complex consisting of the TAP transporter and three chaperones that oversee the binding of β_2-microglobulin and peptide to the MHC class I alpha chain. The other two chaperones in the complex are **ERp57** and **tapasin**. Whereas calnexin and calreticulin are general cellular chaperones, tapasin is a specialized chaperone encoded in the MHC and associated only with the assembly of MHC class I molecules: it binds to both the MHC class I alpha chain and the TAP transporter, stabilizing the MHC molecule and holding it in association with the transporter through which peptides arrive in the endoplasmic reticulum lumen. ERp57 is a cellular thiol oxidoreductase that binds to tapasin and calreticulin and stabilizes the loading complex, where it is also thought to chaperone the formation of the intrachain disulfide bond that stabilizes the peptide-binding pocket of the MHC class I alpha chain. Once a stable complex of MHC class I and peptide has formed, it is released from the chaperones and proceeds to the cell surface. MHC class I molecules that fail to bind peptide are retained in the endoplasmic reticulum and eventually degraded.

The importance of MHC class I molecules in signaling viral infection is attested by the evolution of a remarkable array of viral mechanisms for blocking the MHC class I peptide loading machinery. We shall see in Chapter 10 that herpesviruses in particular have evolved devices for sabotaging almost every step in the process and thus preventing MHC class I expression on infected cells. Because MHC class I molecules are normally expressed on all cells, however, their absence can also be detected as abnormal, and one of the specialized functions of natural killer (NK) cells is the destruction of cells that do not express MHC class I molecules: this important class of cells is discussed in Chapter 8. (Red blood cells do not express MHC class I molecules, as we have mentioned before, and specialized mechanisms exist to protect them from destruction by natural killer cells, a circumstance that is exploited by malaria parasites, which inhabit red blood cells.)

loading onto MHC class I molecules in the lumen of the endoplasmic reticulum.

transporter associated with antigen processing (TAP): a heterodimeric transporter in the endoplasmic reticulum that is encoded in the MHC and transports peptides from the cytosol into the endoplasmic reticulum in an ATP-dependent manner for peptide loading onto MHC class I molecules in the endoplasmic reticulum.

References

Cresswell, P.: **The biochemistry and cell biology of antigen processing** in *Fundamental Immunology* 5th ed. Paul, W.E. ed. (Lippincott Williams & Wilkins, Philadelphia, 2003), 613–629.

Reits, E. *et al.*: **A major role for TPPII in trimming proteasomal degradation products for MHC class I antigen presentation.** *Immunity* 2004, **20**:495–506.

Reits, E.A.J. *et al.*: **The major substrates for TAP *in vivo* are derived from newly synthesized proteins.** *Nature*

2000, **404**:774–778.

Rock, K.L. *et al.*: **Post-proteasomal antigen processing for major histocompatibility class I presentation.** *Nat. Immunol.* 2004, **5**:670–677.

Schubert, U. *et al.*: **Rapid degradation of a large fraction of newly synthesized proteins by proteasomes.** *Nature* 2000, **404**:770–774.

MHC class II molecules display peptides derived from pathogens internalized by specialized immune cells

MHC class II molecules are normally expressed on phagocytes, dendritic cells, B cells and the specialized cells of the thymic epithelium where they participate in the selection of the T cell receptor repertoire (discussed in Chapter 7). Their function is to bind peptides generated from internalized particles in the endosomal–lysosomal systems of these specialized cells and display them for recognition by CD4 helper T cells. In dendritic cells, the processing and display of antigen on MHC class II molecules is subject to specialized regulation, and we discuss the specialized antigen-presenting properties of dendritic cells later. Here we are concerned with the general features of MHC class II antigen processing and presentation in the macrophages and B cells that are the major targets of effector CD4 T cells, which stimulate microbial uptake and destruction by phagocytes (Figure 4-15a) and induce the proliferation and differentiation of B cells to high-affinity antibody-secreting cells (Figure 4-15b).

Antigenic peptides loaded by MHC class II molecules are salvaged from endosomal pathways

Internalized foreign particles and microorganisms delivered to the endosomal–lysosomal systems of B cells and phagocytes are degraded by a battery of enzymes adapted to function in the acidic and reducing environment of the endosomal compartment. Peptide loading by MHC class II molecules therefore requires that the secretory pathway by which they reach the cell surface intersect with the endosomal–lysosomal pathway. Many different enzymes contribute to MHC class II peptide generation, including a large number of **cathepsins**, which are cysteine and aspartyl proteases expressed in all cells, as well as some more specialized cathepsins and other enzymes, many of them induced by interferon-γ, that are expressed only in phagocytes, dendritic cells and B cells (Figure 4-16) and help ensure the generation of a maximal range of potential antigenic peptides. Degradation by these enzymes progresses to destruction with progress through the endocytic pathway, and the open-ended peptide-binding cleft of MHC class II molecules is thought to enable the MHC molecules to capture partly processed antigenic proteins at all stages in the pathway as captured proteins progressively unfold and are degraded, thus ensuring a full sampling of possible antigenic peptides before they are destroyed. The routing of newly synthesized MHC class II molecules into the endosomal pathway, their survival in the hostile endosomal–lysosomal environment and the correct loading of antigenic peptides requires the assistance of a succession of specialized chaperone molecules.

Peptide loading onto MHC class II molecules is chaperoned

The process by which peptides are loaded onto MHC class II molecules is summarized in Figure 4-17. Like the classical MHC class I molecules, newly synthesized MHC class II molecules are unstable in the absence of bound peptide (although they do not require peptide for assembly of the alpha and beta chain into a dimer). This means that their peptide-binding grooves, which project into the endoplasmic reticulum lumen, are exposed to newly synthesized proteins that are not yet fully folded and could become trapped in the open-ended peptide-binding groove. This is prevented in part by a highly specialized chaperone molecule known as the **invariant chain (I chain, or Ii)** which rapidly replaces calnexin, to which the dimer binds transiently to form a nonameric complex containing three I chains and three class II αβ dimers.

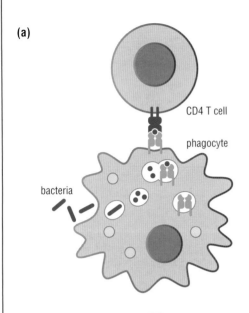

(a)

CD4 T cell

phagocyte

bacteria

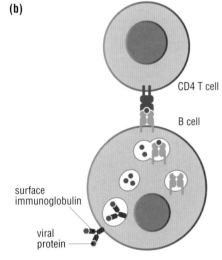

(b)

CD4 T cell

B cell

surface immunoglobulin

viral protein

Figure 4-15 Antigen presentation to helper T cells by phagocytes and B cells (a) Bacteria and other particles ingested by phagocytes are degraded in the specialized lysosomal compartment of these cells and peptides derived from them are loaded onto classical MHC class II molecules which carry the processed antigenic peptides to the cell surface where they are recognized by CD4 T cells. These T cells then activate phagocytes to kill internalized bacteria more efficiently. **(b)** Antigen bound with high affinity to surface immunoglobulin on B cells is internalized by receptor-mediated endocytosis and degraded in the endosomal-lysosomal system, where peptides derived from the intact antigen are loaded onto classical MHC class II molecules and carried to the surface of the B cell. There the peptide–MHC complex is recognized by CD4 T cells, which activate the B cell to proliferate and differentiate into a plasma cell secreting antibodies against the original antigen.

Definitions

cathepsin: any of a family of aspartic and cysteine proteases resident in endocytic vesicles and lysosomes and some of which are implicated in antigen processing for presentation by MHC class II molecules.

class II-associated invariant chain peptide (CLIP): a proteolytic fragment of the **invariant chain** that remains in the peptide-binding groove of newly synthesized MHC class II molecules until they associate with the chaperone **DM**.

CLIP: see **class II-associated invariant chain peptide.**

DM: a non-classical class II MHC molecule of humans that associates with newly synthesized MHC class II molecules after the degradation of the **invariant chain**, resulting in the release of **CLIP** from the peptide-binding groove and chaperoning the loading of peptide. In mouse it is called H-2M.

DO: a non-classical MHC class II molecule of humans that associates with **DM** in B cells and inhibits the chaperone **DM** until the B cell is activated. In mouse it is called H-2O.

The invariant chain not only protects the peptide-binding groove of the MHC class II molecule: it is also responsible for directing the αβ dimers into the endosomal pathway from the secretory pathway after they have passed through the Golgi apparatus from the endoplasmic reticulum. Once in the endosomal compartment the invariant chain is degraded, the process being completed by cathepsin S, which leaves only a fragment, known as **CLIP**, for **class II-associated invariant chain peptide**, in the peptide-binding groove. This complex associates with a third chaperone, **DM**, a non-classical MHC class II molecule encoded in the MHC.

It is thought that classical MHC class II molecules are released from DM only when CLIP is replaced by a peptide with a high enough affinity for the peptide-binding groove to stabilize the molecule, and that this process ensures that the only peptide–MHC class II complexes to reach the cell surface are those that are likely to be long-lived. We shall see in Chapter 5 that the successful activation of CD4 T cells depends upon the stability of such complexes.

In B cells and some dendritic cells, the formation of MHC class II–peptide complexes is doubly chaperoned: a second non-classical MHC class II molecule, **DO**, inhibits DM until the cell is activated, so that loading of MHC class II molecules is deferred until the cell has internalized an activating antigen. Activation of these cells leads to increased acidification of the endosomal compartments containing the MHC class II loading complex, which helps promote peptide loading by activating pH-dependent cathepsins and may also be responsible for inhibiting DO. The regulation of antigen presentation to T cells by B cells is discussed further in Chapter 6. Less specialized mechanisms for regulating antigen presentation are present in most cells, and in the last sections of this chapter we discuss these and the specialized regulation of antigen presentation in dendritic cells.

Enzymes in the Endosomal–Lysosomal Protein Degradation Pathway

Enzyme	Activity	Tissue distribution	IFN-γ inducible
cathepsins B, C, F, H, O, V, Z	cysteine protease	all cells	no
cathepsins D, E	aspartic protease	macrophages, dendritic cells B cells	no
cathepsin L	cysteine protease	thymic epithelial cells	no
cathepsin S	cysteine protease	macrophages dendritic cells B cells	yes
asparaginyl endopeptidase (AEP)	asparaginyl endopeptidase	macrophages dendritic cells B cells	no
IFN-γ inducible lysosomal thiol reductase	thiol reductase	macrophages dendritic cells B cells, IFN-γ inducible in fibroblasts	yes

Figure 4-16 Table of endosomal–lysosomal enzymes implicated in antigen processing for presentation by MHC class II molecules

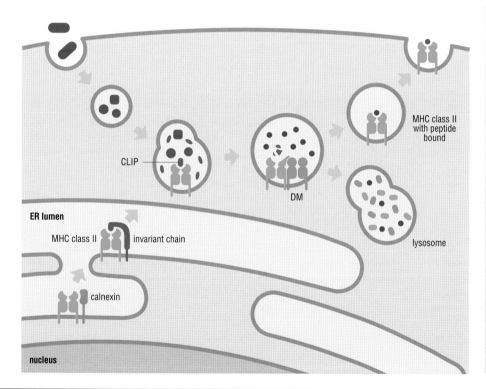

CLIP

DM

MHC class II with peptide bound

ER lumen

MHC class II

invariant chain

lysosome

calnexin

nucleus

Figure 4-17 Peptide loading onto classical MHC class II molecules Immediately on synthesis in the endoplasmic reticulum (ER), the alpha and beta chains of the classical MHC class II molecule dimerize and associate with the chaperone calnexin. Calnexin is rapidly displaced by the invariant chain (I chain, or Ii), which stabilizes the dimer and blocks the peptide-binding groove, preventing it from binding unfolded proteins in the ER. The invariant chain and αβ dimer in fact form nonameric complexes containing three trimers of MHC class II dimer and invariant chain, but for simplicity only one trimer is shown here. Once it has passed through the Golgi apparatus (not shown), the class II molecule is directed by the invariant chain to the endosomal–lysosomal pathway, where the invariant chain is degraded, leaving only a fragment, CLIP, bound in the peptide-binding groove. This is released when the classical class II molecule passes into the care of another chaperone, the non-classical MHC class II molecule DM, which stabilizes it until the peptide-binding groove is filled by a peptide binding with high enough affinity to enable the classical class II molecule to dissociate from DM and proceed to the cell surface.

I chain: see **invariant chain**.

Ii: see **invariant chain**.

invariant chain (I chain or Ii): a membrane-bound protein that associates with MHC class II molecules immediately on synthesis, stabilizing them and filling the peptide-binding groove so that unfolded proteins present in the lumen of the endoplasmic reticulum cannot bind in it.

References

Cresswell, P.: **The biochemistry and cell biology of**

antigen processing in *Fundamental Immunology* 5th ed. Paul, W.E. ed. (Lippincott Williams & Wilkins, Philadelphia, 2003), 613–629.

Glazier, K.S. *et al.*: **Germinal center B cells regulate their capability to present antigen by modulation of HLA-DO.** *J. Exp. Med.* 2002, **195:**1063–1069. Erratum in: *J. Exp. Med.* 2003, **198:**1765.

Lennon-Duménil, A.-M. *et al.*: **A closer look at proteolysis and MHC-class-II-restricted antigen presentation.** *Curr. Opin. Immunol.* 2002, **14:**15–21.

Maric, M. *et al.*: **Defective antigen processing in GILT-free mice.** *Science* 2001, **294:**1361–1365.

Villadangos, J.A. and Ploegh, H.L.: **Proteolysis in MHC class II antigen processing presentation: who's in charge?** *Immunity* 2000, **12:**233–239.

Vogt, A.B. and Kropshofer, H.: **HLA-DM—an endosomal and lysosomal chaperone for the immune system.** *Trends Biochem. Sci.* 1999, **24:**150–154.

Watts, C.: **Antigen processing in the endocytic compartment.** *Curr. Opin. Immunol.* 2001, **13:**26–31.

MHC molecule expression and loading is upregulated by infection

The machinery for generating and loading peptides onto MHC class I molecules is normally present in virtually all cells, while the machinery for generating peptide–MHC class II complexes, as we have seen, is generally confined to the B cells, macrophages and dendritic cells of the immune system. In response to infection, changes occur in the loading and cell-surface expression of both classes of MHC molecules, with consequent increases in their effectiveness in antigen presentation. Increased MHC class I expression makes infected cells better targets for cytotoxic effector T cells, and high cell-surface expression of MHC class II molecules on B cells and macrophages helps ensure recognition by helper effector T cells. The expression and loading of MHC class I and MHC class II molecules are most highly regulated in dendritic cells, whose primary function, as we saw in section 4-1, is to activate the differentiation of naïve T cells into all classes of effector T cells, and which are specialized to display a broad spectrum of the peptide–MHC complexes that T cells are likely to encounter on targets in infected tissues. We shall first describe the regulation of MHC molecule expression on the targets of effector T cells, before turning in the next section to the specialized features of antigen presentation by dendritic cells.

Because neither class of MHC molecule can be expressed on the cell surface until it is stabilized by bound peptide, regulation of MHC expression also entails changes in the expression or activation of the machinery for generating and loading peptides.

MHC class I molecule expression and loading are upregulated by interferons

MHC class I molecules are critical to the detection of viral infection, and their expression is directly upregulated by the type 1 interferons, interferon-α and interferon-β, which are produced by many cells in response to viruses, and coordinate antiviral responses (see section 3-16). Upregulation of the MHC molecules themselves is accompanied by upregulation of peptide loading due to changes in the peptide processing and loading machinery. These are induced by interferon-γ, a pivotal cytokine made only by immune cells and that is produced early in the immune response by natural killer cells and later by T cells and also has important effects on MHC class II expression, as we shall see shortly.

Interferon-γ acts both on the MHC class I loading complex, by increasing the expression of the specialized chaperone tapasin and of the TAP transporter which delivers peptides to the endoplasmic reticulum, and on the production of peptides by the proteasome. Proteasomes occur in two forms: the *constitutive proteasome*, which is present in all cells; and the **immunoproteasome**, which contains specialized subunits that modify both the quantity and the repertoire of peptides generated and that are normally present only in the B cells, macrophages and dendritic cells of the immune system. The specialized subunits and their relationship to the catalytically active site are illustrated schematically in Figure 4-18. The specialized subunits change the cleavage pattern of the proteasome, causing it to produce longer peptides that are thought to be more efficiently transported by the TAP transporter and can be trimmed by cytosolic and endoplasmic reticulum aminopeptidases (also upregulated by interferon-γ) to produce a broader spectrum of antigenic peptides than would be generated from the products of a constitutive proteasome. Interferon-γ induces the production of the specialized subunits in non-immune-system cells, so that in the presence of infection immunoproteasomes are assembled in these cells also, helping to ensure that the repertoire of peptides displayed on potential target cells of effector T cells will match the peptides displayed on the dendritic cells on which the T cells were selected for activation.

Definitions

CIITA: see **class II transactivator**.

class II transactivator (CIITA): a non-DNA-binding transcriptional activator that regulates the expression of MHC class II molecules, both directly by activating the expression of the genes encoding the alpha and beta chains and indirectly by activating transcription of the genes encoding DO, DM and the invariant chain, which are essential for peptide loading.

immunoproteasome: a specialized proteasome that is present in specialized immune cells and induced by interferon-γ in other cells and that processes peptides more effectively than the constitutive proteasome for presentation by MHC class I molecules.

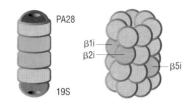

(a) Constitutive proteasome

26S proteasome
20S core proteasome
19S
α
β | catalytically active
β
α
19S

β1
β2
β5

(b) Immunoproteasome

PA28

β1i
β2i
β5i

19S

Figure 4-18 The constitutive proteasome compared with the immunoproteasome (a) The proteasome is a stack of four heptameric rings that form the catalytic chamber of the enzyme complex (the 20S core proteasome), with a regulatory complex at each end. In the immunoproteasome (b), three of the catalytic subunits of the core proteasome (β1, β2 and β5) are exchanged for subunits specific to the immunoproteasome (β1i, β2i and β5i), whereas instead of the 19S regulatory complex, the immunoproteasome is capped at one end of the chamber by a different regulatory complex called PA28.

Immune cytokines modulate the presentation of peptides by MHC class II molecules

MHC class II molecules and the specialized machinery for loading them are under the control of a highly specialized transcriptional activator called **class II transactivator** (**CIITA**) that belongs to the NOD family of proteins (see Figure 3-37) and whose sole function is as a master regulator of MHC class II molecule expression. It is a non-DNA-binding transcriptional regulator that binds to other regulatory proteins at the promoters of the genes encoding the two chains of the MHC class II molecule, the invariant chain, and DO and DM, and activates their transcription.

CIITA is expressed only in the specialized cells of the immune system, and the gene for CIITA is itself regulated by a complex promoter with distinct elements that are active in different cell types. The promoter also includes an element that confers responsiveness to interferon-γ.

The differences in the way in which the expression of CIITA is regulated in different cells of the immune system reflect the distinct functions in immunity of the specialized immune system cells that express MHC class II molecules. B cells, whose primary function is antibody secretion, constitutively express very high levels of MHC class II molecules. They are thereby able to display the antigen internalized by their surface immunoglobulin with high effectiveness to the helper T cells that activate their differentiation into antibody-secreting plasma cells (see Figure 4-15). Macrophages, by contrast, express MHC class II molecules at relatively modest levels until activated by interferon-γ, and never achieve the levels expressed by B cells. This reflects the central functions of macrophages, which have scavenger and phagocytic functions that are independent of T cells and that moreover require the efficient destruction of internalized particles, which may reduce the repertoire of peptides available for loading on MHC class II molecules. In response to interferon-γ, the phagosomes in which antigen is first internalized by the macrophage mature into fully destructive lysosomes, with upregulation of lysosomal proteases (see sections 3-8 and 3-9) that promote rapid degradation, and it is likely that the concomitant upregulation of MHC class II expression through interferon-induced CIITA only partly compensates for the consequent reduction in the quantity of peptides available for loading on MHC class II molecules. In dendritic cells, the sole function of whose phagocytic activities is to enable them to present peptide–MHC complexes to naïve T cells, lysosomal degradation is exquisitely controlled to avoid the premature destruction of antigenic peptides, as we shall see in the next section.

References

Goldberg, A.L. *et al.*: **The importance of the proteasome and subsequent proteolytic steps in the generation of antigenic peptides.** *Mol. Immunol.* 2002, **39**:147–164.

Reith, W. *et al.*: **Regulation of MHC class II gene expression by the class II transactivator.** *Nat. Rev. Immunol.* 2005, **5**:793–806.

The endosomal pathway is highly regulated in dendritic cells

In immature dendritic cells in the absence of infection, the surface expression of MHC class II molecules is kept low in two ways. The first regulatory mechanism is the rapid internalization of those MHC class II molecules that reach the cell surface. The second operates on the loading and release of newly synthesized MHC class II molecules. In the immature cell, protein degradation in the endosomal–lysosomal system is suppressed both by *cystatins*, which inhibit the endosomal proteases, and by downregulation of the proton pumps that regulate the pH of this compartment and generate the acid environment needed to activate these enzymes. This inhibits degradation both of the endosomal–lysosomal contents and of the invariant chain, which holds the MHC class II molecule in the endosomal–lysosomal compartment and protects the peptide-binding groove. Thus in immature dendritic cells, few peptides derived from internalized proteins are generated or loaded, and internalized debris and invariant-chain-bound MHC class II molecules may be held in the endosomal system of the dendritic cell for three or four days.

In the presence of infection, inflammatory cytokines or TLR signaling, or both, induce a burst of phagocytosis that is rapidly followed by shutdown of the phagocytic program, along with the induction of MHC molecules and some endosomal proteases, as well as the activation of the endosomal proton pump. In consequence the pH of the endosomes falls, the invariant chain is degraded, peptides are generated from the recently internalized microbial debris and loaded, and the MHC class II molecules that have been held inside the cell are transported to the cell surface (Figure 4-19). Even in these activating conditions, however, the degradative environment in the endosomal–lysosomal system of dendritic cells is less destructive than in macrophages and allows the survival of a larger repertoire of peptides for loading onto MHC class II molecules.

With maturation of the dendritic cell, MHC class II molecules that reach the cell surface are no longer rapidly endocytosed but can remain on the cell surface for several days; at the same time, synthesis of the transcriptional activator CIITA is shut down, so that production of new

Figure 4-19 Regulation of antigen presentation by MHC class II molecules in dendritic cells In immature dendritic cells (left), antigen is internalized and enters the endosomal–lysosomal pathway but is retained there relatively intact. MHC class II molecules entering the late endosomes are also retained there with invariant chain still protecting their peptide-binding groove. The cell-surface expression of MHC class II molecules in immature dendritic cells is also thought to be regulated by rapid internalization and degradation (not shown). On activation of the dendritic cell by infection, the pH of the endosome rises, lysosmal enzymes are activated leading to the degradation of internalized antigen and invariant chain and to the release of CLIP, peptide is loaded onto the MHC class II molecule, chaperoned by DM, and the peptide–MHC complex travels to the cell surface. The expression of MHC molecules and some endosomal proteases is also increased (not shown).

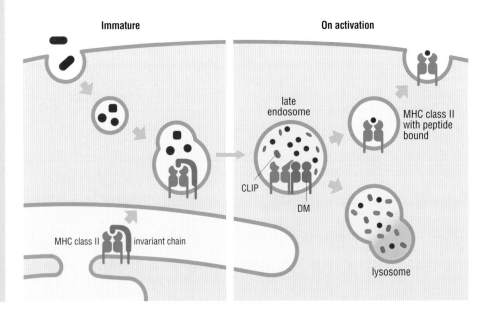

Immature　　　　On activation

late endosome

MHC class II with peptide bound

CLIP

DM

MHC class II　invariant chain

lysosome

Definitions

cross-presentation: the presentation on MHC class I molecules of exogenous peptides derived from proteins internalized by endocytosis, phagocytosis or macropinocytosis.

References

Chow, A. *et al.*: **Dendritic cell maturation triggers retrograde MHC class II transport from lysosomes to the plasma membrane.** *Nature* 2002, **418**:988–994.

Landmann, S. *et al.*: **Maturation of dendritic cells is accompanied by rapid transcriptional silencing of class II transactivator (CIITA) expression.** *J. Exp. Med.* 2001, **194**:379–391.

Shen, L. *et al.*: **Important role of cathepsin S in generating peptides for TAP-independent MHC class I crosspresentation in vivo.** *Immunity* 2004, **21**:155–165.

Trombetta, E.S. and Mellman, I.: **Cell biology of antigen processing in vitro and in vivo.** *Annu. Rev. Immunol.* 2005, **23**:975–1028.

Trombetta, E.S. *et al.*: **Activation of lysosomal function during dendritic cell maturation.** *Science* 2003, **299**:1400–1403.

Villadangos, J.A. *et al.*: **Control of MHC class II antigen presentation in dendritic cells: a balance between creative and destructive forces.** *Immunol. Rev.* 2005, **207**:191–205.

MHC class II molecules ceases. In consequence, the peptides displayed on MHC class II molecules on the dendritic cell surface when it engages with naïve T cells in the secondary lymphoid organs are thought largely to represent a snapshot of the antigenic environment at the time of its activation in the periphery.

Dendritic cells present peptides derived from exogenous proteins on MHC class I molecules to activate naïve CD8 T cells

In the secondary lymphoid tissues, those naïve T cells recognizing antigenic peptides in association with MHC molecules on the dendritic cell are selected for proliferation and differentiation into effector cells that will act on targets bearing the same peptide–MHC complex. This means that dendritic cells must generate the same peptide–MHC complexes that the T cells will encounter in the peripheral tissues. It is clear how this is achieved for CD4 helper T cells, which recognize peptide in association with MHC class II molecules, because both dendritic cells and CD4 T cell targets deliver internalized proteins to the endosomal–lysosomal system. But it is less obvious how dendritic cells activate CD8 cytotoxic T cells, which recognize peptides derived from proteins synthesized in the cytosol by parasites such as viruses and presented on MHC class I molecules. Many viruses do not infect dendritic cells, and although viral components from dead and disintegrated cells can be internalized by these cells during their phagocytic phase, we should expect these to be degraded in the endosomal–lysosomal pathway, where MHC class II and not MHC class I molecules are loaded. Dendritic cells circumvent this constraint by a process known as **cross-presentation**, in which, by a rerouting mechanism that is poorly understood, peptides derived from endosomal contents are loaded onto MHC class I molecules.

There are two main pathways through which cross-presentation is thought to occur. The first and more important depends on the proteasome and the TAP transporter: it is thought to involve the transfer of endosomal contents into the cytosol by transporters whose identity is not yet clear, followed by degradation in the proteasome and transport by TAP into the lumen of the endoplasmic reticulum (Figure 4-20). This pathway thus generates peptides with the same characteristics as those generated by the direct pathway of MHC class I presentation, and that will match those displayed on the infected cells the CD8 effector T cell must recognize. The second pathway is a TAP-independent pathway in which some peptides can be shown to be generated by the endosomal protease cathepsin S and that may involve loading of peptides onto endocytosed MHC class I molecules recycling from the cell surface. Cross-presentation can also occur in macrophages, but only dendritic cells consistently and efficiently cross-present exogenous antigens on MHC class I molecules.

We shall see in the next chapter that dendritic cells not only activate naïve T cells by highly efficient presentation of antigen in association with costimulators, but also profoundly influence their effector properties on differentiation.

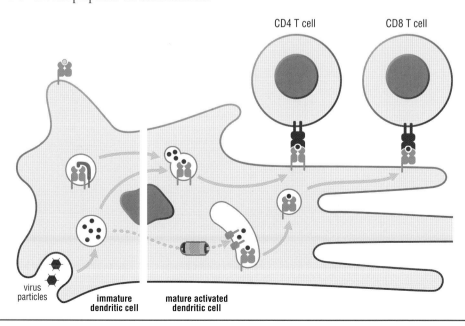

Figure 4-20 Cross-presentation of peptides from internalized proteins by MHC class I molecules Peptides derived from exogenous antigens internalized by immature dendritic cells are transported to the surface of the mature cell both on MHC class II molecules and on MHC class I molecules, which they reach by mechanisms that are incompletely understood but involve proteasomal degradation followed by transport through the TAP transporter into the endoplasmic reticulum, or possibly in some cases recycling of MHC class I molecules from the cell surface through the endosomal pathway. Dendritic cells can thus present antigen derived from the same infectious microorganism to both CD4 and CD8 T cells.

5

Activation and Effector Actions of T Cells

Naïve T cells are activated in secondary lymphoid tissues by recognition of peptide–MHC complexes on the surface of dendritic cells and enter a program of rapid expansion and differentiation to generate effector cells that migrate to B cell follicles, peripheral tissues and epithelial surfaces where they interact with other cells to combat infection. In this chapter, we describe the specialized features of the T cell receptor and signaling pathways that allow sensitive and specific recognition of peptide antigen bound to MHC molecules, and the signals from environmental cytokines and dendritic cells that direct the effector differentiation of naïve T cells to provide a diversity of effector actions adapted to destroy or eliminate different types of infectious agents, and subsequently the emergence of memory T cells that provide lasting protection.

T lymphocytes are central to adaptive immune responses

The functions of T lymphocytes in immunology first began to be understood in the early 1960s with the development of techniques for removing the thymus from newborn mice. The thymus, as we shall see in Chapter 7, is essential for the completion of T cell development, and because most T cell development occurs after birth in mice, removal of the thymus effectively deprives the animals of functional T cells. Thymectomized mice, and humans with **DiGeorge syndrome**, in which the thymus fails to develop, are susceptible to intracellular bacterial, fungal, parasitic and viral infection and cannot make effective antibody responses to many pathogens. This is now known to reflect the part played by T cells in the activation of macrophages and B cells, as well as other immune system cells, and in the destruction of virus-infected cells.

T cells acquire the specialized properties associated with these functions only after their maturation in the thymus and on encounter with antigen on the surface of a dendritic cell. Until then, mature T cells leaving the thymus recirculate continuously through the secondary lymphoid organs, which they enter from the bloodstream, leaving in the efferent lymphatic vessels or blood to rejoin the circulation (see section 1-6). Recirculation is arrested when the naïve T cells recognize antigen in association with costimulatory B7 molecules on the surface of a dendritic cell in the secondary lymphoid tissues. They are then activated to proliferate and differentiate into **effector T cells** equipped on subsequent encounter with the activating antigen rapidly to initiate its destruction or elimination. After differentiation the effector cells leave to migrate to sites of infection, or to the B cell areas of the lymphoid tissue where they activate B lymphocytes, and no longer recirculate through lymphoid tissues.

We begin this chapter with the properties of the variable antigen receptors of T cells that enable them to recognize the variable surfaces presented by peptides bound to polymorphic MHC molecules, and of the intracellular signaling pathways that lead to T cell activation. We then discuss the pivotal events in the secondary lymphoid tissues that lead to the generation of effector T cells, and the role of dendritic cells in directing the differentiation of effector cells with properties appropriate to the infection that initiated the immune response. In the second half of the chapter we describe the functional properties and actions of the different kinds of effector T cells.

T cells are activated by strong and sustained binding to their antigen receptors maintained by adhesive interactions

The antigen receptors of T cells are generated during T cell development by a process involving the assembly of different combinations of distinct gene segments, each present in many copies, that together encode the highly variable polypeptide chains of which the receptor is composed (see section 1-4). It is likely that the genes encoding these receptors have been selected in the course of evolution for recognition of some general structural feature or features of MHC molecules; and we shall see in Chapter 7 that one of the functions of the thymus is to select for maturation those T cells that weakly recognize self-peptide–self-MHC, while eliminating those that bind strongly and thus could be activated against self tissues. In this way, strong binding to a peptide–MHC complex can occur when a foreign peptide bound to a self-MHC molecule increases the contacts between the complex and the T cell receptor, and thus the affinity of the interaction (Figure 5-1). Strong and sustained binding is required for activation of naïve T cells and we shall see that many adhesive and signaling molecules of T cells and dendritic cells contribute to tight and sustained contact that is regulated to tune responses to foreign

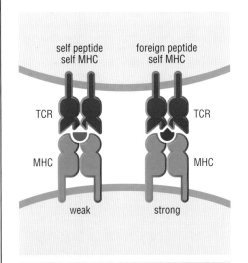

self peptide
self MHC   foreign peptide
self MHC

TCR   TCR

MHC   MHC

weak   strong

Figure 5-1 Recognition of peptide–MHC by the T cell antigen receptor (TCR) T cells are selected during their development in the thymus for weak recognition of self-MHC molecules with self peptides bound: a T cell receptor is schematically shown on the left that makes few contacts with self peptide and thus binds weakly to the complex of self peptide with self MHC. Weak binding is insufficient to activate mature naïve T cells. The replacement of the self peptide with a foreign peptide (right-hand side) can increase the contacts made between the bound peptide and the T cell receptor, thereby increasing binding strength and enabling sustained signaling to occur and the T cell to be activated.

Definitions

DiGeorge syndrome: a genetic disease caused by a deletion in chromosome 22q11 and in which development of several organs is compromised and the thymus fails to develop, leading to T cell immunodeficiency.

effector T cells: T cells that secrete cytokines (in the case of helper T cells) or deliver cytotoxic signals or effector molecules (in the case of cytotoxic T cells) immediately on activation through recognition of peptide–MHC by their antigen receptors.

References

Huseby, E.S. *et al.*: **How the T cell repertoire becomes peptide and MHC specific.** *Cell* 2005, **122**:247–260.

antigens and avoid accidental attack on self. The most important of these regulatory mechanisms is the amplifying signal delivered by the costimulators B7-1 (CD80) and B7-2 (CD86), which signal through CD28 on the T cell and ensure that naïve T cells are normally activated to differentiate into effector cells only in the presence of signals indicating infection.

Most T cell responses depend upon the recruitment or activation of other effector cells

The principal known actions of the T cells of the adaptive immune response, as we have seen, are executed by cells that fall into two major classes: CD8 T cells and CD4 T cells. Of these, the cytotoxic CD8 T cells act primarily by killing infected cells, whereas CD4 T cells act predominantly by the recruitment or activation of other cells through the release of cytokines.

Long-term immunity to most infections depends upon antibodies that are produced by B cells only with T cell help. We discuss the importance of antibodies and the part played by T cells in their induction in detail in Chapter 6, but we shall touch on them in this chapter because distinct functional types of antibodies are induced by different types of T helper cells (see below) and promote the distinct types of immune responses activated by them. The major phagocytic cells, neutrophils and macrophages, are both recruited and activated by CD4 helper T cells. Defense of epithelial surfaces against some important classes of parasites depends upon the activation by helper T cells of eosinophils, basophils and mast cells, which activate expulsive responses to the invaders.

Distinct subtypes of CD4 helper T cells release distinct subsets of cytokines and coordinate different kinds of responses adapted to deal with distinct types of infection. The two best-established helper T cell subtypes are *T$_H$1 cells*, which recruit and activate phagocytic cells in the tissues; and *T$_H$2 cells*, which recruit eosinophils, basophils, mast cells and macrophages in the protection of epithelial barriers. A third helper subset, more recently discovered, seems to be specialized to recruit neutrophils and probably serves to boost the phagocytic response early in an adaptive immune response: these cells are known as *T$_H$17 cells* because they secrete IL-17. Some CD4 T cells contribute to a helper lineage known as *follicular T helper (T$_{FH}$) cells*, which migrate to the B cell follicles of secondary lymphoid tissues and induce the production of antibodies by B cells: these cells include CD4 T cells of the T$_H$1 and T$_H$2 lineages, and possibly also T$_H$17 cells. The generation of the different subsets of T cells is schematically summarized in Figure 5-2.

T cell activation both in the lymph node and in the periphery is regulated by specialized *regulatory T cells (T$_{REG}$ cells)*, which we discuss briefly here because they proliferate in secondary lymphoid tissue as a consequence of the activation of naïve T cells: these cells are of considerable interest because of the potential for manipulating them for therapeutic purposes and will be described in more detail in Chapter 12, where we discuss mechanisms of tolerance.

T cell actions and numbers are controlled by cytokines

The differentiation of the distinct types of T cells, and their recruitment to their sites of action, are controlled by cytokines. At the end of an immune response, the numbers of T cells, which have been expanded by the proliferation of activated naïve cells, return to normal, leaving a legacy of antigen-specific *memory T cells* that provide protection against reinfection. The maintenance of naïve and memory T cell numbers, and proliferation of activated T cells, also all depend upon cytokines, and especially those that signal through the γ_c receptors, so called because they contain a common γ chain through which the signal is transmitted to the cell interior (see section 2-8). T cells constitutively express the common γ chain, but the expression of α and β chains, which determines the cytokine specificity of the response, varies in T cells of different types and at different stages of differentiation, and we shall see in the course of this chapter how T cell survival and proliferation are dynamically controlled by the regulated expression of cytokines, or of the α or β chains of the γ_c receptors, or both.

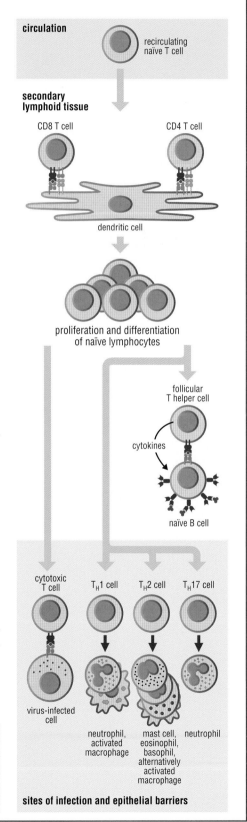

Figure 5-2 Activation and effector functions of T lymphocytes Naïve CD4 and CD8 T lymphocytes circulate through blood and lymph until they encounter the antigen they recognize, with B7 costimulatory molecules, at the surface of a dendritic cell in secondary lymphoid tissue. This stimulates the naïve cells to proliferate and acquire effector functions. CD8 T cells migrate into the tissues as cytotoxic effector cells; CD4 T cells may migrate into lymphoid follicles to activate B cells, or disperse to the periphery as T$_H$1, T$_H$2 or T$_H$17 cells, which coordinate the actions of other cells.

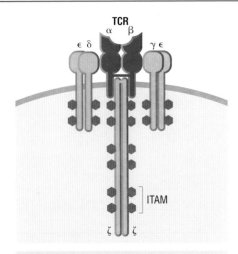

Figure 5-3 Schematic representation of the TCR CD3 complex The TCR antigen-binding αβ heterodimer is associated at the cell membrane with the CD3 complex containing a γ and a δ chain, two ε chains and two ζ chains. Intracellular signaling induced by antigen binding of the TCR is mediated by the ITAMs of the CD3 γ, δ and ε and ζ chains. The signaling complex shown here is that of the αβ receptor: the γδ T cell receptor contains a slightly different combination of signaling chains.

Antigen recognition is mediated by a heterodimeric molecule linked to signaling molecules

The antigen receptors of T cells are adapted to recognize MHC molecules with a cargo of bound peptide and, on binding, to initiate an intracellular signaling cascade that activates the T cell. Each receptor is a complex of eight chains (Figure 5-3). Two of these, the α chain and β chain in the figure, have relatively large and highly variable extracellular domains and form a heterodimer that is responsible for recognition of antigen, but have small cytoplasmic domains with no signaling function. The other six chains are responsible for signaling to the interior of the cell. They are composed of a γ, a δ, two ε and two ζ chains that form three dimers, a γε heterodimer, a δε heterodimer, and a ζ-chain homodimer collectively known as CD3. These chains have cytoplasmic domains containing ITAMs—immunoreceptor tyrosine-based activating motifs—which are sequence motifs that are recognized and phosphorylated by signaling kinases (see section 2-2) and that provide a scaffold for the initiation of a signaling cascade on antigen binding by the αβ heterodimer.

The mechanism whereby antigen binding triggers the intracellular signaling cascade is discussed later in this chapter. In this section and the next, we explain the genetic basis for the variability of the antigen-binding domains of the αβ heterodimer, how the pattern of variation ensures recognition of variability in the peptide–MHC ligand, and how the two coreceptors CD4 and CD8 contribute to T cell recognition of MHC molecules on target cells.

There are two kinds of variable receptors for antigen on T cells

Although all of the established functions of T cells in adaptive immunity are undertaken by T cells bearing variable receptors composed of α and β chains, as shown in Figure 5-3, not all TCRs are αβ heterodimers. Some subsets of T cells express instead an antigen receptor composed of a γ chain and a δ chain (note that these chains form the variable antigen recognition element of the receptor, and should not be confused with the signaling γ and δ chains shown in Figure 5-3). The specialized functions of cells bearing γδ receptors—so-called γδ T cells—are only just beginning to be understood, and will be discussed further in Chapter 8. This chapter is concerned with the functions of αβ T cells and we shall focus largely on the αβ receptor.

The variability of T cell receptors is generated by DNA rearrangements during T cell ontogeny

The receptor whereby T cells recognize antigen is a remarkable feat of natural genetic engineering, surpassed only by the antibody molecule, which is described in Chapter 6. Each chain of the heterodimer is composed of two immunoglobulin domains: an invariant membrane-proximal domain termed the **constant** or **C region**; and a highly variable membrane-distal domain

(a)

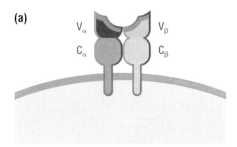

(b)

Figure 5-4 Structure of the αβ TCR heterodimer (a) Schematic representation of the T cell receptor for antigen. There are two chains, the α chain and the β chain, each of which spans the cell membrane and has one membrane-proximal constant region (Cα and Cβ, shown in grey) and one membrane-distal variable domain (Vα and Vβ, shown in red and pink). The hypervariable parts of the variable domains that make contact with the peptide–MHC complex are shown in green. **(b)** Ribbon diagram of the crystal structure of the extracellular domains of the TCR, colored to correspond to the colors in the schematic diagram. The plasma membrane would be below the grey portion of the ribbon diagram. (PDB 1bd2)

Definitions

CDR: see **complementarity-determining region.**

complementarity-determining region (CDR): region of a lymphocyte receptor for antigen that participates in the antigen-binding site and determines its structural complementarity to the antigen.

constant region (C region): region of a lymphocyte receptor for antigen that does not participate in antigen binding and does not vary between cells of different antigen specificities.

C region: see **constant region.**

hypervariable region: region of a lymphocyte receptor for antigen, or the gene encoding it, that contains a large number of amino acids or nucleotides that vary between cells of different antigen specificities and determine the antigen specificity of the cell.

junctional diversity: lymphocyte receptor chain diversity generated by partial digestion of sequences at the junctions of the V, D and J gene segments encoding the **variable region** during their assembly, known as the **variable** or **V region**. Most of the variability is concentrated in three hypervariable loops known as the **hypervariable regions** at the extreme membrane-distal end of the molecule, which makes contact with the peptide–MHC complex; these loops confer on each T cell its distinct specificity for antigen (Figure 5-4): the hypervariable regions are known as CDR1, CDR2 and CDR3, **CDR** standing for **complementarity-determining region**. The T cells of any given individual, taken together, have a repertoire of many millions of different antigen receptors.

The extraordinary diversity of the antigen-binding variable region of the **T cell receptor (TCR)** is achieved by the assembly of the receptor genes during T cell ontogeny from a pool of separate segments; one of these encodes the constant region of each chain, and the variable region is encoded by two or three separate gene segments each of which is present in a number of non-identical copies (Figure 5-5). The separate segments encoding the variable region assemble in different combinations upstream of the constant segment to form the complete gene; this developmentally controlled process is described in Chapter 7.

The DNA encoding the variable region of the α chain is assembled from two segments, a V segment and a J (or joining) segment; the DNA encoding the variable region of the β chain is assembled from three segments—a V and a J segment, and a very short D (for diversity) segment that falls between the V and J segments. The D and J segments encode the third hypervariable region, CDR3. The assembly of lymphocyte receptor genes is called **V(D)J recombination**.

This arrangement contributes four sources of diversity to T cell receptors. First, there are about 42 V segments in the human α pool and 43 in the β pool; the α locus contains about 48 J segments and the β locus about 12 J segments and two D segments. Second, the segments in each pool can assemble in different V-J or V-D-J combinations. Third, although not all α chains are able to associate with every β chain, because of structural constraints that prevent efficient dimerization in some cases, promiscuous dimerization nonetheless makes a substantial additional contribution to receptor diversity. The fourth and largest source of diversity is variation introduced at the junction when two segments join. This variation is known as **junctional diversity** and the mechanism whereby it is generated is discussed in Chapter 7. Its effect is to introduce virtually unlimited variability in CDR3, which, as discussed in the next section, is critical to the recognition of antigenic peptides.

Figure 5-5 Schematic representation of the organization of the gene segments encoding the T cell αβ receptor for antigen The top panel shows the α-chain locus, which has only V, J and C segments, and the bottom panel shows the β-chain locus, which also has D segments. Multiple copies of the variable gene segments are thought to have evolved by duplication in the TCR loci. In the case of the β chain, such an event has led to the duplication of a complete set of D-J-C segments, and a V segment can become joined to the D segment from either set. In both the α and β loci, the *germ-line* DNA is shown—that is, the intact DNA before it has undergone rearrangement. Below each diagram of the germ-line chromo-somal loci for the two chains is shown an example of mRNA derived from separate genomic segments assembled into a coding sequence in T cell DNA. The CDR1 and CDR2 regions are shown in green on the V segments. The mRNA is transcribed from DNA in which the separate segments of the variable region have been joined together (not shown). The process of assembly and expression of mature genes is described in more detail in Chapter 7.

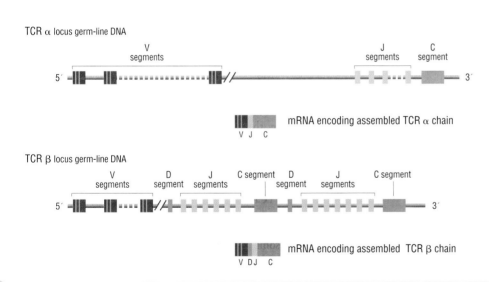

TCR α locus germ-line DNA

V segments   J segments   C segment

5′   //   3′

mRNA encoding assembled TCR α chain
V J C

TCR β locus germ-line DNA

V segments   D segment   J segments   C segment   D segment   J segments   C segment

5′   //   3′

mRNA encoding assembled TCR β chain
V D J C

with insertion of random nucleotide sequences of variable length at the same sites.

T cell receptor (TCR): The complex of variable chains whereby T cells recognize antigen and signaling chains whereby antigen recognition is signaled to the cell interior.

TCR: see **T cell receptor**.

variable region (V region): region of a lymphocyte receptor for antigen that participates in antigen binding and varies between cells of different antigen specificities.

V(D)J recombination: process whereby separate gene segments encoding parts of the chains forming the antigen receptor of a lymphocyte assemble in different combinations in different cells to provide specificity for different antigens.

V region: see **variable region**.

References

Bjorkman, P.J.: **MHC restriction in three dimensions: a view of T cell receptor/ligand interactions.** *Cell* 1997, **89**:167–170.

Davis, M.M. and Bjorkman, P.J.: **T-cell antigen receptor genes and T-cell recognition.** *Nature* 1988, **334**:395–402.

Davis, M.M. and Chien, Y.-H.: **T-cell antigen receptors** in *Fundamental Immunology* 5th ed. Paul, W.E. ed. (Lippincott Williams & Wilkins, Philadelphia, 2003), 227–258.

Glusman, G. *et al.*: **Comparative genomics of the human and mouse T cell receptor loci.** *Immunity* 2001, **15**:337–349.

The mature T cell receptor repertoire is determined by the germ-line pool of gene segments and thymic selection

The theoretical potential T cell receptor repertoire is estimated at 10^{13}, which is at least 10 times the number of T cells in an adult human being. The actual repertoire in any given individual, however, is constrained by three selective processes. Two of these, as we shall see in Chapter 7, occur during T cell development in the thymus. First, all T cells undergo selection in the thymus for weak self-peptide–self-MHC recognition, which is important, as we shall see later, for survival and for T cell activation, as well as for strong recognition of self-MHC plus foreign peptide. Second, to avoid self-reactivity, most cells binding strongly to self-peptide–self-MHC are eliminated in the thymus. The third selective process, which will be the subject of much of the rest of this chapter, is the response to infection. Antigen recognition stimulates vigorous clonal expansion followed at the end of the immune response by the death of most of the responding cells, leaving a small fraction to provide lasting immunity as memory cells. T cell responses are often dominated by relatively few clones of cells, and memory T cells are generally of quite limited diversity: although they may amount to one-third of the total T lymphocytes in an individual, they have been estimated to account for only 1% of total T cell diversity in an adult, with naïve T cells accounting for the rest, and the actual receptor repertoire may be in the region of 2.5×10^7. We shall now examine the features of peptide–MHC recognition by TCRs that determine repertoire selection by infection.

Variable binding orientation and induced fit contribute to the versatility of peptide–MHC binding by αβ TCR

The activation of naïve T cells in the presence of infection requires prolonged signaling from the TCR. This means that the TCR must bind to the peptide–MHC complex with high enough affinity to ensure that signaling is sustained. Structural analysis of 12 TCRs in complexes with MHC and peptide have suggested how the mode of binding of the TCR to the polymorphic surface of the MHC molecule may help to allow this. By contrast with other well understood receptor–ligand interactions, TCRs do not bind to their ligands in fixed orientation, with conserved contacts that contribute most of the binding energy of the interaction, and this is thought to allow accommodation in the face of unpredictably varying peptides. Thus, the TCR can bind at a range of angles to the peptide–MHC complex (Figure 5-6) and with no conserved contacts, which may help to increase the probability of binding with a good enough fit to allow effective signaling.

Of the three hypervariable CDRs, CDR2 of both α and β chains binds predominantly to the MHC molecule, whereas CDR1 and CDR3 bind to both MHC and peptide; thus CDR3, which is encoded by J, or D and J segments and is therefore the most variable part of the TCR V region, contacts the peptide, which is the most variable part of the ligand. The versatility of these binding interactions is increased by the flexibility of the hypervariable loops, which allows them to accommodate themselves to some degree to the surface presented by the peptide–MHC complex. This may play a particularly important part in the case of the CDR3 loop, which often makes the main contacts with the peptide.

The affinity of the interaction between a TCR and a peptide–MHC complex has a profound effect on the response of the T cell. We have seen that strong interactions between the T cell receptor and the peptide–MHC complex are normally required to activate T cells, while weak interactions do not (see section 5-0). In some cases, interactions of intermediate strength can be partly activating—that is, they can activate some functions of T cells, but not all—or

(a)

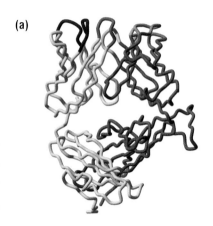

(b)

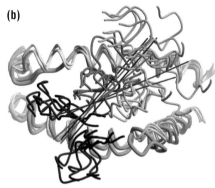

Figure 5-6 Binding orientation of the TCR to MHC molecules (a) The extracellular domains of the αβ TCR are shown for reference, with the CDRs colored as follows: αCDR1, dark blue; αCDR2, purple; αCDR3, green; βCDR1, turquoise; βCDR2, mauve; βCDR3, yellow. (b) Computer graphics derived from crystallographic coordinates have been superimposed for the peptide-binding groove of four MHC class I molecules and two MHC class II molecules binding to six TCRs. Only the complementarity-determining regions of the TCRs are shown: they are numbered for each chain and colored as in panel (a). Note that CDR3 of both chains of the receptor (green and yellow) binds over the middle of the groove where the peptide lies. The binding angles of the TCRs to MHC class I are shown in red, and those to MHC class II in blue. Kindly provided by Markus Rudolph and Ian Wilson. From Rudolph, M.G. and Wilson, I.A.: *Curr. Opin. Immunol.* 2002, **14**:52–65.

Definitions

altered peptide ligands: peptides that when bound to MHC interact with the TCR of a particular T cell with low to intermediate affinity and/or fast dissociation rate, activate the corresponding T cell only partly, or inhibit activation by a peptide–MHC complex that interacts more strongly with the TCR.

coreceptor: (of T lymphocytes) receptor on a T cell that recognizes invariant parts of MHC molecules and forms a recognition complex with the antigen receptor and contributes to intracellular signaling. The coreceptor CD8 binds to MHC class I molecules and is generally expressed on cytotoxic cells; the coreceptor CD4 binds to MHC class II molecules and is generally found on helper cells.

they may even be antagonists and inhibit T cell activation in the presence of higher-affinity peptide–MHC complexes. Peptides with these effects are called **altered peptide ligands**, and we shall see in Chapter 10 that they may be generated naturally during infection with viruses that mutate at a rapid rate, where they may help the virus to erode T cell immunity.

γδ TCRs bind invariant ligands

Unlike αβ T cell receptors, γδ receptors have relatively limited diversity. They do not recognize classical MHC molecules with bound peptide: their ligands are thought to be invariant molecules commonly released by damaged cells, or invariant non-classical MHC class I molecules induced by injury or infection. Consequently, although their germ-line diversity is very high, in practice the γδ receptor repertoire is relatively limited. Structural analysis of a γδ receptor bound to one of the non-classical MHC class I molecules, T22 of mice (see Figure 4-9), has shown a quite different binding mode from that of the classical MHC class I molecule bound to the αβ receptor. Instead of binding to the top of the T22 molecule, the γδ receptor makes contacts chiefly to one side, with the long flexible loop encoded by the D region binding in the non-functional peptide-binding groove of T22. Unlike the D regions in β chains of αβ receptors, which are almost always considerably trimmed during gene rearrangement, the D region of the δ chain of the γδ receptor is often full-length and unchanged from the sequence in the germ-line DNA.

γδ receptors bind with high affinity to their ligands and do not require the cooperation of coreceptors, which are essential to the functioning of αβ receptors.

T cells bind to MHC molecules and signal with the help of coreceptors

Although specific recognition by the αβ TCR is essential for signaling, it is very rarely sufficient for T cell activation. This almost always requires simultaneous binding to the MHC molecule by CD4 or CD8, which are known as **coreceptors** because they make three critical contributions to the response of T cells to MHC molecules bound to foreign peptide.

First, they bind to non-polymorphic regions on MHC molecules, as illustrated schematically in Figure 5-7, and thus contribute to the avidity of binding of T cells to their targets through the TCR: CD4 binds to non-polymorphic regions of MHC class II molecules, and CD8 binds to non-polymorphic regions of MHC class I molecules. Second, because CD4 is expressed on helper T cells and CD8 on cytotoxic T cells, they also focus these cells on antigen presented by the class of MHC molecule appropriate to their function: cytotoxic T cells are specialized for the destruction of infected body cells, which display foreign peptides bound to MHC class I molecules, whereas helper T cells are specialized to activate B cells and macrophages displaying foreign peptide bound to MHC class II molecules. The third critical contribution of the two coreceptors is to T cell signaling on antigen binding: we shall see in the next section that their intracellular domains bind the signaling protein tyrosine kinase Lck and play an essential part in the initiation of the signaling cascade through the TCR.

The CD8 coreceptor of cytotoxic T cells is a heterodimer of two similar chains, CD8α and CD8β. Some dendritic cells and some specialized gut T cells described in Chapter 8 express a homodimer of CD8α chains. The function of the CD8α homodimer on dendritic cells is unknown; on specialized T cells in the gut, as we shall see, it serves as the receptor for the non-classical MHC class I molecule TL (see Figure 4-9).

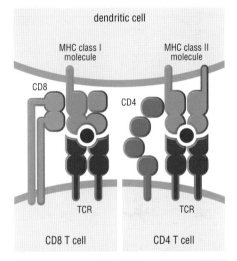

Figure 5-7 T cell coreceptors for antigen
In addition to receptors recognizing a specific complex of MHC molecule and peptide, T cells express coreceptors recognizing invariant features of either the MHC class I molecule or the MHC class II molecule. MHC class I molecules are recognized by the coreceptor CD8, which is a heterodimer containing one α and one β chain (left); MHC class II molecules are recognized by the coreceptor CD4, which is a monomer (right).

References

Adams, E.J. et al.: **Structure of a γδ T cell receptor in complex with the nonclassical MHC T22.** *Science* 2005, **308**:227–231.

Arstila, T.P. et al.: **A direct estimate of the human αβ T cell receptor diversity.** *Science* 1999, **286**:958–961.

Degano, M. et al.: **A functional hot spot for antigen recognition in a superagonist TCR/MHC complex.** *Immunity* 2000, **12**:251–261.

Garcia, K.C. and Adams, E.J.: **How the T cell receptor sees antigens—a structural view.** *Cell* 2005, **122**:333–336.

Hennecke, J. and Wiley, D.C.: **T cell receptor–MHC interactions up close.** *Cell* 2001, **104**:1–4.

Lord, G.M. et al.: **A kinetic differentiation model for the action of altered TCR ligands.** *Immunol. Today* 1999, **20**:33–39.

Nikolich-Žugich, J. et al.: **The many important facets of T-cell repertoire diversity.** *Nat. Rev. Immunol.* 2004,

4:123–132.

Rudolph, M.G. and Wilson, I.A.: **The specificity of the TCR/pMHC interaction.** *Curr. Opin. Immunol.* 2002, **14**:52–65.

Rudolph, M.G. et al.: **How TCRs bind MHCs, peptides, and coreceptors.** *Annu. Rev. Immunol.* 2006, **24**:419–466.

T cells are sensitive to very small amounts of foreign peptide

T cells respond with remarkable sensitivity to the peptide–MHC complexes specifically recognized by their antigen receptors. As few as 1–10 such complexes on the surface of a dendritic cell are sufficient to trigger signaling in a naïve T cell. This is especially remarkable given that there are an estimated 1,000,000 other peptide–MHC complexes on the dendritic cell surface. Two important features of T cell biology are thought to account for this sensitivity. First, although T cells normally respond only to foreign-peptide–MHC complexes recognized with relatively high affinity, all T cells also make weaker interactions with self-peptide–MHC complexes. It is thought that TCRs making such weak interactions with self-peptide–MHC complexes on the surface of dendritic cells associate with TCRs binding to foreign-peptide–MHC complexes and contribute to signaling.

The second factor in the responsiveness of naïve T cells is the contribution of other molecules on the T cell surface to signaling from the TCR. These include adhesion molecules, which stabilize the contact area between the T cell and the dendritic cell, the costimulatory receptor CD28, whose contribution we discuss later, and the coreceptors CD4 and CD8. Because CD4 and CD8, like the TCR, bind to the MHC molecule, receptor and coreceptor are brought together in the membrane and the cytoplasmic tails of these two molecules then cooperate to trigger signaling reactions. The cooperation of the TCR and the coreceptors focuses the T cell on MHC molecules and it is thus central to the ability of T cells to participate in cell-mediated immune responses.

TCR signaling is initiated by phosphorylation of receptor ITAM sequences

TCR signaling is mediated by the immunoreceptor tyrosine-based activation motifs (ITAMs) in the cytoplasmic tails of the CD3 and ζ chains (see Figure 5-3), and in general is similar to ITAM signaling by other Ig superfamily immunoreceptors (see section 2-2), although with some specialized features. One of these is the large number of ITAMs in a TCR, which are thought to amplify the signaling by recruiting multiple kinase molecules when phosphorylated. The ITAM tyrosines are phosphorylated by the protein tyrosine kinase **Lck**, which is constitutively associated with the coreceptors CD4 and CD8. It is attached by lipid groups to the plasma membrane in a specialized region of the membrane called the **lipid raft**, which is rich in glycosphingolipids and cholesterol. Peptide–MHC engagement by the TCR brings the TCR together with its coreceptor, drawing it into the lipid raft where TCR ITAMs are phosphorylated by Lck (Figure 5-8), and the phosphorylated ITAMs serve as binding sites to recruit a second tyrosine kinase, known as **zeta chain associated protein of 70 kDa (ZAP-70)**, which is essential to downstream signaling. ZAP-70 is a cytoplasmic tyrosine kinase with two SH2 domains that bind to the two phosphorylated tyrosines in an ITAM (see Figure 5-8). Once localized to the receptor, ZAP-70 is phosphorylated and activated by Lck.

TCR signaling activates cytoskeletal remodeling and cytokine production

Activation of ZAP-70 is the first step in the assembly of a large protein complex built on a lipid-raft-resident transmembrane scaffold molecule called **linker for activation of T cells (LAT)**. LAT is phosphorylated by ZAP-70 and then assembles a complex containing other adaptor molecules (grey in Figure 5-9) and signaling molecules that generate second messengers and direct cytoskeletal changes promoting prolonged signaling to cytokine production (summarized in Figure 5-9).

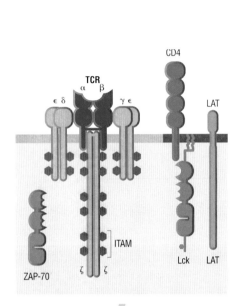

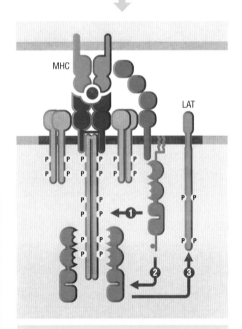

Figure 5-8 Initiation of TCR signaling In the upper panel, the coreceptor, in this case CD4, and the associated kinase Lck are in the lipid raft region of the plasma membrane (red), whereas the TCR is in the other, more fluid portion. When the TCR binds a peptide–MHC complex (lower panel), its ζ and CD3 chains are brought together with CD4 bound to the same MHC molecule and it is drawn into the lipid raft, where Lck phosphorylates ITAMs in the TCR cytoplasmic domains, which thereby recruit the cytoplasmic kinase ZAP-70. ZAP-70 is then phosphorylated and activated by Lck. Note: ZAP-70 may bind to any of the ITAMs in the ζ chain or the CD3 chains. Once activated, ZAP-70 phosphorylates the transmembrane adaptor molecule LAT, which then recruits subsequent signaling components (see Figure 5-9).

Definitions

Itk (IL-2-inducible tyrosine kinase): a TEC-family tyrosine kinase playing a central part in T cell activation through activation of calcium and diacylglycerol signaling.

LAT: see **linker for activation of T cells**.

Lck: a Src-family tyrosine kinase that associates with the coreceptors CD4 and CD8 and plays a central role in TCR signaling.

linker for activation of T cells (LAT): a lipid-raft-localized transmembrane scaffold molecule that plays an essential early part in TCR signaling.

lipid raft: specialized membrane region rich in glycosphingolipids, cholesterol and some membrane proteins. Lipid rafts are dynamic structures in which components are continually exchanged between raft and non-raft regions of the membrane.

phospholipase C (PLC)-γ1: a signaling enzyme that hydrolyzes phosphatidylinositols, generating diacylglycerol and inositol trisphosphate, the latter of which

The receptor-induced assembly of multiprotein complexes serves to localize signaling proteins at the membrane, where many of them act, and to bring together signaling components to facilitate their collaboration. For example, the protein tyrosine kinase **Itk**, a member of a family of kinases called the *TEC kinases*, is recruited to the complex formed by LAT and the other adaptor molecules. Itk is activated by Lck and in turn activates **phospholipase C (PLC)-γ1**, which interacts directly with the membrane. PLC-γ1 then hydrolyzes the membrane-bound lipid phosphatidylinositol 4,5-bisphosphate (PIP_2) to generate the second messengers diacylglycerol (DAG) and inositol trisphosphate (IP_3), leading to activation of cytokine genes.

Localization of the signaling complex adjacent to ligand-bound TCRs is also important for focusing changes in the actin cytoskeleton that are signaled by molecules in the complex and stablilize the area of contact between the T cell and the activating dendritic cell. Actin remodeling is activated by the adaptor *Nck* and a molecule called *Vav*, which is a guanine nucleotide exchange factor important for the activation of Rac, a small GTPase belonging to the Rho family, which plays a part in cytoskeletal changes. Nck recruits **WASP**, which directs the polymerization of new actin filaments. WASP stands for **Wiskott–Aldrich syndrome protein**, so named because it is the protein mutated in **Wiskott–Aldrich syndrome**, an immunodeficiency disease that results in recurrent bacterial, viral and fungal infection reflecting defects in both T and B cell immunity and illustrating the importance of cytoskeletal reorganization at the point of contact established by antigen recognition. We shall see in the next section how this may promote productive signaling.

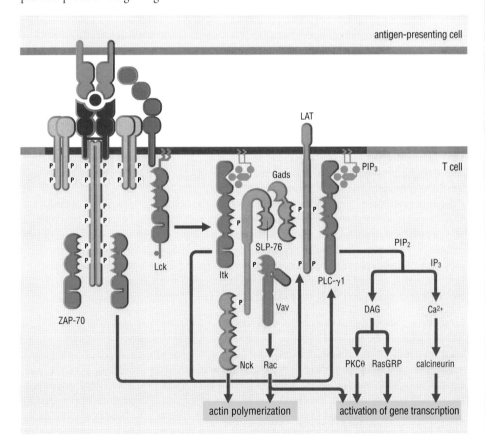

Figure 5-9 Signaling complexes formed during TCR signaling Binding of peptide–MHC ligand to the TCR and coreceptor (CD4 in this case) triggers TCR signaling as described in Figure 5-8. Phosphorylation of the lipid-raft-resident protein LAT leads to assembly of a signaling complex containing two other scaffold/adaptor molecules, SLP-76 and Gads, which recruit phospholipase C-γ1 (PLC-γ1) and Itk in cooperation with specialized domains called *pleckstrin homology (PH) domains*, which bind PIP_3 in the membrane. (SLP-76 and PLC-γ1 also form a direct contact; not shown.) The complex-bound Itk is activated by Lck. Itk and ZAP-70 then phosphorylate and activate PLC-γ1. PLC-γ1 then hydrolyzes a second membrane phosphorylated inositol phospholipid, PIP_2, generating two second messengers, namely IP_3, which causes release of Ca^{2+} from intracellular stores, and diacylglycerol (DAG), which activates protein kinase C (PKC) isoforms, especially PKCθ. Diacylglycerol also stimulates a Ras guanine nucleotide exchange factor, RasGRP, which activates the key signaling molecule Ras. Also present in the LAT–SLP-76 signaling complex are Vav, a guanine nucleotide exchange factor for Rac, and Nck. Activated Rac and Nck direct changes to the actin cytoskeleton that are responsible for plasma membrane movements that promote close contact of the T cell with the dendritic cell; Rac also helps activate cytokine gene transcription via MAP kinases. How these signaling pathways lead to cytokine gene expression is described more fully in section 5-5.

induces elevation of intracellular free calcium.

WASP: see Wiskott–Aldrich syndrome protein.

Wiskott–Aldrich syndrome: X-linked inherited disease in which there are bleeding problems due to platelet defects and also immunodeficiency.

Wiskott–Aldrich syndrome protein (WASP): one of a family of five proteins that promote actin polymerization.

ZAP-70: see zeta chain associated protein of 70 kDa.

zeta chain associated protein of 70 kDa (ZAP-70):

intracellular tyrosine kinase that associates with phosphotyrosines in ITAMs on the ζ chain and CD3 chains of the T cell receptor and is essential in downstream signaling.

References

Acuto, O. and Cantrell, D.: **T cell activation and the cytoskeleton.** *Annu. Rev. Immunol.* 2000, **18**:165–184.

Hořejší, V. *et al.*: **Transmembrane adaptor proteins: organizers of immunoreceptor signalling.** *Nat. Rev. Immunol.* 2004, **4**:603–608.

Huang, Y. and Wange, R.L.: **T cell receptor signaling: beyond complex complexes.** *J. Biol. Chem.* 2004, **279**:28827–28830.

Jordan, M.S. *et al.*: **Adaptors as central mediators of signal transduction in immune cells.** *Nat. Immunol.* 2003, **4**:110–116.

Weiss, A. and Littman, D.R.: **Signal transduction by lymphocyte antigen receptors.** *Cell* 1994, **76**:263–274.

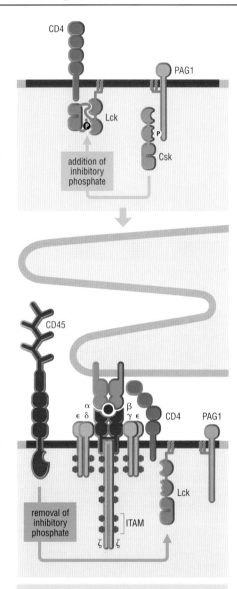

Figure 5-10 Regulation of Lck tyrosine kinase activity by phosphorylation and dephosphorylation Src-family kinases such as Lck have three different states, an inactive state, a partly active state and a fully active state. In the inactive state, the SH2 and SH3 domains of the kinase are clamped to the kinase molecule itself through an inhibitory phosphotyrosine near the carboxyl terminus that binds to the SH2 domain. The inhibitory phosphate is added to the kinase by Csk, which in unstimulated T cells is held in the lipid raft by phosphorylated sites on the lipid-raft-resident scaffold protein PAG1, to which it binds (upper panel). The inhibitory phosphate is removed by the abundant phosphatase CD45, releasing the conformational clamp on the kinase, which is now poised for activation. Activation is facilitated by dephosphorylation of PAG1, which occurs by an unknown mechanism soon after the initiation of TCR signaling and releases Csk (lower panel) so that Lck remains in its unclamped state; clustered Lck molecules (perhaps in TCR microclusters) are then thought to phosphorylate one another at the activating site on the catalytic domain of the kinase for full activation (not shown).

TCR signaling is dynamically regulated to support T cell activation

Productive T cell signaling at early stages after antigen recognition depends on the rapid amplification of the initial signal from the TCR, which must be balanced by mechanisms for ensuring specificity. This balance is maintained in part by events at the cell surface at the initiation of signaling, and in part by properties of the intracellular signaling pathways. Events at the cell surface are regulated at these earliest stages by adhesive interactions. If a small number of antigen receptors on the T cell bind well to peptide–MHC complexes on the dendritic cells, signaling from the TCR induces the high-affinity state of the LFA-1 (αLβ2) integrin on the T cell (see Figure 2-13), allowing it to bind to ICAM-1 on the dendritic cell, allowing a more prolonged sampling of the dendritic cell surface by the TCR. T cell activation is induced only if the presence of antigen is confirmed by continued TCR signaling, helping to ensure specificity for foreign antigens and to avoid responses to transient chance amplification of a weak signal.

Other amplifying mechanisms are thought to include the large numbers of ITAMs in each TCR complex, the cooperativity of coreceptors and their associated Lck with the TCR, and the clustering of signaling components into the lipid raft, as described in the previous section, as well as two additional factors. First, TCRs and their associated coreceptors rapidly associate into micro-clusters of signaling receptors, perhaps including TCRs reacting to self-peptide–MHC complexes; and second, inhibitory events that otherwise restrain TCR signaling are transiently suppressed.

The regulation of Lck provides one example of amplification by suppression of inhibition. Like other Src kinases, Lck exists in three conformational states: an inactive state in which it is clamped by inhibitory phosphorylation of a tyrosine near the carboxyl terminus; a poised but low-activity state in which the inhibitory phosphate has been removed; and a highly active state due to an activating phosphorylation near the active site of the kinase domain. The transition between the inactive and poised states is controlled by the dynamic balance between a kinase called **C-terminal c-Src kinase (Csk)**, which adds the inhibitory phosphate, and the transmembrane protein tyrosine phosphatase **CD45**, an extremely abundant membrane protein, which removes the inhibitory phosphate. In unstimulated T cells, inhibition is favored because Csk is held in the lipid raft region by association with a transmembrane scaffold protein, **PAG1 (for phosphoprotein associated with glycosphingolipid-enriched microdomains)**, and continually phosphorylates and inactivates Lck, which also resides in the raft because of its amino-terminal lipid modification. Csk is tethered to PAG1 by its SH2 domain, which recognizes a tyrosine-phosphorylated site on PAG1 (Figure 5-10). Early after TCR stimulation, PAG1 becomes dephosphorylated (by an unknown mechanism), Csk is released, and inhibition of Lck is thus suspended (see Figure 5-10, bottom panel). Once CD45 has removed the inhibitory phosphate, Lck molecules are in their poised low-activity state. TCR signaling promotes phosphorylation of the activating tyrosine in the Lck kinase domain, resulting in maximal kinase activity. This phosphorylation is thought to result from the action of one Lck molecule on another, and TCR microclusters are likely to promote this process.

Later, when a stable interaction with a single dendritic cell is established, PAG1 is rephospho-rylated by one of the activated signaling molecules (possibly by Lck itself), and Csk is recruited once again into the lipid raft. Csk is in fact found in a complex with a phosphatase, PEP, which removes the activating phosphate from Lck. Thus, addition of the inhibitory phosphate and removal of the activating phosphate both result from recruitment of Csk by PAG1 and provide feedback inhibition of Lck. Feedback inhibition is a common mechanism for countering positive amplification in cell signaling and ensuring that signaling is sustained only in the continued presence of the activating stimulus.

Definitions

Cbl: any of several E3 ubiquitin ligases that bind signaling TCRs or other receptors and target them for ubiquitination and degradation.

CD45: a transmembrane protein tyrosine phos-phatase that is expressed by hematopoietic lineage cells and that dephosphorylates the negative regula-tory site on Src-family tyrosine kinases.

central supramolecular activation complex (c-SMAC): central area of some **immunological synapses** in which TCRs and integrins have segregated into distinct areas.

Csk: see **C-terminal c-Src kinase**.

c-SMAC: see **central supramolecular activation complex**.

C-terminal c-Src kinase (Csk): protein tyrosine kinase that phosphorylates the C-terminal negative regulatory site on Lck and other Src-family tyrosine kinases, thereby inhibiting their activity.

immunological synapse: area of contact between a

A more radical mode of feedback inhibition is contributed by members of a family of ubiquitin ligases, which stop TCR signaling by tagging the receptor complex for destruction. The key ubiquitin ligases in the TCR pathway belong to the **Cbl** family and are recruited to phosphorylated TCR cytoplasmic domains through an adaptor containing an SH2 domain. Once bound to the phosphorylated receptor, the Cbl molecule tags the receptor for internalization and delivery to lysosomes for destruction. This irreversibly terminates signaling by active receptors, so that signaling can continue only if new TCRs become engaged. The importance of this feedback mechanism is illustrated in mice deficient in one of the Cbl-family molecules, Cbl-b, which develop severe autoimmunity. We shall see in Chapter 12 that autoimmunity is a common consequence of the experimental elimination of negative controls on lymphocytes, suggesting that in their absence TCR signaling is not regulated to avoid strong reactions to self.

The contact between T cells and antigen-presenting cells can result in organized arrays of signaling molecules

Once sustained contact is established between a naïve T cell and an antigen-presenting dendritic cell and signaling is initiated, dynamic changes occur in the actin cytoskeleton, promoting a larger area of contact between the two cells where signaling molecules assemble.

The contact zone that forms between the two cells is referred to as the **immunological synapse**, by analogy with the neural synapse, which is the specialized zone of communication between nerve cells. The interactions of the molecules at the immunological synapse can be examined in detail using cultured cells interacting with one another, or with artificial membranes containing key molecules, to give a clearer picture of events at this specialized zone of contact. In these systems, the interacting molecules can be seen to segregate into a central area, called the **central supramolecular activation complex (c-SMAC)** containing the TCR, the coreceptor CD4 or CD8, the costimulatory receptor CD28, the adhesion molecule CD2 and various associated signaling molecules. A ring surrounding the c-SMAC, and therefore called the **peripheral supramolecular activation complex (p-SMAC)**, contains primarily the integrin LFA-1, which is linked to the actin cytoskeleton via a molecule called talin. This organization requires remodeling of the actin cytoskeleton and in part may reflect the differing sizes of the molecules in question: the TCR–MHC interaction is thought to require the membranes of the two cells to be closer to one another than the association between LFA-1 and ICAM-1 (Figure 5-11). Some other surface molecules, including CD45, the phosphatase that, as we saw above, is important for activating Lck, are excluded from the immunological synapse altogether. As this observation implies, signaling by TCRs in the c-SMAC region is poor, reflecting an excess of inhibitory mechanisms over amplifying ones. Although the picture shown in Figure 5-11 appears static, in reality it is highly dynamic. TCR molecules from the c-SMAC are being internalized for degradation in the lysosomes, whereas TCRs toward the edge of the contact area are binding to peptide–MHC ligands and forming highly active signaling microclusters, which are later moved into the c-SMAC as their signaling is attenuated. Thus, this picture of events in the contact zone indicates that signaling continues only if additional TCRs bind to their ligand; and we shall see later that activation of the T cell requires continued signaling for hours.

One of the earliest consequences of T cell activation is the production of the T cell cytokine interleukin 2, which promotes the proliferation of activated naïve T cells. In the next section we shall see how signals from the T cell surface collaborate in the activation of IL-2 gene transcription.

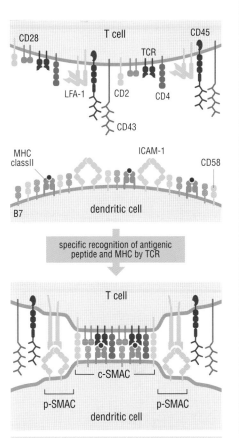

Figure 5-11 The immunological synapse
Top: molecules on the dendritic cell surface are recognized by receptors, coreceptors and adhesive molecules on the T cell surface (in this case a CD4 T cell). CD43 is an abundant cell-surface glycoprotein of T cells that probably helps to prevent adhesion during circulation. Bottom: binding of the peptide–MHC complex on the dendritic cell by the TCR and coreceptor on the T cell leads to T cell signaling, conversion of the integrin LFA-1 to its high-affinity form, and reorganization of the actin cytoskeleton. These events lead to the formation of an organized array of surface molecules consisting of an inner region, known as the central supramolecular activation cluster, or c-SMAC, containing signaling molecules; and an outer ring of adhesive molecules, the peripheral supramolecular activation cluster, or p-SMAC. CD43 and the phosphatase CD45, both abundant tall glycoproteins that might pose a steric obstacle to binding of the TCR to the peptide–MHC complex by virtue of their height, are excluded from the synapse.

T cell and an antigen-presenting cell occurring during antigen recognition by the T cell.

PAG1: see **phosphoprotein associated with glycophospholipid-enriched microdomains**.

peripheral supramolecular activation complex (p-SMAC): peripheral area of an **immunological synapse** in which TCRs and integrins have segregated into distinct areas.

phosphoprotein associated with glycophospholipid-enriched microdomains (PAG1): a transmembrane scaffold protein that when phosphorylated on tyrosine recruits the protein kinase Csk to lipid rafts.

p-SMAC: see **peripheral supramolecular activation complex**.

References

Cloutier, J.-F. and Veillette, A.: **Cooperative inhibition of T-cell antigen receptor signaling by a complex between a kinase and a phosphatase.** *J. Exp. Med.* 1999, **189**:111–121.

Duan, L. *et al.*: **The Cbl family and other ubiquitin ligases: destructive forces in control of antigen receptor signaling.** *Immunity* 2004, **21**:7–17.

Germain, R.N.: **The T cell receptor for antigen: signaling and ligand discrimination.** *J. Biol. Chem.* 2001, **276**:35223–35226.

Grossman, Z. and Paul, W.E.: **Autoreactivity, dynamic tuning and selectivity.** *Curr. Opin. Immunol.* 2001, **13**:687–698.

Lin, J. *et al.*: **The c-SMAC: sorting it all out (or in).** *J. Cell Biol.* 2005, **170**:177–182.

TCR signaling leads to activation of proliferation and clonal expansion through induction of cytokine gene transcription

T cells require cytokines for proliferation and survival, and constitutively express the γ chain that is common to receptors for several important T cell cytokines (see section 2-8). Activation of a naïve T cell at the surface of a dendritic cell induces the expression of both the α chain of the IL-2 receptor, which confers responsiveness to IL-2, and IL-2 itself. IL-2 is a major growth factor for naïve T cells and antigen-stimulated cells are thus driven to proliferate both by autocrine signaling from the antigen-stimulated cell itself, and by paracrine signaling from neighboring T cells also participating in the response. IL-2 production is controlled largely at the level of transcription, and activation of IL-2 gene transcription requires the convergence of many of the signaling pathways illustrated in Figure 5-9 with costimulatory signals that also act on the IL-2 gene.

IL-2 gene transcription requires the activation of four transcriptional regulators by the TCR

The promoter controlling the activation of the IL-2 gene contains binding sites for at least four transcriptional regulators, all of which contribute to maximal transcription of the gene. These transcriptional regulators are activated through pathways initiated by the actions of the signaling enzyme PLC-γ1, which, as we have seen, is activated by signaling through the TCR and hydrolyzes PIP_2 in the membrane (see Figure 5-9). This yields IP_3, which causes an increase in intracellular free calcium, and diacylglycerol (DAG).

Calcium elevation is sufficient to activate a family of transcriptional regulators known as **NFAT**, the **nuclear factor of activated T cells**, which are held in the cytoplasm by phosphorylation on several sites. Calcium, through binding to a protein called calmodulin, activates the protein phosphatase **calcineurin**, which dephosphorylates NFAT molecules, allowing them to translocate to the nucleus and activate transcription (Figure 5-12). NFAT molecules make an essential contribution to the activation of many genes whose products play an important part in T cell activation and function, and we shall see in Chapter 14 that calcineurin is the target of several immunosuppressive drugs. The other product of PIP_2 hydrolysis, diacylglycerol, activates various isoforms of a cytoplasmic protein kinase, protein kinase C, the most important of which in the present context is **protein kinase Cθ (PKCθ)**, which promotes the activation of the transcriptional regulator NF-κB.

A second signaling function of diacylglycerol is to help activate the small GTPase Ras. Ras in turn, through several intermediate steps, activates a kinase, Erk (for extracellular signal-regulated kinase), belonging to the mitogen-activated protein (MAP) kinase family (see section 3-11). MAP kinases are activated by signaling from the cell surface and are of three major types: Erk kinases, JNK kinases and p38 MAP kinases, and they regulate transcriptional regulators as well as other events within cells. Erk MAP kinases act downstream of Ras; the other two major types of MAP kinases are separately controlled and are often activated downstream of another small GTPase, Rac, which is also activated by TCR signaling (see Figure 5-9). Each of these types of MAP kinases plays a unique part downstream of TCR signaling, and we shall describe differential roles of these MAP kinases in T cell development in Chapter 7, and in the differentiation of T helper subsets later in this chapter. Here we are concerned with the induction of IL-2 mRNA, and Erk is important for this process by promoting the activation of a family of transcriptional regulators called *AP-1* (for *activator protein 1*). Finally, Ras is also important for the activity of a fourth type of transcriptional regulator acting at the IL-2 promoter, called Oct1, although less is known about how this is mediated. The regulation of transcriptional

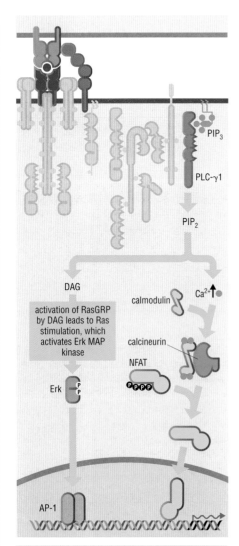

Figure 5-12 Activation of NFAT and AP-1 by Ca²⁺ and Erk pathways TCR signaling leads to the activation of PLC-γ1, which hydrolyzes PIP_2 to generate two second messengers, IP_3 and diacylglycerol (DAG). In addition to activating isoforms of protein kinase C, DAG activates a Ras guanine nucleotide exchange factor, RasGRP, which activates Ras, and Ras in turn activates the MAP kinase pathway leading to the activation of Erk. Erk in turn promotes the activity of the transcriptional regulator AP-1. IP_3 induces the release of calcium from internal stores in the cell. Calcium binds to calmodulin, which then binds to and activates calcineurin, a calcium-dependent protein phosphatase. Calcineurin then removes a number of phosphates from NFAT, allowing it to enter the nucleus, bind to its target sites in DNA and activate transcription. NFAT is subsequently phosphorylated and re-exported to the cytoplasm to terminate signaling.

Definitions

calcineurin: calcium/calmodulin-activated serine/threonine protein phosphatase, an essential component of many calcium-activated signaling pathways. Calcineurin action is essential to activate the **NFAT** family of transcription factors and is the target of the immunosuppressive drugs cyclosporin and FK506.

NFAT: see **nuclear factor of activated T cells**.

nuclear factor of activated T cells (NFAT): any of a family of four closely related transcription factors that

are responsive to calcium elevation.

phosphoinositide-3-kinase (PI3 kinase): a signaling enzyme that adds a phosphate to the 3 position of PIP_2 to generate the membrane-bound second messenger PIP_3.

PI3 kinase: see **phosphoinositide-3-kinase**.

PKCθ: see **protein kinase Cθ**.

protein kinase Cθ (PKCθ): a member of the protein kinase C family. It is important in signaling from the

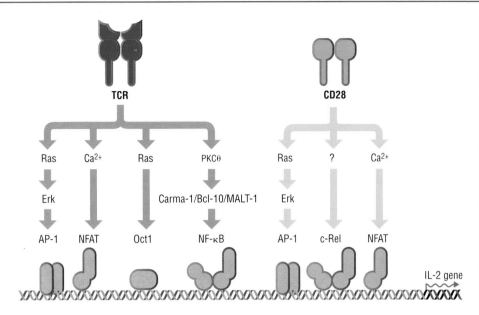

Figure 5-13 Activation of IL-2 transcription by signaling through the TCR and CD28 TCR stimulation leads to elevation of intracellular calcium, activation of Ras, and activation of protein kinase Cθ (PKCθ). All of these events are enhanced by CD28 costimulatory signaling. Calcium elevation activates NFAT and also increases the activation of NF-κB, which is downstream of PKCθ and a complex containing the scaffold and adaptors Carma-1, Bcl-10 and MALT-1. This complex is required for antigen receptors to activate NF-κB and is different from the mechanism of NF-κB activation by TNF receptor family members and Toll-like receptors, as described in section 3-11. Ultimately both types of pathways converge on and activate the same I-κB kinase (IKK) complex, as described in section 2-10 (the classical pathway of NF-κB activation). Ras activation promotes activation of the AP-1 transcription factor through the mitogen-activated protein (MAP) kinase Erk. JNK and p38 MAP kinases also participate in AP-1 activation in some circumstances. Ras also promotes activation of the Oct1 transcription factor. These transcriptional regulators work in concert to drive IL-2 transcription. The amount of IL-2 produced is also subject to other modes of intracellular regulation subsequent to transcription of the IL-2 gene (not shown).

regulators acting at the IL-2 promoter by TCR-induced signaling events is summarized in Figure 5-13. A key feature of transcriptional activation of the IL-2 promoter is the requirement for simultaneous occupancy by different transcription factors responding to different signaling events induced by the TCR, so that IL-2 mRNA is produced only if the cell is undergoing robust and sustained TCR signaling. In addition, maximal activation of IL-2 gene transcription typically requires signaling from the costimulatory receptor CD28.

The costimulatory receptor CD28 amplifies the signal from the TCR

Productive activation of naïve T cells cannot normally occur in the absence of costimulatory signals from B7 molecules (see section 4-1), which are upregulated on the surface of dendritic cells in the presence of infection and bind to the costimulatory receptor CD28 on the surface of the T cell. Thus, while strong signaling through the TCR is required to indicate the presence of foreign-peptide–MHC, signaling through CD28 is required to confirm the presence of infection. Signaling through CD28 is thought to act primarily by amplifying the signal from the TCR to ensure, or increase, cytokine production.

The mechanism of signal amplification by CD28 is thought to be through tyrosines in its cytoplasmic tail that become phosphorylated by the TCR-activated tyrosine kinases and recruit SH2-domain-containing signaling molecules (Figure 5-14). One signaling molecule known to be recruited in this way is the signaling enzyme **phosphoinositide-3-kinase (PI3 kinase)**, which helps with the membrane recruitment of Itk and PLC-γ1 (see Figure 5-9). Although TCR signaling activates PI3 kinase to some degree without CD28, CD28 boosts this critical signaling event, which otherwise might be limiting.

Figure 5-13 shows that the IL-2 promoter contains sites that are bound by transcriptional regulators activated by signaling from both the TCR and CD28. The transcriptional regulators elicited by the two signals are largely overlapping: the dependence of activation on CD28 costimulation may reflect the need for stronger signaling (and thus higher concentrations of transcriptional regulators) to fill lower-affinity DNA sites; or for different family members.

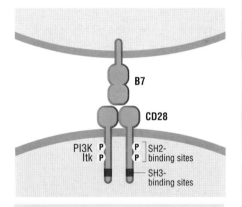

Figure 5-14 Signaling molecules recruited by CD28 The cytoplasmic tail of CD28 contains several potential binding sites for SH2 and SH3 domains of signaling proteins. Some of the intracellular signaling molecules thought to be recruited by CD28 are indicated in the figure. The most important is PI3 kinase (PI3K).

antigen receptor of T cells, where it becomes localized adjacent to the T cell receptor and CD28 and activates NF-κB.

References

Acuto, O. and Michel, F.: **CD28-mediated co-stimulation: a quantitative support for TCR signaling.** *Nat. Rev. Immunol.* 2003, **3**:939–951.

Diehn, M. *et al.*: **Genomic expression programs and the integration of the CD28 costimulatory signal in T cell activation.** *Proc. Natl Acad. Sci. USA* 2002, **99**:11796–11801.

Macian, F.: **NFAT proteins: key regulators of T-cell development and function.** *Nat. Rev. Immunol.* 2005,

5:472–484.

Thome, M.: **CARMA1, BCL-10 and MALT1 in lymphocyte development and activation.** *Nat. Rev. Immunol.* 2004, **4**:348–359.

Activation of naïve T cells occurs when recirculating T cells recognize antigen at the surface of dendritic cells in secondary lymphoid tissues

The signaling events described in the preceding three sections occur in secondary lymphoid tissues when recirculating naïve T cells bind strongly to antigenic peptide–MHC on the surface of a mature activated dendritic cell, and culminate in vigorous proliferation of the cells and differentiation of the expanded clones of cells into fully functional effector T cells.

We shall see shortly that activated mature dendritic cells play an essential part not only in generating effector T cells but also in directing their specialized effector and migratory properties. Before discussing this central role of dendritic cells, however, we revisit the secondary lymphoid organs and the specialized architecture that promotes the critical encounters between T cells, dendritic cells and antigen that determine the subsequent behavior of the T cell. For purposes of this discussion, we specifically describe traffic through the lymph nodes, which are the most extensively studied of the secondary lymphoid tissues.

Naïve T cells are directed into secondary lymphoid tissues by a characteristic combination of homing molecules

Naïve T cells enter the lymph nodes from the bloodstream through the specialized endothelial cells of the high endothelial venules (HEV; see section 1-7), in a process that is described in general terms in section 2-14 and is orchestrated for naïve T cells by L selectin, CCR7 and LFA-1 (Figure 5-15). Although the integrin LFA-1 is essential for the tight binding that leads to entry into tissues from the bloodstream (see the legend to the figure), it is recognition of tissue-specific homing molecules by the selectin and the chemokine receptor that determines the tissue entered; and the combined surface expression of L selectin and CCR7 is characteristic of recirculating T cells. Once in the lymph node, naïve T cells are directed to the T cell zone, also by CCR7, which recognizes the homeostatic chemokine CCL21 that characterizes this zone, and encounter dendritic cells that have also been directed there by CCR7 (see section 4-1). The entering T cells then scan the surface of dendritic cells through low-affinity interactions involving the adhesive molecule ICAM-3, which is expressed on the T cell surface, and a family of C-type lectins on the dendritic cells, as well as molecules of the CD2 family. Scanning is rapid and must be efficient, to ensure the activation of the one T cell in 100,000 that is specific for any given antigen.

While the T cell is exploring the lymph node, the levels of the S1P receptor, which are depressed by high levels of S1P in blood (see section 2-14), recover. Thus if after a period of about 12–18 hours the T cell fails to bind strongly to its specific peptide–MHC ligand, it leaves the lymph node, attracted by S1P in lymph and blood, to rejoin the circulation and explore other secondary lymphoid tissues (Figure 5-16).

T cells that bind strongly to self-peptide–MHC complexes may be inactivated

Although most potentially self-reactive T cells are eliminated in the thymus, as we shall see in Chapter 7, some escape this stringent selection. Some of these become regulatory T cells that suppress immune responses against self; these are discussed later in this chapter. Others enter the recirculating lymphocyte pool but are thought to be prevented from launching autoimmune responses by dendritic cells that have matured spontaneously—that is, in the absence of activation by infection (see section 4-1)—and present self peptides derived from cellular debris ingested in the periphery but without the high levels of B7 required for T cell activation. This leads to abortive signaling from the TCR that induces a state of unresponsiveness to subsequent stimulation: this is known as *anergy*. Maintenance of this state requires persistence of the antigen, so is generally elicited only by self-peptide–MHC complexes that are continuously present. Induction of this unresponsive state is believed to be an important mechanism of *peripheral tolerance*, which we discuss further in Chapter 12.

T cell activation is initiated by lymph-node-resident dendritic cells and stabilized by recruited dendritic cells

While many of the dendritic cells in the lymph node are thought to be cells that have matured spontaneously in the periphery and, as we have just seen, help to maintain tolerance, most are probably immature dendritic cells that have migrated directly from the blood to function as

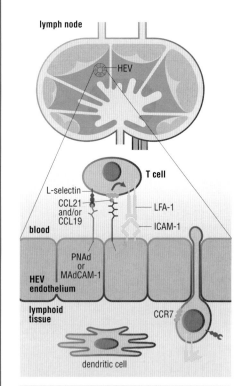

lymph node

HEV

T cell

L-selectin

CCL21 and/or CCL19

blood

LFA-1

ICAM-1

PNAd or MAdCAM-1

HEV endothelium

lymphoid tissue

CCR7

dendritic cell

Figure 5-15 Mechanism of entry of naïve T cells into secondary lymphoid tissue through high endothelial venules The entry of naïve T cells into lymph nodes or mucosal lymphoid tissue requires the sequential interaction of molecules on the T cell surface with their ligands on the endothelial cells of the HEV. Contact is first made by L-selectin on the T cell with carbohydrate determinants of a mucin on the HEV endothelial cells (the mucin is the peripheral node addressin (PNAd) on peripheral lymph node HEV, and mucosal addressin cell adhesion molecule (MAdCAM-1) on mucosal lymphoid tissue HEV). This is followed by recognition of the chemokine CCL21 (and/or CCL19) on the endothelial cell by the chemokine receptor CCR7 on the lymphocyte, which signals to the integrin LFA-1 to undergo a conformational change (not shown) that results in tight binding to ICAM-1, leading to the migration of the lymphocyte through the endothelium into the secondary lymphoid organ. The sequence of events in homing and transmigration through endothelia is described in detail in section 2-14, and conformational changes in integrins in section 2-4. Entry into the spleen is through terminal arterioles and does not require L-selectin, probably because shear forces from the circulation are smaller.

sentinels in the lymph node in the same way that they do in peripheral tissues. Both these and the spontaneous migrants are known as **lymph-node-resident dendritic cells**. The immature resident dendritic cells may play a part in the activation of naïve T cells before the arrival of migrant mature dendritic cells activated by infection in the periphery.

Although most antigens probably reach the lymph node as fragments bound to MHC molecules on activated dendritic cells migrating from peripheral sites of infection, this takes some hours and in the meantime small soluble antigens can be delivered to the lymph node within minutes in the afferent lymph (see section 1-7) and captured by resident dendritic cells. TLR ligands and cytokines entering the lymph node by the same route at the same time activate the cells, which thereby become competent to initiate the activation of circulating naïve T cells. Interactions with resident dendritic cells, however, are transient, and productive activation of the T cells must await the arrival of migrating dendritic cells, which arrive in the lymph node from the afferent lymphatic vessels some 8–12 hours later.

Activated dendritic cells that have matured in the periphery display higher levels of antigen on their surface than the resident dendritic cells, cluster round the HEV through which recirculating T cells enter, and make sustained contacts with several T cells at a time that may last for several hours. These sustained interactions induce vigorous proliferation of the T cells and culminate in the production of fully competent effector cells, which disperse either to the B cell follicles as follicular helper cells or through the bloodstream to sites of infection in the periphery. This sequence of events is schematically illustrated in Figure 5-17.

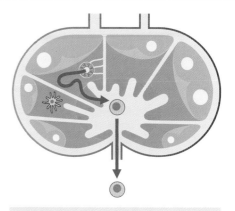

Figure 5-16 Naïve T lymphocyte recirculation through secondary lymphoid tissue T cells enter lymph nodes through high endothelial venules and migrate to the T cell zones (purple) where they encounter dendritic cells that have entered from afferent lymphatic vessels. If they fail to recognize antigen, they leave the lymph node in the efferent lymphatic vessels to be carried back into the bloodstream and sample other secondary lymphoid tissues.

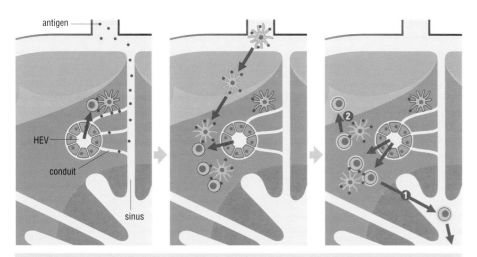

Figure 5-17 Activation of naïve T lymphocytes by resident and migrant dendritic cells in secondary lymphoid tissue Antigen entering directly through the lymphatic vessels crosses the T cell zone in specialized conduits (outlined in blue; first panel), which is taken up by resident dendritic cells (outlined in blue; first panel), which make transient contacts with recirculating naïve T cells and initiate activation. Later, mature activated dendritic cells (outlined in green) migrate into the lymph node through the afferent lymphatic vessels and cluster round the HEV, where they make sustained contacts with T cells (second panel), resulting in the production of effector T cells (third panel: shown with yellow halo), which leave the lymph node for the peripheral tissues (1) or migrate to the follicles where they activate B cells (2).

Definitions

lymph-node-resident dendritic cells: dendritic cells that migrate to the lymph node as immature cells and take up residence there, or that have migrated there from the periphery in the absence of infection.

References

Geijtenbeek, T.B.H. *et al.*: **Identification of DC-SIGN, a novel dendritic cell-specific ICAM-3 receptor that supports primary immune responses.** *Cell* 2000, **100**:575–585.

Germain, R.N. and Jenkins, M.K.: *In vivo* **antigen presentation.** *Curr. Opin. Immunol.* 2004, **16**:120–123.

Itano, A.A. *et al.*: **Distinct dendritic cell populations sequentially present antigen to CD4 T cells and stimulate different aspects of cell-mediated immunity.** *Immunity* 2003, **19**:47–57.

Mempel, T.R. *et al.*: **T-cell priming by dendritic cells in lymph nodes occurs in three distinct phases.** *Nature* 2004, **427**:154–159.

Schwartz, R.H.: **T cell anergy.** *Annu. Rev. Immunol.* 2003,

21:305–334.

Sixt, M. *et al.*: **The conduit system transports soluble antigens from the afferent lymph to resident dendritic cells in the T cell area of the lymph node.** *Immunity* 2005, **22**:19–29.

von Andrian, U.H. and Mackay, C.R.: **T-cell function and migration.** *N. Engl. J. Med.* 2000, **343**:1020–1034.

von Andrian, U.H. and Mempel, T.R.: **Homing and cellular traffic in lymph nodes.** *Nat. Rev. Immunol.* 2003, **3**:867–878.

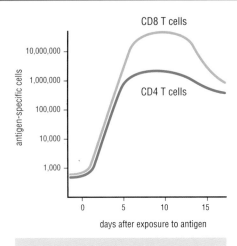

Figure 5-18 Proliferation kinetics of CD4 and CD8 T cells A schematic representation of the expansion in the numbers of antigen-specific CD4 and CD8 T cells based on data from viral infection of mice.

Activation of T cells leads to proliferation accompanied by changes in their surface properties

The activation of naïve T cells at the surface of dendritic cells in secondary lymphoid tissue is known as **priming**, to distinguish it from the activation of effector cells by their targets, and is marked by a succession of changes in the surface molecules of the activated cells that modify their migratory behavior and interactions with other cells, and their responsiveness to the cytokines that drive their proliferation and differentiation and sustain their survival. Two of the first surface molecules to be induced on the activation of naïve T cells are the α chain of the IL-2 receptor (CD25) and a C-type lectin receptor called CD69. The α chain of the IL-2 receptor confers high-affinity binding to IL-2 and, as we saw in section 5-5, drives clonal expansion of activated T cells in response to IL-2, which they also produce. CD69 inhibits $S1P_1$, the receptor that is required to guide recirculating lymphocytes out of the lymphoid tissues and back into the bloodstream (see section 2-14). Activation of naïve lymphocytes causes both downregulation of the S1P receptor and expression of CD69, so that the activated cells are detained in the lymphoid tissue and can engage in sustained interactions with dendritic cells.

Amplification of T cell numbers on activation is rapid and substantial

Activated naïve T cells undergo rapid proliferation during which their numbers can increase by a factor of 10,000; CD8 T cells, which expand even more dramatically than CD4 T cells (Figure 5-18), may increase in number as much as 50,000-fold in response to viral infection. This rapid expansion, with the retention of the activated cells in the lymph node, leads to the enlargement of the lymph nodes draining a site of infection, an effect known as **lymphadenopathy** that is familiar to most people from occasional infections that cause detectable swelling in the lymph nodes of the neck. By the end of this phase of clonal expansion, which peaks at 3–7 days, the T cells have acquired the characteristics of effector cells, so that within a week of encountering antigen, substantial numbers of antigen-specific T cells are ready to migrate to their sites of action to combat infection.

The rapid amplification of the T cell response is promoted and regulated by an exchange of signals between the T cell and the dendritic cell that protect the cells from premature death or anergy and stimulate their vigorous proliferation.

The engagement of the TCR and CD28 by peptide–MHC and B7 on the dendritic cell activates a positive feedback loop

The proliferation of activated T cells is driven by IL-2 and by inflammatory cytokines and type-1 interferons, acting either directly on the T cells or indirectly through the induction of B7 (CD80/CD86) on the dendritic cell. These proliferative signals are rapidly amplified by a feedback loop that is established when signaling through the TCR and CD28 induces the expression of a TNF-family signaling molecule called **CD40L** on the T cell surface. **CD40**, the TNFR-family receptor for CD40L, is expressed at a low level on all dendritic cells but is strongly upregulated on activated cells. The induction of CD40L on the T cell thus initiates signaling to the dendritic cell through CD40. This induces upregulation of B7 on the dendritic cell surface, amplifying the signal to the T cell through CD28 and leading in turn to the upregulation of CD40L and further amplification of the reciprocal signaling between the two cells.

Definitions

CD40: a receptor of the TNF receptor family that is expressed on mature dendritic cells, B cells and activated macrophages and plays an essential part in their interactions with T cells.

CD40L: a TNF-family molecule, also known as CD154, that binds to receptor **CD40** and is induced on activated T cells and plays an essential part in the initiation of T cell responses and in many of the effector actions of CD4 T cells.

CTLA4: see **cytotoxic T lymphocyte antigen 4**.

cytotoxic T lymphocyte antigen 4 (CTLA4): an inhibitory member of the CD28 family of cell surface molecules that is induced after a delay on activated T cells and binds to the B7 costimulator molecules in competition with CD28, reducing or terminating the activating signal to T cells.

lymphadenopathy: enlarged lymph nodes.

priming: activation of naïve T or B lymphocytes on first encounter with antigen.

If this positive feedback loop were left unbroken, lymph nodes would be rapidly overwhelmed with proliferating T cells, and those T cells that arrived first could monopolize the dendritic cell surface, denying access to later-arriving T cells and thereby restricting the repertoire of antigen-specific cells available to counter infection. This is prevented by the induction on the T cell, also by signaling from the TCR and CD28, of **CTLA4** (for **cytotoxic T lymphocyte antigen 4**), which is held in the cytoplasm of T cells and is rapidly directed to the T cell surface on activation through the TCR. It appears after about 24 hours and serves as an activation-dependent reversible brake on T cell proliferation. CTLA4 is closely related to CD28, and also binds B7-1 and B7-2, but with considerably higher affinity than CD28 does and without transmitting an activating signal. It is thought to act largely through competitive binding to the B7 costimulators (particularly B7-1) that prevents binding and thus signaling by CD28 and thereby interrupts the positive feedback loop (Figure 5-19).

Animals from which the gene for CTLA4 has been experimentally deleted die soon after birth from massive lymphoproliferation accompanied by autoimmune disease, reflecting the importance of the braking effect of CTLA4.

CD28 and CTLA4 act mainly during the clonal expansion of T cells in the secondary lymphoid tissues, where an important function of CD28 is to induce antiapoptotic molecules that sustain the survival of the T cells during their rapid proliferation. Once the T cells have differentiated into effector cells, this function is taken over by other molecules that are differentially important for different cell types so that control of T cell numbers can be varied for cells with different functions. These molecules are induced in the lymph node as a downstream consequence of the dialog established between the T cell and the dendritic cell, and may also contribute to the modulation of clonal expansion. We describe these downstream signaling molecules and what is known of their specialized functions in the next section.

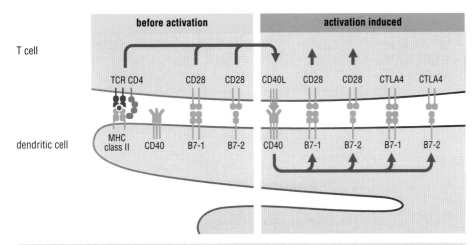

Figure 5-19 The costimulatory dialogue between CD4 T cells and dendritic cells On entry into the lymph node, the activated mature dendritic cell expresses CD40 and the B7 costimulators, which are induced in the presence of infection. The CD4 T cell constitutively expresses CD28, the receptor for B7, and activation through this costimulatory receptor and the TCR induces expression of CD40L, which binds to CD40 on the dendritic cell, causing further upregulation of the B7 costimulators and amplifying the signal to the T cell. This positive feedback circuit is subsequently broken by the induction of CTLA4 on the T cell: although it may also deliver an inhibitory signal (not shown), CTLA4 is thought to act chiefly by competing with CD28 for binding to the B7 costimulators and attenuating the activating signal.

References

Egen, J.G. et al.: **CTLA-4: new insights into its biological function and use in tumor immunotherapy.** Nat. Immunol. 2002, **3**:611–618.

Jenkins, M.K.: **Peripheral T-lymphocyte responses and function** in Fundamental Immunology 5th ed. Paul, W.E. ed. (Lippincott Williams & Wilkins, Philadelphia, 2003), 303–319.

Shiow, L.R. et al.: **CD69 acts downstream of interferon-α/β to inhibit S1P₁ and lymphocyte egress from lymphoid organs.** Nature 2006, **440**:540–544.

Induced modulators of T cell signaling regulate T cell survival and proliferation

The second wave of costimulatory molecules that are induced on T cells as a consequence of signaling from the T cell antigen receptor and CD28 all belong either to the CD28 subfamily of the immunoglobulin superfamily or to the TNFR family. They recognize ligands that belong to the B7 and TNF families and are induced or upregulated on the dendritic cell by signaling through CD40. These ligands are also induced on many cell types, both immune and non-immune, with which effector T cells interact when they disperse to their sites of action. The principal function of most of these molecules is to sustain the survival of T cells by inhibiting apoptosis; some of them, however, like CTLA4, are inhibitory and act to limit the numbers of surviving cells or to limit tissue damage. Because of the structural and functional relationships between these molecules, they are often for convenience collectively called costimulatory molecules, by analogy with B7 and CD28, or in the case of CTLA4 and PD-1, coinhibitory molecules, although there is no rigorous definition of these terms corresponding to this broader usage.

At least 15 costimulatory receptor–ligand pairs are now known, although in many cases their biological roles have yet to be established. Those with known functions are summarized in Figure 5-20. The most important of these molecules for CD4 T cells are the receptors **ICOS** (for **inducible costimulator**) and **OX40** on the T cell, which recognize **ICOSL** and **OX40L** on the dendritic cell; ICOSL and OX40L are also expressed on B cells and other cells in the periphery, and their most significant function is thought to be in the interactions of effector T cells with B cells and in the survival of memory CD4 T cells, which we discuss in later sections of this chapter. The costimulatory molecules thought to be most important for CD8 T cells are 4-1BB, which recognizes 4-1BBL on the dendritic cell, possibly with two other TNF/TNFR-family molecules, OX40/OX40L and the receptor CD27 and its ligand CD70 (see below).

PD-1 (for **programmed death 1**) is an inhibitory receptor that is induced later than CTLA4; it has two ligands, namely **PD-L1**, which is constitutively expressed at low levels on the dendritic cell and is inducible on many cell types in the periphery, and **PD-L2**, which is expressed only on dendritic cells and macrophages. Unlike CTLA4, PD-1 does not act by competitive binding, but through a phosphatase that dephosphorylates and inactivates components of the signaling pathways of antigen receptors and costimulators. Because it is expressed on many cell types, including non-immune cells, it is likely to be important in limiting T cell effector actions in the periphery, which we discuss later, and in maintaining tolerance to self tissue antigens, which we discuss in Chapter 12.

Figure 5-20 Some of the second-wave costimulatory molecules induced by CD28 and CD40 Early costimulatory signals induce a second wave of costimulatory molecules, OX40L and its receptor OX40, and ICOSL and its receptor ICOS, which promote the survival and proliferation of CD4 T cells in secondary lymphoid tissues, although they are more important in later interactions with B cells and in sustaining memory CD4 T cells; 4-1BBL and its receptor 4-1BB, which are important in the maintenance of CD8 T cells; and PD-L1 and PD-L2 and their inhibitory receptor PD-1, which limit T cell responses and may contribute to the maintenance of tolerance to self. All of the ligands for these costimulatory receptors are expressed on targets of T effector cells and/or on peripheral tissues, as well as on dendritic cells.

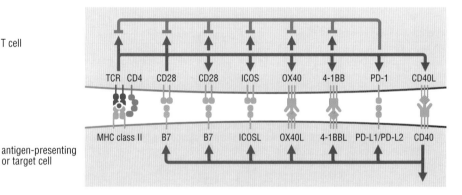

T cell

antigen-presenting or target cell

Definitions

ICOS: see **inducible costimulator**.

ICOSL: the ligand for **ICOS**, belonging to the B7 subfamily of the immunoglobulin superfamily and expressed on the surface of activated dendritic cells, B cells, macrophages and some non-immune cells.

inducible costimulator (ICOS): a receptor belonging to the CD28 subfamily of the immunoglobulin superfamily that is induced on activated and effector T lymphocytes and is important in the maintenance of effector and

memory CD4 T cells and especially those interacting with B cells.

OX40: a TNFR-family molecule that is induced on activated and effector T lymphocytes and is important in the maintenance of effector and memory CD4 T cells, and especially those interacting with B cells. It is called OX40 because it was first identified in an Oxford laboratory as one of a number of molecules induced on activated T cells.

OX40L: the TNF-family ligand for **OX40**, expressed on activated dendritic cells and B cells and some non-

immune cells.

PD-1: see **programmed death 1**.

PD-L1: a B7-family ligand for **PD-1** that is expressed widely on tissue cells and is thought to be important in limiting damage by activated T cells in the periphery.

PD-L2: a B7-family ligand for **PD-1** that is expressed chiefly on dendritic cells and is thought to contribute to T cell tolerance.

programmed death 1 (PD-1): an inhibitory receptor

Pivotal to the activating functions of T cells is the binding of CD40L by CD40. CD40, as we have seen, is upregulated on the activated dendritic cell, where it critically increases the activating potency of the cell; it is also expressed on the macrophages and B cells that are the later targets of CD4 T cell effector actions, and is essential to their activation by T cells. Individuals with defects in the gene encoding CD40L, not surprisingly, have deficient immune responses to many types of infections. The effect on antibody production is particularly severe and is described in more detail in Chapter 6. The importance of CD40L in promoting clonal expansion in the lymph node may in part explain the part played by CD4 T cells in some CD8 T cell responses.

Activation of CD8 T cell immunity sometimes requires help from CD4 T cells

CD8 T cells are less dependent than CD4 T cells on CD28 for activation, and many of them do not express CD40L on activation. The positive feedback loop that drives the proliferation of CD4 T cells therefore does not operate directly on CD8 T cells, and in many cases it is thought that cytokines induced by infection are sufficient to launch vigorous proliferation of these cells on antigen recognition. IL-12 in particular is important in the activation of CD8 T cells in response to bacterial and parasitic infection, and we shall see in the next sections that IL-12 is also critical in the induction of T$_H$1 CD4 T cells, which often participate in the same immune responses as CD8 T cells. In viral infections, type 1 interferons produced by infected cells or by plasmacytoid dendritic cells are the major CD8 T cell-activating cytokines.

CD8 T cell responses to antigens that do not elicit strong cytokine responses (for example transplantation antigens, which we discuss later in the book), as well as low-level chronic infections, for example by latent viruses (which we discuss later in the chapter), require help from CD4 T cells. The exact nature of the help is not fully understood, but may include activation of the dendritic cell through CD40L, and induction on the dendritic cell of ligands for 4-1BB and other TNF-family costimulatory molecules; and it is very likely that the high levels of IL-2 induced by CD4 T cell activation are important in the activation of CD8 T cells (Figure 5-21).

CD4 T cell help is also essential for the generation of CD8 T memory cells, which is programmed during the late clonal expansion of CD8 T cells, and later for their maintenance, and this probably explains why they are needed for chronic viral infections, which are controlled by CD8 memory T cells, as we shall see later. The importance of CD4 T cell help for CD8 T cell-mediated immunity can be seen in the devastating effects of AIDS, in which, as we shall see in Chapter 10, CD4 T cells are gradually destroyed and infections normally controlled by CD8 T cells are among those to which AIDS patients succumb.

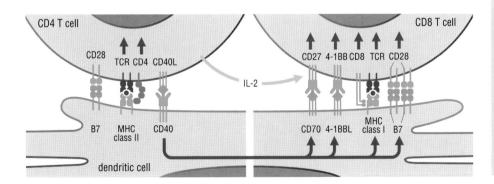

Figure 5-21 Role of CD4 T cells in the activation of CD8 T cells in secondary lymphoid tissue CD4 T cells provide help during the activation of CD8 T cells by signaling to the dendritic cell through CD40 and thereby amplifying the costimulatory signals acting on CD8 T cells, as well as by providing IL-2. The CD4 T cell is sometimes said in this way to be licensing the dendritic cell, because it is thought that it need not bind to the dendritic cell at the same time as the CD8 T cell. The most important CD8 costimulatory molecule is though to be 4-1BB, although CD27 and (not shown) OX40 may also play a part.

belonging to the CD28 subfamily of the immunoglobulin superfamily, with an important part in peripheral T cell tolerance and in limiting immune damage at the effector cell stage. PD-1 has two ligands: **PD-L1**, which is expressed by many stromal cells of tissues, and **PD-L2**, which is expressed by dendritic cells and macrophages.

References

Bevan, M.J.: **Helping the CD8+ T-cell response.** *Nat. Rev. Immunol.* 2004, **4**:595–602.

Castellino, F. and Germain, R.N.: **Cooperation between CD4+ and CD8+ T cells: where, when, and how.** *Annu. Rev. Immunol.* 2006, **24**:519–540.

Kroczek, R.A. *et al.*: **Emerging paradigms of T-cell co-stimulation.** *Curr. Opin. Immunol.* 2004, **16**:321–327.

Peggs, K.S. and Allison, J.P.: **Co-stimulatory pathways in lymphocyte regulation: the immunoglobulin superfamily.** *Br. J. Haematol.* 2005, **130**:809–824.

Song, J. *et al.*: **Sustained survivin expression from OX40 costimulatory signals drives T cell clonal expansion.** *Immunity* 2005, **22**:621–631.

Williams, M.A. and Bevan, M.J.: **Effector and memory CTL differentiation.** *Immunol. Rev.* 2007, **25**:171–192.

Williams, M.A. *et al.*: **Interleukin-2 signals during priming are required for secondary expansion of CD8+ memory T cells.** *Nature* 2006, **441**:890–893.

CD4 T cells differentiate into subsets characterized by distinct patterns of cytokine expression

During their clonal expansion, CD4 and CD8 T cells divide as frequently as every 6–8 hours, completing five to eight divisions. Differentiation occurs during this period of rapid cell division and is accompanied by alterations of the chromatin at specific sets of cytokine genes so that these genes, which are not expressed in naïve T cells, become accessible to key transcriptional regulators, such as NFAT, that are activated after TCR stimulation. Other cytokine genes are reciprocally silenced by a related process that renders them inaccessible to these transcriptional regulators. By the time cells complete their division and differentiation, these changes in chromatin structure become stabilized so that cytokines appropriate to each T cell subset are induced upon TCR stimulation. Because these changes in chromatin do not involve changes in DNA sequence, they are known as **epigenetic** changes. In this way, discrete patterns of cytokines become fixed in effector cells, and cells can be grouped into different CD4 T helper (T_H) or CD8 T cytotoxic (T_C) cell subsets based on the patterns of cytokines that they make.

A number of stable T_H subsets can be distinguished by their cytokine patterns, which in turn determine the types of immunity they mediate (Figure 5-22). Predominant, or canonical, individual cytokines are often used to distinguish subsets. T_H1 cells, identified by secretion of IFN-γ, coordinate immune responses that facilitate the destruction of pathogens, such as many viruses, bacteria and protozoa, by cytotoxic CD8 T cells, NK cells and activated macrophages. T_H2 cells, identified by secretion of IL-4, coordinate immune responses by eosinophils, basophils and mast cells that are dedicated to the attack on pathogens such as parasitic worms and blood-sucking insects at sites of epithelial invasion. T_H17 cells, identified by secretion of IL-17-family cytokines, coordinate local acute inflammatory responses mediated by neutrophils, which are important against rapidly dividing extracellular bacteria and invading fungi. T_{FH}, or T helper follicular cells, are CD4 cells that on activation migrate to follicles, where they undergo further differentiation. T_{FH} cells provide help for B cells and are important for the production of high-affinity antibodies of specific classes, which determines the types of immune responses elicited by these antibodies at peripheral sites of infection. T_{REG}, or T regulatory, cells produce IL-10 and possibly TGF-β, and suppress the immune activation of other effector T cells, thus restraining over-exuberant host immune responses. Although most T_{REG} cells are generated in the thymus (discussed in Chapter 12), induced T_{REG} cells with similar functions can differentiate from naïve CD4 T cells in secondary lymphoid organs under conditions of chronic antigen stimulation. Additional subsets have been described, including T_{R1} cells, which represent IL-10-producing cells particularly prevalent in bowel tissues; and T_H3 cells, which produce TGF-β. The functions of these latter two subsets have not been clearly defined, and they will not be discussed further.

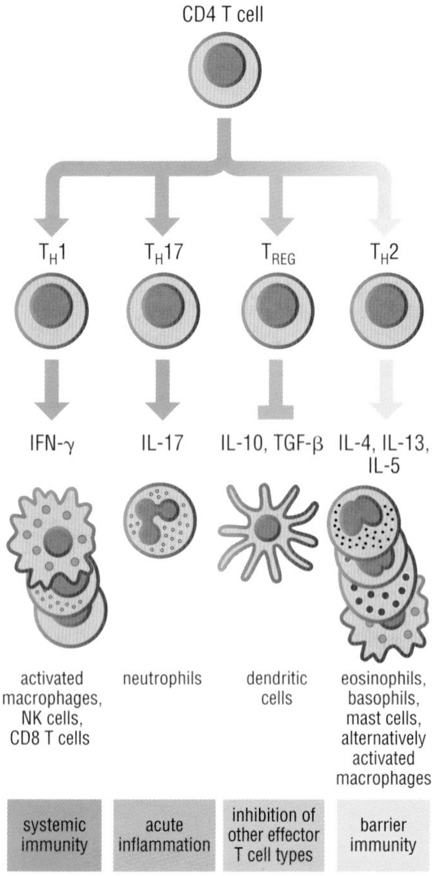

CD8 T cells most commonly differentiate into T_C1 effector cells

Most commonly, naïve CD8 T cells differentiate to T_C1 cells, characterized by IFN-γ production and direct cytotoxicity, and in this way they function in a manner very similar to NK cells, which are also cytotoxic and secrete IFN-γ. Effector CD8 T cells do not move to B cell follicles, but rather disperse widely to peripheral tissues. Although T_C2 cells and T_{C-REG} cells can be generated in various model systems, their contributions to immunity *in vivo* are not understood. Thus, the T_C1 nomenclature is rarely used, and reference to effector CD8 T cells generally specifies the T_C1 phenotype, or cytotoxic T cell.

CD4 T cell

| T_H1 | T_H17 | T_{REG} | T_H2 |

IFN-γ · IL-17 · IL-10, TGF-β · IL-4, IL-13, IL-5

activated macrophages, NK cells, CD8 T cells · neutrophils · dendritic cells · eosinophils, basophils, mast cells, alternatively activated macrophages

systemic immunity · acute inflammation · inhibition of other effector T cell types · barrier immunity

Figure 5-22 The major T_H subsets that arise during differentiation of naïve CD4 T cells The major cytokines produced by each of the subsets is shown, along with primary cells activated by each subset and the major function of the subsets.

Definitions

epigenetic: inherited through mechanisms that are not dependent on DNA sequence. Known epigenetic mechanisms often concern gene regulation and are dependent on modifications of the DNA and associated proteins that affect local chromatin structure.

References

Agace, W.W.: **Tissue-tropic effector T cells: generation and targeting opportunities.** *Nat. Rev. Immunol.* 2006, **6**:682–692.

Ansel, K.M. *et al.*: **An epigenetic view of helper T cell differentiation.** *Nat. Immunol.* 2003, **7**:616–623.

Lee, G.R. *et al.*: **T helper cell differentiation: regulation by *cis* elements and epigenetics.** *Immunity* 2006, **24**:369–379.

Murphy, K.M. and Reiner, S.L.: **The lineage decisions of helper T cells.** *Nat. Rev. Immunol.* 2002, **2**:933–944.

T cell differentiation is driven by dendritic cells and cytokines

The changes in cytokine gene expression that accompany T cell differentiation reflect the integration of signals imparted by environmental cytokines and by activated mature dendritic cells, which themselves have integrated signals from pathogens encountered in the tissue, and thereby induce a network of transcriptional regulators, discussed in the next section, that progressively instruct the changes in gene expression that determine the differentiation pathway of the responding T cells. T cell subset differentiation is influenced both by cytokines released from dendritic cells and other immune cells and by the number and affinity of TCR/peptide–MHC interactions and the strength of costimulation—which combine to contribute what is loosely described as strength of signal, but probably reflects distinct pathways activated downstream of the TCR, as we shall see in the next section. IFN-γ, which can be produced early after infection by NK cells, acts on both the T cell and on the dendritic cell, which is thereby stimulated to secrete IL-12, which promotes T_H1 cell survival. Signals promoting T_H2 differentiation are less well understood. Certain cytokines, such as IL-25, a member of the IL-17 cytokine family, and TSLP (for thymic stromal lymphopoietin), which is produced by epithelial cells during inflammation, are associated with T_H2 differentiation under some conditions, but the mechanisms by which this occurs are not clearly defined. T_H17 and T_{REG} cell differentiation are both promoted by TGF-β, which is produced by many cells, with T_H17 cell differentiation prevailing when the inflammatory cytokine IL-6 is also present, signaling the presence of infection: the differentiation of these two cell types is discussed in more detail later in this chapter.

Dendritic cells, by assimilating signals from cytokines and from microorganisms recognized by Toll-like receptors and C-type lectin receptors, can impart to differentiating T cells an imprint of the type of pathogen encountered in the periphery and are thus critical in linking the differentiation of specific T cell subsets with the distinct types of effector functions required for control of specific kinds of pathogens (Figure 5-23).

Most effector T cells disperse to peripheral tissues and accumulate at sites of infection

The differentiation of effector T cells is accompanied by changes not only in the cytokines they secrete on TCR engagement but in their responsiveness to cytokines and migratory signals and in their sensitivity to signaling through the TCR. Instead of CCR7 and L-selectin, which direct recirculation through the T cell areas of the lymphoid organs, effector T cells express homing molecules that guide them to their sites of action and include the receptor for P-selectin that directs them to sites of inflammation (see section 2-14). Except for T_{FH} cells that express CXCR5, which directs them to the B cell follicles, effector CD4 and CD8 cells are directed out of the lymph nodes and the bloodstream and are distributed throughout peripheral tissues. Their migration is guided by regional signals that are distinct for different tissues and that are recognized by receptors induced by dendritic cells so that effector T cells can be preferentially directed back to the region of the body in which the dendritic cells were activated. For example, activated dendritic cells draining to mesenteric lymph nodes promote more efficient differentiation of intestinal-homing T_H cells (by inducing expression of the receptor for the constitutive small intestinal chemokine, CCL25) than do dendritic cells draining from peripheral skin sites, which promote the differentiation of skin-homing T_H cells (by inducing expression of the receptors for the constitutive dermal chemokines, CCL17 and CCL27). Regional homing molecules are summarized in Figure 5-24.

Unlike naïve T cells, effector cells do not require CD28 costimulation for activation. However, effector T cells, particularly after moving into tissues, also express inhibitory receptors, such as PD-1 (see Figure 5-20), which limit their inadvertent activation until appropriate stimulation is achieved through the antigen receptor. Importantly, inflammatory signals at sites of infection promote the accumulation, retention and activation of effector cells by upregulating further on tissues the ligands for the array of migratory receptors and downregulating the ligands for inhibitory receptors.

The overall result after clonal expansion and differentiation in lymph nodes is to create, from small numbers of naïve precursors, the delivery of large numbers of pathogen-specific cells to sites of inflammation and the dispersal of a pool of pathogen-specific effector cells to many tissues.

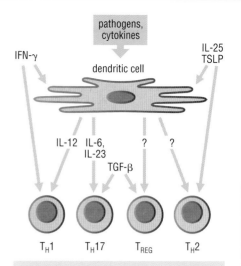

Figure 5-23 Dendritic cells integrate peripheral signals and mediate T_H subset differentiation Pathogens in the tissues activate Toll-like receptors and C-type lectin receptors on DCs, and induce local cytokines, which influence the types of cytokines, such as IL-12, IL-23 and IL-6, that DCs produce during activation of naïve CD4 T cells. Cytokines produced by other cells in the tissues, such as IFN-γ, TGF-β, IL-25 and TSLP, can also contribute to the selection and outgrowth of distinct T_H subsets.

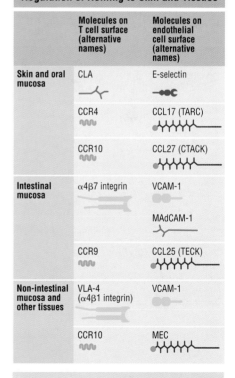

Regulation of Homing to Skin and Tissues

	Molecules on T cell surface (alternative names)	Molecules on endothelial cell surface (alternative names)
Skin and oral mucosa	CLA	E-selectin
	CCR4	CCL17 (TARC)
	CCR10	CCL27 (CTACK)
Intestinal mucosa	α4β7 integrin	VCAM-1
		MAdCAM-1
	CCR9	CCL25 (TECK)
Non-intestinal mucosa and other tissues	VLA-4 (α4β1 integrin)	VCAM-1
	CCR10	MEC

Figure 5-24 Table of homing molecules used by effector T lymphocytes CLA: cutaneous lymphocyte associated antigen; CTACK: cutaneous T cell attracting chemokine; MAdCAM: mucosal addressin cell adhesion molecule; TARC: thymus and activation regulated chemokine; TECK: thymus-expressed chemokine; VCAM: vascular cell adhesion molecule; VLA: very late activation antigen.

Polarization of T cell subsets is accompanied by changes in key transcriptional regulators

The changes in chromatin structure that accompany differentiation of distinct CD4 T cell subsets are required to make specific cytokine genes accessible to transcriptional regulators of two kinds: transcriptional regulators such as NFAT that are activated by TCR signaling and participate in the activation of many genes important for T cell functions (see section 5-5); and transcriptional regulators that are highly specific for the genes that are differentially expressed in the different T cell subsets. For T$_H$1 and T$_H$2 cells, these specific transcriptional regulators are **T-bet** (also known as Tbx21), a member of the *T-box family* of transcriptional regulators that drives T$_H$1 differentiation, and **GATA-3**, a member of the *GATA family* of transcriptional regulators that drives T$_H$2 differentiation. In naïve T cells, both T-bet and GATA-3 are present at low levels. Differentiation occurs as one transcriptional regulator or the other predominates, leading to the expression of genes promoting one cell fate and the repression of those promoting the other, and thus the *polarization* of the cell as a T$_H$1 or T$_H$2 cell. Such polarization of different cell fates can also be seen in the differentiation of T$_H$17 cells and T$_{REG}$ cells, which we discuss later, but it is best understood for the T$_H$1 and T$_H$2 helper subsets.

The differentiation of T$_H$1 cells is mediated by induction of T-bet

T-bet is low in naïve CD4 T cells, and is increased by peptide–MHC interactions from activated dendritic cells that generate TCR signaling resulting in sustained activation of the JNK and p38 MAP kinases and NF-κB. T-bet activates the IFN-γ and the IL-12Rβ2 genes, as well as a second transcription factor, **Hlx**. This initiates a series of auto-reinforcing positive

Figure 5-25 T-bet induction drives the differentiation of T$_H$1 cells TCR activation mediates the induction of T-bet through NFAT, which accumulates until T-bet activates the IFN-γ, Hlx and IL-12Rβ2 genes. IFN-γ, secreted both from the differentiating T cells and from other cells in the environment, such as NK cells recruited to the lymph nodes or spleen, further increases T-bet levels through activation of STAT1. IFN-γ, as well as Toll-like receptor ligation, induces IL-12 release from dendritic cells, which further activates IFN-γ induction in T cells through the activation of STAT4 by stimulation through the upregulated IL-12 receptor. Hlx with T-bet works to establish a transcriptionally accessible chromatin state at the IFN-γ gene (not shown), and Hlx also represses expression of the IL-4Rα gene, making the cells unresponsive to IL-4. IFN-γ also induces the transcription factors IRF-1 and IRF-2, which repress expression of the IL-4 gene. Activated T-bet is modified so that the modified protein can physically interact with and sequester GATA-3 proteins, preventing them from interacting with the IL-4 locus in the differentiating T cells, thus blocking T$_H$2 development.

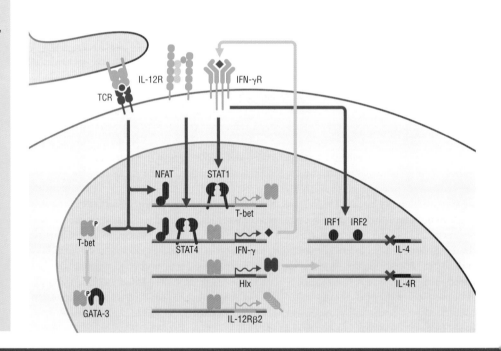

Definitions

chemoattractant receptor homologous molecule expressed on T$_H$2 cells (CRT$_H$2): a G-protein-coupled seven-transmembrane receptor that is expressed on T$_H$2 cells, eosinophils and basophils and that mediates chemotaxis to prostaglandin D$_2$.

c-Maf: a transcriptional regulator implicated in IL-4 and IL-10 expression in T$_H$2 cells and in lens fiber cell differentiation, and the cellular homolog of a proto-oncogene from the avian musculoaponeurotic fibrosarcoma virus.

CRT$_H$2: see **chemoattractant receptor homologous molecule expressed on T$_H$2 cells**.

eomesodermin: a T-box family transcriptional regulator (Tbr2) that directs trophoblast and mesoderm development in vertebrates and is implicated in the differentiation of effector programs in CD8 T cells and NK cells.

GATA-3: a member of a family of six zinc-finger transcription factors that share a conserved DNA-binding domain, which interacts with a core GATA nucleotide sequence that gives the family its name.

Hlx: a member of the homeodomain family of transcriptional regulators expressed in discrete cell types at specific times during development and in adult hematopoietic tissues and cells, as well as in diverse mesodermal tissues.

T-bet: a member of the large T-box family of transcriptional regulators implicated in developmental cell-fate decisions important for vertebrate body plan and organogenesis and named for the mouse *Brachyury* (also known as *T*) and *Drosophila optomotor-blind* genes, which share a common DNA-binding domain designated the T-box.

feedback loops that sustain T_H1 subset differentiation (Figure 5-25). Secreted IFN-γ binds to its receptor, leading to STAT1 activation, which further increases T-bet expression. IL-12Rβ2 pairs with the constitutive IL-12Rβ1 on the naïve CD4 T cells, thus allowing the differentiating T cells to respond to the cytokine IL-12, which is expressed by dendritic cells activated through TLRs and/or the IFN-γR. IL-12 further augments IFN-γ expression by activation of STAT4. Hlx interacts with T-bet to stabilize the processive opening of the IFN-γ gene locus, so that important regulatory areas involved in expression become fully accessible to the core transcriptional regulators, including the NFAT family members, activated after TCR stimulation.

The events that drive a CD4 T cell into the T_H1 subset are coupled with transcriptional events that block differentiation into the T_H2 subset. IFN-γ induces expression of the transcription factors IRF1 and IRF2, which repress IL-4 expression, and Hlx represses IL-4Rα expression, rendering T_H1 cells unresponsive to IL-4. Activated T-bet also interferes with the activity of GATA-3, while other signals repress and silence the GATA-3 gene. Additional cytokine receptors, such as the IL-18R, are expressed on the cells and enable them to respond to inflammatory signals in the tissues that increase IFN-γ secretion after TCR engagement. Finally, the chemokine receptors CCR5, CXCR3, CXCR6 and CX3CR1 become differentially expressed on T_H1 cells, and these are responsive to chemokines induced by IFN-γ, thus promoting the accumulation of T_H1 cells at sites where they are needed.

T-bet also controls competence to secrete IFN-γ in CD8 T cells, but this is in concert with a second T-box family member, **eomesodermin**. These two transcriptional regulators together regulate IFN-γ expression and the cytolytic pathway in both CD8 T cells and NK cells, which share these effector activities. They also induce expression of the IL-15 receptor, a γ_c cytokine receptor required for the survival of CD8 T cells and NK cells. Certain cytokine combinations, such as IL-12 and IL-18, can directly elicit IFN-γ release from CD8 effector T cells and T_H1 cells without the need for TCR stimulation.

The differentiation of T_H2 cells is mediated by induction of GATA-3

Like T-bet, GATA-3 is low in naïve CD4 T cells in part because it is actively degraded. Partial activation of the Erk MAP kinase together with the NF-κB pathway by TCR stimulation leads to impaired GATA-3 degradation, allowing the GATA-3 protein to accumulate and bind to regulatory sites across a cluster of cytokine genes that includes IL-4, IL-13 and IL-5. Although GATA-3 can directly activate the IL-13 and IL-5 genes, its more important role is to mediate changes in chromatin that increase accessibility of the entire cytokine cluster to transcriptional regulators such as NFAT that are activated by TCR engagement, allowing activation of these genes. GATA-3 also feeds back to activate its own gene, thus increasing its own production in a positive feedback loop. IL-4 also promotes GATA-3 expression in an autocrine and paracrine fashion by activating STAT6. ICOSL on dendritic cells, upon binding ICOS on the activated T cells, induces another transcription factor, **c-Maf**, which is required for optimal IL-4 expression (Figure 5-26). Additional pathways, analogous to those operating in T_H1 cells to block T_H2 differentiation, mediate silencing of the IFN-γ and T-bet genes. Finally, T_H2 cells preferentially express receptors, particularly the **chemoattractant receptor homologous molecule expressed on T_H2 cells (CRT_H2)**, but also certain chemokine receptors, such as CCR3, that bind to chemotactic lipids and chemokines induced by IL-4 and IL-13 at sites of epithelial inflammation, thus promoting the accumulation of T_H2 cells at sites where they are needed.

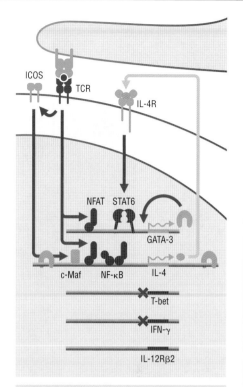

Figure 5-26 GATA-3 induction drives the differentiation of T_H2 cells TCR activation induces the stabilization of GATA-3, allowing the protein to accumulate and bind to sites in the promoter of GATA-3 itself, as well as across sites that span the IL-4 gene, rendering it accessible to transcriptional regulators, such as NFAT, activated by TCR signaling. Ligation of ICOS by receptors on activated dendritic cells activates the expression of c-Maf, a second transcriptional regulator that optimizes expression across the IL-4 gene. IL-4 secreted from T cells binds the constitutively expressed IL-4R and activates STAT6, which further augments the expression of GATA-3. The T-bet and IFN-γ genes become silenced by mechanisms that are not yet fully understood.

References

Ansel, K.M. et al.: **Regulation of TH2 differentiation and Il4 locus accessibility.** Annu. Rev. Immunol. 2006, **24**:607–656.

Dong, C. et al.: **MAP kinases in the immune response.** Annu. Rev. Immunol. 2002, **20**:55–72.

Intlekofer, A.M. et al.: **Effector and memory CD8+ T cell fate coupled by T-bet and eomesodermin.** Nat. Immunol. 2005, **6**:1236–1244.

Mullen, A.C. et al.: **Hlx is induced by and genetically interacts with T-bet to promote heritable T_H1 gene induction.** Nat. Immunol. 2002, **3**:652–658.

Szabo, S.J. et al.: **Molecular mechanisms regulating TH1 immune responses.** Annu. Rev. Immunol. 2003, **21**:713–758.

Yamashita, M. et al.: **Ras-ERK MAPK cascade regulates GATA3 stability and Th2 differentiation through ubiquitin-proteosome pathway.** J. Biol. Chem. 2005, **280**:29409–29419.

T$_H$1 cells coordinate the destruction and control of pathogens by cytolytic cells and activated phagocytes

The function of T$_H$1 cells is the destruction of microorganisms through the activation of phagocytic cells. The actions of T$_H$1 cells on phagocytes are mediated by the characteristic cytokines of T$_H$1 cells, and in particular IFN-γ. Other cytokines produced by T$_H$1 cells include the TNF-superfamily cytokines TNF and lymphotoxin, and also IL-2, GM-CSF, and certain chemokines including MIP family members and RANTES (CCL5). Together, these cytokines coordinate the production, recruitment and activation of immune cells that mediate host defense against many viruses, bacteria, fungi and protozoan parasites. Often, T$_H$1 cells appear in tissues with NK cells and CD8 T cells, the other two major IFN-γ-producing cells of the immune system, together with activated macrophages and dendritic cells.

When T$_H$1 cell-mediated immune responses fail to eradicate organisms, *granulomas* can be induced. Granulomas are characterized by epithelioid cells, activated macrophages, T$_H$1 cells and cytotoxic CD8 T cells, which surround and wall off the organisms from the host tissues. *Mycobacterium tuberculosis*, a lipid-rich organism that is difficult to eradicate, induces chronic granulomas in many tissues, which enclose the persisting bacteria, and the maintenance of these granulomas is dependent on the sustained production of IFN-γ, as we discuss more fully in Chapter 9.

The effects of T$_H$1 cells are mediated by cytokines

The major activities of T$_H$1 cells represent the pleiotropic effects of IFN-γ (Figure 5-27). Although T$_H$1 cells can enter many tissues after leaving the lymph nodes, they are recruited preferentially to sites of inflammation, where they are directed by chemokines and selectins induced on the endothelium of the local blood vessels in response to cytokines produced in the underlying inflamed tissues (see section 3-14). Within inflammatory sites, T$_H$1 cells can be activated either by antigens from pathogens presented on MHC class II molecules at the surface of macrophages or dendritic cells or their precursors recruited from blood, or by the inflammatory cytokines IL-12 and IL-18. Either or both of these stimuli induce T$_H$1 cells to secrete IFN-γ. IFN-γ acts on dendritic cells to promote their maturation and upregulate MHC class II expression and loading (see section 4-8). Major functions of IFN-γ on macrophages are to upregulate MHC class II and the expression of inflammatory cytokines including IL-1β, IL-6,

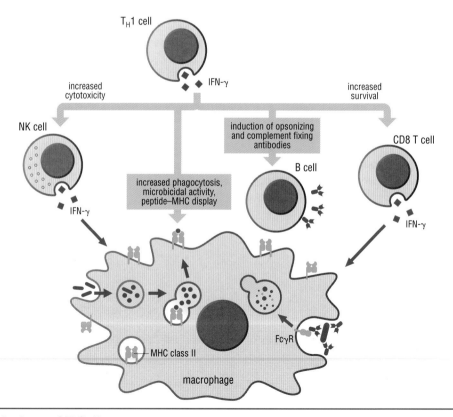

Figure 5-27 The pleiotropic effects of IFN-γ are central to immunity mediated by T$_H$1 cells IFN-γ increases the microbicidal activity of activated macrophages, enhances antigen presentation by macrophages and dendritic cells, induces opsonizing and complement-fixing antibodies, and enhances macrophage phagocytic activity. T$_H$1 cells also enhance the activity of NK cells and the growth and survival of CD8 T cells by the elaboration of IL-2 and other cytokines, which, in turn, allows these cells to contribute additional IFN-γ to the response.

TNF, IL-12 and IL-18, and the IL-12 and IL-18 receptors, and at the same time to activate the oxidative (NADPH oxidase) and nitrosative (iNOS) microbicidal systems and upregulate the Fc receptors and complement receptors that promote phagocytosis. Induction of CCL2 by IFN-γ promotes the release from bone marrow of blood monocyte precursors of a specialized class of cells known as **TipDCs** (for **TNF/iNOS-producing dendritic cells**), which are specialized for the early destruction of intracellular bacteria, such as *Listeria*, as we discuss in Chapter 9. IFN-γ also increases the secretion of chemokines, such as CXCL9, CXCL10, CXCL11 and CXCL16, that bind receptors preferentially expressed on T$_H$1 cells, providing an additional mechanism to polarize the immune response in the corresponding tissue site (Figure 5-28).

Inflammatory cell recruitment and activation is also a property of TNF, which potently activates neutrophils and monocytes; of GM-CSF, which promotes the bone marrow production and subsequent differentiation of blood monocytes to tissue dendritic cells; and of IL-2, which enhances the cytotoxicity of NK cells and CD8 T cells. In turn, cytokines carried to lymph nodes by lymphatics from inflammatory sites, or produced by activated dendritic cells, can induce the entry of NK cells from blood to the lymph node, where IFN-γ released from these cells can contribute to T$_H$1 differentiation by increasing T-bet expression, thus reinforcing polarization.

T$_H$1 cytokines induce B cells to make the types of antibodies that promote the destruction of pathogens by phagocytic cells

IFN-γ also has important effects on B cells (Chapter 6), mediated in part by the induction of T-bet through activated STAT1, as in T$_H$1 cells. After induction during BCR stimulation, T-bet promotes the generation of specific isotypes of antibodies that bind with high affinity to the activating Fc receptors on phagocytes, thus stimulating the ingestion of antibody-coated pathogens. These antibodies are also potent activators of complement (see section 3-5), which can directly destroy microbes as well as facilitating their uptake by phagocytes and recruiting additional inflammatory cells.

T$_H$1 cells coordinate a system for antigen elimination but can cause immunopathology

The overall outcome of T$_H$1 cell activation is the elimination of pathogens largely by innate immune cells such as activated macrophages and NK cells, although CD8 T cells are also recruited. This process is amplified by the enhanced presentation of immunogenic peptides derived from the invading organism on host MHC molecules (see section 4-8). Mice that are IFN-γ-deficient are susceptible to a variety of intracellular and extracellular systemic pathogens, including bacteria, viruses, yeast and protozoa. Organisms that can live within host cells are particularly problematic. In humans, deficiencies of IL-12, the IL-12 receptor and the IFN-γ receptor are associated with persistent *Mycobacteria* and *Salmonella* infections. Complete deficiency of STAT1, which is activated by receptors for both type 1 interferons and IFN-γ, is usually lethal soon after birth, as a result of overwhelming infection.

Inadequately controlled T$_H$1-mediated immunity can result in lethal organ damage, for example overwhelming hepatitis virus infection; or, when present for prolonged periods, it can be associated with chronic inflammation, which has been linked with cancer, autoimmunity and even atherosclerosis.

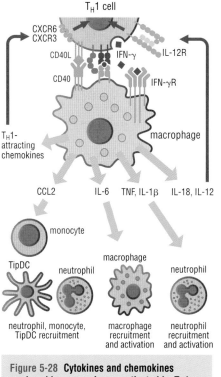

Figure 5-28 Cytokines and chemokines produced by macrophages activated by T$_H$1 cells, and their main actions

Definitions

TipDCs: see **TNF/iNOS-producing dendritic cells**.

TNF/iNOS-producing dendritic cells (TipDCs): monocytes with a specialized function in the early destruction of intracellular bacteria. These cells express the chemokine receptor CCR2 and enter sites of bacterial infection and mature in the presence of IFN-γ and bacterial products into MHC-class-II- and CD11c-positive cells that express high levels of TNF and iNOS.

References

Schroder, K. *et al.*: **Interferon-γ: an overview of signals, mechanisms and functions.** *J. Leukoc. Biol.* 2004, **75**:163–189.

Wald, O. *et al.*: **IFN-γ acts on T cells to induce NK cell mobilization and accumulation in target organs.** *J. Immunol.* 2006, **176**:4716–4729.

Xu, W. and Zhang, J.J.: **Stat1-dependent synergistic activation of T-bet for IgG2a production during early stage of B cell activation.** *J. Immunol.* 2005, **175**:7419–7424.

T$_H$2 cells coordinate barrier immunity against invading pathogens

The function of T$_H$2 cells is to organize a stereotyped set of myeloid cells to prevent blood-sucking worms and insects from flourishing at epithelial barriers. In the absence of T$_H$2 cells, intestinal worms and biting insects feed for prolonged periods, which can lead to epithelial breakdown and bacterial superinfections. The cytokines that mediate the functions of T$_H$2 cells include the canonical T$_H$2 cytokine IL-4, as well as IL-13 and IL-5, which are encoded in the same region of the genome as IL-4 and tend to be regulated coordinately with IL-4, and IL-9 and IL-10. The major functions of these cytokines are to recruit, activate and promote the survival of eosinophils, basophils and mast cells. Additionally, T$_H$2 cells activate macrophages, although in a way that is distinct from the activation of macrophages by T$_H$1 cells, and cause alterations in mucosal epithelial cells and smooth muscle that render it difficult for blood-sucking pathogens to continue to ingest blood. These changes are reversible, and the epithelium reverts to its basal state after invading organisms have been repelled.

T$_H$2 cytokines induce alternatively activated macrophages and activate eosinophils, basophils and mast cells

T$_H$2 cells disperse to many tissues after leaving the lymph nodes, but are retained at inflammatory sites by the actions of adhesive molecules and of chemokines such as CCL1, CCL11, CCL17, CCL22 and CCL24, which are induced by IL-4 and IL-13 and secreted from tissue cells. Like T$_H$1 cells, T$_H$2 cells activate macrophages on recognition of peptide–MHC at their surface; but activation of macrophages by T$_H$2 cells is distinct from classical macrophage activation mediated by IFN-γ, and for this reason macrophages activated by T$_H$2 cells are known as **alternatively activated macrophages** (Figure 5-29). In both cases, antigen presentation and costimulatory molecules are upregulated. Alternative macrophage activation is, however, characterized by the induction of the mannose receptor and dectin, implicated in fungal recognition; chemokines, such as CCL17, CCL18 and CCL22, that attract T$_H$2 cells, eosinophils and basophils; antiinflammatory cytokines, such as IL-1Ra, IL-10 and TGF-β (see section 3-15); arginase, which degrades arginine, the substrate for iNOS, and generates proline for collagen synthesis; chitinases, which can bind and degrade chitin on fungi and insects; and proteases important in degrading the extracellular matrix so that immobilized growth factors and chemokines can be released to promote tissue remodeling. Thus alternatively activated macrophages help recruit cell types distinct from those recruited by classically activated macrophages, and have antiinflammatory and repair functions.

On mucosal epithelial cells, IL-4 and IL-13 increase proliferation and turnover, and upregulate the expression of mucin genes while increasing the secretion of salts and fluids and thus the flow of mucus across the lumen, making it difficult for worms to feed. On smooth muscle cells, IL-4 and IL-13 increase contractility.

IL-5 induces the production of eosinophils in the bone marrow and promotes their survival in tissues. IL-9 similarly promotes the production and survival of mast cells and basophils. These

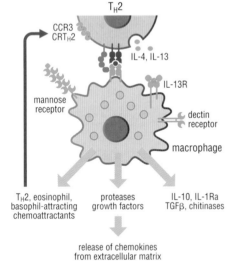

Figure 5-29 Alternatively activated macrophages can be generated by exposure of tissue macrophages to IL-4 and/or IL-13 T$_H$2 cells secrete IL-4 and IL-13, which induce the expression of chemoattractants that recruit additional T$_H$2 cells, as well as eosinophils and basophils, growth factors and proteases involved in tissue remodeling, antiinflammatory cytokines and chitinases involved in the degradation of chitin. Mannose and dectin C-type lectin receptors are also highly expressed.

Definitions

alternatively activated macrophages: macrophages activated by T$_H$2 cells and which promote the recruitment of eosinophils and basophils and more T$_H$2 cells, and secrete antiinflammatory cytokines, as distinct from macrophages activated by T$_H$1 cells and which have inflammatory effects.

Figure 5-30 Mast cell and basophil activation IL-4 induces the production of IgE by B cells. IgE binds to Fc receptors on mast cells and basophils, causing degranulation when the bound IgE is cross-linked by antigen.

cells in turn are not only potent producers of IL-4 and IL-13, they also release vasoactive mediators that lead to tissue edema and constriction of blood vessels, making it difficult for worms and biting insects to feed. IL-10, which is secreted later after activation than the other T$_H$2 cytokines, has an overall antiinflammatory role that is of particular importance in the small intestine, perhaps because of the large numbers of intestinal bacteria in close apposition to immune cells at that site.

T$_H$2 cytokines induce B cells to produce antibodies involved in barrier immunity

Like IFN-γ, IL-4 (and IL-13 in humans) also acts on B cells. IL-4 leads to the production of antibodies with properties that promote the activation of the types of innate immune cells activated by T$_H$2 cells. These antibody isotypes, in contrast to those induced by IFN-γ, activate complement poorly and are less effective in promoting phagocytosis by macrophages. They include the highly specialized IgE isotype, whose production is dependent on IL-4. IgE binds to mast cell and basophil Fc receptors, arming these cells with antibody specific for mucosal pathogens. Upon a second encounter, the cross-linked IgE-bound Fc receptors cause explosive mast cell and basophil degranulation and release of preformed histamine and potent vasoactive lipid derivatives, and induce the cytokines IL-4 and IL-13, which promote more rapid smooth muscle and epithelial changes (Figure 5-30). The second specialized isotype important in barrier immunity is IgA, which is induced by IL-5 and TGF-β. A large proportion of IgA is transported across mucosal epithelial cells to bind and neutralize pathogens and their toxins before they invade the body. Eosinophils express Fc receptors that recognize IgA bound to migrating worms, and activation through these receptors leads to the release of the contents of their toxic granules onto the parasites. Together, IgE and IgA function to augment immune function at epithelial barriers of the body. The main functions of T$_H$2 cells are summarized in Figure 5-31.

Uncontrolled T$_H$2 responses can cause immunopathology

When controlled, T$_H$2-mediated responses protect by inducing smooth muscle contraction, mucus secretion and tissue edema that dislodges worms. They can, however, occur in excess, in response to otherwise innocuous substances such as pollens or peanuts, and in this case such responses result in allergic diseases, which are discussed further in Chapter 13.

When confronted with pathogens that cannot be entirely cleared, T$_H$2 cells, like T$_H$1 cells, can induce tissue granulomas to contain them. T$_H$2 cell-mediated granulomas consist of T$_H$2 cells, epithelioid cells, alternatively activated macrophages and eosinophils, which circumferentially wall off and surround the organisms. Granulomatous responses occur after aberrant worm or egg migration into tissues, and, with time, can result in substantial organ damage. The pathology of schistosomiasis reflects such activities and is discussed in Chapter 11.

Figure 5-31 Actions of T$_H$2 cell cytokines

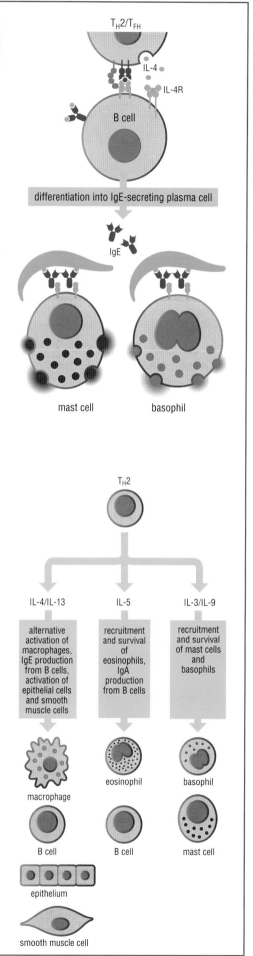

References

Fallon, P.G. *et al.*: **IL-4 induces characteristic Th2 responses even in the combined absence of IL-5, IL-9, and IL-13.** *Immunity* 2002, **17**:7–17.

Gordon, S.: **Alternative activation of macrophages.** *Nat. Rev. Immunol.* 2003, **3**:23–35.

Nair, M.G. *et al.*: **Novel effector molecules in type 2 inflammation: lessons drawn from helminth infection and allergy.** *J. Immunol.* 2006, **177**:1393–1399.

Voehringer, D. *et al.*: **Type 2 immunity reflects orchestrated recruitment of cells committed to IL-4 production.** *Immunity* 2004, **20**:267–277.

T_{FH} cells provide help for germinal center B cells

Follicular T helper cells, or T_{FH} cells, are effector T_H cells that express CXCR5 and are positioned in B cell follicles. It is not known whether these cells arise as a distinct subset, or after further differentiation of T_H1, T_H2 and T_H17 cells that respond to CXCL13, the ligand for CXCR5 constitutively produced by follicular stromal cells. The second possibility, however, would be consistent with the exquisite matching of the functions of antibodies induced to the types of myeloid cells activated by the cytokines released by T helper cells of the different subsets. Thus it seems likely that as cells of each T_H subset clonally expand and differentiate on activated dendritic cells, some cells of each subset respond to CXCL13 and move to the follicles. Once the T cells recognize peptide–MHC on the surface of a B cell, interactions between the T cell and the B cell drive the further differentiation of the T cell. These interactions include, in addition to the signal from the TCR on peptide–MHC class II engagement, the engagement of CD40 with CD40L, which induces ICOS and ICOSL on the interacting cells and promotes the formation of the germinal center: these interactions are discussed in more detail in Chapter 6. Differentiated T_{FH} cells continue to make the specific cytokines induced during their activation by dendritic cells, but remain in follicles and acquire properties that differ from those of peripheral effector T_H subsets. These include the secretion of **IL-21**, an important cytokine for B cell differentiation, and expression of the transcriptional regulator **BCL-6**.

T_H17 cells recruit neutrophils to inflammatory sites

T_H17 cells secrete IL-17A and IL-17F, two closely related members of the **IL-17-family** cytokines, as well as IL-6 and TNF. T_H17 cells are generated from naïve CD4 T cells by priming in the presence of TGF-β and IL-6, which induces the expression of the transcriptional regulator **RORγt** (for **retinoic acid receptor-related orphan receptor-gamma thymus**), which is required for their development. T_H17 cells express the inflammatory chemokine receptor CCR6 and the IL-23 receptor, and their survival is greatly promoted by IL-23 (see section 2-8) released from activated dendritic cells and macrophages. T_H17 cells and dendritic cells that express IL-23 are constitutively present in small bowel tissues, probably reflecting interactions with the large numbers of commensal bacteria in this organ.

IL-17 binds to receptors on stromal, epithelial and endothelial cells, and perhaps some tissue macrophages, to promote the release of IL-6, as well as the inflammatory chemokines CXCL8, CXCL1, CXCL6 and the hematopoietic growth factors G-CSF and GM-CSF. Together, these cytokines promote bone marrow expansion and the release of neutrophils and monocytes, and promote their subsequent recruitment into inflammatory tissue sites, where these potent phagocytic cells engulf and kill invading organisms. IL-12, IFN-γ and IL-4 inhibit T_H17 differentiation, suggesting that once polarized, T_H1 and T_H2 responses downregulate the localized, acute inflammatory processes induced by T_H17 cells.

Small molecules, such as PGE_2 and ATP, that appear extracellularly in areas of acute inflammation and cell necrosis, and certain bacterial cell wall constituents such as those from *Bordetella pertussis*, bind to receptors on activated macrophages and dendritic cells, provoking the release of IL-23. These conditions contribute to the survival and growth of T_H17 cells when they enter inflammatory sites, and amplify the recruitment of neutrophils. Uptake of apoptotic neutrophils in tissues by contrast decreases IL-23 release from macrophages, providing a homeostatic mechanism for decreasing IL-17 signals from the tissues as infection is controlled, thus returning blood neutrophils to normal baseline levels (Figure 5-32).

Definitions

BCL-6: a zinc finger family transcriptional regulator highly expressed in germinal center B cells that prevents differentiation of mature B cells to plasma cells, in part by repressing Blimp-1 transcription while maintaining expression of the machinery required for hypermutation and follicular homing.

FoxP3: a member of the winged helix-forkhead DNA-binding domain protein family highly conserved in vertebrates that acts as a transcriptional repressor. In immunology, FoxP3 is a lineage-specific marker for reg-

ulatory T cells (T_{REG} cells). Genetic deficiencies of FoxP3 in the scurfy mouse and human IPEX syndrome (immunodysregulation, polyendocrinopathy, enteropathy, X-linked syndrome) lead to fatal inflammatory autoimmune diseases.

GITR: see **glucocorticoid-induced TNF-family receptor**.

glucocorticoid-induced TNF-family receptor (GITR): a TNF-family receptor expressed on activated effector T cells and constitutively on regulatory T cells. GITR signals render activated effector T cells refractory to T_{REG}-mediated suppression.

IL-17 family: a six-member family of homodimeric cytokines designated IL-17A to IL-17F; IL-17E is also called IL-25. T-tropic Herpesvirus saimiri encodes an IL-17A-like molecule (viral IL-17).

IL-21: a type I four-helix-bundle-type cytokine that binds to IL-21R/common γ chain receptors on activated B cells, T cells and NK cells.

retinoic acid receptor-related orphan receptor-gamma thymus (RORγt): a largely T cell-specific isoform of the nuclear hormone receptor superfamily expressed predominantly in CD4 CD8 double-positive thymocytes

In the absence of IL-17, mice are highly susceptible to bacterial and fungal infections. When dysregulated, however, IL-17 can cause chronic inflammation. T$_H$17 cells have been implicated in murine models of inflammatory arthritis and autoimmune encephalomyelitis (an animal model of multiple sclerosis that commonly involves pertussis toxin used to induce inflammation), and IL-17 levels are increased in a variety of human inflammatory diseases, including rheumatoid arthritis, inflammatory bowel disease and asthma.

Induced T$_{REG}$ cells suppress activation of effector T$_H$ cells

T regulatory cells, or T$_{REG}$ cells, suppress the actions of other T$_H$ effector cells. Most T$_{REG}$ cells differentiate in the thymus and express a diverse TCR repertoire that is skewed towards recognition of autoantigens. These cells will be discussed in Chapter 12. Some T$_{REG}$ cells, however, differentiate from naïve CD4 T cells in secondary lymphoid organs; they are called induced T$_{REG}$, or iT$_{REG}$, cells. Like T$_{REG}$ cells, iT$_{REG}$ cells express the transcriptional activator **FoxP3** and the α chain of the IL2 receptor, CD25, as well as the TNF family members **GITR** (for **glucocorticoid-induced TNF-family receptor**) and OX40, and inhibitors of TCR activation such as CTLA4 and PD-1.

In vitro, naïve CD4 T cells can be induced by IL-2 and TGF-β to express FoxP3 and differentiate into iT$_{REG}$ cells. It is noteworthy that the further addition of IL-6 blocks FoxP3 induction and leads to the development of T$_H$17 cells, suggesting a mechanism by which T$_{REG}$ differentiation is suppressed under inflammatory conditions that induce T$_H$17 cells, when neutrophils are acutely needed. Although the precise mechanisms by which iT$_{REG}$ cells suppress the activation of naïve T cells are likely to differ under different conditions, suppression requires cell contact, probably with activated dendritic cells in the lymph nodes, and, in some circumstances, the induction of antiinflammatory cytokines such as IL-10 and TGF-β.

Although T$_{REG}$ cells suppress responses to autoantigens, iT$_{REG}$ cells arise *in vivo* in the presence of chronically high levels of antigen. Such conditions are seen experimentally in mice with chronic infectious diseases, such as tuberculosis or schistosomiasis, and in animals with large tumor burdens. In such cases, the differentiation of iT$_{REG}$ cells may serve to limit the pathology that would otherwise be caused by prolonged activation of effector cells. Expansion of thymic T$_{REG}$ cells also occurs under these conditions, and both populations are likely to contribute to the suppression of immunopathology.

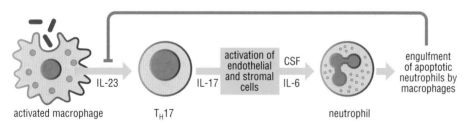

Figure 5-32 T$_H$17 cells amplify acute immune responses mediated by neutrophils T$_H$17 cells are maintained by IL-23 secreted by activated macrophages, and recruit neutrophils by means of secreted IL-17, which acts on endothelial and stromal cells to induce IL-6, chemokines and CSF. Uptake of dead neutrophils by macrophages inhibits IL-23 secretion and thus downregulates the T$_H$17 response. Under resting conditions, levels of IL-17 from γδ T cells and other specialized tissue lymphocytes (discussed in Chapter 8) contribute to basal surveillance by neutrophils.

and also in CD4-positive lymph node inducer cells.

RORγt: see **retinoic acid receptor-related orphan receptor-gamma thymus**.

References

Bettelli, E. *et al*.: **Reciprocal developmental pathways for the generation of pathogenic effector T$_H$17 and regulatory T cells.** *Nature* 2006, **441**:235–238.

Fontenot, J.D. and Rudensky, A.Y.: **A well adapted regu-** latory contrivance: regulatory T cell development and the forkhead transcription factor Foxp3. *Nat. Immunol.* 2005, **6**:331–337.

Ivanov, I.I. *et al*.: **The orphan nuclear receptor RORγt directs the differentiation program of proinflammatory IL-17$^+$ T helper cells.** *Cell* 2006, **126**:1121–1133.

McKenzie, B.S. *et al*.: **Understanding the IL-23–IL-17 immune pathway.** *Trends Immunol.* 2006, **27**:17–23.

Stark, M.A. *et al*.: **Phagocytosis of apoptotic neutrophils regulates granulopoiesis via IL-23 and IL-17.** *Immunity* 2005, **22**:285–294.

Veldhoen, M. *et al*.: **TGFβ in the context of an inflammatory cytokine milieu supports de novo differentiation of IL-17-producing T cells.** *Immunity* 2006, **24**:179–189.

Vinuesa, C.G. *et al*.: **Follicular B helper T cells in antibody responses and autoimmunity.** *Nat. Rev. Immunol.* 2005, **5**:853–865.

Ziegler, S.F.: **FOXP3: of mice and men.** *Annu. Rev. Immunol.* 2006, **24**:209–226.

CD8 T cells are important for the clearance of viruses and bacteria that proliferate in the cytosol

The major established effector function of CD8 T cells is direct cytotoxicity. They are specialized to provide immunity to viruses, which can replicate only inside cells, and to a small but important class of bacteria specifically adapted to enter the cytoplasm of cells and proliferate there. This bacterial class includes *Listeria monocytogenes*, a widespread human pathogen that we discuss in Chapter 9, and *Salmonella* species, which cause typhoid and food poisoning. Such intracellular pathogens are undetectable by macrophages and antibodies during their proliferation, and some spread from cell to cell without ever emerging into the extracellular space: clearance of infection can thus depend upon the elimination of the infected cell by cytotoxic lymphocytes, which detect microbial components displayed on the cell surface by MHC class I molecules.

Differentiation to effector function in CD8 T cells generates cells that on subsequent encounter with an infected target secrete the cytokines IFN-γ and TNF (Figure 5-33), and release cytotoxic mediators that deliver a lethal hit to the target. The actions of IFN-γ have already been discussed in the context of T$_H$1 cells: two important effects of this cytokine are the upregulation of MHC class I molecule expression and the induction of the immunoproteasome, which greatly increase the effectiveness of antigen presentation to CD8 T cells (see section 4-8). TNF can activate apoptosis in some cells, as we have seen in section 2-9 (see Figure 2-27), and is a major inflammatory cytokine. Some chemokines are also released by CD8 T cells. These immune mediators are, however, all also provided by T$_H$ cells, NK cells and activated macrophages that characterize sites of CD8 T cell action, as we have seen in section 5-11; and the most important effector property of CD8 cells is their antigen-specific contact-mediated cytotoxicity, which we describe below.

There are two main mechanisms of CD8 T cell cytotoxicity

CD8 T cells can kill their targets within minutes of contact and are capable of destroying many infected target cells in a few hours while sparing neighboring healthy cells. They also avoid killing themselves.

The two major mechanisms by which CD8 T cells destroy their targets both activate the programmed death of the cell. The more important and faster acts through the release of specialized cytotoxic components, **perforin**, granulysin and **granzymes**, which are held in granules in the CD8 T cell cytoplasm and released by calcium signaling from the TCR. The other operates through the expression of ligands for death receptors (see Figure 2-27a), of which the best-studied is FasL, which is induced on the CD8 T cell on contact with an infected target and binds to Fas on the target cell, activating the apoptotic pathway. FasL cytotoxicity requires transcriptional activation of the gene, and therefore occurs over a period of hours. Granzyme–perforin cytotoxicity operates in minutes. Granzymes (from granule enzyme) comprise many proteases, the most abundant and the best-understood of which are granzyme A and granzyme B; perforin is closely related to the pore-forming complement component C9 (see Figure 3-16). It exists as an inactive monomer in the granule, but on release into the extracellular medium it is exposed to high levels of Ca^{2+}, which activate its assembly into a pore in the target cell membrane. How this leads to the entry of granzymes into the target cell is not clear, but it is thought that Ca^{2+} and small molecules enter through the perforin pore and trigger the uptake of granzymes (Figure 5-34). At the low pH prevailing in the granule, granzymes are inactive: on release from the granule they are activated and on entry into the target they induce the programmed death of the cell through various mechanisms including cleavage of caspase 3

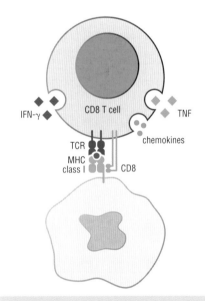

Figure 5-33 Effector actions of CD8 T cells
CD8 T cells secrete the important immune cytokine interferon-γ, as well as the inflammatory cytokine TNF and some chemokines, but their major effector function is the direct killing of infected cells with which they form tight conjugates on recognition of antigen associated with MHC class I molecules.

Definitions

granzymes: proteases that are contained in membrane-bounded granules of cytotoxic lymphocytes and are released on activation of the cytotoxic cell. Many and perhaps all induce programmed cell death on entry into target cells.

perforin: a protein contained in membrane-bounded granules of cytotoxic lymphocytes that is released on activation of the cytotoxic cell and forms pores in the membrane of target cells.

References

Balaji, K.N. *et al.*: **Surface cathepsin B protects cytotoxic lymphocytes from self-destruction after degranulation.** *J. Exp. Med.* 2002, **196**:493–503.

Henkart, P.A. and Sitkovsky, M.V.: **Cytotoxic T lymphocytes** in *Fundamental Immunology* 5th ed. Paul, W.E. ed. (Lippincott Williams & Wilkins, Philadelphia, 2003), 1127–1150.

Keefe, D. *et al.*: **Perforin triggers a plasma membrane-repair response that facilitates CTL induction of**

apoptosis. *Immunity* 2005, **23**:249–262.

Martinvalet, D. *et al.*: **Granzyme A induces caspase-independent mitochondrial damage, a required first step for apoptosis.** *Immunity* 2005, **22**:355–370.

Trambas, C.M. and Griffiths, G.M.: **Delivering the kiss of death.** *Nat. Immunol.* 2003, **4**:399–403.

Wong, P. and Pamer, E.G.: **CD8 T cell responses to infectious pathogens.** *Annu. Rev. Immunol.* 2003, **21**:29-70.

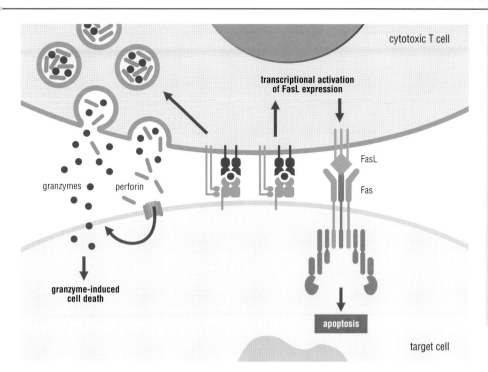

Figure 5-34 Cytotoxic mechanisms of CD8 T cells Engagement of the TCR by an antigen–MHC class I complex induces the expression of FasL, the ligand for the death receptor Fas, and in a relatively slow process induces apoptosis by the caspase-dependent pathway, while at the same time activating a much faster cytotoxic pathway operating through the release of perforin and granzymes from preformed granules. In the environment inside the granule, perforin molecules are stored as inactive monomers. Release into the extracellular space exposes them to higher concentrations of Ca^{2+}, which triggers polymerization so that they form a pore in the membrane of the target cell. By an unknown mechanism, this enables the granzymes to enter the cytoplasm of the target cell. At the zone of contact, interactions between ICAM-1 on the target cell and LFA-1 on the T cell lead to local cytoskeletal reorganization that focuses the secretory apparatus of the T cell at the point of contact (not shown).

and the Bcl-2 family proapoptotic protein Bid (see section 2-12) and a distinct, caspase-independent pathway that directly activates the fragmentation of DNA and inactivates the cellular repair program. We shall see in Chapter 10 that many viruses have evolved mechanisms for blocking apoptosis through caspase-dependent and Bcl-2 family-member-dependent pathways: granzyme-mediated pathways that act directly on DNA, with inactivation of the repair machinery, are independent of either of these pathways and may therefore be important in avoiding viral sabotage.

Conjugate formation between CD8 T cells and their targets has much in common with the formation of the immunological synapse between naïve T cells and dendritic cells (see section 5-4), and indeed is the best-studied example of the immunological synapse. The initial antigen-specific contact between the T cell receptor and the peptide–MHC complex on the infected cell triggers strengthening of the adhesive interactions between the two cells through LFA-1–ICAM-1 interactions, and leads to cytoskeletal reorganization (see section 2-4) that focuses the secretory apparatus of the cytotoxic T cell on its target. This allows the focused and contained delivery of perforin and granzymes onto the surface of the infected target with little leakage onto other cells. But whereas the contacts made by naïve T cells with dendritic cells are protracted, the contact of a cytotoxic T cell with its target is transient and fleeting, and having delivered its cytotoxic cargo the cytotoxic cell quickly passes to another target.

CD8 T cells are protected from their own cytotoxic products

The focused delivery of perforin and granzymes at the target cell surface prevents the destruction of neighboring cells but cannot protect the cytotoxic lymphocyte itself. CD8 T cells are protected from the contents of their own granules by cathepsin B, which is bound to the membrane of the granule containing the perforin and granzyme molecules, and remains attached when the granule fuses with the plasma membrane, so that it is exposed at the immunological synapse between the cells. It can then destroy any perforin molecules at the cytotoxic CD8 T cell membrane before they can assemble: thus the CD8 T cell membrane remains intact while the granule contents are released onto the target cell (Figure 5-35).

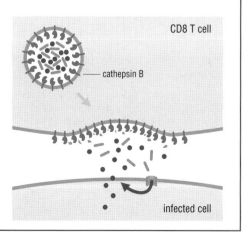

Figure 5-35 Protection of cytotoxic T cells from granule contents by cathepsin B The membrane of the granules containing granzyme and perforin contains the membrane-bound enzyme cathepsin B, which degrades perforin at the membrane surface, preventing it from forming pores. This protects the surface of the cytotoxic T cell by exposing the cathepsin B active site at the cell surface where the granule fuses with the plasma membrane.

T cell numbers are subject to dynamic control

In the course of an immune response, antigen-specific T cell numbers can expand by a factor of 10,000. Yet by the end of the response only about 10% of these cells remain, and in healthy adult humans, numbers of T cells are maintained at a constant level of around 10^{12}. This reflects a dynamic balance between survival and proliferation signals that maintain T cell numbers for effective immune surveillance and rapidly boost them in the face of infection, and proapoptotic mechanisms that prevent them from accumulating and causing permanent congestion of the lymphoid organs.

All normal T cells express proapoptotic members of the Bcl-2 family of cell death regulators (see section 2-12) and are thus programmed to die unless actively protected by survival signals. These signals are provided by cytokines, or by signals from B7- or TNF-family molecules, or both, depending on the activation state of the T cell. In this way T cell numbers are controlled by regulated expression of receptors by the T cell, or regulated production of cytokines and/or other survival signals by other cells, or both. Cytokines that maintain T cell numbers in healthy animals—that is, that maintain homeostasis—are constitutively produced. Those that regulate T cell responses to infection are inducible.

Most of the cytokines required both for maintaining normal T cell numbers and for inducing the proliferation of activated T cells are those that signal through the γ_c receptors (see Figure 2-23): the α and β chains, which determine the cytokine specificity of the receptor, can vary in T cells in different states of activation and differentiation and are thus responsible for the differential control of their survival and proliferation.

Numbers of naïve T cells are maintained by cytokines and stimulation through the TCR

Circulating naïve T cells live for months (in rodents) or for months to years (in humans) and are continuously replaced from bone marrow progenitors that mature in the thymus. They also proliferate and replace themselves at a low level that can be increased in response to depletion of their numbers, or *lymphopenia*. Lymphopenia can occur in response to viral infection (AIDS is an extreme example of this), but it is also a natural consequence of ageing. The thymus gradually ceases to function during the lifetime of an individual and by the age of 60 in humans has virtually stopped producing new T cells. Numbers of circulating naïve cells in these circumstances are maintained almost entirely by proliferation from the existing pool of cells. This mechanism for maintaining T cell numbers results in a reduced repertoire of distinct receptor specificities available to recognize infection; however, this is not necessarily as problematic as might at first appear, partly because of the accumulation of memory cells with useful specificities: we discuss this further in the next section.

At least two signals are thought to be required for the maintenance of naïve T cells. The first of these is through the T cell receptor for antigen, which binds weakly to self-peptide–MHC complexes. T cells experimentally transferred into mice lacking the self-MHC molecules

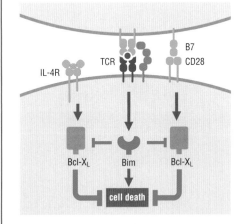

Figure 5-36 Death and survival signals in activated T cells Activation-induced cell death is thought to be due to induction by strong TCR signaling of Bim, which promotes apoptosis by inhibiting antiapoptotic actions of other Bcl-2 family members. Activated cells are protected by the induction of high levels of antiapoptotic regulators (here, Bcl-X$_L$) by cytokines and costimulators (here, IL-4 and B7), sufficient to escape inactivation by Bim.

Definitions

activation-induced cell death: (of T lymphocytes) programmed death of activated T lymphocytes through proapoptotic signals delivered by activating stimuli.

References

Badovinac, V.P. *et al.*: **Programmed contraction of CD8+ T cells after infection.** *Nat. Immunol.* 2002, **3**:619–626.

Boise, L.H. *et al.*: **CD28 costimulation can promote T cell survival by enhancing the expression of Bcl-x$_L$.** *Immunity* 1995, **3**:87–98.

Boursalian, T.E. and Bottomly, K.: **Survival of naïve T cells: roles of restricting versus selecting MHC class II and cytokine milieu.** *J. Immunol.* 1999, **162**:3795–3801.

Jameson, S.C.: **Maintaining the norm: T-cell homeostasis.** *Nat. Rev. Immunol.* 2002, **2**:547–556.

Marrack, P. and Kappler, J.: **Control of T cell viability.** *Annu. Rev. Immunol.* 2004, **22**:765–787.

recognized by their antigen receptors survive poorly. The other signal is from IL-7, which is thought to be the most important cytokine for the survival of naïve T cells. IL-7 is constitutively produced by many cell types, and it is believed that the control of numbers of naïve T cells depends in part on competition for this cytokine. Both IL-7 and signals from the T cell receptor are thought to contribute to the induction of the antiapoptotic regulator Bcl-2, which inhibits the action of the apoptotic effector molecules Bax and Bak and thereby protects the cell from death (see Figure 2-34). IL-7 is thought also to be responsible for inducing the homeostatic proliferation of naïve T cells, which occurs continuously at a low level and is increased in conditions of lymphopenia.

Effector T cells are programmed to undergo apoptosis

The expansion of numbers of activated T cells is believed to be driven initially by autocrine signaling through the induction of IL-2 along with the IL-2 receptor α chain (see section 5-7) and by inflammatory cytokines. These powerful proliferative signals are accompanied by the induction through the T cell receptor of the proapoptotic regulator Bim, which counteracts the antiapoptotic regulators and provides an automatic brake on T cell expansion, an effect known as **activation-induced cell death**. The programmed death of activated T cells is inhibited during the expansion phase of the immune response both by the costimulatory B7 molecules, signaling through CD28, and through inflammatory cytokines produced in response to infection. They are thought to act by inducing high levels of the antiapoptotic regulator Bcl-X_L that overwhelm the proapoptotic action of Bim (Figure 5-36). ICOSL, OX40L and 4-1BBL are thought to be essential for maintaining protective induction of Bcl-X_L in T effector cells in the B cell follicles and the periphery.

Inhibitory signals help to limit T cell responses in the periphery

Most effector T cells die by about 15 days after activation by antigen even if the infection is not cleared, although survival can be prolonged by the presence of inflammatory cytokines. This probably reflects the failure of survival signals to provide sustained protection from activation-induced cell death, as well as the action of inhibitory signals in the periphery. These inhibitory signals are provided by the induction on peripheral tissues of the negative costimulator PD-L1, which is recognized by PD-1, an inhibitory receptor that is induced on activated effector cells (see section 5-8). PD-1 recruits cytoplasmic phosphatases that dephosphorylate tyrosine residues in the cytoplasmic tails of costimulatory receptors, thereby abrogating their protective effect and exposing the cells to activation-induced death.

As a result of these intrinsic and extrinsic factors, T cell numbers decline steeply and the 10% of cells that remain survive as a pool of antigen-specific memory cells with new properties that are sustained by IL-7 and IL-15. The course of a T cell response, with the cytokine dependence of the T cells at each stage, is illustrated schematically in Figure 5-37. The properties of memory T cells are discussed in the next section.

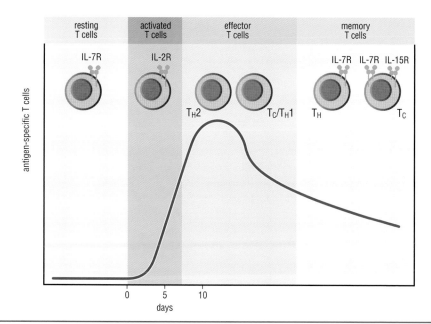

Figure 5-37 Cytokine dependence of T cells at different stages of an immune response The cytokine receptors shown schematically on the cells are those that are thought to play a major part in survival and/or proliferation at each stage. It is not clear how effector T cells are kept alive, although inflammatory cytokines contribute to their survival and they express the constitutive γ_c and β chains of the IL-2 receptor, which confer low-affinity recognition of IL-2 and IL-15. TNF- and B7-family signals from peripheral tissue and target cells are also important. Other less well understood signals are likely to operate at most stages.

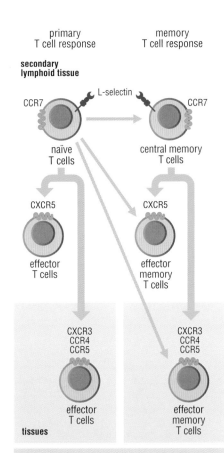

primary
T cell response

memory
T cell response

Figure 5-38 T cell activation leads to the production of effector cells and two subsets of memory cells Naïve T cells are dependent for survival on IL-7 and express L-selectin and CCR7, which direct their recirculation through secondary lymphoid tissue and their migration through the T cell areas. The two subsets of memory T cells depend on IL-7 (for CD4 T cells) and IL-7 with IL-15 (for CD8 T cells) for survival and have distinct migratory properties determined by the homing receptors they express. Effector memory cells, like effector T cells, express chemokine receptors that direct them to sites in the periphery (see Figure 5-24) and quickly express effector functions on reexposure to antigen. Central memory cells, like naïve T cells, express L-selectin and CCR7 and recirculate through secondary lymphoid organs, where they can be stimulated to proliferate and differentiate into effector cells on reexposure to antigen. Many of these cells express the follicular homing chemokine receptor CXCR5.

Two types of memory T cells are generated during immune responses

Most infections generate memory T cells that are capable of effector responses within hours of reinfection and provide a reservoir of antigen-specific cells that can expand and rapidly differentiate into new effector cells. These two functions of memory responses, often called *recall responses*, reside in two distinguishable subsets of cells: memory T cells capable of immediate effector activation are known as **effector memory T cells (T_EM)**; those that can proliferate to provide new effector cells are known as **central memory T cells (T_CM)**. The two subsets of memory T cells are distinguished by their migratory properties, state of differentiation, and proliferative capacity.

Effector memory T cells have migratory properties typical of effector T cells: L-selectin and CCR7 are downregulated and although they can circulate through secondary lymphoid organs, they are found mainly in peripheral sites. The homing molecules expressed on these cells direct them predominantly to epithelia, especially of the skin and gut, perhaps because epithelia are the main portal of entry for microorganisms and thus repay constant surveillance. Internal organs and tissues, by contrast, may recruit memory effectors chiefly through inflammatory chemokines produced in response to infection or damage.

On recurrence of infection, T_EM cells are armed for immediate deployment. CD8 effector memory cells contain cytotoxic granules, and Fas ligand and cytokine secretion are rapidly induced on TCR activation; CD4 effector memory cells on reencounter with antigen express the chemokine receptors and cytokine repertoires characteristic of the major CD4 T cell subsets, usually reflecting the polarizing conditions in which they were generated. Effector T memory cells can be immediately deployed in the face of reinfection; but they have relatively limited capacity for proliferation and decline more rapidly with time than do central memory cells.

Central memory T cells do not express the differentiated functions of effector cells but proliferate rapidly and differentiate into effector cells on reencounter with antigen. These cells are activated to proliferate more rapidly, and require less costimulation than naïve T cells. Moreover because there are many more memory cells specific for a given antigen than there are naïve T cells, large numbers of effector cells are much more rapidly produced. Central memory T cells express L-selectin and CCR7 and recirculate through secondary lymphoid organs, although bone marrow is an important reservoir of memory CD8 T cells expressing additional chemokine receptors that direct them to this site. A high proportion of CD4 T_CM cells express CXCR5, which recognizes the follicular chemokine CXCL13 and thereby directs the cells to B cell follicles: these serve as a reservoir of T_FH cells. It is thought that central memory cells, like naïve T cells, are activated by dendritic cells in secondary lymphoid tissues through which they recirculate. Thus they act as a reserve of already clonally expanded cells of appropriate specificity that can expand further and boost the numbers of available T_EM cells, as well as providing a pool of cells that can be activated on subsequent recurrence of infection. The two subsets of memory cells and their migratory properties are schematically illustrated in Figure 5-38.

Memory T cells are maintained by cytokines and stimulation through the TCR, and possibly by antigen

Memory T cells can persist for a decade or more in humans; CD8 memory T cells are longer-lived than CD4 memory T cells. CD8 memory T cells specific for vaccinia virus for example have been found more than 30 years after vaccination. Like naïve T cells, memory T cells are dependent for survival on cytokines, or stimulation through the T cell receptor by self-peptide–MHC, or both. The maintenance requirements for memory CD4 and memory CD8 seem to be distinct.

Definitions

central memory T cells (T_CM): long-lived circulating T lymphocytes that are generated from antigen-specific T cells during a primary immune response and can be activated to proliferate and differentiate into effector cells on reencounter with antigen. Central memory cells provide a reservoir of antigen-specific cells that can be rapidly activated on recurrence on an infection.

effector memory T cells (T_EM): long-lived effector T lymphocytes that are generated from antigen-specific T cells activated during a primary immune response and

that disperse to the tissues or the bone marrow and can provide immediate effector action on recurrence of infection.

T_CM: see **central memory T cells.**

T_EM: see **effector memory T cells.**

References

Ahmed, R. *et al.*: **Immunological memory and infection** in *Immunology of Infectious Diseases* Kaufman, R. H. *et al.* ed. (ASM Press, Washington, D.C., 2002), 175–189.

Doherty, P.C. *et al.*: **Influenza and the challenge for immunology.** *Nat. Immunol.* 2006, **7**:449–455.

Fearon, D.T. *et al.*: **The rationale for the IL-2-independent generation of the self-renewing central memory CD8+ T cells.** *Immunol. Rev.* 2006, **211**:104–118.

CD4 T cells, like naïve CD4 T cells, require IL-7 and peptide–MHC. CD8 memory T cells are maintained by IL-7 and especially by IL-15, which like IL-7 is constitutively available, and are thought not to require peptide–MHC; but they require CD4 T cell help, although the mechanism is not yet clear (see section 5-8).

As well as keeping memory cells alive, cytokines can stimulate their homeostatic proliferation at a low level. This is thought to be particularly important in individuals who cannot produce new naïve T cells either because the thymus has been surgically removed for therapeutic reasons or, as we have seen, because of the decline of thymic function with age. In these individuals, the T cell repertoire is provided by the existing homeostatically proliferating naïve T cells, along with the accumulated repertoire of memory T cells.

Although memory implies persistence without restimulation, it is not clear how far memory T cells survive in the complete absence of antigenic stimulation. Not all memory cells are long-lived, and immunity after infection can wane over time. Moreover, in many cases it is difficult to rule out persistent antigenic stimulation, either from peptide–MHC complexes that cross-react with the original stimulating antigen–MHC complex or through reencounter with circulating pathogens. For some viruses, which we discuss below, the persistence of the original infection ensures the restimulation of memory CD8 T cells. The issue of maintenance of immunity in the absence of antigen is of growing practical importance now that vaccination is largely eliminating many diseases from the populations of industrialized countries, so that immunity is no longer periodically boosted by natural reinfection. We discuss this issue in Chapter 14.

T cell memory is particularly important for immunity to latent viruses

In general, antibodies are the most important component of immune memory, and we shall see in Chapter 6 that these are produced by plasma cells that persist for years after infection. However, although very little is known in practice about how the different components of immune memory collaborate, it seems reasonable to suppose that microorganisms that replicate inside cells may sometimes escape protective antibodies in small numbers, and T cell-mediated immunity then becomes important for their elimination. More subtly, CD8 T cells may provide a degree of immunity to highly mutable viruses whose coat proteins change rapidly so that they are no longer recognized by antibodies, which see only the surface molecule of pathogens. T cells, by contrast, are often specific for peptides derived from internal proteins of the pathogen, many of which are likely to be conserved proteins that cannot readily accumulate mutations without loss of function, and thus CD8 T memory cells may in some cases arrest infection by destroying infected cells where antibodies have failed to prevent infection. This is thought to account for limited immunity, for example, in some cases of infection with new antigenic variants of influenza virus, which is notoriously mutable and will be discussed in Chapter 10.

For protection from persistent or latent viral infection, CD8 T cell memory is thought to be critical. Latent viruses may remain for most of the time in a state of dormancy from which they periodically emerge to cause overt infection. A rapid CD8 T cell response is then required to control the infection, which generally cannot be eliminated: most people in developed countries are chronically infected with one or more persistent viruses, of which the commonest are the herpesviruses CMV and EBV (which causes mononucleosis and in immunosuppressed individuals leads to the development of lymphomas and nasopharyngeal carcinoma). Because of the recurrent activation of these viruses, memory CD8 T cell numbers are constantly boosted, and in the elderly the CD8 T cell repertoire can in consequence come to be dominated by cells specific for latent viruses. This is thought to contribute to the deterioration of immunity with age.

Hataye, J. et al.: **Naïve and memory CD4+ T cell survival controlled by clonal abundance.** Science 2006, **312**:114–116.

Klenerman, P. and Hill, A.: **T cells and viral persistence: lessons from diverse infections.** Nat. Immunol. 2005, **6**:873–879.

Robertson, J.M. et al.: **Not all CD4+ memory T cells are long lived.** Immunol. Rev. 2006, **211**:49–57.

Sallusto, F. et al.: **Central memory and effector memory T cell subsets: function, generation, and maintenance.** Annu. Rev. Immunol. 2004, **22**:745–763.

Sallusto, F. et al.: **Two subsets of memory T lymphocytes with distinct homing potentials and effector functions.** Nature 1999, **401**:708–712.

Tough, D.F. and Sprent, J.: **Immunologic memory** in Fundamental Immunology 5th ed. Paul, W.E. ed. (Lippincott Williams & Wilkins, Philadelphia, 2003), 865–899.

Zinkernagel, R.M. and Hengartner, H.: **Protective 'immunity' by pre-existent neutralizing antibody titers and preactivated T cells but not by so-called 'immunological memory'.** Immunol. Rev. 2006, **211**:310–319.

6

B Cells and Humoral Immunity

The antibodies produced by B lymphocytes are the most important effectors of lasting immunity for many viral and microbial infections. They contain highly diverse binding sites for recognizing antigens, linked to specialized domains of more limited diversity that couple the antigen-binding domains to mechanisms for destroying the bound infectious agent. During an immune response, there is rapid production of low affinity antibody accompanied by extensive mutation of immunoglobulin genes and selection for higher affinity variant antibodies, which have greater protective potency and are then secreted in large amounts.

6-0 B Cells and the Importance of Antibodies for Immune Defense

Humoral immunity is complementary to cell-mediated immunity

B cells are responsible for producing antibodies, which serve a role complementary to that of the cell-mediated immunity provided by T cells. Whereas T cells act primarily through direct cellular interactions at the location of infection, B cells provide systemic protection through antibodies carried rapidly throughout the body by the blood and lymph or secreted through epithelial layers to protect the interface between the organism and its environment. Antibodies also provide a mechanism by which adaptive immunity can be passed from a mother to a fetus or infant, conferring adaptive immune protection during the critical early period of life. Antibodies are the secreted form of the B cell antigen receptor: the more general term for both the membrane and the secreted form of the molecule is **immunoglobulin (Ig)**.

Whereas the function of the variable membrane-bound receptors of both B and T cells is to recognize antigen and to signal that recognition to the interior of the cell, secreted immunoglobulin must engage other mechanisms that mediate the destruction of the bound antigen. This is achieved through binding to other molecules such as complement, or to specialized receptors known as Fc receptors (see Figure 2-5) on the surface of effector cells such as phagocytes, which are thereby activated. Antibodies are made in five structurally distinct types, or classes, known as *IgM, IgD, IgG, IgA* and *IgE*, which have different functional properties.

Antibodies produced by distinct types of B cells are important in early protection from infection and in preventing reinfection

There are three main types of B cells with distinct roles in the control of infection. Two of them, *B1 cells* and *marginal zone B cells*, belong to a specialized group of lymphocytes that share some features with innate immunity. These B cells often do not require T cell help in order to differentiate into plasma cells and do not circulate through the lymph nodes, but migrate directly to the sites of their effector action: B1 cells are directed to the peritoneal and pleural cavities and protect these body cavities, while marginal zone B cells monitor bloodborne antigens in the marginal zone of the spleen and protect against infections of the bloodstream. In addition, B1 cells produce low-affinity antibodies of limited diversity in the absence of infection. These so-called *natural antibodies* are reminiscent of innate immune opsonins such as the collectins. The specialized properties of B1 and marginal zone B cells will be described in Chapter 8.

In this chapter we shall be concerned primarily with the third type of B cell, the *follicular B cells*. These B cells require activation by T cells followed by clonal expansion and differentiation, and, like T cells, they recirculate through blood and lymph until they encounter antigen. It is because they reside mainly in the B cell follicles of the lymph node and other secondary lymphoid organs that these cells are called follicular B cells. They produce highly diverse antibodies recognizing an effectively limitless range of molecules, and a major function of the antibodies produced by these B cells is to prevent reinfection with microorganisms that have been encountered previously. Like the other two types of mature B cells, all follicular B cells start out making IgM. Signals from helper T cells direct them to switch to the production of other antibody isotypes, allowing the effector properties of the antibodies to be customized depending on the nature of the immune response. During the immune response, the average affinity of the antibodies produced increases, a process called **affinity maturation**, due to somatic mutation of the immunoglobulin genes and selection for B cells with beneficial mutations. Low levels of high-affinity antibodies are sufficient for example to prevent infection by viruses, by blocking

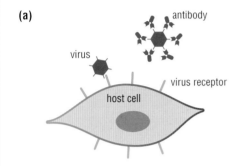

(a)

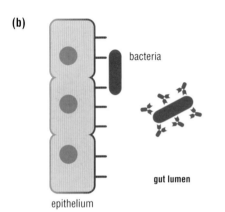

(b)

Figure 6-1 Neutralization by antibodies
(a) Viruses bind through specific viral coat proteins to one or more host cell surface proteins as a prelude to infection. Antibodies that bind to the virus and block binding to host cells block this step and prevent infection. **(b)** Enteric bacteria colonize the gut by attaching themselves to the mucosal wall via specific interactions. Antibodies against the bacterial cell surface components may block this attachment and thereby interfere with colonization and subsequent infection. Not shown, antibodies against bacterial toxins may block their interactions with cells and thereby prevent them from causing disease. Antibodies that block binding of pathogens or toxins to cells are said to be neutralizing antibodies.

Definitions

affinity maturation: the property of an immune response in which the average affinity of antibodies produced against an antigen increases as the response continues. This occurs over a period of several weeks.

Ig: see **immunoglobulin**.

immunoglobulin (Ig): a class of proteins produced by B lymphocytes of the immune system and that recognizes and binds to foreign antigens. Also called an antibody. The most common form of an immunoglobulin in the blood has a dimeric structure with two antigen-binding sites.

neutralizing antibodies: antibodies that can directly block infection by viruses or attachment by bacteria or toxins, without need of complement or Fc receptors.

serotype: variant of a pathogenic microorganism or virus in which the major antigen recognized by protective antibodies has been changed to the extent that it is not recognized by the antibodies directed to the other serotypes.

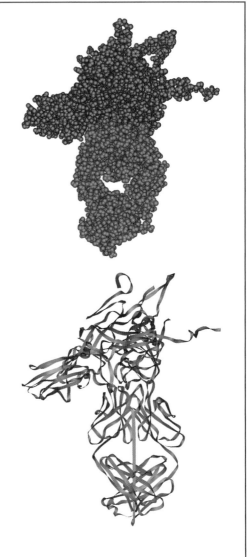

Figure 6-2 Complementarity of a neutralizing antibody to its viral antigen The structure of a complex of a fragment of a neutralizing antibody with a viral capsid protein from foot and mouth disease virus, as determined by cryoelectron microscopy. The viral protein is shown in red and the antibody fragment is shown in purple. On the top is the space-filling model structure, showing the close complementarity between the antibody and its protein antigen. On the bottom is shown the ribbon diagram of the same structures. The paired immunoglobulin V domains are contacting the viral antigens with the loops in the immunoglobulin domains. The heavy chain and light chain constant domains are paired at the lower end of the structure. (PDB 1qgc)

entry into cells. Antibodies that prevent infection by blocking the recognition of host cells by a pathogen or a toxin are called **neutralizing antibodies** (Figures 6-1 and 6-2). Most and possibly all currently effective vaccines work on this principle, as does much of the protective effectiveness of maternal antibody transmitted to offspring.

Antibody is also critical to immune defense against many primary infections. Viruses with surfaces characterized by highly repetitive arrays of viral proteins induce rapid IgM antibody responses that are typically important for control of these infections. Examples include vesicular stomatitis virus (VSV), polyoma virus, rotavirus, Sindbis virus and flavivirus. Such viruses typically exhibit distinct **serotypes**, variants that are not recognized by the same neutralizing antibodies, so that exposure to a virus of one serotype will not confer protection from another. The existence of many distinct serotypes of certain viruses testifies to the importance of antibody for protection against these viruses, which have presumably evolved such variants under selective pressure to avoid recognition. Not all viruses, however, are susceptible to control by antibodies. Hepatitis B virus and HIV, for example, often establish persistent infections, and cytotoxic T cells are thought to be more important in controlling infection with these viruses.

Often viruses are not completely cleared by the immune response to a primary infection and instead establish a long-lived latent state. In at least some cases, antibodies participate in keeping such latent infections under control.

Although early protection against bacterial infections relies to a high degree on innate immune mechanisms, antibodies also have a role in controlling infections by most bacterial pathogens. Extracellular bacteria can become coated with antibodies that greatly stimulate phagocytosis via Fc receptors. Antibodies again have a role in preventing reinfection. Enteric bacteria, for example, must attach to the gut wall in the early stages of infection, and this can be prevented by antibodies secreted into the gut (see Figure 6-1). Although antibodies are important in controlling most bacterial infections, an exception is *Listeria monocytogenes*, which grows in the cytoplasm of cells and spreads from cell to cell without passage through the extracellular milieu. Cytotoxic T cells are necessary for defense against *Listeria*, although antibodies can protect against reinfection. Several other important bacterial pathogens, including *Salmonella* and mycobacteria largely escape antibody control by invading host cells. Neutralizing antibodies can also protect against bacterial toxins, and this is the basis of the tetanus and diptheria vaccines.

Parasites, as we shall see in Chapter 11, pose a particularly difficult problem for immunity; but antibodies do help to control infection by malaria and trypanosomes and these organisms have evolved elaborate mechanisms for varying their surface antigens to evade humoral immunity. The principal effector actions of the five classes of antibodies are summarized in Figure 6-3 and are discussed in more detail throughout this chapter.

References

Bachmann, M.F. and Kopf, M.: **The role of B cells in acute and chronic infections.** *Curr. Opin. Immunol.* 1999, **11**:332–339.

Bachmann, M.F. and Zinkernagel, R.M.: **The influence of virus structure on antibody responses and virus serotype formation.** *Immunol. Today* 1996, **17**:553–558.

Casadevall, A.: **Antibody-mediated protection against intracellular pathogens.** *Trends Microbiol.*

1998, **6**:102–107.

Hangartner, L. *et al.*: **Antiviral antibody responses: the two extremes of a wide spectrum.** *Nat. Rev. Immunol.* 2006, **6**:231–243.

Major Functional Properties of Antibodies

Antibody class	Major functional properties
IgM	neutralization, complement activation, antigen trapping, antigen receptor of naïve and some memory B cells
IgG	neutralization, complement activation, induction of phagocytosis, ADCC, transfer of adaptive immunity to offspring, regulation of antibody production
IgA	neutralization, protection of mucosa, induction of phagocytosis, protection of newborn mucosa via milk
IgE	activation of mast cells and basophils to promote barrier immunity
IgD	antigen receptor on naïve follicular B cells

Figure 6-3 Table of antibody functions ADCC: antibody-dependent cellular cytotoxicity, a process in which antibody-coated cells are killed by Fc receptor-expressing immune system cells (see Figure 6-11).

6-1 Structure of Antibodies

Antibodies contain an antigen-binding region and a region that connects to effector mechanisms

Antibodies, the first recognized members of the immunoglobulin superfamily (see section 2-2), are glycoproteins composed of two types of chains, both of which contain multiple immunoglobulin domains. Unlike those of the TCR, however, the two chains of antibodies are of very different sizes: one, known as the light chain, is similar in size to a TCR chain; the other is known as the heavy chain and is much larger. It is the heavy chain that has a transmembrane region and in the membrane-bound receptor form of the molecule anchors the molecule in the membrane; in the secreted form—the antibody—the heavy chain is responsible for engaging with the cells and molecules that destroy the microorganisms to which the antibody binds. The basic design of antibodies is illustrated for IgG in Figure 6-4, with other details that we discuss below. The structure contains two identical heavy chains and two identical light chains: each light chain associates with a heavy chain, creating two identical antigen-binding units. The two heavy chains also pair with each other. The chains are linked together by disulfide bonds.

The specificity of antibodies derives from the paired amino-terminal immunoglobulin domains of each chain. These variable (V) domains, like those of TCR chains, can have many different amino-acid sequences and are generated during B cell development by the assembly of multiple gene segments along with variation introduced at the joints, a process known as V(D)J recombination. This process has already been briefly discussed for T cell antigen receptors and is outlined for B cells in Figure 6-5: it will be discussed in detail in Chapter 7. Antigen-binding specificity derives from the loops at one end of the V domains of both the light chain and the heavy chain (Figure 6-6), and these loops contain most of the variability from one antibody to the next. Like those of the TCR, they are known as hypervariable regions. As they form the basis for binding to antigen, which is in part determined by shape complementarity, they are also referred to as complementarity-determining regions (CDRs).

The structure of the complementarity-determining regions can vary considerably. The antigen-binding site of some antibodies is a relatively flat surface which is good for binding to native protein antigens, whereas in other antibodies it can contain a deep pocket or a cleft, in which small organic molecules, oligosaccharides, and so on, can bind.

The rest of the antibody molecule has much less variation than the V domains, and therefore the immunoglobulin domains that comprise these parts of the molecule are called constant (C) domains. Light chains have one C domain, whereas heavy chains have either three or four, depending on the type of heavy chain. It is the C domains of the heavy chains that are responsible for the effector functions of antibodies.

Figure 6-4 Basic structure of antibodies
Antibody structure is illustrated by IgG, and consists of paired heavy and light chains in a H_2L_2 structure. The names of individual domains are indicated on the right half of the molecule **(a)**. Disulfide bonds (represented as yellow lines) covalently link light chains to heavy chains and also heavy chains to each other in the hinge region. The sites of papain and pepsin proteolytic cleavage are shown. Papain cleavage gives rise to two Fab fragments and one Fc fragment. Pepsin cleavage gives rise to primarily a single $F(ab')_2$ fragment; the Fc fragment is cleaved further by pepsin. The antigen binds to a site contributed by the V_H and V_L domains, whereas C1q and Fc receptor-binding sites are in the Fc part of the molecule. **(b)** Ribbon diagram of IgG in which the light chains and the heavy chains are shown in the same colors as in (a). Note that the domains are not arranged in a strictly linear array, but rather there are generally bends between adjacent domains. In the Fc region this is in part due to carbohydrate attached to C_H2, which sterically separates these domains, in contrast to the C_H3 domains, which make close contact with one another. (PDB 1igy)

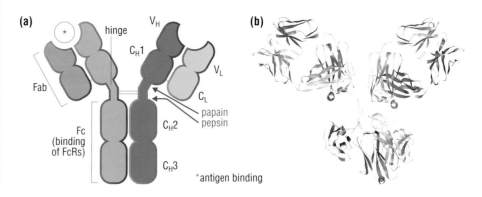

Definitions

Fab fragment: fragment of an antibody generated by limited digestion with the protease papain. The Fab fragment contains the light chain plus the V and C_H1 domain of the heavy chain, and represents a monomeric antigen-binding fragment without effector functions other than neutralization.

Fc fragment: fragment of an antibody generated by limited digestion with the protease papain and containing the C_H2 and C_H3 domains that connect to complement and Fc receptors.

hinge region: flexible stretch of amino acids between two domains, C_H1 and C_H2, of an IgG molecule, that allows the antigen-binding sites mobility relative to one another.

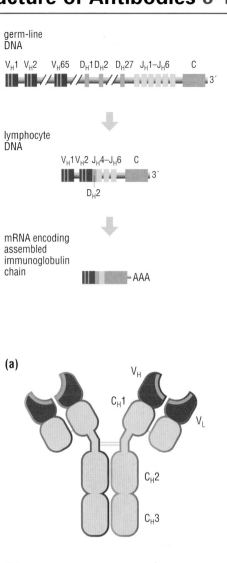

germ-line
DNA

V_H1 V_H2 V_H65 D_H1D_H2 D_H27 J_H1–J_H6 C

lymphocyte
DNA

V_H1V_H2 J_H4–J_H6 C

D_H2

mRNA encoding
assembled
immunoglobulin
chain

AAA

Figure 6-5 Functional immunoglobulin genes of B cells are generated by somatic recombination between gene segments Shown is a representation of the human IgH locus. At the top is the configuration of the germ-line chromosomal DNA. In B cell precursors, one D_H gene segment is moved next to one J_H segment by deletion of the intervening DNA and then one V_H gene segment is juxtaposed to the DJ combination, again by deleting intervening DNA, resulting in a single exon composed of V, D and J segments (middle). The D and J segments contribute amino-acid sequence to the third loop in the immunoglobulin domain, which corresponds to the third complementarity-determining region of the heavy chain (CDR3; also called hypervariable region 3; see Figure 6-6). If the open reading frame of the D and J segments is maintained, then this is a functional rearrangement that can encode an immunoglobulin heavy chain. Transcription of the rearranged immunoglobulin gene allows the production of functional mRNA after RNA splicing, which eliminates introns including any J segments between the one that is joined to the D segment and the C segment (bottom). An analogous process occurs at the light-chain loci, except that there are only V and J regions. These genetic mechanisms permit the encoding of approximately 10^7 different specificities by relatively little genetic information. In addition, there is variability in the exact sites of recombination between the segments, and additional bases can be inserted or deleted at the junctions, creating additional junctional diversity in the CDR3 loops. These features are described in more detail in Chapter 7.

The structure of antibodies was originally deduced in part from studies of the fragments of antibodies that could be generated by treatment with proteases such as papain and pepsin, and the mode of cleavage by these enzymes and the fragments they generate are still instructive. Limited proteolysis cleaves antibodies between immunoglobulin domains and the most sensitive site in IgG molecules is between the first and second constant domains (numbered from the amino terminus). Between these domains is an extended sequence of amino acids called the **hinge region**. The hinge has a flexibility that allows the two antigen-binding sites to have considerable mobility relative to one another, facilitating the binding of antibodies to antigens in fixed positions on virions, bacterial surfaces, and so on. Cleavage of the hinge with papain generates two fragments, one containing the light chain bound to the V and C_H1 domains of the heavy chain, called the **Fab fragment**, and a second containing the C_H2 and C_H3 domains of the two heavy chains. This essentially represents the cleavage of the molecule into its two functional parts: the antigen-binding regions comprising the Fab fragments, and the effector region, comprising the **Fc fragment**. (Fc stands for fragment crystallizable, and reflects the fact that in early studies of antibodies, only the structurally uniform Fc regions of antibodies could be crystallized: because there was at that time no way of obtaining large quantities of antibody with the same V region, Fab fragments were always mixtures and were impossible to crystallize.) Pepsin cleaves at a slightly different site that leaves the two Fab fragments (called Fab′, since they are slightly different from papain-generated Fab fragments) connected by a disulfide bond. Thus, pepsin treatment generates $F(ab')_2$ fragments. Fab and $F(ab')_2$ fragments can be used to block molecular interactions experimentally and therapeutically, without incurring effector actions that may have unwanted effects.

(a)

V_H

C_H1

V_L

C_H2

C_H3

(b)

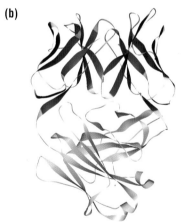

Figure 6-6 Hypervariable regions of the V domains The variability between different antibodies largely clusters in the variable domains in three regions of each chain called hypervariable regions, the positions of which are shown in green on the schematic diagram **(a)** and on a ribbon diagram (PDB 1aj7) **(b)** showing just one Fab fragment (equivalent to one arm of the antibody shown schematically in panel (a)) with its paired V_H and V_L domains (in red with the hypervariable loops shown in green) and paired constant domains (in gray at the bottom of the structure). The hypervariable regions (shown in green) correspond to loops in the immunoglobulin domains that are at the end of the molecule. These six loops are all located near one another and comprise the antigen-binding site. For this reason, these loops are also called the complementarity-determining regions (CDRs). The other parts of the V domains are shown in red. Note the similarity to the T cell receptor in Figure 5-4.

References

Davies, D.R. *et al.*: **Antibody–antigen complexes.** *Annu. Rev. Biochem.* 1990, **59**:439–473.

Edelman, G.M.: **Antibody structure and molecular immunology.** *Science* 1973, **180**:830–840.

Porter, R.R.: **Structural studies of immunoglobulins.** *Science* 1973, **180**:713–716.

There are five main kinds of antibodies with distinct functional properties

Figure 6-7 Oligomeric structures of antibodies Shown are the relative structures of the major types of antibodies. Human and mouse antibodies exist in two major types: those with heavy chains consisting of four immunoglobulin domains and a hinge (IgG, IgD and IgA) **(a)**, and those consisting of five immunoglobulin domains, but without a hinge (IgM and IgE) **(b)**. The number of disulfide bonds (yellow lines) between heavy chains varies between different isotypes. IgA and IgM exist in polymeric forms. The most common form of IgA is a dimer held together by J chain and bound to secretory component **(c)**, which is derived from the polymeric Ig receptor, a specialized receptor whereby IgA is transported into mucosal secretions. IgA can also form a tetramer (not shown). Secreted IgM exists primarily as a pentamer **(d)**, held together by disulfide bonds (yellow lines) between C_H3 and C_H4 domains and by J chain. In the absence of J chain, IgM forms a hexamer.

We have seen that prevention of infection by viruses depends on blocking their entry into cells, whereas protection from extracellular bacteria depends on delivery to phagocytes for destruction. While the effectiveness of an antibody in blocking the entry of a virus into a cell depends only on the rapidity and strength of binding to viral proteins, delivery of a bacterium to a phagocyte also requires recognition of the antibody by Fc receptors on the phagocyte surface (see Figure 3-23). The different requirements for protection from different infectious agents is reflected in the distinct properties of the five major antibody types, or **immunoglobulin classes**, also referred to as **isotypes**, a general term for distinct variants of a given structure. The effector actions of the five classes of antibodies are discussed in detail in the next section. Here we describe the distinctions in structure and oligomeric state that underlie them, and the specialized mechanisms whereby some of them are transported to their sites of action.

The five antibody classes are **immunoglobulin M (IgM)**, **immunoglobulin G (IgG)**, **immunoglobulin A (IgA)**, **immunoglobulin E (IgE)** and **immunoglobulin D (IgD)**. Their distinctive features, which are illustrated schematically in Figure 6-7, are conferred by their heavy chains, of which the five different isotypes are named with the Greek letters μ, γ, α, ϵ and δ corresponding to the antibody classes IgM, IgG, IgA, IgE and IgD, respectively. IgD exists only as a cell-surface receptor involved in B cell activation and has no effector properties. IgM is found mainly in the blood. It exists in a pentameric state in which there are five of the standard H_2L_2 units linked together by disulfide bonds and by binding to a distinct polypeptide called J chain (see Figure 6-7); less commonly it can be found in a hexameric state. The higher-order

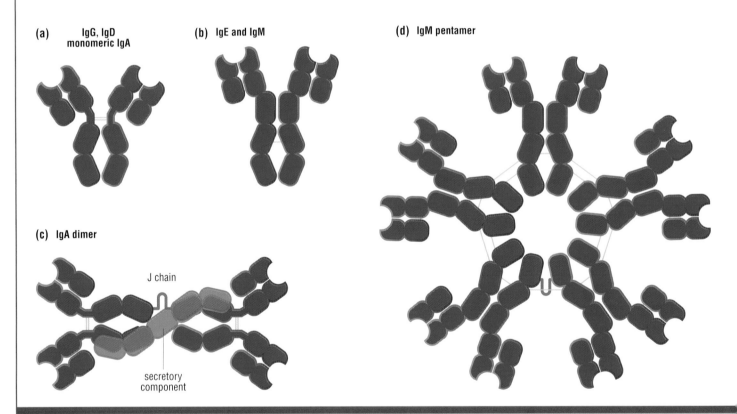

(a) IgG, IgD monomeric IgA

(b) IgE and IgM

(d) IgM pentamer

(c) IgA dimer

J chain

secretory component

Definitions

FcRn: neonatal FcR, involved in transport of **IgG** across endothelial cells.

immunoglobulin A (IgA): a class of antibody that is secreted across epithelial layers.

immunoglobulin classes: different **isotypes** of antibodies characterized by the type of heavy chain.

immunoglobulin D (IgD): a class of antibody that is primarily an antigen receptor found on naïve B cells.

immunoglobulin E (IgE): a class of antibody that binds with high affinity to mast cells and basophils and participates in immune defense against helminthic worms and also in the symptoms of allergies and asthma.

immunoglobulin G (IgG): the class of antibody that is the major antibody in the blood and extravascular fluids.

immunoglobulin M (IgM): the class of antibody that is made initially by newly generated B cells as a membrane protein and serves as a component of the B cell antigen receptor. Later, after antigen stimulation, IgM is the first class of antibody to be secreted, in which form

it is a pentamer.

isotype: any of several highly related forms of a molecule.

polymeric Ig receptor: receptor mediating transcytosis of **immunoglobulin A** across epithelial cells to mucosal surfaces.

secretory component: a chain of secreted **IgA** that is derived from the **polymeric Ig receptor** and remains bound to polymeric IgA after proteolytic cleavage of the transport receptor.

Figure 6-8 Genetic organization of human IgH locus Following the six J_H gene segments are exons encoding the different immunoglobulin heavy-chain constant regions in the order shown. V(D)J recombination generates a variable-domain-encoding segment near the Cμ exons, allowing the synthesis of mRNA encoding μ heavy chains. (The V and D gene segments to the left of the J segments are not shown.) Heavy chains of the δ type can also be made following alternative splicing that occurs if transcription proceeds beyond the Cμ locus and through the Cδ locus. Production of downstream isotypes of heavy chains requires class switch recombination, which is described later in this chapter. Note also that each heavy-chain constant-region-encoding locus is more complex than shown here, with a separate exon encoding each immunoglobulin domain and typically two additional exons encoding the transmembrane and cytoplasmic domains of the membrane form of the heavy chain (see Figure 6-19).

structure of IgM has the important advantage of increased binding strength through increased avidity: that is, the linked binding sites can cooperate and thus hold together a complex more strongly than would be the case with single binding interactions. We shall see later in this chapter that IgM is generally of relatively low affinity and the increased avidity of the pentamer helps to compensate for this. The other class of immunoglobulin that exists in a higher oligomeric state is IgA (see Figure 6-7), which protects mucosa and is secreted in sweat, tears, saliva, milk and colostrum, is held in a dimeric complex by J chain, and also contains *secretory component*, a protein derived from a specialized receptor that transports IgA across epithelial cells to the locations where it operates (see below). Humans have two isotypes of IgA, both of which can form polymeric structures and be secreted across epithelium. Monomeric IgA is primarily found in serum, where it constitutes 10–15% of antibody.

IgG, the main immunoglobulin class in blood and tissues, is produced in multiple isotypes; in humans these are IgG1, IgG2, IgG3 and IgG4, named in order of their prevalence in human blood. The different isotypes have distinct functional properties which we discuss in the next section. Mice also have four isotypes of IgG, called IgG1, IgG2a, IgG2b, and IgG3, with similarly distinct functional properties, although there is not a simple one-to-one correspondence between mouse and human IgG subclasses. The genetic organization of the IgH constant-region gene segments is shown schematically in Figure 6-8. Later in this chapter we shall discuss how the production of the different types of antibodies is controlled.

As well as mediating their oligomerization, the constant regions of immunoglobulin molecules also determine their interactions with other molecules and cells of the immune system as described in the next section. These actions depend on the interaction of the heavy-chain constant region with complement and with Fc receptors on the surface of immune effector cells. Fc receptors are specialized to recognize particular classes of antibodies, and there are Fc receptors recognizing all classes.

Specialized Fc receptors transport antibodies to their sites of action

Specialized Fc receptors are also responsible for transporting some classes of antibodies to the locations in which they are adapted to operate. The **polymeric Ig receptor** transports polymeric IgA (dimer or higher-order forms) across epithelial cells of mucosal surfaces or exocrine glands. This molecule is primarily responsible for transporting IgA to mucosal sites or to secretions such as saliva, sweat and milk. The polymeric Ig receptor binds a polymeric IgA–J chain complex (and, less well, an IgM–J chain complex) on the basolateral side of the epithelial cell and transports it to the apical side, where a protease cleaves the polymeric Ig receptor, releasing IgA bound to part of the transport receptor referred to as **secretory component** (Figure 6-9). Secretory component has two functions: it protects IgA from the proteases that may be present and it binds to mucus, keeping IgA at the desired location on the mucosal surface.

A second transporting Fc receptor is called **FcRn** (n for neonatal): unlike the other Fc receptors, all of which belong to the immunoglobulin superfamily, it is one of the non-classical MHC class I molecules (see Figure 4-9) and it transports IgG of all isotypes across the placenta and into the mammalian fetus. It also is responsible for maintaining a long half-life of IgG in the blood, where it persists for 14–21 days.

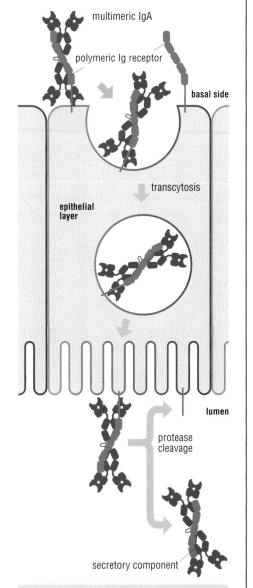

Figure 6-9 Transepithelial transport of IgA Polymeric IgA with J chain attached binds to the polymeric Ig receptor, a transmembrane protein expressed at the basal surface of epithelial cells of the gut, airways and various secretory glands. Binding of IgA induces transcytosis of the polymeric Ig receptor, delivering the complex to the apical surface of the epithelial cell and thus into the lumen of the organ, where proteases cleave the polymeric Ig receptor near the membrane, releasing the majority of the extracellular domain still bound to polymeric IgA, and now called secretory component.

6-3 Effector Functions of Antibodies

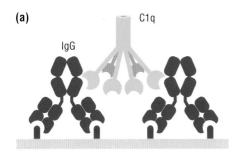

(a)

C1q

IgG

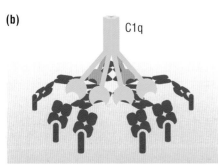

(b)

C1q

Antibodies can act by neutralization and complement activation

Some functions of antibodies are mediated purely by binding, as we have seen; these are known as neutralizing antibodies. For example, antibodies may bind to virions, preventing them from infecting cells (see Figure 6-1). Similarly, antibodies at mucosal surfaces may bind to bacteria and prevent them from attaching to molecules on the epithelia they need to colonize or invade.

Binding of IgG or IgM antibodies to particles can also trigger the complement cascade, as described in Chapter 3, through C1q, a hexameric protein that is a soluble component of plasma. C1q can bind to IgG only in clusters of at least two or three IgG molecules, and thus complement is activated only when IgG is clustered by virtue of binding to antigen, as can occur on the surface of a microorganism (Figure 6-10a). It then initiates the activation of the proteolytic reactions of complement to generate transmembrane pores that contribute to defense against Gram-negative bacteria, fungi, and enveloped viruses; to tag particles for phagocytosis via complement receptors; and to release peptides that induce inflammation and attract neutrophils (see Figure 3-16). Complement can also be activated by a single IgM molecule, which as a pentamer of H_2L_2 structures functions analogously to multiple IgG molecules. Activation of complement by IgM also requires bound antigen, however, because in its absence the H_2L_2 units of IgM are in a planar configuration that cannot bind C1q: antigen binding to the multiple binding sites of IgM distorts its configuration in a way that allows C1q to bind and activate complement (Figure 6-10b).

Figure 6-10 Activation of complement by IgM and IgG Clustered binding of two or three IgG molecules, for example on a bacterial surface, permits the binding of multiple heads of C1q to the Fc regions of multiple IgG molecules. Multipoint binding distorts the hexameric structure of C1q, which triggers the proteolytic action of bound subunits C1r and C1s (see Figure 3-13), which then act on C2 and C4 to start the classical complement cascade (see Chapter 3 for details). C1q can also bind to IgM when it is bound to antigens in a way that distorts its structure. In solution, IgM is in a planar pentagon configuration that does not allow multipoint binding of C1q. However, binding of IgM to surfaces such as virions or bacterial cell surfaces often distorts the IgM structure such that the bound Fab portions are taken out of the plane of the pentamer. This allows C1q to bind to IgM Fc regions and trigger complement activation.

Fc receptors mediate many of the effector functions of antibodies

The third general way in which antibodies can contribute to immune defense is through cellular receptors. Such receptors are selective for the class of antibody they recognize by virtue of their binding to the Fc regions, and hence are called Fc receptors. The isotype of antibody bound by each is indicated by the appropriate Greek letter: hence the Fc receptor recognizing IgA is called FcαR, and so on. Most Fc receptors activate immune system cells for phagocytosis, release of inflammatory mediators, and/or killing reactions. The exception is a receptor for IgG. IgG has several Fc receptors, one of which, designated FcγRIIB, inhibits cell activation, reflecting a role of IgG in regulating immune responses, briefly mentioned in Chapter 5 (sections 5-11 and 5-12) and discussed in more detail below.

In phagocytosis, the complement-activating functions of antibodies potentiate the effects of Fc receptor binding, because as we saw in Chapter 3, complement receptors cooperate with Fc receptors to induce the vigorous uptake of particles coated with IgG and complement (see section 3-8).

The biological actions of the activating Fc receptors depend largely on the cells expressing them. In macrophages and neutrophils the principal effect is to activate phagocytosis and the production of reactive oxygen compounds (see section 3-9), although the production of cytokines can be induced as well. Mast cells, basophils and eosinophils are stimulated through Fc receptors to degranulate, and natural killer cells are stimulated to secrete perforin and granzymes to induce the apoptosis of target cells, as described in section 5-13 for cytotoxic T cells. This process is

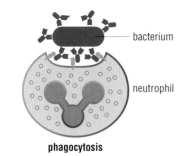

bacterium

neutrophil

phagocytosis

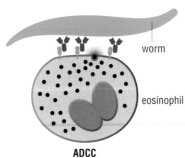

worm

eosinophil

ADCC

Figure 6-11 Effector functions performed by the activating Fc receptors Activating Fc receptors mediate the phagocytosis of IgG-coated particles by neutrophils and macrophages (upper panel). IgG-induced deposition of C3b on the particle can further enhance phagocytosis by complement receptors on the phagocytes (not shown). Fc receptors also mediate release of granule contents by eosinophils, mast cells, basophils and natural killer cells. When the granules contain toxic compounds that can kill cells, as is true of eosinophils and natural killer cells, this process is called antibody-dependent cellular cytotoxicity (ADCC) (lower panel).

Definitions

ADCC: see **antibody-dependent cellular cytotoxicity**.

antibody-dependent cellular cytotoxicity (ADCC): a process whereby FcR-bearing cells encounter an antibody-coated target cell and degranulate, releasing contents that kill the antibody-coated cell.

References

Nimmerjahn, F. and Ravetch, J.V.: **Fcγ receptors: old friends and new family members.** *Immunity* 2006, **24**:19–28.

Ravetch, J.V. and Bolland, S.: **IgG Fc receptors.** *Annu. Rev. Immunol.* 2001, **19**:275–290.

van Egmond, M. *et al.*: **IgA and the IgA Fc receptor.** *Trends Immunol.* 2001, **22**:205–211.

Woof, J.M. and Burton, D.R.: **Human antibody–Fc receptor interactions illuminated by crystal structures.** *Nat. Rev. Immunol.* 2004, **4**:89–99.

called **antibody-dependent cellular cytotoxicity (ADCC)** (Figure 6-11). Eosinophils and macrophages are also capable of ADCC, although the mechanisms by which they kill cells differ.

The structures of the human Fc receptors are illustrated schematically in Figure 6-12, with the exception of the receptor for IgM, for which no Fc receptor has been clearly characterized in humans; there is a receptor for IgM in mice that is shared with IgA. Most Fc receptors have relatively low affinity for their corresponding antibody ligand. These receptors primarily bind to antibodies attached to particles or to soluble immune complexes containing multiple bound antibodies as the simultaneous binding to multiple Fc receptors compensates for low intrinsic affinity, and Fc receptor cross-linking in this way triggers receptor signaling and cellular responses. An important exception to this general rule is FcεRI, which has a very high affinity for IgE. Most of the IgE in the body is bound to mast cells through FcεRI and remains bound to mast cells in tissues for a very long time. Binding of bivalent or oligovalent antigen to the IgE bound to mast cells clusters FcεRI molecules and thereby triggers degranulation and release of histamine and other inflammatory mediators. Mast cells are often involved in immune defense at mucosal surfaces (see section 5-12) and are responsible for many of the symptoms we associate with allergies (rhinitis, watering eyes, sneezing and so on). Basophils also express FcεRI and bind IgE in the blood. They can be stimulated by antigen binding to their bound IgE to release inflammatory mediators (see section 1-2).

The activating Fc receptors signal through ITAMs (see section 2-2), which for FcγRIIA are in the cytoplasmic domain of the receptor, while for the other activating Fc receptors the ITAMs are contained in a separate transmembrane chain, the γ chain. The inhibitory receptor FcγRIIB has an immunoreceptor tyrosine-based inhibitory motif (ITIM) in its cytoplasmic tail, which when brought adjacent to signaling ITAM-containing receptors inhibits their signaling (the mechanism of this is described in the context of B cell signaling in section 6-8).

Several bacteria have evolved mechanisms for interfering with Fc-receptor-mediated phagocytosis. For example, various streptococci evade IgA-based protection by making IgA-binding proteins that block binding to FcαRI and thereby interfere with IgA-induced phagocytosis by neutrophils and macrophages. Protein A of *Staphylococcus aureus* likewise binds to IgG and may block its effector functions. Protein A has also been extraordinarily useful for the biochemical purification of antibodies.

FcγR effector function is regulated by the balance of activating and inhibitory receptor action

The four different isotypes of IgG fall into two functional groups. In human, IgG2 and IgG4 and in mouse IgG1 neither activate complement well nor bind well to most of the Fcγ receptors, and are therefore thought to function primarily as neutralizing antibodies. IgG1 and IgG3 in human and IgG2a, IgG2b and IgG3 in mouse by contrast both activate complement and bind to activating Fc receptors and are thus important in promoting phagocytosis and inflammatory responses. This function of phagocytes is regulated by cytokines released by helper T cells, which alter the levels of expression of the activating and inhibitory FcRs, as described in section 5-11. Thus interferon-γ secreted by T$_H$1 cells upregulates activating Fc receptors on phagocytes and downregulates expression of the inhibitory FcγRIIB, promoting uptake and killing of microbes, whereas IL-4 secreted by T$_H$2 cells has the opposite effects. Control of immune cell activation by the balance of activating and inhibitory receptor function is also important in other immune cells, most prominently in natural killer cells, as we shall see in Chapter 8.

Figure 6-12 Activating and inhibitory human Fc receptors Activating Fc receptors have immunoreceptor tyrosine-based activating motifs (ITAMs) either in the associated FcR γ chain or in their own cytoplasmic tail (FcγRII A and FcγRII C). Mast cells and basophils uniquely express the FcεRI β chain, which is thought to amplify receptor signaling. FcγRIIIA also pairs with the β chain in mast cells and basophils and lacks this chain in other cells, including natural killer cells. The inhibitory FcγRIIB contains an immunoreceptor tyrosine-based inhibitory motif (ITIM) in its cytoplasmic tail, and can inhibit signaling by activating receptors when brought into proximity to them, for example when both are bound to the same immunoglobulin-coated particle. Finally, neutrophils express FcγRIIIB, which has neither an ITAM nor an ITIM associated with it but is held in the plasma membrane through a glycosylphosphoinositide (GPI) tail. GPI-anchored proteins are highly enriched in lipid rafts, which are subdomains of the plasma membrane that support ITAM signaling. It is thought that the activating FcRs are outside lipid rafts before ligand binding and then enter lipid rafts to initiate signaling. Therefore FcγRIIIB may facilitate signaling by activating FcRs bound to the same particle by promoting their entry into lipid rafts (neutrophils also express the activating receptors FcαRI, FcγRIIA and FcγRI). The Fc receptors shown here are those found on human cells: mice have an array of Fc receptors that are broadly similar but distinct in detail. Mice lack the activating FcγRIIs and the FcγRIIIB and instead have FcγRIII and FcγRIV. Mouse FcγRIV is similar to human FcγRIIIA.

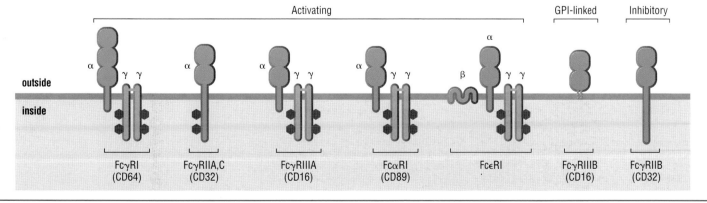

| Activating | | | | | GPI-linked | Inhibitory |

| FcγRI (CD64) | FcγRIIA,C (CD32) | FcγRIIIA (CD16) | FcαRI (CD89) | FcεRI | FcγRIIIB (CD16) | FcγRIIB (CD32) |

Monoclonal antibodies can be produced to capture a single antibody specificity

In addition to their natural role in immune defense, antibodies are also extremely useful reagents for biological research, and in medicine for diagnosis and therapy. Historically, antiserum or purified antibodies were used for these purposes, including for example antitoxins to treat tetanus or snake bites, and antibodies from immune individuals to prevent hepatitis A virus infection. Antiserum, however, contains a mixture of different antibodies derived from distinct precursor cells: such mixtures of antibodies are known as **polyclonal antibodies** because they are the product of clones of cells originating from many different precursors. The use of such mixtures is limited in three ways. First, reproducibility is limited in some cases, particularly if the antigen used is poorly immunogenic. Second, the quality of the reagent is limited by the purity of the antigen used in immunization, because of the presence of antibodies reacting against contaminating proteins. Finally, the therapeutic use of polyclonal antibodies from animals is limited to a single treatment because individuals often have a strong response to serum proteins from other species (horse, rabbit and so on) in which the antibody is raised. Nonetheless, polyclonal antibodies have been extremely useful and are still used in a variety of ways: for example individuals with genetic B cell deficiencies that result in failure to make antibodies can be treated by injecting them with antibody pooled from multiple human donors.

An alternative to the use of polyclonal antibodies was developed in the late 1970s with the discovery of techniques for immortalizing plasma cells of predetermined specificity. Such immortalized cells produce essentially limitless quantities of a homogeneous single antibody: such antibodies are called **monoclonal antibodies**. They are made by fusing spleen or lymph-node cells from an immunized animal with an immortalized cell line derived from a **plasmacytoma**, which is a cancerous version of a plasma cell. The hybrid cell that produces the monoclonal antibody is called a **hybridoma**. This procedure is illustrated in Figure 6-13.

Monoclonal antibodies have permitted the identification of most specialized surface molecules of lymphocytes

One of the most wide-ranging contributions of monoclonal antibodies to our understanding of immunology has been their use in identification of the cell-surface molecules of lymphocytes and leukocytes. A systematic effort to make monoclonal antibodies against cell-surface molecules of different subsets and differentiated states of the cells of the immune system is the basis of the CD nomenclature for describing these molecules. CD stands for cluster of differentiation,

Figure 6-13 Method for making monoclonal antibodies Activated B cells from immunized mice are taken around 3 days after immunization with the desired antigen and fused to a mouse plasmacytoma cell line variant that has lost its rearranged Ig loci and also cannot make the enzyme hypoxanthine-guanine phosphoribosyl transferase (HGPRT). The fused cells (known as hybridomas) are then selected for their ability to grow in medium in which HGPRT expression is required (called HAT medium because it contains hypoxanthine, aminopterin and thymidine). The antibiotic aminopterin in HAT medium blocks *de novo* purine synthesis and therefore makes cell growth dependent on the use of the salvage pathway for nucleoside utilization, which requires HGPRT. The medium is supplemented by hypoxanthine and thymidine to facilitate the salvage pathway and allow the rapid growth of HGPRT-expressing cells in the presence of aminopterin. The unfused plasmacytoma cells die because they cannot maintain purine levels needed for DNA replication. The unfused B cells cannot grow continuously in culture, but the hybridoma cells have the ability to grow continuously because they have the HGPRT gene from the B cell partner and the continuous growth and plasma cell differentiation state derived from the plasmacytoma parent. Hybridomas are screened for production of the desired antibody and then cloned so that a single antibody is produced. Fortuitously, activated B cells fuse with myeloma cells much more efficiently than do resting B cells, so hybridomas of the desired specificity are relatively easy to obtain after most immunizations. PEG: poly(ethylene glycol).

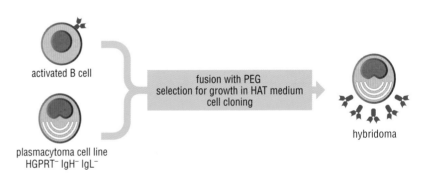

activated B cell

plasmacytoma cell line
HGPRT⁻ IgH⁻ IgL⁻

fusion with PEG
selection for growth in HAT medium
cell cloning

hybridoma

Definitions

HAMA: see **human anti-mouse Ig antibody**.

human anti-mouse Ig antibody (HAMA): antibody produced by an immune response by humans to determinants on mouse Ig, which limits the therapeutic usefulness of mouse **monoclonal antibodies**.

humanized antibodies: monoclonal antibodies made by using mouse cells and subsequently manipulated by recombinant DNA technology to produce a molecule most of which is derived from human sequences, with only the antigen-binding loops of the mouse molecule.

hybridoma: cell produced by fusing an activated B cell to a **plasmacytoma** cell line.

monoclonal antibodies: antibodies produced by a single clone of cells and therefore with a single specificity.

phage display: a technique for isolating proteins with specific binding characteristics whereby the desired protein is expressed on the surface of a bacteriophage. This allows phage to be selected for binding to a tissue culture dish coated with the ligand.

plasmacytoma: cancerous plasma cell. Human cancers of this type are often called multiple myelomas, from the propensity of these cells to grow in the bone marrow.

polyclonal antibodies: heterogeneous antibodies against an antigen that are obtained by immunizing an individual.

and it reflects the identification of key molecules through the clustering of a number of different monoclonal antibodies recognizing the same molecule. Molecules identified in this way are known as CD antigens. Monoclonal antibodies have been useful in determining the function of these molecules as well.

Human monoclonal antibodies can be made for therapeutic uses

Originally, monoclonal antibodies could be made only from rodent species (mice, rats and hamsters). This limited their therapeutic use to situations in which a single treatment was sufficient because individuals generally make a very strong immune response to determinants in the constant domains of the rodent immunoglobulins, or even to determinants in the variable domains. This is called the **human anti-mouse Ig antibody (HAMA)** response. In recent years a variety of approaches have largely overcome this limitation. For example, it is possible to use recombinant DNA technology to replace the constant domains of a mouse monoclonal antibody with human constant domains. This, however, still leaves the possibility of responses to mouse variable domains. Still less immunogenic antibodies can be made by grafting onto human V domain frameworks the mouse complementarity-determining regions, the loops in the structure of the V_H and V_L domains (see Figure 6-6). Such antibodies are known as **humanized antibodies**. This approach is generally sufficient to remove the main immunogenicity of the monoclonal antibody and permit multiple treatments although the affinity of the humanized antibody can be weaker than that of the original antibody. It is possible, of course, to make an antibody response to the antigen-binding site, but in most individuals this is not a problem.

Another approach that has been successful in creating human monoclonal antibodies for therapy involves using transgenic technology to create mice with much of the IgH and IgL loci of humans in place of the mouse Ig loci. Such mice produce antibodies that are essentially identical to human antibodies, but allow experimental immunization and hybridoma technology.

Finally, it is also possible to isolate human monoclonal antibodies by the creation of libraries of bacteriophage expressing combinations of human V_H domains and V_L domains and screening these libraries for binding to the antigen of choice. Once isolated from the **phage display** libraries, these clones can be grafted back onto the desired constant domains to generate antibodies with long half-lives and effector functions characteristic of natural antibodies.

Some of the monoclonal antibodies used in human disease therapy are listed in Figure 6-14.

Monoclonal Antibodies Used in Therapies

Monoclonal antibody	Target	Disease
trastuzumab	HER2	breast cancer
infliximab	TNF	rheumatoid arthritis, Crohn's disease
rituximab	CD20	non-Hodgkin's lymphoma
abciximab	GPIIb/IIIa	coronary disease
OKT3	CD3	graft rejection
alemtuzumab	CD52	B cell chronic lymphocytic leukemia

Figure 6-14 Table of monoclonal antibodies used in therapy

References

Berger, M. *et al.*: **Therapeutic applications of monoclonal antibodies.** *Am. J. Med. Sci.* 2002, **324**:14–30.

Köhler, G. and Milstein, C.: **Continuous cultures of fused cells secreting antibody of predefined specificity.** *Nature* 1975, **256**:495–497.

Mendez, M.J. *et al.*: **Functional transplant of megabase human immunoglobulin loci recapitulates human antibody response in mice.** *Nat. Genet.* 1997, **15**:146–156.

Winter, G. *et al.*: **Making antibodies by phage display technology.** *Annu. Rev. Immunol.* 1994, **12**:433–455.

Antibodies have many uses in diagnostic techniques and research

The specificity of antibodies makes them very useful for a wide range of diagnostic and research purposes. For example, the presence of antibodies against a particular infectious agent is often an excellent way of determining whether the person is or has been infected with that agent and is used to screen blood to make sure it is safe to use in transfusion. In this section we describe some of the basic features of antibody–antigen interactions that are useful in this regard, and in this and the next section we illustrate how those properties are used in the more important applications of antibodies in these areas.

Immune complexes of antibodies and complex antigens can generate large insoluble aggregates

The interaction of polyclonal antibodies derived from the serum of a person or an animal with a macromolecular antigen in the correct proportions will typically lead to the formation of a large insoluble aggregate or **immune precipitate**. This is because antibodies are at least bivalent and typically can interact with the antigen at more than one site, or **epitope**. When the antibody or the antigen is in substantial excess, then smaller aggregates are formed, which are soluble, but in between these two extremes, a lattice of antibody-antigen interactions forms and this large complex is insoluble and precipitates (Figure 6-15a). Such complexes can form pathologically in some autoimmune diseases in which large quantities of polyclonal antibodies are produced against soluble self components, such as DNA released from dying cells; these are called **immune complexes** and can cause tissue damage as discussed in Chapter 13. Immune precipitates are the basis for a variety of laboratory tests for antibodies and/or antigens put into semi-solid agar gels, in which the precipitate forms a visible white line. Most of these laboratory tests have, however, been superseded by more sensitive methods.

Use of antibodies in the analysis and measurement of biological molecules

Antibodies can be used to characterize the antigens with which they interact by **immuno-precipitation**, a technique in which the antibody is mixed with a soluble biological preparation containing the antigen and then the immune complexes are collected, typically by binding to beads that are covalently attached to either an antibody that will react to the antibody being used (for example, a rabbit antibody that reorganizes mouse antibody constant regions) or to protein A, an IgG-binding protein from some strains of the bacterial pathogen *Staphylococcus aureus*, which uses this protein to evade humoral immunity. In either case the beads collect the immunoprecipitate, and components of the preparation that do not bind to the antibody directly or indirectly (by association with the antigen) are removed by washing. The molecules bound to the beads can then be analyzed to identify them, to see whether they have become phosphorylated, and so on.

The high degree of specificity of antibodies can be used to measure the amount of an antigen, such as a hormone in the blood. One common method of this type is the **radioimmunoassay (RIA)**. A typical RIA employs a radioactively labeled antigen plus antibodies specific to the antigen. These two components are mixed together along with a biological sample containing an unknown amount of the antigen. The unlabeled antigen in the sample displaces the radioactive antigen from the antibody, which is then precipitated by a high salt concentration

Definitions

epitope: molecular feature of an antigen that is specifically recognized by a lymphocyte or an antibody. Epitopes recognized by T cell receptors are peptides. Epitopes for antibodies on proteins can be linear, as for T cells, or can be made up of different regions of a polypeptide that come together in the three-dimensional structure of the antigen. The latter kind of epitope is called a discontinuous or conformational epitope.

hemagglutination: the property by which antibodies binding to red blood cells cause them to agglutinate.

immune complex: noncovalent molecular complex formed when an antibody binds to the antigen it specifically recognizes.

immune precipitate: precipitate formed by the complex of a polyclonal antibody with a complex antigen, when the ratio of the components is favorable for forming a large lattice.

immunoprecipitation: isolation of a protein or other macromolecule through the use of specific antibodies that bind to the molecule of interest and are usually collected by beads covalently coupled to an antibody

that reacts to the antibody being used or to protein A from *Staphylococcus aureus*, a protein that binds to antibodies.

radioimmunoassay (RIA): a technique to measure the amount of an antigen of interest in a solution by its ability to inhibit the incorporation of radiolabeled antigen into an **immunoprecipitation**.

RIA: see **radioimmunoassay**.

or collected on beads as in immunoprecipitation (Figure 6-15b). The amount of radioactivity in the precipitate is measured and the decrease in radioactivity is a measure of the amount of antigen in the biological sample. The RIA is performed in solution.

A variation on this kind of test is the use of a cell that has the antigen on its surface, such as red blood cells and blood type antigens. In this case, at the appropriate concentration the antibodies will form bridges between cells. As a result, the cells all stick together. In the **hemagglutination** test, red blood cells are used in this way to test for the presence on them of blood group antigens or for the presence of antibodies against blood groups in the serum of an individual (Figure 6-15c). Related assays put known antigens onto latex beads and then mix them with patient's sera to test for antibodies reactive to particular infectious agents that, if present, will agglutinate the beads. This assay has the advantage of high sensitivity.

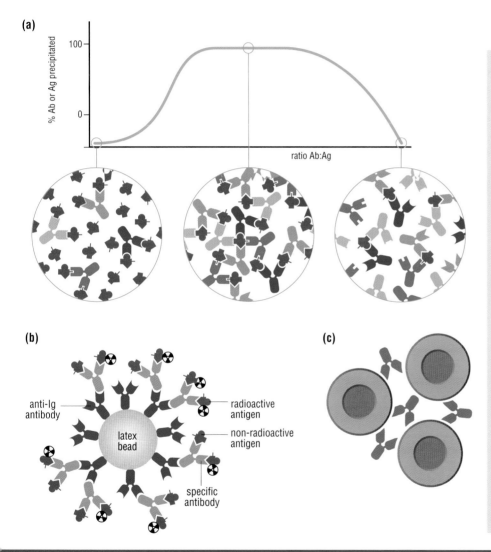

(a)

% Ab or Ag precipitated

100

0

ratio Ab:Ag

(b)

anti-Ig antibody

latex bead

radioactive antigen

non-radioactive antigen

specific antibody

(c)

Figure 6-15 Diagnostic assays based on immune precipitation (a) The effect of the ratio of antibody (Ab) to antigen (Ag) on the formation of an immune precipitate. Examples of the equilibrium binding between antigen and antibody at three places along the curve are illustrated. When antigen is in excess, no lattices are formed because most antigen is bound to only single antibody molecules, and when antibody is in excess, lattices do not form because immune complexes are very small, containing one antigen and a small number of antibodies bound to different epitopes on it, with none of these antibodies bridging to other antigen molecules. **(b)** Illustration of the principle of the RIA. Radiolabeled antigen (indicated by the radioactive symbol) is mixed with a biological sample containing an unknown amount of an unlabeled antigen (such as a hormone in blood) and with antibody specific to the antigen. The antibody and bound antigen are then collected on a latex bead that has attached to it a protein that binds to antibodies (in this case a different antibody that recognizes the specific antibody being used) and the amount of radioactive antigen collected is measured. **(c)** In the hemagglutination test, red blood cells are mixed with an antibody sample such as a patient's serum and are put in a round-bottomed (U-shaped) well. Agglutination occurs when there are enough antibodies present to form bridges between the cells. If the red blood cells are agglutinated, then the agglutination keeps them more spread out against the sides of the well, but if they are not agglutinated, then gravity causes them to form a tight dot in the bottom of the well (not shown).

References

Berzofsky, J.A. *et al.*: **Antigen–antibody interactions and monoclonal antibodies** in *Fundamental Immunology* 5th ed. Paul, W.E. ed. (Lippincott Williams & Wilkins, Philadelphia, 2003), 69–105.

Antibodies bound to a solid substrate can be detected by the production of colored products

Diagnostic techniques such as the RIA that depend on antibody–antigen interaction in solution are not always sufficiently sensitive for the purpose at hand. Often this problem can be solved by putting the antigen on a solid support. For this type of diagnostic test, the antibody is chemically linked to a marker such as a fluorophore or to an enzyme that can generate a colored product, and then added to the solid-phase sample. Typically, the binding phase takes only a few minutes, given the small volumes used and the fast on-rate of binding of most antibodies. The unbound material is then washed away, leaving only the specifically bound antibody to be detected. This method is readily adapted for high-throughput analysis, for example by the use of 96-well plates.

In the **enzyme-linked immunosorbent assay (ELISA)** an antigen is bound to a solid support, typically the plastic wells of a flat-bottomed 96-well plate, and antibodies are added that are specific for the antigen and are chemically linked to an enzyme, such as horseradish peroxidase or alkaline phosphatase, that can generate a colored product when incubated with relevant substrates. The amount of color produced is related to the amount of enzyme and hence to the amount of antibody that is bound to the plate. This type of ELISA is very commonly used to test for the presence of antibodies specific for the antigen put on the plate, for example in screening units of blood to avoid those from people infected with HIV, hepatitis C virus, and so on. In this type of ELISA, serum is added to the antigen-coated plate, unbound antibody is washed away, and then a second antibody, in this case specific for human IgG or human IgM, is added to detect the antibody of interest. The second antibody is conjugated to the color-producing enzyme. Alternatively, if the goal is to measure the amount of antigen in a biological sample, one monoclonal antibody (the capture antibody) is put on the plate, and the biological sample is incubated with the antibody-coated plate. Next, the unbound antigen is washed away, and a second monoclonal antibody that binds a separate epitope is added and used to detect the amount of antigen that was captured by the first antibody. This is commonly referred to as a sandwich ELISA (Figure 6-16), with the two antibodies being the bread and the antigen being the middle part of the sandwich.

A variation of the ELISA, called the **ELISPOT** assay, is used to determine the number of cells secreting a particular product, such as plasma cells secreting antibodies or effector T cells secreting cytokines. In this assay, a cell population is placed in a plastic dish that had been previously coated with the relevant capture antibody or antigen, and the cells are incubated to allow secretion to occur. Then the cells are removed, the enzyme-linked secondary antibody is added and a substrate is added that is converted to an insoluble colored product instead of a soluble one. Where there were cells secreting the protein of interest, discrete colored spots are generated, which are counted.

A protein or glycoprotein antigen can also be identified or quantitated by resolving a protein mixture by SDS-polyacrylamide gel electrophoresis, transferring (blotting) the proteins in the gel to a solid support such as a nitrocellulose membrane, and then incubating this with the specific enzyme-linked antibody (**immunoblotting** or western blotting). As in the ELISPOT assay, a substrate is used that is converted to an insoluble product or alternatively a substrate and enzyme pair is used that gives off light to expose film. Similarly, the location of an antigen (or antibody) within a tissue section can be determined by chemically fixing a tissue section in some way that does not destroy the antigen, and then adding antibody that is either fluorescently labeled, to produce **immunofluorescence**, or linked to an

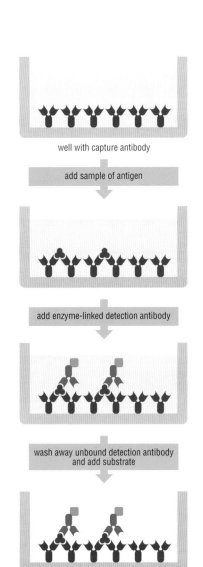

well with capture antibody

add sample of antigen

add enzyme-linked detection antibody

wash away unbound detection antibody and add substrate

Figure 6-16 A capture or sandwich ELISA to test for the amount of an antigen present A monoclonal antibody is purified and is allowed to stick to the plastic wells of a 96-well flat-bottomed plate. Unbound antibody is removed and remaining binding sites on the plastic are filled with an irrelevant protein. Next the biological sample containing an unknown amount of antigen is added. After allowing a short time for binding (minutes to a few hours), the unbound material is again washed away and the detection antibody is added. This enzyme-linked antibody must be able to bind to a different site on the antigen than the capture antibody, so typically it is a different monoclonal antibody or a polyclonal antibody. After unbound material has again been washed away, the substrate is added and the plate is incubated for a set time before the amount of color in each well is measured.

Definitions

ELISA: see **enzyme-linked immunosorbent assay**.

ELISPOT: an assay similar to a capture **ELISA** in which cells are incubated in a plastic dish coated with the capture antibody, followed by incubation of the dish with an enzyme-conjugated detection antibody and finally incubation with a substrate that is converted to an insoluble colored product, so that colored spots appear where cells secreting the molecule of interest were present.

enzyme-linked immunosorbent assay (ELISA): a technique for measuring the amount of a molecule in a sample in which the antigen is put or trapped on a solid support and then an antibody with an attached enzyme is used to bind to the trapped antigen. Alternatively, an ELISA can be used to measure the amount of a specific antibody in a sample. The amount of enzymatic activity is directly related to the amount of the antigen or antibody on the plate.

flow cytometry: a technique used to enumerate and analyze a sample of cells by incubating them with one

enzyme that catalyzes the production of a colored insoluble product that is deposited where the antibody is localized. This technique is called **immunohistochemistry**.

Antibodies can be used to characterize the properties of subsets of cells

It is often important, in both the laboratory and the clinic, to be able to measure the number of immune cells of different types that are present in tissues or in blood, for example to identify the type of cell that is expanded in a leukemia patient, or the missing populations in a person with an immunodeficiency disease. This can be achieved with fluorescently labeled monoclonal antibodies recognizing proteins on the surface of cells, in conjunction with **flow cytometry**, a method that measures the amount of fluorescence of individual cells (Figure 6-17). A cell suspension is reacted with several monoclonal antibodies, each covalently attached to a different fluorescent molecule, or *fluorophore*, and the flow cytometer determines the amount of each color of fluorescence bound to each cell in the mixture. In Figure 6-17, an example is shown in which cells in the blood are stained with antibodies against CD3 and CD19 (markers for T cells and B cells, respectively), but the use of more than two colors is now routine for many applications. Recent advances in fixation and permeabilization of the plasma membrane allow analysis of internal molecules as well as cell surface molecules, for example making it possible to identify the cytokines being produced by an activated T cell at the single-cell level.

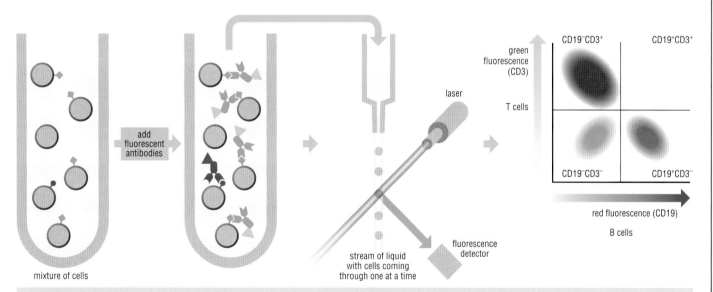

Figure 6-17 Flow cytometry to characterize cells Cells are removed from blood or tissue and disaggregated so that each cell is free from other cells to make what is called a single-cell suspension. Fluorescently labeled antibodies are added, each specificity marked with a different color fluorophore, and the sample is then put through narrow tubing that creates a stream of liquid containing one cell at a time. The stream goes past a laser beam that excites the fluorophores to give off fluorescence, which is detected with multiple detectors using different wavelength filters that distinguish the fluorophores being used. The flow cytometer also measures forward and side light scatter properties of the cells, which can also distinguish between some types of cells (granulocytes have high side light scatter due to their cytoplasmic granules whereas lymphocytes have low; forward light scatter is generally related to cell size). A computer collects all of the data obtained on each cell and is used to show the results. Shown is an idealized version of CD3 and CD19 expression profiles of lymphocytes from blood (CD3 is a component of the TCR and is only on T cells; CD19 is a coreceptor expressed only on B cells and is discussed later in this chapter). Note that cell types can often be distinguished on the basis of the amount of a particular cell surface protein, not simply whether the cell expresses the protein or not, and many different colored fluors can be used. This technique can also be used to separate cells, a process called cell sorting. In this related method, the stream of cells is broken into small droplets after the laser beam, the droplets are given electric charge based on which type of cell they contain, and the droplets are separated by their charge into different containers.

or more fluorescently labeled antibodies and/or other molecules that can bind to cellular components and measuring the fluorescence intensity of each fluor for each cell.

immunoblotting: also called western blotting. A technique for analyzing the proteins in a sample in which the sample is fractionated by SDS-polyacrylamide gel electrophoresis, which resolves denatured polypeptides by their approximate molecular mass; the resolved proteins are transferred to a solid support such as a flexible membrane made out of nitrocellulose, which is then reacted with labeled antibody specific to the protein of interest.

immunofluorescence: a technique in which fluorescently labeled antibodies are used to determine the location of the corresponding antigen in a tissue section or in cells.

immunohistochemistry: a technique in which enzyme-linked antibodies and a substrate that is converted to a colored insoluble product are used to determine the location of the corresponding antigen in a tissue section.

Membrane forms of immunoglobulin comprise the ligand-binding part of the B cell antigen receptor

Lymphocyte development (described in Chapter 7) generates large numbers of B cells, each of which makes an immunoglobulin with a unique antigen-binding site. As enunciated by the clonal selection theory (see section 1-4), infections induce the selective activation of B cells making antibodies that can recognize components of the infecting agent. This can be achieved because immunoglobulins are made both as secreted antibodies and as membrane-bound receptors for antigen displayed on the surface of B cells. Therefore, antigen can interact with the membrane immunoglobulin (mIg) and selectively induce the activation of those B cells, which later secrete antibodies with the identical antigen-binding site (see Figure 1-24). Membrane Ig exists on the cell surface as a complex with a disulfide-linked heterodimer of two other polypeptides, **Igα** and **Igβ**, which are made only in B-lineage cells (Figure 6-18). This complex comprises the **B cell antigen receptor (BCR)**. Igα and Igβ each contain an amino-terminal extra-cellular domain including one Ig domain, a transmembrane domain, and a cytoplasmic domain ~50 amino acid residues long containing the ITAM sequence that provides the signaling function.

The BCR binds native proteins, glycoproteins and polysaccharides

As described in Chapter 4, αβ T cells recognize antigens only after they have been processed to peptides and bound to MHC molecules. In contrast, antibodies and the BCR can recognize native conformations of proteins, glycoproteins and polysaccharides, although they are also capable of recognizing denatured proteins and peptides. Although immunoglobulins are extremely versatile and can recognize virtually any molecule the B cell encounters, recognition of a molecule by the BCR alone is often not sufficient to induce a response on the part of the B cell and the consequent production of antibodies, but rather depends on additional signals from other immune cells such as T cells, or from innate immune receptors, as we see in the next sections.

The type of Ig heavy chain produced is dependent on the differentiated state of the B cell

The transition from membrane to secreted immunoglobulin marks a key step in the transition from resting B cell to **plasma cell**, in which the synthetic machinery of the cell is dedicated to the production of large quantities of secreted antibody, with a concomitant increase in the membrane of the endoplasmic reticulum to accommodate this activity. In this section and the next we describe the mechanism of the membrane–secreted switch and of signaling through membrane Ig in B cell activation. In subsequent sections we describe the nature of the antibody responses of B lymphocytes and how plasma cells providing long-lived protection are generated by cell–cell interactions in the lymphoid tissues.

The membrane and secreted forms of immunoglobulins differ at the carboxy-terminal end of the heavy chain. The secreted forms of immunoglobulin heavy chains contain a few unique amino acids at the carboxyl terminus, but lack an obvious transmembrane domain. In contrast, the membrane forms of immunoglobulin include at their carboxyl terminus a transmembrane domain and a short cytoplasmic domain. The transmembrane and cytoplasmic sequences are encoded by two exons (μm1 and μm2 in the DNA shown for the μ heavy chain in Figure 6-19) that lie beyond the coding region of the secreted form (indicated as s in the μ heavy-chain DNA in Figure 6-19). All heavy chains can be made as either membrane-bound or secreted forms with the exception of the heavy chain of IgD (δ), which is made only as a membrane form.

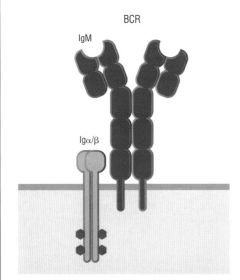

BCR

IgM

Igα/β

Figure 6-18 Structure of the B cell antigen receptor The BCR is a complex of membrane immunoglobulin with a heterodimer of Igα and Igβ, which have the signaling function contained in their cytoplasmic domains. Shown is a BCR containing membrane IgM, but all five isotypes of antibodies can be expressed as membrane forms that also form BCRs in complex with Igα and Igβ.

Definitions

B cell antigen receptor (BCR): the complex of membrane immunoglobulin that recognizes antigen and a hetero-dimer of **Igα** and **Igβ** that signals antigen recognition in B lymphocytes.

BCR: see **B cell antigen receptor**.

Igα/Igβ: the disulfide-linked heterodimer that serves as the signaling component of the antigen receptor of B lymphocytes. Both proteins have an amino-terminal single Ig-like domain as an extracellular domain, a trans-membrane domain and a moderate length (~50 amino-acid residues) cytoplasmic domain containing an immunoreceptor tyrosine-based activation motif (ITAM).

μm mRNA: mRNA encoding the membrane form of the μ heavy chain of immunoglobulin.

μs mRNA: mRNA encoding the secreted form of the μ heavy chain of immunoglobulin.

plasma cell: terminally differentiated B lineage cell secreting large quantities of antibody.

The process whereby the two forms of immunoglobulin are produced from the heavy-chain gene is outlined in Figure 6-19. Naïve B cells make only the membrane-bound form of immunoglobulin (Figure 6-19a). This is achieved by preferential use of the second of two polyadenylation sites (pA2 in the figure): the short sequence encoding the tail of the secreted form is included in the primary transcript but is eliminated by RNA splicing, along with stop codons that would otherwise prevent translation of the membrane-specific sequences. This results in synthesis of the membrane form of the heavy-chain mRNA (**μm mRNA** in the case of the μ heavy chain). Secreted Ig is produced when the first polyadenylation site is used: transcription then terminates before the membrane exons, giving rise to a mRNA that encodes the secreted form (**μs mRNA** in the case of the μ heavy chain). Thus, the terminal differentiation of B cells into plasma cells, which secrete antibody at a high rate, is accompanied by a change in transcriptional termination and RNA processing of the Ig heavy-chain mRNA. The net result is that the cells produce large amounts of the secreted form of the heavy chain and secrete large quantities of antibody.

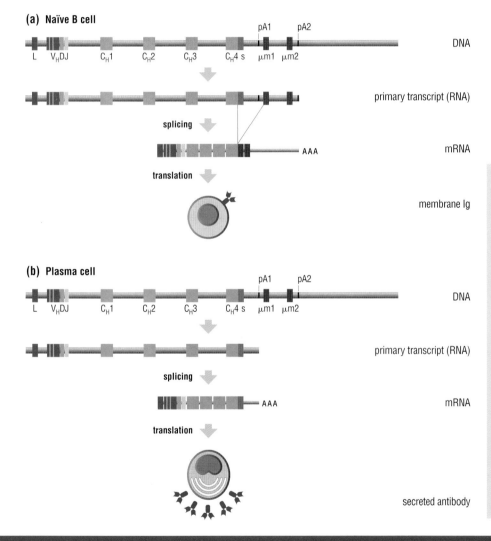

Figure 6-19 Production of membrane and secreted Ig (a) For the membrane form of Ig, transcription of the IgH locus (μ chain in the example shown) terminates downstream of the two exons encoding the transmembrane domains (μm1 and μm2). This permits the use of the poly(A) addition site that is located after μm2 (pA2), which then allows RNA splicing between the 3′ end of the C_H4 exon and the 5′ end of the μm1 exon (indicated by the blue lines). This splicing cuts out the RNA encoding the secretory tail (s), and joins the C_H4 exon to the exons encoding the transmembrane and cytoplasmic domains. In plasma cells, which produce secreted Ig **(b)**, most transcription stops in the intron between the secretory poly(A) addition site (pA1) and the μm1 exon. Thus, the only poly(A) addition site that is available is pA1 and there is no splicing of the mRNA after the C_H4 exons because there is no acceptor splice site. The s-encoding sequences remain in the mRNA. Some Ig heavy chains do not have added secretory tail sequences, but otherwise the principles are maintained. L: exon that encodes signal sequence/leader peptide directing the nascent polypeptide to the lumen of the endoplasmic reticulum.

References

Matsuuchi, L. and Gold, M.R.: **New views of BCR structure and organization**. *Curr. Opin. Immunol.* 2001, **13**:270–277.

Venkitaraman, A.R. *et al.*: **The B-cell antigen receptor of the five immunoglobulin classes**. *Nature* 1991, **352**:777–781.

The BCR signals via ITAM sequences in the Igα and Igβ subunits

Antigen engagement by the BCR triggers signaling through the associated Igα and Igβ chains (Figure 6-20) in a manner that is similar to other ITAM signaling events (sections 2-2 and 5-3). Antigen binding to more than one BCR induces these receptors to move into the lipid raft region of the plasma membrane, which contains required signaling components. Src-family tyrosine kinases in the lipid rafts phosphorylate the two tyrosines of the ITAMs, which triggers binding of the cytosolic tyrosine kinase **Syk**. Syk becomes activated by phosphorylation and then phosphorylates adaptors and signaling enzymes, triggering signaling reactions very similar to those that the TCR triggers in T cells (section 5-3). These signaling events promote the activation of B cells.

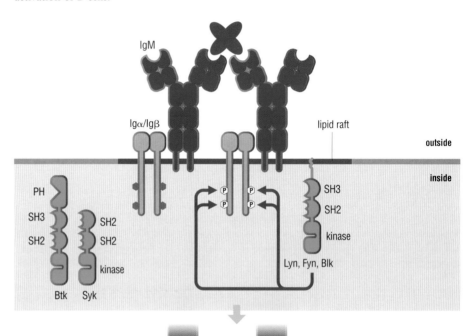

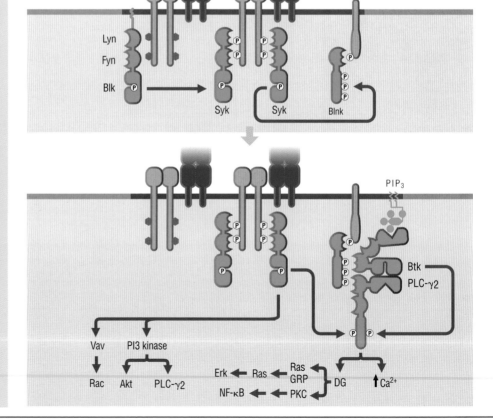

Figure 6-20 Signaling by the BCR In the unstimulated state, the BCR is not located in lipid rafts, whereas the Src-family tyrosine kinases Lyn, Fyn and Blk are (not shown). For reasons that are not understood, antigen-induced oligomerization of BCRs causes them to associate with the plasma membrane lipid raft domains, and Src-family kinases phosphorylate the ITAMs on oligomerized receptors (top panel). BCRs with phosphorylated ITAMs recruit the cytosolic tyrosine kinase Syk. Binding of Syk promotes its activation, which involves an activating phosphorylation of a tyrosine near the catalytic site (middle panel). Among the targets of Syk are adaptor molecules such as Blnk. Blnk is recruited to lipid rafts via non-ITAM sequences in the cytoplasmic domain of Igα or perhaps via an unidentified apaptor (as shown). Blnk serves as a platform for assembling a complex of signaling components including Btk and phospholipase (PL) C-γ2 (bottom panel). BCR signaling also activates PI3 kinase, which makes PIP_3. PIP_3 promotes the activation of the PLC-γ2 pathway by aiding the recruitment of PLC-γ2 and Btk. PIP_3 also activates the protein kinase Akt, which has been implicated in several cell survival pathways. PLC-γ2 hydrolyzes PIP_2, generating inositol (1,4,5)-trisphosphate, which elevates calcium in the cytoplasm, and diacylglycerol, which activates protein kinase C as well as Ras guanine nucleotide releasing protein (Ras GRP), which activates the key signaling molecule Ras. Protein kinase C promotes activation of the transcription factor NF-κB, as in T cells.

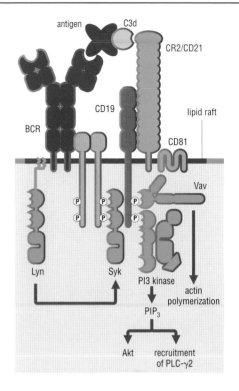

Figure 6-21 Coreceptor function of the complement receptor 2 complex Binding of the BCR to an antigen that has bound one of the small C3b-derived complement fragments brings the CR2 complex (including CD19 and CD81) adjacent to the BCR. This leads to the phosphorylation of several tyrosines in the cytoplasmic domain of CD19 and the recruitment of Vav and the signaling enzyme PI3 kinase, both of which have an integral role in BCR signaling (see Figure 6-20). Src-family tyrosine kinases such as Lyn also bind to tyrosine phosphorylated CD19 (not shown) and may act to amplify signaling.

BCR signaling is modulated by positive and negative coreceptors

Just as the responses of T cells are directed and amplified by the coreceptors CD4 and CD8, which focus T cells on the appropriate type of MHC molecule, B cells have coreceptors that provide important additional information about the nature of molecules that bind to the BCR or the type of cell on which the antigen is found. Some of these coreceptors amplify B cell responses; others inhibit them.

The positively acting coreceptor of B cells is the complement receptor 2 (CR2, also called CD21) in complex with two other molecules, called CD19 and CD81. If complement fragments are deposited on an antigen, then the antigen is much more effective at inducing an antibody response, because the CR2/CD19/CD81 complex can boost suboptimal BCR signaling (Figure 6-21). Since complement activation often occurs in the presence of microorganisms and is inhibited by negative regulators expressed on the surface of host cells (see section 3-6), CR2 coreceptor function helps to ensure that B cells are activated in the presence of infection and not by cell surface autoantigens. (It should be noted, however, that CD19 is thought to participate as an adaptor molecule in BCR signaling even without complement involvement, although less efficiently.)

A second coreceptor of B cells is the negatively acting Fc receptor, FcγRIIB, which is important in limiting B cell responses. It recognizes complexes of IgG with antigen and blocks the activation of naïve B cells (see Figure 6-21), thus mediating a negative feedback loop. Once sufficient IgG has been produced to complex with free antigen, this acts to inhibit further antibody production. As with other immunoreceptor tyrosine-based inhibitory motif (ITIM)-containing receptors, FcγRIIB acts by recruiting phosphatases that counter positive signaling events (Figure 6-22).

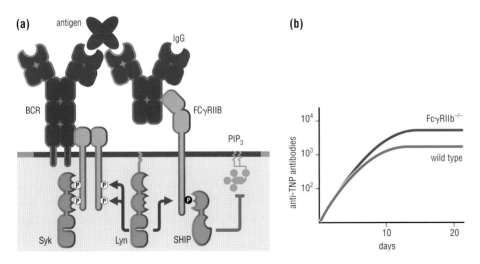

Figure 6-22 Negative coreceptor function of FcγRIIB When soluble IgG is bound to antigen, this inhibits further B cell activation through coreceptor action of the inhibitory Fc receptor, FcγRIIB (FcγRIIb in mouse). **(a)** IgG-containing immune complexes bring the BCR and FcγRIIB adjacent to one another, which leads to the phosphorylation of the ITIM of FcγRIIB by kinases activated by the BCR (primarily Lyn). The phosphorylated ITIM of FcγRIIB then recruits SHIP, an SH2 domain-containing inositol phosphatase that hydrolyzes PIP₃. Given the important role of PIP₃ in BCR signaling (see Figure 6-20), this effectively inhibits BCR signaling. Red membrane: lipid raft microdomain of plasma membrane. **(b)** Mice deficient in FcγRIIb because of targeted gene disruption have an enhanced IgG antibody response to many antigens. Shown here is the production of anti-trinitrophenyl (TNP) antibodies in response to immunization with TNP-keyhole limpet hemocyanin in complete Freund's adjuvant (mineral oil plus mycobacterial extract containing TLR ligands) on day zero. Adapted from Takai, T. *et al.*: *Nature* 1996, **379**:346–349.

Takai, T. *et al.*: **Augmented humoral and anaphylactic responses in FcγRII-deficient mice.** *Nature* 1996, **379**:346–349.

Definitions

Syk: an intracellular protein tyrosine kinase of most hematopoietic cell types that is recruited to phosphorylated ITAMs and has an essential role in ITAM-based signaling. ZAP-70 of T cells is a close relative.

References

Fearon, D.T. and Carroll, M.C.: **Regulation of B lymphocyte responses to foreign and self-antigens by the CD19/CD21 complex.** *Annu. Rev. Immunol.* 2000, **18**:393–422.

Kelly, M.E. and Chan, A.C.: **Regulation of B cell function by linker proteins.** *Curr. Opin. Immunol.* 2000, **12**:267–275.

Kurosaki, T.: **Regulation of B cell fates by BCR signaling components.** *Curr. Opin. Immunol.* 2002, **14**:341–347.

Protective antibodies are preexisting and induced

Three kinds of antibody responses can be distinguished by the time course of their appearance; and before any of these responses is initiated, a primitive level of protection is provided by natural antibody, as briefly mentioned earlier (see section 6-0). Natural antibodies are produced by B1 cells in the absence of infection by pathogens, although it is thought that they may be stimulated by commensal microorganisms; they are of low affinity and include antibodies that are widely cross-reactive, but can be shown to be important for defense against some pathogens, as we shall see in Chapter 8. All other antibodies are produced in response to antigen, and the quantity of antibody produced as well as its specificity and affinity for antigen are higher than for natural antibody.

The most rapid type of antigen-induced antibody response is made by B1 cells and marginal zone B cells in response to infectious agents without T cell help, and is known as the **T cell-independent (TI) antibody response**: these antibodies start to appear within 48 h of infection and can include secretory IgA in addition to IgM. The next response is an early response made by marginal zone and **follicular B cells** that depend on T cell help, and is thus part of the **T cell-dependent (TD) antibody response**. This response requires clonal expansion of both antigen-specific T cells and antigen-specific B cells but nonetheless is rapid, with substantial antibody production by three to four days after infection. The antibodies produced in these two responses, as well as preexisting natural antibodies, are predominantly IgM, and like natural antibodies, they are often of relatively low affinity. The microbial components that stimulate these early responses, however, are often multivalent antigens such as a viral or bacterial surface, and the multivalency of the pentameric IgM complex and the consequent enhancement of avidity helps to compensate for the low affinity of these antibodies. Some IgG is also produced in these early responses and in the case of some viruses the antibody produced is very effective at neutralization.

The third type of antibody response, and the key contribution of the T cell-dependent follicular B cells, is the *germinal center reaction*, so called because it gives rise to the follicular germinal center, a site of vigorous B cell proliferation (see Figure 1-19). This component of the humoral response takes one to three weeks to develop fully, but produces IgG, IgA and/or IgE antibodies with much higher affinity for antigen. The B lineage cells that synthesize these antibodies are known as **long-lived plasma cells** and produce antibody for months and years. The antibodies produced in these responses also recruit a wider range of effector mechanisms than IgM and may prevent subsequent infections with the same microorganism. The time course and nature of the four types of antibody responses are summarized in Figure 6-23.

The germinal center reaction and the high-affinity antibodies generated by it are the main focus of the remainder of this chapter: the B1 cells and marginal zone B cells that produce early low-affinity IgM responses belong to a class of lymphocytes that have many features of innate immunity and are described in more detail, along with other lymphocytes that fall into this group, in Chapter 8. Here we briefly discuss the properties of these T cell-independent responses before turning to the T cell-dependent responses that evolve in the germinal center.

TI antibody responses are induced by two different mechanisms and are critical to the responses to some pathogens

T cell-independent antibody responses are typically seen in response to microbial surface antigens or virus-coat proteins and are predominantly IgM antibodies whose principal effector action is the activation of complement. This antibody is not only directly destructive to the

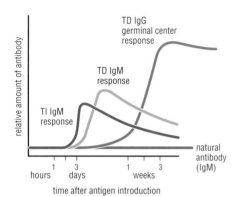

Figure 6-23 Time course of antibody production Four types of antibody responses are compared. Natural antibody is made before antigen encounter by B1 cells using Ig genes encoded by the more 3' V_H elements and lacking junctions with extensive nucleotide additions. These antibodies therefore vary little from the germ-line sequences of the gene segments and have relatively limited diversity. TI IgM responses are the most rapid responses induced by antigen but do not give rise to memory B cells or long-lived plasma cells. The T cell-dependent response bifurcates. There is first an IgM response of unmutated Ig genes that again does not give rise to memory or long-lived plasma cells. This response is followed by the slower germinal center response, which gives rise to high-affinity, isotype-switched antibodies such as IgG and long-lived plasma cells. Note that there is some IgG produced in the T cell-dependent non-germinal center responses at the same time as the IgM (not shown).

microbe but also helps to activate other cells of the immune system. In particular, complement components initiate inflammation, attract neutrophils and monocytes, and may activate dendritic cells. Dendritic cells not only help to initiate T cell responses but also have a role in stimulating B cell responses through the TNF superfamily cytokine BAFF. BAFF is secreted by activated dendritic cells (and by other immune cell types) and, as we shall see later, is required for the survival of mature marginal zone and follicular B cells; it also enhances TI B cell responses, perhaps through its pro-survival effects.

Antigens that induce T cell-independent responses are traditionally divided into two classes: *TI-1* and *TI-2*, although they are in fact of three kinds. **TI-1 antigens** comprise repetitive and ordered viral protein coats such as those of polio virus, and antigens that contain ligands for both the BCR and a Toll-like receptor (TLR), as exemplified by bacterial lipopolysaccharide (LPS). LPS induces a very strong TI antibody response primarily against the diverse O-antigen portion of the molecule, which is distinct from the lipid A portion that engages TLR4. Antigens such as LPS are thought to induce strong B cell responses because they stimulate both through the BCR and through a TLR. **TI-2 antigens** comprise repetitive polysaccharides, such as those found in bacterial capsules. The repetitive antigens (highly ordered viral protein coats and bacterial cell wall polysaccharides) are able to activate B cells in the absence of T cell help or a TLR signal probably because of the strength and duration of BCR signaling that they can sustain through their repetitive nature.

The division of TI antigens into two classes is based on the effects of mutations in the B cell signaling component Btk (see Figure 6-20). Defects in Btk in the mouse cause an X-linked immune deficiency (**X-linked immunodeficiency**, or **XID**) but in the human cause the more severe B cell immunodeficiency, X-linked agammaglobulinemia, in which B cell development is blocked because the BCR signaling pathway is also needed for B cell development, as described in Chapter 7. In both species the gene for Btk is on the X chromosome. In mice, whereas B1 cells and responses to TI-2 antigens are eliminated in XID mice, responses to TI-1 and T cell-dependent antigens are intact. The reason for the difference in severity between human and mouse Btk mutant phenotypes is not known, nor is the reason for the differential sensitivity of TI-1 and TI-2 responses to mutations in Btk.

The distinctive features of TI-1 and TI-2 antigens are summarized in Figure 6-24. TI-2 responses, although they arise early in the immune response to infection, do not develop early in ontogeny, and in consequence infants fail to make early responses to bacterial polysaccharides. The importance of these responses is seen in the susceptibility of children under two years old to severe bacterial meningitis caused by *Haemophilus influenzae* type b, although they can be protected from *H. influenzae* by a vaccine designed to elicit a T cell response to the bacterial polysaccharide, as we shall see in Chapter 14.

Comparison of T Cell-Independent Antibody Responses

Antigen type	Response of XID mice; neonatal mammal	Involvement of TLRs	Example
TI-1	+	–	repetitive, paracrystalline virions
TI-1	+	+	O-antigen of Gram-negative bacteria; intact bacteria (?)
TI-2	–	–	purified polysaccharides

Figure 6-24 Table comparing T cell-independent antibody responses XID mice have a mutation in the Btk signaling component. TLR: Toll-like receptor.

mice but do in T cell-deficient mice. In general, these antigens also do not induce robust antibody responses in young individuals.

XID: see X-linked immunodeficiency.

X-linked immunodeficiency (XID): a genetic immunodeficiency disease of mice caused by a mutation in the Btk gene leading to a partly immunodeficient phenotype characterized by failure to respond to a subset of T cell-independent antigens (**TI-2 antigens**).

References

Baumgarth, N. *et al.*: **B-1 and B-2 cell-derived immunoglobulin M antibodies are nonredundant components of the protective response to influenza virus infection.** *J. Exp. Med.* 2000, **192**:271–280.

Hangartner, L. *et al.*: **Antiviral antibody responses: the two extremes of a wide spectrum.** *Nat. Rev. Immunol.* 2006, **6**:231–243.

Ochsenbein, A.F. and Zinkernagel, R.M.: **Natural antibodies and complement link innate and acquired immunity.** *Immunol. Today* 2000, **21**:624–630.

Vos, Q. *et al.*: **B-cell activation by T-cell-independent type 2 antigens as an integral part of the humoral immune response to pathogenic microorganisms.** *Immunol. Rev.* 2000, **176**:154–170.

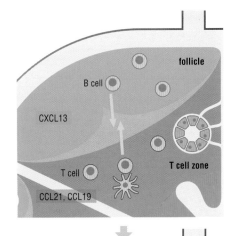

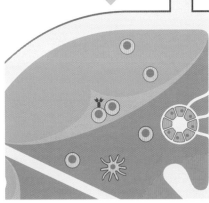

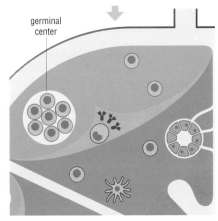

germinal
center

Figure 6-25 B cell–T cell interactions in the lymph node Antigen activation of follicular B cells (blue outline) induces increased expression of CCR7 and migration to the boundary of the T cell zone and the follicle (top panel). This occurs within several hours of antigen contact. Some antigen-activated CD4 T cells (red outline) also migrate to this location and scan B cells for expression of MHC class II–peptide complexes that react with their TCR (middle panel). If B cells and T cells capable of recognizing the same antigen come into contact, this leads to a stable contact promoting B cell activation. After two days, some B cells differentiate into IgM-secreting or, less commonly, IgG-secreting plasma cells where they are, whereas others migrate back to the follicle along with some specific helper T cells and start a germinal center. Note that lymph nodes are essentially full of lymphocytes.

Antigen binding induces follicular B cells to migrate to the boundary between the B cell follicles and the T cell zone

Follicular B cells recirculate between blood and secondary lymphoid organs, homing to the lymphoid follicles by means of CXCR5, which is the receptor for the main follicular chemokine, CXCL13, and departing again after about 16 hours to sample another location unless they meanwhile encounter the antigen they recognize. Contact with antigen induces a rapid 2–3-fold increase in expression of the chemokine receptor CCR7 (see section 2-13), which recognizes the two predominant chemokines expressed in the T cell zone of the spleen or lymph node, CCL21 and CCL19 (see section 1-6), and causes the B cell to migrate from the follicle to the edge of the T cell zone, a position determined by the competing responsiveness to CCL21, CCL19 and CXCL13 (Figure 6-25). B cells arrive at this new location several hours after antigen introduction and await interaction with antigen-specific helper T cells, which arrive soon after activation by antigen presented by dendritic cells in the central part of the T cell zone. Their migration is the result of the complementary induction of CXCR5. If T cell help fails to arrive, these B cells will, like T cells that encounter antigen without costimulation (see section 5-6), enter a state of anergy: we discuss this further in Chapter 12 where we describe mechanisms of peripheral tolerance. Prompt arrival of antigen-specific helper T cells, in contrast, will lead to B cell activation and antibody production.

Antigen bound to the BCR is taken up efficiently, processed and presented to helper T cells

Helper T cells initially stimulated by antigen presented by dendritic cells may migrate to the boundary of the T cell zone, where they scan B cells for antigen displayed as peptide bound to MHC class II molecules. These T cells have a distinctive phenotype and are called T_{FH} (see section 5-9). If the T cell recognizes the MHC–peptide complexes for which it is specific and receives costimulation, the helper T cell establishes a sustained monogamous interaction with the presenting B cell that leads to activation of the B cell as well as further activation of the helper T cell.

On the B cell side, this interaction involves the efficient uptake of antigen that is bound to the BCR. B cells can take up antigen by nonspecific pinocytosis, but this is generally about 10,000-fold less efficient than for antigen bound to the BCR. Some of this increased efficiency is simply the concentrative effect of specific binding of the antigen to the surface of the B cell, combined with a slow constitutive rate of internalization of the BCR. BCR signaling events triggered by receptor oligomerization by antigen, however, greatly enhance the rate of internalization and also serve to target the BCR–antigen complex and MHC class II molecules to the late endosomes, where peptides are loaded onto MHC class II molecules (Figure 6-26). BCR signaling also promotes the acidification of these endocytic compartments, which further promotes peptide loading due to the pH-dependent activation of cathepsins, proteases that promote antigen presentation (see section 4-7). BCR signaling additionally promotes the ability of B cells to interact with and activate helper T cells by inducing additional expression of MHC class II molecules, which can be loaded with the internalized antigen, by increasing the adhesiveness of the LFA-1 ($\alpha L\beta 2$) integrin, which binds to ICAM-1 on the T cell, and by inducing expression of the B7-2 costimulatory molecule. In addition, B cells constitutively express ICOS ligand, which, together with B7-2, contributes to the induction of CD40L and cytokines in the T cell (Figure 6-27).

Definitions

X-linked hyper-IgM syndrome: a human immunodeficiency characterized by elevated levels of serum IgM, but very low levels of serum IgG, IgA or IgE. This disease is caused by loss-of-function mutations in the gene encoding CD40L (CD154). Much less common are mutations of the autosomal genes encoding CD40 or activation-induced cytidine deaminase (AID), an enzyme required for class switch recombination and somatic hypermutation, which give rise to similar diseases.

References

Bishop, G.A. and Hostager, B.S.: **B lymphocyte activation by contact-mediated interactions with T lymphocytes.** Curr. Opin. Immunol. 2001, **13**:278–285.

Clark, M.R. et al.: **B-cell antigen receptor signaling requirements for targeting antigen to the MHC class II presentation pathway.** Curr. Opin. Immunol. 2004, **16**:382–387.

Leonard, W.J. and Spolski, R.: **Interleukin-21: a modulator of lymphoid proliferation, apoptosis and dif-**

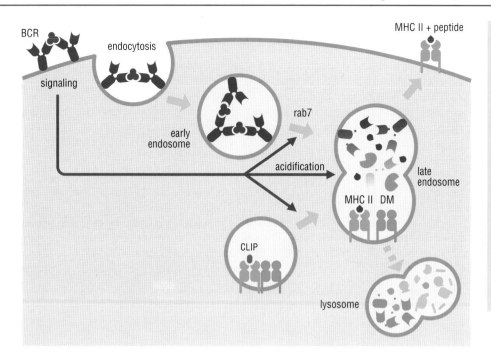

Figure 6-26 Specific uptake of antigen by the BCR strongly promotes antigen presentation BCR binding promotes the efficient endocytosis of antigen and antigen presentation to CD4 T cells. BCR signaling promotes the movement of internalized BCR molecules to the late endosomes, which are the major compartment where MHC class II loading takes place in the B cell. Unstimulated B cells have low levels of the small GTPase rab7, which mediates movement from early endosomes into late endosomes, and various means of stimulating B cells increase its expression, which may facilitate antigen presentation. BCR signaling also promotes MHC class II molecule transit to late endosomes and acidification of that compartment, both of which promote antigen presentation. Antigenic peptide loading onto MHC class II, facilitated by DM and DO (see section 4-7 and Figure 4-17; not shown), is followed by transit of the loaded MHC molecules to the cell surface.

Helper T cells induce rapid antibody production and the germinal center reaction

The helper T cell receiving activation signals from a B cell presenting the correct antigen remains in contact with that B cell for an hour. During this time, the helper T cell synthesizes and secretes cytokines, of which IL-6, IL-10 and IL-21 are thought to be particularly important, and also synthesizes the membrane-bound TNF family member CD40L (CD154), which interacts with CD40 on the B cell. This is an essential signal for both rapid antibody production and the *germinal center reaction*, which is described in the next section. Individuals with loss-of-function mutations of CD40L have a disease called **X-linked hyper-IgM syndrome**, characterized by severe lack of IgG, IgA and IgE and recurrent infections. Elevated levels of IgM in these individuals probably reflect poorer immune defense, more and longer infections, and consequently greater T cell-independent stimulation to make IgM. Interestingly, individuals with this deficiency in addition to having increased infections with extracellular bacteria, characteristic of antibody deficiencies, also have increased numbers of opportunistic infections, as seen in T cell immunodeficiencies. This reflects the important role of CD40L in cell-mediated immune responses (see section 5-8).

Some B cells interacting with helper T cells proliferate and then differentiate into antibody-secreting plasma cells either in the T cell area of the lymph node or, in the case of the spleen, in the adjacent red pulp area. These are the B cells, described in the preceding section, that provide the early T cell-dependent IgM and IgG response. Other B cells migrate back to the B cell follicle along with antigen-specific T_{FH} cells and initiate a germinal center reaction, which we describe in the next two sections.

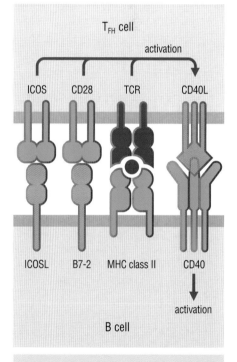

Figure 6-27 Cell–cell interaction molecules involved in helper T cell activation of B cells For a helper T cell to provide help for an antigen-activated B cell, its TCR must recognize the specific peptide–MHC class II complex on the surface of the antigen-presenting B cell. In addition, the helper T cell must receive costimulatory signals, provided by B7-2 and ICOS ligand (ICOSL) expressed on the B cell. Naïve B cells constitutively express ICOSL but not B7-2. The latter is upregulated by BCR and/or TLR stimulation. T cell activation results in the expression of CD40L, which provides an essential signal for T cell-dependent activation of B cells, and cytokines (not shown), of which IL-6, IL-10 and IL-21 are thought to be especially important.

ferentiation. *Nat. Rev. Immunol.* 2005 **5**:688–698.

MacLennan, I.C.M. *et al.*: **Extrafollicular antibody responses.** *Immunol. Rev.* 2003, **194**:8–18.

Okada, T. and Cyster, J.G.: **B cell migration and interactions in the early phase of antibody responses.** *Curr. Opin. Immunol.* 2006, **18**:278–285.

Reif, K. *et al.*: **Balanced responsiveness to chemoattractants from adjacent zones determines B-cell position.** *Nature* 2002, **416**:94–99.

Snapper, C.M. *et al.*: **Distinct types of T-cell help for the induction of a humoral immune response to** *Streptococcus pneumoniae*. *Trends Immunol.* 2001, **22**:308–311.

The germinal center is a transient structure that supports the production of high-affinity antibodies

Those B cells that do not differentiate into antibody-secreting plasma cells on their initial encounter with helper T cells at the edge of the T cell zone migrate into the B cell-rich follicle, where they proliferate very rapidly (6–8 h doubling time). Likewise, some of the antigen-activated T cells migrate into the same follicular region as the antigen-stimulated B cells. This region soon assumes the unique appearance of the germinal center, characterized by the presence of a **dark zone** full of the very rapidly dividing B cells, also called **centroblasts**, and an adjacent **light zone**. This contains the progeny of centroblasts that have stopped proliferating, which are known as **centrocytes**, as well as antigen-specific helper T cells, and a third type of cell, **follicular dendritic cells (FDCs)**, which trap intact antigen on their surface (see section 1-7) (Figure 6-28). The interactions of B cells with antigen on the surface of follicular dendritic cells and subsequently with helper T cells in the germinal center sustain the **germinal center reaction**, which produces a mature T cell-dependent antibody response.

The major results of the germinal center reaction are (1) rapid mutation of the antibody-encoding genes of the B cells followed by selection of centrocytes for those that can make high-affinity antibodies, (2) **class switch recombination** at the immunoglobulin heavy-chain gene locus so that the antibody secreted is IgG, IgA or IgE rather than IgM, (3) generation of long-lived plasma cells that secrete this high-affinity antibody of the appropriate class for a long time, and (4) generation of memory B cells that allow a more rapid adaptive immune response if a new infection can breach the protective barrier provided by the prolonged secretion of high-affinity antibody. The first and second outcomes represent the key events of the germinal center, whereas the third and fourth outcomes represent alternative cellular fates for putting this high-quality antibody to use.

Activation-induced cytidine deaminase mediates class switch recombination and hypermutation of Ig genes

The rapid mutation of the antibody genes in germinal center B cells is a specialized process known as **somatic hypermutation**. Somatic hypermutation and class switch recombination are molecular genetic processes that alter the immunoglobulin genes generated during B cell development by the assembly of separate gene segments (see Figure 6-19). But while somatic hypermutation generates point mutations in the gene segments encoding the variable regions of the antibody, class switch recombination is a breakage and rejoining reaction that replaces one constant-region gene segment with another. Surprisingly, these two disparate processes are under the control of a single gene, **activation-induced cytidine deaminase (AID)**. AID participates in both events directly by deaminating cytidine residues in the V region (for somatic hypermutation) or in repetitive sequences found upstream of each heavy-chain constant region, called **S regions** (for class switch recombination). Cytosine deamination converts cytosine to uracil, which on DNA replication pairs with adenine not guanine and thus gives rise to a point mutation or can lead to error-prone DNA repair or to DNA strand breakage (see Figure 3-52). Genetic defects in AID, like those in CD40L, give rise to a hyper-IgM syndrome in which there is almost no IgG, IgA or IgE (except maternally derived IgG).

AID expression in germinal center B cells leads to a very high rate of mutation of sequences in and near the part of the Ig genes encoding the heavy and light chain V domains, which determine antigen-binding properties of the antibody. Mutations approach a frequency of ~1 in 1000 base pairs per generation, compared with a normal mutation rate below 1 in 10^7 base

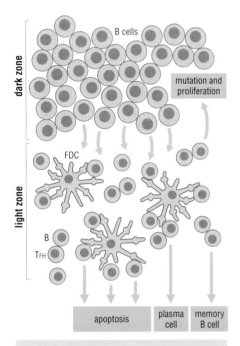

Figure 6-28 Anatomy of the germinal center The germinal center can be seen with the microscope to contain a dark zone and a light zone, represented schematically here. The dark zone has extensive proliferation of B cells, referred to as centroblasts. These are thought to be the cells in which somatic hypermutation occurs. After proliferation, the B cells, now called centrocytes, are found in the adjacent light zone, which has in addition antigen-specific helper T cells and follicular dendritic cells (FDCs). Centrocytes that cannot bind antigen die by apoptosis. Those that can bind antigen and receive survival signals from BCR signaling and from interactions with the helper T cells can either migrate back to the dark zone for another round of proliferation, mutation and selection, or can assume one of two cell fates, that of the memory B cell or that of the plasma cell, which migrates to the bone marrow for long-term antibody secretion.

Definitions

activation-induced cytidine deaminase (AID): a cytidine deaminase necessary for both **class switch recombination** and **somatic hypermutation**. AID is expressed in germinal center B cells and is related to APOBEC3G.

AID: see **activation-induced cytidine deaminase**.

centroblast: rapidly proliferating B cell in the **dark zone** of the germinal center.

centrocyte: post-mitotic germinal center B cell that is

programmed to die by apoptosis unless it receives survival signals from the BCR and helper T cells.

class switch recombination: genetic recombination between switch regions that are found 5' of the constant exons of each Ig heavy-chain gene. Usually this involves the switch region upstream of the C_μ exons (S_μ) and a downstream **S region**.

dark zone: (of lymphoid follicles) region of the germinal center full of rapidly dividing **centroblasts**.

FDC: see **follicular dendritic cell**.

follicular dendritic cell (FDC): specialized non-hematopoietic cell found in B cell areas of secondary lymphoid tissue and specialized for collecting antigen bound to antibody or complement components and presenting it to **centrocytes**, which can thereby be selected for high-affinity binding.

germinal center reaction: the T cell-dependent antibody response of follicular B cells in which there is rapid proliferation combined with **somatic hypermutation** and **class switch recombination**, resulting in the production of high-affinity IgG, IgA and/or IgE antibodies,

Figure 6-29 Class switch recombination The IgH locus is shown with a functional VDJ recombination adjacent to the μ locus. This is the configuration that is generated during B cell development in the bone marrow. The μ locus is transcribed under the control of an enhancer (E_μ) in the intron between the J and the first C_μ segment and gives rise to mRNA encoding the μ heavy chain. Upon receipt of the appropriate signals, including IL-4 in this case, there is transcriptional activity at one or more unrearranged downstream heavy-chain constant-region-encoding locus (γ1 is shown in this example: it is responsive to IL-4 in the mouse). Class switch recombination machinery, involving AID, then recombines the S_μ and $S_\gamma 1$ sequences, deleting the intervening DNA and juxtaposing the functional VDJ sequences encoding the V_H domain with the exons encoding a downstream constant region (γ1). Transcription of this locus leads to the production of mRNA encoding the new heavy chain. Note that not all of the heavy-chain constant exons for μ and γ1 are shown.

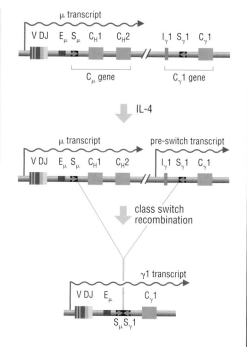

pairs per generation. Mutations are targeted by the transcriptional enhancers that lie in the introns between the exons encoding the V domains and the C domains. The enhancers seem to be the only element of the Ig genes that are required, because heterologous genes with active promoters will undergo mutation in B cells if either the heavy-chain enhancer or the light-chain enhancer is inserted into the gene. Interestingly, there are a few other genetic loci that also seem to attract the hypermutation machinery in AID-expressing B cells, and these presumably adventitious mutations may contribute to the formation of B cell lymphomas.

AID expression also makes a germinal center B cell competent for class switch recombination, which is a genetic recombination between a region of repetitive sequence in the intron between the V domain-encoding exon and the C_μ-encoding exons (called S_μ for switch region of the IgH μ locus) and a corresponding region in front of the exons encoding the constant domains of the isotypes other than δ (called $S_\gamma 1$, $S_\gamma 2$, and so on) (Figure 6-29). The cytosine deaminase activity of AID also is important for generating breaks in the DNA at the S regions, but the exact molecular mechanism is not understood. The result of switch recombination is that the recombined exon encoding the V domain is moved adjacent to the constant domains encoding a different heavy-chain isotype, so that the B cell now produces an antibody of the original specificity but different class.

Although AID is necessary for all class switch recombination, the isotype selected is determined by environmental cues that the B cell receives. The best understood of these are the cytokines IL-4, IFN-γ and TGF-β. The first two are the canonical T_H2 and T_H1 cytokines, respectively. IL-4 or IL-13 promote isotype switching to form IgG1 (in mouse; IgG4 in humans) or IgE, whereas IFN-γ induces isotype switching to IgG2a (in mouse). Thus, the type of helper T cell greatly influences the nature of antibody produced. TGF-β induces isotype switching to IgA (Figure 6-30). The situation in humans is not completely analogous. IL-4, as in the mouse, stimulates class switch recombination to IgE and to one isotype of IgG, namely IgG4, which is poor at activating complement or interacting with activating Fcγ receptors. T_H1 cells do stimulate isotype switching to the complement-activating IgG isotypes (IgG1, IgG3), but the particular cytokines responsible are not established. The mechanism for isotype switching depends on transcriptional activation of the downstream IgH locus, in some way directing the class switch recombination machinery to use the S region at that site.

AID expression is also required for gene conversion events, which are analogous to somatic hypermutation in that they change the specificity of the antibody produced, although the mechanism is distinct in that the changes are templated by small regions of other V genes in the genome. This process is not important for creating diversity and specificity of antibodies of human or mouse, but is important in other species including rabbit and chicken (see Chapter 7).

Direction of Class Switch Recombination of Mouse B Cells by Cytokines	
Cytokine	Isotype favored
IL-4	IgG1, IgE
IFN-γ	IgG2a, IgG3
TGF-β	IgA, IgG2b

Figure 6-30 Table of direction of class switch recombination by cytokines Shown are isotypes of mouse antibodies. For human antibody responses, IL-4 induces class switch recombination to IgE and IgG4 and T_H1 cells induce class switch to IgG1 and IgG3, which are effective at activating complement and engagement of activating Fc receptors.

long-lived plasma cells, and memory B cells.

light zone: (of lymphoid follicles) region of the germinal center containing **centrocytes, follicular dendritic cells** and antigen-specific helper T cells.

somatic hypermutation: a process whereby mutations are introduced into and near the V-domain-encoding exons of the IgH and IgL genes.

S regions: highly repetitive sequences upstream of each immunoglobulin heavy-chain constant gene locus. **Class switch recombination** occurs between two S regions.

References

Ahonen, C.L. et al.: **The CD40-TRAF6 axis controls affinity maturation and the generation of long-lived plasma cells.** Nat. Immunol. 2002, **3**:451–456.

Bishop, G.A.: **The multifaceted roles of TRAFs in the regulation of B-cell function.** Nat. Rev. Immunol. 2004, **4**:775–786.

Durandy, A. et al.: **Hyper-immunoglobulin M syndromes caused by intrinsic B-lymphocyte defects.** Immunol. Rev. 2005, **203**:67–79.

Maizels, N.: **Immunoglobulin gene diversification.** Annu. Rev. Genet. 2005, **39**:23–46.

Neuberger, M.S. et al.: **Immunity through DNA deamination.** Trends Biochem. Sci. 2003, **28**:305–312.

Tew, J.G. et al.: **Follicular dendritic cells: beyond the necessity of T-cell help.** Trends Immunol. 2001, **22**:361–367.

Vinuesa, C.G. et al.: **Follicular B helper T cells in antibody responses and autoimmunity.** Nat. Rev. Immunol. 2005, **5**:853–865.

Centrocytes are selected for the ability of their antigen receptors to bind effectively to antigen

Germinal center B cells proliferate rapidly and mutate their Ig genes at a high rate. After multiple rounds of replication, centroblasts stop proliferating and migrate to the adjacent light zone, which contains FDCs and antigen-specific helper T cells. The B cells, now called centrocytes, are subjected to stringent competition based on affinity for antigen. The result of this competition is the selection of cells with the best affinity for antigen, some of which return to the dark zone for another round of amplification, mutation and selection. Figure 6-31 shows a reconstruction of the evolution of high-affinity antibodies generated in this way.

Germinal centers are initiated by a small number of founder B cells, and hence are oligoclonal in nature. Thus, competition is set up between a few different clones of cells that mutate rapidly, generating many variants. Most of these will bind antigen less effectively: they will be unable to compete for binding to the available antigen, which is necessary for their survival (see below), and will die by apoptosis. Over time, antigen levels fall, selection becomes more stringent and B cells expressing mutated Ig with increasingly higher affinity for antigen survive preferentially.

This selection is driven by antigen held on the surface of FDCs for long periods by their complement receptors (CR2) and Fc receptors (FcγRIIb). Centrocytes are programmed to die by apoptosis unless they receive a survival signal, which comes in part from BCR signaling by B cells recognizing antigen bound to the FDC and in part from survival signals provided by the antigen-specific helper T cells. For the latter, the B cell must take antigen from the FDC, internalize it, process it and present it, as described above. This is thought to be aided by a propensity of the FDC to release small cell fragments in the form of membrane vesicles, referred to as immune-complex-coated bodies or **iccosomes**. The nature of the helper T cell involvement in selection of centrocytes by antigen is not well understood, but disruption of CD40–CD40L interactions, for example with anti-CD40L antibodies, leads to the disruption of germinal centers, indicating that CD40L is needed to maintain the germinal center reaction.

Centrocytes give rise to long-lived plasma cells and memory B cells

As mentioned above, the process of mutation and selection can be iterative and multiple cycles can occur. Indeed the germinal center reaction often continues for at least six to seven weeks. During this period the centrocytes are directed down one of two alternative paths. Some become memory cells: these are described below. The others differentiate into plasma cells. This is a terminal differentiation step ending in the eventual death of the cell by apoptosis. Plasma cell differentiation occurs throughout the six or seven weeks of the germinal center reaction and results in part from downregulation of the transcriptional repressor BCL-6, which allows the expression of Blimp-1, a transcription factor that drives the plasma cell differentiation program. The transcription factor network involved in plasma cell fate is summarized in Figure 6-32. These plasma cells are longer-lived than those generated before the germinal center response or in T cell-independent antibody responses. Whereas the other types of plasma cells undergo apoptosis in about a week, plasma cells generated in a germinal center response migrate to the bone marrow or other appropriate locations (for example, lamina propria of gut for IgA-secreting plasma cells), where they may live for years. Migration to these locations is mediated by changes in chemokine receptor expression: decreased expression of CXCR5 and CCR7 together with increased expression of either CXCR4, which binds CXCL12 in bone marrow and splenic red pulp (for IgG-secreting plasma cells), or of CCR9, which binds CCL25 in gut lamina propria (for IgA-secreting plasma cells).

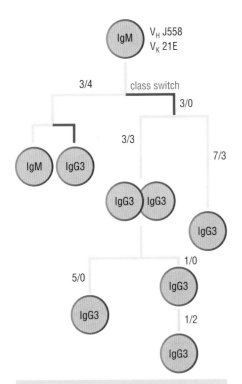

Figure 6-31 Evolution of high-affinity antibodies Shown is a reconstructed genealogical tree of the relationship of a group of hybridomas recovered from a single autoimmune MRL/lpr mouse spleen, all of which exhibited rheumatoid factor (anti-IgG2a, which is similar to an autoantibody frequently seen in rheumatoid arthritis patients) specificity. The tree is based on the assumption that shared mutations occurred before unique mutations. The number of mutations in the heavy-chain and light-chain genes are shown as H/L associated with each line. Two identical hybridomas were isolated in one case (shown as overlapping cells). The switch region DNA indicated that all hybridomas except the IgG3 hybridoma with the same Ig gene sequence as the IgM hybridoma (top left of tree) were the progeny of a cell with the same class switch recombination event. Data are from Shan, H. et al.: *J. Exp. Med.* 1990, **172**:531–536.

Definitions

iccosomes: cell fragments released by follicular dendritic cells and containing immune complexes bound to their surface. Also called immune-complex-coated bodies.

Long-lived plasma cells are largely responsible for the high levels of serum IgG and for the continuous secretion of IgA into the gut and act to provide protection against reinfection. They can first be detected 7–10 days after exposure to antigen and accumulate over the next several weeks, reaching a plateau level by 6–7 weeks. Such cells are responsible for the protective action of vaccines directed at bacterial toxins (for example tetanus toxin, for which protective antibody levels are maintained for about 10 years). Presumably there is a limited capacity of bone marrow and other locations to support long-lived plasma cells, limiting their overall number in the body. In mouse, bone marrow has been estimated to hold about 400,000 long-lived plasma cells. This represents about 0.4% of mononucleated cells, and a similar fraction of plasma cells is found in human bone marrow. Each plasma cell secretes about 2000 antibody molecules per second.

Memory B cells can be rapidly activated to produce higher levels of antibodies

The alternative fate of germinal center B cells that survive selection is to become memory B cells. Memory B cells are generated somewhat later than the first long-lived plasma cells and therefore have a somewhat higher average affinity for antigen. Once generated, memory B cells are maintained for a long time without requiring contact with the initiating antigen. If a subsequent infection overwhelms the levels of circulating antibody produced by the long-lived plasma cells, these memory B cells can become activated by unbound antigen and memory T cells. Memory B cells can again initiate a germinal center reaction, although they are more likely to differentiate into plasma cells. Often a secondary response will produce about ten times more long-lived plasma cells than were generated in the initial response. This increases the long-term titer of the specific antibody in question, a phenomenon that is very important in the design and efficacy of vaccines and is also exploited in the experimental production of antibodies.

How the alternative cell fates of surviving centrocytes is determined is not well understood. CD40L inhibits terminal differentiation to plasma cells, whereas IL-6 promotes plasma cell differentiation, as do several other cytokines. Interestingly, IL-6 is a growth and survival factor for many multiple myeloma cells, which are cancerous cells of a plasma cell phenotype.

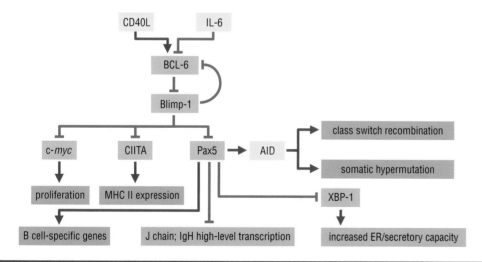

Figure 6-32 Control of differentiation to plasma cells Inputs from CD40 and other signals impinge on the expression of the BCL-6 gene, which encodes a transcriptional repressor necessary for germinal center development. BCL-6 acts to repress another transcriptional repressor, Blimp-1, which acts to promote terminal differentiation to the plasma cell state. Therefore when BCL-6 is on, terminal differentiation is blocked, but when BCL-6 is turned off, Blimp-1 can induce terminal differentiation both by turning on functions required for high-level antibody secretion (shown in red) and by turning off functions required by germinal center B cells but dispensable in plasma cells (shown in blue). Among the genes repressed by Blimp-1 are c-myc, which drives proliferation, CIITA, the master regulator of MHC class II expression and Pax5. Pax5 is a transcriptional activator of B cell-specific genes and a transcriptional repressor of the gene encoding J chain and of high-level transcription of the IgH locus, and thus represses functions required for the plasma cell state. It is not known how the high-level production of Ig light chains is controlled. Plasma cells are also characterized by high levels of the components of the secretory pathway, including the endoplasmic reticulum (ER). These components are controlled by the transcription factor XBP-1, which is repressed by Pax5. In addition, Blimp-1 blocks the expression of AID, which is required for somatic hypermutation and class switch recombination, functions of the germinal center B cell before terminal differentiation.

References

Manz, R.A. *et al.*: **Maintenance of serum antibody levels.** *Annu. Rev. Immunol.* 2005, **23**:367–386.

Maruyama, M. *et al.*: **Memory B-cell persistence is independent of persisting immunizing antigen.** *Nature* 2000, **407**:636–642.

Randall, T.D. *et al.*: **Arrest of B lymphocyte terminal differentiation by CD40 signaling: mechanism for lack of antibody-secreting cells in germinal centers.** *Immunity* 1998, **8**:733–742.

Shan, H. *et al.*: **Heavy-chain class switch does not terminate somatic mutation.** *J. Exp. Med.* 1990, **172**:531–536.

Shapiro-Shelef, M. and Calame, K.: **Regulation of plasma-cell development.** *Nat. Rev. Immunol.* 2005, **5**:230–242.

Tarlinton, D.M. and Smith, K.G.C.: **Dissecting affinity maturation: a model explaining selection of antibody-forming cells and memory B cells in the germinal centre.** *Immunol. Today* 2000, **21**:436–441.

7

Development of Lymphocytes and Selection of the Receptor Repertoire

The tremendous diversity of antibodies and T cell receptors is created during specialized genetic recombination events that occur during the development of lymphocytes. Developing B cells and T cells are also assorted into a handful of different subtypes, including both the major adaptive immune cells, the helper T cell, the cytotoxic T cell, and the follicular B cell, and also the specialized, more innate-like T cell and B cell populations described in the next chapter. Finally, the use of highly diversifying mechanisms for creating antigen-specific receptors on lymphocytes means that generation of self-reactive lymphocytes capable of inducing autoimmune disease is an inevitable byproduct, and several mechanisms operate during lymphocyte development to remove these cells or limit their ability to cause autoimmunity.

Lymphocyte development generates cells of many different specificities

Lymphocytes, like other blood cells, develop from hematopoietic stem cells in a series of maturational steps dependent on specialized microenvironments. What is unique about lymphocyte development is that the key receptors of these cells, their antigen receptors, must be generated by genetic rearrangements of germline DNA. This occurs by V(D)J recombination, a process we touched on in Chapter 5 and Chapter 6, and which is directed by the proteins encoded by the **recombination-activating genes-1 and -2** (*RAG-1* and *RAG-2*). Ig genes are rearranged in a stereotypical order in developing B cells in the bone marrow, and TCR genes are rearranged in a stereotypical order in developing T cells, which are known as **thymocytes** because they develop in the thymus. The result is two sets of remarkably diverse cells with at least 10^7 different possible antigen receptors, of which each individual cell expresses only one (or possibly two in the case of T cells).

Developing lymphocytes make multiple lineage choices

B cells, T cells and natural killer (NK) cells all develop from early lymphoid progenitors, which in turn arise from pluripotent hematopoeitic stem cells. Progenitor cells that commit to the B cell lineage continue to develop in the bone marrow, progressing from a **pro-B cell**, which does not have any functionally rearranged Ig genes, to a **pre-B cell**, which has a functional Ig heavy chain, and then to a B cell, which expresses membrane IgM (Figure 7-1).

Alternatively, progenitor cells may migrate to the thymus, or less commonly to the gut, and develop into T cells. This lineage choice is reinforced by signaling by members of the **Notch** family of receptors, which are often involved in determining cell fate during development in vertebrates and invertebrates. The thymus and specialized locations in the gut express Notch ligands and promote T cell development in these locations.

Once in the thymus, cells of the T lineage have two further decisions to make. First, they must become either αβ T cells (those with αβ TCRs) or γδ T cells (those with γδ TCRs). Second, the αβ T cells must become either CD8 cells or CD4 cells and moreover coreceptor expression must be coordinated with MHC recognition so that CD4 cells recognize antigenic peptides associated with MHC class II molecules and CD8 cells recognize antigenic peptides associated with MHC class I molecules. The differentiated fate of γδ T cells is determined when their γ and δ genes have been successfully rearranged. αβ T cells, not surprisingly, undergo a more complex developmental sequence. Their β-chain genes are rearranged before their α-chain genes, and at this intermediate stage the cells express both CD4 and CD8. Indeed the stages of development of thymocytes are often defined by expression of these coreceptors, so that the earliest T cell precursors, which express neither CD4 nor CD8, are called *double-negative (DN) thymocytes*, and those cells expressing both coreceptors are called *double-positive (DP) thymocytes*. Further maturation leads to downregulation of either CD4 or CD8 coordinated with adoption of the cytotoxic or helper T cell lineage. While still in the thymus, the relatively

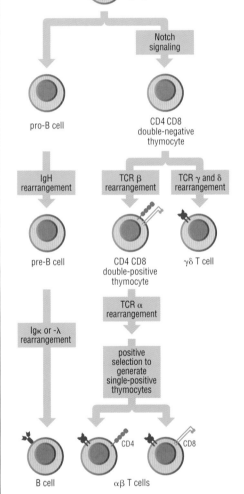

Figure 7-1 Rearrangements of antigen receptor genes and lineage choices during lymphoid development Early lymphoid progenitors can develop into either B cells or T cells, with Notch1 signaling reinforcing the latter choice. Further lineage choices are made between different types of B cells and T cells, directed by stage of development (fetal versus adult), environmental cues (cytokines, Notch ligands), or antigen receptor specificity (CD4 versus CD8 T cells). In addition, Ig and TCR rearrangements are required for developmental progression. Early lymphoid progenitors can also give rise to natural killer cells, and to a subset of dendritic cells (not shown).

Definitions

negative selection: a process during the development of lymphocytes in which cells whose antigen receptors bind strongly to self antigens die.

Notch: any of a family of cell-surface receptors that often participate in cell fate determination during development.

positive selection: a process during the development of T cells in which cells that have a low but positive response to a self antigen are selected to continue

development. Positive selection is generally coupled to differentiation into either the CD8 cytotoxic T cell lineage or the CD4 helper T cell lineage. B cells are thought to undergo antigen receptor-dependent positive selection as well.

pre-B cell: the stage of B cell development at which a functional Ig heavy chain protein is expressed, but a functional Ig light chain is not expressed.

pro-B cell: the stage of B cell development at which no functional Ig chains are produced but the cell is committed to the B cell lineage and DNA at the IgH locus is

undergoing rearrangement.

RAG-1: see **recombination activating gene-1**.

RAG-2: see **recombination activating gene-2**.

recombination activating gene-1 (*RAG-1*): a gene encoding one of two proteins that mediate initial DNA cleavages involved in V(D)J recombination.

recombination activating gene-2 (*RAG-2*): a gene encoding one of two proteins that mediate initial DNA cleavages involved in V(D)J recombination.

mature T cell precursors expressing only one coreceptor are referred to as *single-positive thymocytes* (see Figure 7-1). During the transition to the single-positive stage, the thymocytes undergo selection for the appropriate MHC recognition properties.

Lymphocytes undergo positive and negative selection during their development

Both B cells and T cells are selected during their development so that those lymphocytes able to respond to antigen are retained and those lymphocytes that are self-reactive are eliminated. The processes whereby this is achieved are known as **positive selection** and **negative selection**. For both B cells and T cells, positive selection depends upon a weak response to self antigen that promotes the maturation and survival of the cell. But whereas in B cells this may be little more than a test of the functional integrity of the BCR, in T cells it also plays an important part in ensuring that cells with the correct combination of TCR and coreceptor are selected for recognition of either MHC class I or MHC class II, in either case complexed with self peptides. Thus double-positive thymocytes that recognize self-peptide–MHC complexes mature to the single-positive stage corresponding to the type of MHC they recognize (that is, cells recognizing self peptides + class II MHC become CD4 cells, and those recognizing self peptides + class I MHC become CD8 cells).

Positive selection ensures that mature T cells have some ability to recognize MHC molecules, but the threshold for this recognition is set relatively low, so that the same TCR may recognize the complex of a foreign peptide bound to MHC molecules with higher affinity. The threshold for activation of mature T cells requires this higher affinity. Positive selection is thought to select only about 10% of developing thymocytes for further maturation, and therefore it appears that the function of positive selection is to increase the frequency of potentially useful T cells in the peripheral T cell pool by requiring some reactivity with the type of MHC corresponding to the coreceptor expressed. Mature naïve B and T lymphocytes require continual low-level antigen receptor engagement for survival, and this is thought to increase the effectiveness of the lymphocyte populations in the periphery.

Although a low-level self-reactivity appears to be beneficial for lymphocytes, clearly a strong affinity for self antigens presents the danger of autoimmunity, especially given the considerable element of chance introduced by the V(D)J recombination mechanism into the antigen-binding sites of the antigen receptors, as described in detail later in this chapter. Indeed, specialized mechanisms exist for removing these cells from the repertoire of lymphocytes before they finish their development. In the T cell lineage this involves the death of cells: this is the process of negative selection, also known as clonal deletion. In B cells, self-reactivity above a certain threshold reinduces V(D)J recombination and light-chain genes are further rearranged in a process called *receptor editing* that can serve to eliminate the self-reactivity. If receptor editing fails or if the developing B cell leaves the bone marrow and encounters the self antigen in the spleen soon after arriving at that site, then this cell also undergoes clonal deletion.

In this chapter we describe the process of V(D)J recombination and how it is controlled, as well as reviewing some of the alternative mechanisms that operate in non-mammalian species to generate antigen receptor diversity. This is followed by a discussion of the mechanisms of positive and negative selection. Genetic defects in components of the machinery of V(D)J recombination, or the signaling pathways that direct lymphocyte development, cause severe immunodeficiencies and these are discussed at the end of the chapter, along with genetic accidents of V(D)J recombination that in later life cause cancers of lymphoid cells.

thymocytes: developing T cells in the thymus.

References

Pelayo, R. *et al.*: **Lymphoid progenitors and primary routes to becoming cells of the immune system.** *Curr. Opin. Immunol.* 2005, **17**:100–107.

Pillai, S.: *Lymphocyte Development: Cell Selection Events and Signals During Immune Ontogeny* (Springer-Verlag, New York, 1997).

Antigen receptor loci have multiple gene segments that are rearranged to generate functional genes

The variable domains of antibody heavy and light chains and of TCR α, β, γ and δ chains are genetically encoded in two or three types of gene segments, as we have already seen in Chapters 5 and 6, and these segments must be recombined at the DNA level in order to generate functional Ig- or TCR-encoding genes. These gene segments are called V (variable), D (diversity) and J (joining) genes, and each is encoded in multiple copies in most of the antigen receptor loci. IgH, TCR β and TCR δ loci contain all three types of gene segments, whereas Igκ, Igλ, TCR α and TCR γ loci contain only V and J segments (Figure 7-2).

For the IgH and TCR β loci, one of multiple D segments first recombines with one of multiple J segments. This is followed by recombination of one V gene to the rearranged DJ site (Figure 7-3). In contrast, the TCR δ locus segments can recombine in either order. The process of V(D)J recombination is simpler for the other loci; they just need to rearrange a V region to a J region, although the resulting diversity coming from choice of segments—that is, the **combinatorial diversity** of the assembled sequences—is correspondingly lower. For example, in the IgH locus of humans there are ~40 functional V_H segments, 27 D_H segments, and six J_H segments, permitting approximately 6,480 different H chains to be generated. For the human Igκ locus, there are 40 Vκ genes and five Jκ segments, giving rise to 200 different combinations. Additional diversity comes from the combination of two different chains (heavy and light chains of antibodies, and

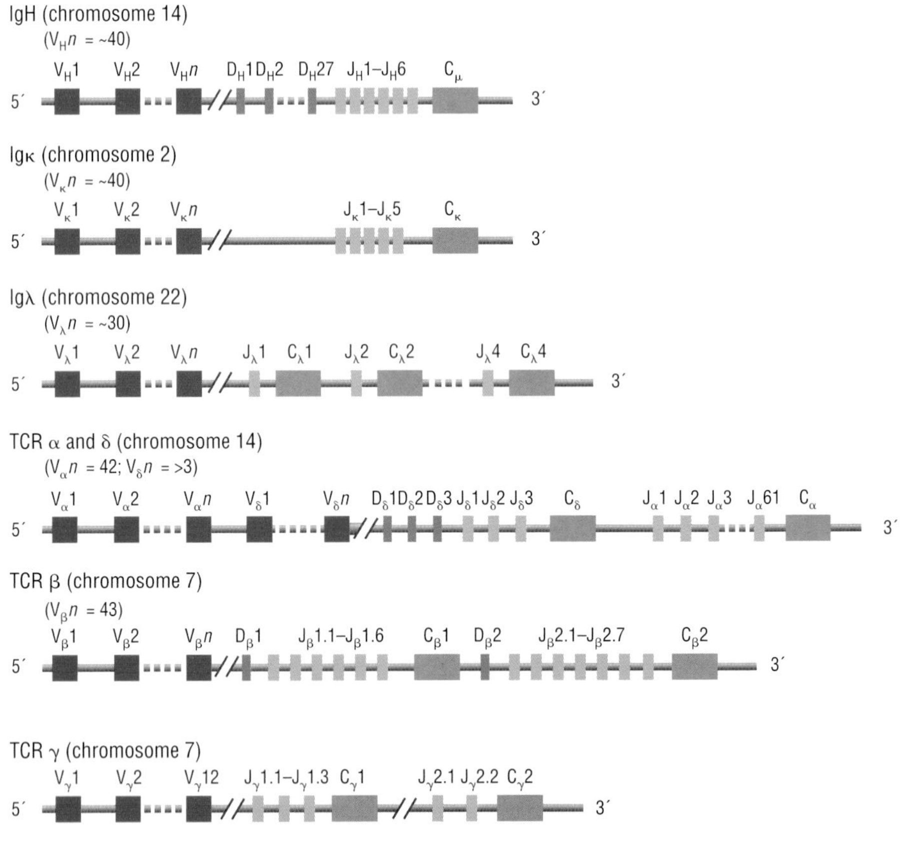

Figure 7-2 Germline structure of immunoglobulin and TCR genes Current understanding of Ig and TCR genetic loci is indicated, with the numbers of gene segments shown. Note that the TCR α locus is embedded within the TCR α locus. Most TCR α rearrangements delete the TCR δ locus. Some V gene segments can contribute to assembled genes for either α or δ chains. Inverted V genes occur in several loci, including Igκ and TCR β (not indicated in the figure). Recombination involving these gene segments results in inversions of the DNA, rather than deletions of the DNA between the two RSSs being used. Although the C_μ locus is in fact made up of multiple exons (see Figure 6-19), only one is shown here for simplicity. Also not shown are short exons upstream of V genes that encode the signal sequences for these transmembrane proteins.

Definitions

12/23 rule: a property of V(D)J recombination whereby gene segments that are recombined are limited to combinations of one **RSS** with a 12-bp spacer and one **RSS** with a 23-bp spacer. For example, gene segments with 12-bp spacer **RSS**s never combine with each other. This is one of the mechanisms that limits V(D)J recombination to avoid undesirable combinations.

combinatorial diversity: the diversity of the antigen receptors of lymphocytes resulting from different combinations of V, D and J gene segments.

non-homologous end joining: mechanism for repairing double-strand breaks in DNA in which the broken ends are rejoined directly, usually with the loss of nucleotides at the join.

recombination signal sequence (RSS): the sequence elements adjacent to lymphocyte antigen receptor gene segments that direct V(D)J recombination. RSSs are composed of a palindromic heptamer adjacent to the coding region, a non-conserved spacer of 12 or 23 base pairs in length, and an AT-rich nonamer.

RSS: see recombination signal sequence.

References

Bassing, C.H. *et al.*: **The mechanism and regulation of chromosomal V(D)J recombination.** *Cell* 2002, **109**:S45–S55.

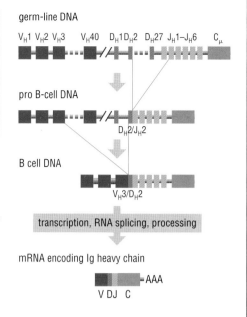

germ-line DNA

pro B-cell DNA

D_H2/J_H2

B cell DNA

V_H3/D_H2

transcription, RNA splicing, processing

mRNA encoding Ig heavy chain

V DJ C

Figure 7-3 Ordered rearrangement of gene segments to create a functional Ig heavy-chain-encoding gene First a single D_H gene segment recombines with one of the J_H segments. Next a V_H gene segment recombines with the DJ segment. Note that the reading frames of the V_H segment and of the J_H segment are fixed and only about one-third of the time will the number of nucleotides of the D region and of the two junctions be an even multiple of three such that the same reading frame is maintained between V and J (and ultimately to the C region). D regions can typically be read in two or even all three of the possible reading frames without having any stop codons. This potentially increases combinatorial diversity by a factor of two or three, although in the case of antibodies a single reading frame of the D segment is most commonly seen.

so on) and from nucleotide variability at the site where the junctions are made between the gene segments: this is known as junctional diversity, and is already mentioned in the chapters on T cell and B cell receptors (sections 5-1 and 6-1). Junctional diversity is generated by a specialized mechanism in the lymphocytes of adaptive immunity; we describe this in the next section. Some antigen-receptor loci have single gene segments of the V, D or J type, and in those cases all of the diversity contributed by that segment comes from junctional diversity or subsequent diversification mechanisms such as somatic mutation in the Ig loci (see section 6-12) or, in the Ig loci of certain species, gene conversion. The processes generating junctional diversity create variation not only in nucleotide sequence but also in the numbers of nucleotides inserted or deleted at the junctions. As codons consist of three nucleotides, and the number of nucleotides inserted or deleted is essentially random, the number of nucleotides at the junctions maintains the alignment of the reading frames between the V and C regions in only one-third of cases. Since the reading frames must be aligned to encode a functional antigen receptor protein, approximately two-thirds of the time rearrangements result in a non-functional allele of the gene.

Antigen receptor genes are recombined by a site-specific recombinase that generates antigen receptors of great diversity

Rearrangement of all of the antigen receptor loci is mediated by the same machinery, involving two proteins, RAG-1 and RAG-2, which are the products of the recombination-activating genes-1 and -2 and are expressed exclusively in developing lymphocytes. Recombination is completed by ubiquitously expressed DNA repair proteins that join chromosome ends, by a process known as **non-homologous DNA end-joining**. Although all these proteins are expressed both in T cells and in B cells, Ig genes rearrange in B cell precursors and rarely in T cell precursors, and the reverse is true for TCR loci. Moreover, as described in more detail below, the antigen receptor genes recombine in a specific order. Thus V(D)J recombination is a highly regulated process.

All of the functional V, D and J gene segments are flanked on the side that will be joined by a characteristic **recombination signal sequence (RSS)** that contains a conserved palindromic heptamer, a non-conserved spacer of 12 nucleotides or 23 nucleotides, and an AT-rich nonamer (Figure 7-4). Each type of gene segment for a particular locus contains RSSs with a spacer of the same length. The significance of the spacer is that each recombination event bringing two of these elements together occurs between one RSS with a 12-bp spacer and another RSS with a 23-bp spacer, a phenomenon called the **12/23 rule**. This restricts the rearrangements that can occur and in this way prevents rearrangements that are less useful. For example, at the IgH locus, V_H to J_H rearrangements cannot occur because both of these elements have RSSs of the 23-bp spacer type; rather, both elements must recombine with D_H elements, which have RSSs on both sides with 12-bp spacers. Clearly, however, there are other mechanisms that dictate specificity for V(D)J recombination, since V_H elements recombine only with DJ elements and not with unrearranged D_H segments, despite the fact that the upstream RSSs are the same for both. Similarly, at the TCR β-chain locus, the 12/23 rule would not prevent Vβ to Jβ rearrangements because the former segments have 23-bp spacer RSSs and the latter have 12-bp spacer RSSs; nonetheless, such rearrangements occur infrequently.

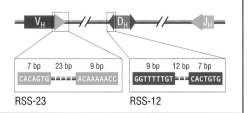

Figure 7-4 Recombination signal sequences RSSs are of two types, one with a 12-bp spacer (represented by a purple arrowhead) and another with a 23-bp spacer (represented by a blue arrowhead). These differ by almost exactly one turn of the DNA double helix, so the heptamer and nonamer sequences are oriented similarly with respect to each other, although farther apart in the case of the 23-bp spacer RSS. The heptamer is always adjacent to the coding sequence, and the AT-rich nonamer is toward the intervening DNA. Shown are consensus sequences; actual sequences vary to a limited degree.

V_H D_H J_H

7 bp 23 bp 9 bp
CACAGTG----ACAAAAACC
RSS-23

9 bp 12 bp 7 bp
GGTTTTTGT--CACTGTG
RSS-12

RAG-1 and RAG-2 proteins are lymphocyte-specific components of the V(D)J recombinase

The nature of V(D)J recombination was originally inferred from the analysis of rearranged and germ-line antigen receptor genes, and much was subsequently learned from more detailed sequence analysis. But a major breakthrough in the understanding of this process came with the identification of the *RAG-1* and *RAG-2* genes. These recombination-activating genes were cloned by transfer of genomic DNA fragments into a fibroblast cell line containing an artificial recombination substrate followed by selection for cells in which the substrate had recombined (Figure 7-5). The DNA responsible for this activity turned out to encode two closely linked genes, *RAG-1* and *RAG-2*, that together mediate this process. The tandem location of these two genes in all vertebrate species and their biochemical properties indicate that they were originally derived from a transposable element that was captured and evolved to acquire the function of generating diverse antigen receptors. Indeed, the RAG-1 and RAG-2 proteins can still cause DNA transposition *in vitro*.

Initial steps in V(D)J recombination can be accomplished *in vitro* with purified RAG-1 and RAG-2 proteins and template DNA. RAG-1 and RAG-2 proteins act together in V(D)J recombination to bind to RSS elements. They prefer to bind to two RSS elements in a DNA molecule, one with a 12-bp spacer and another with a 23-bp spacer, following the observed 12/23 rule. The RAG-1–RAG-2 protein complex initiates the recombination reaction by cleaving one strand of the DNA at the border between the RSS heptamer and the antigen receptor-coding region (Figure 7-6). The free 3′ OH group then serves to attack a phosphodiester bond on the other DNA strand, resulting in cleavage of that strand and generation of a covalently sealed hairpin DNA strand on the coding side of the cleavage. The RSS side of the cleavage is a double-stranded DNA end. Both ends of both RSS cleavages are held together in this complex.

Non-homologous end joining recombination components also participate in V(D)J recombination

Upon cleavage of the RSSs, components of the non-homologous DNA end joining machinery are recruited to the site. These include the DNA end-binding Ku70 and Ku80 molecules, DNA-protein kinase catalytic subunit (DNA-PKcs), XRCC4, DNA ligase IV and a nuclease called Artemis. **Terminal deoxynucleotidyl transferase (TdT)**, an enzyme important for creating junctional diversity, is also recruited to this complex. The double-stranded breaks at the heptamers of the RSSs are then simply joined to generate what is known as the **signal joint**, since it contains the RSSs. The two hairpin-containing sites are also joined to generate the **coding joint**, but this reaction is more complicated, as described below.

In most cases, the signal joint circularizes the DNA that was between the two coding regions being joined. The circular DNA is ultimately lost from the cell. This is what happens when

(a)

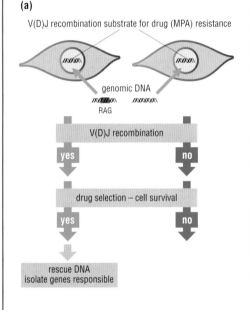

V(D)J recombination substrate for drug (MPA) resistance

genomic DNA

RAG

V(D)J recombination

yes   no

drug selection – cell survival

yes   no

rescue DNA
isolate genes responsible

(b)

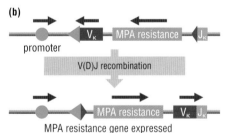

promoter

V_κ   MPA resistance   J_κ

V(D)J recombination

MPA resistance   V_κ   J_κ

MPA resistance gene expressed

Figure 7-5 Discovery of RAG-1 and RAG-2 RAG-1 and RAG-2 were discovered by experiments in which genomic DNA was introduced into a fibroblast cell line that had an artificial V(D)J recombination substrate present such that recombination would permit the expression of a drug resistance marker (the mycophenolic acid (MPA) resistance gene) **(a)**. The introduced DNA fortuitously contained both *RAG-1* and *RAG-2* genes (because they are located near each other in the genome) and expressed them ectopically, even though fibroblasts do not normally express *RAG-1* or *RAG-2*. **(b)** DNA configuration of V(D)J recombination induced an inversion of the DNA between the two RSSs, allowing the mycophenolic acid resistance gene to be expressed as it was now in the correct orientation with regard to the promoter element.

Definitions

coding joint: region of the DNA where V and D, D and J or V and J segments of lymphocyte antigen receptor genes are joined. Often this occurs imprecisely with the inclusion of **P regions** or **N regions**, and/or the removal of a few bases. This property of coding joints generates junctional diversity.

N region: nucleotides introduced into the coding joints of antigen receptor genes of lymphocytes by **TdT**.

P region: palindromic nucleotides introduced into the

coding joints of antigen receptor genes of lymphocytes by asymmetric cleavage of the hairpin intermediate in V(D)J recombination and subsequent copying of the resulting overhanging bases.

signal joint: DNA in which two recombination signal sequences (RSSs) in gene segments of lymphocyte antigen receptor genes have been joined, usually in a precise manner. Most signal joints are found in circles of excised DNA, but some V(D)J recombination events result in inversion of the DNA between two RSSs, in which case the signal joint remains in the chromosomal

DNA of the developing lymphocyte.

TdT: see **terminal deoxynucleotidyl transferase**.

terminal deoxynucleotidyl transferase (TdT): an enzyme expressed in developing lymphocytes that inserts random nucleotides into the sites of recombination between V and D segments, D and J segments, or V and J segments of antigen receptor genes in an untemplated fashion to generate **N regions** and much junctional diversity at the CDR3 loop, particularly of the Ig heavy chain and TCR β chain.

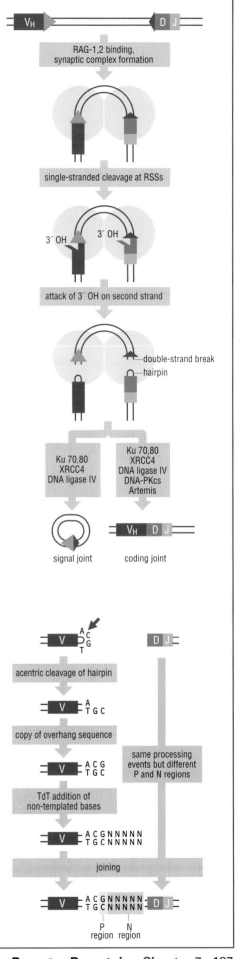

Figure 7-6 Mechanism of V(D)J recombination *In vitro* studies with recombinant purified RAG-1 and RAG-2 proteins have revealed the mechanism of V(D)J recombination. RAG-1 and RAG-2 bind directly to RSS sequences, and binding is favored by the presence of a 12-bp spacer RSS and a 23-bp spacer RSS, in accordance with the 12/23 rule. Cleavage of the DNA at the edge of the RSS occurs in two steps and generates a flush double-strand break on the RSS side and a covalently closed hairpin on the coding side of the cleavage. The RSSs are joined in the signal joint, and the hairpins are further processed to give junctional diversity and form the coding joint, as illustrated in greater detail in Figure 7-7. Components of the cellular non-homologous end joining machinery are involved in the formation of the signal joint and the coding joint.

the two gene segments are in the same orientation with respect to one another, as shown in Figure 7-6. Less frequently, one gene segment starts out in an inverted orientation relative to the other (for example, approximately one-half of all V_κ segments are inverted relative to the J_κ segments). In this configuration, the two segments can still be recombined, but now the signal joint and the coding joint result in reinserting the DNA between the two RSSs, where it originated, but in an inverted orientation (Figure 7-5b).

Several mechanisms contribute to the generation of junctional diversity

After initial RAG action, the coding sequences are covalently sealed hairpins that must be opened up by nucleolytic action. Artemis is thought to be responsible for this cleavage. A hairpin can be cleaved in the middle or a few nucleotides to one side of the center. The latter action creates a staggered break in the DNA. Replication of the overhanging nucleotides generates a short inverted repeat or palindrome, which is called a **P region**, as illustrated in Figure 7-7. Before joining of the two coding regions, with or without P regions, there can be further processing of the ends by removal of nucleotides and/or addition of random nucleotides, an untemplated reaction catalyzed by TdT, giving rise to **N regions**. The ends then need to be made flush by DNA synthesis or nuclease trimming and joined. Rearrangements during fetal life often occur without TdT involvement, so that junctional diversity is restricted. Note that this process will generate both functional and non-functional Ig or TCR genes, depending on the number of nucleotides present in the junctions, since a multiple of three is required to maintain the reading frame of the downstream coding sequences.

Artemis and TdT are involved only in the coding joint formation, not in the signal joint formation, and the role of DNA-PKcs is also primarily in coding joint formation. People with mutations in Artemis or RAG-1 or RAG-2 exhibit *severe combined immunodeficiency (SCID)* due to an absence of T cells and B cells (we describe this deficiency in Chapter 12). Interestingly, partial loss-of-function mutations in this process give rise to a rare genetic disease called *Omenn syndrome*, characterized by a decreased number of lymphocytes, immune deficiency and autoimmune phenomena. Evidently, a small number of lymphocytes with normal antigen receptors is generated by compromised V(D)J recombination, but the lymphocytes are insufficiently diverse to provide good immune protection and their small numbers may contribute to dysregulation of autoreactive cells.

Figure 7-7 Processing of coding ends to generate junctional diversity The hairpin can be cleaved, probably by Artemis, either centrally (not shown), or asymmetrically. In the latter case, as shown, this creates two or more unpaired nucleotides, which can then be copied to generate P regions. TdT then adds non-templated nucleotides (represented as N; these nucleotides can be A, T, C or G), which are called N regions. There can also be removal of nucleotides at the junction by nucleases (not shown).

References

Brandt, V.L. and Roth, D.B.: **A recombinase diversified: new functions of the RAG proteins.** *Curr. Opin. Immunol.* 2002, **14**:224–229.

Fugmann, S.D. *et al.*: **The RAG proteins and V(D)J recombination: complexes, ends, and transposition.** *Annu. Rev. Immunol.* 2000, **18**:495–527.

Gellert, M.: **V(D)J recombination: RAG proteins, repair factors, and regulation.** *Annu. Rev. Biochem.* 2002, **71**:101–132.

Grawunder, U. and Harfst, E.: **How to make ends meet in V(D)J recombination.** *Curr. Opin. Immunol.* 2001, **13**:186–194.

Schatz, D.G. and Baltimore, D.: **Stable expression of immunoglobulin gene V(D)J recombinase activity by gene transfer into 3T3 fibroblasts.** *Cell* 1988, **53**:107–115.

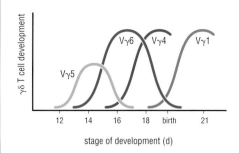

Figure 7-8 Ontological control of V gene usage in γδ T cell development Generation of γδ T cell subpopulations with distinct receptor chains occurs in waves at specific times during fetal development during which particular Vγ genes are favored for recombination. Similarly, in B cell development, VH genes close to DH (more 3′) are preferentially used during fetal B cell development and are used primarily by B1 cells, whereas postnatal B cell development uses the entire VH repertoire, with more 5′ VH genes being heavily used by follicular B cells. Data shown are for the mouse.

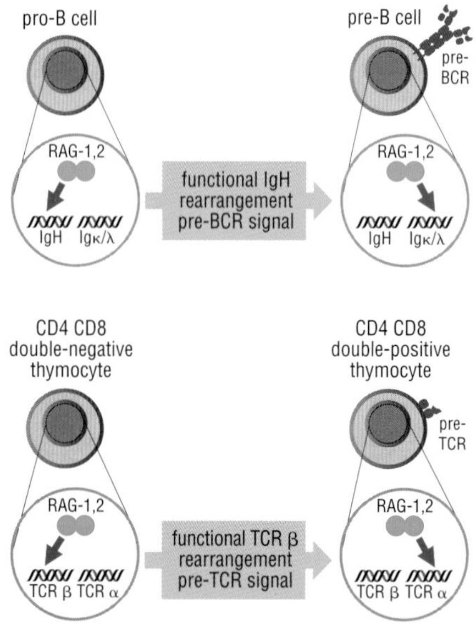

Figure 7-9 Feedback control of V(D)J recombination in B cell and T cell development In pro-B cells, RAG-1 and RAG-2 proteins target the IgH locus. Once a functional Ig heavy-chain protein can be made, it combines with the surrogate light chain (λ5 and VpreB) and the Igα/Igβ signaling heterodimer to form a pre-BCR, which signals feedback control of V(D)J recombination, so that RAG-1 and RAG-2 are now targeted to the Ig light-chain loci instead of to the IgH loci. In a parallel fashion, generation of a functional TCR β gene leads to feedback control of V(D)J recombination in developing T cells by formation of a signaling pre-TCR containing TCR β, pre-TCR α and the CD3 signaling subunits.

V(D)J recombination is regulated by modifications to chromatin structure affecting DNA accessibility

Although the RSS sequences and the RAG-1 and RAG-2 proteins are common to all V(D)J recombination events, there is lineage and developmental stage specificity to which gene segments in which loci are recombined. In the B cell lineage, IgH rearranges before light-chain loci; within IgH, DH recombines with JH and this is followed by VH to DJ recombination. Similarly, in developing T cells, the β, γ and/or δ loci rearrange before TCR α rearrangements. In addition, the gene segments that recombine change during fetal development, so that particular rearrangements are restricted in time (Figure 7-8). This is important for the functional properties of γδ T cells and of B1 cells, which we discuss in Chapter 8.

In most cases this specificity in V(D)J recombination is thought to be determined by the 12/23 rule and by accessibility of the gene segments in question to the recombination machinery. Indeed, the presence of DNase hypersensitivity sites correlate with gene rearrangement, as do other features of chromatin that has been opened up for transcription, including histone acetylation, DNA demethylation, and transcription itself. These features are in turn controlled by enhancers and promoters of the loci in question. Whereas chromatin structure and consequent accessibility probably controls many aspects of which gene segments rearrange and which do not, the relative efficiency of different recombinations that can occur at the same time (for example Igκ and Igλ rearrangements) may be determined by the effectiveness of individual RSS sequences. In the case of the TCR β1 locus, the ability of the Vβ segments to recombine with Dβ and not Jβ (both of which have RSSs with 12-bp spacers) is due to the RSS elements themselves, rather than chromosomal position.

Often, only one allele of an Ig or TCR locus recombines productively

In the vast majority of mature B cells, there is only one functionally rearranged heavy-chain allele and only one functionally rearranged light-chain allele. This property is referred to as **allelic exclusion**. Since antibodies have a H2L2 structure (or multiples of that), allelic exclusion ensures that all of the antibodies produced by a single cell have the same specificity—that is the two antigen–binding sites in each antibody are identical. If, for example, a B cell produced two different light chains, then many of the antibodies produced by that cell would have one of each of the two types of light chains and therefore would have different specificities for the two antigen-binding sites. Such antibodies might be poor effectors, particularly where several antigen-binding sites are required, for example for activating complement or forming large enough immune complexes to promote phagocytosis via Fc receptors (see Chapter 6). Moreover, because they could no longer be crosslinked by antigen, such antibodies might be ineffectual antigen receptors, leading to poor activation of the B cells and poor antibody production. It is clear, therefore, that allelic exclusion of Ig loci is an important feature of the control of V(D)J recombination.

How allelic exclusion is managed is incompletely understood, but part of the explanation at the IgH locus is that a functional Ig heavy-chain protein forms a complex with pre-existing **surrogate light chain** subunits that signals its presence in B cell precursors (Figure 7-9). Note that only roughly one-third of V(D)J recombinations will maintain the same reading frame between the V segment and the J segment and hence potentially produce a functional protein. The complex of a functional heavy chain and the surrogate light chains is called the **pre-BCR** (Figure 7-10). Signaling by the pre-BCR promotes developmental progression from the pro-B

Definitions

allelic exclusion: the property that a single lymphocyte expresses only one functionally rearranged allele of an antigen receptor gene. Both IgH and IgL loci show allelic exclusion, as does the TCR β locus, whereas TCR α and δ loci do not.

pre-BCR: receptor complex formed in pre-B cells between the μ chain and the **surrogate light-chain** subunits λ5 and VpreB, as well as Igα and Igβ. The pre-BCR signals analogously to engaged BCRs to promote B cell developmental progression to the

pre-B cell stage.

pre-TCR: receptor complex found in double-negative thymocytes between the TCR β chain, pre-TCRα and CD3 chains. The pre-TCR signals to promote maturation to the double-positive thymocyte stage.

surrogate light chain: an immunoglobulin light chain-like complex of the VpreB and λ5 chains and which bind to μ heavy chains analogously to the light chain to form the **pre-BCR**. Surrogate light-chain subunits are expressed only in developing B cells.

Figure 7-10 Pre-BCR structure and function The pre-BCR is composed of μ heavy chain bound to the surrogate light-chain subunits λ5 and VpreB, which together take the place of light chain, and to the Igα/Igβ signaling components. This molecule has obvious similarities to the BCR and it is thought to signal by essentially the same mechanism. Mutations that prevent formation of the pre-BCR lead to strong blocks in B cell development at the pro-B cell to pre-B cell transition, with the exception of surrogate light-chain mutations, which lead to a partial block. Similarly, B cell development is blocked in mice mutated in genes encoding several proteins involved in BCR signaling. In humans, a severe block in B cell development occurs for this reason in individuals with X-linked agammaglobulinemia, which is caused by mutations in the X-linked gene encoding Btk. Btk is a protein tyrosine kinase that participates in BCR signaling (see Figure 6-20).

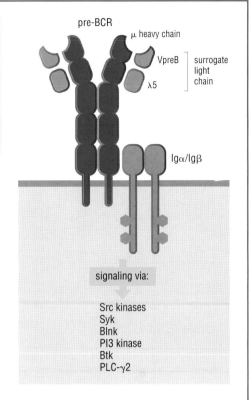

to the pre-B cell stage and a short burst of clonal expansion, and in the process in some way targets the V(D)J recombinase away from the other IgH locus and toward the light-chain loci (see Figure 7-9), perhaps via changes in accessibility. In addition, allelic exclusion may be facilitated by making one allele relatively accessible compared with the other, so that really only one allele is competent to recombine at any one time. However, it should be noted that B cells with one non-productively rearranged IgH locus and one productively rearranged IgH locus are frequently observed, so there must be a mechanism for opening up the second allele if no heavy chain protein is produced in a limited period because of a non-functional rearrangement. For Ig light chains, it is currently thought that production of a functional light chain that can pair with the heavy chain in some way sends a signal that terminates RAG-1 and RAG-2 expression. Thus, allelic exclusion of the light chain is facilitated by cessation of rearrangements.

Allelic exclusion is also a prominent feature of the TCR β locus in αβ T cells. In a process analogous to that in developing B cells, a functional β chain pairs with a molecule called pre-TCRα and this **pre-TCR** sends a signal that mediates β chain allelic exclusion (see Figure 7-9). The TCR α and δ chains do not exhibit allelic exclusion. In developing αβ T cells, TCR α gene rearrangements continue until positive selection occurs, whereupon RAG-1 and RAG-2 expression is silenced. The TCR α locus is especially well configured to permit many rearrangements, since it has 61 J segments, allowing continued recombination of upstream Vα segments with downstream Jα segments, excising earlier VJ rearrangements (Figure 7-11). Indeed, rearrangements seem to begin at the 5′ end of the Jα locus and then over time progress toward the 3′ end.

Figure 7-11 Successive rearrangements at the TCR α locus allow multiple α chains to be tested in conjunction with a single TCR β chain until positive selection occurs Initial rearrangements at the TCR α locus favor the use of more 5′ Jα segments. RAG-1 and RAG-2 expression is not turned off in the double-positive thymocyte until a positive selection signal has been received. If positive selection fails to occur when the current α chain pairs with the β chain present in that cell, then successive V(D)J recombination can occur at the same TCR α locus, because downstream V genes can be rearranged to upstream J elements, resulting in the removal of the earlier functional V/J unit and the creation of a new one (lowest line). The new TCR α chain can pair with the β chain to generate a new TCR, which may be able to induce positive selection.

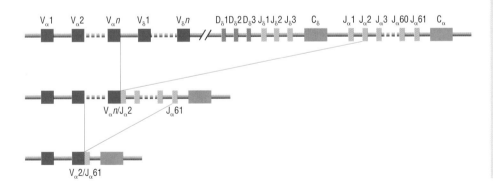

References

Bergman, Y. and Cedar, H.: **A stepwise epigenetic process controls immunoglobulin allelic exclusion.** *Nat. Rev. Immunol.* 2004, **4**:753–761.

Jung, D. *et al.*: **Mechanism and control of V(D)J recombination at the immunoglobulin heavy chain locus.** *Annu. Rev. Immunol.* 2006, **24**:541–570.

Khor, B. and Sleckman, B.P.: **Allelic exclusion at the TCRβ locus.** *Curr. Opin. Immunol.* 2002, **14**:230–234.

Melchers, F.: **The pre-B-cell receptor: selector of fitting immunoglobulin heavy chains for the B-cell repertoire.** *Nat. Rev. Immunol.* 2005, **5**:578–584.

Roth, D.B. and Roth, S.Y.: **Unequal access: regulating V(D)J recombination through chromatin remodeling.** *Cell* 2000, **103**:699–702.

The key elements of the adaptive immune system are found in all jawed vertebrates

All six types of antigen receptor loci found in mammals (IgH, IgL, TCR α, TCR β, TCR γ, and TCR δ) have been found in other jawed vertebrate species, including the **cartilaginous fish** (sharks and skates), but they have not been found in jawless fish or invertebrates. Moreover MHC class I and MHC class II loci are also found in these species, as are *RAG-1* and *RAG-2*. Thus, the essential features of adaptive immunity as currently known were achieved approximately 500 million years ago, early during vertebrate evolution (Figure 7-12).

Detailed analysis of antigen receptor loci indicate that the basic elements of V(D)J recombination were also present early, including recognizable RSSs with 12- and 23-bp spacers. Presumably, these specificity elements were inherent in the ancestral transposon/transposase system that was captured and exploited for antigen receptor diversification. TdT contribution of N regions to junctional diversity is also seen broadly, again going back to cartilaginous fish.

Although the study of TCR loci in divergent vertebrate species is less advanced, clearly recognizable TCR α, β, γ and δ loci have been found in many species, including cartilaginous fish. These loci show rapid sequence diversification, but the overall organization of TCR loci seems to be broadly similar to that in human and mouse. Similarly, IgM-like heavy chains are seen in all jawed vertebrates, as are light chains and somatic hypermutation. Strikingly, however, the organization of Ig gene loci and the genetic mechanisms largely responsible for generating diverse repertoires of antibodies vary dramatically between species.

Sharks have a tandem array of V/D/J/C IgH cassettes

The basic architecture of human and mouse IgH and κ loci, in which there are multiple V, D and J or V and J elements upstream of a single C region supports combinatorial diversity, and this architecture is seen in many vertebrates, starting from the ray-finned fish. In contrast, the Ig loci of cartilaginous fish are arranged primarily as repeating cassettes that greatly limit combinatorial diversity but do retain some junctional diversity (Figure 7-13). In some species, there is even a substantial number of preassembled IgH or IgL genes. For example, in the horned shark IgM genes, there are approximately 100 cassettes; about one-half of these are already rearranged in the germ line. The gene organizations of sharks suggest that there must be novel mechanisms of gene expression to account for allelic exclusion. In addition to IgM,

Figure 7-12 Evolution of major features of adaptive immunity Many of the major elements of adaptive immunity appeared soon after the appearance of vertebrates and are seen in all jawed vertebrates. IgH class switch arose either in lobed-finned fish or early amphibians. Cartilaginous fish and birds have evolved distinctive arrangements of their Ig loci and mechanisms of diversification. Lamprey and hagfish, the jawless fish, have evolved rearranging antigen-specific receptors from leucine-rich repeat (LRR) structures, rather than from Ig domains.

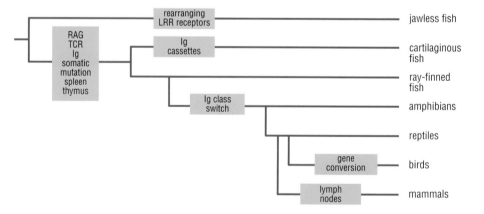

Definitions

bursa of Fabricius: a specialized primary lymphoid organ of chickens that does not have a strict counterpart in mammals. In the bursa, newly formed B cells proliferate and undergo gene conversion events at their IgH and IgL loci before antigenic stimulation.

cartilaginous fish: primitive jawed fish with cartilage instead of true bone. Includes sharks and skates.

gene conversion: the nonreciprocal transfer of information between homologous sequences.

pseudogene: a non-functional gene in the genome. Pseudogenes provide genetic information for **gene conversion** mechanisms that provide much of the diversity of Ig genes in chickens.

V_H1 $D_H1.1$ $D_H1.2$ J_H1 C_μ V_H2 $D_H2.1$ $D_H2.2$ J_H2 C_μ V_H3 D J C_μ V_H4 D J_H4 C_μ

cartilaginous fish have two other types of heavy chains, which are not particularly similar to any mammalian IgH isotypes, although one, called NAR, exhibits somatic mutation after antigenic stimulation. Sharks and skates have two divergent types of light chains, also arrayed in cassette form, although they are typically not more κ-like or λ-like. In ray-finned fish, the IgM heavy-chain locus is more like the mammalian locus; light-chain loci are either mammalian tandem loci or shark-like cassettes, depending on the species.

The most primitive fish are the jawless fish, consisting of lamprey and hagfish. Although phenomenological evidence indicates that these organisms have both humoral and cell-mediated adaptive immunity, they appear to lack *RAG-1* and *RAG-2* genes, Ig genes and TCR genes. Recently, an entirely distinct set of diverse and rearranging genes has been discovered in these species. The protein products consist largely of leucine-rich repeat structures, similar to the extracellular domains of Toll-like receptors. Moreover, these molecules are expressed by lymphocyte-like cells of the blood. Their roles in adaptive immunity of jawless fish remain to be elucidated, but they appear to represent an alternatively evolved mechanism of adaptive immune recognition.

Chickens and rabbits make extensive use of gene conversion to generate antibody diversity

Whereas amphibians and reptiles in general have Ig loci that resemble those of mammals, including the sequences necessary for the heavy-chain class switch and a secreted IgA-like antibody called IgX in amphibians, birds have a distinct primary mechanism for diversification of their antibody genes. Chickens have only single functional VH and VL genes, but many VH and VL **pseudogenes**—that is non-functional copies of VH and VL genes—that are templates for **gene conversion** events that copy small numbers of nucleotides from a pseudogene and convert the corresponding regions of the functional and rearranged V gene (Figure 7-14). Some of the pseudogenes contain J-like sequences, which also participate in gene conversion events. This diversification occurs in a unique bird organ called the **bursa of Fabricius**, located near the gut, and occurs before antigenic stimulation. Gene conversion appears to be dependent on AID, the cytidine deaminase involved in somatic hypermutation and switch recombination (see section 6-11).

Surprisingly, rabbits use a strategy very similar to that of chickens to diversify their antibody repertoire before antigen encounter. Sheep have a similar but distinct mechanism in which somatic hypermutation rather than gene conversion is used, and in these mammals the pre-antigen-exposure diversification occurs in the ileal Peyer's patch.

Figure 7-13 Shark IgH locus organization Cartilaginous fish have IgH and IgL loci containing repeated cassettes of recombining Ig loci with very limited numbers of gene segments in each cassette. Shown is a partial representation of the horned shark IgM locus, which has about 100 cassettes, about 50% of which have cassettes that are rearranged or partially rearranged in the germ line. It is presumed that there are mechanisms to control the expression of the pre-rearranged loci to permit allelic exclusion. Note that this organization minimizes combinatorial diversity, but junctional diversity is possible except for the pre-rearranged loci.

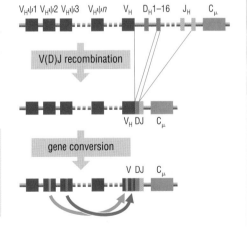

$V_H\psi1$ $V_H\psi2$ $V_H\psi3$ $V_H\psi n$ V_H D_H1–16 J_H C_μ

V(D)J recombination

V_H DJ C_μ

gene conversion

V DJ C_μ

Figure 7-14 Chicken Ig gene diversification by gene conversion The chicken IgH locus is shown. There are 80–100 upstream pseudo-VH genes. These genes are truncated at the 5' end and lack RSSs at the 3' end. Some contain DH and JH sequences rearranged to them. These pseudogenes serve as template material for gene conversion events in which short stretches of nucleotides are copied into the corresponding position in the functional and rearranged VH gene. A similar organization and diversification mechanism is seen for the IgL locus, which is λ-like. There are 26 pseudo-VLs.

References

Hsu, E. *et al.*: **The plasticity of immunoglobulin gene systems in evolution.** *Immunol. Rev.* 2006, **210**:8–26.

Lanning, D. *et al.*: **Development of the antibody repertoire in rabbit: gut-associated lymphoid tissue, microbes, and selection.** *Immunol. Rev.* 2000, **175**:214–228.

Litman, G.W. *et al.*: **Evolution of antigen binding receptors.** *Annu. Rev. Immunol.* 1999, **17**:109–147.

Pancer, Z. *et al.*: **Somatic diversification of variable lymphocyte receptors in the agnathan sea lamprey.** *Nature* 2004, **430**:174–180.

Reynaud, C.-A. *et al.*: **Formation of the chicken B-cell repertoire: ontogenesis, regulation of Ig gene rearrangement, and diversification by gene conversion.** *Adv. Immunol.* 1994, **57**:353–378.

Reynaud, C.-A. *et al.*: **Hypermutation generating the sheep immunoglobulin repertoire is an antigen-independent process.** *Cell* 1995, **80**:115–125.

Commitment to the B cell lineage involves expression of transcriptional regulators and the RAG-1 and RAG-2 proteins

Pluripotent hematopoietic stem cells differentiate into early lymphoid progenitor cells, which can then become committed to the B cell lineage. This lineage choice is reinforced by the absence of Notch signaling, which promotes T cell development, as described in section 7-0. Commitment to the B cell lineage is controlled by sequential expression within the cell of the transcription factors E2A, EBF and Pax5, which direct the expression of B-lineage-specific gene products such as Igα, Igβ and surrogate light-chain subunits λ5 and VpreB, and of RAG-1 and RAG-2 (Figure 7-15).

B cell developmental stages are defined primarily by Ig gene rearrangement and Ig chain expression

Progress through the B cell developmental program is defined in two ways: by the status of Ig gene rearrangements and by the cell surface expression of proteins characteristic of various stages of B cell development. We shall refer to these stages here as B-lineage-committed progenitors, pro-B cells, pre-B cells and immature B cells—all of which are found in the bone marrow—and, in the periphery, transitional B cells and mature B cells of the three types described in Chapter 6 (B1 cells, marginal-zone B cells and follicular B cells). Some of the distinguishing phenotypic characteristics of the developmental intermediate cell types are shown in Figure 7-16; those of the three mature B cells are shown in Figure 8-18.

Expression of Ig heavy chain is responsible for driving B cell maturation in the bone marrow

As described in section 7-3, once a functional IgH rearrangement has occurred in the pre-B cell, heavy-chain protein is produced and pairs with the surrogate light chain to form the pre-BCR (see Figure 7-10). Signaling from the pre-BCR signifies that the developing B cell has produced a satisfactory rearrangement of the IgH locus. The pre-BCR induces signaling reactions similar to those induced by the BCR (see section 6-8).

It is unclear whether the pre-BCR generates a signal spontaneously or whether it must interact with a ligand, for example expressed on bone marrow stromal cells. Despite considerable effort, evidence for a pre-BCR ligand is lacking. In any case, this signal has three effects on the pro-B cell: it induces phenotypic transition to the pre-B cell stage, it makes the cell responsive to IL-7 for expansion, and it stops further V(D)J recombination at the IgH locus, which thereby contributes to allelic exclusion of the heavy-chain locus.

Pro-B cells proliferate in the bone marrow, and studies *in vitro* indicate that this proliferation requires contact with bone marrow stromal cells. In contrast, pre-B cells proliferate in response to the cytokine IL-7; this proliferation does not require stromal cell contact. Therefore, it has been suggested that pro-B cells proliferate in the bone marrow until they fill up the niche of contact with the appropriate stromal cells. Division of these pro-B cells would require that one daughter cell lose contact with the growth-stimulating stromal cell. If this cell fails to make a productive IgH rearrangement, it fails to express the pre-BCR and cannot continue to grow. These cells die and are lost from the bone marrow. In contrast, if the cell does make a productive IgH rearrangement, it expresses the pre-BCR and acquires responsiveness to the IL-7 that is present in the microenvironment. This leads to an expansion of these pre-B cells. This

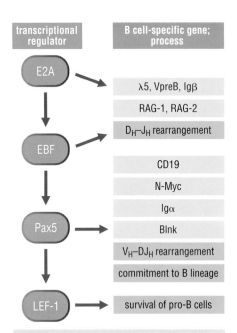

transcriptional regulator	B cell-specific gene; process
E2A	λ5, VpreB, Igβ
	RAG-1, RAG-2
EBF	D$_H$–J$_H$ rearrangement
	CD19
	N-Myc
	Igα
Pax5	Blnk
	V$_H$–DJ$_H$ rearrangement
	commitment to B lineage
LEF-1	survival of pro-B cells

Figure 7-15 Control of B cell development by a cascade of transcription factors E2A, a basic helix–loop–helix family transcription factor forms homodimers in the early lymphoid progenitor cells. If E2A action is not blocked by Notch signaling to give T cells or by signals that induce NK-cell development by promoting the expression of Id3, which forms inactive dimers with E2A, then it induces B cell development. E2A induces the expression of EBF (early B cell factor). E2A and EBF then collaborate to induce the expression of some key components needed for development down the B cell lineage, including the opening up of the IgH locus for D to J rearrangements and probably also RAG-1 and RAG-2 expression. EBF induces the expression of Pax5, also called BSAP (B cell-specific activating protein). Pax5 activates the expression of additional genes required for B cell development and also represses the expression of genes, such as that encoding the M-CSF receptor, that are required for other differentiated states in the hematopoietic lineage (see Figure 1-4). Among the targets of Pax5 is the transcription factor LEF-1 (lymphoid enhancer factor 1), which is required for the survival of pro-B cells.

Definitions

receptor editing: a process whereby immature B cells in the bone marrow that express antigen receptors recognizing self arrest maturation and undergo additional DNA rearrangements at the Ig light-chain loci resulting in a new antigen receptor that may not recognize self antigens.

References

Hardy, R.R.: **B-cell commitment: deciding on the players.** *Curr. Opin. Immunol.* 2003, **15**:158–165.

Hardy, R.R. and Hayakawa, K.: **B cell developmental pathways.** *Annu. Rev. Immunol.* 2001, **19**:595–621.

Kurosaki, T.: **Regulation of B cell fates by BCR signaling components.** *Curr. Opin. Immunol.* 2002, **14**:341–347.

Matthias, P. and Rolink, A.G.: **Transcriptional networks in developing and mature B cells.** *Nat. Rev. Immunol.*

2005, **5**:497–508.

Meffre, E. *et al.*: **Antibody regulation of B cell development.** *Nat. Immunol.* 2000, **1**:379–385.

Monroe, J.G.: **ITAM-mediated tonic signaling through pre-BCR and BCR complexes.** *Nat. Rev. Immunol.* 2006, **6**:283–294.

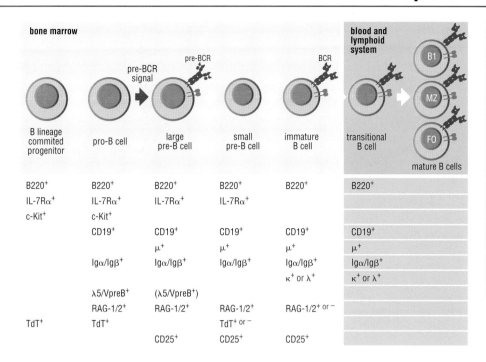

Some important phenotypic characteristics of B-lineage precursors are shown. The pre-BCR signal that drives pro-B cell to pre-B cell development also leads to a turning off of surrogate light-chain ($\lambda 5$ and VpreB) expression, so pre-B cells initially express the pre-BCR, but stop expressing it during the proliferative burst caused by IL-7 stimulation in the absence of stromal cell contact. Immature B cells may not express RAG-1 and RAG-2, or they may express it if they are undergoing receptor editing. For the phenotypic properties of the mature B cell subsets, see Figure 8-18. FO: follicular; MZ: marginal zone; TdT: terminal deoxynucleotidyl transferase. B220 is a B cell-specific isoform of the transmembrane protein tyrosine phosphatase CD45. c-Kit is a growth factor receptor that is important for growth of early hematopoietic stem cells.

proliferative response is thought to be limited by the fact that pre-B cells also stop making surrogate light chains, which will ultimately result in the cessation of pre-BCR expression. The proliferative response to IL-7 is thought to be a response to pre-BCR signaling plus IL-7, so the loss of pre-BCR signaling limits the expansion of pre-B cells; indeed, a limited burst of expansion of pre-B cells is followed by the generation of post-proliferative small pre-B cells. In the mouse, B cell development is highly dependent on IL-7, but in the human the dependence is much less, suggesting the existence of an alternative cytokine.

Small pre-B cells, those that have stopped proliferating, are competent to rearrange IgL loci. In humans and mice, the κ loci generally recombine first, perhaps because of intrinsically higher efficiency of their recombination signal sequences. Successful rearrangement at a light-chain locus leads to the production of light chain, which combines with μ chain to generate the BCR. BCR expression turns off expression of RAG-1 and RAG-2 and induces further maturation to the immature B cell stage. How this happens is poorly understood, but the BCR induces a low level of constitutive signaling in the absence of binding to an antigen, and this low-level tonic signaling may be responsible. BCR and pre-BCR signaling require the intracellular protein tyrosine kinase Btk, as illustrated in Figure 6-20. Btk is encoded by a gene on the X chromosome and defects in this gene lead to a severe defect in B cell development, resulting in a disease called *X-linked agammaglobulinemia*, in which antibody levels are very low.

Self-reactive immature B cells in the bone marrow arrest maturation and undergo receptor editing

In the absence of a stronger signal than that provided by tonic BCR signaling, immature B cells leave the bone marrow via the blood and enter the next stage of maturation, which we describe in the next section. However, if the immature B cell in question is self-reactive and if the relevant self antigen is present in the bone marrow, then stronger BCR signaling is generated. This stronger signaling arrests maturation and keeps the immature B cell in the bone marrow. Expression of RAG-1 and RAG-2 is reactivated (or maintained), and mediates further rearrangements, primarily at the light chain loci (Figure 7-17). This process is called **receptor editing**, because the B cell already has a functional BCR. If receptor editing results in changes to the light chain that eliminate self-reactivity, then the edited cell resumes the maturation program, turns off RAG-1 and RAG-2 and leaves the bone marrow. Thus, the selection against self-reactivity at this stage is a molecular selection rather than a cellular selection, presumably for efficiency reasons. We shall see in the next section that the bone marrow supports receptor editing because it provides survival signals to the immature B cell that prevent BCR-induced apoptosis. In contrast, transitional B cells in the periphery lack this support and they rapidly undergo deletion if they encounter antigen before maturing.

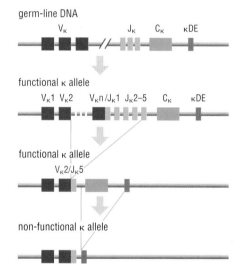

Figure 7-17 Receptor editing of the Igκ locus Immature B cells in the bone marrow with strong reactivity for self components arrest maturation, stay in the bone marrow and re-express RAG-1 and RAG-2, permitting additional rearrangements at the light-chain loci. The top line shows a germ-line Igκ locus. On the second line is an initial functional κ re-arrangement, typically involving a 5' Jκ element. This configuration permits a nested rearrangement between an upstream Vκ and a downstream Jκ that deletes the original functional κ rearrangement (third line). In addition, there is a less efficient rearrangement element downstream of the Cκ gene, called κ deleting element (κDE), which can recombine with a site in the intron between Jκ and Cκ, with deletion of the Cκ region and hence inactivation of that particular allele. Such deletions are followed by rearrangements of other light chain alleles until either a light chain is generated that when paired with the heavy chain of that cell results in a BCR that is not autoreactive or, if this does not occur in time, the cell dies.

Transitional B cells are deleted or anergized by antigen contact in the periphery

Immature B cells that leave the bone marrow and enter the periphery are termed **transitional B cells**. These transitional B cells will mature into one of the three types of mature B cells within several days or they will die, either from failure to be positively selected or, if they are self-reactive cells not eliminated in the bone marrow, by negative selection for self tolerance.

Although immature B cells with strong reactivity to self antigens cease maturation in the bone marrow and undergo receptor editing, some self-reactive B cells do reach the periphery. These B cells may have a weaker reactivity for self antigen than the cells that undergo receptor editing, or they may be specific for a self antigen that is preferentially expressed in the periphery rather than in the bone marrow.

The fate of newly formed self-reactive peripheral B cells that encounter their antigen is dependent on the degree of BCR signaling that is induced by the antigen (Figure 7-18). A moderate or high level of signaling, as is typically seen with a membrane-bound form of antigen, leads to rapid cell death, and thus clonal deletion. A lower but still significant level of BCR signaling induces the B cell to move to the T cell zone of the spleen or lymph node and, if it does not receive T cell help within a short period, to adopt an anergic phenotype, characterized by downregulation of surface IgM and attenuated signaling through the antigen receptor. These anergic B cells are hard to activate because of poor signaling from the antigen receptor, and they are also readily killed if they interact with activated helper T cells, through FasL on the helper T cell which delivers a death signal to the B cell via Fas (see Figure 2-30). These properties contribute to immune tolerance to self antigens.

Transitional B cells are positively selected into different types of mature B cells

As described in Chapter 6, there are three types of mature B cells: B1 cells, marginal zone B cells and recirculating follicular B cells. B1 cells make natural antibody without antigenic stimulation, and also participate in T cell-independent antibody responses, particularly in the peritoneal and pleural cavities. Marginal zone B cells are located in the marginal zone of the spleen, where they come into contact with bloodborne antigens and make T cell-independent and rapid T cell-dependent IgM antibody responses; and follicular B cells recirculate between blood and lymphoid follicles and participate in T cell-dependent germinal-center antibody responses.

All three types of mature B cells are thought to require some sort of stimulus through their antigen receptors to enter the mature long-lived stage, a process analogous to the positive selection of T cells. Surprisingly, genetic alterations in signaling components have very different effects on the appearance of these three populations, suggesting that the selective events are distinctive in some ways (Figure 7-19). These and other observations have led to the hypothesis that strong signaling from the antigen receptor is needed for transitional B cells to enter into the long-lived B1 cell population. Weaker signals seem to be sufficient to drive transitional B cells into the follicular B cell pool, and perhaps even weaker BCR signals induce the maturation of transitional B cells into the marginal-zone population.

It is thought that marginal zone and follicular B cell types derive from a single uncommitted transitional B cell precursor, whereas B1 cells develop early in life from precursors committed to this type of B cell. B1 cells are distinct in that their functional IgH alleles have almost no

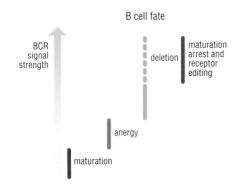

B cell fate

BCR signal strength

maturation arrest and receptor editing

deletion

anergy

maturation

Figure 7-18 Effect of BCR signaling strength on fate of immature and transitional B cells
Strong signaling from the antigen receptor, as occurs with high-affinity binding to multivalent or membrane-bound antigen, induces maturation arrest (preventing exit from the bone marrow) and receptor editing. If receptor editing is unable to remove strong signaling from the antigen receptor, or if strong signaling occurs in the periphery, then rapid cell death by apoptosis (deletion) occurs. A lower-level of signaling from the antigen receptor may induce an anergic phenotype, which generally precludes participation in immune responses unless a more potent form of antigen is sensed. Still lower signaling from the antigen receptor promotes the maturation of transitional B cells. Some signaling from the antigen receptor is required for maturation and also for continued survival in the periphery. B1 cell maturation is promoted by intermediate levels of signaling, whereas a lower level of signaling promotes follicular and marginal zone B cell maturation. Therefore, precursors of B1 cells, which are generated during fetal life, are likely to have different thresholds for anergy and deletion from those of precursors of follicular B cells and marginal zone B cells, which are generated in the adult bone marrow.

Transitional B cells can be phenotypically divided into two subtypes, called T1 and T2. T1 cells have the property of being very sensitive to apoptosis induced through the antigen receptor and T2 cells seem to be further advanced toward the mature state.

Figure 7-19 Impact of antigen receptor signaling mutations on maturation of B cell subsets Effect of mutations of the Igα ITAM of the antigen receptor, or of signaling components, on the maturation of different mature B cell types (FO: follicular B cell; MZ: marginal zone B cell). Blue areas with minus signs indicate that there is a substantial decrease in the numbers of these cells.

Mutation	MZ	FO	B1
none	+	+	+
Igα ITAM	–	–	–
Btk	+	–	–
PLC-γ2	+	–	–
PKC-β	+	+	–

N-region additions between V, D and J segments, indicating that V(D)J recombination occurred in a pro-B cell that did not express terminal deoxynucleotidyl transferase (TdT). These cells are primarily generated early in life, during fetal development, and it is clear that TdT is not expressed at this stage. In contrast, marginal zone B cells start to appear later, after 2–3 weeks of life in rodents or after about 1 year in humans. The choice between the marginal zone and follicular B cell fate depends on two factors: the level of signaling from the antigen receptor in response to self antigens, and Notch2 signaling. It is thought that low levels of signaling from the antigen receptor in response to self antigens favor follicular B cell development, whereas still lower levels favor maturation into the marginal zone B cell type. Superimposed on the effects of antigen receptor signaling are effects of Notch2 signaling. Marginal zone B cells require Notch2 signaling for their development. It has been proposed that signaling from the antigen receptor inhibits Notch2 signaling, which therefore favors follicular B cell development. Often in development Notch signaling determines both which of two cell fates is adopted, and the ratio of the two differentiated cell types produced. By analogy, one hypothesis is that the involvement of Notch in this developmental decision ensures that both follicular B cells and marginal zone B cells are produced from transitional B cells in appropriate numbers.

BAFF is an important survival factor for transitional and mature B cells

Survival of follicular B cells and marginal zone B cells is strongly dependent on a soluble TNF family member called **BAFF (B cell-activating factor of the TNF family)**, acting through one of its receptors, called the BAFF receptor. Mice defective in the BAFF receptor or in BAFF have greatly decreased numbers of marginal zone B cells and follicular B cells, and conversely, injection of BAFF increases B cell numbers and antibody responses to T cell-independent and T cell-dependent antigens. B1 cells are relatively less dependent on BAFF; their numbers are normal in the absence of BAFF and overexpression of BAFF induces only a modest increase in B1 cell numbers. *In vitro*, BAFF enhances the survival of mature B cells, promotes the maturation of immature and transitional B cells, and enhances proliferative responses via the BCR.

How this powerful system for regulating B cell responses is controlled is an important issue that is not yet fully understood. Lymphoid stromal cells constitutively produce BAFF in a manner that is independent of B cell numbers. In addition, various myeloid cells (monocytes, dendritic cells and neutrophils) make BAFF after innate immune stimulation and can thereby support enhanced B cell survival locally during an immune response. Like TNF, BAFF is made as a membrane-bound form that can be released as a soluble form by proteolytic cleavage. B cells express three different receptors, BAFF receptor, BCMA and TACI, the last two of which also bind another TNF family member called APRIL. Each of these receptors mediates distinct responses of B cells. A strain of mice (A/WySnJ), previously recognized to have a phenotype similar to that of mice in which the gene for BAFF has been deleted, turns out to have a deletion in the cytoplasmic domain of BAFF receptor. Mice in which the BCMA gene is deleted have long-lived plasma cells that have shorter survival times in the bone marrow than do those of normal mice. Mice from which the TACI gene is deleted have a defect in T cell-independent type 2 antibody responses, so TACI appears to have positive roles in B cell activation and acts analogously to CD40 in promoting isotype switching (Figure 7-20).

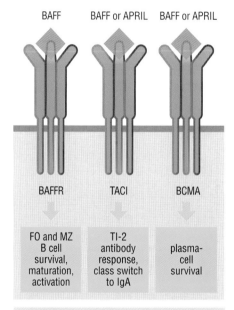

Figure 7-20 BAFF and its receptors on B cells BAFF binds to three TNF receptor family members and all three of these receptors are expressed by B cells. Evidence for the role of the individual receptors in the effects of BAFF (and possibly another TNF family member APRIL) on B cells is from experiments with mice mutated in individual receptors. In addition, some B cell immunodeficiency patients have been found to have defects in BAFF receptor or TACI.

References

Cariappa, A. and Pillai, S.: **Antigen-dependent B-cell development.** *Curr. Opin. Immunol.* 2002, **14**:241–249.

Martin, F. and Kearney, J.F.: **B1 cells: similarities and differences with other B cell subsets**. *Curr. Opin. Immunol.* 2001, **13**:195–201.

Rolink, A.G. and Melchers, F.: **BAFFled B cells survive and thrive: roles of BAFF in B-cell development.** *Curr. Opin. Immunol.* 2002, **14**:266–275.

Schneider, P.: **The role of APRIL and BAFF in lymphocyte activation.** *Curr. Opin. Immunol.* 2005, **17**:282–289.

7-7 Organization of the Thymus and Early T Cell Development

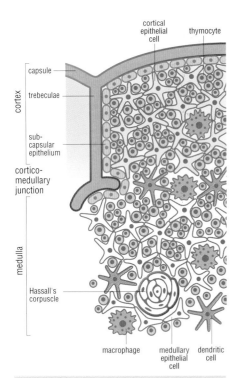

cortical epithelial cell
thymocyte
capsule
cortex
trebeculae
sub-capsular epithelium
cortico-medullary junction
medulla
Hassall's corpuscle
macrophage
medullary epithelial cell
dendritic cell

Figure 7-21 Organization of the thymus
Shown is a lobule of the thymus, defined by the invaginations of connective tissue, which is lined by the subcapsular epithelial layer. Inside this layer is the cortex, which has a meshwork of cortical epithelial cells that express MHC class I and class II molecules bound to self peptides. There are also some macrophages in the cortex, along with a high density of thymocytes. Rapid proliferation of double-negative thymocytes occurs primarily in the subcapsular region of the cortex. The central region of the thymus, the medulla, also contains at least two specialized medullary epithelial cell types, along with dendritic cells of bone marrow origin. A striking feature of the medulla is the presence of Hassall's corpuscles, cell aggregates of unknown function composed of concentric layers of one of the types of epithelial cells. Single-positive thymocytes are found primarily in the medulla. At the cortico-medullary junction is a meshwork of macrophages and dendritic cells of bone marrow origin, as well as blood vessels (not shown).

The thymus provides a specialized environment for T cell development

Most T cell development requires a specialized lymphoid organ in the chest, the thymus. Neonatal thymectomy or mutations that affect formation of the thymus (nude in mice; DiGeorge syndrome in humans) result in individuals lacking most recirculating T cells (there is a specialized subset of T cells, discussed in Chapter 8, that seems to develop in the gut rather than the thymus).

The thymus consists mostly of T-lineage cells but also contains a variety of specialized cell types that support T cell development. These include non-hematopoietic thymic epithelial cells, which express MHC class II molecules as well as MHC class I molecules and have roles in positive and negative selection, and hematopoietic cells including macrophages and dendritic cells.

The thymus is morphologically divided into a cortex and a medulla (Figure 7-21), and these regions support distinct phases of T cell development. Stem cells enter the thymus from the blood at the junction between these two regions, and early stages in development are characterized by movement of the thymocytes toward the outer cortex and their expansion there. As maturation proceeds, thymocytes move toward and into the medulla, where the more mature cells are located prior to their exit to the periphery. The developmental events occurring between the arrival of an early lymphocyte progenitor in the thymus and the export of a mature CD4 or CD8 T cell involve several checkpoints extending over a period of about 12 days.

Notch is important for commitment to the T cell lineage

Early lymphoid progenitors with a propensity toward entry into the T cell lineage leave the bone marrow and migrate to the thymus. There they encounter ligands for Notch1, which induces commitment to the T cell lineage, as shown by the fact that inhibition of Notch1 function leads to B cell development rather than T cell development in the thymus (see Figure 7-1). There are at least five ligands for Notch, but which one is important for T cell development is not established.

Stages in T cell development are marked by expression of coreceptors

The stages of T cell development are largely defined by the expression of the two major coreceptors of T cells, CD4 and CD8 (Figure 7-22). The early lymphoid progenitors express neither of these coreceptors, and the same is true for the earliest T-committed cells, collectively referred to as **double-negative (DN) thymocytes** because they express neither coreceptor. DN thymocytes can be further subdivided into four stages (DN1–4) based on the expression of CD44 and CD25 (the IL-2 receptor α chain). Once a TCR β chain has been produced, it combines with the pre-TCRα chain to form the pre-TCR, which induces maturation to the CD4 CD8 **double-positive (DP) thymocyte** stage. Next, TCR α chain genes are rearranged and if the resulting αβ TCR can react with self peptide–MHC complexes in the thymus, then positive selection occurs and the thymocyte matures into a CD4 or CD8 **single-positive (SP) thymocyte**, which after several days leaves the thymus and becomes a mature αβ T cell if it does not die as a result of negative selection.

Definitions

double-negative (DN) thymocytes: immature T cells at the earliest stage of T cell development in the thymus, characterized by the absence of CD4 or CD8.

double-positive (DP) thymocyte: immature T cells at an intermediate stage in the development of αβ T cells, during which the cells express CD4 and CD8.

pre-TCRα: a transmembrane Ig superfamily member that is expressed during the development of T lymphocytes and pairs with TCR β chain to form the pre-TCR. The gene encoding pre-TCRα does not undergo rearrangement.

single-positive (SP) thymocytes: thymocytes that have undergone positive selection successfully and express either CD4 or CD8 but not both.

References

Borowski, C. et al.: **On the brink of becoming a T cell.** Curr. Opin. Immunol. 2002, **14**:200–206.

MacDonald, H.R. et al.: **T cell fate specification and αβ/γδ lineage commitment.** Curr. Opin. Immunol. 2001, **13**:219–224.

Robey, E.A. and Bluestone, J.A.: **Notch signaling in lymphocyte development and function.** Curr. Opin. Immunol. 2004, **16**:360–366.

Takahama, Y.: **Journey through the thymus: stromal guides for T-cell development and selection.** Nat. Rev. Immunol. 2006, **6**:127–135.

von Boehmer, H.: **Unique features of the pre-T-cell receptor α-chain: not just a surrogate.** Nat. Rev. Immunol. 2005, **5**:571–577.

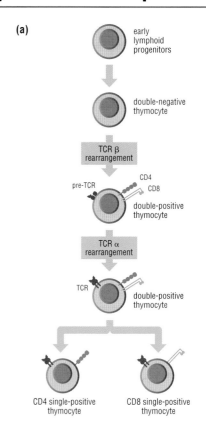

(a)

Figure 7-22 Stages in the development of αβ T cells (a) Early lymphoid progenitors enter the thymus, become committed to the T cell lineage and undergo expansion. At this stage they do not express either coreceptor: that is, they are double negative (DN) for CD4 and CD8. TCR β chain rearrangement and pre-TCR signaling is required to progress past this stage. This progression includes induced expression of CD4 and CD8, leading to the double-positive (DP) stage, and shifting of V(D)J recombination from the TCR β locus to the TCR α locus. Once an α chain has been produced, the DP thymocyte may receive a signal inducing positive selection to the single-positive (SP) stage as described in the next section. **(b)** An example of flow cytometry data defining the stages of thymocyte development by the expression of CD4 and CD8 is shown, with arrows depicting the developmental progressions.

Lineage choice between αβ T cells and γδ T cells depends on functional TCR rearrangements

During the DN stage both TCR β and TCR γ and δ rearrangements can occur, and both types of rearrangements are frequently seen in individual precursors, indicating that there is not a complete lineage split before rearrangements start.

The choice between αβ and γδ T cell lineages is determined by which of the appropriate TCR loci successfully undergo rearrangement (Figure 7-23). If a cell acquires functionally rearranged γ and δ TCR chains, then it will become a γδ T cell. Conversely, if a functional TCR β chain is made and expressed, then it will pair with **pre-TCRα**, a non-rearranged α-like chain that pairs with β chain to form the pre-TCR. Analogously to the pre-BCR, the pre-TCR sends a signal that promotes progression to the next stage of thymocyte development (see Figure 7-22), although in T cell development the process is complicated by the existence of the two possible cell fates, the αβ T cell fate and the γδ T cell fate. Thus, the developing T cell must distinguish between the signal provided by the pre-TCR and the signal provided by the γδ TCR. Evidence so far indicates that the discrimination involves quantitative differences (the γδ TCR provides a stronger signal than the pre-TCR), but it has also been proposed that there are qualitative differences in signaling.

Pre-TCR signaling induces commitment to the αβ TCR lineage

The pre-TCR signals the DN thymocyte to proceed down the developmental pathway leading to αβ T cells. This includes (1) proliferation, expanding each precursor approximately 100-fold, (2) expression of the coreceptors of αβ T cells, CD4 and CD8, and (3) re-targeting of the V(D)J recombinase away from the TCR β locus and toward the TCR α locus. The pre-TCR mediates these events via signaling reactions we have already discussed in Chapter 5, including activation of the tyrosine kinases Lck and ZAP-70, elevation of intracellular free Ca^{2+} and activation of Erk MAP kinase (see sections 5-5 and 5-6). Activation of Ras or Raf, which are upstream of Erk, is sufficient for the proliferative response and induced expression of coreceptors, but not for mediating β-chain allelic exclusion. It is thought that the pre-TCR can signal spontaneously, without the need for a ligand to trigger its signaling. IL-7 is required for the survival of thymocytes at this stage, and a defect in IL-7 or its receptor leads to a strong defect in T cell development with corresponding immunodeficiency.

Once the newly generated CD4 CD8 double-positive thymocytes have stopped proliferating in response to pre-TCR signaling, the RAG-1 and RAG-2 proteins are re-induced and targeted to the TCR α locus. Rearrangement of the TCR α locus begins, as described above, with rearrangements of Vα segments to more 5′ Jα regions, leaving the potential for additional productive TCR α gene rearrangements at that locus (see Figure 7-11). This process enables thymocytes to make several attempts at achieving a TCR that can mediate positive selection, and can continue for several days. It is stopped by positive selection, as described in the next section.

(b)

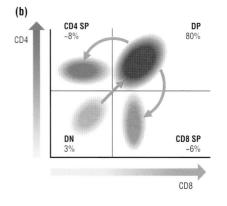

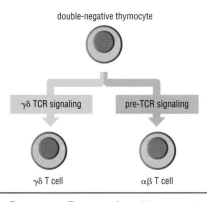

Figure 7-23 αβ versus γδ T cell lineage choice Double-negative thymocytes can rearrange the TCR β, γ and δ loci. Functional rearrangement of the TCR β locus leads to formation of the pre-TCR, which signals the cell to become a CD4 CD8 double-positive thymocyte and to proceed with development in the αβ T cell lineage. Conversely, functional rearrangement at both TCR δ and γ loci results in the formation of the γδ TCR, which signals to induce the cell to enter the γδ lineage.

αβ TCR specificity determines cell fate and lineage choice between helper and cytotoxic T cells

Once a functional TCR α chain is expressed in a DP thymocyte, there are several possible cell fates, depending on the recognition properties of the resulting αβ TCR (Figure 7-24). The first possibility, and apparently the most common, is that the TCR will not have significant reactivity for any the complexes of endogenous peptides and MHC found in the thymus. In this case, the cell will continue to rearrange TCR α gene segments, either by recombination involving the same allele (see Figure 7-11), which will excise the previous TCR α rearrangement, or by rearranging the other allele. If the second allele is rearranged, this could result in two functional α chains in one cell: this sometimes occurs and does not seem to be detrimental. Rearrangements will continue until a TCR with sufficient affinity for self peptide + MHC is produced, or until time runs out, in 3–4 days, after which the cell will die, a fate referred to as **death by neglect**.

If the TCR expressed by the DP thymocyte has some reactivity for self peptides and MHC, then it will either be selected into the population of mature lymphocytes (positive selection) or will rapidly die (negative selection). The latter fate will be discussed in detail later in this chapter.

Positive selection of a thymocyte can occur if its TCR recognizes a self-peptide–MHC complex with low or intermediate affinity, whereas strong binding is required for negative selection in the thymus or for activation in the periphery during an immune response. The selective events in the thymus discriminate primarily on the basis of affinity for the ligand rather than on the amount of the self-peptide–MHC ligand that is available for binding. This discrimination is thought to reflect qualitative differences in signaling between TCRs engaged by lower-affinity ligands with faster dissociation rates and those engaged by higher-affinity ligands with slower dissociation rates.

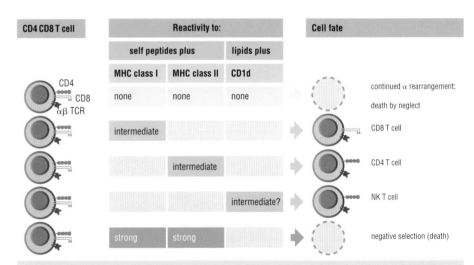

Figure 7-24 Different fates of double-positive thymocytes The fate of CD4 CD8 αβ TCR DP thymocytes is determined by their reactivity for ligand–MHC complexes. In the absence of a positive TCR signal, TCR α locus rearrangement continues, allowing the generation of new αβ TCR pairs for possible positive selection. Low or intermediate reactivity to a self peptide and MHC class I leads to adoption of the CD8 T cell differentiated state, whereas low or intermediate reactivity to a self peptide and MHC class II leads to the CD4 state. Some sort of reactivity (intermediate or possibly strong) to CD1d complexed to certain lipids (not shown) leads to development of an NK T cell. Finally, too-strong reactivity for self peptide + MHC class I or MHC class II leads to negative selection by apoptosis.

Definitions

death by neglect: the fate of double-positive thymocytes that fail to receive a positive selection signal. This occurs after 3–4 days of residence at the DP stage.

References

Santori, F.R. *et al.*: **Rare, structurally homologous self-peptides promote thymocyte positive selection.** *Immunity* 2002, **17**:131–142.

Starr, T.K. *et al.*: **Positive and negative selection of T cells.** *Annu. Rev. Immunol.* 2003, **21**:139–176.

Possible ways in which this may work in the elimination of self-reactive cells are described in the section on negative selection. A similar distinction in biological response is seen in mature T cells, which are activated in response to strong binding of the TCR to foreign-peptide–MHC complexes but may be activated partially or not at all in response to altered peptide ligands with lower affinity for the TCR (see section 5-2).

In a few cases it has been possible to define the nature of the self peptides required for positive selection by DP thymocytes expressing a particular $\alpha\beta$ TCR. In some cases there is structural similarity between the positive selecting self peptide and the antigenic peptide that will induce a strong response by the mature T cell bearing that $\alpha\beta$ TCR (Figure 7-25). These observations support the concept that positive selection can be mediated by a weak interacation between the TCR and self-peptide–MHC complexes, whereas a stronger interaction is required for activation of T cells in an immune response.

A low-level TCR signal for a DP thymocyte induces developmental maturation, and this involves a decision between two major lineages, the CD4 helper T cell lineage and the CD8 cytotoxic T cell lineage, and a less common type of unconventional T cell, the NKT cell, which we discuss in Chapter 8. As described in Chapter 4, CD4 helper $\alpha\beta$ T cells recognize antigenic peptides together with MHC class II molecules and CD8 cytotoxic $\alpha\beta$ T cells recognize peptide together with MHC class I molecules. Positive selection also involves determination of this lineage choice so that those DP thymocytes that receive a positive selection signal via an interaction with MHC class II molecules turn off the expression of CD8 and assume the developmental program of helper T cells, whereas those DP thymocytes that receive a positive selection signal via an interaction with MHC class I molecules turn off the expression of CD4 and assume the developmental program of cytotoxic T cells. In this way, each cell retains the coreceptor recognizing the type of MHC molecule preferentially bound by the TCR of that cell, and adopts the functional properties appropriate to recognition of peptides loaded on to that type of MHC molecule. This feature of T cell development is critical for the proper functioning of the immune system. For example, if some of the CD4 T cells had a cytotoxic T cell potential instead of a helper T cell potential, then these cells would kill B cells during an immune response, which would interfere with antibody production.

The linkage between expression of CD4 or CD8 and differentiated functional properties of the mature T cell is determined by transcriptional regulators in a way that is only partly understood. One key regulator of T cell differentiation state is the transcription factor Th-POK, which is related to the Kruppel family of transcriptional regulators, first identified as participating in development in fruit flies. In the absence of Th-POK, all positively selected T cells become CD8 T cells. Conversely, if Th-POK expression is experimentally enforced in developing thymocytes, all positively selected T cells become CD4 T cells. Exactly how TCR signaling determines whether Th-POK will be expressed is not known.

The third possible differentiated cell type that can arise at this stage occurs when DP thymocytes recognize antigen in conjunction with the non-classical MHC class I-like molecule CD1d. These thymocytes assume the fate of NKT cells, the properties of which are further described in Chapter 8. How CD1d recognition leads these cells to become NKT cells and not CD4 T cells or CD8 T cells is unknown.

Figure 7-25 Self peptides that cause positive selection An extensive effort was made to identify self peptides that could cause positive selection of thymocytes expressing a transgenic TCR obtained from a cell specific for a peptide from chicken ovalbumin (shown in the top line). Two endogenous self-peptides were identified and are shown in the single-letter amino-acid code. The first, representing residues 329–336 or β-catenin, was more effective at positive selection than the second, representing residues 92–99 of F-actin capping protein A. Forty-three other peptides that can bind to the same MHC class I molecule (K^b) were ineffective at inducing positive selection of thymocytes expressing this TCR. Note the sequence similarities of the two positively selecting peptides to the antigenic peptide, particularly in the carboxy-terminal half of the peptides (F and Y both have benzene rings; E and D are negatively charged amino acids; and K and H are both positively charged amino acids). It should be noted that in a few cases self peptides responsible for positive selection have been identified that lack this structural similarity to the foreign peptide that activates the T cell. In those cases, it is presumed that the self-peptide–MHC complex binds to the TCR with a moderate affinity due to distinct atomic contacts, rather than structurally similar ones. Color coding: red amino acids are the anchor residues that mediate binding to the MHC class I molecule; blue amino acids have side chains that can be contacted by the TCR. Data are from Santori, F.R. *et al.*: *Immunity* 2002, **17**:131–142.

foreign peptide   S I I N F E K L

self peptide   R T Y T Y E K L

self peptide   I S F K F D H L

Double-positive thymocytes may mature by a stochastic/selection or by an instructive mechanism

The choice between the CD4 and CD8 lineages is a central event in T cell development; but its mechanism is still incompletely understood. Two major hypotheses were originally developed to explain this process: the **stochastic/selection model** and the **instruction model** (Figure 7-26). Currently favored ideas are combinations or variations of these two extremes.

In the stochastic/selection model, it is postulated that either CD4 or CD8 is randomly turned off. A positive selection event is then needed to prevent death by neglect. Since in most cases signaling requires binding both of the TCR to the polymorphic region of the MHC molecule and of the coreceptor to the non-polymorphic region, cells will be positively selected only if the MHC molecule recognized by the TCR corresponds to that recognized by the remaining coreceptor (that is, class I for CD8 and class II for CD4). According to this model therefore lineage choice is coupled to which coreceptor is still expressed.

In contrast, the instruction model posits that DP thymocytes receive a signal from the TCR while they express both coreceptors and moreover that they can distinguish whether the signal is from TCR + CD4 or TCR + CD8. Signaling through TCR + CD4 instructs the downregulation of CD8, and signaling though TCR + CD8 downregulates CD4.

Many experiments have been done to test these models. For example, transgenic mice have been created that force the expression of CD4 or CD8 in order to test whether this results in any cells of the inappropriate phenotype: for example, in the presence of forced expression of CD4, do any MHC class II-specific T cells develop that express endogenous CD8? The appearance of such cells is predicted by the stochastic/selection model, since the transgene-encoded CD4 would rescue cells

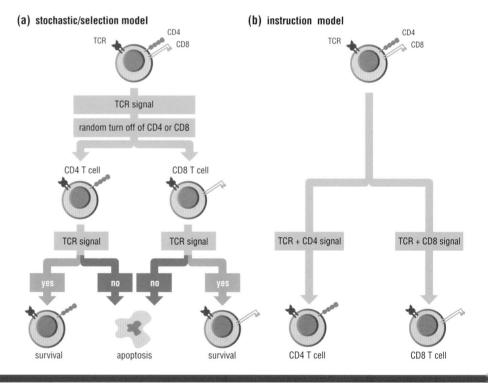

(a) stochastic/selection model

(b) instruction model

Figure 7-26 Original models to explain CD4/CD8 lineage choice Initially, two opposite hypotheses were proposed to explain CD4/CD8 lineage choice. According to the stochastic/selection model **(a)** a TCR signal in a DP thymocyte induces random downregulation of one of the coreceptors. Subsequently, receipt of a TCR signal leads to survival of the cell, whereas lack of this signal (for example due to absence of the coreceptor needed for a response to the recognized self-peptide–MHC complex) leads to apoptosis. The alternative hypothesis, the instruction model **(b)**, postulates that the DP thymocyte can distinguish the signal provided by TCR + CD4 from that provided by TCR + CD8 and this allows the cell to turn off the coreceptor that is not contributing to the signal.

Definitions

instruction model: hypothesis to explain the mechanism of positive selection of developing T lymphocytes in which a DP thymocyte can recognize the type of MHC molecule that it has interacted with and downregulates the coreceptor that is not needed to interact with that class of MHC.

kinetic signaling model: a recent variation of the **instruction model** for positive selection of developing T lymphocytes in which the fate of the cell is determined by whether or not TCR signaling declines when

CD8 expression is downregulated. If it does, then the cell turns off CD4, turns CD8 back on and adopts the CD8 lineage; if signaling is maintained, the cell chooses the CD4 lineage. This model incorporates features of the **stochastic/selection model**, but there is no stochastic element.

stochastic/selection model: hypothesis to explain the mechanism of positive selection of developing T lymphocytes in which one of the two T cell coreceptors is randomly turned off at the start of positive selection; additional signaling must be received once the cell

only has one coreceptor, or the cell dies by neglect.

strength of signal model: a form of the **instruction model** for positive selection of developing T lymphocytes in which TCR + CD4 recognition sends a stronger signal whereas TCR + CD8 recognition sends a weaker signal. This model generally incorporates a checking mechanism after downregulation of one coreceptor that is analogous to the selection step in the stochastic model. This is a way of ensuring that the instructional step has provided accurate information.

that made the wrong choice and downregulate endogenous CD4. Conversely, mice have been manipulated to alter the signaling properties of thymocytes and the effects on CD4 versus CD8 T cell development have been assessed. Some evidence for both models has been obtained. This has led to two more complicated models that are generally thought to be likely explanations for positive selection (Figure 7-27). The first of these is the **strength of signal model**. This model holds that TCR + CD4 inherently sends a stronger signal than TCR + CD8, possibly because Lck associates much better with CD4 than with CD8. According to this model, a weak positive selection signal instructs the DP thymocyte to downregulate CD4, whereas a stronger positive selection signal instructs the DP thymocyte to downregulate CD8. After this, when the cell expresses only a single coreceptor, there must be an additional TCR signal of comparable magnitude to complete positive selection. This feature is included to explain the observations that forced expression of one of the two coreceptors can rescue cells that are inappropriately expressing the other, and that positive selection requires prolonged signaling. Thus, this model is really a hybrid of the two original models. Downregulation of one coreceptor is instructed, but is followed by selection to make sure that the inference from strength of signal was correct.

The other current model, the **kinetic signaling model**, also has features of both models. It posits that the recognition of self-peptide–MHC complexes always causes downregulation of CD8 in DP thymocytes. If signaling continues, the cell becomes a CD4 helper T cell. If signaling stops while CD8 levels are decreasing, the cell turns CD4 off and CD8 back on and becomes a CD8 cytotoxic T cell. In other words, the cell is instructed by the kinetics of signaling. As experimental manipulations that increase or decrease TCR signaling may also affect the kinetics of TCR signaling, many experiments do not definitively distinguish between the two newer models. Additional experiments will be required to determine which of these two models more closely approximates the mechanism of positive selection.

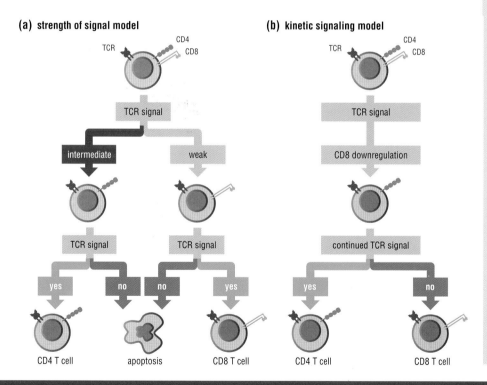

(a) strength of signal model

(b) kinetic signaling model

Figure 7-27 Recent models to explain CD4/CD8 lineage choice Experimental tests of the stochastic/selection and instruction models have provided evidence for both hypotheses and directed the formulation of two more recent models that include elements of both models but are largely instructional in nature. In the strength of signal model **(a)** TCR + CD4 is postulated to send a stronger signal than TCR + CD8 and this biases the selection of the coreceptor that is turned off. After one or other coreceptor has been turned off, TCR signals must be received or the cell dies by apoptosis (death by neglect). This is an element of the stochastic/selection model, but note that there is a bias to which coreceptor is turned off, which is an element of the instruction model. The kinetic signaling model **(b)** is an instruction model in which the cell is able to discriminate kinetically which coreceptor is participating in providing the positive-selection TCR signal by first turning off the expression of CD8. If this turns off the TCR signal, then the cell recognizes that the TCR signal was dependent on CD8 and it therefore chooses the CD8 T cell fate, turning off CD4 and turning CD8 back on. On the other hand, if the TCR positive-selection signal does not decline over this period, the cell chooses the CD4 T cell fate. This model is purely instructional—there is no stochastic element—but is similar to the stochastic/selection model in that whether or not the TCR signals after downregulation of a coreceptor (always CD8 in this model) is a key determinant of cell fate. Instead of dying, the cell that has lost its positive selection signal reverses coreceptor expression.

References

Germain, R.N.: **T-cell development and the CD4–CD8 lineage decision.** *Nat. Rev. Immunol.* 2002, **2**:309–322.

Hogquist, K.A.: **Signal strength in thymic selection and lineage commitment.** *Curr. Opin. Immunol.* 2001, **13**:225–231.

Singer, A.: **New perspectives on a developmental dilemma: the kinetic signaling model and the importance of signal duration for the CD4/CD8 lineage decision.** *Curr. Opin. Immunol.* 2002, **14**:207–215.

Mutations Causing Defects in Negative Selection and Leading to Autoimmunity

Gene product	Function
Aire	transcriptional regulator
PTEN	counteracts PI3 kinase
LAT(Y136F)	TCR signaling adaptor
ZAP-70(W163C)	TCR signaling kinase

Figure 7-28 Table of mutations causing defects in negative selection and leading to autoimmunity LAT(Y136F): point mutation in LAT changing a tyrosine at position 136 to a phenylalanine. ZAP-70(W163C): point mutation in ZAP-70 changing amino acid 163 from tryptophan to cysteine. These two mutations are partial loss of function mutations; complete loss of function mutations in ZAP-70 or LAT cause blocks in T cell development in the thymus.

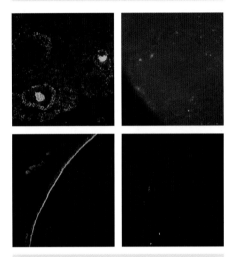

Figure 7-29 Autoimmunity to multiple organs in mice deficient in Aire Mice deficient in Aire exhibit spontaneous autoimmunity in multiple organs, as evidenced by inflammation and the deposition of autoantibodies. Shown are indirect immunofluorescence micrographs of ovary (top panels) and retina (bottom panels), showing autoantibodies reactive to these tissues. Tissue sections were incubated with serum from Aire-deficient (left) or normal (right) mice and then with fluorescently labeled anti-mouse IgG antibodies. Note that the autoantibodies are specific for particular cell layers in each organ, suggesting that a small number of self antigens is recognized. Courtesy of Mickie Chang, Jason DeVoss and Mark Anderson.

Thymocytes encountering high-affinity peptide–MHC ligands in the thymus die rapidly

Whereas contacts of moderate affinity between a developing T cell and peptide–MHC complexes induce the differentiation of CD4 and CD8 T cells from DP thymocytes, high-affinity contacts with peptide–MHC complexes lead to rapid death by apoptosis, a process called *negative selection* or *clonal deletion*. This process is essential to achieving tolerance to self and avoiding autoimmunity, as revealed by the fact that various mutations that compromise negative selection lead to autoimmune phenomena (Figure 7-28).

Negative selection can occur in either the cortex at the DP stage or in the medulla at the single-positive stage, although it is generally thought that it occurs most often in the medulla, after positive selection. Cells that can mediate negative selection include the cortical epithelial cells that mediate positive selection, dendritic cells and macrophages at the cortico-medullary junction and in the medulla, and medullary epithelial cells. All of these cells express MHC class II molecules, as well as the ubiquitously expressed MHC class I molecules, and are therefore able to present self-peptide–MHC complexes for developing T cells to recognize. In addition, cells in the medulla express B7 molecules, which bind CD28 and thereby promote TCR signaling. In this way, the signals received by the T cell at this stage are matched to the signals it will later encounter in the secondary lymphoid tissues, where antigen-presenting dendritic cells present antigen to mature T cells, as described in Chapter 5. By engaging the costimulatory signaling of CD28 in the thymic medulla as well as in the secondary lymphoid tissues, tolerance to self peptides can be effectively achieved, in combination with other mechanisms for achieving self tolerance, that we describe in Chapter 12.

Special mechanisms exist to promote self-peptide presentation in the thymus

Medullary epithelial cells have the remarkable property that they express many self proteins, such as insulin, myelin basic protein and thyroglobulin, that are otherwise expressed only in specific organs. Thus, medullary epithelial cells are thought to be important in inducing negative selection of T cells that recognize organ-specific self antigens. How medullary epithelial cells express these otherwise highly restricted proteins is not entirely understood, but part of the explanation lies in a transcriptional regulator called **autoimmune regulator (Aire)**. Aire was discovered as the product of a gene mutated in an inherited multiorgan autoimmune disease called **autoimmune polyendocrine syndrome type 1**. Mice in which this gene has been deleted also exhibit autoimmunity to several organs, which is consistent with a loss of self tolerance to some organ-specific self antigens (Figure 7-29). They also lose expression of some self antigens in the thymic medulla. Thus, it is thought that medullary epithelial cells express several transcriptional regulators (Aire and others not yet identified) responsible for the thymic expression of otherwise tissue-restricted self proteins and function to induce negative selection of autoreactive thymocytes.

Thymocytes discriminate between positive selection and negative selection based in part on TCR off-rate

A central question in T cell development is how thymocytes can distinguish positively selecting peptide–MHC ligands from negatively selecting ones. Although positive selection often occurs before negative selection, this is not obligatory and negative selection can even occur in the

Definitions

Aire: see **autoimmune regulator**.

autoimmune polyendocrine syndrome type 1: a rare inherited disease due to mutation of the gene encoding **Aire** and characterized by multiple-organ-specific autoimmune disease.

autoimmune regulator (Aire): transcriptional regulator expressed in thymic medullary epithelial cells and responsible for expression in these cells of a number of proteins otherwise expressed only in one particular

organ. Originally discovered as the mutant protein responsible for autoimmune polyendocrine syndrome (APS) type 1 in humans; mice in which the gene encoding Aire is disabled have a deficiency in central tolerance and develop diverse solid-organ autoimmune diseases.

immunodysregulation, polyendocrinopathy, enteropathy, X-linked (IPEX) syndrome: an X-linked multiorgan autoimmune disease caused by mutations in FoxP3, a transcription factor required for the development of regulatory T cells.

IPEX: see **immunodysregulation, polyendocrinopathy,**

enteropathy, X-linked syndrome.

regulatory T cells (T$_{REG}$): T cells that act to inhibit responses of other T cells. Regulatory T cells are dependent upon a differentiation program driven by the transcriptional regulator FoxP3 and usually express CD4 and CD25.

T$_{REG}$: see **regulatory T cells**.

References

Anderson, M.S. *et al.*: **Projection of an immunological self shadow within the thymus by the Aire protein.**

©2007 New Science Press Ltd

absence of positive selection. Thus, thymocytes can respond in either way and must be able to distinguish between these two types of ligands with high fidelity. This requirement comes in the face of a great variety in the level of expression of various self peptides in the thymus, so it is unlikely that thymocytes distinguish between a high-affinity peptide–MHC ligand and a moderate-affinity one simply by the number of occupied receptors, which would be determined by the number of complexes of a particular peptide with an MHC molecule. Measurements of peptide–MHC affinity of TCRs indicate that the major difference between positively selecting ligands and negatively selecting ones is often that the latter have slower dissociation rates for the TCR (recall that higher affinity can result from a slower off-rate, a faster on-rate or a combination of these). Therefore, it has been hypothesized that these two types of ligands induce qualitatively different signaling events. For example, if the TCR-induced signaling events include some events that are rapidly triggered and others that are triggered more slowly, then a slow off-rate ligand would induce both classes of signaling events whereas a faster off-rate ligand might induce only the more rapidly generated signaling events. The cell would then be positively selected if only the fast signaling events occurred, whereas it would be induced to die if the slow signaling events also occurred (Figure 7-30). So far, no signaling event has been identified that fits this simple idea. One striking difference is that negatively selecting ligands induce a short, strong activation of Erk MAP kinase, whereas positively selecting ligands induce a slow but sustained activation of Erk (see Figure 7-30). Erk has been shown to be critical for positive selection, as has calcineurin, which activates the transcription factor NFAT downstream of calcium elevation (as described in section 5-5). In contrast, these two signaling events seem not to be required for negative selection, for which the other two types of MAP kinases, JNK and p38 MAP kinase, are important instead.

Key downstream mediators of negative-selection-induced death seem to include two related transcriptional regulators, Nur77 and Nor1, which are induced in a calcium-dependent manner. Also implicated in negative selection is inhibition of the transcriptional regulator NF-κB, which has pro-survival actions (see Figure 3-49). Death is dependent on the BH3-only Bcl-2 family member Bim, the synthesis of which is induced by negative-selection-inducing peptides.

Some thymocytes encountering high-affinity ligands survive and become regulatory T cells

Although cell death is the predominant fate of thymocytes encountering high-affinity peptide–MHC ligands in the thymus, some autoreactive T cells instead survive and become **regulatory T cells (T$_{REG}$)**. These cells are also called central T$_{REG}$ or cT$_{REG}$, to distinguish them from T$_{REG}$ cells that differentiate in the periphery as part of an immune response (induced T$_{REG}$ or iT$_{REG}$; see section 5-13). The action of both types of T$_{REG}$ cells in the periphery is to inhibit T cell immune responses, and hence they act to enforce immune tolerance or limit tissue damage in a chronic infection.

Thus, high-affinity peptide–MHC ligands induce both clonal deletion and the production of cells that can inhibit immune responses. Both of these events contribute to immune tolerance. At this stage it is not known why a particular self-reactive thymocyte dies or becomes a central T$_{REG}$, although the transcriptional regulator FoxP3 has been shown to be important for acquisition of the T$_{REG}$ differentiation state. Interestingly, this gene is mutated in patients with **immunodysregulation, polyendocrinopathy, enteropathy, X-linked (IPEX) syndrome**, another multiorgan autoimmune disease. Regulatory T cells and their role in immune tolerance are discussed in more detail in Chapter 12.

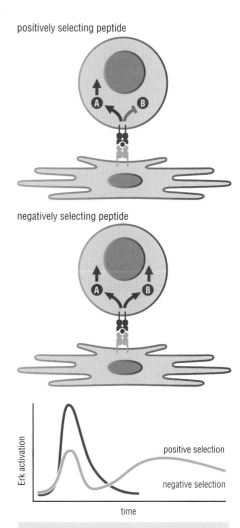

positively selecting peptide

negatively selecting peptide

Figure 7-30 Kinetic signaling model for discrimination between negatively selecting peptides and positively selecting peptides It is postulated that the positively selecting peptide–MHC complex has a faster off-rate and therefore TCR signaling stimulates signaling reaction A but not signaling reaction B, which is activated only on longer receptor occupancy. A negatively selecting peptide, in contrast, supports a longer interaction with the TCR and hence stimulates both signaling reactions. Experimental examination of signaling reactions indicates that activation of Erk MAP kinase shows a strong but transient activation with negative selection and a slow but sustained activation with positive selection (lower panel), a complex behavior that does not fit this model, particularly as Erk activity is necessary for positive selection but not for negative selection. So far, no single signaling event with activation characteristics corresponding to signaling event B in the upper two panels has been described, but the JNK and p38 MAP kinases have been shown to be important for negative selection and hence may contribute to postulated signaling event B.

Science 2002, **298**:1395–1401.

He, Y.-W.: **Orphan nuclear receptors in T lymphocyte development.** *J. Leukoc. Biol.* 2002, **72**:440–446.

Hogquist, K.A. *et al.*: **Central tolerance: learning self-control in the thymus.** *Nat. Rev. Immunol.* 2005, **5**:772–782.

Hori, S. *et al.*: **Control of regulatory T cell development by the transcription factor Foxp3.** *Science* 2003, **299**:1057–1061.

Kyewski, B. *et al.*: **Promiscuous gene expression and central T-cell tolerance: more than meets the eye.** *Trends Immunol.* 2002, **23**:364–371.

Palmer, E.: **Negative selection — clearing out the bad apples from the T-cell repertoire.** *Nat. Rev. Immunol.* 2003, **3**:383–391.

Siggs, O.M. *et al.*: **The why and how of thymocyte negative selection.** *Curr. Opin. Immunol.* 2006, **18**:175–183.

Sprent, J. and Kishimoto, H.: **The thymus and negative selection.** *Immunol. Rev.* 2002, **185**:126–135.

Some Genetic Causes of Immunodeficiency Diseases

Gene defect	Result
RAG-1 or RAG-2	T⁻B⁻ NK⁺ SCID
Artemis	T⁻B⁻ NK⁺ SCID
cytokine-receptor γc chain	T⁻B⁺ NK⁻ SCID (X-SCID)
Jak3	T⁻B⁺ NK⁻ SCID
Btk	X-linked agammaglobulinemia

Figure 7-31 Table of genetic defects in human immunodeficiencies Approximately 3–10% of SCID patients have defects in V(D)J recombination, about half due to defects in RAG-1 or RAG-2 and about half due to defects in Artemis. The most common cause of SCID is a defect in the cytokine-receptor γc chain, reflecting the importance of the γc-containing receptor for IL-7 in lymphocyte development (particularly for T cell development). Less frequently, SCID can result from mutation of the gene encoding Jak3, which is required for signaling from the γc chain. Another relatively frequent genetic immunodeficiency is X-linked agammaglobulinemia, which is due to a defect in Btk, a protein tyrosine kinase involved in signaling from the BCR and its pre-B cell counterpart, the pre-BCR. In this disease almost no B cells develop.

Defects or aberrations related to lymphocyte development can lead to immunodeficiency or cancers

Serious inherited deficiencies in adaptive immunity are rare, at an estimated 1 in 100,000 live births, but they have made a significant contribution to the understanding of the mechanisms of lymphocyte development. Collectively they are known as *severe combined immunodeficiency (SCID)*, but they represent a broad spectrum of defects in different specialized molecules essential to the development of lymphocytes.

Leukemias or lymphomas arising from genetic accidents to somatic cells in children or adults are more frequent than SCID, at approximately 140 per 100,000, and most of these arise directly from the specialized mechanisms we have described for lymphocyte diversification.

In this section we describe some of the defects underlying SCID and explain how aberrations in the operation of the specialized genetic mechanisms that generate lymphocyte receptor diversity can sometimes lead to cancerous changes in cells.

Genetic defects in lymphocyte development lead to immunodeficiency

Severe combined immunodeficiency, or **SCID**, is an inherited condition in which both T cells and B cells are absent or non-functional. Individuals with SCID are always strongly deficient in T cell function and may have defects in B cell function ranging from defects in T cell-dependent antibody production due to the T cell deficiency to complete loss of antibody production. Some of the mutations that lead to SCID affect V(D)J recombination. Defects in the signaling pathways necessary for lymphocyte development are also frequent causes of SCID.

Between 3 and 10% of cases of SCID are caused by mutations in RAG-1, RAG-2, the recombinases that mediate V(D)J recombination, or in Artemis, a nuclease that is part of the normal cellular machinery recruited for V(D)J recombination (see Figure 7-6). In individuals with these mutations, T cells and B cells completely fail to mature because they cannot make antigen receptors, which deliver essential developmental signals.

The majority of SCID cases, however, are due to defects in components of signaling pathways. Of these, the commonest is an X-linked condition, known as *X-SCID*, that is due to a mutation in the cytokine-receptor γc chain. In X-SCID, which accounts for about 45% of severe combined immunodeficiencies, T cell development is blocked but B cells are found in the blood. These cells are not fully functional because the γc chain is required for responses to a variety of cytokines (see section 2-8). The developmental defect in T cell development is largely due to a loss of responsiveness to IL-7, which promotes the survival of double-positive thymocytes. In B cell development, IL-7 promotes the expansion of pre-B cells after the pre-BCR signal has been conveyed, but is not absolutely required for some B cell development (in humans). NK cell survival is also affected because the cytokine-receptor γc chain is required for the

Figure 7-32 Chromosome translocation activating Myc Example of a translocation occurring between chromosome 8 (containing the *MYC* gene) and the IgH locus on chromosome 14. This translocation fuses the three exons of the *MYC* gene to the IgH constant-region locus, where the enhancer (Eμ) upstream of the IgH constant-region exons drives Myc expression strongly.

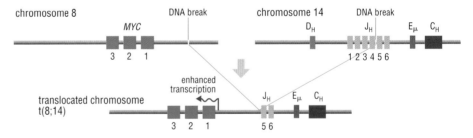

response to IL-15. A rarer form of SCID with the same phenotype results from mutation in Jak3, an intracellular tyrosine kinase required for γ_c signal transduction (see section 2-7). These forms of SCID are treated with considerable success by bone marrow transplantation, as normal hematopoietic stem cells can give rise to functional lymphocytes in the treated patient. SCID can also result from defects in purine metabolism (adenosine deaminase or purine nucleoside phosphorylase) that act to inhibit lymphocyte development.

More selective immunodeficiencies can result from mutation of components required for the development of B cells or T cells but not both. The most common immunodeficiency of this type is **X-linked agammaglobulinemia**, where mutation of the gene encoding Btk, an intracellular tyrosine kinase that participates in signaling by the BCR and pre-BCR (see section 6-8), results in a block of B cell development caused by inability of the pre-BCR to signal. Some common causes of immunodeficiency are listed in Figure 7-31.

DNA breaks introduced by V(D)J recombination can lead to lymphomas and leukemias

DNA breakage and rejoining is potentially hazardous to the functional integrity of cells, and indeed the actions of the RAG-1 and RAG-2 proteins can in aberrant cases lead to malignancies of B and T lineage cells. This occurs when the DNA breakages lead to **chromosomal translocations** in which a gene controlling cellular growth or survival (a **proto-oncogene**) becomes juxtaposed to an active transcriptional control region in an Ig or TCR locus. The translocation causes dysregulated expression of the proto-oncogene, which can thus contribute to the development of malignancy (Figure 7-32). Often such translocations occur at at a J or D segment in the Ig or TCR locus, presumably reflecting DNA breakage at that position by the action of RAG-1 and RAG-2. In such cases, sometimes an RSS-like sequence can be found at the end of the translocation where the dysegulated gene is situated, whereas in other cases no such sequence can be identified. These findings indicate diversity in mechanisms of translocations. In either case, breaks in DNA at the two loci are followed by fusion of the newly created chromosomal ends to one another to produce the chromosomal translocation. In some cases, V(D)J recombination may generate both DNA breaks, whereas in other cases it may produce the break at the Ig or TCR locus, while some other mechanism produces the DNA break next to the proto-oncogene. In B cell malignancies, translocations also occur in IgH switch regions, presumably reflecting the action of AID and class switch recombination-mediated DNA breaks.

Genes whose dysregulation promotes lymphoid malignancies include regulators of cell proliferation (Myc, cyclin D1), regulators of cell survival (Bcl-2), regulators of differentiation (BCL-6) and other genes with less well understood functions. Some examples of common translocations associated with lymphoid malignancies and resulting from V(D)J recombination are listed in Figure 7-33.

Oncogenic Translocations Caused by V(D)J Recombination Errors

Tumor	Cell type	Translocation	Oncogene product	Ig/TCR locus
Burkitt's lymphoma	germinal center B cell	t(8;14)	Myc	IgH
follicular lymphoma	memory B cell	t(14;18)	Bcl-2	IgH
mantle cell lymphoma	naïve B cell	t(11;14)	cyclin D1	IgH
pre-T acute lymphocytic leukemia	thymocyte	t(10;14) t(8;14)	Tcl-3 Myc	TCR α TCR α

Figure 7-33 Table of oncogenic translocations caused by V(D)J recombination errors The chromosome numbers are shown for the two chromosomes that have been fused together. Note that, in Burkitt's lymphoma, fusions of Myc to the Ig κ or λ loci are also seen, although less frequently than to IgH.

References

Antoine, C. *et al.*: **Long-term survival and transplantation of haemopoietic stem cells for immunodeficiencies: report of the European experience 1968–99.** *Lancet* 2003, **361**:553–560.

Fischer, A. *et al.*: **Gene therapy of severe combined immunodeficiencies.** *Nat. Rev. Immunol.* 2002, **2**:615–621.

Leonard, W.J.: **Cytokines and immunodeficiency diseases.** *Nat. Rev. Immunol.* 2001, **1**:200–208.

Marculescu, R. *et al.*: **V(D)J-mediated translocations in lymphoid neoplasms: a functional assessment of genomic instability by cryptic sites.** *J. Exp. Med.* 2002, **195**:85–98.

Rabbitts, T.H.: **Chromosomal translocations in human cancer.** *Nature* 1994, **372**:143–149.

8

Specialized Lymphocytes in Early Responses and Homeostasis

The lymphocytes of adaptive immunity become effective in defense against infection only after undergoing clonal expansion and differentiation, which take several days. Different types of specialized lymphocytes exist, however, that are equipped to produce effector molecules or actions immediately on encounter with antigen. These cells have relatively invariant receptors adapted to detect damaged cells, or conserved elements shared by a variety of bacteria or viruses, and play an important part in the early response to infection.

Specialized types of lymphocytes can be grouped functionally by their capacity to mediate rapid immune responses

Naïve T and B lymphocytes that populate secondary lymphoid organs express receptors that can recognize the vast array of antigenic shapes presented by pathogens. The requirements for clonal expansion and effector differentiation, however, delay their participation in host defense for some days. Immediate protection is provided by the molecules and phagocytic cells of the innate immune system, which are activated within minutes to hours of infection (see Chapter 3). Populations of specialized lymphocytes are present in tissues and can be functionally grouped by their capacity to respond to infection during the period between activation of the phagocytic cells of innate immunity and the T and B cells of adaptive immunity (Figure 8-1). Under resting conditions these lymphocytes do not circulate through the lymph nodes but are localized in blood, including the spleen, and tissues, particularly near epithelia. These specialized effector populations include *natural killer (NK) cells*, which lack the variable antigen receptors of T and B cells; *NKT cells*, *γδ T cells*, intestinal *intraepithelial lymphocytes* (IEL) and *mucosal-associated invariant T (MAIT) cells*, which express T cell receptors for antigen; and *B1 cells* and splenic *marginal zone B cells*, which express B cell receptors for antigen. The TCRs and B cell receptors for antigen expressed by these antigen receptor-bearing cells are frequently generated early in development, often in the fetal and perinatal periods. These T cell populations, unlike conventional T cells, do not interact with classical MHC class I or class II molecules, but commonly interact with non-classical MHC class I molecules. The activation of NK cells is determined by summation of activating and inhibitory receptors that interact with both classical and non-classical MHC molecules, both of which are commonly modulated after inflammation or injury. Unlike conventional B cells, B1 and marginal zone B cells are poised for rapid differentiation into antibody-secreting plasma cells in response to a restricted range of microbial antigens, even in the absence of T cell help. Although these various cells arise from distinct lineages, they share with innate effector cells the capacity for rapid effector function, and with adaptive effector cells the use of rearranged antigen receptors and/or interactions with MHC and MHC-like molecules for their activation.

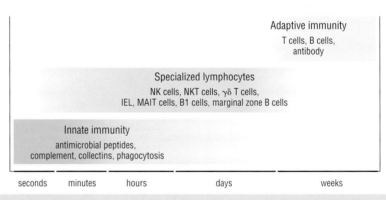

Figure 8-1 Specialized lymphocytes bridge innate and adaptive immune responses The innate response to infection is initiated within seconds to minutes by serum opsonins and phagocytes, whereas adaptive immunity mediated by specific T and B cells is initiated after 4–7 days. Specialized lymphocytes localized in epithelia and blood respond to a different range of signals of infection over a period of hours to days, thus bridging the gap between the innate and adaptive arms of immunity.

References

Bendelac, A. *et al.*: **Autoreactivity by design: innate B and T lymphocytes.** *Nat. Rev. Immunol.* 2001, **1**:177–186.

Gleimer, M. and Parham, P.: **Stress management, MHC class I and class I-like molecules as reporters of cellular stress.** *Immunity* 2003, **19**:469–477.

Activation of specialized lymphocytes can be mediated by self or foreign ligands

In general, the diversity of ligands that stimulate these various lymphocyte populations is substantially less than that of the ligands that activate conventional T and B cells. This reflects the limited diversity of receptors expressed on these cells. Ligands that have been shown to stimulate these cells, however, often consist of both conserved microbial antigens and common self ligands (Figure 8-2). Many of the activating ligands seen by these specialized cells are carbohydrates, glycolipids and small non-peptide antigens that are not recognized by conventional T cells. Some of these self ligands can be induced by cell injury or inflammation, as occurs during viral infection, DNA damage, exposure to inflammatory cytokines or modifications of the stromal microenvironment.

In this way, specialized lymphocytes can respond directly to microbial invasion and also to cell activation or injury arising as a consequence of invasion, and by serving as sentinels for injury as well as for infection, may amplify the response. Perhaps not surprisingly for cells reactive to common self ligands, these cells constitutively express a number of cell surface markers more typically induced on activated cells, and in this way function in a manner more typical of memory lymphocytes, which can also reside in tissues and express their effector capacity rapidly upon activation. These cells commonly express various inhibitory receptors that modulate their threshold for activation, thus guarding against their inadvertent activation in the absence of tissue injury or pathogen invasion.

Specialized Lymphocyte Populations with Limited Receptor Diversity

Cells	Tissue distribution	Self ligands	Foreign ligands	Function
NK cells	blood, BM, spleen, liver, LN*	MHC class I, MHC class I-like	viral MHC-like proteins	cytolytic, IFN-γ release
NKT cells	blood, BM, spleen, liver	iGb3/CD1d	α-proteobacteria glycosylceramides, mycobacterial PIM	cytolytic, rapid cytokine release
γδ T cells	squamous epithelia	unknown, stress-induced	?	cytolytic, epithelial growth factor release
	blood, spleen (human Vδ2)	cell phosphoantigens	bacterial phosphoantigens, environmental alkylamines	cytolytic, IFN-γ release
	blood, spleen (human non-Vδ2)	?	unknown, CMV-infected cells	cytolytic, IFN-γ release
	liver/intestine (human Vδ1)	MICA/MICB/UBLP	?	antimicrobial
intestinal IEL	polarized epithelia	? (mouse TL, costimulatory)	?	cytolytic, epithelial growth factor release
MAIT cells	intestinal lamina propria	unknown, MR1-presented ligand	unknown, MR1-presented ligand	unknown, restrains intestinal IgA
B1 cells	serosal cavities	oxidized lipoproteins	bacterial capsular polysaccharide derivatives	antibacterial, apoptotic cell removal
marginal zone B cells	spleen	oxidized lipoproteins	bacterial capsular polysaccharide derivatives	antibacterial, apoptotic cell removal

Figure 8-2 Table of specialized lymphocytes that recognize self and foreign ligands
BM: bone marrow; CMV: cytomegalovirus; iGb3: isoglobotrihexosylceramide; LN: lymph nodes; MICA and MICB: MHC class I chain related molecules A and B; PIM: phosphatidylinositol mannoside; TL: thymic leukemia antigen; ULBP: UL16-binding protein. *after activation

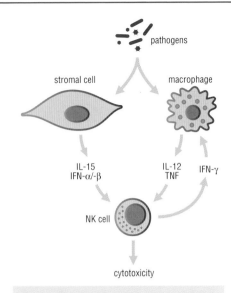

Figure 8-3 Activation of NK cells and their role in immunity Pathogens entering the tissues cause the release of cytokines IL-15 and IFN-α/β from stromal cells and IL-12 and TNF from macrophages, and these in turn activate NK cells, inducing their cytotoxic function and activating them to secrete IFN-γ, which in turn activates macrophage microbicidal functions.

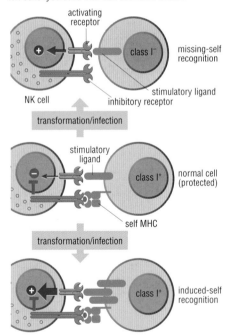

NK cells lyse cells that lack self MHC class I

NK cells lyse MHC class I⁺ cells that upregulate stimulatory ligands

Figure 8-4 Activation of NK cells by a balance of activating and inhibiting ligands expressed on target cells The use of a dual receptor system for activation allows NK cells to become activated by cells that have lost ligands for inhibitory receptors, such as occurs with MHC class I downregulation, or have gained ligands for activating receptors, such as occurs during cell damage.

Natural killer cells provide early protection from intracellular pathogens, particularly viruses

Natural killer (NK) cells arise from the same bone marrow progenitor cells as T cells and share with CD8 T cells two of their most important effector functions: they are cytotoxic and they secrete the proinflammatory cytokine IFN-γ. They are identified by their expression of families of transmembrane proteins that regulate their activation, as well as IL-2Rβ chain (CD122), which is required for IL-15 responsiveness for their generation and survival. The molecular families that regulate NK cell activation are described below: in mice, they are C-type lectins and one of them, known as **NK1.1 (CD161)**, is widely used as a marker for NK cells. Unlike T cells, NK cells do not have variable receptors for antigen and instead of maturing in the thymus they migrate directly to the bloodstream, where they are believed to provide an innate immune defense, known as **natural cytotoxicity**, which may be particularly important against viruses and tumors. They can also be recruited later in an immune response by the antibodies of adaptive immunity to kill antibody-coated cells by antibody-dependent cellular cytotoxicity (ADCC) (see section 6-3).

NK cells were discovered for their ability to kill some tumor cells *in vitro* and were thought to mediate natural immunity to tumors, a possibility in which there is still considerable interest. It is now clear, however, that NK cells are important in immunity to infections, although it is only in viral infections that they have been unequivocally implicated in protection from natural disease. In humans, a rare immunodeficiency disease results in a selective lack of NK cells; affected individuals, despite having normal B and T cells, suffer severe viral infections, particularly by **herpesviruses**. Herpesviruses are large DNA viruses that have acquired a wide array of mechanisms for avoiding recognition by CD8 T cells and are a common cause of latent infections that can be reactivated when immunity wanes (see Chapter 10). In some cases the receptors of NK cells may have evolved to recognize cells infected with this class of pathogenic viruses.

NK cells are recruited and activated by cytokines released by other cells

NK cells are found in the circulation and in blood-filtering organs, including the spleen, liver, lungs and bone marrow, but usually are not present in lymph nodes and do not enter other tissues unless recruited during an inflammatory response. During an infection, NK cells are recruited and activated by cytokines and chemokines secreted by myeloid cells or infected non-hematopoietic tissues at the site of infection. The type 1 interferons (IFN-α/β), interleukin-15 (IL-15) and interleukin-12 (IL-12) are particularly important. The type 1 interferons, which are produced by many cell types in response to viral infection and have direct antiviral activity (see sections 3-16 and 3-17), are potent stimulators of NK cell activation, augmenting both cell-mediated cytotoxicity and cytokine production. IL-15 is required for the development of NK cells from bone marrow progenitors and also induces the activation and promotes the survival of mature NK cells. IL-15 and IL-12 (produced by activated dendritic cells and macrophages) together induce substantial IFN-γ secretion by NK cells. Other proinflammatory cytokines produced at the site of infection by macrophages and stromal cells (for example IL-1, IL-6, IL-18, TNF, and others) may serve to amplify an NK cell-mediated response. In turn, secretion of IFN-γ by NK cells serves to activate macrophages and to promote an antiviral state in the local site and, after the recruitment of NK cells into lymph nodes in response to inflammation, can influence the differentiation of T cells toward a T$_H$1 phenotype. The activation and actions of NK cells are schematically summarized in Figure 8-3. IFN-γ secretion by NK cells is negatively regulated by TGF-β, which can be activated from a latent state in the extracellular matrix in response to inflammation.

Definitions

CD161: see **NK1.1.**

herpesvirus: any member of a family of more than 100 large, enveloped, double-stranded DNA viruses that cause widespread infections that are commonly asymptomatic in healthy individuals.

KIR family: family of receptors belonging to the immunoglobulin superfamily that are expressed on human NK cells. Most are inhibitory and recognize classical MHC class I molecules, thereby preventing activation of NK cytotoxicity by normal cells.

Ly49 family: family of C-type lectin receptors expressed on mouse NK cells. Most are inhibitory and recognize classical MHC class I molecules, thereby preventing activation of NK cytotoxicity by normal cells.

natural cytotoxicity: the process whereby NK cells kill virus-infected cells and some tumors without requiring prior immunization of the host.

natural killer (NK) cells: cytotoxic lymphocytes lacking

NK cells express both activating and inhibitory receptors

Cytotoxicity to virus-infected or tumor cells requires direct contact of the attacking NK cell with its target and demands that NK cells have a reliable mechanism for focusing their cytotoxic machinery on abnormal cells and sparing normal healthy ones. This dual requirement is met by two general sets of receptors, one inhibitory and one activating. The inhibitory receptors suppress killing by binding to self MHC class I molecules or other constitutive surface molecules that are expressed by normal healthy cells. In contrast, the activating receptors promote killing by binding to surface receptors that are induced by cell damage or that are encoded by infecting viruses. Many of these receptors are encoded by genes that belong to one or other of two major gene families (see below) and are clustered in the genome. Activation of killing by NK cells depends upon the balance between the activating and inhibitory signals delivered by these receptors. In this way, NK cells can kill tumor or virus-infected cells that have down-regulated MHC class I from their surface ("missing-self") or cells that have upregulated damage-induced targets of activating receptors ("induced-self" or virus-encoded antigens) (Figure 8-4).

The major activating and inhibitory receptors of NK cells belong either to the immunoglobulin superfamily (see section 2-2) or to the C-type lectin family (see section 2-5). The main families of receptors are described in the next section. Here we outline the features of two major families, the **KIR family** of human NK cell receptors and the functionally equivalent **Ly49 family** of murine NK receptors. Signaling by the inhibitory receptors is mediated by cytoplasmic ITIMs, and by the activating receptors by their association with membrane adaptor proteins (Figure 8-5).

NK cells express an array of polygenic and polymorphic receptors specific for host MHC class I

The inhibitory KIR and Ly49 family receptors recognize different variants of polymorphic MHC class I molecules. Since virtually all cells express MHC class I molecules, inhibition of killing by these receptors is one of the most important mechanisms for preventing the destruction of normal cells by NK cells. Remarkably, inhibitory KIR and Ly49 molecules share almost equivalent expression, ligand-binding and regulatory properties, although structurally they are members of two different superfamilies, immunoglobulin for human receptors, and C-type lectin for mouse receptors, and are thus an example of convergent evolution.

Although most cells of the body express all of the inherited MHC class I variants, each NK cell expresses only a subset of the MHC class I-binding inhibitory receptors, and these subsets vary on different cells. In general, individual NK cells express one to five different KIR or Ly49 receptors, creating a repertoire of different MHC-binding specificities. The precise mechanisms that regulate this diversity remain unknown, but they develop on pre-NK cells during development in the bone marrow. Surprisingly, up to 10% of NK cells may lack the expression of any KIR or Ly49 inhibitory receptors that bind host MHC class I molecules, although these cells are functionally tolerized and toxicity to the host is avoided. Additional receptors, as discussed in the next section, also restrain NK cell activation. Importantly, the activation threshold of NK cells is finely tuned by the levels of host MHC class I molecules, thus allowing NK cells to respond rapidly to viruses that block host MHC class I expression and thereby avoid activating cytotoxic CD8 T cells, which comprise the major cellular adaptive immune response to viruses.

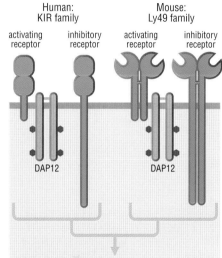

polygenic and polymorphic,
expressed on overlapping NK subsets,
inhibitory and activating receptors coexpressed,
receptors bind polymorphic MHC

Figure 8-5 The KIR and Ly49 families of receptor molecules Whereas the KIR family is immunoglobulin-related, the Ly49 molecules are C-type lectins. Nonetheless, the two families share all their important functional properties. Each family contains both activating and inhibitory receptors: the activating receptors signal through the associated DAP12 molecule (and DAP10 in the mouse). The KIR molecules shown here have two immunoglobulin domains; other KIR family molecules (not shown) have three.

antigen-specific receptors but with invariant receptors that detect infected cells and some tumor cells and activate their destruction.

NK1.1: a member of the polymorphic NKR-P1 family of C-type lectins encoded in the NK complex on mouse chromosome 6. In the mouse, both activating and inhibitory members are present, and bind to C-type lectin ligands also encoded in the NK complex. In humans, a single NKR-P1 inhibitory receptor binds a homologous C-type lectin.

References

Biron, C.A. et al.: **Natural killer cells in antiviral defense: function and regulation by innate cytokines.** *Annu. Rev. Immunol.* 1999, **17**:189–220.

Cerwenka, A. and Lanier, L.L.: **NK cells, viruses and cancer.** *Nat. Rev. Immunol.* 2001, **1**:41–49.

Di Santo, J.P.: **Natural killer cell developmental pathways: a question of balance.** *Annu. Rev. Immunol.* 2006, **24**:257–286.

Kim, S. et al.: **Licensing of natural killer cells by host major histocompatibility complex class I molecules.** *Nature* 2005, **436**:709–713.

Raulet, D.H. and Vance, R.E.: **Self-tolerance of natural killer cells.** *Nat. Rev. Immunol.* 2006, **6**:520–531.

Activating NK cell receptors recognize a variety of ligands

The main activating receptors of NK cells are thought to recognize either damage-induced self molecules or viral products. One important group of damage-induced ligands is composed of a subset of non-classical MHC class I molecules (see section 4-4) that are recognized by a C-type lectin receptor known as **NKG2D**, that is expressed on NK cells. Expression of these ligands is regulated so that a variety of cellular stress pathways, particularly DNA breakage, induce their surface expression. Certain members of the KIR and Ly49 families, which make up the main inhibitory receptors on human and mouse NK cells, are also activating. Some of these bind MHC class I with lower affinity than their inhibitory counterparts and may be affected by alterations in the peptides displayed in the MHC binding groove that occur with infection. Others, however, are likely to bind directly to viral gene products that mimic MHC molecules. These mimic molecules are believed to have evolved to bind inhibitory receptors on the NK cell, thus preventing its activation, but to have been countered by mutation of the inhibitory receptor into an activating one. Another class of activating receptors, NKp30, NKp44 and NKp46, particularly the latter, has been implicated in the recognition of virus hemagglutinins, and in protection against influenza. As noted previously, CD16, the low-affinity FcγRIII receptor, is activated by antibody bound to target cells to mediate ADCC. Finally, activating receptors such as CD94/NKG2C, CD94/NKG2E and CD244, recognize more widely expressed ligands whose expression is likely to be altered by damage or stress.

Activating NK cell receptors share conserved signaling pathways

As in other signal transduction pathways, activating receptors couple ligand binding outside the cells via adaptor proteins to robust signaling cascades within the cell (Figure 8-6). As in the T cell receptor, charged amino acids in the transmembrane domains of the receptors assemble with the adaptors during their transport to the cell surface. The activating receptors of the KIR and Ly49 families signal through the adaptor protein DAP12, which has cytoplasmic ITAMs. CD16, NKp30, NKp44 and NKp46 also signal through ITAM-bearing adaptor proteins for signaling. ITAM-based receptors activate NK cells through the Syk or ZAP-70 tyrosine kinase pathways. Since mice rendered Syk- and ZAP-70-deficient by gene disruption still have NK cells with certain functions intact, it is clear that other pathways must exist. One of these operates through the activating receptor NKG2D. NKG2D signals through DAP12 and a second adaptor protein, DAP10, in mouse NK cells, but only through DAP10 in human NK cells.

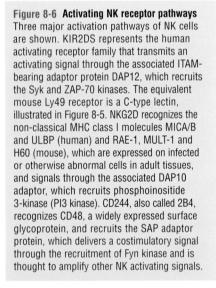

Figure 8-6 Activating NK receptor pathways Three major activation pathways of NK cells are shown. KIR2DS represents the human activating receptor family that transmits an activating signal through the associated ITAM-bearing adaptor protein DAP12, which recruits the Syk and ZAP-70 kinases. The equivalent mouse Ly49 receptor is a C-type lectin, illustrated in Figure 8-5. NKG2D recognizes the non-classical MHC class I molecules MICA/B and ULBP (human) and RAE-1, MULT-1 and H60 (mouse), which are expressed on infected or otherwise abnormal cells in adult tissues, and signals through the associated DAP10 adaptor, which recruits phosphoinositide 3-kinase (PI3 kinase). CD244, also called 2B4, recognizes CD48, a widely expressed surface glycoprotein, and recruits the SAP adaptor protein, which delivers a costimulatory signal through the recruitment of Fyn kinase and is thought to amplify other NK activating signals.

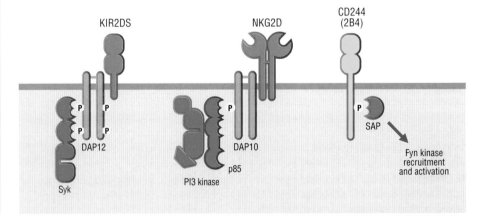

Definitions

NKG2D: monogenic activating receptor of the C-type lectin receptor family that is expressed on NK cells as well as on some γδ and effector CD8 T cells. NKG2D signals through noncovalent association with DAP10 (and, in the mouse, with DAP12), a transmembrane signaling molecule lacking an ITAM motif but containing a cytoplasmic phosphorylatable YXXM motif shared by costimulatory molecules, such as CD28 and ICOS.

References

Gazit, R. *et al.*: **Lethal influenza infection in the absence of the natural killer receptor gene *Ncr1*.** *Nat. Immunol.* 2006, **7**:517–523.

Lanier, L.L.: **NK cell recognition.** *Annu. Rev. Immunol.* 2005, **23**:225–274.

Tangye, S.G. *et al.*: **Functional requirement for SAP in 2B4-mediated activation of human natural killer cells as revealed in the X-linked lymphoproliferative syndrome.** *J. Immunol.* 2000, **165**:2932–2936.

Vilches, C. and Parham, P.: **KIR: diverse, rapidly evolving receptors of innate and adaptive immunity.** *Annu. Rev. Immunol.* 2002, **20**:217–251.

NKG2D is expressed by all NK cells, as well as some γδ T cells and CD8 T cells. It binds to the non-classical MHC class I RAE-1 proteins H-60 and MULT1 (murine ULBP-like transcript-1) in mice, and MICA/B and ULBP proteins (orthologs of the mouse RAE-1 molecules) in humans. NKG2D binds with high affinity to these ligands and transmits signals through DAP10, with one NKG2D homodimer assembling with four chains of DAP10. DAP10 contains a YXXM motif in the cytoplasmic domain that recruits the p85 subunit of phosphoinositide 3-kinase (PI3K) and activates the enzyme.

A third activation pathway of NK cells is thought to be important for amplifying NK cell-mediated cytolytic activity and cytokine secretion in response to other activating signals. The best-known receptor for this pathway is CD244 (also called 2B4), one of a family of six similar proteins. CD244 is expressed on all NK cells and binds to CD48, a broadly distributed cell surface glycoprotein of the immunoglobulin superfamily (section 2-3). A motif in the cytoplasmic domain of CD244 (and related receptors), TXYXXV/I, recruits members of a family of SH2-containing adaptor proteins, the best known of which is SH2D1A (also known as SAP) expressed from the X chromosome. Details of SAP-mediated signaling are not fully understood, but loss-of-function mutations in the SAP gene in humans cause an X-linked lymphoproliferative disorder and patients frequently succumb to fatal Epstein–Barr virus (EBV) infection or develop EBV-induced lymphomas.

Additional inhibitory signals regulate NK cell activation

As discussed previously, each NK cell expresses only a subset of all the inhibitory receptors, and therefore not all NK cells recognize the polymorphic MHC class I variants on a given cell. To compensate for this, a more promiscuous system for recognition of host MHC class I alleles by NK cells is mediated by the inhibitory CD94/NKG2A receptor. CD94/NKG2A recognizes the relatively non-polymorphic non-classical HLA-E molecule or its mouse equivalent, Qa-1, which binds peptides from the leader sequences that target classical MHC class I α chains to the plasma membrane. Disruption of MHC class I expression thus deprives HLA-E (or Qa-1) of the peptides necessary for stability and export for expression at the cell surface, removing the ligand for CD94/NKG2A. Like KIR and Ly49 members, CD94/NKG2A signals through an ITIM motif in its cytoplasmic tail to recruit cellular tyrosine phosphatases, typically SHP-1 or SHP-2, to dampen NK cell activation (Figure 8-7). A more ubiquitous inhibitory signal is represented by the KLRG1 receptor on mouse NK cells, which is engaged by cadherins, molecules frequently downregulated during cell transformation that accompanies oncogenesis. Thus the magnitude of the NK cell response is determined by the dynamic balance of activating and inhibitory signals. The major families of these receptors are summarized in Figure 8-8.

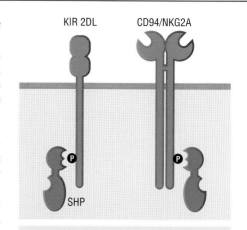

Figure 8-7 Inhibitory NK cell receptor pathways KIR2DL represents the human ITIM-bearing receptor family that recognizes polymorphic MHC class I molecules and transmits an inhibitory signal through tyrosine phosphatases. The equivalent mouse Ly49 receptor is a C-type lectin, illustrated in Figure 8-5. CD94/NKG2A of both human and mouse recognizes the non-classical MHC molecules HLA-E (human) and Qa-1 (mouse), which are expressed when these molecules bind leader peptides from classical MHC class I molecules; these molecules also signal through tyrosine phosphatases that bind to ITIMs.

Major Receptor Families of Human and Mouse NK Cells

Receptor	Species	Structural	Polymorphic	Polygenic	Activating	Inhibitory	Ligand family
KIR[*1]	human	Ig	yes	yes	yes	yes	MHC class I (?MHC-like viral products)
Ly49	mouse	C-type	yes	yes	yes	yes	MHC class I, MHC-like viral products
CD94/NKG2A	human/mouse	C-type	no	no	no	yes	HLA-E (human), Qa-1b (mouse)
CD94/NKG2C and NKG2E	human	C-type	no	no	yes	no	HLA-E with stress or pathogen-derived peptides[*2]
NKG2D	human/mouse	C-type	no	no	yes	no	MICA/B, ULBPs (human), RAE-1, H60, MULT-1 (mouse)
CD16	human/mouse	Ig	yes	no	yes	no	IgG complexes
CD244	human/mouse	Ig	no	no	yes	yes[*3]	CD48
LILR	human	Ig	no	yes	no	yes	MHC class I
NKp30, NKp44, NKp46	human/mouse[*4]	Ig	no	no	yes	no	influenza hemagglutinins?
CD226	human	Ig	no	no	yes	no	CD155, CD112
NKR-P1	human[*5]/mouse	C-type	yes	yes	A/C/F	D	C-type lectins
KLRG1	mouse	C-type	no	no	no	yes	cadherins

[*1] KIR: KIRs occur in two subsets containing either two or three extracellular immunoglobulin domains. The cytoplasmic tails can be either short, abbreviated S, which usually occurs in the activating receptors that use adaptors, or long, abbreviated L, which usually occurs in inhibitory receptors. Thus, a given KIR is represented as KIR3DL.

[*2] Peptides from cellular stress proteins and from some pathogens, including *Mycobacterium tuberculosis* and some viruses, have been identified.

[*3] CD244 (also known as 2B4), can be inhibitory under some circumstances, perhaps related to its ability to recruit different adaptors.

[*4] Mouse NK cells expresses only NKp46 in this family.

[*5] Human NK cells express only a single inhibitory NKR-P1 family member.

Figure 8-8 Table of activating and inhibitory receptors of NK cells C-type: C-type lectin superfamily; Ig: immunoglobulin superfamily; LILR: leukocyte immunoglobulin-like receptors.

NKT cells are specialized for detection of glycolipid antigens

Natural killer T (NKT) cells derive their name from their co-expression of NK cell receptors and T cell receptors. Many conventional lymphocytes acquire expression of various NK cell-associated receptors during maturation to tissue effector cells, but do not express these receptors when naïve. In contrast to conventional T cells, canonical NKT cells are positively selected in the thymus on CD1d, a β2-microglobulin-associated non-classical MHC I molecule implicated in the presentation of glycolipids (see section 4-4). The study of canonical NKT cells in human and mouse was greatly aided by their fortuitous identification after staining with CD1d tetramers loaded with the marine sponge glycosphingolipid α-galactosylceramide (αGalCer), which presumably mimics the endogenous natural ligand. Less well characterized are small populations of non-canonical NKT cells that are CD1d-restricted but not reactive to CD1d/αGalCer tetramers, and non-canonical, non-CD1d-restricted NKT cells. Unless otherwise noted, the term NKT cells will be used in reference to canonical NKT cells recognized by CD1d/αGalCer tetramers.

NKT cells are selected in the thymus and populate the liver, spleen and bone marrow

NKT cells derive from bone marrow-derived conventional T cell precursors that enter the thymus but undergo selection on CD1d presented by CD4 CD8 (double-positive) cortical thymocytes. The endogenous selecting ligand is thought to be **isoglobotrihexosylceramide (iGb3)**, a lysosomal degradation product of glycosylated membrane lipids, the **glycosphingolipids**. Thus the development of NKT cells requires transport of CD1d to lysosomes, lysosomal cathepsins, lipid transfer proteins (saposins) and glycosphingolipid-degrading enzymes (hexosaminidase) (Figure 8-9). This selects for the rare cells that randomly express a single conserved Vα-Jα combination (Vα14-Jα18 in mouse, homologous to Vα24-Jα18 in human), which pairs with any of a limited number of conserved Vβ chains (predominantly Vβ8.2 in mouse, homologous to Vβ11 in human). CD4 CD8 double-positive cells expressing the canonical receptor undergo intra-thymic expansion as CD4, NK1.1-negative precursors. After emigration to tissues, canonical NKT cells typically express CD4 and NK1.1 (CD161), as well as CD94/NKG2A and various receptors of the Ly49 (mouse) or KIR (human) families.

In mice, NKT cells are absent at birth but accumulate to reach adult levels in tissues by 5–6 weeks. NKT cells comprise up to 1 million cells in each of liver, spleen, bone marrow and thymus in inbred mice, and make up to 30% of total liver lymphocytes. Liver accumulation of NKT cells reflects interactions between CXCR6 on NKT cells and CXCL16, which is expressed on hepatic sinusoidal endothelial cells. The development, distribution and numbers of NKT cells are unaffected by germ-free conditions. The maintenance of peripheral NKT cells does not require CD1d expression, just as the maintenance of conventional memory/effector T cells does not

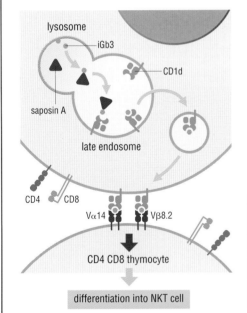

Figure 8-9 Degradation of lysosomal membrane glycosphingolipids generates the endogenous NKT ligand for presentation by CD1d Degradation of membrane isoglobo-glycosphingolipids in the lysosome of CD4 CD8 double-positive cortical thymocytes by β-hexosaminidase generates iGb3 (isoglobotrihexosylceramide) by removal of the terminal *N*-acetylgalactosamine from iGb4. Saposin A extracts iGb3 from the membrane, and assists loading onto CD1d molecules in late endosomal compartments. iGb3-loaded CD1d is shuttled to the membrane, where interactions with CD4 CD8 thymocytes that by chance have made the canonical Vα14/Vβ8.2 rearrangement directs them into the NKT lineage.

Definitions

α-proteobacteria: large, metabolically diverse and widely dispersed, evolutionarily ancient group of Gram-negative bacteria that express an altered lipid A and fatty-acid side-chain pattern, resulting in an unusual LPS that can be several thousand-fold less stimulatory for TLRs than LPS derived from typical Gram-negative bacteria; some members of the family, such as *Sphingomonas*, totally lack LPS and utilize instead a family of **glycosphingolipids** in the outer membrane; α-proteobacteria include a large number of organisms that establish chronic infections in plants (*Agrobacterium*, *Rhizobia*), insects (*Wolbachia*) and vertebrates, including humans (*Bartonella*, *Brucella*, *Rickettsia*), and probably contains the mitochondrial ancestor.

glycosphingolipids: amphiphilic plasma membrane molecules consisting of one or more sugar residues linked with sphingosine to long-chain fatty-acid side chains.

isoglobotrihexosylceramide (iGb3): lysosomal degradation product of **glycosphingolipids**. Thought to be the CD1d ligand on which NKT cells are selected in the thymus.

natural killer T (NKT) cells: T lymphocytes that express both αβ T cell receptors and NK cell receptors and recognize glycolipids presented on CD1d non-classical MHC class I molecules.

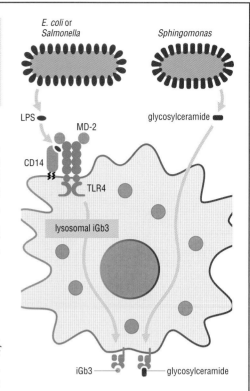

Figure 8-10 NKT cells are activated by Gram-negative bacteria through indirect and direct pathways Typical Gram-negative bacteria activate TLR4 via LPS, which stimulates lysosomal glycosphingolipid degradation, resulting in the generation of iGb3 and loading onto CD1d for expression at the surface. Some α-proteobacteria, such as *Sphingomonas*, do not express LPS but instead utilize glycosphingolipids, some of which can be loaded directly onto CD1d. Both complexes generate activating ligands for the canonical NKT cell receptor.

require classical MHC molecules, but is dependent on IL-15 and signals from the IL-15 receptor. Activated dendritic cells and macrophages express CD1d, as do hepatocytes and marginal zone B cells. Whereas tissue NKT cells typically express NK1.1, recent thymic NKT emigrants in blood or spleen do not, and nor does a small population of canonical NKT cells in Peyer's patches that uniquely express CD49d (α4 integrin). Mice lacking CD1d or Jα18 are NKT cell deficient.

Less is known about human NKT cells. Although they express homologous highly conserved TCR, canonical NKT cells are less prevalent than in mice, and CD8 expression is common. Even less is known regarding non-canonical NKT cells.

NKT cells release cytokines rapidly after activation and can mediate defense against bacteria not detected by Toll-like receptors

Tissue NKT cells have an activated surface phenotype typical of effector/memory T cells. Isolated cells are inherently autoreactive on CD1d *ex vivo*, suggesting that combinations of inhibitory and activating NK cell-like receptors may regulate their activation *in vivo*. Stimulation by injection of αGal/Cer or anti-CD3 antibody *in vivo* leads to a rapid secretion of IL-4 and IFN-γ over 90 minutes that does not occur in NKT cell-deficient mice. The capacity to secrete both cytokines is acquired during thymic differentiation and does not require priming, in contrast to conventional T cells. Activation through the TCR results in profound downregulation of the receptor, and strong signals can cause activation-induced cell death. Re-expression of receptors and/or repopulation from bone marrow-derived precursors may take days to restore baseline numbers, suggesting a process that regulates the duration of their effector function. NKT cells are particularly potent in activating NK cells by the direct and indirect effects of IFN-γ. Human CD4 NKT cells and double-negative NKT cells are enriched for IL-4-producing and IFN-γ-producing capacities, respectively.

In macrophages and dendritic cells, LPS-stimulated TLR4 signals induce the degradation of lysosomal membrane glycosphingolipids, generating endogenous iGb3, which is loaded onto CD1d, thus generating activating ligands for NKT cells. Intriguingly, some α-**proteobacteria**, a family of Gram-negative bacteria whose cell walls do not contain canonical LPS, are instead characterized by a family of cell-wall glycosylceramides, some of which powerfully activate mouse and human NKT cells upon direct presentation of these bacterial glycosphingolipids by CD1d. NKT cell-deficient mice are unable to clear infections by such bacteria, suggesting that these cells may have evolved to defend against microbes that evade TLR-mediated recognition (Figure 8-10). NKT cells have been implicated in tumor defense, perhaps through the activation of NK cell immunity. An inverse relationship has been noted in the numbers of NKT cells and autoimmune diseases such as diabetes. In the diabetes-prone *NOD mouse*, NKT cell function is compromised, and diabetes is delayed when it is restored. Research continues on the possibility that the rapid effector function of NKT cells might be therapeutically exploited in various human diseases, particularly cancer, although the relatively small numbers of NKT cells in humans may make this difficult.

References

Brigl, M. *et al.*: **Mechanism of CD1d-restricted natural killer T cells activation during microbial infection.** *Nat. Immunol.* 2003, **4**:1230–1237.

Kinjo, Y. *et al.*: **Recognition of bacterial glycosphingolipids by natural killer T cells.** *Nature* 2005, **434**:520–525.

Kronenberg, M.: **Toward an understanding of NKT cell biology: progress and paradoxes.** *Annu. Rev. Immunol.* 2005, **23**:877–900.

Mattner, J. *et al.*: **Exogenous and endogenous glycolipid antigens activate NKT cells during microbial infections.** *Nature* 2005, **434**:525–529.

Zhou, D. *et al.*: **Lysosomal glycosphingolipid recognition by NKT cells.** *Science* 2004, **306**:1786–1789.

8-4 γδ T Cells

γδ T cells express antigen receptors of limited functional diversity

γδ T cells express products of the γ and δ TCR genes on their surface rather than the α and β TCR gene products expressed by conventional CD4 and CD8 T cells. γδ T cells are thought to provide an immediate inflammatory and cytotoxic response to tissue invasion, with a later regulatory role in which they may function to limit tissue damage. They comprise only 1–5% of circulating blood lymphocytes but selectively populate epithelia, where they make up over 50% of total T cells. Most peripheral blood γδ T cells are CD4 CD8 double-negative; γδ intestinal epithelial lymphocytes typically express CD8αα homodimers rather than the CD8αβ heterodimers expressed on conventional CD8 T cells. Intestinal intraepithelial cells, or IEL, are discussed in the next section. Although γδ T cells, like αβ T cells, are theoretically capable of expressing a large combinatorial repertoire of TCRs, their functional γδ TCR repertoire is in practice highly constrained, for reasons that are not entirely understood. They recognize various microbial and damage-induced molecules, including non-classical MHC class I molecules.

γδ T cells develop early in ontogeny and expand as oligoclonal populations in response to diverse environmental antigens

Although derived from common bone marrow cell precursors, γδ and αβ T cells diverge early in thymic development during the immature double-negative stage. Whereas in αβ T cells TCR-β chains initially pair with the pre-TCRα chain to produce a receptor through which the cells are selected for maturation, in γδ T cells commitment and differentiation are directed by the expression of a complete γδ heterodimer with no intermediate stage. In mouse and human, the TCR-δ genes are embedded between the TCR Vα and Jα genes within the TCR-α locus (see Figure 7-2). Although successful Vα rearrangement usually results in deletion of the TCR-δ genes, this does not affect γδ T cell differentiation because Vα rearrangements generally occur late during thymic T cell differentiation, whereas Vβ, Vγ and Vδ gene rearrangements occur earlier, before such deletions have occurred.

The potential for γδ or αβ T cell fate correlates with the numbers of available gene segments. Mice and humans have relatively few Vγ (7 commonly expressed in mice, 4 in one family; 6 commonly expressed in human, 5 in one family) and Vδ genes (about 16 in mouse, 8–10 in human), whereas sheep, cattle and chickens have many more TCR γδ genes than αβ genes and have greater numbers of γδ T cells. γδ T cells depend on signals from IL-7 and the IL-7Rα chain for development.

Generation of γδ T cells precedes that of αβ T cells in the fetal thymus and, as best studied in the mouse, begins with a series of developmentally programmed waves of cells that express specific V genes (Figure 8-11). Most of these γδ TCRs are largely germ-line-encoded, reflecting rearrangements that occur before the expression of terminal deoxynucleotidyl transferase (TdT) in development. The first subsets, expressing TCRs that contain Vγ5/Vδ1 or Vγ6/Vδ1, populate epithelia of the skin or oral/uterine cavities, respectively. Vγ5 cells in the epithelial layer of skin of mice are referred to as dendritic epidermal T cells, or DETCs, because of their characteristic dendritic appendages *in situ*. Humans lack a direct equivalent of murine DETCs, and skin γδ T cells exist predominantly in the dermis rather than the epidermis.

In mice, Vγ5 and Vγ6 rearrangements cease before birth, and subsequent γδ T cells that express different V gene segments emigrate to populate the intestines, blood and lymphoid organs. In mice and humans, the circulating peripheral γδ T cell repertoire becomes focused by antigen exposure, narrowing from its most diverse at birth to a few functional specificities in maturity.

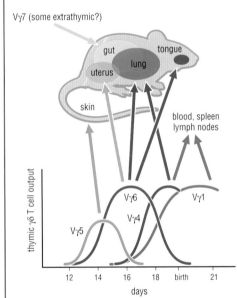

Figure 8-11 Waves of γδ T cells in mammalian development The highly stereotyped, programmed, appearance of distinct classes of γδ T cells at mucosal barriers during late embryonic and perinatal life suggests that these cells may provide protection at epithelial and systemic sites until adaptive immunity matures more fully.

Definitions

γδ T cells: T lymphocytes expressing receptors containing variable γ and δ chains rather than α and β chains, and thought to be specialized to respond to a limited range of damage-induced and environmental antigens.

phosphoantigens: small, phosphorus-containing non-peptides that collectively represent antigens capable of activating human Vγ9Vδ2 T cells. Several isoprenoid pyrophospate derivatives represent natural phospho-antigens produced by bacteria, including mycobacteria.

References

Adams, E.J. *et al.*: **Structure of a γδ T cell receptor in complex with the non-classical MHC T22.** *Science* 2005, **308**:227–231.

Allison, T.J. *et al.*: **Structure of a human γδ T-cell antigen receptor.** *Nature* 2001, **411**:820–824.

Carding, S.R. and Egan, P.J.: **γδ T cells: functional plasticity and heterogeneity.** *Nat. Rev. Immunol.* 2002, **2**:336–345.

Jameson, J. *et al.*: **A role for skin γδ T cells in wound repair.** *Science* 2002, **296**:747–749.

Prinz, I. *et al.*: **Visualization of the earliest steps of γδ T cell development in the adult thymus.** *Nat. Immunol.* 2006, **7**:995–1003.

In humans, Vγ9/Vδ2 T cells, whose formation is largely confined to the perinatal period, expand to represent the vast majority of blood γδ T cells in adults. This is thought to reflect the activation of these cells by common environmental alkylamines and by small phosphorus-containing nonpeptides known as **phosphoantigens**. The most potent activating phosphoantigens, such as isopentenyl pyrophosphate, are abundant in bacteria and mycobacteria but are also formed during host metabolism. How Vγ9/Vδ2 T cells recognize these antigens and distinguish between foreign and self ligands is not known. Human tissue γδ T cells, including intestinal γδ T cells, frequently express Vγ2/Vδ1 TCRs.

γδ T cells have immediate effector function

γδ T cells are activated by antigen presented on cell surfaces, but, unlike αβ T cells, they do not recognize peptide antigen presented by classical MHC molecules. The γδ TCR complex, unlike the αβ TCR complex, does not contain CD3δ, but rather consists of two dimers of CD3γ and CD3ε, along with TCR-ζ or FcεRIγ homodimers. Stimulation through γδ TCRs results in more robust signaling than occurs through αβ TCRs, and consequently γδ T cells can be activated in the absence of CD28 costimulation. Thus γδ T cells function in a manner more closely resembling memory/effector T cells than naïve T cells.

Although γδ T cell activation is not entirely understood, it is likely that these cells are activated *in vivo* through summation of signals generated through the TCR and activating receptors, such as the NKG2D/DAP10 pathway (see section 8-3), which recognize non-classical MHC molecules induced by damage, inflammation or cell activation. For example, some of the prominent Vγ2/Vδ1-bearing cells in human tissues can be activated by bacterial and mycobacterial antigens presented to the TCR by CD1c, in combination with NKG2D activation by the damage-induced MICA and MICB ligands. This has led to the suggestion that epithelial γδ T cells are tuned to the expression of diverse environmental antigens released from both bacterial and host cells in conjunction with the inflammation-induced expression of self ligands, which together trigger their immediate effector function. In this way, γδ T cells are activated when bacterial and host antigens reach levels that trigger a host response (Figure 8-12).

In general, activation of epithelial γδ T cells elicits proinflammatory effector function, including the production of cytokines, particularly IFN-γ, and perforin-mediated cytotoxicity, which are induced by TCR and NKG2D stimulation, respectively. The cytolytic protein granulysin is also secreted. Gene expression microarrays have revealed that these cells exist with many effector transcripts poised for rapid translation, with a number of concurrently expressed inhibitory receptors, presumably engaged to maintain a high activation threshold necessary to avoid unprovoked tissue injury.

γδ T cells may have a role in tissue repair later in the immune response

Animal studies of γδ T cell responses in infectious disease models have suggested a bipartite role for tissue and recruited cells. Early tissue γδ T cell responses peak before αβ T cell responses, and consist largely of proinflammatory cytokines and tissue growth factors that contribute to initial host defense but also tissue integrity. Conversely, a second γδ T cell response consists of a wave of recruited cells that peaks after αβ T cells enter tissue. The recruited blood γδ T cells produce regulatory cytokines, such as IL-10, and may serve to limit inflammation after the pathogenic antigen load is cleared. Deletion of the initial γδ T cell response leads to a delay in pathogen clearance, whereas deletion of the secondary γδ T cell response leads to enhanced tissue injury. A homeostatic role for γδ T cell responses is suggested by the impaired epithelial integrity and enhanced risk for epithelial tumors that occur in TCR γδ-deficient mice.

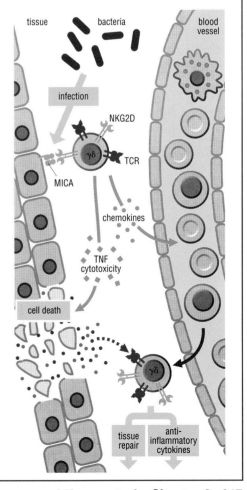

Figure 8-12 Suggested sequence of events in γδ T cell surveillance Epithelial γδ T cells are poised to react immediately to ubiquitous bacterial and host intracellular metabolites, as well as to cell damage-induced non-classical MHC molecules, such as MICA, MICB and ULBP, which are ligands for the activating receptor NKG2D. Damaged cells from microbes or host release additional intracellular products, such as low-molecular-weight phosphoantigens and heat shock proteins (red and blue dots), which are presented by tissue macrophages and dendritic cells and help to activate circulating γδ T cells that have been recruited to inflammatory sites (lower γδ T cell). After migration into tissues, circulating γδ T cells can also engage NKG2D ligands to enhance their activation (not shown), although, in general, the response of these recruited cells contributes to tissue repair and control of inflammation.

Intraepithelial lymphocytes populate the epithelia of bowel and lung

Pathologists have long noted the presence of lymphocytes resident in close proximity to the lamina propria between polarized epithelial cells of the bowel and lung. These specialized lymphocytes, designated **intraepithelial lymphocytes (IEL)**, comprise two major cell populations. The first represents conventional αβ T cells that are activated by antigen in association with classical MHC molecules in draining lymph nodes and emigrate into intestinal or respiratory epithelia, presumably awaiting a second exposure to antigen, akin to memory effector cells in other tissues. They are particularly prevalent in areas of high antigen load, such as the colon, where they are dependent on colonic bacterial colonization. They will not be further discussed here. The second IEL population is made up of TCR γδ and TCR αβ cells. TCR αβ IEL of this specialized type express CD8αα on their surface; TCR γδ IEL are CD4 CD8 double-negative or, like TCR αβ IEL, express CD8αα homodimers. These cells are not dependent on classical MHC molecules for selection or activation but instead are often activated by non-classical MHC molecules. They do not require bacterial colonization of the bowel for their localization. These specialized IELs are prevalent in small bowel mucosa, where they are best characterized, but also in lung and other polarized epithelia, where they may have a role in sustaining epithelial barriers in response to tissue injury.

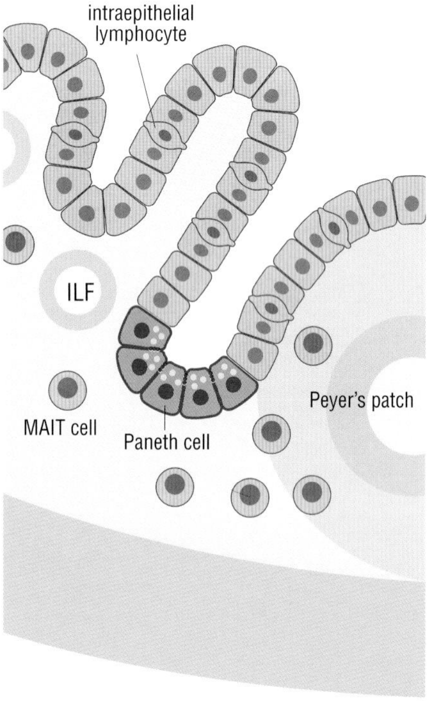

IEL populate the bowel and lung in late fetal and early postnatal life. CD8αα TCR αβ IEL arise in the thymus as a branch from thymic CD8 single-positive αβ T cells (which express CD8αβ coreceptors in the resting state). The precise lineage of γδ IEL remains unclear, as some of these cells can arise in mice in the absence of a thymus. IEL development is normal in mice genetically engineered so that normal secondary lymphoid tissues fail to develop. Despite the absence of lymphoid tissue-inducer cells, and hence intestinal Peyer's patches and isolated lymphoid follicles, in such animals, IEL are present. Development of the various IEL populations is dependent on some combination of IL-2, IL-7 and IL-15. In consequence, mice deficient in γ_c or JAK3 lack IEL, as these elements are required for signal transduction by receptors for each of these cytokines. β2-microglobulin-deficient mice are deficient in αβ TCR IEL but not γδ IEL; deficiencies in classical MHC class I (or II) genes are without effect in both populations. Interventions that impair homing to the intestine, including blockade of β7 integrins, CCR9/CCL25 or TGF-β function, or that affect hematopoietic stem cell niches, such as c-Kit mutations, selectively deplete IEL compared with conventional T cells.

In mouse and human small intestine, IEL densities approach 10–20 cells per 100 villous epithelial cells (enterocytes), with approximately 70% and 50% expressing CD8αα and γδ TCR, respectively. Homing and differentiation in these sites presumably reflect local constitutive chemokine gradients and integrin/cadherin expression. Less is known regarding the maintenance of IEL in other mucosal tissues.

Intraepithelial lymphocytes are cytotoxic

Freshly isolated IEL resemble effector CD8 T cells in function; both populations are cytotoxic by perforin/granzyme- and FasL-mediated mechanisms, and both are induced rapidly to secrete IFN-γ. CD8αα γδ and αβ IEL demonstrate similar gene expression profiles, suggesting that they have a common functional role. In addition to their activated profile, resting IEL express a number of inhibitory receptors, such as PD-1 and Ly49 family members. Tonic inhibition may serve to maintain cells in a state poised for rapid activation in response to epithelial injury. Epithelial activation or injury results in the expression on epithelia of costimulatory ligands

Figure 8-13 Small-intestine T cells
Gut-associated T cells include individual IEL positioned outside the basement membrane of polarized epithelia, as well as cells scattered in the lamina propria (including MAIT cells) or organized into Peyer's patches and isolated lymphoid follicles (ILFs). Mutations that ablate fetal CD4 lymphoid inducer cells lead to a loss of Peyer's patches and ILFs but have no effect on IEL.

Definitions

H2-M3 restricted T cells: T lymphocytes recognizing the non-classical MHC class I molecule H2-M3.

IEL: see **intraepithelial lymphocytes**.

intraepithelial lymphocytes (IELs): T lymphocytes localized between the epithelial cells in gut and lung mucosa. Some are conventional memory T cells; others are specialized cells some of which express γδ receptors instead of αβ receptors for antigen and recognize non-classical MHC molecules.

MAIT cells: see **mucosal-associated invariant T (MAIT) cells**.

mucosal-associated invariant T (MAIT) cells: T lymphocytes of unknown function that are found in mucosa and express invariant receptors recognizing the non-classical MHC class I molecule MR1.

that release IEL effector function. Induced epithelial ligands include non-classical MHC I molecules, such as TL and MICA/MICB/ULBP proteins, which serve as ligands for CD8αα and NKG2D, respectively, and which, when stimulated, lower the threshold for cell activation. Human MICA/MICB not only serves as a ligand for NKG2D, but may also bind directly to the Vδ1-containing γδ TCR expressed on 70–90% of human IEL. The Rae-1 and H60 proteins may subserve similar roles in the mouse. γδ TCR activation leads to secretion of chemokines and enhances expression of activating receptors, such as NKG2D, serving to amplify the response.

Infection of mice genetically deficient in IEL typically leads to inability to restrain the local entry and spread of pathogens. As noted above, epithelial injury may also be increased in the absence of recruitment of systemic γδ T cells that may work to dampen the initial inflammatory response. Human *celiac disease*, an intestinal inflammatory response to gluten proteins in wheat and other grains, is characterized by massive infiltration of γδ IEL which are thought to contribute to the destruction of the intestinal epithelium. We discuss celiac disease in Chapter 13.

Additional specialized lymphocyte populations exist that recognize non-classical MHC class I molecules that may present microbial components

In addition to those discussed in this and the preceding sections, other T cells in mouse and humans, some relatively rare, are activated by various non-classical MHC molecules. **Mucosal-associated invariant T (MAIT) cells**, which express canonical TCRs in mice and humans, are selected by the β2-microglobulin-dependent, monomorphic MHC I-like molecule MR1. Structural studies suggest that MR1 presents a ligand necessary for the activation of MAIT cells, possibly related to the requirement for ligand in stabilizing the transport and expression of MR1 on the cell surface. MAIT cells depend on intestinal commensal flora and activated B cells (which express MR1) for their expansion in the lamina propria. Mice deficient in MAIT cells have elevated intestinal IgA levels, suggesting that these cells have a regulatory role. Although the precise role of MAIT cells is unknown, the conserved MHC–ligand–TCR recognition preserved from mouse to humans suggests an evolutionarily significant biologic function. Intriguingly, at least in the mouse, several of these unusual lymphocyte populations, like invariant NKT cells, are selected in the thymus by ligands presented on hematopoietic cells rather than on stromal cells, which present ligands involved in selecting conventional T cells. The localization of IEL and MAIT cells in intestinal epithelium is illustrated in Figure 8-13.

H2-M3, a non-classic MHC I molecule, is present in only some strains of inbred mice. H2-M3 presents peptides with *N*-formylated methionine at the amino termini in its large, hydrophobic, ligand-binding pocket (see section 4-4). Formylated peptides from mitochondrial proteins have been implicated in thymic selection of T cells specifically recognizing H2-M3, or **H2-M3-restricted T cells**. H2-M3 is retained intracellularly in macrophages but rapidly mobilized to the surface when it binds formylated peptides. *Listeria* infection of macrophages leads to the production of H2-M3 binding peptides, including a dominant hexamer derived from the extracellular domain of a transmembrane protein designated lemA (*Listeria* epitope with M3). After infection of C57BL/6 mice with *Listeria*, lemA-specific, H2-M3-restricted CD8 T cells show earlier activation and effector function than conventional CD8 T cells, which recognize peptides bound to classical MHC class I molecules (Figure 8-14). Like their conventional counterparts, H2-M3-restricted CD8 T cells are cytolytic against *Listeria*-pulsed targets and secrete proinflammatory cytokines, particularly IFN-γ.

(a) primary infection

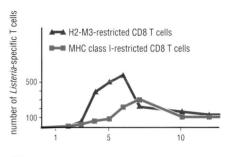

(b) secondary infection

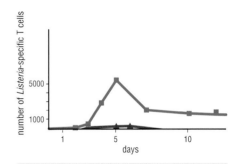

Figure 8-14 Activation of H2-M3-restricted CD8 T cells after *Listeria* infection of C57BL/6 mice Activation kinetics of conventional CD8 cytotoxic T cells and H2-M3-restricted CD8 cytotoxic T cells after primary and secondary infection with *Listeria*. In the secondary infection, the response of H2-M3-restricted CD8 T cells is similar in magnitude to that in the primary response (note different scales), whereas the response of conventional αβ CD8 T cells exhibits immunological memory and is much stronger in response to the second infection as a result of the presence of large numbers of memory CD8 T cells. Adapted from Kerksiek, K.M. *et al.*: *J. Exp. Med.* 1999, **190**:195–204.

References

Busch, D.H. *et al.*: **Processing of *Listeria monocytogenes* antigens and the *in vivo* T cell response to bacterial infection.** *Immunol. Rev.* 1999, **172**:163–169.

Eberl, G. and Littman, D.R.: **Thymic origin of intestinal αβ T cells revealed by fate mapping of RORγt+ cells.** *Science* 2004, **305**:248–251.

Hayday, A. *et al.*: **Intraepithelial lymphocytes: exploring the Third Way in immunology.** *Nat. Immunol.* 2001, **2**:997–1003.

Kerksiek, K.M. *et al.*: **H2-M3–restricted T cells in bacterial infection: rapid primary but diminished memory responses.** *J. Exp. Med.* 1999, **190**:195–204.

Treiner, E. *et al.*: **Selection of evolutionarily conserved mucosal-associated invariant T cells by MR1.** *Nature* 2003, **422**:164–169.

Urdahl, K.B. *et al.*: **Positive selection of MHC class Ib-restricted CD8+ T cells on hematopoietic cells.** *Nat. Immunol.* 2002, **3**:772–779.

B1 cells produce natural antibody, which is present before infection, and T-independent antibody that protects body cavities

The mammalian immune system contains three types of mature B cells, each with specialized functions: recirculating follicular B cells, which are the majority population responsible for adaptive germinal center antibody responses and are discussed in Chapter 6; marginal zone B cells, which reside in the spleen and rapidly make antibody against bloodborne infections and are discussed in the next section; and **B1 cells**, which secrete antibody before exposure to antigen. B1 cells thus behave like a component of the innate immune system, and the antibody they produce, which is of limited diversity, is known as **natural antibody**. These antibodies provide a measure of early protection against infections before the production of higher-affinity antibodies, which are often necessary for full protection (Figure 8-15). In response to infection, B1 cells can also make rapid T cell-independent antibody responses and have an important role in immune defense against infectious agents that invade the body cavities around the lungs and gut.

Natural antibody is synthesized by B1 cells from a preprogrammed repertoire of immunoglobulin genes

Consistent with their role in producing antibody before infection, B1 cells develop from their hematopoietic precursors primarily during fetal and early postnatal life, after which their numbers are maintained by self-renewal of mature B1 cells. B cell precursors at this time are committed to producing B1 cells, the development of which differs from that of other B cells in several key ways. One key difference is that these precursors produce antibody of much less diversity than that of other B cells, although not restricted to the extremes of some of the T cell populations described in earlier sections of this chapter. This reduced diversity, particularly of immunoglobulin heavy chains, derives from two factors: use of only the most D-proximal V_H gene segments and an almost complete lack of junctional diversity. The former is thought to reflect limited accessibility to the RAG-1/RAG-2 recombinase of the more distal V_H gene segments in the B1 precursors, whereas the latter results in large part from a failure of these cells to express terminal deoxynucleotidyl transferase, the enzyme responsible for N-region addition (see section 7-2). These two features of V(D)J recombination at the fetal and early postnatal period limit the diversity of the B1 cell immunoglobulin repertoire to specificities hard-wired by the evolution of the D-proximal V_H genes, giving natural antibody a limited but useful diversity. Indeed, these natural antibodies are important in immune defense against a variety of common pathogens, including *Streptococcus pneumoniae* and influenza virus.

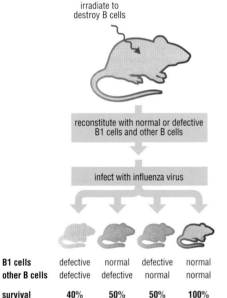

irradiate to
destroy B cells

reconstitute with normal or defective
B1 cells and other B cells

infect with influenza virus

B1 cells	defective	normal	defective	normal
other B cells	defective	normal	normal	normal
survival	40%	50%	50%	100%

Figure 8-15 Experiment examining the role of natural antibody in defense against influenza virus infection in mice Mice were lethally irradiated and reconstituted with adult bone marrow, as a source of hematopoietic cells including follicular and marginal zone B cells but not B1 cells, and with peritoneal B cells, which reconstitute B1 cells because these cells can self-renew as mature B cells. Mice were reconstituted in various combinations with B1 cells or bone marrow either from wild-type mice or from mice genetically mutated so that secretory IgM could not be produced but membrane IgM and secretory IgG, IgA and IgE could all be produced (referred to in the figure as defective). The resulting mice were then infected with a dose of influenza virus that is not lethal for wild-type mice. Mice either lacking natural IgM (produced only by B1 cells) or lacking antigen-induced IgM (coming from follicular and/or marginal zone B cells) exhibited increased susceptibility to influenza virus, demonstrating the protective and complementary nature of both types of IgM. Data from Baumgarth, N. *et al.*: *J. Exp. Med.* 2000, **192**:271–280.

Definitions

B1 cells: B cells generated early in life and responsible for most **natural antibody** and some T cell-independent antibody production.

natural antibody: antibody, primarily IgM and IgA, from individuals previously unexposed to antigen, that is produced mainly by **B1 cells** and binds to microbes or virus particles and participates in host defense.

B1 cells also participate in first-line IgM and IgA responses to bacteria in the peritoneal and pleural cavities

In addition to producing natural antibody, B1 cells also make IgM, IgA, and to a smaller extent IgG3 (in mouse) antibody responses to repeating epitopes of bacterial cell wall carbohydrates and virus particles. These antibody responses do not require helper T cells and are of the T-independent type (see section 6-9). B1 cells also respond vigorously to TLR ligands, which are thought to promote antibody responses by these B cells. B1 cells are localized primarily to the peritoneal and pleural cavities, which puts them in position to respond to antigens in the gut or lungs or to infectious agents entering the body from these locations. B1 cells are guided to and are held in these cavities by expression of the $\alpha M\beta 2$ integrin (CD11b/Mac-1—normally found on macrophages) and CXCR5, which binds to the chemokine CXCL13, which is produced in these locations as well as in the B cell follicles of secondary lymphoid organs. The importance of induced antibody responses by B1 cells is illustrated by the observations that approximately 40% of IgA-secreting cells in the gut are of B1 cell origin and that examination of fecal bacteria from mice has revealed that about two-thirds of them are coated with IgA produced exclusively by B1 cells. This IgA is the result of immune stimulation; it is not natural antibody. B1 cells are also present in small numbers in the spleen and do contribute to immune responses to bloodborne bacteria or viruses, although marginal zone B cells, described in the next section, are typically more important for these responses.

B1 cells require antigen recognition for their development and are restrained by the inhibitory receptor CD5

As mentioned above, the diversity of antibodies produced by B1 cells is restricted by the mechanisms of V(D)J recombination in their precursors. Diversity is further limited by the developmental requirement that immature B cells committed to this lineage must be positively selected by reactivity to self-antigens in order to mature and enter into the long-lived pool of B1 cells (Figure 8-16). In this regard, B1 cells resemble NKT cells (see section 8-3), except that there are multiple self-antigens that can positively select for B1 cells, whereas canonical NKT cells are selected to react to just one related lipid or a small family of such lipids. Interestingly, the level of self-reactivity that is required for the maturation and survival of a B1 cell results in clonal deletion in immature B cells committed to becoming follicular or marginal zone B cells, illustrating that B cell precursors are pre-committed to producing either B1 cells or follicular and marginal zone B cells and have different properties as a result.

B1 cells are characterized by the expression of the inhibitory receptor CD5, high levels of membrane IgM and low levels of membrane IgD. The cell surface proteins distinguishing B1 cells from follicular B cells and marginal zone B cells are summarized in the next section (see Figure 8-18). CD5 is an inhibitory receptor that decreases signaling from the BCR and it is necessary to prevent mature B1 cells from responding to the same self antigens that were required for their positive selection; in this regard CD5 is analogous to the inhibitory receptors present on NK cells, described earlier in this chapter. Presumably, the highly repetitive nature of carbohydrate epitopes on bacterial cell walls or of protein epitopes on virus particles induces sufficiently strong BCR signaling to overcome the inhibitory effects of CD5 and other inhibitory molecules of the B1 cell and allow a rapid antibody response to occur. In addition to receiving a strong BCR signal, the B1 cell, like most other lymphocytes, requires second signals for activation; these can come in the form of certain cytokines made in response to innate recognition or by direct recognition of innate ligands by the TLRs on the B1 cells.

Specificities that are Selected into the B1 Cell Population

Transgene	Specificity
VH12/Vκ4	phosphatidylcholine
SM610	Thy-1
4C8	mouse red blood cell antigen
2-12H	Smith antigen (snRNP)

Figure 8-16 Table of specificities that are selected into the B1 population
Immunoglobulin transgenic mice often give rise primarily to one of the three major types of B cells. This observation suggests that the nature of the specificity directs development to one or another B cell type. Shown here are some of the specificities that lead to a strong preference for B lineage cells to develop into B1 cells, all of which represent self-reactive antibodies. Antibodies against Smith antigen are found in a subset of patients with the immune complex autoimmune disease systemic lupus erythematosus (described in Chapter 12); snRNP: small nuclear ribonuclear particle. Some of these specificities found in B1 cells are likely to have roles in immune defense. For example, phosphatidylcholine is chemically similar to phosphorylcholine, which is found in the cell walls of some bacterial pathogens.

References

Ansel, K.M. *et al.*: **CXCL13 is required for B1 cell homing, natural antibody production, and body cavity immunity.** *Immunity* 2002, **16**:67–76.

Baumgarth, N. *et al.*: **B-1 and B-2 cell–derived immunoglobulin M antibodies are nonredundant components of the protective response to influenza virus infection.** *J. Exp. Med.* 2000, **192**:271–280.

Baumgarth, N. *et al.*: **Inherent specificities in natural antibodies: a key to immune defense against pathogen invasion.** *Springer Semin. Immunopathol.* 2005, **26**:347–362.

Berland, R. and Wortis, H.H.: **Origins and functions of B-1 cells with notes on the role of CD5.** *Annu. Rev. Immunol.* 2002, **20**:253–300.

Hardy, R.R. and Hayakawa, K.: **B cell development pathways.** *Annu. Rev. Immunol.* 2001, **19**:595–621.

Martin, F. and Kearney, J.F.: **B1 cells: similarities and differences with other B cell subsets.** *Curr. Opin. Immunol.* 2001, **13**:195–201.

Martin, F. and Kearney, J.F.: **B-cell subsets and the mature preimmune repertoire. Marginal zone and B1 B cells as part of a "natural immune memory".** *Immunol. Rev.* 2000, **175**:70–79.

Wang, H. and Clarke, S.H.: **Regulation of B-cell development by antibody specificity.** *Curr. Opin. Immunol.* 2004, **16**:246–250.

Spleen

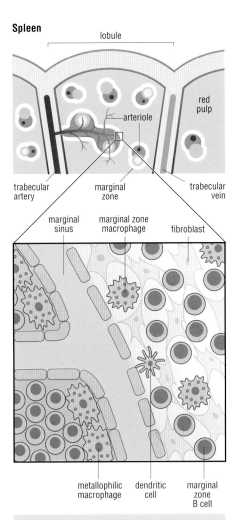

Figure 8-17 The marginal zone of the spleen
The spleen serves as a filter for the blood, promoting the efficient removal of senescent red blood cells and the efficient detection of infectious agents in the blood. The anatomy of a section of the spleen is shown in the top panel. Blood flows through central arteries, which are surrounded by white pulp containing lymphocytes. Side branches of these arteries terminate in the red pulp or empty into marginal sinus, which is adjacent to the marginal zone and separated from it by a leaky endothelial layer. The splenic marginal zone contains numerous macrophages and some dendritic cells that enter from the blood, along with specialized marginal zone B cells.

Blood is filtered in the spleen via the specialized cells of the marginal zone

Blood filtered through the spleen percolates past the cells of the marginal zone, whose function is to allow immune cells to remove debris and detect infectious agents. In contact with the blood as it is being filtered are the specialized cells of the marginal zone including two types of macrophages and a specialized type of B cell, the **marginal zone B cell** (Figure 8-17) (see section 1-7). Marginal zone B cells represent the third type of naïve mature B cell, in addition to the recirculating follicular B cells and B1 cells. The characteristic cell-surface phenotypes of these three types of B cells are summarized in Figure 8-18.

Marginal zone B cells make antibody responses to bloodborne infectious agents

Marginal zone B cells are responsible for making rapid antibody responses to infectious agents that have entered the blood and typically produce IgM, although they can also produce IgG3 (in mouse). As we have seen (section 6-3), IgM and IgG3 are efficient activators of complement, an action that is critical to the importance of these cells, as we shall see below. To support their ability to produce antibody early in an immune response, marginal zone B cells have a partially activated phenotype before they encounter antigen, in common with many of the other cells described in this chapter and in contrast to the resting appearance of follicular B cells and most $\alpha\beta$ T cells. Despite this activated phenotype, many marginal zone B cells are naïve B cells, whereas some are memory B cells; the former are more prevalent in the mouse and the latter are more prevalent in the human. Whereas B1 cells arise from fetal hematopoietic precursor cells, marginal zone B cells and follicular B cells arise from a common intermediate, referred to as transitional B cells (see section 7-6). The lineage choice between these two mature B cell types is controlled by Notch2, which promotes the marginal zone fate, and also by BCR signaling. How the latter contributes is not well established, but most evidence favors the hypothesis that a moderate level of reactivity to self components promotes the follicular B cell fate, probably by inhibiting Notch signaling, whereas a lower level of reactivity to self antigens is needed to allow Notch2 signaling to proceed and induce the marginal zone B cell fate. Once they develop, marginal zone B cells are held in the marginal zone by the expression of αLβ2 and α4β1 integrins, ligands for which are present there, and by poor responsiveness to the follicle-attracting chemokine CXCL13.

The location of marginal zone B cells makes them ideally suited to make responses to bloodborne antigens and they do so by both T-independent and T-dependent mechanisms, depending on the nature of the antigen (see section 6-9). Marginal zone B cells may contact antigen in its free form, as an immune complex with IgG (see below), or bound to the surface of an immune cell that captured it in the blood and transported it to the marginal zone. Bacteria or viruses introduced into the blood can rapidly become bound by immature dendritic cells of the blood. These cells can deliver their bound cargo to the marginal zone for activation of rapid antibody responses by marginal zone B cells. In addition to carrying the antigen to the B cells, these dendritic cells also secrete BAFF and/or APRIL, TNF family members that greatly promote B cell activation in response to particulate antigens such as those present in the cell wall of bacteria or the exterior of a virus particle (Figure 8-19).

Marginal zone B cells are primarily responsible for the T-independent antibody that is produced against the polysaccharides of bloodborne encapsulated bacteria (see section 3-2). Interesting in this regard is the fact that the number of marginal zone B cells is small in a neonate and does

Definitions

marginal zone B cells: B cells resident in the marginal zone of the spleen and providing rapid T cell-independent antibody responses to capsular polysaccharides of bacteria, as well as early T cell-dependent responses to bloodborne antigens.

not reach a substantial number until one to two years of age in humans or two to three weeks in rodents, and a similar time course is seen for the ability to make antibody responses to vaccines that are purely polysaccharide. The inability of such polysaccharide vaccines to induce an antibody response in young children has led in recent years to the development of *conjugate vaccines*, in which capsular polysaccharides of bacteria are coupled to an immunogenic foreign protein, which permits an effective T cell-dependent antibody response by follicular B cells: these vaccines are described in more detail in Chapter 14.

As well as making T-independent antibody responses, marginal zone B cells can make T-dependent responses. Marginal zone B cells express TLRs, which can contribute to their activation. The binding of antigen either to the BCR or to both BCR and TLR induces the expression of B7 and migration to the T cell zone of the spleen. In the absence of T cells recognizing the same antigen, a T-independent IgM or IgG3 response results. If T cells specific for the same antigen are present, these cells can stimulate T-dependent production of a broader range of IgG isotypes that may exhibit some somatic mutation, although marginal zone B cells do not make the highly developed germinal center immune responses that give rise to high-affinity isotype-switched antibody. These interactions with T cells can, however, occur with naïve T cells that are activated by peptide–MHC complexes and B7 on the B cells themselves: this can be initiated within four hours of stimulation by antigen, whereas activation of follicular B cells by previously activated T cells takes eight hours to begin.

Most marginal zone B cells differentiate into plasma cells and secrete IgM by the third day after exposure to bloodborne particulate antigens. In the rapidity of their response they resemble B1 cells and memory B cells. For this reason, B1 and marginal zone B cells have been referred to as natural memory cells.

As well as providing an antibody response that is much faster than the follicular B cell response, marginal zone B cells contribute indirectly to the germinal center response in two related ways. First, the antibodies they produce bind antigen and fix complement, leading to the deposition of C3b and further processed forms of it (especially C3d) attached to the immune complexes. Second, marginal zone B cells bind immune complexes in the blood via complement receptor CR2 (CD21), which is expressed at high levels by these cells. This binding is sufficient to trigger their migration to the B cell follicle, where the complement-coated immune complexes are transferred to follicular dendritic cells, the cells that are necessary for antigen selection in the germinal center reaction. Follicular dendritic cells then hold the antigen in this form on their cell surface for a prolonged period and interact with somatically mutating follicular B cells to promote selection for B cells with increased affinity (see section 6-9). In this latter role, the marginal zone B cell is behaving like an antigen-presenting cell rather than an antigen-specific lymphocyte; all that is required is for the marginal zone B cell to bind the immune complex via its complement receptors—it does not need to recognize the antigen within the immune complex with its BCR. The key here is the initial location of this cell in a place where immune complexes in the blood will come into contact with it.

Figure 8-19 Dendritic cells aid in the activation of marginal zone B cells Bacteria in the blood are rapidly bound by neutrophils and blood immature dendritic cells. The dendritic cells make contact with marginal zone B cells in the spleen and promote their activation by presenting antigen and secreting BAFF or APRIL, which are necessary for marginal zone B cell responses to antigen. After the initial activation, the marginal zone B cells move to the T cell zone of the spleen and make either a T-independent or a T-dependent antibody response, depending on the nature of the antigen.

Phenotypes of Naïve Mature B Cell Subtypes

	Type of B cell		
Properties	B1	MZ	Follicular
location	PC	MZ	recirc.
diversity	low	moderate	high
Surface molecule			
IgM	+++	+++	+
IgD	+/–	+/–	+++
CD1d	–	+	–
CD21	+/–	+++	++
CD23	–	–	++
CD5	+ or –	–	–
αMβ2 integrin (MAC-1)	+ or –	–	–
CD9	+	+	–

Figure 8-18 Table of phenotypes of naïve mature B cell subtypes Some of the cell surface molecules useful for distinguishing the three types of naïve mature B cells in the mouse are shown. CD1d is a non-classical MHC class I molecule that presents lipid antigens to NKT cells (see section 8-3). CD21 is the complement receptor CR2; CD23 is the low-affinity Fc receptor for IgE, FcεRII, the function of which is not well established. For αMβ2 integrin, + or – refers to the fact that in the peritoneum, B1 cells express this molecule, whereas in the spleen they do not. MZ: marginal zone of spleen; PC: peritoneal and pleural cavities.

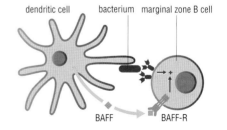

References

Balázs, M. *et al.*: **Blood dendritic cells interact with splenic marginal zone B cells to initiate T-independent immune responses.** *Immunity* 2002, **17**:341–352.

Lopes-Carvalho, T. *et al.*: **Marginal zone B cells in lymphocyte activation and regulation.** *Curr. Opin. Immunol.* 2005, **17**:244–250.

Lu, T.T. and Cyster, J.G.: **Integrin-mediated long-term B cell retention in the splenic marginal zone.** *Science* 2002, **297**:409–412.

Martin, F. and Kearney, J.F.: **Marginal-zone B cells.** *Nat. Rev. Immunol.* 2002, **2**:323–335.

Martin, F. *et al.*: **Marginal zone and B1 B cells unite in the early response against T-independent blood-borne particulate antigens.** *Immunity* 2001, **14**:617–629.

Pillai, S. *et al.*: **Marginal zone B cells.** *Annu. Rev. Immunol.* 2005, **23**:161–196.

Tanigaki, K. *et al.*: **Notch–RBP-J signaling is involved in cell fate determination of marginal zone B cells.** *Nat. Immunol.* 2002, **3**:443–450.

9

The Immune Response to Bacterial Infection

Bacteria comprise a large group of organisms resident on normal skin and mucosal surfaces and that are important in sustaining a healthy epithelial barrier. Pathogenic bacteria are distinguished by genes that enable them to invade body tissues. Innate phagocytic cells and adaptive T cells and antibodies are all called into play in immune responses against pathogenic bacteria. Such responses, when excessive, can lead to damage and even death.

Commensal bacteria colonize microenvironmental niches and contribute to a healthy epithelial interface

Bacteria are ubiquitous in the environment. Indeed, there are more bacterial cells on the epithelial surfaces of the skin and bowel (by at least tenfold) than there are in the vertebrate host. Different epithelial surfaces support different bacterial communities, reflecting adaptation by the organisms to physicochemical features and nutrient access that sustain their growth. These normal colonizing bacteria, or **commensal bacteria**, become established rapidly after birth. The great majority of vertebrate commensal bacteria reside in the intestines, where bacterial densities increase from 10^3/ml of intestinal contents in the proximal small bowel to 10^{11}/g of intestinal contents in the colon. Although many of these bacteria still cannot be grown in the laboratory, the identifiable flora is quite stereotyped among different individuals, suggesting long-standing adaptation (Figure 9-1).

Increasing evidence suggests that the relationship between the host and the commensal flora reflects host–microbe mutualism, or **symbiosis**, by which both host and microbe benefit. This is more readily apparent in instances of colonization by a monospecific flora, as for example by certain light-emitting bacteria in the luminescent organs of squid and other marine animals. In return for providing the squid with luminescence, bacteria grow in a relatively noncompetitive microenvironment, free from other microbes. In like manner, some prevalent intestinal anaerobes of humans, such as *Bacteroides thetaiotamicron*, actively participate in regulating the development of the bowel epithelium to a mature, energy-absorbing surface. Intestinal cells provide specific nutrients that support these bacterial communities, while the microorganisms, in turn, contribute metabolic pathways that expand our capacity to extract calories and micronutrients from a polysaccharide plant diet. Compared with animals reared in a bacteria-free environment, normally colonized mice require 30% less caloric intake to maintain body weight. Thus commensal bacteria are not solely passive colonizers but are active contributors to a healthy epithelial interface. This healthy interface, consisting of a continuously renewed epithelial cell layer and, on mucosal surfaces, a fluid matrix containing antimicrobial peptides (see section 3-1), retains commensals at surfaces by sustaining the niche that favors their growth.

Disrupting the commensal bacterial flora creates an opportunity for pathogenic bacteria

Pathogenic bacteria are acquired from the environment and must compete with the normal commensal bacteria to establish a viable community. In healthy individuals, some pathogens can be recovered from among the normal flora, but this usually reflects transient colonization. Organisms like *Staphylococcus aureus* and *Pseudomonas aeruginosa*, which can be important human pathogens, are examples of bacteria that can be recovered from normal epithelial surfaces for days to weeks, but then become dispersed back to the environment (so-called *transients* or *tourists*). The probability that such organisms will remain and expand into a community of organisms capable of attacking the host is greatly increased by disruption of the commensal flora or the epithelial barrier that sustains it. Broad-spectrum antibiotics can be a two-edged sword: although necessary to kill pathogenic invaders, these drugs deplete the normal bacterial species that contribute to the maintenance of a healthy epithelium. In this way, transient, potentially pathogenic species can colonize the host and set up invasion.

Predominant Human Commensal Bacteria	
Tissue	**Major commensal bacteria**
skin and hair follicles	*Staphylococcus epidermidis*
	Propionibacteria species
	Corynebacteria species
mouth	viridans group
	Streptococcus species
	Staphylococcus epidermidis
	Neisseria species
	Corynebacteria species
	anaerobes (gingival crevices)
upper small intestine	Gram-positive anaerobic species
	Streptococcus species
	Enterococcus species
large intestine	*Bacteroides* species
	Bifidobacterium species
	Streptococcus species
	Clostridium species
	Gram-negative facultative species (predominant *Escherichia coli*)
	Staphylococcus epidermidis
	Lactobacillus species
	Enterococcus species
	anaerobic methanogenic species (Archaea)
vagina	*Lactobacillus* species

Figure 9-1 Table of predominant human commensal bacteria

Definitions

commensal bacteria: bacteria that colonize a niche of the host. From the Latin for one who eats at the same table.

symbiosis: a mutually beneficial relationship that sustains a stable interaction between different organisms.

Pathogenic bacteria exploit epithelial dysfunction and subvert innate immunity

Pathogens deploy a range of specialized mechanisms to gain entry into the tissues. In most cases, these specialized mechanisms are utilized most efficiently when the normal epithelial interface is disrupted or dysfunctional, thus promoting colonization. Not unexpectedly, bacterial infections are the most common complications of traumatic wounds, viral respiratory infections (which damage surface epithelial cells) or acquired or genetic deficiencies that compromise the mucociliary clearance apparatus of mucosal surfaces (smoking or cystic fibrosis). Having established a beachhead, pathogenic bacteria must also avoid destruction by the innate immune response. Many can avoid cytolytic killing by mannose-binding lectin (MBL), complement and natural antibodies through physical properties of their cell walls. Thick-walled, polysaccharide capsules or secreted toxins are common mechanisms for overcoming phagocytosis and killing by neutrophils. Finally, a large number of bacterial pathogens exist wholly or predominantly as intracellular organisms, sheltered within cytoplasmic compartments altered by bacterial proteins to allow their growth, and shielded from the immune system.

Control of pathogenic organisms requires both innate and adaptive immune responses

Different organisms have evolved different strategies to cause disease, and virtually any organ of the body can be targeted by pathogens of different sorts. Common infectious disease syndromes are caused most often by a relatively small number of bacterial types (Figure 9-2). Importantly, many of the symptoms and pathologic consequences of infectious diseases are due to activation of the host immune response. The recognition of foreign molecules by complement, Toll-like receptors, NOD proteins and other innate immune moieties initiates the induction of cytokines, chemokines and other inflammatory mediators that provoke the host response. Genetic deficiencies that attenuate either innate cell functions or adaptive cell functions are associated with increased numbers of infections by bacterial pathogens. Loss of both arms of immunity, as occurs with bone marrow ablative therapy for certain hematologic malignancies, inevitably results in fatal infections without the use of broad antimicrobial therapy and without the eventual recovery of marrow-derived immune cells. In general, high-affinity antibodies and inflammatory T_H17, T_H1 and CD8 T cells are required to activate phagocytes to an antimicrobial state or to clear intracellular residents and sustain immunity against pathogenic bacteria.

Figure 9-2 Table of bacteria associated with common infectious disease syndromes

Bacteria associated with Common Infectious Disease Syndromes	
Disease	**Microorganism**
pneumonia	*Streptococcus pneumoniae, Chlamydia pneumoniae, Mycoplasma pneumoniae, Hemophilus influenzae, Legionella pneumoniae, Mycobacterium tuberculosis*
gastrointestinal infection	*Escherichia coli, Shigella, Campylobacter, Helicobacter pylori, Salmonella, Vibrio cholerae, Listeria monocytogenes, Yersinia enterocolitica*
meningitis	*Neisseria meningitidis, Streptococcus pneumoniae, Listeria monocytogenes*, Group B *Streptococcus*
sexually transmitted diseases	*Neisseria gonorrhoeae, Chlamydia trachomatis, Treponema pallidum, Hemophilus ducreyi*
epithelial infection, septicemia	*Staphylococcus aureus, Streptococcus pyogenes* (Group A *Streptococcus*)
urinary tract infections	*Escherichia coli*

References

Hooper, L.V. and Gordon, J.I.: **Commensal host-bacterial relationships in the gut.** *Science* 2001, **292**:1115–1118.

Hooper, L.V. *et al.*: **How host-microbial interactions shape the nutrient environment of the mammalian intestine.** *Annu. Rev. Nutr.* 2002, **22**:283–307.

Nyholm, S.V. and McFall-Ngai, M.J.: **The winnowing: establishing the squid–*Vibrio* symbiosis.** *Nat. Rev. Microbiol.* 2004, **2**:632–642.

Sonnenburg, J.L. *et al.*: **Getting a grip on things: how do communities of bacterial symbionts become established in our intestine?** *Nat. Immunol.* 2004, **5**:569–573.

Epithelia present both a physical and an antimicrobial barrier

Epithelial cells establish barriers at the cutaneous and mucosal interfaces between host and environment. Stratified epithelial cells of the skin and oral cavity secrete glycolipoproteins that maintain a permeability barrier; keratinization of the outermost skin cells reinforces the cutaneous barrier. The simple polarized epithelia of the intestine and airways are sealed by intercellular tight junctions (Figure 9-3).

Epithelial cells of the eye, hair follicles, oral, intestinal, respiratory and urogenital surfaces are coated by a fluid matrix that physically restricts access of organisms to the underlying epithelia. The fluid is constantly replenished and removed by mechanical means by blinking the eyes or by intestinal motility, or by ciliary motion in the lungs, which make it difficult for bacteria to adhere. In addition to complex glycosaminoglycans and mucins that bind bacterial outer membrane constituents, the fluid matrix contains antibacterial peptides and proteins that are constitutively secreted and further induced in response to bacterial cell wall components such as lipopolysaccharide. Antimicrobial peptides, such as β-defensins in the lung and skin, α-defensins in Paneth cells of the intestinal tract, and antimicrobial proteins, such as lysozyme and secretory leukocyte protease inhibitor (SLPI), in fluids bathing the cornea of the eye, are toxic to diverse groups of microbes. Antimicrobial peptides have optimal activities that reflect the distinctive physiology of given sites, such as the acid pH of the stomach, the alkaline pH of the vagina or the salinity of the respiratory tract. Epithelial antimicrobial peptides are complemented by α-defensins, which are a major constituent of neutrophil granules. Cathelicidins are another group of antimicrobial defensins produced by activated neutrophils and epithelial cells. In addition to their antimicrobial activity, some cathelicidins trigger signaling through the formyl peptide receptor (see section 3-14), and thus recruit neutrophils and monocytes to inflammatory sites.

Infections are common when epithelial barriers are disrupted. Wounds, burns, intravenous access devices and insect vectors allow organisms on the skin to gain access to the tissues. Cancer chemotherapy or irradiation can also compromise epithelial integrity by killing dividing cells that maintain this dynamic barrier. Cystic fibrosis compromises the volume and salinity of respiratory tract fluids, attenuating the activity of antimicrobial peptides and leading to overgrowth of colonizing bacteria and the eventual destruction of lung tissue. Antacids, by neutralizing stomach pH, can allow bacteria that would normally be destroyed to reach the intestines and cause disease. Antibiotics, by destroying susceptible nonpathogenic commensals, can allow the outgrowth of organisms, such as *Clostridium difficile* in bowel, that secrete toxins that cause disease.

Failure to maintain barriers against normal microbial flora can be demonstrated in certain genetic deficiencies associated with innate immunity (Figure 9-4). Animals with these genetic deficiencies die early in life with multiple intestinal, hepatic, cutaneous and lung microabscesses. Such outcomes reflect the loss of the homeostatic interactions between commensal flora and a healthy epithelial barrier. Conversely, overexpression of innate response proteins can protect animals. For instance, expression of human defensin-5 in murine Paneth cells protects mice

(a) cutaneous barrier

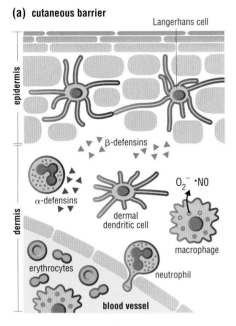

(b) mucosal barrier

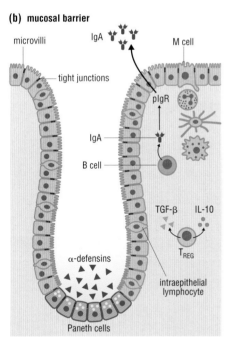

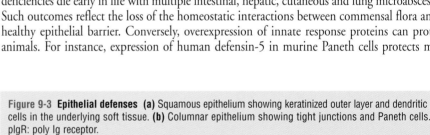

Figure 9-3 Epithelial defenses (a) Squamous epithelium showing keratinized outer layer and dendritic cells in the underlying soft tissue. **(b)** Columnar epithelium showing tight junctions and Paneth cells. pIgR: poly Ig receptor.

References

Frenette, P.S. *et al.*: **Susceptibility to infection and altered hematopoiesis in mice deficient in both P- and E-selectins.** *Cell* 1996, **84**:563–574.

Macpherson, A.J. *et al.*: **A primitive T cell-independent mechanism of intestinal mucosal IgA responses to commensal bacteria.** *Science* 2000, **288**:2222–2226.

Mellman, I. and Steinman, R.M.: **Dendritic cells: specialized and regulated antigen processing machines.** *Cell* 2001, **106**:255–258.

Salzman, N.H. *et al.*: **Protection against enteric salmonellosis in transgenic mice expressing a human intestinal defensin.** *Nature* 2003, **422**:522–526.

Shibuya, A. *et al.*: **Fcα/μ receptor mediates endocytosis of IgM-coated microbes.** *Nat. Immunol.* 2000, 1:441–446.

Shiloh, M.U. *et al.*: **Phenotype of mice and macrophages deficient in both phagocyte oxidase and inducible nitric oxide synthase.** *Immunity* 1999, **10**:29–38.

van Egmond, M. *et al.*: **IgA and the IgA Fc receptor.** *Trends Immunol.* 2001, **22**:205–211.

Wilson, C. L. *et al.*: **Regulation of intestinal α-defensin activation by metalloproteinase matrilysin in innate host defense.** *Science* 1999, **286**:113–117.

Yang, D. *et al.*: **β-defensins: linking innate and adaptive immunity through dendritic and T cell CCR6.** *Science* 1999, **286**:525–528.

from *Salmonella typhimurium* invasion. The epithelial barrier can also be disrupted when the adaptive immune response to bacterial commensals becomes dysregulated. Such dysregulated responses have provided highly instructive models for human inflammatory bowel diseases, such as Crohn's disease and ulcerative colitis, which we discuss in Chapter 13.

Specialized cell types in epithelia initiate and execute adaptive immune responses

Antigen sampling from epithelia occurs by the close association of an interdigitating dendritic cell network in the stratified epithelia of the skin or within the simple columnar epithelia in the mucosa of the lung and intestinal tissues, or through specialized lymphoid organs associated with the gastrointestinal organs, such as the tonsils or intestinal mucosa-associated lymphoid tissues (MALT) (see section 1-6). M cells comprise a unique population of cells in intestinal and respiratory epithelia that promote the transport of luminal antigens to abluminal dendritic cells and macrophages (see Figure 1-22). In each case, however, dendritic cells represent the sentinel network that escorts bacterially derived antigens to the adaptive immune system in the draining or specialized lymphoid organs. Some epithelial β-defensins can bind to CCR6, the CCL20 chemokine receptor, which is expressed on immature dendritic cells and memory T cells, and thus serve to link innate with adaptive immunity by attracting these cells to sites of epithelial injury.

The mechanisms that maintain the integrity of the epithelial barriers are not known for certain, but γδ T cells are believed to play a part. Their localization to epithelial basement membranes in both mouse and man is suggestive, and some loss of intestinal epithelial integrity can be demonstrated in γδ T cell-deficient mice. Gene expression profiling of γδ intraepithelial lymphocytes, or IELs (see section 8-5), reveals expression of an array of immune defense genes, including those encoding the antimicrobial peptide cryptdin and granzymes A and B, as well as transcripts for inhibitory receptors, including CTLA4, PD-1 and various NK cell inhibitory receptors, that may keep these cells poised for activation. In humans, intestinal Vδ1 T cells express NKG2D and are activated by the stress-induced nonpolymorphic MHC class I molecules MICA and MICB (see section 4-4).

IgA is the most prevalent immunoglobulin in mucosal secretions and is believed to contribute to epithelial defense by binding to antigens, including bacteria and toxins, and thus preventing their uptake. Epithelial cells in the intestine and hepatocytes of the liver mediate the transfer of dimeric serum IgA to the intestinal lumen via the polymeric Ig receptor. After engagement at the basolateral surface by IgA, the polymeric IgR–IgA complex undergoes transcytosis to the apical membrane. Proteolytic cleavage of the ectodomain of the polymeric Ig receptor with IgA releases the complex into the lumen. The released complex, which is termed secretory component, is believed to protect luminal IgA from degradation by bacterial proteases. IgA deficiency is a common immunodeficiency in humans. Although some individuals suffer from recurrent sinus, pulmonary and/or intestinal infections, most remain asymptomatic, probably because IgM, which also binds the polymeric Ig receptor, can compensate for IgA in many cases.

After opsonization with IgA or IgM, organisms are endocytosed by Fcα/μR (see section 6-3) expressed on mature B cells, macrophages and dendritic cells. In humans, a second IgA receptor, FcαRI, mediates the uptake of organisms coated with serum IgA by phagocytic cells, such as neutrophils.

Genetic Deficiencies Affecting Responses to Commensal Bacterial Flora

Primary deficiency	Examples	Mouse phenotype	Human disease
neutrophil defects	P selectin E selectin C/EBPε	skin, mucocutaneous, conjunctival abscesses	human secondary granule deficiency
	Phox NOS2	abdominal abscesses	CGD (Phox only)
intestinal mucosa	SMAD3	abdominal abscesses	
	matrilysin	Paneth cell α-defensins absent with exuberant bacterial colonization	
	IL-10	inflammatory bowel disease	
	N-cadherin	inflammatory bowel disease	
	IL-2, IL-2Rα IL-2Rβ, TCRα	inflammatory bowel disease	
	NOD2		Crohn's disease
	MyD88	impaired response to bowel injury	

Figure 9-4 Table of representative gene deficiencies with abnormal responses to commensal bacterial flora C/EBPε: myeloid-specific CCAAT/enhancer binding transcription factor family, in which mutation leads to loss of neutrophil-specific granules that contain cathelicidins in mice and humans; Phox: phagocyte oxidase (disrupted gp91 is used in this example); NOS2: inducible nitric oxide synthase; SMAD3: transcriptional regulator downstream of TGF-β family serine/threonine receptor kinases (homologs of *C. elegans* Sma and *Drosophila* Mad proteins); matrilysin: MMP-7, a Paneth cell metalloproteinase required to process pro-α-defensins to active α-defensin; NOD2: intracellular innate sensor for peptidoglycan fragments of bacterial cell wall, and mutations have been associated with increased susceptibility to Crohn's disease (see Chapters 3 and 13); MyD88: TLR and IL-1R superfamily adaptor.

9-2 Evasion of Epithelial Defenses by Pathogenic Bacteria

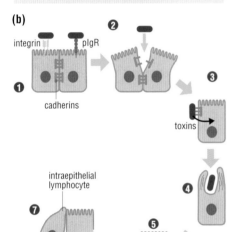

(a)

IgA protease

IgA

pathogen

α-defensins

altered charge on pathogen surface makes them resistant to defensins

(b)

integrin   pIgR

cadherins

toxins

intraepithelial lymphocyte

apoptosis

Figure 9-5 Pathogens overcome epithelial barriers (a) Pathogens alter their membranes to become less susceptible to defensins, and secrete IgA proteases that cleave luminal IgA. **(b)** (1) Specialized receptors target either M cells or epithelial cells themselves, integrins, cadherins and pIgRs being common examples of host receptors exploited by pathogens. (2) Secreted toxins and/or cell binding disrupt intercellular tight junctions, allowing pathogens to adhere firmly to exposed cadherins. (3) Type III secretory systems introduce bacterial toxins into cells. (4) These proteins target cellular cytoskeletal systems to induce membrane ruffling and uptake of organisms into the cell. (5) Other toxins activate the cell apoptotic pathway, (6) disrupting the basilar membrane and allowing invasion to proceed. (7) Epithelial cell disruption may be aided by stress-induced expression of MICA, which is a ligand for NKG2D, a receptor on γδ intraepithelial lymphocytes, leading to activation of the lytic effector pathways.

Pathogenic bacteria have evolved to traverse epithelial barriers

Pathogenic bacteria express adhesins through which they attach to epithelial cells. Attachment often involves fimbriae, hairlike appendages that, in some cases, are capable of extensive variation, thus circumventing antibody-mediated blockade. Structural studies provide evidence for molecular mimicry of host proteins by these bacterial appendages, which are thereby able to interact with cell surface adhesion proteins such as cadherins and integrins. Pathogens secrete enzymes capable of degrading extracellular matrix and creating sites for firm adhesion, allowing deeper penetration into tissues. Some organisms secrete toxins capable of loosening epithelial tight junctions; *zonula occludens toxin (ZOT)* of *Vibrio cholerae* disrupts these junctions, allowing the bacterium access to regions between epithelial cells of the intestine. Some pathogenic *E. coli* responsible for diarrhea encode their own receptor, secreting it into host epithelial cells to which they adhere to enable them to make tighter interactions with the cells. Pathogens also have mechanisms for increasing their resistance to innate defense responses that include biochemical alterations in their outer membranes, which affect their overall charge characteristics or make it more difficult for the insertion of the antimicrobial peptides. IgA proteases that inhibit the deposition of IgA on the bacterial surface are widely prevalent among mucosal pathogens.

In contrast to commensal bacteria, pathogens share conserved genes that program invasion of normal cells. These genes are typically coordinately regulated from a single large cluster, suggesting that they have been selected in evolution as a unit. In many Gram-negative organisms, such genes, which can number 40 or more, are clustered on single large insertion elements, termed **pathogenicity islands**. Best studied is the **type III secretion system**. This genetic element directs the production of a multi-component structure, the *needle complex*, which traverses the bacterial membrane and physically inserts itself into the target cell, allowing the direct transfer of bacterial proteins into the host cytosol (Figure 9-5). Parts of the structure are believed to have evolved from genes that mediate flagellar assembly. In *Yersinia*, *Pseudomonas* and enteropathogenic *E. coli*, some of the transferred proteins target the host cytoskeletal machinery and inhibit uptake by professional phagocytes. Conversely, intracellular organisms such as *Salmonella* and *Legionella* target aspects of the same, or related, machinery to induce their uptake, but subsequently modify the phagolysosome to generate a supportive niche. Other effector molecules target protein synthetic or apoptotic pathways, cell signaling molecules and proteins involved in localization to distinct intracellular compartments. The type IV secretory system, which is related to the type III secretory system, serves in *Helicobacter* to deliver a toxin, *vacuolating cytotoxin A (VacA)*, that disrupts tight junctions in the gastric epithelium but also induces host inflammatory responses that have been linked to peptic ulcers and carcinoma. Gram-positive organisms can target similar processes using **cytolysin-mediated translocation**, a mechanism by which secreted cytolysins can induce pores that allow ingress of molecules that interdict critical intracellular processes in host cells. Host targets for these bacterial virulence factors are broadly conserved during evolution—the same virulence genes in *Pseudomonas aeruginosa*, a Gram-negative pathogen of humans, are necessary for invasion of plants, worms and mice.

Many intestinal pathogens target M cells, specialized antigen sampling cells that protrude into the bowel lumen; M cells also exist in the respiratory tract. M cells take up antigens from their apical surface and release them into a basal pocket in close apposition to macrophages, dendritic cells and lymphocytes. These cells have been repeatedly implicated as sites of invasion by intestinal pathogens, which may exploit unique antigen-sampling receptors or the thinning of the mucus layer over the luminal dome. Some bacteria, such as *Shigella*, traverse M cells to reach the basolateral surface of intestinal epithelial cells for invasion, whereas others, such as *Salmonella*, can apically invade and lyse M cells to reach deeper tissues.

Definitions

CRIg: complement receptor of the immunoglobulin superfamily that is highly expressed on liver macrophages (Kupffer cells) and that mediates the clearance of pathogens bound to complement components C3b and iC3b from the blood.

cytolysin-mediated translocation: mechanism by which Gram-positive pathogens introduce host-modifying proteins into target cells through cytolysin-mediated pores.

pathogenicity islands: DNA encoding contiguous virulence genes and found in pathogenic bacteria.

phase variation: phenotypic variation within a species of bacteria.

type III secretion system: mechanism by which Gram-negative bacteria translocate host-modifying proteins into the cytoplasm of target cells through a syringe-like structure penetrating the host cell membrane. The type III secretion system is encoded in **pathogenicity islands**.

Pathogenic bacteria evade innate defense mechanisms

Having traversed the epithelial barrier, bacteria are confronted by components of the complement system and mannose-binding lectin (MBL); each serves to activate the complement cascade, generating components that coat the organisms and lead to the recruitment of phagocytic cells. Collectins, primarily in lung (surfactant protein-A; SP-A) but also at other mucosal sites (surfactant protein-D; SP-D), also function as opsonins (see section 3-2). Pathogenic organisms typically have large or complex polysaccharide capsules that obstruct efficient activation of complement or MBL and that impede phagocytosis by neutrophils. Low-affinity IgM antibodies and trapping by specialized splenic and alveolar macrophages that express scavenger receptors such as MARCO may be critical for the early sequestration of these organisms. Humans with diminished levels of C3 or MBL, or who have hypogammaglobulinemia or have been splenectomized, are particularly prone to early dissemination of encapsulated organisms. A similar immune deficiency is seen in mice deficient in early complement proteins or complement receptors, SP-A, T cell-independent IgM antibody, MARCO or **CRIg**, a specialized complement receptor that is expressed in the Kupffer cells of the liver, which are important in defense against bloodborne organisms (see section 1-3).

Several important pathogenic bacteria have specialized mechanisms for varying their cell wall proteins to evade immune surveillance. This is called **phase variation**. *Neisseria meningitidis* undergoes extensive phase variation in capsule formation during its adaptation from a nasal colonizing organism to a pathogen that invades the blood and meninges. Phase variation can occur by several mechanisms and in some species is enhanced by mutations in DNA mismatch repair enzymes, which increase the likelihood of alterations in bacterial surface molecules. Increased prevalence of mutations arising in this way is found in enteric pathogens, such as *E. coli* and *Salmonella*, and in *Pseudomonas* species that persistently colonize the lungs of patients with cystic fibrosis.

Adaptive immunity is required for protection against pathogenic bacteria

Humans with severe immunodeficiencies in adaptive immunity, or mice rendered deficient in B and T lymphocytes, are able to restrict commensal organisms at epithelial barriers but are vulnerable to substantial morbidity and mortality from pathogenic bacteria, indicating the critical role of adaptive immunity against these organisms. Many pathogens, as well as having specialized mechanisms for overcoming innate immunity, are able to establish niches within the host and/or release toxins that facilitate their persistence. Niches include the cytoplasm or phagosomes of cells, or abscesses, areas of focal coagulative necrosis where the access of serum factors or phagocytic cells may be limited. Toxins, typically cell wall components or secreted proteins, can generate severe systemic responses and obscure the focus of infection. Some pathogens persist in sites that may be poorly surveyed by the adaptive immune system, such as the synovia of the joints, the central nervous system or the prostate gland.

Successful resolution of infections caused by pathogenic bacteria usually requires T cell help for the production of high-affinity opsonizing antibodies or neutralizing anti-toxin antibodies, and, for intracellular organisms, the induction of cytotoxic CD8 T cells capable of killing infected cells. T_H1 and T_H17 immune responses that recruit and activate phagocytes are generally involved. IFN-γ, TNF and IL-6 have been implicated in host defense against bacterial pathogens in various human and murine infections.

References

Coombes, B.K. *et al.*: **Evasive maneuvers by secreted bacterial proteins to avoid innate immune responses.** *Curr. Biol.* 2004, **14**:R856–R867.

Ganz, T.: **Fatal attraction evaded: how pathogenic bacteria resist cationic peptides.** *J. Exp. Med.* 2001, **193**:F31–F33.

Helmy, K.Y. *et al.*: **CRIg: a macrophage complement receptor required for phagocytosis of circulating pathogens.** *Cell* 2006, **124**:915–927.

Madden, J.C. *et al.*: **Cytolysin-mediated translocation (CMT): a functional equivalent of type III secretion in Gram-positive bacteria.** *Cell* 2001, **104**:143–152.

Mahajan-Miklos, S. *et al.*: **Molecular mechanisms of bacterial virulence elucidated using a *Pseudomonas aeruginosa-Caenorhabditis elegans* pathogenesis model.** *Cell* 1999, **96**:47–56.

Stebbins, C.E. and Galán, J.E.: **Structural mimicry in bacterial virulence.** *Nature* 2001, **412**:701–705.

van der Woude, M.W. and Bäumler, A.J.: **Phase and anti-** genic variation in bacteria. *Clin. Microbiol. Rev.* 2004, **17**:581–611.

Vogelmann, R. *et al.*: **Breaking into the epithelial apical–junctional complex—news from pathogen hackers.** *Curr. Opin. Cell Biol.* 2004, **16**:86–93.

(a) bacterial cell surface

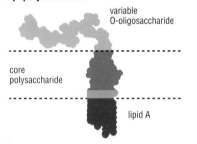

(b) lipopolysaccharide

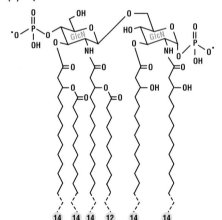

(c) lipid A

Figure 9-6 **Structure of LPS** **(a)** Simplified diagram of Gram-negative bacterial cell surface showing LPS and some other major features. Approximately 3×10^6 LPS molecules decorate 75% of the surface of Gram-negative bacteria. **(b)** The LPS molecule consists of an outer O-specific oligosaccharide that is highly variable among different species, a conserved inner core, and a lipid—lipid A—that forms part of the bacterial outer membrane. **(c)** Lipid A is a phosphoglycolipid consisting of a core hexosamine disaccharide with ester- and amide-linked acylated fatty acid tails arranged in either asymmetric or symmetric arrays that anchor the structure in the membrane. The asymmetric structure of *E. coli* lipid A is shown here. The numbers refer to the number of carbon atoms in the tail.

Sepsis syndrome is a systemic response to invasive pathogens

Sepsis syndrome, or sepsis, is an adverse systemic response to infection that includes fever, rapid heartbeat and respiration, low blood pressure and organ dysfunction associated with compromised circulation. Approximately 250,000–750,000 cases of sepsis occur annually in the United States of America, with mortality ranging from 20% to 50% overall and as high as 90% when shock develops. Sepsis can occur through infection with Gram-positive bacteria and even fungi and viruses, or as a consequence of secreted toxins, which we discuss in the next section. However, the sepsis syndrome occurs commonly in response to lipopolysaccharide (LPS) from Gram-negative bacteria, which will be illustrated here.

Lipopolysaccharide recognition occurs via the innate immune system

LPS is a major constituent of Gram-negative bacterial cell walls (see section 3-0) and is essential for membrane integrity. The portion of LPS that causes shock is the innermost and most highly conserved phosphoglycolipid, lipid A (Figure 9-6), which acts by potently inducing inflammatory responses that are life-threatening when systemic (see below), and is known as bacterial **endotoxin**. Multicellular organisms from horseshoe crabs and fruit flies to humans have evolved proteins specialized for the recognition of LPS. These proteins are found both on the surface of phagocytic cells and as soluble proteins in blood.

LPS is removed by macrophages through scavenger receptors (for example, SR-A) that are highly expressed in the liver and are thus positioned to remove LPS from portal blood draining the intestines, and by neutrophils through the primary granule protein, bactericidal permeability-increasing protein (BPI), which is toxic to Gram-negative bacteria (see section 3-9). The homologous LPS-binding protein, LBP, transfers LPS to membrane-bound or soluble CD14, enabling interactions with Toll-like receptors (TLRs) on the phagocyte membrane (see section 3-10), and to high-density lipoprotein (HDL) particles for removal. In mice and humans the LPS receptor includes CD14, an LPS-interacting moiety, TLR4, the signal transducing element, and MD-2, a small extracellular protein tightly bound to TLR4 (see Figure 3-34). Mice deficient in any of the LPS receptor components are more susceptible to Gram-negative bacterial infection but, at the same time, are less susceptible to the sepsis syndrome.

Sepsis results from the activation of lipopolysaccharide-responsive cells in the bloodstream

TLRs have a lethal function in the septic shock syndrome. The physiological function of signaling through phagocyte TLRs is to induce the release of the cytokines TNF, IL-1, IL-6, IL-8 and IL-12 and trigger the inflammatory response, which is critical to containing bacterial infection in the tissues. However, if infection disseminates in the blood, the widespread activation of phagocytes in the bloodstream is catastrophic.

Humans injected with purified LPS develop a cytokine cascade in the serum (Figure 9-7). The early cytokine response (TNF, IL-6 and IL-8) coincides with the onset of fever and the activation of blood neutrophils, monocytes and lymphocytes. A subsequent increase in the numbers of circulating neutrophils, or **neutrophilia**, is driven by effects of colony stimulating factors, such as G-CSF, whereas the decreased numbers of circulating lymphocytes and monocytes, designated **lymphopenia** and **monocytopenia**, is sustained by their activation-induced exit and retention in peripheral sites. This is followed by a pituitary response and a regulatory or antiinflammatory response (see section 3-15).

Definitions

endotoxin: (of bacteria) a non-secreted toxin inherent in the cell membrane and that induces strong innate inflammatory responses that can be fatal when systemic.

LD$_{50}$: the dose at which 50% of treated individuals die.

lymphopenia: a decrease in the numbers of circulating lymphocytes.

monocytopenia: a decrease in the numbers of circulating monocytes.

neutrophilia: a rise in the number of circulating neutrophils.

protein C: a vitamin K-dependent plasma serine protease that is synthesized in liver and activated by a complex of thrombin bound to thrombomodulin on membranes and that then cleaves coagulant factors VIIIa and Va, inactivating them.

Figure 9-7 Time course of sepsis The clinical manifestations of sepsis are shown above the successive waves of the serum cytokine cascade. (Cytokine concentrations are not shown to scale.) In humans injected with purified LPS, TNF rises almost immediately and peaks at 1.5 h; the sharp decline of TNF may be due to modulation by its soluble receptor sTNFR. A second wave of cytokines that peaks at 3 h activates the acute-phase response in the liver and the systemic pituitary response (via IL-6 and IL-1) and the activation and chemotaxis of neutrophils (via IL-6, IL-8 and G-CSF). Neutrophil activation results in the release of lactoferrin from neutrophil secondary granules; the activation of endothelial procoagulants is shown by the rise of tissue plasminogen activator (t-PA). Pituitary-derived adreno-corticotropic hormone (ACTH) and migration inhibition factor (MIF) peak at 5 h and coincide with peak levels of the regulatory cytokines IL-Ra and IL-10 that counteract the release or activity of inflammatory cytokines. Diffuse endothelial activation is shown by the appearance of soluble E-selectin that peaks at about 8 h and remains elevated for several days.

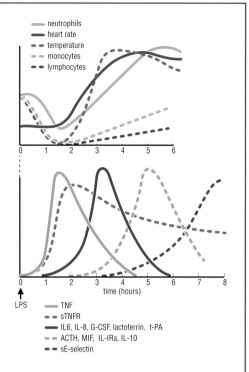

neutrophils
heart rate
temperature
monocytes
lymphocytes

time (hours)

LPS

TNF
sTNFR
IL6, IL-8, G-CSF, lactoferrin, t-PA
ACTH, MIF, IL-IRa, IL-10
sE-selectin

Inflammation leads to widespread endothelial cell activation and organ dysfunction

Cytokine production in the bloodstream results in widespread endothelial cell activation, with expression of adhesion molecules, activation of the coagulation cascade and the production of chemokines and cytokines by the endothelial cells themselves, with consequent amplification of the inflammatory cascade. The adhesion and activation of circulating neutrophils at the endothelium results in both oxidative and elastase-mediated damage, resulting in the loss of vascular integrity and failure to maintain adequate blood pressure. TNF and IL-1 also depress myocardial function directly. Refractory shock, with leakage of edema fluid, and the failure of organs with large capillary beds, such as the lung and kidney, leads to death.

Levels of circulating TNF, IL-6, IL-1 and LPS are directly correlated with the probability of death in humans with sepsis. Despite this, anti-LPS and anti-TNF antibodies, soluble TNF receptors, IL-1Ra and corticosteroids have all failed to alter the outcome of septic shock. Greater success has been achieved with activated **protein C**, an antithrombic, antiinflammatory serine protease activated by thrombin and consumed during sepsis. Levels of activated protein C and antithrombin III are inversely correlated with the probability of death from sepsis, and replacement of activated protein C can reduce the relative risk of death during severe sepsis by almost 20%.

Animal models have helped clarify mechanisms of sepsis

Two widely used models are commonly referred to as the high-dose and low-dose models (Figure 9-8). High-dose LPS challenge involves injection of mice intraperitoneally or intravenously with doses typically between 25 and 100 μg per animal. The **LD50** approximates 150 μg, with death occurring in about 35 hours. Mortality is due to proinflammatory cytokines and widespread endothelial cell injury. Mice with defects in neutrophil adhesion or activation demonstrate a higher LD50 in this assay. The low-dose LPS model relies on concurrent administration (from 1 h before to 4 h after) of D-galactosamine, a potent and specific inhibitor of hepatic macromolecular synthesis. Mice given 300 mg galactosamine/kg (typically 20 mg) have an LD50 of 0.5 ng LPS, with death occurring in about 7 h. In the low-dose model, death is due to massive hepatic necrosis in response to LPS by a process dependent upon TNF and IFN-γ. Mice deficient in the production or recognition of these cytokines demonstrate a higher LD50 in the low-dose model. Mice deficient in the clearance or recognition of LPS itself demonstrate an altered LD50 in both models.

Susceptibility to LPS Toxicity in Gene Knockout Mice

Defect	Protein	High LPS	Low LPS/D-Gal
LPS recognition	CD14	decreased	decreased
	LBP	no change	decreased
	TLR4	decreased	decreased
	MD-2	decreased	decreased
	MyD88	decreased	decreased
	SR-A	enhanced	not done
phagocyte function	Hck/Fgr	decreased	no change
	ICAM-1	decreased	no change
	L-selectin	decreased	not done
	GM-CSF	decreased	not done
	TNFR1	no change	decreased
inflammation	TNFR2	no change	no change
	IL-1Ra	enhanced	not done
	IL-1β	no change	no change
	IFN-γR	decreased	decreased
	caspase 1	decreased	not done

Figure 9-8 Table of susceptibility to LPS toxicity in gene knock-out mice The proteins encoded by the deleted genes are listed. SR-A is scavenger receptor A; Hck and Fgr are Src-family kinases with an essential role in integrin-mediated migration of neutrophils out of the bloodstream. D-Gal: D-galactosamine.

References

Arbour, N.C. et al.: **TLR4 mutations are associated with endotoxin hyporesponsiveness in humans.** Nat. Genet. 2000, **25**:187–191.

Bernard, G.R. et al.: **Efficacy and safety of recombinant human activated protein C for severe sepsis.** N. Engl. J. Med. 2001, **344**:699–709.

Casey, L.C. et al.: **Plasma cytokine and endotoxin levels correlate with survival in patients with the sepsis syndrome.** Ann. Intern. Med. 1993, **119**:771–778.

Faust, S.N. et al.: **Dysfunction of endothelial protein C activation in severe meningococcal sepsis.** N. Engl. J. Med. 2001, **345**:408–416.

Kuhns, D.B. et al.: **Increased circulating cytokines, cytokine antagonists, and E-selectin after intravenous administration of endotoxin in humans.** J. Infect. Dis. 1995, **171**:145–152.

Mira, J.-P. et al.: **Association of TNF2, a TNF-α promoter polymorphism, with septic shock susceptibility and mortality: a multicenter study.** JAMA 1999, **282**:561–568.

Suffredini, A.F.: **Promotion and subsequent inhibition of plasminogen activation after administration of intravenous endotoxin to normal subjects.** N. Engl. J. Med. 1989, **320**:1165–1172.

van der Poll, T. et al.: **Activation of coagulation after administration of tumor necrosis factor to normal subjects.** N. Engl. J. Med. 1990, **322**:1622–1627.

9-4 Sepsis Syndrome: Bacterial Superantigens

Bacterial pyrogenic exotoxins can cause sepsis syndrome

In the early 1980s, an outbreak of fever and hypotension that resembled endotoxin-mediated shock occurred among menstruating young women, predominantly in the United States of America, and was called toxic shock syndrome (TSS). Frequently accompanied by a peeling, sunburn-like rash, vomiting, diarrhea and multiorgan failure, TSS was epidemiologically linked to the use of hyperabsorbent tampons. Although infrequently recovered from blood, *Staphylococcus aureus* was present in the vagina. Biochemical and genetic studies established the presence of a secreted toxin, or **exotoxin**, TSS toxin-1 (TSST-1), that was pyrogenic (fever-inducing) and capable of causing a shock-like syndrome when injected into laboratory animals. Seven related staphylococcal exotoxins—SEA, SEB, SEC1, SEC2, SEC3, SED and SEE—were associated subsequently with diarrheal syndromes and are therefore classified as **enterotoxins**; the SEB and SEC1 enterotoxins also cause nonmenstrual TSS (for example in post-surgical wounds).

From 5% to 25% of clinical *S. aureus* isolates contain the *tst* gene that encodes TSST-1 as part of a larger virulence element. Expression of the exotoxin, however, is controlled by a regulatory protein produced during late-logarithmic growth in response to environmental conditions—including low Mg^{2+}, aerobic conditions with elevated CO_2, high serum proteins and neutral pH—encountered during menses and particularly sustained by use of hyperabsorbent tampons. The discontinuation of hyperabsorbent tampons and public awareness ended the outbreak. Susceptible cases occur among the approximately 20% of individuals who lack neutralizing antibody against the toxin and that is presumably acquired during low-level colonization.

In the late 1980s, invasive group A streptococcal infections with several features of TSS and a mortality approaching 50% appeared in the USA and Europe. Disease was linked to streptococcal pyrogenic exotoxins—usually SPE-A but also -B, -C and SSA. The streptococcal pyrogenic exotoxin A gene, *speA*, is homologous to the staphylococcal *entB* and *entC1* genes.

Subsequent studies have identified similar exotoxins in other organisms (Figure 9-9), although the relationship of exotoxin production with disease remains less clear than with the staphylococcal and streptococcal toxins.

Bacterial pyrogenic exotoxins are superantigens and activate T cells expressing unique Vβ T cell receptors

Bacterial pyrogenic exotoxins are secreted proteins of approximately 200 amino acids and have a common structure consisting of an N-terminal β barrel domain and a C-terminal β-grasp domain. These proteins all have profound effects on T cells which have led to the designation as **superantigen** for these molecules. Unlike conventional peptides or lipids, superantigens require neither antigen processing nor TCR specificity for the superantigen itself. Instead, they bind to regions shared by distinct subsets of TCR Vβ proteins and simultaneously to conserved regions of MHC class II molecules. Optimal T cell activation requires presentation by MHC class II on antigen-presenting cells. Different superantigens accomplish this by different mechanisms (Figure 9-10). MHC class II binds superantigens at two independent sites—a low-affinity site ($K_d \sim 10^{-5}$ M) on the conserved α chain, and a zinc-coordinated high-affinity site ($K_d \sim 10^{-7}$ M) on the polymorphic β chain. SEA binds to both sites and effectively cross-links two MHC molecules. Whereas SPE-C binds to the high-affinity site, SEB, SEC and TSST-1 bind to the low-affinity site. These latter superantigens serve to create a wedge between the TCR and the MHC α chain, displacing the TCR antigen-binding site from the

Microbial Superantigens and the Diseases in which they are Implicated

Superantigen	Disease
Staphylococcal exotoxins	
SEA enterotoxin	food poisoning
SEB enterotoxin	food poisoning; TSS
SEC1 enterotoxin	food poisoning; TSS
SEC2 enterotoxin	food poisoning
SEC3 enterotoxin	food poisoning
SED enterotoxin	food poisoning
SEE enterotoxin	food poisoning
SEA G-L	food poisoning
Toxic-shock-syndrome toxin (TSST-1)	toxic shock syndrome (TSS)
Exfoliative toxins A and B (ETA and ETB)	scalded-skin syndrome
Mycoplasma arthritidis superantigen (MAS)	arthritis, shock
Streptococcal erythrogenic exotoxins	
SPE-A, -B, -C	scarlet fever, strep toxic shock
Streptococcal mitogenic exotoxins	
SPEF, SSA, SPM, SPM-2, SMEZ, SPEG, SPEH, SPEJ, SMEZ-2	unknown
Clostridium perfringens enterotoxin	food poisoning
Yersinia pseudotuberculosis mitogen (YPM)	enteritis, mesenteric adenopathy

Figure 9-9 Table of microbial superantigens and the diseases in which they are implicated

Definitions

enterotoxins: secreted toxins that target intestinal mucosal cells and induce diarrhea by interfering with normal water resorption and/or secretion. Some enterotoxins are **superantigens** and cause diarrhea through secondary effects on mucosal cells of the local release of high levels of host cytokines.

exotoxin: molecule secreted from pathogens that has toxic effects on host cells.

superantigen: protein produced by bacteria or viruses that bind both to the antigen receptor chains of T lymphocytes and to the MHC molecules that they recognize and thereby strongly activate T cells. Most known superantigens are bacterial **exotoxins** that cause septic shock or diarrhea.

References

Bonfoco, E. *et al.*: **Inducible nonlymphoid expression of Fas ligand is responsible for superantigen-induced peripheral deletion of T cells.** *Immunity* 1998, **9**:711–720.

Choi, Y. *et al.*: **A superantigen encoded in the open reading frame of the 3′ long terminal repeat of mouse mammary tumour virus.** *Nature* 1991, **350**:203–207.

Cone, L.A. *et al.*: **Clinical and bacteriologic observations**

©2007 New Science Press Ltd

conventional peptide–MHC complex. Whereas SEB and SEC bind to the side of the peptide-binding site of MHC, TSST-1, SEA and SPE-C make extensive contacts with the buried peptide, so that peptides affect the binding of these toxins to MHC molecules.

Superantigen-mediated interactions between CD4 T cells and antigen-presenting cells results in T cell proliferation and activation with the secretion of large amounts of T cell-derived and antigen-presenting-cell-derived cytokines and inflammatory mediators. Because superantigens bypass the need for antigen specificity and instead bind to β chains shared by many T cells (see Figure 9-10), superantigens can activate from 2% to 15% of T cells in the periphery, a number several orders of magnitude greater than occurs in antigen-specific stimulation. Optimal T cell activation by superantigens requires antigen-presenting-cell-derived costimulatory signals, such as B7, but some activation can occur with TCR cross-linking alone. Activation of T cells by superantigens is often followed by anergy and eventual deletion, perhaps mediated by activation in the absence of costimulation, by superantigen-induced expression of FasL on peripheral tissues, predominantly the liver and small intestine, that initiates the apoptosis of activated, Fas-expressing T cells, and by cytokine deprivation.

Mouse minor lymphocyte stimulating genes are endogenous retroviral superantigens derived from mouse mammary tumor viruses

The *Mls*, or minor lymphocyte stimulating, genes of mice were named for their capacity to stimulate marked CD4 T cell activation in mixed lymphocyte cultures between inbred strains matched at the MHC genes. Mls antigens activate distinct Vβ-expressing CD4 T cells by an MHC class II-dependent mechanism (Figure 9-11). During development, mice expressing a given Mls delete T cells expressing the relevant Vβ subsets. *Mls* genes have been mapped to sites of endogenously integrated mouse mammary tumor proviruses. These integrated retroviruses derive from exogenous mouse mammary tumor viruses (MMTVs), viruses vertically transmitted from mother to offspring in milk and capable of inducing mammary cancer in mice. The mystery of T cell activation by *Mls* was resolved with the discovery that the 3′ long terminal repeats of the endogenous proviruses encode superantigens that are expressed in activated B cells. Intriguingly, immune activation by the viral superantigen is required for the retrovirus to establish productive infection of the breast and accomplish transmission, and T cell deletion by the encoded superantigen abrogates infection by the same retrovirus.

Mice are susceptible to shock induced by superantigens binding to closely similar Vβ chains

Mice and other animals develop shock upon challenge with staphylococcal or streptococcal superantigens. As with LPS, sensitivity to shock is increased by the previous administration of D-galactosamine. The LD_{50} after D-galactosamine is approximately 2 μg within 8 h. The Vβ proteins targeted by superantigens are those most closely related in sequence to the corresponding human T cell Vβ genes. Although sensitivity to superantigen-mediated shock is less in mice with gene deficiencies that attenuate TNF and IFN-γ, deletions in CD14, LBP and TLR4 do not affect sensitivity to superantigens, indicating that the response to LPS does not contribute to this form of sepsis. In contrast, deletion of the relevant Vβ-bearing T cell subsets (or MHC class II) abrogates sensitivity.

Figure 9-11 **Table of representative endogenous MMTV-encoded superantigens**

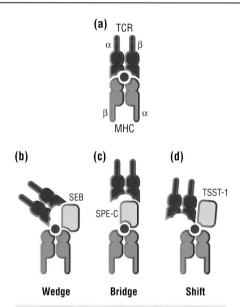

Figure 9-10 Diverse modes of superantigen binding All superantigens bind to both the MHC class II molecule and the TCR so that they are held together, but they bind in quite different ways. **(a)** TCR bound to the peptide–MHC complex in the absence of superantigen; **(b)** schematic representation of SEB binding as a wedge, resulting in the tilting of the TCR relative to the peptide–MHC complex; **(c)** schematic representation of the binding of SpeC, which binds between the TCR and the peptide–MHC complex, forming a bridge connecting the two; **(d)** schematic representation of TSST-1 binding so as to displace the TCR from the peptide, resulting in a shift, with the TCR bound to the MHC β chain.

Representative Endogenous Mouse Mammary Tumor Virus-encoded Superantigens

Strain	H-2	MMTV	Vβ Deleted
BALB/c	d	6, 8, 9	3, 5, 11, 12
C57BL/6	b	8, 9, 17	5, 7, 12 (partial)
C3H/HeJ	k	3, 5, 11, 12	1, 8, 11
DBA/2	d	1, 6, 13, 17	3, 5, 6, 7, 8.1, 9, 11, 12

of a toxic shock-like syndrome due to *Streptococcus pyogenes*. *N. Eng. J. Med.* 1987, **317**:146–149.

Fitzgerald, J.R. *et al.*: **Evolutionary genomics of *Staphylococcus aureus*: insights into the origin of methicillin-resistant strains and the toxic shock syndrome epidemic.** *Proc. Natl Acad. Sci. USA* 2001, **98**:8821–8826.

Held, W. *et al.*: **Superantigen-induced immune stimulation amplifies mouse mammary tumor virus infection and allows virus transmission.** *Cell* 1993, **74**:529–540.

Li, H. *et al.*: **Three-dimensional structure of the complex between a T cell receptor β chain and the superantigen staphylococcal enterotoxin B.** *Immunity* 1998, **9**:807–816.

Li, Y. *et al.*: **Crystal structure of a superantigen bound to the high-affinity, zinc-dependent site on MHC class II.** *Immunity* 2001, **14**:93–104.

Shands, K.N. *et al.*: **Toxic-shock syndrome in menstruating women: association with tampon use and *Staphylococcus aureus* and clinical features in 52 cases.** *N. Engl. J. Med.* 1980, **303**:1436–1442.

Streptococcus pneumoniae is a common cause of bacterial disease with significant mortality

Streptococcus pneumoniae, the pneumococcus, is probably the single most common cause of bacterial pneumonia, meningitis, septicemia and otitis media (middle ear infection). In the USA, seven million cases of childhood otitis media, 500,000 cases of pneumonia, 50,000 cases of bacteremia and 3,000 cases of meningitis occur annually. Despite effective antibiotics, mortality of invasive disease remains over 20% in adults. The emergence of widespread penicillin resistance in the pneumococcus has refocused attention on vaccine strategies to control disease better.

Pneumococci disseminate widely in susceptible hosts

The pneumococcus is a Gram-positive streptococcus covered by a complex capsular polysaccharide (see section 3-2): over 90 antigenically distinct types, or *serotypes*, have been distinguished. Organisms intermittently colonize the human pharynx, usually for periods up to six weeks. Persistent colonization is associated with conversion of colonies from opaque to transparent appearance, reflecting phase variation in capsule formation. The genomic sequence reveals a capacity to use a wide range of polysaccharide and hexosamine nutrients with a substantial commitment to sugar transporters, consistent with the organisms' niche in the oral cavity.

Infections occur when pneumococci are microaspirated into the lungs, where they bind to alveolar type II cells, or when they are refluxed into contiguous, normally sterile spaces, such as the sinuses or inner ear. Under most circumstances the mucous layer overlying ciliated epithelia at these sites removes potentially pathogenic pneumococci. However, conditions that compromise the structural integrity of the mucociliary apparatus, including previous upper respiratory tract viral infections (which damage epithelial cells), smoking and emphysema, or that create access to normally sterile tissues, including head trauma or brain surgery, allow the rapid growth of organisms to levels at which host inflammatory responses occur.

Pneumococci attach to and cross nasopharyngeal epithelium by means of the polymeric Ig receptor. Phosphocholine in the cell wall tethers choline-binding proteins, including choline-binding protein A, CbpA, which binds the secretory component domain in the receptor and initiates reverse transport from the apical membrane. Pneumococcal cell wall and membrane constituents activate a vigorous innate immune response, with the release of inflammatory cytokines and chemokines, and initiation of the procoagulant cascade. Inflammatory cytokines are believed to upregulate the expression of the polymeric Ig receptor, which is expressed normally in a decreasing gradient on epithelium from the upper to the lower respiratory tract, but also **platelet-activating factor receptors (PAF receptors)** on lung epithelium and endothelium. Phosphocholine mediates the binding of pneumococcal cell wall constituents to PAF receptors, further enhancing inflammation. Lysis of pneumococci leads to the release of toxic proteins, including pneumolysin, a pore-forming cytotoxin, that damage endothelial and epithelial barriers and lead to leakage of edema fluid and extensive fibrin deposition. Pneumococci can thence gain access to the bloodstream and disseminate widely through the body, a complication that occurs in approximately 30% of cases.

Control of pneumococci requires anti-capsular antibody and phagocytes

The control of progressive infection depends upon the phagocytosis and destruction of pneumococci by alveolar macrophages and neutrophils. The capsular polysaccharide, which is required for virulence, impedes efficient phagocytosis. Innate recognition, both by opsonins

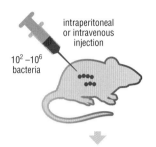

intraperitoneal or intravenous injection

10^2–10^6 bacteria

dissemination through blood and CSF and death of half of the mice in three days

Manipulations affecting susceptibility	
Gene manipulation	Effect on susceptibility
NF-κB p50 knockout	increase ↑
TNFR1 knockout	increase ↑
CRP transgene	decrease ↓

Figure 9-12 Mouse model 1 Mice are inoculated intraperitoneally or intravenously and death occurs within three days with dissemination through the blood and cerebral spinal fluid. Mice with deletion of NF-κB p50 are susceptible to *Streptococcus pneumoniae* but not to various Gram-negative bacteria, indicating differences in the responses to these distinct classes of pathogens. TNFR1 mice are also susceptible. Mice susceptible to other streptococcal species and likely to be susceptible to pneumococci include animals deficient in complement factors C3 or C4, and SP-A. Mice expressing a human CRP transgene had lower mortality when challenged with *S. pneumoniae*. Mice deficient in the polymeric Ig receptor or its signal-transducing kinase, P62^{yes}, are resistant to colonization and invasion after intranasal inoculation. CRP: C-reactive protein (see section 3-15).

Definitions

PAF receptor: see **platelet-activating factor receptor.**

platelet-activating factor receptor (PAF receptor): receptor for platelet-activating factor that is expressed on neutrophils and can enhance neutrophil–endothelial adhesion in response to endothelial activation and production of platelet-activating factor (PAF), a substituted glycerol phosphate derivative.

References

Eskola, J. *et al.*: **Efficacy of a pneumococcal conjugate vaccine against acute otitis media.** *N. Engl. J. Med.* 2001, **344**:403–409.

Jansen, W.T.M. *et al.*: **Fcγ receptor polymorphisms determine the magnitude of in vitro phagocytosis of Streptococcus pneumoniae mediated by pneumococcal conjugate sera.** *J. Infect. Dis.* 1999, **180**:888–891.

Martin, F. *et al.*: **Marginal zone and B1 B cells unite in the early response against T-independent blood-**

borne particulate antigens. *Immunity* 2001, **14**:617–629.

Mi, Q.-s. *et al.*: **Highly reduced protection against Streptococcus pneumoniae after deletion of a single heavy chain gene in mouse.** *Proc. Natl Acad. Sci. USA* 2000, **97**:6031–6036.

Musher, D.M.: **Infections caused by Streptococcus pneumoniae: clinical spectrum, pathogenesis, immunity and treatment.** *Clin. Infect. Dis.* 1992, **14**:801–807.

including C-reactive protein (CRP) and SP-A and by T-independent IgM secreted by serosal B1 cells and splenic marginal zone B cells, is believed to be critical for the restriction of early invasion. Although interaction of these opsonins with constituents of the cell wall, particularly phosphocholine, results in complement activation, the thick Gram-positive peptidoglycan cell wall keeps the bacterial cell membrane at a safe distance from the complement membrane-attack complex. Additionally, the pneumococcal surface protein A, PspA, interferes with the rate of C3b deposition on the bacterial surface, thus slowing the generation of iC3b. In the absence of efficient phagocytosis, infection proceeds, resulting in the continued recruitment and activation of neutrophils with ongoing inflammation and tissue destruction, particularly in a confined space like the meninges and brain. The inflammatory response to pneumococci is thus a two-edged sword responsible both for the limitation of infection and for the destruction of tissue that leads to morbidity and mortality.

Infection is ultimately cleared by activation of the adaptive immune system, resulting in the generation of high-affinity IgG recognizing exposed determinants of the polysaccharide capsule that are unique to each serotype. The external deposition of IgG and subsequent decoration with C3b allows effective recognition by phagocyte complement and FcγRIIA (CD32) receptors; the latter provide particularly strong signals for phagocytosis (see section 3-8). In humans, IgG2 antibodies constitute the major pneumococcal opsonins and mediate successful bacterial clearance, which occurs predominantly in the liver. A polymorphism (H131R) in the membrane-proximal Ig domain of the FcγRIIA receptor reduces affinity for IgG2 and may influence the course of human infection.

Adults at particular risk of disease are those with impaired mucociliary clearance (post-viral pneumonia, smoking, lung cancer, primary ciliary dysfunction), attenuated bacterial clearance (inherited MBL or complement deficiencies, liver disease) or abrogated antibody production (HIV infection, malnourishment, alcoholism, hypogammaglobulinemia, multiple myeloma). Splenectomy due to surgery or autoinfarction (as in sickle-cell disease) confers risk for high-grade pneumococcal bacteremia due to the combination of impaired bacterial clearance and deficient antibody production. Rare genetic deficiencies in IRAK4, a TLR signaling component, and components of the IκB-kinase (IKK) complex are also accompanied by increased susceptibility to pneumococcal infections.

The observation that passive antibody is effective in both human and animal infections led to the development of a pneumococcal vaccine. Consisting of 23 prevalent polysaccharide types, the vaccine is recommended for individuals in high-risk groups as well as for persons over 65 years of age. Although the polysaccharide vaccine is poorly immunogenic in young children, a protein-conjugate polysaccharide vaccine containing seven prevalent serotypes is recommended for all children less than 2 years of age and for high-risk children to age 5. These vaccines are discussed in Chapter 14.

Mice are susceptible to *Streptococcus pneumoniae* and can be used to analyze factors in susceptibility and to test therapeutic strategies: two mouse models are illustrated in Figures 9-12 and 9-13. Isolates vary widely in virulence, with LD$_{50}$ ranges from fewer than 100 to over 10^6 bacteria. Pathogenic human isolates must be titered within individual laboratories, although some reference strains have been characterized.

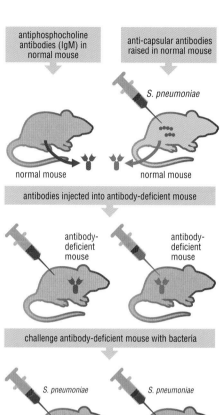

antiphosphocholine antibodies (IgM) in normal mouse

anti-capsular antibodies raised in normal mouse

S. pneumoniae

normal mouse   normal mouse

antibodies injected into antibody-deficient mouse

antibody-deficient mouse   antibody-deficient mouse

challenge antibody-deficient mouse with bacteria

S. pneumoniae   *S. pneumoniae*

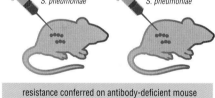

resistance conferred on antibody-deficient mouse

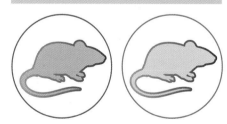

Figure 9-13 Mouse model 2: antibody-deficient mice Mice deficient in germline-encoded IgM (for example *btk*-deficient) are reconstituted with preimmune or serotype-specific antisera from normal mice immunized with heat-killed *S. pneumoniae*. In this model, either anti-capsular IgG or high-titer, germline-encoded antiphosphocholine antibodies are capable of conferring resistance. Indeed, mice deficient in the canonical heavy-chain gene, *V1*, used by germline-encoded antiphosphocholine antibodies, fail to produce high-affinity antibody after immunization and remain more susceptible to *S. pneumoniae*, suggesting evolutionary pressure to maintain the germline-encoded repertoire produced by B1 cells.

Ren, B *et al.*: **The virulence function of *Streptococcus pneumoniae* surface protein A involves inhibition of complement activation and impairment of complement receptor-mediated protection.** *J. Immunol.* 2004, **173**:7506–7512.

Stein, M.-P. *et al.*: **C-reactive protein binding to FcγRIIa on human monocytes and neutrophils is allele-specific.** *J. Clin. Invest.* 2000, **105**:369–376.

Tettelin, H. *et al.*: **Complete genome sequence of a virulent isolate of *Streptococcus pneumoniae*.** *Science* 2001, **293**:498–506.

Tuomanen, E.I. *et al.*: **Pathogenesis of pneumococcal infection.** *N. Engl. J. Med.* 1995, **332**:1280–1284.

Zhang, J.-R. *et al.*: **The polymeric immunoglobulin receptor translocates pneumococci across human nasopharyngeal cells.** *Cell* 2000, **102**:827–837.

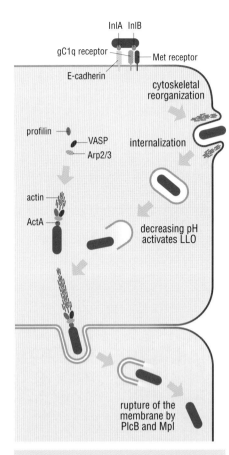

Figure 9-14 Intracellular and intercellular life-cycles of *L. monocytogenes* Beginning at the top, *Listeria* uses the internalin A and B (InlA, InlB) receptors to invade through E-cadherin, and the Met receptor kinase/gC1qR receptors to activate cytoskeletal reorganization leading to uptake within a phagocytic vacuole. Activation of LLO expression by acidification leads to escape to the cytosol, with LLO targeted for degradation via PEST sequences. Cytosolic *Listeria* asymmetrically binds ActA protein at one end, which docks host VASP and profilin complexes, leading to the directional assembly of actin filaments that propel the bacteria through the cell. At the membrane, pseudopods drive *Listeria* into neighboring cells, thus bypassing the extracellular milieu. Rupture of the double membrane is accomplished after enzymatic activation of phospholipases (predominantly plcB) by the metalloproteinase Mpl, thus re-establishing intracellular infection.

Listeria monocytogenes can be lethal to individuals with lowered cell-mediated immunity

Listeria monocytogenes is a Gram-positive bacillus that is environmentally widespread but particularly associated with farm and dairy animals. Epidemiologic evidence has repeatedly linked disease to ingestion of foods or beverages that have become contaminated during their preparation or storage. In normal adults, *Listeria* causes asymptomatic infection or self-limited gastrointestinal disease with low-grade fever, cramps and diarrhea that resolves without therapy. However, individuals with compromised cell-mediated immunity, including those on steroids or receiving cancer chemotherapy, or with AIDS, can suffer devastating bacteremia, frequently complicated by life-threatening meningitis and meningoencephalitis. *Listeria* is particularly dangerous to neonates and to pregnant women in late gestation, presumably reflecting the incompletely developed and suppressed cellular immunity that respectively accompany these conditions.

Listeria is a facultative intracellular organism

The cell biology of *Listeria* is a remarkable example of the adaptation of pathogens to the host. The organism directly penetrates epithelial cells of the small bowel, and infects primarily hepatocytes and macrophages; epithelial and endothelial cells can also be infected. Attachment and invasion are mediated by at least two **internalins**, proteins belonging to a large family of proteins characterized by a region made of leucine-rich repeats (LRRs) and that are covalently linked to the peptidoglycan. The receptor for internalin A is E-cadherin, a cell adhesion molecule and a major constituent of *adherens junctions* in polarized epithelial cells. *Listeria* may gain access to E-cadherin exposed during the normal turnover of aged cells from the tips of the intestinal villae. Rodent E-cadherin does not bind internalin A, and mice are thus not good models for orally acquired infection unless engineered to express human E-cadherin, which contains a critical proline, rather than glutamic acid, at position 16. A second ligand, internalin B, binds Met, a receptor tyrosine kinase whose natural ligand is hepatocyte growth factor, as well as gC1qR, a ubiquitous receptor for diverse innate ligands, including C1q.

Engagement of the internalin LRRs by these receptors induces host cell cytoskeletal reorganization and internalization of *Listeria* into a vacuole (Figure 9-14). *Listeria* escape from phagocytic vacuoles using a pore-forming, acid-activated hemolysin, **listeriolysin O (LLO)**, a member of the cholesterol-dependent cytolysin (CDC) family. In the cytosol, a sequence known as the **PEST sequence** in LLO targets the cytolysin for degradation, so that the cell is not killed and the organisms can replicate. After division, the asymmetrically distributed surface protein ActA on the cytosolic bacteria activates the actin-nucleating **Arp2/3 complex** which rapidly nucleates polymerization of actin filaments that propel organisms through cells at rates up to 1 µm/s. Contact with the host cell membrane creates protrusions into neighboring cells, from which bacteria escape using metalloproteinase and two lecithinase enzymes, together with LLO. Thus, bacteria traverse two membranes and establish infectivity in adjoining cells without ever leaving the cell interior (see Figure 9-14). The genes encoding the proteins required for intracellular residence are expressed coordinately from a single pathogenicity island, LIPI-1 (*Listeria* pathogenicity island 1). Intriguingly, activation of these genes may be signaled through the recognition of glucose phosphates present within the host cytosolic compartment. Although *Listeria* is capable of extracellular growth, it is likely that the pathogenesis reflects prolonged residence in intracellular compartments and the ability to establish niches within the central nervous system, placenta and fetus. The expression of E–cadherin and Met receptors at the blood-brain and fetoplacental barriers probably underlies this tropism.

Definitions

Arp2/3 complex: a complex including two actin-related proteins, Arp2 and Arp3 and that initiates the polymerization of branched-chain actin filaments. It is normally activated by proteins of the Wiskott–Aldrich syndrome (WASP) and Scar/WAVE families, thus linking cytoskeletal remodeling with various signaling pathways. *Listeria* ActA mimics these proteins, establishing bacterial control of the actin activating pathway.

internalins: protein products of a *Listeria* gene family containing leucine-rich repeats; most are anchored covalently in the peptidoglycan cell wall; inlA and inlB are invasins sufficient to trigger uptake by cells expressing the appropriate receptors.

listeriolysin O (LLO): member of the cholesterol-dependent cytolysin family that includes streptolysin O from *Streptococcus pyogenes*.

LLO: see **listeriolysin O**.

PEST sequence: amino acid sequence (P, proline; E, glutamate; S, serine; T, threonine) exposed on cytosolic proteins and constituting one mechanism for targeting proteins for degradation by proteosomes.

Host immunity requires a type 1 response with T$_H$1 cells and cytotoxic T cells

Defense against *Listeria* involves a wide range of innate and adaptive responses. Neutrophils and NK cells are important in limiting early infection through phagocytosis and the production of IFN-γ, respectively. T cell-deficient mice are ultimately unable to control infection, however, reflecting the strict requirement for T cell-mediated immunity. The organism potently induces IL-12 from macrophages and dendritic cells, which serves to promote the development of effector T$_H$1 cells and cytotoxic T cells. A complex multicellular response ensues in which, as established by experiments with mice deficient in different components of the cellular immune response, CD8 T cells are more important than CD4 T cells, and αβ T cells are more important than γδ T cells, which are more important than NK cells. γδ T cells are required for efficient granuloma formation that is mediated by αβ T cells and that serves to sequester organisms before their destruction by cytotoxic T cells, activated macrophages and TipDCs. TipDCs (see section 5-11) are a population of specialized monocytes mobilized and recruited to sites of bacterial invasion, where they produce large amounts of TNF and nitric oxide. In some strains of mice, cytotoxic T cells that recognize the nonpolymorphic H2-M3 MHC class I molecule can also be activated early in the response (see section 8-5). Humans lack H2-M3. The development of effective cytotoxic T cells is dependent on cross-priming by dendritic cells (see section 4-9). Perforin is required for optimal cytotoxic T cell effector function, and IFN-γ is required for the optimal activation of infected macrophages to a microbicidal state, and perhaps in regulating optimal MHC molecule expression on infected cells.

The importance of different components of the immune response to intracellular pathogens can be evaluated in mice infected with *Listeria*

Listeria is an excellent model for evaluating the breadth of immunity to an intracellular pathogen. Mice with defects in neutrophil function, macrophage microbicidal systems or T cell-dependent immunity show distinct deficiencies in their response to *Listeria*. Antibody has only a minor role. Efficient oral infection of mice requires transgenic expression of human E-cadherin in the intestine.

Mice inoculated intraperitoneally or intravenously with a sublethal dose of *Listeria*, usually 5×10^3 bacteria, support the replication of organisms in macrophages throughout the body, but particularly in the spleen and liver, as well as in hepatocytes (Figure 9-15). Replication peaks after 5–7 days. Defects in neutrophil function (adhesion, chemotaxis, killing) or NK cells, or in the production of IL-1, IL-6 or TNF, lead to increased early growth and dissemination of bacteria. Subsequently, bacterial load decreases and organisms are cleared by 14 days. Defects in CD8 or CD4 T cells, in αβ and, to a lesser extent, γδ T cells, or in IL-12, IFN-γ and NOS2 impair immunity. For reasons incompletely known, mice deficient in the production or response to type 1 interferons are resistant to *Listeria*, suggesting that the bacterium exploits activation of this innate antiviral pathway to facilitate its growth. On recovery, mice are immune to rechallenge with a lethal dose of bacteria, typically 10^5 organisms: immunity can be transferred to other mice through CD8 T cells; co-transfer of immune CD4 T cells increases the protection. Listeriolysin O peptides have been identified as immunodominant targets of protective cytotoxic T cells in mice. Macrophages doubly deficient in NOS2 and Phox enzymes are completely unable to restrain intracellular growth of virulent *Listeria* and animals with these deficiencies die of disseminated infection.

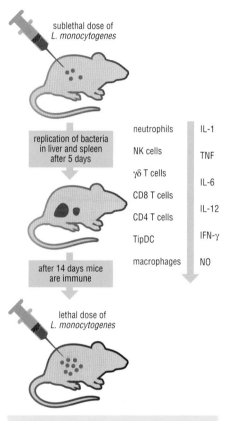

Figure 9-15 The immune response to *L. monocytogenes* Mice inoculated intravenously with a sublethal *Listeria* dose support peak replication in the liver and spleen after 5 days. After 14 days, mice are immune to challenge with an otherwise lethal dose. The cells (left column) and effector pathways (right column), in the approximate temporal sequence in which they become activated in the infected mice, are depicted next to the arrow. Note that cell types and effector molecules are displayed temporally and do not necessarily represent production by that cell or only that cell.

References

Chico-Calero, I. *et al.*: **Hpt, a bacterial homolog of the microsomal glucose-6-phosphate translocase, mediates rapid intracellular proliferation in *Listeria*.** *Proc. Natl Acad. Sci. USA* 2002, **99**:431–436.

Decatur, A.L. and Portnoy, D.A.: **A PEST-like sequence in listeriolysin O essential for *Listeria monocytogenes* pathogenicity.** *Science* 2000, **290**:992–995.

Hsieh, C.-S. *et al.*: **Development of T$_H$1 CD4$^+$ T cells through IL-12 produced by *Listeria*-induced**

macrophages. *Science* 1993, **260**:547–549.

Lecuit, M. *et al.*: **A transgenic model for listeriosis: role of internalin in crossing the intestinal barrier.** *Science* 2001, **292**:1722–1725.

Mombaerts, P. *et al.*: **Different roles of αβ and γδ T cells in immunity against an intracellular bacterial pathogen.** *Nature* 1993, **365**:53–56.

Pamer, E.G.: **Immune responses to *Listeria monocytogenes*.** *Nat. Rev. Immunol.* 2004, **4**:812–823.

Schlech, W.F 3rd.: **Foodborne listeriosis.** *Clin. Infect. Dis.* 2000, **31**:770–775.

Vázquez-Boland, J.A. *et al.*: **Listeria pathogenesis and molecular virulence determinants.** *Clin. Microbiol. Rev.* 2001, **14**:584–640.

Tuberculosis is a major world health problem

M. tuberculosis, the cause of human tuberculosis, is estimated to infect one-third of the world's population and to be alone the eighth most common cause of death. Mycobacteria are widely prevalent soil organisms characterized by the addition of a complex array of lipidoglycans to the peptidoglycan cell wall that protect them from environmental desiccation. *M. tuberculosis* represents the outgrowth of a clonal variant that became highly adapted for human parasitism some 20,000 years ago, an event that is likely to have occurred in Africa.

M. tuberculosis is transmitted by respiratory aerosol from a diseased patient to noninfected recipients. Organisms remain viable in small airborne respiratory droplets and are inhaled into terminal bronchioles and alveoli. The infectious dose is believed to be fewer than 100 bacteria. Most infected individuals remain asymptomatic or develop a self-limited respiratory illness— only 2–4% of immunocompetent contacts will develop disease within a year. By comparison, more than one-third of persons infected with HIV develop disease within five months. Similarly, any individuals with compromised cellular immunity—steroids or cancer chemotherapy, pregnancy, newborns—are at increased risk of developing severe infection with widespread dissemination (so-called *miliary TB*) at the time of initial infection. The success of *M. tuberculosis* as a pathogen resides in its capacity to establish chronic infection complicated by reactivation and disease later, thus sustaining transmission to new hosts (Figure 9-16).

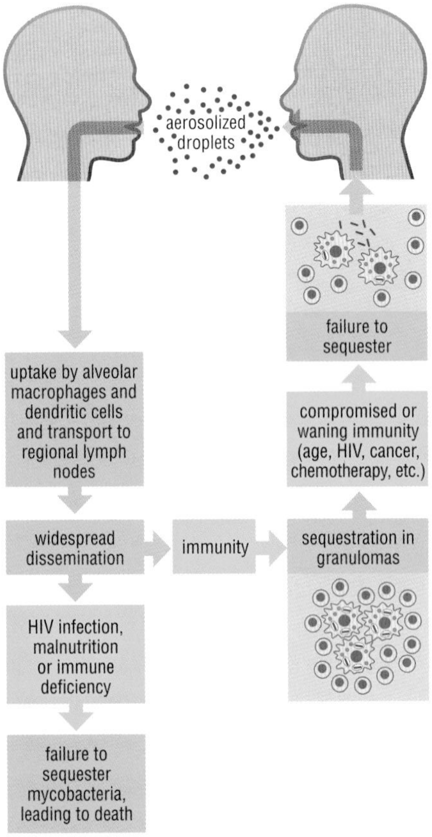

Mycobacterium tuberculosis establishes chronic infection

After lodging in the respiratory tract, organisms translocate across epithelia at M cells, drain to the pulmonary lymph nodes and thence disseminate widely through the bloodstream. *M. tuberculosis* activates the complement system leading to opsonization and ingestion by macrophages, primarily via CR3. After ingestion, mycobacteria block the normal maturation of the phagolysosome, replicate within this unique intracellular compartment and eventually destroy cells by activating endogenous autophagic and apoptotic pathways. The mycolic acid- and lipid-rich cell wall of mycobacteria is a powerful adjuvant that drives proinflammatory responses in areas of high bacterial load. Mycobacterial lipoproteins activate IL-12 and NOS2 expression in macrophages via TLR2. The organism is microaerophilic, persisting in areas of relatively high blood flow and oxygen tension, and relatively poor lymphatic drainage which may drive antigen accumulation and persistent inflammation. The hallmark lesion of persistence is the granuloma, induced by chronic T cell-mediated stimulation of macrophages so that the organisms are circumferentially walled off by plump, epithelioid macrophages and multinucleated giant cells surrounded by a collar of proliferating fibroblasts and activated T cells. Viable bacteria survive but grow poorly within the necrotic center of these granulomas; caseous necrosis is the term used to indicate their cheesy, lipid-rich, nature. Bacteria remain dormant for years within mature granulomas, or *tubercles*, primarily in the apical regions of the lung, in the kidney and in areas of the central nervous system. Classically, latent infection is detected by an inflammatory response known as *delayed-type hypersensitivity* to cutaneously injected soluble purified protein derivative of *M. tuberculosis*, which we discuss in Chapter 13): this test is known as the **PPD skin test**.

Control of tuberculosis requires T$_H$1-mediated immunity, and disease occurs when immunity is impaired

The restriction of organisms to macrophages and the sequestration within granulomas requires complex cellular interactions that are incompletely understood but are known to involve T$_H$1-mediated immunity (see section 5-11). Activated cells within granulomas express TNF and

Figure 9-16 Dynamics of immunity in control of *Mycobacterium tuberculosis* Infectious droplets containing *M. tuberculosis* are inhaled into the lungs, taken up by macrophages and dendritic cells, and transported to the regional lymph nodes. Bacteremia and dissemination occur until the organisms are sequestered within granulomas by a process dependent upon T$_H$1-mediated immunity. Individuals with suppressed immunity, including malnourishment and HIV infection, are unable to sequester organisms and die of widespread miliary TB. In immune individuals, sequestered granulomas containing viable organisms persist in well oxygenated sites, such as the apices of the lungs, the kidneys and the meninges. With waning immunity, in the face of aging, malnourishment, HIV infection or immuno-suppression, granulomas break down and viable bacilli are released to resume growth. In the lung, inflammation promotes the access of bacilli into the airspaces, triggering cough, aerosolization of organisms, and transmission to new hosts.

Definitions

bacille Calmette–Guerin (BCG): attenuated *M. bovis* used to vaccinate against tuberculosis. The attenuated bacterium was derived by passage of a virulent strain 230 times between 1908 and 1921; seed lots from original isolates have been separately propagated under varying conditions for use in childhood BCG vaccines.

BCG: see **bacille Calmette–Guerin**.

PPD skin test: a test for infection with *M. tuberculosis* using purified protein derivative from the bacterium. A standardized preparation, PPD-S, is injected intracutaneously on the arm using 5 tuberculin units (0.0001 mg of protein); delayed-type hypersensitivity is quantified after 48–72 h by measuring the width of the indurated (hardened infiltrate) response.

IFN-γ, and mice defective in these cytokines are unable to restrain growth of *M. tuberculosis*. Among healthy infected humans, the risk for reactivation of latent tuberculosis infection is less than 1% per year. Loss of immunity markedly increases the reactivation risk, which reaches 10% per year among HIV-infected individuals. Patients with rheumatoid arthritis or Crohn's disease treated with anti-TNF reagents are at increased risk for reactivation of latent TB and progressive primary infections. Transmission occurs primarily with cavitary pulmonary disease, when coughing releases organisms in large numbers into the environment. Cavitary disease reflects the breakdown of pulmonary granulomas and the strong proinflammatory components of mycobacterial cell walls. The ensuing tissue destruction leads to invasion of the airspaces and triggering of a strong cough response.

Immune control is not absolute—even patients receiving curative therapy after infection have been reinfected with newly acquired exogenous organisms. A vaccine based on attenuated bovine mycobacterium, **bacille Calmette–Guerin (BCG)**, is widely used as a childhood vaccine in countries with a high incidence of tuberculosis but convincing evidence that it confers protection against *M. tuberculosis* infection is lacking.

Tuberculosis antigens stimulate γδ T cells and T cells recognising CD1

Adult blood γδ T cells express Vγ2/Vδ2 antigen receptors, which react strongly to mycobacterial non-peptide antigens, particularly prenyl pyrophosphates and certain phosphorylated nucleotide conjugates. Cord blood cells react with these antigens, and the number of such cells increases slowly over adult life, presumably reflecting expansion in response to nonpathogenic mycobacteria that are widespread in water and soil. Similarly, some human γδ CD4 CD8 double-negative and CD8 αβ T cells respond to mycobacterial lipoarabinomannan and glycolipids presented on CD1a, b or c molecules (see sections 8-4 and 8-5). Such activated γδ T cells and T cells that recognize CD1 and CD1-restricted cells generally produce inflammatory cytokines, such as IFN-γ, or kill targets infected with mycobacteria.

Inherited deficiencies in T$_H$1 immunity predispose to mycobacteria

Investigations of children unable to control the attenuated *M. bovis* used in the BCG vaccine have uncovered inherited immunodeficiencies in defense against mycobacteria. The most severe are mutations in either chain of the IFN-γR. The phenotype is usually recessive, reflecting a requirement that both alleles of the gene be defective; less frequently, the defect is caused by a frameshift at a mutation hotspot in the IFN-γR1 gene that disables the receptor even in the presence of one normal copy of the gene because the normal chain forms dimers with the mutant chains and these dimers are defective (this is known as a *dominant negative* effect). Affected children are unable to control BCG or common environmental mycobacteria. Intestinal *Salmonella* and *Listeria* infections are prolonged and difficult to control, and granuloma formation is defective. Proportionally less severe are recessive mutations in either the IL-12 p40 gene or the IL-12Rβ1 chain. Disease occurs at a later age in such children and is more readily controlled. Granuloma formation tends to be normal. Recombinant IFN-γ has been an effective treatment. Rare mutations in the gene for STAT1 that exert a dominant negative effect on the nuclear localization of STAT1 dimers in response to IFN-γ and IFN-α/β are also associated with susceptibility to mycobacterial disease. Patients with mutations in NEMO or IκBα associated with anihidrotic ectodermal dysplasia with immunodeficiency also have an increased risk. Mice deficient in IFN-γ, IFN-γR, TNFR1, IL-12 or NOS2 are unable to control *M. tuberculosis* or BCG; larger granulomas and enhanced bacterial load occur in IL-18-deficient mice.

References

Brosch, R. *et al.*: **A new evolutionary scenario for the *Mycobacterium tuberculosis* complex.** *Proc. Natl Acad. Sci. USA* 2002, **99**:3684–3689.

Casanova, J.-L. and Abel, L.: **Genetic dissection of immunity to mycobacteria: the human model.** *Annu. Rev. Immunol.* 2002, **20**:581–620.

Cole, S.T. *et al.*: **Deciphering the biology of *Mycobacterium tuberculosis* from the complete genome sequence.** *Nature* 1998, **393**:537–544.

Flynn, J.L. and Chan, J.: **Immunology of tuberculosis.** *Annu. Rev. Immunol.* 2001, **19**:93–129.

Gutierrez, M.G. *et al.*: **Autophagy is a defense mechanism inhibiting BCG and *Mycobacterium tuberculosis* survival in infected macrophages.** *Cell* 2004, **119**:753–766.

Keane, J. *et al.*: **Tuberculosis associated with infliximab, a tumor necrosis factor α-neutralizing agent.** *N. Engl. J. Med.* 2001, **345**:1098–1104.

Russell, D.G.: ***Mycobacterium tuberculosis*: here today,** and here tomorrow. *Nat. Rev. Mol. Cell Biol.* 2001, **2**:569–586.

Schaible, U.E. *et al.*: **Apoptosis facilitates antigen presentation to T lymphocytes through MHC-I and CD1 in tuberculosis.** *Nat. Med.* 2003, **9**:1039–1046.

van Rie, A. *et al.*: **Exogenous reinfection as a cause of recurrent tuberculosis after curative treatment.** *N. Engl. J. Med.* 1999, **341**:1174–1179.

The Immune Response to Viral Infection

Viruses are the commonest recurring infections of humans. Immunity to viruses is marked by induction of the type 1 interferons and the activation of NK cells. Cytolytic T cells and neutralizing antibodies are major adaptive immune responses that confer long-term protection from viruses. The high mutation rate among viruses, particularly RNA viruses and retroviruses, and the capacity of pathogenic viruses to evade host immunity, are primarily responsible for our inability to establish lasting protection against reinfection by some viruses.

Viruses are ubiquitous and highly diverse

Viruses are responsible for the commonest recurring infections of humans. Although most viral infections are transient and cause only relatively minor discomfort, the economic consequence is enormous. As with other infectious diseases, morbidity and mortality are greatest at the extremes of life, reflecting the combined effects of immature or attenuated immunity and limited physiologic reserves. Increases in the human population, changes in social behavior, the spread and accessibility of rapid transportation and even climate variations have facilitated the dissemination of viruses that have jumped from animal species, such as HIV from chimpanzees in Africa, Hantaviruses from rodents in North and South America, and severe acute respiratory syndrome (SARS) coronavirus, likely to have spread from bats in China. In general, such viruses cause substantially greater mortality than endemic viruses, such as human herpesviruses, that have adapted over greater evolutionary time to their hosts. Type 1 interferons, complement and natural killer (NK) cells provide the major innate, and antibody and cytotoxic T cells (CTL) the major adaptive, protection against viruses.

Despite their existence as obligate intracellular pathogens dependent on host proteins for replication, viruses display enormous diversity. They may contain as few as three or four to up to several hundred genes expressed from single- or double-stranded RNA or DNA genomes. Viral genomes are enclosed within protein coats, or *capsids*, encoded by viral genes and, in the case of enveloped viruses, can be further surrounded by a lipid membrane acquired from the host while budding from infected cells. Different viruses have different effects on cells: *cytopathic (lytic) viruses*, such as poliovirus and influenza virus, lyse host cells by inducing apoptosis or autophagy after completing their replication; non-cytopathic (non-lytic) viruses, such as hepatitis B virus and lymphocytic choriomeningitis virus (LCMV), replicate without destruction of cells. Some cytopathic viruses, such as herpesviruses, can infect cells without producing infectious virions until immunity wanes or until reactivated by various physiologic signals, and are designated *latent viruses*. Despite their dormancy, latent viruses express a limited array of viral genes that function to mask recognition of infected cells by the immune system. Predictably, viruses have evolved several strategies for confounding detection by the immune system and for facilitating their own replication. These include the capture of genes from the host they infect: such captured genes have provided considerable insights into normal cell biology.

Tissue-specific receptors underlie viral tropism

Like bacteria, viruses must overcome epithelial barriers of skin and mucosa to establish infection. Intestinal viruses, such as reovirus in mice or poliovirus in humans, exploit the mucosal antigen-sampling M cells to enter the host. Respiratory viruses, such as influenza viruses and the rhinoviruses that cause colds, establish infection in epithelial cells of the nose and airways. Some viruses, such as dengue virus and West Nile virus, are transmitted by biting insects that serve as vectors. Bloodborne viruses, such as HIV and hepatitis B virus, can invade through mucosa or at sites of epithelial damage due to trauma or injury.

Viruses exploit specific molecules on cells as receptors for invasion. While some viruses bind widely expressed surface molecules, and hence can infect virtually any mammalian cell, most viruses invade using receptors with restricted tissue expression patterns, and are said to display **tissue tropism**. The coevolution of receptor–ligand pairs on the host and the virus contributes in part to the species specificity of most viruses. In some cases, expression of human receptors from transgenes in the mouse has made possible the development of models for the study of

Examples of Viral Receptors Expressed in the Immune System

Receptor family	Host receptor	Virus
Ig superfamily	CD4	HIV
	CD150	measles
complement receptors	CD21	EBV
	CD46	measles, HHV6
	CD55	echoviruses
chemokine receptors	CXCR4, CCR5, etc.	HIV
TNF superfamily	TNFRS14 (HveA)	HSV
C-type lectin receptors	CD209 (DC-SIGN)	HIV, dengue virus
	CD209L (L-SIGN)	SARS coronavirus
Toll receptors	TLR4/CD14	RSV

Figure 10-1 Table of examples of viral receptors expressed in the immune system
HHV6: human herpes simplex virus; HSV: herpes simplex virus; RSV: respiratory syncytial virus.

Definitions

poxvirus: any member of a family of large, enveloped, double-stranded DNA viruses distinguished by replication in the cytoplasm, rather than in the nucleus as occurs for other DNA viruses. Variola (smallpox) and vaccinia (cowpox) viruses are poxviruses.

tissue tropism: (of a virus) restriction of viral growth to a specific tissue, often reflecting tissue-restricted expression of the viral entry receptor, although numerous intracellular processes can further restrict viral infection. An example is hepatitis B virus, which grows only in liver cells.

References

Bomsel, M. and Alfsen, A.: **Entry of viruses through the epithelial barrier: pathogenic trickery.** *Nat. Rev. Mol. Cell Biol.* 2003, **4**:57–68.

Dimitrov, D.S.: **Virus entry: molecular mechanisms and biomedical applications.** *Nat. Rev. Microbiol.* 2004, **2**:109–122.

Goulder, P.J.R. et al.: **Evolution and transmission of stable CTL escape mutations in HIV infection.** *Nature* 2001, **412**:334–338.

Gupta, A. et al.: **Anti-apoptotic function of a microRNA encoded by the HSV-1 latency-associated transcript.** *Nature* 2006, **442**:82–85.

Li, W. et al.: **Bats are natural reservoirs of SARS-like coronaviruses.** *Science* 2005, **310**:676–679.

Seet, B.T. et al.: **Poxviruses and immune evasion.** *Annu. Rev. Immunol.* 2003, **21**:377–423.

Sidorenko, S.P. and Clark, E.A.: **The dual-function CD150 receptor subfamily: the viral attraction.** *Nat. Immunol.* 2003, **4**:19–24.

human viruses that otherwise do not infect rodents. Often, however, additional intracellular interactions required for efficient viral replication can be exquisitely species-specific, and even cell-specific, thus limiting virus promiscuity. In many instances, viruses target cells or receptors of the immune system as a mechanism for achieving immunosubversion or for disseminating systemically (Figure 10-1).

Viruses evade immune responses by diverse mechanisms

Viruses have several strategies for avoiding destruction by host cells, schematically summarized in Figure 10-2. We have already mentioned viral latency, a state in which viral gene expression is restricted to a small number of viral proteins, designated *latent proteins*. Latent proteins sustain the permissive state for viral activation and impede immunologic recognition of infected cells. In herpes simplex virus, latency is sustained by a single viral transcript that encodes a microRNA, which downregulates TGF-β signaling involved in the induction of apoptosis in neurons, where the virus remains. Herpesviruses are latent viruses in humans, with examples including herpes simplex and varicella (chickenpox) viruses, which maintain latency in neuronal cells, and Epstein–Barr virus (EBV), which maintains latency in memory B cells. Latency reflects immunologic détente: CD8 T cells sustain surveillance that limits viral reactivation to episodic bursts (see section 5-16). Severe immunosuppression, as occurs with transplantation, is complicated by the predictable reactivation of latent herpesviruses.

A second important mechanism of viral evasion is mutation. Viruses can accumulate mutations in genes encoding viral coat proteins, thus avoiding detection by antibodies induced by the original, non-mutated proteins. Influenza virus, a common cause of endemic respiratory disease which we discuss in this chapter, evades neutralization through mutation of coat proteins. Viruses can also accumulate mutations that change the repertoire of peptides capable of binding to the host MHC molecules. In this way, the virus can evade detection by CTL or even generate altered peptide ligands (see section 5-2) that anergize CTL reactive to the original virus peptide–MHC complex. Viruses with RNA genomes replicate by means of RNA polymerases, which are more error-prone than DNA polymerases and generate approximately one mutation in the RNA genome with each replication cycle. Hepatitis C virus and HIV are examples of viruses that can generate altered peptide ligands during infection, a process that may contribute to their ability to establish chronic infections.

The third important mechanism of viral evasion is direct interference with or subversion of host antiviral immune mechanisms. Viruses must sustain the machinery of the host cells until replication and assembly have been completed, and often block common pathways that have evolved to prevent their replication. These pathways include apoptosis, or programmed cell death, which effectively destroys the virus factory (see section 3-18); type I interferon signaling, which induces a multitude of pathways that interfere with viral replication (see section 3-16); and MHC class I expression (see section 4-6), which is fundamental to the recognition of cells by cytotoxic CD8 T cells and by NK cells. Viral pathogens must confront these fundamental cellular processes to be successful, although the degree to which different viruses interfere with each pathway varies. For instance, RNA viruses not uncommonly induce apoptosis after completing replication, whereas DNA viruses block apoptosis to ensure the timely completion of replication. In general, viruses with large genomes, such as herpesviruses and **poxviruses**, encode additional proteins that target the host immune response, suggesting a direct relationship between the overall amounts of foreign antigen in the pathogen (size of the genome) and the need to block immune detection in order to replicate. Thus, herpesviruses and poxviruses encode many proteins, so-called *virokines*, which serve as ligands for host cytokines and cytokine receptors. Virokines bind to host receptors for cytokines, chemokines and growth factors, and thus generate agonist signals that modify the host response to promote the survival of infected cells. Similarly, so-called *viroreceptors* function as soluble decoys that bind host cytokines and chemokines to achieve the same outcome (Figure 10-3).

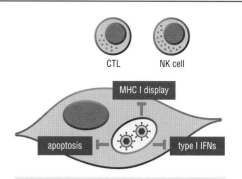

Figure 10-2 Common cellular targets of viral infection Viruses block cellular pathways that activate apoptosis, generate or respond to type 1 interferons, and alter MHC class I expression in ways that abrogate CTL or NK cell recognition.

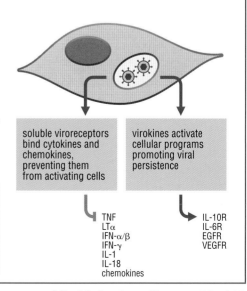

Figure 10-3 Large DNA viruses of the herpesvirus and poxvirus families encode viroreceptors and virokines Viroreceptors are soluble, virus-encoded receptors that bind to key mediators of inflammation, thus preventing them from binding to host receptors. Common targets of viroreceptors are shown. Virokines are virus-encoded cytokines, chemokines and growth factors that bind to host receptors and initiate signals that favor viral persistence. Host cytokines mimicked by common virokines are shown. EGFR: epidermal growth factor receptor; VEGFR: vascular endothelial growth factor receptor.

RNA and DNA sensors activate cells to achieve an antiviral state and to recruit host immune cells

Cells have cytosolic and membrane-associated nucleic acid sensors that serve to activate the antiviral state by producing type 1 interferons (Figure 10-4). Double-stranded RNA (dsRNA), which accumulates in the cytosol during viral infection, binds to the RNA helicases RIG-I and Mda5 (see section 3-16), which have overlapping specificity for different types of viruses. Thus, RIG-I is activated by paramyxoviruses and influenza, whereas Mda5 is activated by picornaviruses. Activation of the helicases facilitates a CARD domain (see section 3-12) interaction with a mitochondrial outer membrane adaptor protein, **mitochondrial activating signaling (MAVS) protein**, which serves to recruit kinase-activating complexes resulting in the activation of NF-κB and IRF3. An unknown cytosolic DNA sensor also activates these transcriptional regulators by a pathway independent of the helicases and MAVS.

A class of Toll-like receptors (TLRs), including TLR3 (dsRNA), TLR 7 and 8 (ssRNA) and TLR9 (CpG DNA, which in herpesviruses is highly enriched) is also specialized for nucleic acid detection (see section 3-10). These TLRs are predominantly in intracellular compartments (see Figure 3-32), and are thought to detect viral nucleic acids in materials phagocytosed from infected, apoptotic cells. TLR activation also initiates a signaling cascade that results in the nuclear translocation of NF-κB and IRF3 (in the case of TLR3) or IRF7 (in the case of TLRs 7, 8 and 9) in cells that express these transcriptional regulators (see below). Components of the MAPK cascade are also activated (see section 3-11). Activated IRF3 localizes to the nucleus, where it directs the transcription of IFN-β, IFN-α4 and the chemokine RANTES. RANTES recruits NK cells and effector T cells, as well as immature dendritic cells, from the blood. The IFNs bind to the type 1 IFN receptor in an autocrine manner, leading to activation and translocation of STAT1–STAT2 complexes to the nucleus, where they direct the transcription of another latent transcriptional regulator, IRF7. Upon further TLR activation, IRF7 becomes phosphorylated and translocates to the nucleus, where it initiates the transcription of the many additional IFN-α genes, thus greatly amplifying the amount of interferon produced. NF-κB (see section 2-10) directs the transcription of a large number of genes involved in inflammation and immunity, including chemokines and cytokines, such as IL-1β, IL-6, TNF, IL-12p40 and IL-15. Mice deficient in MAVS, RIG-I and Mda5, type 1 interferon receptors, STAT1 and STAT2, or IRF3 and IRF7 demonstrate marked viral susceptibility.

Plasmacytoid dendritic cells are specialized producers of IFN-α

Plasmacytoid dendritic cells (pDC), which are specialized to produce type 1 interferons (see section 3-16), constitutively express TLR7, 8 and 9, and IRF7, in addition to IRF3. pDC do not require the RNA helicases, MAVS or priming by IFN-β to generate large amounts of IFN-α, in part because they can hold ingested nucleic acids in endosomes for prolonged

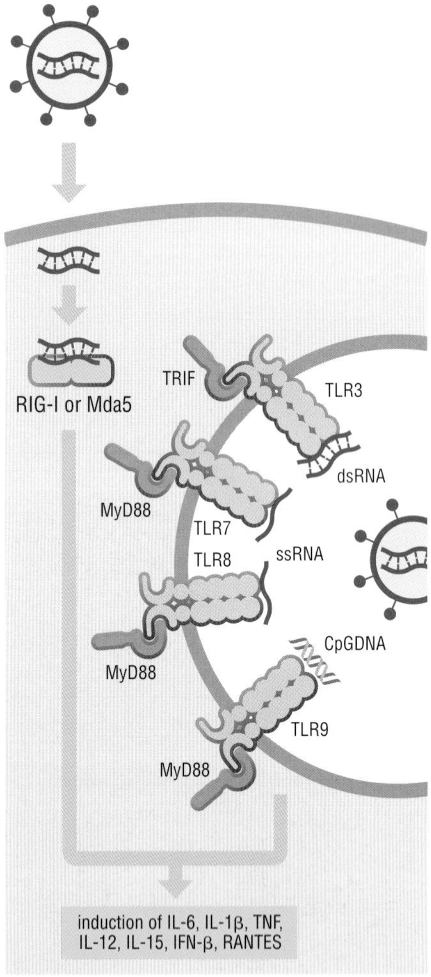

TRIF

RIG-I or Mda5

TLR3

dsRNA

MyD88

TLR7

TLR8

ssRNA

MyD88

CpGDNA

TLR9

MyD88

induction of IL-6, IL-1β, TNF, IL-12, IL-15, IFN-β, RANTES

Figure 10-4 Cell nucleic acid sensors activate the antiviral response In tissue cells, RNA and DNA are bound by cytosolic sensors as well as specialized TLRs to activate pathways that converge on activation of the NF-κB and IRF family of transcriptional regulators, leading to the generation of inflammatory cytokines, IFN-β and RANTES. IFN-β regulates the autocrine and paracrine induction of the many IFN-α family members by activation of the type 1 interferon receptor and induction of IRF-7. Plasmacytoid dendritic cells constitutively express IRF-7 and transcribe the IFN-α genes immediately after the detection of viral infection. TLR stimulation also activates components of the MAP kinase pathway (not shown).

Definitions

MAVS: see **mitochondrial activating signaling protein**.

mitochondrial activating signaling (MAVS) protein: a CARD domain-containing adaptor protein in the mitochondrial outer membrane that links upstream activation of the RNA helicases RIG-I and Mda5 with downstream kinase complexes responsible for activation of the transcriptional regulators NF-κB, IRF3 and IRF7; also known as IPS-1 (IFN-β promoter stimulator-1), VISA (virus-induced signaling adaptor), and Cardif (CARD adaptor inducing IFN-β).

periods, thus facilitating TLR signaling. pDC are the major circulating cell type that produces type 1 interferons after systemic viral infection. After activation, pDC migrate to regional lymphoid tissues to occupy niches surrounding high endothelial venules or to the spleen to localize at areas where T and B cells interact, and play an important part in their activation and survival. Induction of CD69 by IFN-α inhibits egress of the activated lymphocytes from the lymphoid tissues (see section 5-7) and because CD8 T cell proliferative responses are so vigorous, the accumulation of the activated cells in the lymphoid tissues gives rise to the swollen lymph nodes, or lymphadenopathy, often associated with viral infections. Type 1 interferons also promote the migration and maturation of myeloid dendritic cells from peripheral tissues to draining lymph nodes, thus augmenting the adaptive immune response.

NK cells contribute to early antiviral immunity

NK cells respond to virus infection within hours to days, and thus provide protection during the period before the adaptive immune system can respond. Infection with viruses induces NK cell cytotoxicity and the production of cytokines, particularly IFN-γ and TNF superfamily members. NK cell depletion experiments have demonstrated contributions by NK cells to antiviral immunity in a number of experimental systems. Humans with selective NK cell deficiencies suffer severe viral infections, particularly from herpesviruses (see section 8-1).

NK cells constitutively express receptors for, and become activated in response to, the cytokines induced by virus infection, including type 1 interferons, IL-15, TNF, IL-12 and IL-18. These cytokines also induce ligands for NK cell activating receptors on target cells. In this way, the integration of multiple activating signals serves to overcome signals from NK inhibitory receptors that otherwise prevent NK cell activation (see section 8-1). After infection with herpesviruses, human cells express MHC class I-like MIC and ULBP molecules (see section 4-4) that serve as ligands for activating NKG2D receptors. Mouse cells display the comparable RAE molecules after infection with murine cytomegalovirus (MCMV).

Herpesviruses may target NK cell receptors directly in order to evade NK cell killing of infected cells. MCMV infection leads to the expression of a virally encoded MHC-like protein, m157, on the surface of infected cells that binds to Ly49I, a mouse inhibitory NK cell receptor, leading to failure to control viral infection (Figure 10-5). Resistant mice express the homologous Ly49H receptor on NK cells, which is an activating, rather than inhibitory, NK cell receptor (see section 8-2). It is thought that the existence of activating and inhibitory receptors with similar specificities may reflect the evolution of NK cell receptors under pressure from herpesviruses. The viral protein m157 presumably evolved to mimic the natural ligands for Ly49I, and in some mouse strains this has been followed by recombination between the *Ly49I* gene and a gene encoding an activating receptor, resulting in the substitution of activating cytoplasmic domains for inhibitory ones, to generate the activating Ly49H.

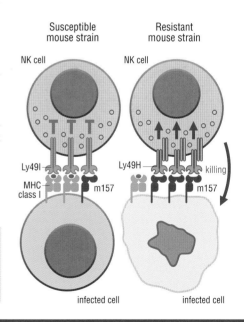

Figure 10-5 Viral proteins can interact directly with NK cell receptors Mouse CMV protein m157 interacts with the NK inhibitory receptor Ly49I, thus attenuating NK cell cytotoxicity in inbred strains that express this receptor. Resistant strains express the related receptor Ly49H, which binds to m157 but no longer binds self MHC and has become an activating receptor, so that it activates lysis of infected cells.

References

Arase, H. *et al.*: **Direct recognition of cytomegalovirus by activating and inhibitory NK cell receptors.** *Science* 2002, **296**:1323–1326.

Cella, M. *et al.*: **Plasmacytoid monocytes migrate to inflamed lymph nodes and produce large amounts of type I interferon.** *Nat. Med.* 1999, **5**:919–923.

Groh, V. *et al.*: **Costimulation of CD8αβ T cells by NKG2D via engagement by MIC induced on virus-infected cells.** *Nat. Immunol.* 2001, **2**:255–260.

Honda, K. *et al.*: **Spatiotemporal regulation of MyD88–IRF-7 signaling for robust type-I interferon induction.** *Nature* 2005, **434**:1035–1040.

Karst, S.M. *et al.*: **STAT1-dependent innate immunity to a Norwalk-like virus.** *Science* 2003, **299**:1575–1578.

Kawai, T. and Akira, S.: **Innate immune recognition of viral infection.** *Nat. Immunol.* 2006, **7**:131–137.

Siegal, F.P. *et al.*: **The nature of the principal type 1 interferon-producing cells in human blood.** *Science* 1999, **284**:1835–1837.

Taniguchi, T. *et al.*: **IRF family of transcription factors as regulators of host defense.** *Annu. Rev. Immunol.* 2001, **19**:623–655.

Viral infections can activate massive oligoclonal expansion of CD8 T cells

Activation of CD8 T cells in response to viral infection results in massive expansion of antigen-specific cells, which reach frequencies as high as 1:3 or 1:4 CD8 T cells at the peak of some anti-viral responses. This has become clear now that the CD8 T cell response to viruses can be recorded in intact animals and in people using antigen-induced cytokine assays (enzyme-linked immunospot assay, or ELISPOT, see section 6-6), and soluble recombinant **MHC tetramers** linked to or loaded with peptides from viral proteins and that bind to and thus identify antigen-specific T cells. Expansion of CD8 T cells on activation is rapid, with the cells dividing every 6–8 h to achieve overall increases of 10^4–10^5-fold above resting levels (Figure 10-6). This expansion is accompanied by differentiation to effector cytotoxic CD8 T cells, which release cytokines, particularly IFN-γ and TNF, kill cells directly through perforin and granzymes, and express receptors that induce migration into non-lymphoid tissue (see section 5-14). IFN-γ and TNF promote inflammatory cell recruitment and activation, but also directly inhibit the replication of some viruses, such as hepatitis B virus, by non-cytopathic mechanisms. In some cases, inflammatory actions of cytotoxic CD8 T cells contribute to pathology by inducing extensive tissue damage. In most instances the virus is eliminated, and the sudden competition for survival factors initiates a massive apoptotic collapse in 90–95% of antigen-specific cytotoxic CD8 T cells (see section 5-15). Thereafter, a small proportion of the clonally expanded CD8 cells, which differ from effector cytotoxic CD8 T cells by expression of distinct survival receptors, such as IL-7R, persists as memory CD8 T cells (see section 5-16). Memory CD8 T cells develop only from cells that have received help from CD4 T cells during their initial activation, and such CD8 T cells can be sustained for many years. These cells reexpress the lymph-node homing receptors CD62L and CCR7, and no longer have immediate effector function upon activation but must first proliferate and differentiate again. Frequencies of 1–5% of peripheral blood CD8 T cells can be demonstrated for T cells specific for dominant epitopes from common recurrent or latent human viruses.

Antibodies are critical for the control of many viruses

Although helper and cytotoxic memory T cells accelerate the secondary response to virus infection by 2–3 days, the appearance of effector cells of the adaptive immune system still takes up to 5 days because of the needs for antigen processing, dendritic cell maturation and migration, and T cell clonal expansion and migration to sites of infection. Immunity against reinfection is mediated by neutralizing antibodies, which are believed to underlie most successful human vaccines. In general, acute, cytolytic viruses, such as influenza, elicit protective neutralizing antibodies relatively quickly after infection. Neutralizing antibodies generally target proteins on the virion surface or on the surface of infected cells. Opsonization by antibody, often with complement, can prevent viruses from binding to cellular

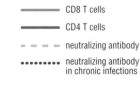

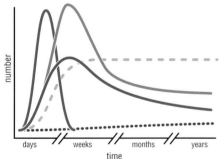

Figure 10-6 The adaptive immune response to acute virus infection Appearance of virus is followed rapidly by a massive expansion of virus-specific CD8 T cells. Upon viral clearance, the vast majority of cytotoxic CD8 T cells undergo apoptosis. CD4 T cell expansion and contraction are of less magnitude. Memory cells represent long-lived cytotoxic CD8 T cells that differentiate after the peak cytotoxic CD8 T cell response and persist at frequencies proportional to the extent of the initial clonal burst by a process dependent upon CD4 T cells. Neutralizing antibodies are produced relatively early after acute infection with cytolytic viruses. In contrast, chronic viruses induce neutralizing antibodies only with prolonged infection, and often these are of low affinity. Adapted from Kaech, S.M. *et al.*: *Nat. Rev. Immunol.* 2002, **2**:251–262.

Definitions

MHC tetramers: recombinant soluble MHC molecules either physically linked or subsequently loaded with a peptide epitope of interest and chemically tetramerized to generate a four-valent binding molecule with sufficient avidity to bind to TCRs on the T cell membrane in an antigen-specific manner. The method has been widely used to study the frequency of antigen-specific T cells to epitopes from viruses and other pathogenic organisms.

receptors, or can lead to damage to virions or infected cells that targets them for removal by phagocyte Fc and complement receptors. Antibody bound to the surface of infected cells (usually IgG1 and IgG3 in humans) displays Fc regions that can bind to the low-affinity IgG receptor, FcγRIII (CD16), expressed on NK cells. Activation of NK cell FcγRIII activates cytotoxicity against the virus-infected cell, a process known as ADCC, or antibody-dependent cell-mediated cytotoxicity (Figure 10-7; and see section 6-3). Mucosal viruses elicit IgA, which is released into mucosal secretions to enable early encounter with the virus upon reexposure. Finally, maternal IgG crosses the placenta, thus providing the newborn with substantial immunity against viral infection during the neonatal period, when protection is also provided by the transfer of maternal immunoglobulins through breast milk to the neonatal intestine.

For reasons not completely understood, viruses that cause chronic infections, such as hepatitis C virus or HIV in humans, or persisting lymphocytic choriomeningitis (LCMV) infections in mice, lead to the production of neutralizing antibodies much more slowly and of much lower affinity and, hence, potency. This may underlie the current difficulty in developing vaccines against these types of viruses. Although the reasons for this failure to generate neutralizing antibodies during chronic infections are unclear, several things may contribute. Thus, critical epitopes tend to be conformationally hidden, and exposed only transiently during the infectious cycle, such as during the membrane fusion process between the viral and host membranes; critical epitopes are commonly surrounded by sites of glycosylation, creating a glycan shield; accessible residues near functional epitopes can sustain mutations without compromising their biologic function; potent CD4 helper T cell responses may enhance the polyclonal expansion of B cells producing low-affinity antibody at the expense of high-affinity neutralizing antibody; cytotoxic CD8 T cells may kill B cells producing high-affinity antibody because of capture and presentation of viral antigens; and such viruses may have evolved to escape interactions with natural antibodies. We discuss some of these issues further in Chapter 14 in connection with vaccines.

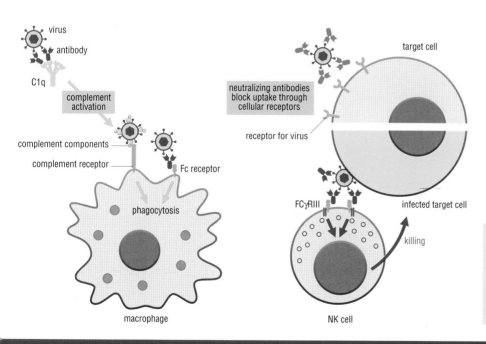

Figure 10-7 Mechanisms of protective neutralizing antibody responses to viruses Antibodies can directly target the virion surface or antigens on the surface of infected cells, thus mediating their destruction by complement, phagocyte-mediated clearance or NK cell-mediated ADCC.

References

Kaech, S.M. *et al.*: **Effector and memory T-cell differentiation: implications for vaccine development.** *Nat. Rev. Immunol.* 2002, **2**:251–262.

Pinschewer, D.D. *et al.*: **Kinetics of protective antibodies are determined by the viral surface antigen.** *J. Clin. Invest.* 2004, **114**:988–993.

Planz, O. *et al.*: **Specific cytotoxic T cells eliminate cells producing neutralizing antibodies.** *Nature* 1996, **382**:726–729.

Recher, M. *et al.*: **Deliberate removal of T cell help improves virus-neutralizing antibody production.** *Nat. Immunol.* 2004, **5**:934–942.

Robbins, J.B. *et al.*: **Perspective: hypothesis: serum IgG antibody is sufficient to confer protection against infectious diseases by inactivating the inoculum.** *J. Infect. Dis.* 1995, **171**:1387–1398.

Wherry, E.J. and Ahmed, R.: **Memory CD8 T-cell differentiation during viral infection.** *J. Virol.* 2004, **78**:5535–5545.

10-3 Subversion of Immune Mechanisms by Viruses

Viral Inhibition of Apoptosis

Cellular target/homolog	Viruses
Fas	adenovirus
death-domain receptors	HHV8, HVS, EHV-2, MCV, BHV4
caspases	cowpox, vaccinia, baculoviruses, adenovirus
Bcl-2 homologs	EBV, HHV8, AFSV, HVS, MHV, adenovirus
p53	adenovirus, SV40, HPV
transcription (inhibits TNF-induced apoptosis)	CMV, Marek's disease virus
oxidative stress	MCV

Figure 10-8 Table of targets of viral inhibition of apoptosis A variety of viruses, most prominently poxviruses and herpesviruses, have developed many mechanisms for interfering with apoptosis of infected cells. The targets of the viral interference mechanisms are listed on the left, the viruses that produce the interfering proteins on the right. Virus names: AFSV: African swine fever virus; BHV4: bovine herpesvirus 4; CMV: human cytomegalovirus; EBV: Epstein–Barr virus; EHV-2: equine herpesvirus 2; HHV8: human herpesvirus 8 (causative agent of Kaposi's sarcoma); HPV: human papillomavirus; HVS: herpesvirus saimiri; MCV: molluscum contagiosum virus (a human poxvirus); MHV: murine herpesvirus 68; SV40: simian virus 40.

Pathogenic viruses have captured genes that target cellular apoptosis, interferon and innate inflammation, and the display of peptide–MHC class I complexes

Just as bacterial pathogens have captured clusters of genes on pathogenicity islands that confer virulence (see section 9-2), pathogenic viruses have captured genes that are not required for growth in tissue culture cells but are essential for infection of animals. In general, these virulence determinants target the cellular apoptotic machinery (see section 3-18), cytokines critical to innate immunity, particularly the type 1 interferons, and the MHC class I pathway involved in the immune recognition of virally infected cells (Figures 10-8 and 10-9). Blocking the cell apoptotic response allows the virus time to complete replication and assembly into virions, and exit from the cell.

Type 1 interferons represent a common target of pathogenic viruses. Hepatitis C virus, an RNA virus responsible for chronic liver infections, encodes a protease that cleaves MAVS from the mitochondrial membrane, thus blocking signaling from the RNA helicases RIG-1 and

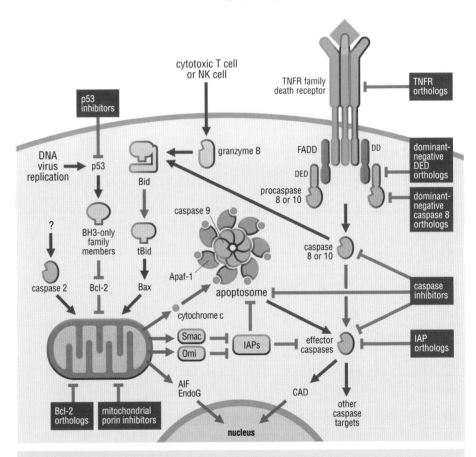

Figure 10-9 Viral antiapoptotic strategies Viruses block the activation of apoptosis in the cell by numerous strategies (highlighted in enclosed rectangles) that interrupt both the extrinsic (death receptor) and intrinsic (mitochondrial) cell death pathways. Adapted from Benedict, C.A. *et al.*: *Nat. Immunol.* 2002, **3**:1013–1018. For a description of vertebrate cell apoptotic pathways, see sections 2-11 and 2-12.

Definitions

Kaposi's sarcoma: multiple, highly vascular, nodules consisting of proliferating spindle-shaped cells, endothelial cells and infiltrating lymphocytes in the skin, mucous membranes and internal organs that occur most frequently in immunocompromised individuals as a consequence of human herpesvirus 8 infection.

Figure 10-10 An IL-6-like virokine from HHV8 promotes escape from effects of type I interferons
Activation of the type I IFN receptor initiates cell-cycle arrest and downregulates surface expression of the human IL-6 receptor, thus allowing cells to escape the negative regulation of IFN signaling by IL-6 through the IL-6 receptor acting through the signal-transducing gp130 subunit. The viral IL-6, vIL-6, is activated by IFN signaling in the cell, and is capable of signaling solely by binding to the gp130 subunit, thus blocking signaling from the type I IFN receptor. Adapted from Chatterjee, M. *et al.*: *Science* 2002, **298**:1432–1435.

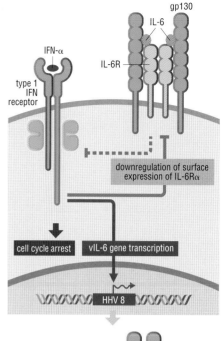

Mda5. Ebola and influenza viruses encode proteins that bind viral RNA, thus masking them from cysolic RNA sensors, and a number of viruses target the STAT1 signaling pathway, thus blocking amplification of the type 1 interferon response. In some cases, virally encoded molecules, so-called virokines and viroreceptors (see section 10-0), interdict similar pathways. A striking example is provided by human herpesvirus 8, the cause of **Kaposi's sarcoma**, which in response to IFN-α signaling secretes an IL-6-like molecule that inhibits host antiviral responses, so that the viral defense response is activated by the signal that launches the host antiviral program (Figure 10-10). Finally, interference with the MHC class I pathway delays the presentation of viral antigens while the simultaneous production of class I decoys or other mechanisms that engage inhibitory receptors abrogates recognition by NK cells. Examples of viral mechanisms for blocking apoptosis are shown in Figures 10-8 and 10-9. Some of the multiple viral strategies for abrogating class I recognition by the immune system are illustrated using cytomegalovirus as an example in Figure 10-11.

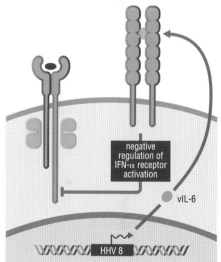

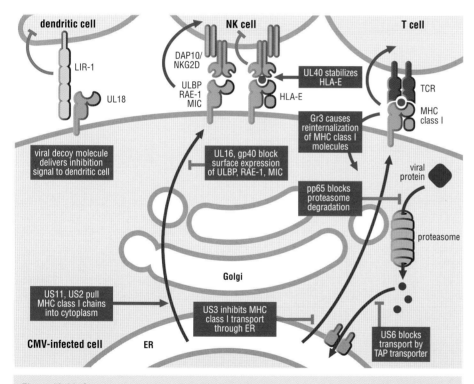

Figure 10-11 Cytomegalovirus targets the MHC class I pathway at multiple sites Human and murine (mCMV) cytomegaloviruses encode a number of molecules that affect MHC antigen expression at the cell surface, thus altering recognition by cytotoxic CD8 T cells and NK cells.

References

Benedict, C.A. *et al.*: **To kill or be killed: viral evasion of apoptosis.** *Nat. Immunol.* 2002, **3**:1013–1018.

Chatterjee, M. *et al.*: **Viral IL-6-induced cell proliferation and immune evasion of interferon activity.** *Science* 2002, **298**:1432–1435.

Collot-Teixeira, S. *et al.*: **Human tumor suppressor p53 and DNA viruses.** *Rev. Med. Virol.* 2004, **14**:301–319.

Gewurz, B.E. *et al.*: **Virus subversion of immunity: a structural perspective.** *Curr. Opin. Immunol.* 2001, **13**:442–450.

Levy, D.E. and García-Sastre, A.: **The virus battles: IFN induction of the antiviral state and mechanisms of viral evasion.** *Cytokine Growth Factor Rev.* 2001, **12**:143–156.

Lodoen, M. *et al.*: **NKG2D-mediated natural killer cell protection against cytomegalovirus is impaired by viral gp40 modulation of retinoic acid early inducible 1 gene molecules.** *J. Exp. Med.* 2003, **197**:1245–1253.

Meylan, E. *et al.*: **Cardif is an adaptor protein in the RIG-I antiviral pathway and is targeted by hepatitis C virus.** *Nature* 2005, **437**:1167–1172.

AIDS originated as a zoonotic disease introduced into humans from primates

Acquired immunedeficiency syndrome (AIDS) is a devastating disease caused by the human immunodeficiency virus (HIV), which infects and eventually destroys CD4 T cells. First recognized clinically in the early 1980s, HIV infection has since grown to an estimated 42 million infected people and 29 million dead. Worldwide approximately 6,000 new infections occur daily, half of which are in women. Two-thirds of infected persons live in Africa, and another one-fifth are in Asia. The vast majority of human AIDS is caused by infection with HIV-1, one of the two known types of HIV. The other, HIV-2, is largely restricted to countries of West Africa and is much less virulent.

HIV-1 and HIV-2 are derived from retroviruses indigenous to African primates. Thus, HIV was originally a **zoonotic infection**, transmitted from animals to humans, although the virus is now spread person-to-person by body fluids. Sequence analysis shows that HIV-1 is derived from a chimpanzee simian immunodeficiency virus (SIV_{cpz}) that occurs in *Pan troglodytes*, the chimpanzee, in Central Africa. Similarly, HIV-2 evolved from SIV_{sm}, which infects sooty mangabey monkeys in West Africa. Both viruses seem highly adapted to their primate hosts, and cause little obvious disease. Central Africa is an area with an active bushmeat trade in which primates are butchered for food, with the risk of human infection by primate viruses.

Sequence analysis of HIV-1 isolates suggests that SIV_{cpz} was introduced stably into humans on at least three occasions, with the last common ancestor entering the human population from chimpanzees around 1930. Molecular diversity remains greatest in regions of Africa where the epidemic originated and where the virus has had longest to diversify, whereas isolates around the world represent selective outgrowth of the virus introduced from different points in its evolutionary adaptation to humans. Three major genetic groups of HIV-1 have been identified, with group M (major) accounting for most worldwide infections; groups N (Cameroon) and O (Gabon) remain largely confined to Africa. HIV-2 has also been introduced into humans on more than one occasion and can be divided into six lineages, designated A to F.

HIV is a retrovirus in the lentivirus subfamily

HIV is the most important human *retrovirus*, one of a diverse group of enveloped RNA viruses called retroviruses because they transcribe viral RNA into DNA, thereby reversing the usual sequence in which DNA is transcribed into RNA. Reverse transcription of the viral RNA generates a double-stranded DNA provirus, which is then stably integrated into the host genome. Among the retroviruses, HIV is in the *lentivirus* subfamily, a group distinguished by a relatively complex genome and specializations allowing them to integrate into non-dividing cells. Other animal lentiviruses often infect immune cells and kill their hosts slowly through effects of anemia, immune deficiency or invasion of the central nervous system. The integrated provirus HIV-1 genome is 10 kilobases and encodes 16 different proteins (Figure 10-12).

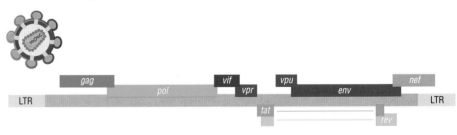

Figure 10-12 The integrated HIV-1 proviral genome LTR; long terminal repeat; retrovirus-associated *gag* (group-specific antigen: the gene encodes the capsid proteins of the viral coat), *pol* (polymerase: the gene encodes the reverse transcriptase that reads the viral RNA into cDNA, the integrase that inserts the cDNA copy into the host genome, and the protease that cleaves the precursor into the separate enzymes), and *env* (envelope; encodes membrane-exposed proteins on the virion that mediate attachment and fusion) genes; the regulatory genes, *tat* (transcriptional transactivator necessary for expression of the viral genes) and *rev* (regulator of virion gene expression exports unspliced and singly spliced viral mRNA into the cytoplasm); and the accessory genes, *nef* (negative regulator; alters endosomal recycling with downregulation of surface molecules such as CD4 and MHC class I), *vif* (viral infectivity factor), *vpr* (viral protein r: important for nuclear import) and *vpu* (viral protein u: required for viral assembly and budding). Overlapping boxes indicate overlapping transcripts; thin connecting lines indicate differentially spliced transcripts from the same gene.

Definitions

endosomal sorting complex required for transport (ESCRT): cellular machinery for sorting ubiquitinated cargo proteins to endosomal membranes for degradation; mediates intra-luminal budding and extrusion to generate a multivesicular body, or MVB.

ESCRT: see **endosomal sorting complex required for transport.**

long terminal repeat (LTR): repeated sequences at the 5′ and 3′ ends of a proviral or transposon genome that, when integrated into the host genome, act as promoters or polyadenylation and termination sites, respectively.

LTR: see **long terminal repeat.**

multivesicular body (MVB): vesicles of the endosomal pathway created when sections of endosome membrane are pinched off into vesicles inside the endosome. The internalized material is targeted for degradation and also becomes sequestered away from the cytoplasm, which therefore terminates signaling reactions by ligand-bound receptors in that membrane. MVBs facilitate antigen degradation within a structured membrane-bounded compartment, and in antigen-presenting cells are a site of peptide–MHC complex assembly.

MVB: see **multivesicular body.**

zoonotic infection: an infection that crosses species, from an animal, to infect humans.

Entry of the virus into cells is mediated by the viral envelope glycoprotein gp120, which binds to CD4 and chemokine receptors (primarily CCR5 or CXCR4; see section 2-13) on the host cell surface and is thus the principal determinant of the tropism of the virus: that is, the cells it can productively infect—in this case, cells of the immune system. Binding to its receptors leads to a conformational change in gp120 that exposes the transmembrane component of the envelope protein gp41, which mediates fusion of the virus envelope with the host cell membrane, primarily at sites of lipid rafts. The viral inner core enters the cytoplasm as an intact pre-integration complex (PIC). In the cytosol, viral RNA is reverse transcribed to double-stranded cDNA, and the PIC is transported to the nucleus. There, the cDNA is integrated through specialized sequences at its ends, the **long terminal repeat (LTR)** sequences, into transcription-ally active regions of the host genome (Figure 10-13). The product of the viral *tat* gene is necessary for productive expression of the proviral genome. Membrane-associated envelope proteins recruit structural proteins and two copies of the full-length viral RNA genome for assembly of virions at host cell membranes. Budding is mediated by sequences in a *gag*-derived protein which serves to attract and assemble the cellular machinery, designated the **endosomal sorting complex required for transport (ESCRT)**, that normally drives invagination and extrusion of membrane vesicles into endosomes. In infected T cells, budding occurs from the surface. In macrophages, mature virions assemble on endosomal membranes and bud intra-cellularly into **multivesicular bodies (MVBs)**. Viral release occurs when MVBs fuse with the plasma membrane, a process believed to be accelerated by contact with activated T cells. In dendritic cells, which are more resistant to infection, viruses are internalized and recycled to the surface after receptor-mediated endocytosis triggered by binding to C-type lectin receptors such as DC-SIGN. As in macrophages, this process seems to be accelerated by T cell contact.

Figure 10-13 HIV-1 life-cycle The trimeric gp120 subunit binds CD4 and undergoes a conformational change that promotes interactions with chemokine receptors and exposes the trimeric transmembrane fusion protein gp41. After envelope fusion, the inner core acts as a template for the reverse transcription of the two copies of single-stranded viral RNA into double-stranded cDNA, and the complex is transported via nuclear localization signals to the nucleus, where integration of the proviral genome occurs. Expression is greatly augmented by host transcriptional regulators activated by cell stimulation, such as NF-κB, and by the viral transactivator protein, Tat. Mature viral RNAs are transported to the cytoplasm by Rev, and virions are assembled at sites where envelope proteins become incorporated into membranes at the cell surface (T cells) or in MVBs (macrophages). Budding is mediated by the cellular ESCRT machinery.

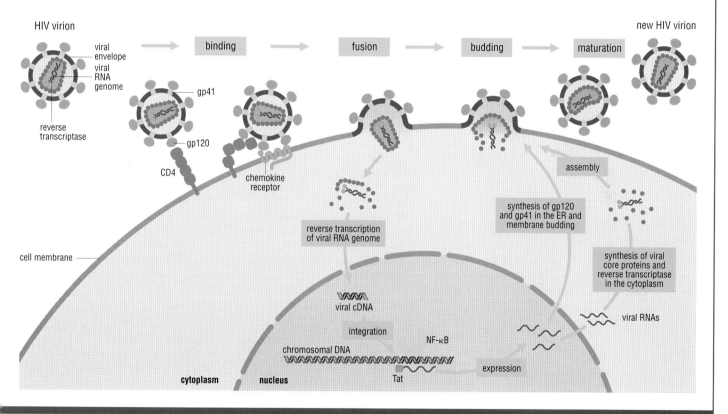

References

Amara, A. and Littman, D.R.: **After Hrs with HIV.** *J. Cell Biol.* 2003, **162**:371–375.

Fauci, A.S.: **Twenty-five years of HIV/AIDS.** *Science* 2006, **313**:409.

Freed, E.O.: **HIV-1 and the host cell: an intimate association.** *Trends Microbiol.* 2004, **12**:170–177.

Hahn, B.H. *et al.*: **AIDS as a zoonosis: scientific and public health implications.** *Science* 2000, **287**:607–614.

Heeney, J.L. *et al.*: **Origins of HIV and the evolution of resistance to AIDS.** *Science* 2006, **313**:462–466.

Keele, B.F. *et al.*: **Chimpanzee reservoirs of pandemic and nonpandemic HIV-1.** *Science* 2006, **313**:523–526.

HIV crosses mucosal barriers, establishes early viremia, and targets CD4 T cells

HIV-1 can cross mucosal surfaces through M cells, which are prevalent in tonsil and rectal epithelia, or can gain access through epithelia damaged by concurrent ulcerative infections or by trauma or injection (Figure 10-14), as occurs with transfusion of contaminated blood or during intravenous drug abuse. Although vaginal epithelium lacks M cells, HIV may gain access to interdigitating processes on Langerhans cells or, with microtrauma, to subepithelial dendritic cells. These dendritic cell populations express diverse C-type lectin receptors (CLRs), including Langerin, which is expressed on Langerhans cells, DC-SIGN (see section 5-6) and the mannose receptor, which bind high-mannose N-linked oligosaccharides on gp120. Interactions with these receptors initiate dendritic cell migration to regional lymph nodes, where intact virus, internalized with CLRs, is reexposed at the surface and thus capable of infecting T cells interacting with the dendritic cells. Despite being relatively resistant to productive viral infection themselves, dendritic cells are likely to be crucial in bringing the virus to lymphoid tissues. On the basis of studies of SIV in rhesus monkeys, which are highly susceptible to AIDS-like disease (in contrast to the natural sooty mangabey monkey host for the virus), dendritic cells translocate virus from mucosa within 30 minutes, initiating a wave of viral proliferation in lymph nodes that peaks 4–7 days after infection. Viremia typically peaks at 14 days and all lymphoid tissues in the body are infected by three weeks. In human infection, the route and dose of infection, as well as viral and host genotypes, affect the timing and course of disease.

Primary mucosal HIV-1 infection is usually established by viruses that require CCR5 as a coreceptor, and these are designated **R5 viruses**. CCR5 is expressed on dendritic cells and macrophages (which in humans frequently express CD4), but also on activated T cells. Epithelial cells in the human small intestine express CCR5 and galactosylceramide, which can substitute for CD4 in tethering HIV for binding to the chemokine receptor, and efficiently transcytose virus. Regardless of the route of infection, the predominant site of virus replication early in disease is in the small intestine, perhaps reflecting the large numbers of activated T cells that express CCR5 in that organ.

As CD4 T cell numbers decline, HIV-1 envelope variants commonly emerge that infect by means of CD4 with CXCR4 coreceptors, which are expressed on many cell types, including CD4 T cells. These variants are designated **X4 viruses**. In some instances, the emergence of X4 variants has been associated with steep declines in CD4 T cells and rapid progression to AIDS. Natural ligands for CXCR4—such as CCL21 and CCL19—are constitutively expressed in lymph nodes and may delay the outgrowth of pathogenic X4 variants through competition for the receptor. In advanced HIV, the normal lymphoid architecture becomes ablated and the subsequent decrease in CCL21 levels may contribute to accelerated disease.

Even when X4 variants emerge, mucosal transmission occurs primarily via R5 viruses, perhaps for the reasons listed above. The best evidence for this is the resistance to infection of individuals

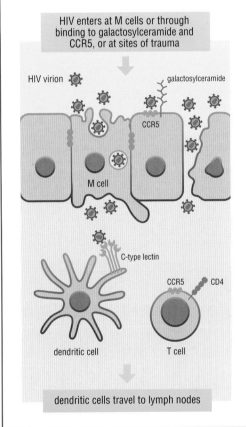

HIV enters at M cells or through binding to galactosylceramide and CCR5, or at sites of trauma

HIV virion

galactosylceramide

CCR5

M cell

C-type lectin

CCR5   CD4

dendritic cell   T cell

dendritic cells travel to lymph nodes

Figure 10-14 HIV transport to lymph nodes HIV crosses mucosal epithelial barriers by transcytosis through M cells or upper small-intestinal epithelia, or access to epithelial or subepithelial dendritic cells, which capture virus using a variety of C-type lectin receptors. Virus is internalized and transported by dendritic cells to regional lymph nodes, where intact virus is presented to activated T cells clustering against dendritic cell membranes.

Definitions

APOBEC family: family of cytidine deaminases including both RNA- and DNA-editing enzymes, members of which are responsible for the somatic hypermutation of Ig genes as well as viral defense through the introduction of catastrophic mutations into retroviral DNA. APOBEC3G, 3F and 3B, and perhaps others, have antiviral activity and are likely to have evolved to protect against viruses, including retroviruses and hepatitis B virus, that replicate through reverse transcription, and to suppress the transposition of endogenous retroelements in the genome.

R5 virus: HIV-1 isolates with gp120-binding affinity for CD4 and CCR5; most primary mucosal HIV-1 infections are due to R5 viruses.

TRIM5α: see **tripartite motif 5α**.

tripartite motif 5α (TRIM5α): tripartite motif 5α, a protein that contributes to species-specific blocking of retroviral infection through interactions with capsid proteins, interfering with uncoating and reverse transcription.

X4 virus: HIV-1 isolates with gp120-binding affinity for CD4 and CXCR4. X4 viruses can emerge late in HIV infection and be associated with rapid immune deterioration.

homozygous for a 32 amino-acid residue CCR5 deletion (CCR5Δ32), which abrogates receptor expression despite an intact CXCR4. Heterozygous carriage of CCR5Δ32, which occurs in 20% of Caucasians of west European heritage, is associated with a reduced risk of infection and slower disease progression. Viral variants that arise during infection develop an affinity for other CCR and CXCR types that may influence the eventual tissue distribution of the virus and the rate of infection. CCR3 and CCR5 are primary receptors by means of which HIV infects microglia in the central nervous system. The primary HIV-2 receptors *in vivo*, CCR5 and CXCR4, are shared with HIV-1, but a wider spectrum of chemokine receptors bind HIV-2 *in vitro*.

Activated CD4 T cells drive viral replication while HIV subverts immune detection

HIV has adapted exquisitely to abrogate host immunity. Viral replication is driven by occupation of NF-κB and NFAT binding sites in the retrovirus long terminal repeats (LTRs). HIV is quiescent in resting CD4 T cells, but on T cell activation NF-κB and NFAT translocate to the nucleus and bind to the viral LTRs, activating virus production, which results in the eventual death of infected cells. Over 99% of virus is produced by newly infected CD4 T cells, which have a half-life of about 1.6 days; approximately 10^7–10^8 CD4 T cells are producing virus at a given time. The immunological consequence is initial targeting of helper T cells that become activated in response to HIV-1 itself, leading to their destruction and loss of helper activities necessary to maintain cytotoxic CD8 T cell and antibody responses. The related primate retroviruses do not cause comparable immune activation in their natural hosts, despite active viral replication. In primate T cells, the viral protein Nef downregulates surface TCR expression, thus suppressing T cell activation and limiting viral replication, but this mechanism has been lost during infection of human T cells, and hence the greater pathogenicity of HIV.

Viral proteins target key components of innate cellular defense and adaptive responses in order to sustain persistent replication and viremia. Nef mediates the downregulation of surface MHC class I, while sparing expression of the NK inhibitory receptor ligands HLA-C and HLA-E, thus masking recognition by cytotoxic CD8 T cells and avoiding activation of NK cells. Although Old World monkeys, such as macaques, are susceptible to SIV (or HIV-2), they are resistant to HIV-1 because of a block early after viral entry. Resistance is due to a protein known as **tripartite motif 5α (TRIM5α)**, a cytoplasmic protein that interacts with capsid proteins and causes improper uncoating of the viral core and failure of reverse transcription. Despite high homology with monkey TRIM5α, human TRIM5α is unable to bind HIV-1 capsid proteins, although it does restrict infection by murine retroviruses. Finally, another viral accessory protein, vif (virion infectivity factor), that is incorporated into the virion, binds and inactivates cellular RNA-editing enzymes, which constitute important cellular defenses against retroviruses. These enzymes, known as the **APOBEC family,** introduce C to U transitions in the first (minus)-strand cDNA during reverse transcription, leading to lethal mutations when the proviral genome is transcribed (see section 3-19). APOBEC3G is the family member that is most prominent in protection from HIV. Vif excludes APOBEC3G during virion formation, primarily through binding and directed degradation (Figure 10-15). Resting CD4 T cells express high levels of APOBEC3G, accounting for their relative resistance to HIV infection, but T cell activation sequesters the active protein, allowing unimpaired reverse transcription and proviral integration. HIV-1 vif binds human, not primate, APOBEC3G, whereas SIV vif binds primate, but not human, APOBEC3G. Thus, vif demonstrates species-specific evolution in targeting APOBEC3G.

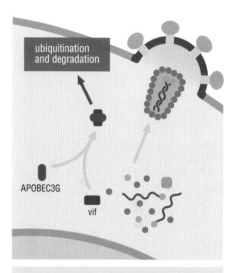

Figure 10-15 Vif protects HIV virions from incorporating APOBEC proteins APOBEC3G is a host cytidine deaminase that is incorporated into budding virions through interactions with capsid proteins; vif binds and degrades human APOBEC proteins, thus preventing incorporation into virions and the actions of these enzymes during reverse transcription during subsequent infection, which otherwise leads to hypermutation of the proviral cDNA and to non-functional virus.

References

Bieniasz, P.D.: **Restriction factors: a defense against retroviral infection.** *Trends Microbiol.* 2003, 11:286–291.

Chiu, Y.-L. *et al.*: **Cellular APOBEC3G restricts HIV-1 infection in resting CD4+ T cells.** *Nature* 2005, 435:108–114.

Harris, R.S. and Liddament, M.T.: **Retroviral restriction by APOBEC proteins.** *Nat. Rev. Immunol.* 2004, 4:868–877.

Liu, R. *et al.*: **Homozygous defect in HIV-1 coreceptor accounts for resistance of some multiply-exposed individuals to HIV-1 infection.** *Cell* 1996, 86:367–377.

Meng, G. *et al.*: **Primary intestinal epithelial cells selectively transfer R5 HIV to CCR5+ cells.** *Nat. Med.* 2002, 8:150–156.

Navarro, F. and Landau, N.R.: **Recent insights into HIV-1 Vif.** *Curr. Opin. Immunol.* 2004, 16:477–482.

Schindler, M. *et al.*: **Nef-mediated suppression of T cell activation was lost in a lentiviral lineage that gave rise to HIV-1.** *Cell* 2006, 125:1055–1067.

Stevenson, M.: **HIV-1 pathogenesis.** *Nat. Med.* 2003, 9:853–860.

Turville, S.G. *et al.*: **Diversity of receptors binding HIV on dendritic cell subsets.** *Nat. Immunol.* 2002, 3:975–983.

Veazey, R.S. *et al.*: **Gastrointestinal tract as a major site of CD4+ T cell depletion and viral replication in SIV infection.** *Science* 1998, 280:427–431.

Progression to AIDS is determined by the viral set-point

By three to six months of infection, the balance between viral destruction of immune cells and the host response results in steady-state levels of virus, as measured by plasma HIV RNA, which remain stable for months to years. This is designated the **viral set-point**. APOBEC-driven mutagenesis occurs, despite the presence of vif, and neutralizing antibodies and cytotoxic CD8 T cells can be demonstrated. Numerous viral variants, termed quasi-species, appear, because of the infidelity of the viral reverse transcriptase and the absence of viral nucleotide repair systems, and over time, **escape variants** emerge with mutations selected to evade immune recognition. Indeed, large numbers of virus circulating during infection are replication-incompetent, presumably because of mutations to proteins essential for replication. Escape mutants allowing evasion of neutralizing antibodies can be achieved without any fitness cost to the virus however because the critical CD4 binding site on gp120 is masked by highly variable loops whose precise structure is not important to the function of the protein. Escape from CD8 T cells does involve a fitness cost to the virus because T cells recognize many peptides from internal viral proteins, changes to which may result in loss of function.

The level of plasma virus at the viral set-point is a general predictor of the rate of CD4 T cell depletion and the progression to AIDS (Figure 10-16). A number of host determinants has been linked to the viral set-point. Mutant chemokine receptor genes have been linked with resistance and slower progression of disease (see section 10-5). Individuals homozygous for MHC class I alleles progress more rapidly to AIDS, presumably because of the potential for escape variants to arise more quickly when there is less host variation. Conversely, relatively uncommon MHC class I alleles may slow disease progression because viral peptides selected

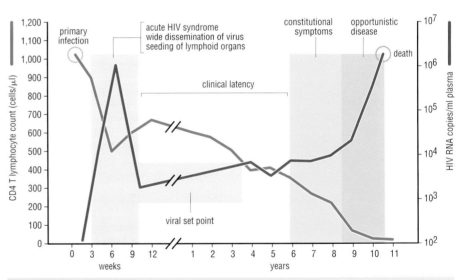

Figure 10-16 The natural history of untreated HIV infection Infection typically results in a flu-like illness from days to weeks after exposure, associated with a drop of CD4 T cells in the blood and the appearance of HIV RNA in virions in plasma. The CD4-dependent antiviral cytotoxic CD8 T cell response develops and is responsible for a prolonged period of stable viremia, designated the viral set point. Patients are generally asymptomatic until the CD4 T cell counts decline further. Opportunistic infections begin when CD4 T cells reach levels less than 200/μl, and death typically ensues in about two years.

Definitions

escape variants: viral progeny that differ from the parental virus through the acquisition of mutations that prevent recognition by antibody or binding to MHC class I so that the virus is no longer bound by antibody or presented to cytotoxic CD8 T cells in a given host.

HAART: see **highly active antiretroviral therapy**.

highly active antiretroviral therapy (HAART): a combination of an HIV aspartyl protease inhibitor with at least two reverse transcriptase inhibitors used to treat AIDS. Combination therapy targeting multiple enzymes is necessary to avoid the rapid selection of resistance mutants.

viral set-point: the relatively steady-state level of plasma HIV reached after acute infection becomes controlled by CD8 T cell responses. Although viral and CD4 T cell levels are relatively stable for prolonged periods, these levels reflect dynamic turnover in both compartments.

References

Belyakov, I.M. and Berzofsky, J.A.: **Immunobiology of mucosal HIV infection and the basis for development of a new generation of mucosal AIDS vaccines.** *Immunity* 2004, **20**:247–253.

Brander, C. and Walker, B.D.: **Gradual adaptation of HIV to human host populations: good or bad news?** *Nat. Med.* 2003, **9**:1359–1362.

Gao, X. *et al.*: **Effect of a single amino acid change in MHC class I molecules on the rate of progression to**

against population-wide alleles may be less able to accumulate further mutations without compromising viral fitness. Concurrent infections may enhance progression by increasing the pool of activated CD4 T cells and thus increasing rounds of viral replication. Eventually, variant viruses infect developing lymphocytes in the bone marrow and thymus, resulting in failure to replace lost T cells and increasingly accelerated disease. The increasing loss of T cells exposes the individual to opportunistic infections—infections due to organisms that are normally suppressed by the immune system. Such infections become inevitable when CD4 T cells decline below 200/µl. At this stage, the normal lymphoid architecture is lost, reflecting the loss of the normal interactions between immune cells that are required to sustain it (see section 1-5).

Infections, together with the propensity of the virus to generate variants with the capacity to infect additional cell types, lead to death in 10–12 years, usually from opportunistic infection or damage to the nervous system. It is estimated that up to 10^{10} virions are produced per day, with a latent reservoir in resting memory T cells comprising about 10^6 cells. Additional latent virus may be maintained in dendritic cells, monocytes, the astrocytes and microglia of the central nervous system, and seminal cells of the testes.

Control of HIV is mediated by CD4 T cell-dependent cytotoxic CD8 T cells

Although neutralizing antibodies against HIV may contribute to immunity by blocking uptake at mucosal surfaces, increasing evidence has implicated CD4 T cell-dependent cytotoxic CD8 T cells in host control. In mice, and probably in humans, CD4 T cells are required for secondary expansion and memory in CD8 T cells. In infected patients, the level of the viral set-point is inversely proportional to the antiviral cytotoxic lymphocyte response, which is lost with the eventual decline in CD4 T cells. During acute infection, up to 10% of CD8 T cells reflect oligoclonal responses to HIV-specific epitopes, and levels of 1–2% of all CD8 T cells remain specific for HIV during the viral set-point period. Chemokines that compete for HIV binding to chemokine receptors are produced by cytotoxic T effector cells. The loss of antiviral cytotoxic CD8 T cells correlates with the rapid progression of disease and with increases in plasma virus. Disease progression has been linked with the appearance in plasma of virus expressing escape variants for major viral epitopes. Non-neutralizing antibodies can opsonize virus and enhance uptake by macrophage Fc and complement receptors, but this results in increased infection of myeloid cells, rather than clearing infection, and is termed *immune enhancement*. Documented second infections (superinfections) with HIV from different subtypes, or clades, suggest that little protective immunity is acquired during natural infection. Treatment with an HIV aspartyl protease inhibitor combined with at least two reverse transcriptase inhibitors (**highly active antiretroviral therapy**, or **HAART**) can clear viral RNA from blood and maintain CD4 T cells, although latent virus becomes rapidly reexpressed when therapy is stopped.

In primate studies, depletion of CD8 T cells abrogates the control of infection, and some success has been achieved in vaccination studies against SIV by using constructs designed to expand cytotoxic CD8 T cells with specificity against gag and env epitopes. Viral escape variants ultimately emerge, however. The predominant transmission of HIV by mucosal routes will undoubtedly require a multi-pronged approach: it is likely that effective vaccination will require the elicitation of CD4 T cells capable of providing help for antiviral cytotoxic CD8 T cells, but also of providing help for neutralizing IgA and IgG mucosal antibodies. Targeting of effector CD8 T cells to mucosal tissues may also be important. We discuss these issues in Chapter 14.

AIDS. *N. Engl. J. Med.* 2001, **344**:1668–1675.

Jost, S. *et al.*: **A patient with HIV-1 superinfection.** *N. Engl. J. Med.* 2002, **347**:731–736.

Klausner, R.D. *et al.*: **The need for a global HIV vaccine enterprise.** *Science* 2003, **300**:2036–2039.

McCune, J.M.: **The dynamics of CD4⁺ T-cell depletion in HIV disease.** *Nature* 2001, **410**:974–979.

McMichael, A.J. and Rowland-Jones, S.L.: **Cellular immune responses to HIV.** *Nature* 2001, **410**:980–987.

Perelson, A.S. *et al.*: **HIV-1 dynamics in vivo: virion clearance rate, infected cell life-span, and viral generation time.** *Science* 1996, **271**:1582–1586.

Schmitz, J.E. *et al.*: **Control of viremia in simian immunodeficiency virus infection by CD8⁺ lymphocytes.** *Science* 1999, **283**:857–860.

Stebbing, J. *et al.*: **Where does HIV live?** *N. Engl. J. Med.* 2004, **350**:1872–1880.

Zolla-Pazner, S.: **Identifying epitopes of HIV-1 that induce protective antibodies.** *Nat. Rev. Immunol.* 2004, **4**:199–210.

Influenza viruses cause recurrent epidemics and occasional pandemics

There are three classes of influenza virus—influenza A, B and C. Influenza A and B are major causes of human disease, but only influenza A is responsible for pandemic outbreaks. Virus is spread by aerosol droplets and establishes cytolytic infection in epithelial cells of the upper and lower respiratory tract. Influenza is typically a self-limited febrile illness characterized by fever, muscle aches and cough that last approximately a week. Epidemics occur every 1–3 years in human populations and are characterized by attack rates above 10% during the winter months. Pandemics occur less frequently and their consequences can be devastating: the Spanish influenza pandemic of 1918–19 killed over 20 million people worldwide. This pattern of recurrent epidemics and occasional pandemics reflects the pattern of mutations of the influenza A virus, which, as discussed below, undergoes continuous minor antigenic changes with an occasional major shift that causes pandemics. Mortality due to influenza is greatest at the extremes of age and is commonly due to complications of the viral infection, particularly bacterial pneumonia. Adaptive immunity against influenza virus is mediated by neutralizing antibody and cytotoxic CD8 T cells.

Influenza A viruses are avian viruses that adapt to humans

Influenza viruses are single-strand, negative-sense, segmented RNA viruses of the orthomyxovirus family. The viral genome has eight RNA segments that encode 10 or 11 proteins, depending on the strain. The segmented character of the genome is thought to provide the basis for major *antigenic shifts* that underlie pandemics, by allowing the exchange of segments that differ between two viruses.

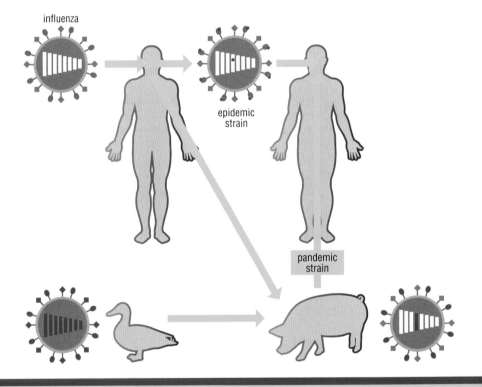

influenza

epidemic strain

pandemic strain

Figure 10-17 Antigenic drift and shift underlie influenza epidemics and pandemics, respectively Epidemics occur when point mutations (red dots) accumulate in surface hemagglutinin (HA, shown as ovals), or neuraminidase (NA, shown as diamonds), leading to antigenic drift so that in many previously immune individuals these surface molecules (in this case the HA) are no longer recognized by antibodies produced in response to earlier infections. Pandemics can occur when avian influenza viruses and human influenza viruses infect pigs and reassortment of the two genomes leads to the expression of avian HA or NA in a virus otherwise adapted to infect humans. The eight-segmented RNA genome is depicted by boxes. Adapted from Ito, T. *et al.*: *J. Virol.* 1998, **72**:7367–7373.

Definitions

antigenic drift: (of influenza virus) point mutations, predominantly in **hemagglutinin** and **neuraminidase**, that affect recognition by neutralizing human antibodies. Antigenic drift gives rise to epidemic infections.

antigenic shift: (of influenza virus) reassortment of independent RNA segments from two different influenza genomes to generate recombinant virus with new antigenic subtypes. Antigenic shift gives rise to pandemic outbreaks.

HA: see **hemagglutinin**.

hemagglutinin (HA): influenza homotrimeric envelope protein that binds to widely distributed host-cell-surface glycoproteins and glycolipids that are modified by sialic acid additions and is required for viral entry into host cells. Binding by HA induces internalization and transport of the sialylated receptors to endosomal compartments, where low pH induces a marked conformational change in HA that mediates fusion of the viral and host membranes, allowing translocation of the influenza nucleocapsid to the cytosol.

highly pathogenic avian influenza (HPAI): influenza viruses with high virulence for chickens and other birds. These viruses commonly express H5, H7 or H9 **hemagglutinin** proteins with an expanded basic amino-acid repeat that increases the ease of proteolytic digestion required to activate the fusion domain.

HPAI: see **highly pathogenic avian influenza**.

NA: see **neuraminidase**.

neuraminidase (NA): influenza homotetrameric sialidase expressed on the virion that cleaves terminal sialic acid

Influenza A is a spherical or rod-shaped enveloped virus covered with two spike-like glycoproteins: trimeric **hemagglutinin (HA)** and tetrameric **neuraminidase (NA)**. After cleavage by a trypsin-like host enzyme, HA mediates binding to the cell surface and internalization, and induces fusion with cell membranes upon pH-induced conformational change within host endosomes. NA cleaves sialic acid and promotes viral release from infected cells. NA inhibitors reduce the duration of symptomatic infection in humans and prevent disease when administered prophylactically to family members of infected individuals.

Influenza genotypes are diverse in birds, particularly migratory waterfowl, and express a variety of HA (H1 to H15) and NA (N1 to N9) subtypes. Trans-species adaptation to humans is determined in part by HA receptor specificity, as well as by internal proteins that also contribute to virulence. Avian viruses prefer receptors, primarily on intestinal cells, that have the α-2,3 sialic acid linkage to galactose, whereas human viruses prefer the α-2,6 linkage expressed on respiratory epithelia. Pigs express both types of receptors on respiratory epithelial cells. Influenza viruses enter the human population in regions where humans, pigs and waterfowl, as well as domestic chickens, are in close proximity, as occurs throughout Southeast Asia (Figure 10-17). Experimental evidence suggests that pigs can be infected with both human and avian influenza, and exchange of segments between viral genomes in the pig can give rise to variants expressing novel surface proteins together with human-adapted virulence determinants.

Human infections are generally limited to viruses that express H1, H2 or H3 and N1, N2 and possibly N8 subtypes. **Highly pathogenic avian influenza (HPAI)** viruses, which express H5, H7 or H9, have become established in poultry throughout China and Southeast Asia. HPAI H5N1 viruses were transmitted to individuals exposed to poultry during outbreaks in Hong Kong in 1997 and have since spread, primarily through migratory birds, to infect poultry and humans in the Middle East and northern Africa. Although human transmission of H5N1 viruses appears infrequent, mortality in infected humans approaches 60%. Ominously, pigs in China have been infected with avian H5N1 and human H3N2 viruses, which may facilitate the reassortment of viral genome segments to produce a more transmissible virus. Thus, a new pandemic analogous to the 1918–19 outbreak may arise from such viruses.

Antigenic drift and shift in HA and NA underlie influenza A epidemics and pandemics

Influenza-neutralizing antibodies target HA and block receptor attachment, but do not impede the fusion apparatus. Point mutations in HA can allow the virus to escape antibody recognition without compromising the fusion function of the molecule. Antibodies against NA reduce virus release from infected cells, and decrease spread and transmission. Escape variants of HA and NA due to point mutations in the original virus cause disease every three to four years among susceptible individuals; these continuously circulating variants are said to be due to **antigenic drift**. Occasionally, a new virus with different HA and NA genes emerges, usually from Southeast Asia for the reasons explained above, and a worldwide pandemic can result as disease spreads into non-immune populations. The reassortment of viral genomes leads to **antigenic shift**, and underlies the increased morbidity and mortality associated with influenza pandemics, of which there have been three in the past century: the 1918 Spanish flu (H1N1), the 1957 Asian flu (H2N2) and the 1968 Hong Kong flu (H3N2). The mutational basis of epidemics and pandemics is illustrated schematically in Figure 10-17. Worldwide surveillance for influenza genotypes that circulate continuously is used to define the composition of the influenza vaccines for use in the autumn in temperate climates.

residues from host glycoconjugates to allow the release of infectious progeny virus.

References

Baigent, S.J. and McCauley, J.W.: **Influenza type A in humans, mammals and birds: determinants of virus virulence, host-range and interspecies transmission.** *Bioessays* 2003, **25**:657–671.

Centers for Disease Control and Prevention (CDC): **Update: Influenza activity—United States and worldwide, 2003–04 season, and composition of the 2004–05 influenza vaccine.** *MMWR Morb. Mortal. Wkly Rep.* 2004, **53**:547–552.

Fauci, A.S.: **Emerging and re-emerging infectious diseases: influenza as a prototype of the host-pathogen balancing act.** *Cell* 2006, **124**:665–670.

Ito, T. *et al.*: **Molecular basis for the generation in pigs of influenza A viruses with pandemic potential.** *J. Virol.* 1998, **72**:7367–7373.

Li, K.S. *et al.*: **Genesis of a highly pathogenic and potentially pandemic H5N1 influenza virus in eastern Asia.** *Nature* 2004, **430**:209–213.

Neuzil, K.M. *et al.*: **The effect of influenza on hospitalizations, outpatient visits, and courses of antibiotics in children.** *N. Engl. J. Med.* 2000, **342**:225–231.

Nicholson, K.G. *et al.*: **Influenza.** *Lancet* 2003, **362**:1733–1745.

Tran, T.H. *et al.*: **Avian influenza A (H5N1) in 10 patients in Vietnam.** *N. Engl. J. Med.* 2004, **350**:1179–1188.

10-8 Influenza Virus: Innate and Adaptive Immunity

Innate immunity to influenza depends on type 1 interferons

Resistance to influenza, like resistance to most viruses, requires the induction of type 1 interferons (see section 3-16). Although interferons are strongly induced with heat-killed influenza, viable viruses do not efficiently activate the cellular expression of interferon. This is because influenza viruses avoid inducing interferon through a viral protein called **non-structural protein-1 (NS1)** which is produced by the replicating virus. NS1 binds RNAs formed during the viral life cycle, sequestering these elements from host RNA receptors and thus preventing recognition by cellular RNA helicases (particularly RIG-I), PKR (double-stranded RNA-activated kinase), TLR3 and Dicer-mediated RNA-silencing pathways (see sections 3-17 and 3-19). The result is effectively to bypass the innate antiviral immune response (Figure 10-18). One consequence of activating these pathways is to induce apoptosis in cells, and abrogation of the apoptotic cell response enables the virus to complete its life cycle and spread. Through incompletely defined mechanisms, NS1 from highly pathogenic avian H5N1 viruses also confers resistance to the antiviral effects of exogenous interferons and TNF *in vitro*. Viruses with NS1 deleted are avirulent (that is, do not cause disease) in normal mice but are fully virulent in mice lacking either PKR, which recognizes double-stranded RNA, or STAT1, which is an essential signaling component in pathways activated by interferon.

In mice, key downstream targets of type 1 interferon signaling include PKR and Mx1, a small nuclear GTPase that blocks influenza virus replication (see section 3-17). Mice with naturally occurring Mx1 deficiency can be protected by a human *MxA* transgene. Protection is correlated with the ability to accumulate threshold levels of Mx1 protein at sites of infection. Humans express two *Mx* genes, encoding the proteins MxA and MxB, which also demonstrate antiviral activity against some tick-borne influenza-like viruses.

Viral replication and cell lysis during infection lead to the induction of inflammatory cytokines, including IL-1, IL-6, IL-8, IL-15, IL-18, TNF and IFN-γ, as well as chemokines that serve to recruit lymphoid and myeloid effector cells. It is likely that many of the symptoms of influenza are due, in part, to immunopathological effects of the inflammatory host response.

Influenza virus infection is cleared by cytotoxic CD8 T cells

Viral clearance correlates with the appearance of virus-specific cytotoxic CD8 T cells in the lung. Common targets of the antiviral cytotoxic T cell response in humans include the viral proteins HA and NP, as well as a protein called PB2, which is a subunit of the viral RNA polymerase, and M2, a pore-forming channel protein and a target of the antiviral drug amantadine. Cytotoxic T cells specific for M2 are almost universally found after infection. Cytotoxic CD8 T cells specific for M2 commonly recognize amino acids 58–66 presented by a specific allelic variant of the MHC class I HLA-A2 molecule, designated HLA-A2 (A*0201). These cytotoxic cells often express a TCR with a β chain containing a conserved arginine motif that binds to a notch in the otherwise featureless M2 peptide–HLA-A2 structure, along with an α chain that positions the motif over the notch (Figure 10-19). Since this HLA-A2 is the most common human HLA allele,

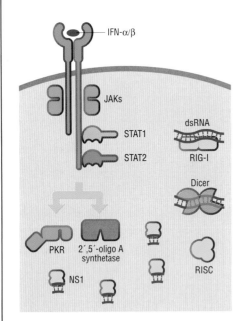

Figure 10-18 Abrogation of innate immunity by influenza virus NS1 Viral RNA is detected by the RIG-I helicase, and by protein kinase PKR and by 2′,5′-oligoadenylate synthetase, which are induced by type 1 interferons and prevent viral replication (see Figure 3-46): and by siRNA generated by RISC from fragments produced by Dicer and that target viral mRNA for degradation (see Figure 3-53). The viral protein NS1 binds to viral RNA, preventing it from being detected and thereby aborting the innate immune defense mechanisms activated by these proteins.

Definitions

non-structural protein-1 (NS1): a protein produced by influenza virus that downregulates host cell antiviral innate immune responses by binding to viral RNAs and sequestering them from cytosolic nucleic-acid-sensing proteins that trigger type 1 interferon production.

NS1: see **non-structural protein-1.**

References

Arnheiter, H. *et al.*: **Mx transgenic mice—animal models of health.** *Curr. Top. Microbiol. Immunol.* 1996, **206**:119–147.

Arulanandam, B.P. *et al.*: **IgA immunodeficiency leads to inadequate Th cell priming and increased susceptibility to influenza virus infection.** *J. Immunol.* 2001, **166**:226–231.

Bizebard, T. *et al.*: **Structure of influenza virus haemagglutinin complexed with a neutralizing**

antibody. *Nature* 1995, **376**:92–94.

Doherty, P.C. and Christensen, J.P.: **Accessing complexity: the dynamics of virus-specific T cell responses.** *Annu. Rev. Immunol.* 2000, **18**:561–592.

Flynn, K.J. *et al.*: **Virus-specific CD8+ T cells in primary and secondary influenza pneumonia.** *Immunity* 1998, **8**:683–691.

García-Sastre, A. *et al.*: **Influenza A virus lacking the NS1 gene replicates in interferon-deficient systems.** *Virology* 1998, **252**:324–330.

expressed by over a billion people, and influenza viruses are recurrent throughout life, some have speculated that anti-M2 cytotoxic T cells carry the most prevalent TCR in the world.

Viral protection is mediated by neutralizing antibody

Although cytotoxic CD8 T cells can develop in the absence of CD4 T cells, efficient viral clearance and protective immunity require helper CD4 T cells and T cell-dependent neutralizing antibody. Antibodies are primarily IgG and IgA. Secreted IgA serves both to neutralize virus in the lumen of the respiratory tract and other mucosa and to target virions to antigen-presenting cells through the Fcα/μ receptor. IgA-deficient mice are thus more susceptible to infection by influenza virus. Systemic IgG induced by influenza vaccine is sufficient to confer protection against infection: this is discussed in Chapter 14.

Protective, neutralizing anti-HA antibodies are made after natural influenza infection and prevent infection but are subtype specific. Anti-NA antibodies do not prevent infection but restrict viral spread. Influenza vaccine induces neutralizing antibodies that confer protection against infection, but small changes in HA incurred by antigenic drift can enable escape from vaccine-induced protection. The 2003–04 influenza season was moderately severe because of the emergence of an antigenic drift variant that had accumulated two mutations at sites of antibody binding to HA that were distinct from those in the H3N2 virus used to formulate the vaccine (Figure 10-20).

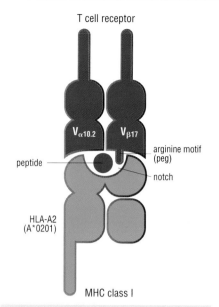

Figure 10-19 The most common TCR in the world? The worldwide prevalence of influenza and the expression of HLA-A2 (A*0201) by millions of people have driven the expansion of a TCR containing an arginine motif (the so-called peg) that binds tightly into a notch in the peptide–MHC complex, thus forming a high-affinity complex. The α chain binds MHC and is required to position the TCR optimally over the notch.

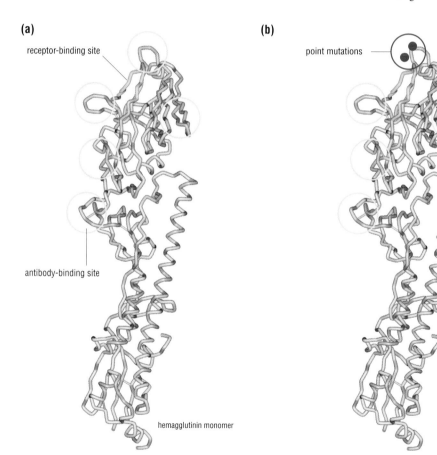

(a) receptor-binding site, antibody-binding site, hemagglutinin monomer

(b) point mutations

Figure 10-20 Antigenic drift variants in circulating H3N2 influenza A viruses The HA1 subunit from the A/Panama/2007/99 virus **(a)** used in the 2003–04 influenza vaccine has five neutralizing antibody-binding sites (yellow circles), one of which contains two mutations (**(b)**; red dots mark the mutations) at the tip of the receptor-binding site (green) in A/Fujian/411/2002 viruses that came to dominate the influenza season. Such antigenic drift underlies the seasonal endemicity of influenza. Adapted from Treanor, J.: *N. Engl. J. Med.* 2004, **350**:218–220. (PDB 1qfu)

Geiss, G.K. *et al.*: **Cellular transcriptional profiling in influenza A virus-infected lung epithelial cells: the role of nonstructural NS1 protein in the evasion of the host innate defense and its potential contribution to pandemic influenza.** *Proc. Natl Acad. Sci. USA* 2002, **99**:10736–10741.

Li, W.-X. *et al.*: **Interferon antagonist proteins of influenza and vaccinia virus are suppressors of RNA silencing.** *Proc. Natl Acad. Sci. USA* 2004, **101**:1350–1355.

Seo, S.H. *et al.*: **Lethal H5N1 influenza viruses escape host anti-viral cytokine responses.** *Nat. Med.* 2002,

8:950–954.

Stewart-Jones, G.B.E. *et al.*: **A structural basis for immunodominant human T cell receptor recognition.** *Nat. Immunol.* 2003, **4**:657–663.

Subbarao, K. *et al.*: **Development of effective vaccines against pandemic influenza.** *Immunity* 2006, **24**:5–9.

Treanor, J.: **Influenza vaccine – outmaneuvering antigenic shift and drift.** *N. Engl. J. Med.* 2004, **350**:218–220.

Lymphocytic choriomeningitis virus (LCMV) is extensively used in mouse models of immunity to viruses

LCMV is an enveloped, single-strand, segmented RNA virus of the arenavirus family. Although the virus can infect humans and cause self-limited meningitis, it is a natural virus of rodents, and because of the range of immune responses it can elicit in laboratory mice it has had a central role in the history of immunology. The importance of MHC class I molecules in viral immune responses was originally clarified in experiments using LCMV infection of inbred mice. LCMV has also provided experimental models for acute systemic infection, immune tolerance and chronic infection leading to loss of virus-specific T cell responses, a state known as **immune exhaustion,** which we discuss below. Host defense against LCMV is mediated by cytotoxic CD8 T cells in primary infection, and by neutralizing antibody and CD4-dependent memory cytotoxic CD8 T cells in secondary infection.

LCMV is a noncytopathic virus that establishes persistent infection in mice

LCMV enters cells by means of the G1 and G2 surface glycoproteins of the virion which bind to a cell-surface glycoprotein called **α-dystroglycan** and perhaps other host receptors through which it is internalized. Membrane fusion delivers the virus to an endocytic compartment where the acidic pH induces uncoating. LCMV infects many cell types, including dendritic cells which rapidly transport the virus to lymph nodes. Infectious virions are produced within 3–10 h of infection of a cell, and progeny virus spreads rapidly throughout the tissues and circulation.

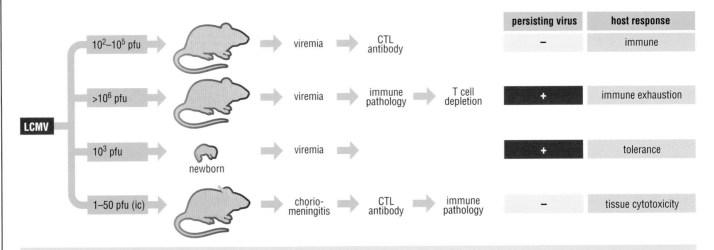

Figure 10-21 Models of viral immunity using LCMV infection of inbred mice Depending on the virus inoculum and the age of the mouse, model systems using LCMV have been established to examine states of antiviral immunity, immune exhaustion, tolerance and persistent infection, and tissue-specific immunity. ic: intracerebral inoculation; pfu: plaque-forming units.

Definitions

α-dystroglycan: a heavily glycosylated protein associated noncovalently on the cell surface with the integral membrane protein β-dystroglycan; the two dystroglycans are derived by post-transcriptional processing of a single gene product and contribute to interactions of extracellular matrix ligands, including laminin and agrin, with the cell actin cytoskeleton. α-Dystroglycan is implicated in the binding of arenaviruses, including LCMV and Lassa fever virus, but also *Mycobacterium leprae*, the causative agent of leprosy, to cell surfaces.

immune exhaustion: a state defined by loss of virus-specific T cell responses despite persistent viral infection and more commonly seen using large viral inocula with retroviruses or RNA viruses with high rates of mutation, which allow escape variants to arise through evasion of immune detection.

Most naturally infected mice are infected *in utero*, and are tolerant to the virus. Persistent life-long viremia ensues with no ill effects on the mouse. High titers of virus are excreted in urine and saliva, and humans can become infected when exposed to infectious aerosols or on being bitten. Some 5–10 days later, viremia in humans is accompanied by fever, aching, swollen lymph glands, headache and rash. Although the disease is usually self-limiting, some individuals develop the meningitis that gives the virus its name. Symptoms of meningitis begin at the time that cytotoxic CD8 T cell responses appear and start to eliminate virus-infected cells. In a few cases, infected individuals can develop immunopathological complications at the time of resolution of infection: inflammation of the testes, joints and heart, and loss of hair can occur. Infection during pregnancy can result in congenital infection and fetal abnormalities.

Acute infection, immune exhaustion and immune tolerance can be studied using LCMV in inbred mice

Infection of inbred mouse strains with LCMV has proven an exceptional model for studies of viral immunity. Although this varies somewhat depending on the strain of virus, infection with 10^2–10^5 virions generally results in systemic infection of all organs except the brain (Figure 10-21). This leads to massive virus-specific CD8 T cell expansion (~1,000-fold) that peaks around 8 days with frequencies of cells specific for dominant epitopes as high as 1 in 3–4 CD8 T cells. CD4 T cells peak around 10 days, although expansion is less (7- to 35-fold to individual epitopes) than that in CD8 T cells. Antibodies can be detected as early as 7 days, although neutralizing antibodies do not appear for three to four months. Virus is completely cleared by 14 days by virus-specific cytotoxic CD8 T cells. T cell-deficient, CD8-deficient, and perforin-deficient mice are unable to clear acute infection. Type 1 IFN-α,β receptor-deficient mice are also unable to clear LCMV. In the absence of CD4 T cells, cytotoxic CD8 T cells develop and acutely eliminate virus normally, but viral replication recurs and cannot be cleared. CD4 T cells are required for the production of neutralizing antibody and memory cytotoxic CD8 T cells that mediate protection from reinfection.

With larger infecting doses—more than 10^6 virions—persistent infection is established in immunocompetent mice. Collapse of the adaptive response to LCMV ensues through a number of causes, including massive depletion of heavily infected dendritic cells, emergence of escape variants and loss of T cells: this is the state known as immune exhaustion. Animals are persistently viremic but can develop immunopathology before the immune response collapses. Immunopathology wanes as T cell responses become attenuated. LCMV-specific T cells become unresponsive to stimulation, and eventually undergo apoptosis and die. When infected mice have reached the state of immune exhaustion, virus can be cleared only by adoptive transfer of LCMV-specific immune cells. Before the loss of lymphocytes, immunity can also be restored by blocking the inhibitory CD28 family receptor PD-1 (see section 5-8), which is upregulated on exhausted CD8 T cells. Blockade of PD-1 restores cytotoxic and cytokine-mediated antiviral activity of these cells.

Tolerance and persistent infection that mimic the outcome of natural infection can be established by injecting newborn mice with 10^3 virions of LCMV. Mice remain viremic for life but have no ill effects. Finally, the direct intracerebral inoculation of adult mice with 1–50 virions of LCMV causes choriomeningitis and death within 7–9 days. Death is due to host immune responses. LCMV-immune mice have a much earlier and attenuated inflammatory response and survive intracranial inoculation. This model has been used as a stringent test for vaccine efficacy.

References

Barber, D.L. *et al.*: **Restoring function in exhausted CD8 T cells during chronic viral infection.** *Nature* 2006, **439**:682–687.

Binder, D. *et al.*: **Aplastic anemia rescued by exhaustion of cytokine-secreting CD8⁺ T cells in persistent infection with lymphocytic choriomeningitis virus.** *J. Exp. Med.* 1998, **187**:1903–1920.

Butz, E.A. and Bevan, M.J.: **Massive expansion of antigen-specific CD8⁺ T cells during an acute virus infection.**
Immunity 1998, **8**:167–175.

Cao, W. *et al.*: **Identification of α-dystroglycan as a receptor for lymphocytic choriomeningitis virus and Lassa fever virus.** *Science* 1998, **282**:2079–2081.

Murali-Krishna, K. *et al.*: **Counting antigen-specific CD8 T cells: a reevaluation of bystander activation during viral infection.** *Immunity* 1998, **8**:177–187.

Sevilla, N. *et al.*: **Infection of dendritic cells by lymphocytic choriomeningitis virus.** *Curr. Top. Microbiol. Immunol.* 2003, **276**:125–144.

Sourdive, D.J.D. *et al.*: **Conserved T cell receptor repertoire in primary and memory CD8 T cell responses to an acute viral infection.** *J. Exp. Med.* 1998, **188**:71–82.

Whitmire, J.K. *et al.*: **Long-term CD4 Th1 and Th2 memory following acute lymphocytic choriomeningitis virus infection.** *J. Virol.* 1998, **72**:8281–8288.

11

The Immune Response to Fungal and Parasitic Infection

Fungi and parasites comprise a large group of genetically complex organisms that can infect humans. For both groups of pathogens, the immune system responds by organizing specific types of innate and adaptive host immune cells, which act together to mediate host defense. In turn, evolution has resulted in adaptation by many of these types of organisms, resulting in chronic or recurrent infections that can last many years.

Illustrative Spectrum of Human Fungal Infections

Endemic mycoses

Histoplasma	Ohio, Mississippi River valleys, other areas worldwide; enriched in soil with bird and bat guano
Coccidiodes	arid desert areas of New World
Blastomyces	areas abutting St Lawrence River, Great Lakes, Mississippi and Ohio River valleys
Paracoccidioides	tropical and subtropical forests of South America

Environmental saprophytes

Aspergillus species	decaying vegetable matter
Rhizopus, Absidia, Mucor	decaying organic material
Cryptococcus	soil enriched with pigeon guano (serotypes A, D); flowering eucalyptus trees (serotypes B, C)

Human commensal fungi

Candida albicans	normal bowel, pharyngeal flora
dermatophytes	keratinized tissues (skin, hair, nails)

Figure 11-1 Table showing the illustrative spectrum of human fungal infections

Pathogenic fungi can be commensals, ubiquitous environmental molds, or geographically endemic species

Many species of fungi are unicellular organisms that are ecologically important for their capacity to degrade complex organic constituents to simple carbon sources. They are widespread in the environment, although some, termed endemic mycoses, reside in geographic areas with distinct climatic conditions that favor fungal growth and dispersal (Figure 11-1). Many fungi share small, spore-like, stages that allow them to become airborne and reach terminal airspaces in the lung. Others are introduced in food or by injuries that disrupt the skin barrier. Although many are transmitted from the environment, fungi that cause disease can also be part of the normal human microbial flora, such as *Candida albicans*.

Phagocytes and dendritic cells express receptors for fungal constituents

The unicellular fungi are aerobic eukaryotes with a rigid cell wall that enables their survival over a range of environmental conditions. Rigidity of the cell wall is maintained by chitin, polymers of N-acetyl-D-glucosamine, embedded within a complex polysaccharide matrix of glucans (glucose polymers), mannans (mannose polymers) and galactans (galactose polymers). Innate recognition is mediated by interactions with these conserved fungal constituents. Dectin-1 is highly expressed on dendritic cells, and in lesser amounts on macrophages, where its C-type lectin carbohydrate recognition domain mediates the recognition of $\beta1,3$- and $\beta1,6$-linked glucans on the surface of intact yeast and **zymosan** particles. Engagement of dectin-1 results in tyrosine phosphorylation of its cytoplasmic ITAM and subsequent recruitment of TLR2 to the phagosome, and these two innate receptors synergistically induce inflammatory cytokines, including TNF, and the production of reactive oxygen species. The macrophage/dendritic cell mannose receptor (CD206) mediates phagocytosis of yeast and zymosan through recognition of α-mannan. Finally, many fungi activate complement or mannan-binding lectin (MBL), or bind fibronectin or vitronectin in serum, and interact with phagocytic cells via complement and integrin receptors. The long pentraxin PTX3 facilitates the recognition of *Aspergillus* by macrophages and dendritic cells. In *Drosophila*, fungal infection leads to the generation of Spätzle dimers, which are ligands for Toll. Toll signaling via the NF-κB-like transcription factor dorsal results in the expression and secretion of drosomycin, an antifungal defensin.

Host defense against fungi relies on epithelial barriers and phagocyte activation facilitated by T$_H$1 and T$_H$17 cells

In mice and humans, epithelial disruption and quantitative or qualitative deficiencies in neutrophils are associated with an increased risk of fungal infections. Predisposing genetic deficiencies in neutrophils include both those in oxidative radical formation (chronic granulomatous disease) as well as various granule defects (myeloperoxidase, elastase or cathepsin G deficiency). CD4 T cells are required to maintain optimal defense against fungi. Studies in mice show that the production of IFN-γ and TNF by T$_H$1 cells, and of IL-17A and TNF by T$_H$17 cells, is particularly important. Cases of fungal infection among patients with mutations of the IFN-γ receptor and among patients treated with anti-TNF antibodies for autoimmune diseases support these mechanisms in humans as well.

Protozoa are a diverse group of systemic and intestinal pathogens

Protozoa constitute a diverse group of unicellular pathogens that typically cause chronic infections, indicating a high degree of evolutionary adaptation. This is perhaps best illustrated with malaria,

Definitions

antigenic variation: (in parasites) clonal expression of members of families of proteins among parasite progeny; examples include the major surface glycoproteins of trypanosomes and the red cell adhesins encoded by malaria parasites; variant expression allows parasites to evade immune recognition.

concomitant immunity: immunity to reinfection that coexists with a pathogen that remains viable in an otherwise asymptomatic host; common to many chronic infections, including latent tuberculosis, leishmaniasis

and intestinal worms; regulatory T cells may contribute to concomitant immunity.

hemoglobinopathies: genetic mutations affecting the structure of hemoglobin and, hence, its capacity to bind and release oxygen; examples include sickle cell disease and thalassemia. They are evolutionarily selected in areas of high malaria transmission as a result of a restricted ability of parasites to grow in red cells heterozygous for the abnormal hemoglobin.

sterile immunity: a state of immunity in which a pathogen is completely eradicated from the host.

zymosan: cell wall preparation of *Saccharomyces cerevisiae* composed of β-glucans, mannans, mannoproteins and chitin.

which has selected for a series of devastating **hemoglobinopathies** while reaching immunologic detente with its human host. **Antigenic variation** is a common theme that underlies the capacity of these organisms to persist in the host. The impact of protozoal infections on human health is enormous, and the true incidence of these widespread diseases can only be estimated (Figure 11-2). Protozoa can be divided into organisms introduced systemically, usually by insect bites or, less commonly, by tissue injury, and organisms introduced by the ingestion of contaminated food or water, or by sexual transmission.

Protozoan pathogens induce disease-protective, but not sterile, immunity

Humans in endemic areas are infected many times with protozoa, but are unable to develop an immune response, termed **sterile immunity**, that completely eliminates the organism. Immunity is critical, however, in protecting from immunopathology and death. *Plasmodium falciparum* causes devastating infections in young children, who develop high levels of parasitized red cells and suffer the most deadly tissue damage, including cerebral malaria. Many of the pathologic consequences of primary protozoan infections reflect robust inflammatory immune responses, accompanied by high levels of TNF, IL-6 and IFN-γ. Among children who survive, sequential infections become more mild, such that young adults with parasitized red cells are asymptomatic. The development of immunity against disease in the setting of persistent infection with viable organisms is designated **concomitant immunity**. Although incompletely understood, protection against disease reflects an interplay of both cellular and humoral responses to the parasite, including the expression of modulatory cytokines, such as IL-10 and TGF-β, the appearance of regulatory T cells, and the development of antibodies against disease-promoting moieties of the organisms. Some (for example, *Toxoplasma*, *Cryptosporidium*), but not all (for example, malaria, ameba), protozoa are important opportunistic infections among patients with advanced HIV infection, although the distinguishing features are not readily apparent.

Helminths constitute a diverse group of systemic and intestinal pathogens

Helminths are large, multicellular organisms, and generally do not replicate within the mammalian host; passage through intermediate hosts or through soil or water is required. Organisms reproduce sexually to create larval stages that mediate transmission. Helminths consist of nematodes (roundworms), trematodes (flukes) and cestodes (flatworms, or tapeworms), and can cause intestinal or systemic infections. These worms are responsible for substantial morbidity, particularly in developing areas of the world, where they infect enormous numbers of persons (Figure 11-3).

Helminths cause chronic infections and incomplete immunity

Helminths cause chronic infections, with a duration measured in years. Most induce strong T$_H$2-associated immune responses, characterized by eosinophilia, mucosal mastocytosis and raised IgE. Immunity is incomplete, and reinfections are common. Over time, the burden of organisms becomes less, reflecting the acquisition of partial immunity by mechanisms similar to those involved against protozoa. Intestinal pathogens can bring about anemia from chronic blood loss. Disease caused by systemic pathogens frequently results from immune responses to eggs or worms over many years, with the end result being substantial end-organ damage, often from fibrosis. The pathology is diverse. In various parts of the world, helminths are major causes of blindness (*Onchocerca*), liver failure (schistosomiasis), epilepsy (*Taenia solium*) and elephantiasis (filariasis).

Estimated Prevalence of Selected Protozoan Infections

Organism	Prevalence
malaria (*Plasmodium* species)	>300 million
ameba	500 million
giardia	many millions
Trypanosoma cruzi	17 million
African trypanosomes	0.1 million
Leishmania species	12 million
Toxoplasma gondii	3–90% in various populations
cryptosporidia	65–75% seroprevalence in developing countries
cyclospora	many millions
Trichomonas	many millions

Figure 11-2 Table of estimated worldwide prevalence of selected protozoan infections

Estimated Prevalence of Helminth Infections

Nematodes	
Intestinal	
Hookworms	900 million
Ascaris lumbricoides	1 billion
Strongyloides stercoralis	75 million
Trichuris trichiura	many millions
Tissue-dwelling	
Wuchereria bancrofti	80 million
Brugia malayi	10 million
Onchocerca volvulus	13 million
Loa loa	13 million
Dracunculus medinensis	30 million
Cestodes	
Taenia solium	1 million
Echinococcus granulosis	1.5 million
Trematodes	
Schistosoma mansoni	150 million
Schistosoma japonicum	2 million
Schistosoma hematobium	50 million

Figure 11-3 Table of estimated prevalence of helminth infections

References

Brown, G.D. and Gordon, S.: **Fungal β-glucans and mammalian immunity.** *Immunity* 2003, **19**:311–315.

Loukas, A. and Prociv, P.: **Immune responses in hookworm infections.** *Clin. Microbiol. Rev.* 2001, **14**:689–703.

Maizels, R.M. and Yazdanbakhsh, M.: **Immune regulation by helminth parasites: cellular and molecular mechanisms.** *Nat. Rev. Immunol.* 2003, **3**:733–744.

Miller, L.H. *et al.*: **The pathogenic basis of malaria.** *Nature* 2002, **415**:673–679.

Romani, L.: **Immunity to fungal infections.** *Nat. Rev. Immunol.* 2004, **4**:1–13.

Sacks, D. and Sher, A.: **Evasion of innate immunity by parasitic protozoa.** *Nat. Immunol.* 2002, **3**:1041–1047.

Candida albicans is a commensal and opportunistic pathogen of humans

Candida albicans is a commensal fungus that colonizes the normal human oral, gastrointestinal and urogenital mucosa. Disruption of mucosal barriers can lead to persistent superficial infection, which, in the setting of compromised immunity, can disseminate to cause life-threatening infections of virtually any organ. Antibiotics, which kill commensal bacteria but not *Candida*, allow proliferation of the organism to high densities on epithelial surfaces, and constitute a major predisposing factor for *Candida* infection. *Candida* is the most common fungal infection of humans. In the USA, *Candida albicans* is the fourth most common bloodstream infection acquired in the hospital, and is estimated to add over $1 billion per year in medical costs. Thrush, a superficial *Candida* infection of the oral mucosa, and *Candida* esophagitis are common opportunistic infections in patients with HIV infection. The organism contains a 16-megabase haploid genome comprising approximately 6,600 genes. Aside from the role of an intact epithelium, defense against *Candida* infection is mediated by phagocytic cells, including neutrophils and macrophages, and CD4 T cells, which activate phagocyte anti-fungal mechanisms.

Candida albicans virulence is linked to the ability to grow vegetatively in morphologically distinct forms

Although it is a commensal on normal mucosa, *Candida* grows primarily as oval yeast, like the baker's yeast *Saccharomyces cerevisiae*, and reproduces asexually by budding. In response to a wide variety of microenvironmental signals that accompany tissue invasion, including changes in pH, cell density, exposure to serum and iron deprivation, *Candida* switches from the yeast form to filamentous forms (Figure 11-4). Filamentous growth occurs by two different processes that result in elongated, polarized forms. Pseudohyphae are elliptical yeast forms that remain attached at constricted septa and grow as a microcolony in branching patterns. True hyphae are elongated forms with parallel sides and interspersed septa that distinguish daughter cells. The cytoskeletal organization of the hyphae creates a rigid structure capable of generating forces that disrupt normal tissue and cellular barriers. The ability of *Candida albicans* to switch between yeast and filamentous forms has been linked with pathogenicity, and both forms are likely to contribute. Yeast forms are likely to disseminate more easily through the bloodstream, whereas the filamentous forms facilitate the invasion and evasion of phagocytosis. Mutants unable to undergo morphogenesis between yeast and filamentous forms are commonly non-pathogenic in animal models.

Phagocyte oxidative killing is important in *Candida* host defense

Candida expresses surface adhesins and mannoproteins that activate the alternative and lectin pathways of complement (see section 3-3), resulting in opsonization and phagocytosis by neutrophils and macrophages. Phagocytosis triggers the assembly of the NADPH oxidase and the production of reactive oxygen species catalyzed, in part, by myeloperoxidase (MPO). Interactions of fungal cell wall components, such as mannoproteins and β-glucan, with phagocyte receptors, such as DC-SIGN, TLR2 and dectin-1, facilitate attachment and stimulate the release of cytokines including IL-1, IL-6 and TNF. The production of cytokines by activated neutrophils, together with direct interactions of neutrophils with dendritic cells within inflammatory sites, promotes the maturation of tissue dendritic cells, such that adaptive immunity becomes activated. Humans and mice with genetic or acquired **neutropenia** or

---

Definitions

glyoxylate cycle: the reactions catalyzed by the enzymes isocitrate lyase and malate synthase, which provide a by-pass in the tricarboxylic acid (TCA) cycle to allow the assimilation of two-carbon compounds, allowing organisms to synthesize glucose in glucose-deficient environments.

neutropenia: low neutrophil numbers in blood; in humans, susceptibility to infectious diseases increases when neutrophil counts fall below 1,000 cells/μl and control of endogenous microflora is impaired when

counts fall below 500 cells/μl.

Sabouraud dextrose agar: a selective medium developed for the cultivation of fungi; antibiotics are frequently added to suppress growth of contaminating bacteria.

genetic defects in components of the NADPH oxidase or MPO are predisposed to *Candida* infection. In mice, deficiencies of TNF or IL-17A, potent neutrophil-activating cytokines, markedly attenuate resistance to *Candida*. Relatively common deleterious variant alleles of the gene encoding MBL (see section 3-2) have been correlated with recurrent vulvovaginal candidiasis in some human populations.

Uptake of *Candida* yeasts into the glucose-deficient environment of the phagolysosome triggers the induction of *Candida* genes for the **glyoxylate cycle** that can be used to generate energy for macromolecular synthesis. Organisms managing to survive the initial oxidative response undergo filamentous transformation and erupt from the phagolysosome to proliferate in the cytosol. Unlike the yeast forms, which interact with immature dendritic cells via receptors that induce IL-12 expression, filamentous forms interact with distinct dendritic cell receptor subclasses that trigger inefficient IL-12 induction. In this way, filamentous forms attenuate the adaptive T_H1 response that promotes phagocyte activation and the resolution of infection. T_H17 cells also have a role in protection. Adaptive immunity is required to restrain *Candida* invasion in the mouth and esophagus. Patients with HIV infection and fewer than 200 CD4 T cells per μl have recurrent mucocutaneous *Candida* infections. Chronic mucocutaneous candidiasis is a genetic disease characterized by an inability to clear mucosal *Candida* infection. Some patients with this disease have mutations in Aire, a protein implicated in establishing T cell tolerance (see section 7-10). How dysfunction of the gene is related to the specific defect in *Candida* immunity remains unknown.

Mouse models exist for both systemic and mucosal *Candida* infection

A commensal highly adapted to humans, *Candida* can also be used to infect mice in models of mucocutaneous and systemic infection. For systemic infection, inbred mice are injected via the tail vein with 10^4–10^6 *Candida albicans* yeast cells. Virulence can vary depending on the *Candida* strain, with clinical isolates typically having an LD_{50} of about 5×10^4 organisms. Death usually occurs in the first week. Quantitation is achieved by removing the spleen and kidneys and determining the numbers of viable *Candida* by plating serial dilutions of the homogenized tissue on **Sabouraud dextrose agar**. Histologic examination of these organs reveals the presence of both yeast and filamentous forms. Resolution of infection is accompanied by T_H1 and T_H17 responses to *Candida* antigens.

Mucocutaneous infection is commonly examined using DBA/2 mice, which are genetically C5 deficient. Animals are pretreated with cyclophosphamide to reduce T cell responses, and with a cocktail of oral antibiotics to deplete the competing gastrointestinal bacterial microflora. Up to 10^8 yeast cells are administered by oral gavage. Colonization and evidence for local and systemic invasion is documented after 3–7 days by histologic examination of the esophagus and kidneys, which reveals both yeast and filamentous forms. Quantitative cultures of the homogenized tissues can also be done.

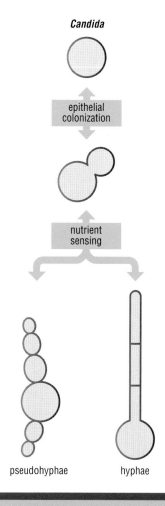

Candida

epithelial colonization

nutrient sensing

pseudohyphae   hyphae

Figure 11-4 *Candida albicans* **grows in three distinct forms** Yeast cells colonizing epithelial surfaces usually divide by budding. Under stress and nutrient-deficient conditions, including those associated with tissue invasion, *Candida* undergoes filamentous transformation. Pseudohyphae create microcolonies of elongated daughter cells that remain attached at sites of constriction. True hyphae are highly polarized extensions with parallel sides. Daughter cells are demarcated by septa within the hyphal extension. Filamentous forms can revert to oval yeast forms when transforming conditions are alleviated.

References

Berman, J. and Sudbery, P.E.: *Candida albicans*: a molecular revolution built on lessons from budding yeast. *Nat. Rev. Genet.* 2002, **3**:918–930.

d'Ostiani, C.F. *et al.*: Dendritic cells discriminate between yeasts and hyphae of the fungus *Candida albicans*: implications for initiation of T helper cell immunity in vitro and in vivo. *J. Exp. Med.* 2000, **191**:1661–1674.

Farah, C.S. *et al.*: T cells augment monocyte and neu-trophil function in host resistance against oropharyngeal candidiasis. *Infect. Immun.* 2001, **69**:6110–6118.

Gantner, B.N. *et al.*: Collaborative induction of inflammatory responses by dectin-1 and Toll-like receptor 2. *J. Exp. Med.* 2003, **197**:1107–1117.

Huang, W. *et al.*: Requirement of interleukin-17A for systemic anti–*Candida albicans* host defense in mice. *J. Infect. Dis.* 2004, **190**:624–631.

Lorenz, M.C. and Fink, G.R.: The glyoxylate cycle is required for fungal virulence. *Nature* 2001, **412**:83–86.

Romani, L. *et al.*: Adaptation of *Candida albicans* to the host environment: the role of morphogenesis in virulence and survival in mammalian hosts. *Curr. Opin. Microbiol.* 2003, **6**:338–343.

Staib, P. *et al.*: Differential activation of *Candida albicans* virulence gene family during infection. *Proc. Natl Acad. Sci. USA* 2000, **97**:6102–6107.

Whiteway, M. and Oberholzer, U.: *Candida* morphogenesis and host–pathogen interactions. *Curr. Opin. Microbiol.* 2004, **7**:350–357.

Pneumocystis is an important cause of pneumonia in immunocompromised hosts

Pneumocystis, originally described as a protozoan on the basis of morphologic criteria, is now placed in the Kingdom Fungi in a unique Order (Pneumocystidales). Previously recognized as an uncommon cause of pneumonia in immunocompromised patients, *Pneumocystis* became widely recognized after the HIV epidemic, where it is the most common opportunistic infection among patients with AIDS. Before the development of prophylactic antimicrobials, up to 85% of HIV-infected individuals were estimated to have at least one episode of *Pneumocystis* pneumonia, with mortality up to 20%.

Pneumocystosis is a common childhood respiratory infection. Symptoms are typically mild, resolve without specific therapy, and may occur unrecognized with other common childhood respiratory infections. Most humans have antibody before four years of age. Humans are believed to be the natural reservoir of the fungus, and transmission is airborne. CD4 T cells, B cells and antibody are required to control *Pneumocystis* and eliminate it from the lung. HIV-infected patients with CD4 T cells less than 200/μl are at high risk for infection and require antibiotic prophylaxis. The recognition of *Pneumocystis* infection in immunodeficient mice has facilitated study of the organism, although it is still not possible to grow *Pneumocystis* *in vitro* and this remains an obstacle to research.

Pneumocystis comprises diverse host-specific strains of unusual fungi

Pneumocystis infects a wide spectrum of mammals, but strains from one host animal do not infect other animal species. Thus, mouse, rat, ferret, human and monkey *Pneumocytis* strains are different and do not establish infections in other than the parent host. Genetic variation is

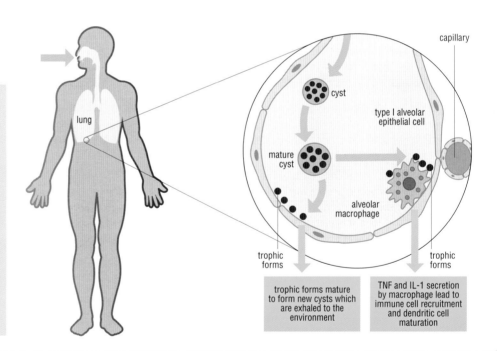

Figure 11-5 Simplified representation of the presumed life cycle of *Pneumocystis jirovecii* Inhaled organisms mature in alveoli to generate trophic forms that adhere to type I alveolar epithelial cells. Interactions with alveolar macrophages through β-glucan and mannose receptors leads to the production of cytokines, including TNF and IL-1, that promote the clearance of organisms by phagocyte recruitment and activation of innate antimicrobial mechanisms until adaptive CD4 T cells and antibody can provide definitive clearance. Trophic forms that escape immune destruction either undergo mitosis to generate more trophic forms in an asexual cycle (not shown) or undergo conjugation to generate precysts (not shown). Whether the cysts or trophic forms or both are exhaled to sustain transmission to new hosts is unknown.

Figure labels: capillary; cyst; type I alveolar epithelial cell; mature cyst; alveolar macrophage; lung; trophic forms; trophic forms; trophic forms mature to form new cysts which are exhaled to the environment; TNF and IL-1 secretion by macrophage lead to immune cell recruitment and dendritic cell maturation

Definitions

Gomori methenamine silver stain: stain used to detect reactive aldehydes in fungal cell walls by the precipitation of black silver salts at aldehyde groups exposed by treatment with chromic acid.

Pneumocystis jirovecii: the human-adapted species of *Pneumocystis*, and preferred over the designation *Pneumocystis carinii* f. sp. *hominis*; honors Otto Jirovec, the Czech parasitologist who published the first description of human *Pneumocystis* infection; *Pneumocystis carinii* refers to the rat-adapted species.

type 1 alveolar epithelial cells: cells covering approximately 90% of the alveolar surface of the lung and participating in gas exchange between the air space and the capillaries; type 2 alveolar epithelial cells secrete hydrophobic surfactant proteins (surfactants B and C) involved in moderating alveolar surface tension.

less among isolates from one host, but substrains are common. The original organism, *Pneumocystis carinii*, was derived from rats, and human isolates are now designated *Pneumocystis carinii* f. sp. *hominis*, or, more correctly, as **Pneumocystis jirovecii**.

Organisms are acquired by the airborne route. The life cycle of *Pneumocystis* is illustrated in Figure 11-5 and described in the legend. Cysts mature in the lung alveoli, and rupture to release eight haploid vegetative, or trophic, forms. Trophic forms interact with the collectins SP-A and SP-D, which facilitate phagocytosis by immune cells, and bind to **type I alveolar epithelial cells** using the mannose receptor or by bridging interactions with fibronectin and vitronectin. β-Glucan in the fungal cell wall activates the alveolar macrophage dectin-1 and TLR2 receptors, causing a release of TNF and IL-1 that contributes to protective immunity. In most cases, the trophic forms remain extracellular in the alveolar air spaces. Organisms are eliminated by a process that depends on the recruitment of CD4 T cells and B cells and on the production of antibody against surface proteins. In the absence of appropriate immune effector cells, organisms proliferate and provoke the accumulation of fluid in the alveoli. Progressive hypoxia ensues. Severe infection is treated both with antimicrobial drugs against *Pneumocystis* and also with corticosteroids to blunt the inflammatory response that contributes to the air-space disease. In patients coinfected with HIV, institution of anti-retroviral agents can worsen disease, a phenomenon termed immune reconstitution syndrome and associated with increased numbers of CD4 T cells that exacerbate inflammation. In severe immunodeficiency, organisms can disseminate widely through the body.

In immunocompetent hosts, *Pneumocystis* is typically eliminated without establishing a latent state. However, it is likely that this microorganism can evade lasting immunity and reinfect an individual many times. The major surface glycoprotein, or MSG, comprises a family of 100 genes dispersed over the organism's genome. Like those encoding trypanosome coat proteins, MSG genes switch by recombination between a telomeric site at which they are expressed and other silent MSG gene loci elsewhere in the chromosome. Although antibody and T cell responses to a given MSG gene are protective, variant organisms can evade prior immunity and reestablish infection. Immunocompetent humans are probably reinfected many times with *Pneumocystis jirovecii*. Such infections, although mild and unrecognized, ensure the maintenance of *Pneumocystis* in its mammalian reservoir.

Pneumocystis infection of mice comprises an experimental model for studying the immune response

Pneumocystis outbreaks were recognized as a common complication among colonies of SCID (severe combined immunodeficient) and B cell-deficient mice, and led to the routine use of prophylactic antimicrobial drugs in drinking water for such colonies. Asymptomatic airborne transmission occurs among immunocompetent mice, as in humans. In order to study *Pneumocystis* experimentally, mice are treated with anti-CD4 antibodies to deplete CD4 T cells, and then inoculated intratracheally twice, 1 week apart, with approximately 2×10^5 cysts prepared from the lungs of infected, immunodeficient mice (Figure 11-6). The numbers of organisms in the lung, determined using special stains that highlight the cyst cell wall (**Gomori methenamine silver**), increase rapidly by 7 days, but then decline in normal mice, in which *Pneumocystis* is cleared by 28 days. In contrast, immunodeficient mice continue to support the proliferation of organisms up to 40 days, when mice begin to die. Reconstitution with CD4 T cells or with hyperimmune anti-*Pneumocystis* serum results in a rapid clearance of organisms.

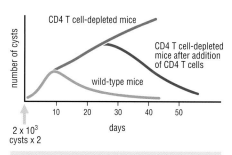

Figure 11-6 Mouse model of *Pneumocystis* infection Wild-type mice or mice depleted of CD4 T cells are inoculated with *Pneumocystis* purified from the lungs of immunodeficient mice. Organisms are cleared after 28 days in wild-type mice but continue to proliferate in CD4 T cell-deficient mice. Administration of CD4 T cells or hyperimmune antisera facilitates the clearance of organisms.

References

Bishop, L.R. and Kovacs, J.A.: **Quantitation of anti-*Pneumocystis jiroveci* antibodies in healthy persons and immunocompromised patients.** *J. Infect. Dis.* 2003, **187**:1844–1848.

Chen, W. *et al.*: **Interleukin 1: an important mediator of host resistance against *Pneumocystis carinii*.** *J. Exp. Med.* 1992, **176**:713–718.

Edman, J.C. *et al.*: **Ribosomal RNA sequence shows *Pneumocystis carinii* to be a member of the Fungi.**

Nature 1988, **334**:519–522.

Shellito, J. *et al.*: **A new model of *Pneumocystis carinii* infection in mice selectively depleted of helper T lymphocytes.** *J. Clin. Invest.* 1990, **85**:1686–1693.

Thomas, C.F. Jr and Limper, A.H.: **Pneumocystis pneumonia.** *N. Engl. J. Med.* 2004, **350**:2487–2498.

Totet, A. *et al.*: **Pneumocystis jiroveci genotypes and primary infection.** *Clin. Infect. Dis.* 2003, **36**:1340–1342.

Wada, M. *et al.*: **Antigenic variation by positional control of major surface glycoprotein gene expression in *Pneumocystis carinii*.** *J. Infect. Dis.* 1995, **171**:1563–1568.

Leishmania major is a protozoan pathogen transmitted by sandflies

Leishmania species are **kinetoplastid protozoa** estimated to infect over 12 million persons in tropical and subtropical areas of the world. The organisms have a dimorphic life cycle, and live as motile, extracellular promastigotes in the gut of the sandfly vector and as obligate intracellular amastigotes in the vertebrate host. *Leishmania* species infect a wide variety of forest and desert rodents. Human encroachment into areas of high endemicity among animals results in transmission from female sandflies, which require blood for egg maturation, to both humans and their domestic pets. Although species of *Leishmania* differ in their propensity to disseminate from the site of inoculation in the skin to distant organs, all forms establish prolonged infectious forms in the skin that enable transmission during subsequent insect bites, thus sustaining the reservoir (Figure 11-7).

Leishmania major is used widely in immunology to probe T_H subset differentiation, as discussed below. The organism contains a 33.6 Mb haploid genome and expresses some 8,000 genes from 36 chromosomes. *L. major* is widespread across Northern Africa and the Middle East, and through Afghanistan and Pakistan, where infection is endemic in desert sand rats and gerbils. A cause of classic Old World cutaneous leishmaniasis, the infection begins as a nodule at the site of the insect bite, which gradually enlarges until the central crust cavitates, exposing a shallow ulcer covered with a serous exudate. Saliva from sandflies promotes infectivity *in vitro*. Without therapy, lesions heal slowly over many weeks, leaving a slightly depressed central hypopigmented scar. Individuals with healed scars are resistant to the development of new lesions.

Macrophages are the major cell type sustaining infection by the intracellular amastigotes, although dendritic cells and even fibroblasts may contribute to the infectious reservoir, which persists in a latent state for years in asymptomatic individuals. Control of *Leishmania* requires activation of macrophages by parasite-specific T_H1 cells, although sterile immunity, defined as clearance of persistent organisms (see section 11-0), is uncommon. Many cases occur among otherwise well individuals who leave areas of endemic leishmaniasis, only to be afflicted with the disease years later at the time of an immunosuppressing event such as HIV infection or organ transplantation.

Leishmania parasites invade macrophages without inducing inflammatory cytokines and reside in a phagolysosomal compartment

A variety of receptors have been implicated in the interaction of promastigotes with the macrophages that are their main reservoir (for example, integrin and scavenger family receptors) and dendritic cells (for example, DC-SIGN). In contrast to many organisms, *Leishmania* parasites do not activate IL-12 expression during macrophage internalization, thus diminishing the tendency to develop a curative T_H1 response.

Ingested organisms reside in parasitophorous (literally, parasite-bearing) vacuoles that become acidified by lysosomal recruitment. The parasitophorous vacuole remains in contiguity with materials endocytosed via fluid-phase pinocytosis and contains MHC class II antigens and other constituents of the class II loading pathway. Organisms differentiate into oval, non-flagellated, amastigotes, the obligate intracellular forms that characterize the vertebrate cycle. Amastigotes are impervious to killing by tissue macrophages, and replicate by binary fission. Eventually, infected cells rupture, releasing amastigotes to spread to uninfected macrophages.

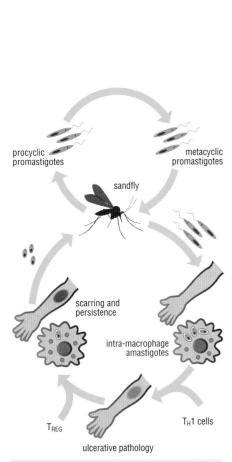

Figure 11-7 Life cycle of *Leishmania*
Maturation in the sandfly is associated with modifications of the surface lipophosphoglycan, or LPG, that allow the multiplying immature organisms, termed procyclics, to be released from the gut epithelia and migrate to the proboscis. The modified LPG on the mature organisms, termed **metacyclic promastigotes**, confers resistance to complement and macrophage killing mechanisms. Upon inoculation in the skin, dermal macrophages and dendritic cells are infected, and organisms replicate within macrophages. Initiation of T_H1-mediated immune responses corresponds with tissue pathology, resulting in an ulcer. Activation of regulatory T cells serves to moderate pathology, but viable parasites are maintained for indefinite periods. Uninfected sandflies probing sites of infection take up amastigotes, which transform back to the motile promastigotes and multiply in the insect gut.

Labels in figure: procyclic promastigotes; metacyclic promastigotes; sandfly; scarring and persistence; intra-macrophage amastigotes; T_{REG}; T_H1 cells; ulcerative pathology

---

Definitions

arginase: arginase I and arginase II enzymes metabolize L-arginine to urea and L-ornithine, a polyamine precursor; in macrophages, arginase I is induced by IL-4, IL-13, IL-10 or TGF-β, and serves to remove arginine available for oxidation by NOS2 in the formation of L-citrulline and reactive nitric oxide.

kinetoplastid protozoa: flagellated protozoa containing a single mitochondrion with mitochondrial DNA massed into a region called the kinetoplast; medically important kinetoplast parasites belong to the genera

Leishmania or *Trypanosoma*, while *Phytomonas* species cause disease in plants.

LACK: see *Leishmania* homolog of receptor for activated C kinase.

***Leishmania* homolog of receptor for activated C kinase (LACK):** a WD-repeat protein containing an immunodominant peptide for the CD4 T cell response.

metacyclic promastigotes: infectious forms of *Leishmania* that represent terminally differentiated promastigotes released from the intestinal epithelia and

that migrate to the sandfly proboscis; metacyclic forms are covered with a lengthened form of surface lipophosphoglycan molecules that renders parasites more resistant to complement and macrophage oxidant killing.

Immunity to *Leishmania* requires T_H1 cells, and persistence is maintained by regulatory T cells

Activation of macrophages, principally by IFN-γ, enables infected cells to inhibit amastigote replication through mechanisms dependent upon activated oxygen and nitrogen radicals. T_H1 cells are necessary for *Leishmania* control. T cell-deficient mice or animals deficient in IFN-γ, IL-12 or NOS2 (inducible nitric oxide synthase) are therefore highly susceptible to *L. major* infection, as are mice deficient in ligand–receptor pairs important for T_H1 cell activation, such as CD40 and CD40L. Long-term immunity is correlated with the appearance of IFN-γ-producing CD8 T cells as well, although the mechanisms by which CTL become recruited into the reservoir of memory cells remain unknown.

Small numbers of viable organisms persist in infected hosts without causing overt disease. Persistence of parasites in the absence of overt symptoms together with resistance to reinfection, or concomitant immunity, requires active immune surveillance: depletion of CD4 T cells or neutralization of IFN-γ or nitric oxide production leads to the reappearance of destructive lesions in immune animals. In *L. major*-infected mice, concomitant immunity is dependent upon regulatory T cells, a CD4 CD25 T cell subset capable of suppressing the activation of effector T cells specific for parasite antigens. Suppression by regulatory T cells in *L. major* infection is due in part to IL-10, which induces macrophage **arginase**, thus attenuating nitric oxide production. In the absence of IL-10 or regulatory T cells, *Leishmania* parasites are cleared rapidly and completely, but protection against reinfection fails to develop, perhaps reflecting the lack of persistent antigenic stimulation. Thus, regulatory T cells promote sustained antigen exposure required for concomitant immunity (see section 11-0), which the parasite exploits to achieve persistence and efficient transmission to biting insects.

Some strains of mice develop an aberrant and fatal T_H2 response to *L. major*

Some inbred strains of mice (for example, those belonging to strains collectively known as BALB, and in particular one known as BALB/c), are unable to control *L. major* infection and die with extensive cutaneous lesions and disseminated visceral disease. In contrast to the dominant T_H1 response that develops in most inbred mouse strains, BALB/c mice develop an aberrant T_H2 response to parasite antigens. The result is high levels of IL-4 and IL-13, and the failure to achieve macrophage activation that is required to restrict amastigote replication (Figure 11-8). Experiments using *L. major* infection of BALB/c and resistant C57BL/6 mice were among the first demonstrations of the importance of T_H subsets to the outcome of infectious diseases.

The early CD4 T cell response in infected BALB/c mice is directed at a peptide from the parasite *Leishmania* **homolog of receptor for activated C kinase (LACK)** antigen, which induces an oligoclonal IL-4 response among T cells expressing Vα8/Vβ4 T cell receptors. Depletion of the LACK-reactive T cell repertoire before infection attenuates disease in BALB/c mice. Despite these observations, the early expansion of IL-4-expressing, LACK-reactive CD4 T cells is comparable among MHC-matched susceptible and resistant mice, and the mechanisms underlying the pathogenic effects of LACK-dependent responses in BALB/c mice remain unclear. In the parasite, deletion of three of the four identical LACK genes was associated with a marked loss of virulence *in vivo*, despite its having little effect on promastigote growth *in vitro*. Genetic studies of the susceptibility of BALB/c mice have suggested contributions from genes expressed both in helper T cells and in antigen-presenting cells.

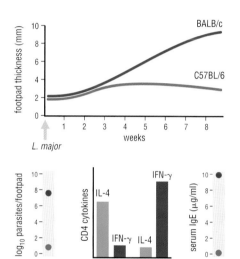

Figure 11-8 Infection of inbred mouse strains with *L. major* Susceptible BALB/c and resistant C57BL/6 mice are inoculated in the footpads with promastigotes selected from stationary-phase cultures by their loss of binding to the lectin peanut agglutinin, which binds to immature forms of LPG. The course of infection is plotted by weekly measurements of the footpad lesions, and evaluated terminally by serial dilutions of the footpads for recovery of viable promastigotes in liquid medium. Restimulation of isolated lymph-node CD4 T cells with *L. major* antigens reveals cytokine patterns typical of T_H1 and T_H2 responses, with the latter accompanied by high levels of IgE in BALB/c mice.

References

Belkaid, Y. *et al.*: **CD4⁺CD25⁺ regulatory T cells control *Leishmania major* persistence and immunity.** *Nature* 2002, **420**:502–507.

Julia, V. *et al.*: **Resistance to *Leishmania major* induced by tolerance to a single antigen.** *Science* 1996, **274**:421–423.

McConville, M.J. *et al.*: **Developmental modification of lipophosphoglycan during the differentiation of *Leishmania major* promastigotes to an infectious stage.** *EMBO J.* 1992, **11**:3593–3600.

Melby, P.C.: **Recent developments in leishmaniasis.** *Curr. Opin. Infect. Dis.* 2002, **15**:485–490.

Noben-Trauth, N. *et al.*: **The relative contribution of IL-4 receptor signaling and IL-10 to susceptibility to *Leishmania major*.** *J. Immunol.* 2003, **170**:5152–5158.

Stenger, S. *et al.*: **Reactivation of latent leishmaniasis by inhibition of inducible nitric oxide synthase.** *J. Exp. Med.* 1996, **183**:1501–1514.

Stetson, D.B. *et al.*: **Rapid expansion and IL-4 expression by *Leishmania*-specific naive helper T cells in vivo.** *Immunity* 2002, **17**:191–200.

Nippostrongylus brasiliensis is a roundworm that matures during migration through the lung to the small intestines

Nippostrongylus brasiliensis is an intestinal roundworm, or **nematode**, parasite of rats that has been adapted for use in mice. The life cycle (Figure 11-9) resembles that of hookworm in humans and is divided into five stages, designated L1–L4 (larvae) and adults. Eggs in soil release motile, free-living larval worms that molt twice to become infective L3 larvae. L3 larvae can penetrate intact skin or can be injected subcutaneously. In the host, organisms invade the venous circulation and are swept to the lung, where they are trapped in the capillaries. The worms molt to the L4 stage, rupture the capillaries to escape into the alveoli, and are coughed up and swallowed, finally reaching the small intestines 3–4 days after infection. After a final molt, organisms mature to the adult males and females and on day 6 begin laying eggs. Organisms are adapted to rats and continue to lay eggs for prolonged periods. In mice, however, immune responses lead to cessation of egg-laying by day 8, and expulsion of adult worms from the intestines by day 10. Reinfection of immune mice with L3 larvae results in trapping of most migrating larvae in the skin and lungs, with few worms ever maturing to adult forms.

Nippostrongylus induces a protective T_H2-associated immune response

Nippostrongylus invokes the full spectrum of T_H2-associated immune responses, including recruitment of T_H2 cells, eosinophils and basophils into tissues, proliferation of mucosal mast cells, differentiation of epithelial **goblet cells**, secretion of mucus, and production of IgE. CD4 T cells and T_H2 cytokines are necessary for the expulsion of worms through the activation of overlapping patterns of effector immunity.

T cell deficiency or depletion of CD4 T cells by using antibodies renders infected mice unable to expel *N. brasiliensis*. Studies in cytokine and cytokine receptor knockout mice have revealed complex interactions among the cardinal T_H2 cytokines, IL-4, IL-13, IL-5 and IL-9. Obligate

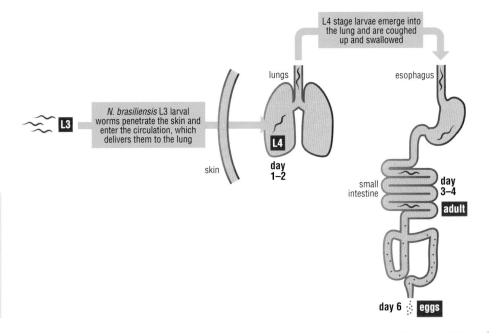

Figure 11-9 Life cycle of *N. brasiliensis* L3 larvae penetrate skin or are injected experimentally and migrate to the lungs, where they molt once to become L4 larvae and ascend to the oral cavity. After descending through the esophagus, worms migrate to the small intestine, where they undergo further maturation to become adults. Eggs begin to appear in the stool by day 6.

Definitions

goblet cells: mucus-producing epithelial cells; induced *in vitro* and *in vivo* by the differentiation of epithelial cells with IL-4 or IL-13.

nematode: unsegmented roundworm. Most nematodes are free living, but together they are the largest cause of helminth infections of humans; hookworms are an example of a nematode parasite of humans.

roles for IL-4 in IgE production, IL-9 and IL-4 in intestinal mastocytosis, IL-5, IL-4 and IL-13 in eosinophilia, and for all of these cytokines in epithelial goblet cell differentiation have been demonstrated, emphasizing the overlapping functions shared by this group of cytokines. Although the absence of IL-4 does not affect time to worm expulsion, the successive deletion of IL-13, IL-5 and IL-9 results in progressively impaired immunity (Figure 11-10). In the absence of IL-13, IL-5, and IL-9, however, IL-4 can provide immunity to worms, emphasizing the primary role of IL-4 signaling in mucosal immunity, although the time course of expulsion is slower than in the presence of the other interleukins. Mice deficient in IL-4Rα or STAT6 (and hence unresponsive to IL-4 and IL-13) are unable to expel worms. Induction and activation of T$_H$1 cells at the time of infection by administration of exogenous IL-12, IFN-γ or IFN-α impairs the expulsion of worms by interfering with the production and activities of the T$_H$2 cytokines.

Recruitment of immune cells to tissue and activation of worm expulsion reflect compartmentalized signals via the IL-4Rα chain and STAT6

Immunity to *Nippostrongylus* involves the recruitment of IL-4-producing innate and adaptive cells to the tissue. At the time that worms are expelled, the lung contains three populations of IL-4-producing cells. Eosinophils comprise 70% of the cells, with T$_H$2 cells and basophils making up equivalent populations of the remaining cells. Recruitment of eosinophils and T$_H$2 cells requires the STAT6-dependent production of the chemokines CCL11 (eotaxin), CCL24 (eotaxin-2) and CCL17 (TARC) by a tissue-resident hematopoietic cell, perhaps macrophages. Although depletion of CD4 T cells abrogates primary immunity, depletion of CD4 T cells during secondary infection of immune mice does not compromise immunity. Although incompletely understood, this probably reflects the accumulation of innate cells, such as eosinophils, in subcutaneous tissues in response to the first infection. Eosinophils attack migrating larvae in the tissue and lung, preventing their ability to reach the intestines and mature.

Expulsion of worms requires the expression of the genes encoding IL-4Rα and STAT6 by tissue gastrointestinal cells. Actions of IL-4 and IL-13 on type II IL-4 receptors (see section 2-8) in various tissue cell types of the small intestine include stimulation of smooth muscle contractility, enhanced epithelial permeability and increased mucus secretion. Together, these effects may interfere with the ability of the worms to sustain contact with the intestinal mucosa, facilitating their passage in the stool.

Cytokines in Immunity to *Nippostrongylus brasiliensis*

mice	days to 50% worm reduction	eosinophils	mucosal mast cells	IgE	epithelial goblet cells
wild type	7	+++	+++	+++	+++
IL-4 ko	7	+++	+++	–	+++
IL-4/-13 ko	25	+++	+++	–	+++*
IL-4/-13/-5 ko	35	–	+++	–	+++*
IL-4/-13/-5/-9 ko	60	–	–	–	+*
IL-9 ko	7	+++	++*	+++	+++*
IL-13/-5/-9 ko	25	++	+++*	+++	+++*

*intact but delayed response

Figure 11-10 **Table of T$_H$2 cytokines essential for immunity to *N. brasiliensis*** Various combinations of cytokine knockout (ko) mice have been used to reveal the overlapping roles for this class of cytokines in mediating the stereotyped phenotypes of immunity. Note that IL-4 can alone drive all of the cardinal features of T$_H$2-associated immunity, although responses are delayed. Data from Fallon, P.G. *et al.*: *Immunity* 2002, **17**:7–17.

References

Fallon, P.G. *et al.*: **IL-4 induces characteristic Th2 responses even in the combined absence of IL-5, IL-9, and IL-13.** *Immunity* 2002, **17**:7–17.

Shinkai, K. *et al.*: **Helper T cells regulate type 2 innate immunity *in vivo*.** *Nature* 2002, **420**:825–829.

Urban, J.F. Jr *et al.*: **IL-13, IL-4Rα, and Stat6 are required for the expulsion of the gastrointestinal nematode parasite *Nippostrongylus brasiliensis*.** *Immunity* 1998, **8**:255–264.

Voehringer, D. *et al.*: **Type 2 immunity reflects orchestrated recruitment of cells committed to IL-4 production.** *Immunity* 2004, **20**:267–277.

Zhao, A. *et al.*: **Dependence of IL-4, IL-13 and nematode-induced alterations in murine small intestinal smooth muscle contractility on Stat6 and enteric nerves.** *J. Immunol.* 2003, **171**:948–954.

Schistosomes are a widespread human trematode infection of the abdominal vasculature

Schistosoma, the causative agents of schistosomiasis, are the most important parasitic **trematode**, or fluke, infection of humans. Over 200 million persons in tropical and subtropical countries are believed infected by any of six species of schistosomes. Infection occurs by exposure to fresh water contaminated with motile, fork-tailed, cercariae, which penetrate skin using proteases stored within glands in the head. Organisms shed their tails, and the immature schistosomula migrate via the circulation to reach the liver microcirculation, where they mature. Worms do not multiply in the host. Adult females reside intimately within a canal in the ventral body of the male, and pairs migrate against the portal venous blood flow to release eggs into veins draining various abdominal sites, including the intestines (*S. mansoni*, *S. japonicum*, *S. intercalatum*, *S. mekongi* and *S. malayi*) and bladder (*S. haematobium*). Egg-laying begins after 4–6 weeks, with estimated rates of release varying from 300 (*S. mansoni*) to 3,000 (*S. japonicum*) eggs per day. Eggs elicit a granulomatous inflammatory response that is believed to assist their transit into the intestines or bladder, and excretion from the body in the stool or urine, respectively. In fresh water, eggs hatch to yield motile miracidia, which infect snails. In their intermediate snail hosts, the miracidia undergo maturation and multiplication to release many infectious cercariae, thus completing the life cycle (Figure 11-11).

Host immune responses are modulated over the course of infection and culminate in a highly polarized type 2 granulomatous response to eggs

In animal studies, schistosomes fail to mature fully in the liver microcirculation in the absence of hepatic CD4 T cells, although the signals required to promote worm development are unknown. Immune responses during the early stages of infection are directed against antigens of schistosomula, and demonstrate a T_H1 profile. Similar responses occur in mice vaccinated with irradiated cercariae that fail to survive and complete maturation. Although usually asymptomatic, acute schistosomiasis (**Katayama fever**) can provoke inflammatory symptoms, including fever, particularly in adults from nonendemic countries. Despite residence in the bloodstream, worms evade innate immunity through a variety of mechanisms, including adsorption on their tegumental surface of complement regulatory proteins from the host that enable them to avoid activating complement.

With the onset of egg-laying, T_H1 responses are replaced by vigorous T_H2 responses directed against egg antigens. Characterized in humans and mice by tissue eosinophilia, elevated IgE and a T_H2 pattern of cytokines, the cardinal feature of schistosomiasis is the presence of tissue granulomas surrounding eggs (Figure 11-12). The granulomas are characterized by an organized, circumferential infiltrate of T_H2 cells, eosinophils, macrophages and fibroblasts within a dense collagen-rich matrix, and are designated **type 2 granulomas** to distinguish them from granulomas associated with T_H1 responses (see sections 3-9, 5-10 and 5-11), as occurs in tuberculosis (which is described more fully in section 9-7).

Although infected individuals develop T cell and B cell responses, adult worms live for 5–10 years with no evidence that host immunity impedes survival. Epidemiologic studies in areas endemic for human schistosomiasis show a correlation between IgE levels and acquired immunity that develops over many years but is incomplete. The intensity of chronic infection has been linked to the region of human chromosome 5q that encodes the T_H2 cytokine genes.

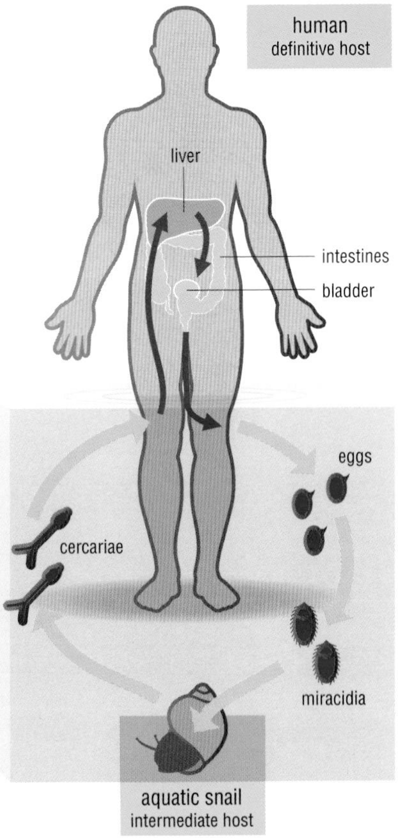

Figure 11-11 Life cycle of schistosomes Cercariae released from infected freshwater snails penetrate exposed skin, shed their tails to become schistosomula, and migrate via the circulation to the vessels of the liver. After maturation, males and females pair and begin to lay eggs after migrating out into the abdominal vessels draining the intestines or bladder (*S. haematobium*). Eggs that access fresh water hatch to release the motile miracidia. Miracidia invade snails, maturing and multiplying over 25–40 days to produce the infectious cercariae and complete the cycle.

Definitions

cirrhosis: irreversible chronic injury to the liver manifested pathologically by extensive fibrosis.

Katayama fever: febrile illness marked by fatigue and tissue inflammatory responses to schistosome egg antigens; most commonly occurs in nonimmune adult travelers who encounter heavy schistosome infection; typically due to *S. japonicum* or *S. haematobium* with onset at the time of egg-laying, perhaps reflecting a serum sickness reaction.

portal hypertension: elevated blood pressure in the portal vein draining from the intestines to the liver; normally low because of the distensible hepatic sinusoids, portal blood pressure is raised by any obstruction to flow, including the fibrosis induced by chronic egg deposition of schistosomiasis; the high blood pressures result in retrograde flow, leading to splenomegaly, ascites (accumulation of peritoneal fluid) and **varices**.

trematode: flatworms or flukes; parasitic trematodes of humans typically have oral and ventral suckers to assist anchoring while feeding, and alternate between an adult sexual cycle in humans and a larval asexual multi-

plication in an intermediate host.

type 2 granuloma: granulomatous response orchestrated by T_H2 cells, and including T_H2 cells, eosinophils, alternatively activated macrophages and collagen-secreting mesenchymal cells organized in a dense spherical structure around a central inciting agent, such as a schistosome egg.

varices: collaterals that develop in the portal circulation to by-pass liver obstruction: the absence of valves in the portal blood system allows retrograde flow, which leads to the formation of varices in the lower esophagus and rectum.

Type 2 granulomatous responses benefit the host and the parasite until fibrosis compromises the function of critical organs of the body

Viable eggs secrete antigens through pores in the tough outer membrane that are believed to orchestrate the intense T_H2 immune response. Eggs have spines that help them to lodge in the endothelium, but the host granulomatous response is necessary for their translocation from the microcirculation of the small bowel or bladder to the intestinal or bladder lumen. T cell-deficient mice or HIV-infected patients with low CD4 T cell counts have diminished egg excretion in stool. Thus, the immune response facilitates the entry of eggs into the feces and urine, which enables their return to fresh water to complete the life cycle.

Eggs that become dislodged from the endothelium are swept with the blood draining the abdomen back into the liver and become trapped in the liver sinusoids. A granulomatous response surrounds the eggs, sequestering egg antigens that are otherwise toxic to liver cells (see Figure 11-12). Over time, entrapped eggs die, and the granulomas dissipate, leaving a fibrous scar. After years, the coalescence of collagenous scars leads to **cirrhosis**, obstruction of blood flow causing **portal hypertension**, and the creation of aberrant vascular by-pass channels called **varices**. Death is usually due to bleeding from varices. Fibrosis also occurs in response to eggs trapped in the bowel wall, or, in *S. haematobium* infection, in the bladder, where disease is complicated by blood in the urine and obstruction of urine flow from the kidneys. Approximately 4–8% of individuals with chronic schistosomiasis reach this stage, which is believed to reflect underlying modifying genes that influence whether pro-fibrotic cytokines such as IL-13, or anti-fibrotic cytokines such as IFN-γ, are predominantly produced. Predisposition to hepatic fibrosis has been linked to a region near the *ifnγr1* (IFN-γ receptor-1) gene on chromosome 6 in humans infected with *S. mansoni*.

Mouse models for *S. mansoni* infection and for type 2 granulomatous responses reveal distinct roles for T_H2 cytokines

The best-characterized murine model for schistosomiasis involves percutaneous infection with *S. mansoni*. Tails of immobilized mice are exposed to cercariae collected from infected snail colonies. Cercariae can also be injected subcutaneously. Animals develop a T_H1 response to larval antigens that is replaced by a T_H2 response to egg antigens at the time that egg-laying begins after 5–6 weeks. With heavy inoculations, death can occur from liver fibrosis by 9–12 weeks. In the absence of IL-4, granuloma formation is compromised. As a consequence, egg excretion in the stools is diminished, but animals succumb by 6–8 weeks with hepatic toxicity due to non-sequestered egg antigens. Hepatotoxicity may also reflect compromised bowel integrity, as the marked increase in eggs trapped in the intestinal wall leads to mucosal lesions and the appearance of bacterial endotoxin in portal blood. In the absence of IL-13, fibrosis is attenuated and survival improves, implicating IL-13 as a major fibrogenic cytokine. Failure to achieve alternative activation of macrophages by IL-4 and IL-13 also results in immunopathology. IL-10 modulates both T_H1 and T_H2 responses in the murine model.

In the synchronous granuloma model, *S. mansoni* eggs (~5,000) purified from the livers of infected rodents are injected into the tail vein and become trapped in the lung. After 14 days, mice are killed and the granuloma volumes and cell composition are determined by histology. Tissue and systemic T_H2 responses are quantitated by histology, cytokine production by T cells stimulated with egg antigens *in vitro*, and elevated IgE levels. Pulmonary collagen is quantitated to assess the extent of fibrosis. Studies in mice have shown that IL-4 and IL-13 are important for granuloma formation, and that IL-13 is important for goblet-cell hyperplasia and collagen deposition.

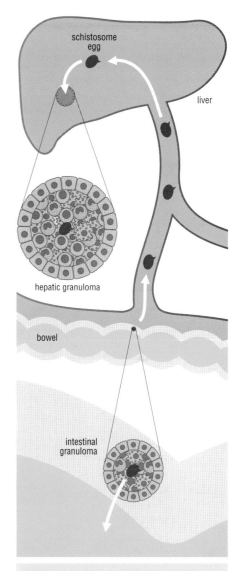

Figure 11-12 Type 2 granulomatous responses to eggs Eggs deposited in the intestinal vasculature elicit a granulomatous response that is necessary for translocation into the intestinal lumen for excretion. Eggs that are swept into the liver become trapped in the hepatic sinusoids. The granulomatous response sequesters toxic egg antigens and kills the eggs, but resolves leaving fibrotic scars that distort the liver architecture. In the absence of IL-4, type 2 granuloma formation is compromised. Eggs are not excreted, but hepatic toxicity results from damage to the liver and intestine.

References

Davies, S.J. *et al.*: **Modulation of blood fluke development in the liver by hepatic CD4+ lymphocytes.** *Science* 2001, **294**:1358–1361.

Fallon, P.G. *et al.*: **Schistosome infection of transgenic mice defines distinct and contrasting pathogenic roles for IL-4 and IL-13: IL-13 is a profibrotic agent.** *J. Immunol.* 2000, **164**:2585–2591.

Herbert, D.R. *et al.*: **Alternative macrophage activation is essential for survival during schistosomiasis and** downmodulates T helper 1 responses and immunopathology. *Immunity* 2004, **20**:623–635.

Hoffmann, K.F. *et al.*: **IL-10 and the dangers of immune polarization: excessive type 1 and type 2 cytokine responses induce distinct forms of lethal immunopathology in murine schistosomiasis.** *J. Immunol.* 2000, **164**:6406–6416.

Pearce, E.J. and MacDonald, A.S.: **The immunobiology of schistosomiasis.** *Nat. Rev. Immunol.* 2002, **2**:499–511.

Ross, A.G.P. *et al.*: **Schistosomiasis.** *N. Engl. J. Med.* 2002, **346**:1212–1220.

Wynn, T.A. *et al.*: **Immunopathogenesis of schistosomiasis.** *Immunol. Rev.* 2004, **201**:156–167.

12

Tolerance and Autoimmunity

The stochastic nature of the recombination process by which the antigen receptors of adaptive immunity are generated means that autoreactive lymphocytes are generated at a high frequency during lymphocyte development. In this chapter, we first discuss the various mechanisms by which robust immune tolerance to self components is achieved and then consider the mechanisms by which this process fails for limited numbers of self components in the 3% of people who develop autoimmune diseases. These diseases result from a combination of multigenic susceptibility and environmental triggers, which are not understood for most autoimmune diseases, but which may include particular infections.

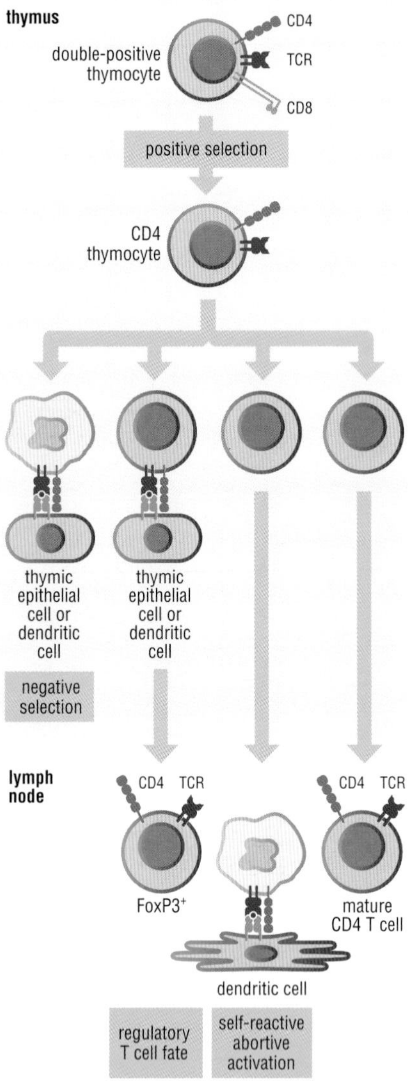

12-0 Overview: Tolerance and Autoimmunity

Immune tolerance is established and maintained by both central and peripheral mechanisms

Whereas innate immune recognition mechanisms have arisen through evolution and have been selected to avoid recognition of self components, adaptive immune recognition receptors are generated by genetic diversification mechanisms that introduce an element of randomness into the junctions between Ig and TCR gene segments, generating what is known as junctional diversity (see section 7-1), and therefore cannot evolve to avoid self-reactivity. Indeed, self-reactivity and the potential for attacking healthy tissues pose a major challenge for the immune system. The need to establish **immune tolerance** to self is met in two broad ways. The first is through the existence of developmental programs in which self-reactive lymphocytes are deleted or converted into regulatory cell types. These mechanisms are responsible for **central tolerance**, so called because they operate in central lymphoid organs where lymphocytes develop, namely the bone marrow for B cells and the thymus for T cells (Figure 12-1). The second is through mechanisms that operate in the peripheral lymphoid organs to prevent or limit activation of self-reactive lymphocytes that have not been eliminated by central tolerance. These mechanisms thus contribute to **peripheral tolerance**. In some cases this involves abortive activation, which occurs when a mature B cell or T cell encounters antigen without additional signals that are needed for its activation: T cell help in the case of B cells or costimulation in the case of T cells. As innate immune recognition induces costimulators on antigen-presenting cells, these peripheral tolerance mechanisms are engaged when antigen is recognized in the absence of innate immune cell recognition of infection or tissue damage.

These mechanisms act to limit self-reactivity of B cells and T cells but do not fully contain it. Indeed, immune tolerance to self is rarely absolute, but if it is limited to a level below that which causes pathological consequences, then the system is viewed as achieving tolerance.

A key element of the immune system that limits auto-reactive immune responses and also limits the pathology of immune responses to severe or chronic infections is the dominant suppression of immune responses by a specialized type of CD4 T cell, the regulatory T cell (T_{REG}). These cells are generated in two ways: they arise from a proportion of self-reactive immature T cells in the thymus as an alternative to negative selection (see section 7-10), and less frequently they also arise in the periphery during chronic T cell immune responses (see section 5-13). Although they are generated chiefly in the thymus, self-reactive T_{REG} cells act in the periphery and are thus considered to be one of the components of peripheral tolerance.

The action of T_{REG} cells is also controlled by strong innate immune signals, so that their suppressive action is suspended during a response to an infection, at least during the early stages. Thus, suppression by T_{REG} cells shares with other peripheral tolerance mechanisms the property that in the absence of signals from the innate immune system, antigen recognition by naïve lymphocytes leads to their elimination or silencing. In this way, the adaptive immune system is heavily dependent upon recognition by the innate immune system for initiating a T cell immune response, as discussed in Chapter 5.

Failures in self tolerance lead to autoimmune disease

Autoimmune diseases represent a failure of immune tolerance and collectively affect approximately 3% of the population in developed countries. With the rare exception of individuals with genetic diseases in which a major mechanism of immune tolerance is compromised, individuals with autoimmune disease have lost immune tolerance to a limited number of self antigens that are usually associated through their pattern of expression or cellular location

thymus

double-positive thymocyte — CD4, TCR, CD8

positive selection

CD4 thymocyte

thymic epithelial cell or dendritic cell

thymic epithelial cell or dendritic cell

negative selection

lymph node

CD4 TCR

CD4 TCR

FoxP3⁺

mature CD4 T cell

dendritic cell

regulatory T cell fate

self-reactive abortive activation

Figure 12-1 Effect of self-reactivity on fates of thymocytes and T cells Shown are different possible fates for a developing T cell with a TCR that reacts to self-peptide–MHC class II complexes well enough to be positively selected into the CD4 single-positive thymocyte population (upper arrow). If the reactivity for self-peptide–MHC class II complexes is high and the peptide is present in the thymus, TCR recognition leads either to death of the cell in the thymus (negative selection) or to induction of the FoxP3 transcriptional regulator and adoption of the T_{REG} differentiation state. This differentiation state is compatible with export to the periphery and long-term survival there. It is thought that stronger self-reactivity probably favors negative selection over T_{REG} development, but this is not clearly established. Some self-reactive T cells leave the thymus and enter the periphery as mature CD4 T cells, as do cells positively selected as a result of weak reactivity to self-peptide–MHC class II complexes. It is thought that, while recirculating between different secondary lymphoid organs, the self-reactive T cells encounter a high enough concentration of self-peptide–MHC complexes in one or another location to induce an abortive activation, leading to death of the cell. This occurs if the dendritic cells present the self antigen with low levels of costimulators (B7-1, B7-2 and so on) after maturation in the absence of an infection or tissue damage and cell necrosis. Naïve CD4 T cells with lower reactivity to self-peptide–MHC II class complexes can survive in the periphery for long periods until a foreign peptide–MHC class II complex is presented along with costimulators to induce an immune response.

(for example, the self antigens targeted in autoimmune diabetes are all expressed in the insulin-producing pancreatic beta cells). Thus, even in individuals with autoimmune disease, immune tolerance to the vast majority of self components remains effective. Therefore, it is clear that the multiple mechanisms of immune tolerance constitute a robust system for protecting the individual from autoimmune attack. What is less clear is why this robust system breaks down in restricted cases in a substantial number of people. Indeed, for almost every autoimmune disease, the mechanisms by which autoimmunity arises are mysterious. Genetic susceptibility is a major contributor to the development of autoimmunity: genetically identical twins typically develop the same autoimmune disease in between 10% and 50% of cases, which is a much higher percentage than seen in other siblings. Except in rare cases, autoimmunity also seems to require a precipitating event, which is generally thought to be an infection of some sort but is not defined in the vast majority of autoimmune diseases.

Some autoimmune diseases are treated by blocking molecules of the immune system

Even without an understanding of the mechanisms by which autoimmune diseases arise, there has recently been impressive progress in developing new and effective treatments for some of these conditions (Figure 12-2). Especially effective has been blockade of the inflammation-inducing cytokine TNF, which is now being used to treat rheumatoid arthritis, Crohn's disease (a prominent form of inflammatory bowel disease), and two other forms of arthritis, ankylosing spondylitis and psoriatic arthritis. In contrast, blocking TNF exacerbates disease in multiple sclerosis, a disease which is thought to result from autoimmune attack on the myelin sheaths surrounding nerve cells in the central nervous system. For this disease, blocking T cell entry into sites of inflammation with antibodies that block the $\alpha 4$ integrins was found to be effective, although its use has been restricted because of severe viral infection of the brain in a few treated individuals. This illustrates one feature of therapies that block immune responses at these common levels: they provide benefit for the autoimmune disease at the cost of increasing the risk of severe infection. For this reason, the ultimate aim of research in this field is to develop means of reinstating immune tolerance to specific antigens. The goal is to transiently enhance one or more mechanisms of peripheral tolerance acting only on antigen-responding lymphocytes and in this way silence the self-reactive lymphocytes causing the autoimmunity, without reemergence of the disease when the therapy is discontinued (as occurs in current therapies) and without continued suppression of other lymphocytes. Many exciting efforts are underway in this area. One example of an immune tolerance-based therapeutic agent may be an agent that blocks the B7 costimulators (*CTLA4–Ig*). This agent has recently been approved for use in rheumatoid arthritis, but it remains to be seen whether it induces specific tolerance and whether its use can be discontinued after a short period.

Figure 12-2 Table of therapeutic agents for treating human autoimmune diseases Shown are therapeutic agents approved for treatment of autoimmune diseases in the USA as of the end of 2005. Note, however, that at the time of writing the use of the $\alpha 4$ integrin blocking monoclonal antibody has been associated with an increased risk of severe virus infection. The TNF-blocking agents are highly successful, as described further in section 12-9. The IL-1-blocking agent has a short half-life in serum, which is a disadvantage for a therapy that must be given chronically. However, recent evidence indicates that this agent is more effective than blocking TNF for a subset of rheumatoid arthritis patients (those that contract the disease as juveniles), so improvements in this approach may be useful. Interferon-β has been used for a number of years to treat multiple sclerosis and it is effective in slowing progression of the disease in a subset of patients, but the mechanism is not known. Finally, in 2005, CTLA4–Ig was approved for use in rheumatoid arthritis. It acts by binding to and blocking B7-1 and B7-2, thereby preventing CD28 costimulatory action for T cell activation.

Therapeutic Agents for Human Autoimmune Diseases

Therapeutic agents	Target	Function	Disease
TNF-specific Ab	TNF	block inflammation	rheumatoid arthritis, Crohn's disease
TNFR–Ig	TNF	block inflammation	rheumatoid arthritis
IL-1Ra	IL-1 receptor	block inflammation	rheumatoid arthritis
$\alpha 4$ integrin-specific antibody	$\alpha 4$ integrins	block effector T cell entry to tissue	multiple sclerosis
interferon-β	?	anti-viral agent, immune regulation	multiple sclerosis
CTLA4–Ig	B7-1, B7-2	block T cell costimulation	rheumatoid arthritis

Definitions

central tolerance: immune non-responsiveness due to mechanisms operating during lymphocyte development to eliminate self-reactive cells.

immune tolerance: non-responsiveness of the adaptive immune system to an antigen. Tolerance is usually to self components but can also occur to transplanted tissue, infectious agents, therapeutic proteins, and so on.

peripheral tolerance: immune non-responsiveness

due to mechanisms operating on mature self-reactive lymphocytes not eliminated during development.

References

Bluestone, J.A. *et al.*: **CTLA4Ig: bridging the basic immunology with clinical application.** *Immunity* 2006, **24**:233–238.

Feldmann, M. and Steinman, L.: **Design of effective immunotherapy for human autoimmunity.** *Nature* 2005, **435**:612–619.

Rioux, J.D. and Abbas, A.K.: **Paths to understanding the genetic basis of autoimmune disease.** *Nature* 2005, **435**:584–589.

Highly self-reactive B cells and T cells are eliminated during their development

As we have seen in Chapter 7, many self-reactive T cells and B cells are eliminated during development and are thereby prevented from maturing and entering the peripheral pool of lymphocytes. However, the mechanisms whereby central tolerance is established in this way differ somewhat between T cells and B cells.

In B cells, maturation of immature cells in the bone marrow is arrested if antigen receptor signaling indicates reactivity with self antigens in the environment of the cell. This keeps self-reactive B cells in the bone marrow while RAG-1 and RAG-2 proteins act on their Ig light-chain loci. The functionally rearranged light-chain gene is inactivated, as described in section 7-5, and a new light chain is generated, in a process known as receptor editing. If the new BCR is no longer self-reactive, then BCR signaling will stop and the cell can leave the bone marrow for the periphery. If receptor editing fails to release the immature B cell from self-reactivity, then its BCR will continue to signal and the cell will die in 1–2 days, a process called clonal deletion (Figure 12-3). The main advantage of receptor editing is thought to be efficiency: it is more efficient to rescue self-reactive immature B cells than to generate new ones. Indeed, in mice at least 25% of mature B cells contain Ig light-chain gene rearrangements that indicate receptor editing occurred during their development.

Tolerance can also be enforced for B cells that leave the bone marrow and encounter self antigens upon their arrival in the spleen. Newly emerging B cells leaving the bone marrow are referred to as transitional B cells. They go to the spleen, where further maturation and selective events are required for the cell to become either marginal zone or follicular B cells. If the transitional B cell encounters antigen, it will die rapidly: receptor editing is not an option for these cells, as only the bone marrow environment will support cell survival during this process. The negative response of transitional B cells does not abrogate the response to foreign antigens entering the blood or spleen because of the presence of B cells that matured before the start of the infection and are able to respond positively and make antibodies.

It is worth noting that B cell central tolerance appears to require that the self component be accessible either in the bone marrow or in the spleen, which filters the blood. Therefore, it is thought that self antigens that are highly tissue restricted do not induce effective central tolerance in the B cell compartment. As we shall see in the next section, mechanisms of peripheral tolerance for B cells are probably important for preventing production of autoantibodies directed against these self proteins.

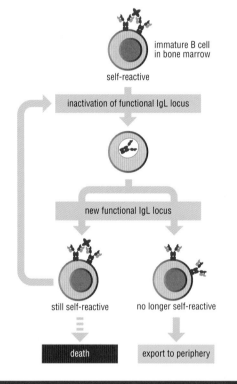

immature B cell in bone marrow

self-reactive

inactivation of functional IgL locus

new functional IgL locus

still self-reactive

no longer self-reactive

death

export to periphery

Figure 12-3 Receptor editing and clonal deletion of developing B cells After successful rearrangement of the Ig heavy chain and Ig light chain (IgL) loci, an immature B cell in the bone marrow expresses membrane IgM together with Igα and Igβ as the B cell antigen receptor (BCR). If this BCR is engaged by self antigens present in its local environment, maturation of the B cell is arrested and it stays in the bone marrow. RAG-1 and RAG-2, the critical components of the V(D)J recombination machinery, are reexpressed and act on the Ig light-chain loci. Their action can destroy the functional IgL gene (see section 7-5) and generate a new functionally rearranged IgL gene, resulting in a BCR with the original heavy chain but a new light chain, a process called receptor editing. If this BCR is still autoreactive, then it can try again for a limited time before dying (clonal deletion).

References

Goodnow, C.C. et al.: **Cellular and genetic mechanisms of self tolerance and autoimmunity.** Nature 2005, **435**:590–597.

Kyewski, B. and Klein, L.: **A central role for central tolerance.** Annu. Rev. Immunol. 2006, **24**:571–606.

Pelanda, R. and Torres, R.M.: **Receptor editing for better or for worse.** Curr. Opin. Immunol. 2006, **18**:184–190.

Villaseñor, J. et al.: **AIRE and APECED: molecular insights into an autoimmune disease.** Immunol. Rev. 2005, **204**:156–164.

The third type of mature B cell, the B1 cell, appears to be controlled primarily by peripheral tolerance mechanisms rather than by clonal deletion. Indeed, some degree of self-reactivity is a requirement for maturation by precursors of this type of B cell.

The thymus has specialized mechanisms for expressing self antigens

Central tolerance of self-reactive T cells is similar to that of B cells in some ways and distinct in others. A major result of strong reactivity to a peptide–MHC complex presented by antigen-presenting cells in the thymus is death of the developing T cell, a process called negative selection that leads to clonal deletion (see section 7-10) (Figure 12-4). Receptor editing is thought to occur infrequently for self-reactive T cells in the thymus. The reason for this difference between B cells and T cells is not known. Peptides bound to MHC on the cells that are thought to induce negative selection—that is, on thymic dendritic cells and thymic medullary epithelial cells—are primarily self peptides, because there is no special mechanism for bringing foreign peptides to the thymus as there is to the lymph nodes; tissue dendritic cells picking up antigens during infections migrate to lymph nodes but not to the thymus. For this reason, negative selection of thymocytes is heavily biased toward self-antigens, although foreign and self antigens in the blood at sufficient concentrations are presented in the thymus. Thus, in the T cell compartment as in the B cell compartment, a key feature of central tolerance to self antigens is that self antigens are continually present and induce clonal deletion of the higher affinity self-reactive lymphocytes on a continual basis. Foreign antigens may induce clonal deletion of newly developing lymphocytes at the time of an infection, but the pool of mature lymphocytes that developed prior to the infection includes many cells that can recognize antigens of the infecting agent and therefore these lymphocytes can mount an effective immune response. In addition to this timing mechanism, it is likely that many foreign antigens fail to reach the bone marrow or the thymus and therefore do not induce clonal deletion even during an infection.

One might imagine that central T cell tolerance would be limited to those self proteins expressed in the thymus and would thus exclude many proteins that are expressed only in specific specialized peripheral tissues. It turns out, however, that the thymus expresses many organ-specific proteins that are not expressed in most tissues. The mechanism of this expression is not well understood, but in the case of a number of self antigens of endocrine tissues this expression requires a nuclear protein called Aire, which is thought to be a transcriptional regulator. Rare individuals with genetic defects in this protein have been identified as suffering from an autoimmune disease in which multiple endocrine organs are attacked, autoimmune polyendocrine syndrome type I (see section 7-10), demonstrating the critical nature of central tolerance of T cells for maintaining immunological tolerance. So far, we have discussed mechanisms that remove self-reactive B cells and T cells from the repertoire of mature lymphocytes. A second critical mechanism for ensuring tolerance is the generation in the thymus of self-reactive regulatory T cells (T$_{REG}$) that can act to block immune responses in the periphery. These cells differentiate from CD4 T cells, and as we shall see in section 12-3 they have a critical role in the maintenance of immune tolerance.

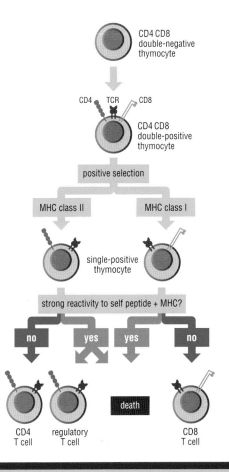

Figure 12-4 Central T cell tolerance Thymocyte development goes through stages in which cells are first positively selected for responsiveness to self-peptide–MHC complexes and differentiate as CD4 T cells specific for peptide–MHC class II or CD8 T cells specific for peptide–MHC class I. If this reactivity is too strong, however, it can lead to death by apoptosis (called negative selection or clonal deletion) or to adoption of the regulatory T cell differentiation state. These cells differentiate from CD4 T cells and function in suppressing T cell responses in the periphery, as described in section 12-3.

Central tolerance is insufficient to generate fully effective immune tolerance

Although some individuals who develop autoimmune disease (for example, people with Aire deficiencies) have defects in central tolerance, it is thought that this is the exception and that most autoimmune disease arises from failure to maintain tolerance in those self-reactive lymphocytes that are not removed by central tolerance mechanisms.

Peripheral tolerance involves multiple mechanisms, each of which contributes to maintaining immune tolerance to self. These fall into two categories: attenuation of immune responses by T_{REG} cells, as described in the next section, and mechanisms that are intrinsic to the self-reactive B cell or T cell. In these latter mechanisms, the general principle is that antigen stimulation of a mature B cell or T cell is not sufficient to generate an immune response. The antigen-stimulated lymphocyte must receive additional signals that ultimately indicate that innate immune recognition mechanisms have been engaged. In the absence of these additional signals, the antigen-stimulated lymphocyte becomes inactivated. For a self-reactive lymphocyte, self antigen is always present in some location, so when a recirculating lymphocyte arrives at a location where the self antigen is prevalent, it will initiate activation. But if there is no infection at that point in time in that location, the activation will be abortive and the cell will either die or enter an anergic state from which activation is more difficult. This will be the fate of most self-reactive lymphocytes, but by chance some lymphocytes may arrive at the antigen-expressing secondary lymphoid organ for the first time while an infection or inflammatory reaction is occurring. These cells may become activated. After the infection is cleared, they will need to be silenced by another mechanism, which may involve either the action of T_{REG} cells or the depletion of survival cytokines.

B cells are anergized by antigen contact in the absence of T cell help

As described in Chapter 6, mature follicular B cells that contact antigen in secondary lymphoid organs migrate to the boundary of the follicle and the T cell zone, where some helper T cells also migrate after initial activation by antigen-presenting dendritic cells. These T cells rapidly migrate within this region, sampling B cells for the presence of the peptide–MHC class II complexes that they recognize. B cells that can present antigen to helper T cells and receive T cell help become activated to make antibodies; those that do not receive T cell help enter a refractory state referred to as **anergy**, from which activation becomes increasingly difficult. Anergic B cells die much more rapidly than mature follicular B cells, so this state is an intermediate on the way to cell death. Anergy typically occurs when the B cell contacts antigen in a form that stimulates antigen receptor signaling to a modest degree. Mature B1 cells, which are often self-reactive, appear to be in an anergic state, but unlike anergic follicular B cells, B1 cells are long-lived. Initial recognition of antigen in a form that induces strong antigen receptor signaling, such as an antigen expressed on a cell surface, either induces antibody production by a T cell-independent type II route (see

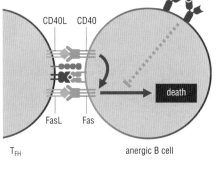

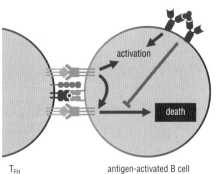

Figure 12-5 Prevention by Fas of helper T cell-induced antibody responses by anergic B cells CD40L stimulation of B cells induces expression of Fas on their surface, making them susceptible to apoptosis induced by FasL, which, like CD40L, is induced on the surface of helper T cells by antigen recognition. Thus, an anergic B cell presenting antigen to a helper T cell will be induced to undergo apoptosis (top panel). In contrast, a B cell that is not anergic, but is binding antigen via its antigen receptor, will have active signaling by the antigen receptor, which prevents Fas-induced apoptosis, allowing cell activation induced by antigen receptor signaling, CD40 signaling and cytokine receptor signaling (lower panel).

Definitions

ALPS: see **autoimmune lymphoproliferative syndrome**.

anergy: (of lymphocytes) an unresponsive state resulting from stimulation by antigen in the absence of additional signals, such as those provided by helper T cells (for B cells) or by costimulation (for T cells) that are generally necessary for activation.

autoimmune lymphoproliferative syndrome (ALPS): a genetic disease characterized by splenomegaly, lymphadenopathy, and autoantibody production, and resulting from missense mutations in genes encoding Fas, FasL or caspase 10. These mutations result in defective Fas signaling. The increased cellularity of the secondary lymphoid organs in these individuals is due primarily to the accumulation of an unusual type of T cell lacking the expression of CD4 or CD8, and an accumulation of activated B cells.

section 6-9) by B1 cells or marginal zone B cells, or induces death more rapidly for a follicular B cell in the absence of T cell help, giving rise to clonal deletion rather than anergy.

Until they die, anergic B cells are located in the region of the secondary lymphoid organ where activated helper T cells are present if they are responding to an infection. Thus, they may receive some helper T cell-derived signals (CD40L and cytokines) without presenting antigen to these T cells. This bystander activation is prevented by the expression on antigen-activated helper T cells of not only CD40L but also FasL, another cell-bound member of the TNF family. Whereas CD40 on the B cell is essential to its activation by a T cell-dependent antigen, Fas on the B cell can induce its death, as Fas is a death-domain-containing TNF receptor family member (see section 2-9). This mechanism is thought to contribute to the silencing of anergic B cells, as CD40 induces Fas expression on the B cell. Anergic B cells are more sensitive to Fas killing than are non-anergic B cells, because the former have strongly attenuated antigen receptor signaling and antigen receptor signaling can prevent Fas-induced death of a B cell interacting with a helper T cell (Figure 12-5). The significance of this mechanism of peripheral tolerance is suggested by the elevated autoantibody production, including antibodies to nuclear self components, in people who have a genetic disease caused by defects in Fas, called **autoimmune lymphoproliferative syndrome (ALPS)**.

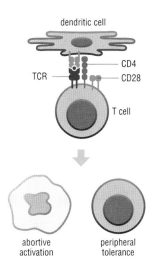

Dendritic cells constitutively present self antigen without costimulation or cytokines and induce abortive activation of self-reactive T cells

In the absence of specific mechanisms to promote peripheral tolerance, antigen presentation during infections would be expected to induce activation of self-reactive T cells that had escaped central tolerance. This is because, during an infection, dendritic cells in the infected tissue upregulate expression of costimulators (B7-1, B7-2 and other costimulators) and migrate to the nearest secondary lymphoid organ bearing on their surface peptide–MHC complexes containing peptides from both the infectious agent and self molecules from cellular debris at the infected site. Thus, naïve recirculating T cells may recognize self peptides presented by these dendritic cells in association with costimulators and thereby be activated to proliferate and differentiate into effector cells. This is prevented by prior tolerizing interactions of newly generated self-reactive T cells with dendritic cells that have matured in the absence of infection. These dendritic cells migrate from tissues to the draining lymph node and present self-peptide–MHC complexes but with low levels of costimulation and without secretion of cytokines that support effector T cell development. In this way, dendritic cells induce an abortive activation of recently generated self-reactive T cells, resulting in a short-lived proliferation followed by a rapid decline in T cell numbers (Figure 12-6). A few self-reactive T cells do survive, but they are thought to be refractory to subsequent activation.

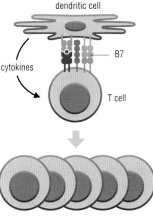

A final checkpoint for peripheral T cell tolerance is provided by the inhibitory receptor PD-1 (programmed death 1), which is induced on the T cell surface upon effector cell differentiation (see section 5-8), and one ligand for which (PD-L1) is expressed on many stromal cells of the tissues. When PD-L1 engages PD-1 on effector T cells, it inhibits their activation. The exact nature of this tolerance checkpoint is still poorly understood, but currently it is thought that the more strongly the T cell is stimulated in the lymph node, the less well PD-1 can subsequently block its activation in peripheral tissue sites. How PD-L1 levels might be regulated or this inhibition otherwise modulated is not yet understood, but experiments with mice in which PD-1 is mutated have demonstrated that PD-1 plays an important part in preventing organ-specific autoimmunity.

Figure 12-6 Abortive activation of self-reactive T cells in the periphery Some self-reactive cells escape the induction of central tolerance and enter the periphery as naïve CD4 or CD8 T cells. Often, these cells encounter the self-peptide–MHC complexes they recognize in elevated amounts in one or more secondary lymphoid organs presented by dendritic cells that have matured in the absence of infection and migrated to the draining lymph node. These dendritic cells present antigen with low levels of costimulators (B7-1, B7-2, and so on), and hence cannot induce strong T cell activation. The result is an abortive clonal expansion that is rapidly terminated and associated with high levels of apoptosis. The surviving antigen-specific T cells are poorly responsive to subsequent antigenic stimulation. In contrast, naïve T cells stimulated by antigen presented by dendritic cells that matured in the presence of infection are exposed to higher levels of costimulators during antigen presentation, and this stronger stimulation promotes high-level secretion of cytokines such as IL-2 that promote T cell expansion. In addition, dendritic cells that receive direct stimulation from their TLRs during their maturation produce cytokines such as IL-12 that promote effector T cell differentiation.

References

Bidère, N. et al.: **Genetic disorders of programmed cell death in the immune system.** Annu. Rev. Immunol. 2006, **24**:321–352.

Bluestone, J.A. et al.: **CTLA4Ig: bridging the basic immunology with clinical application.** Immunity 2006, **24**:233–238.

Noelle, R.J. and Erickson, L.D.: **Determinations of B cell fate in immunity and autoimmunity.** Curr. Dir. Autoimmun. 2005, **8**:1–24.

Regulatory T cells are a distinct type of effector T cells that suppress T cell immune responses

The mechanisms of immune tolerance described above are intrinsic to the T cell or B cell. They result either from the developmental timing of recognition of antigen (central tolerance) or from the effect of recognizing antigen in the periphery in the absence of second signals that are needed to mount an immune response (peripheral tolerance). In addition to these mechanisms for eliminating or inactivating self-reactive lymphocytes, there is also a suppressive mechanism of tolerance imposed by a separate type of CD4 T cell, the regulatory T cell (T_{REG}). These cells are potent inhibitors of T cell immune responses, but their activity is prevented during an immune response when dendritic cells release inflammatory cytokines while presenting antigen to naïve T cells. Innate immune recognition by dendritic cells, for example via TLRs, programs them to release cytokines, especially IL-6, which interfere with T_{REG}-mediated suppression. So in this respect T_{REG} function is similar to other peripheral tolerance mechanisms of T cells, which also are largely restricted to situations in which antigen is recognized without evidence of infection.

Most T_{REG} cells are self-reactive CD4 T cells induced to differentiate as regulatory cells upon recognition of peptide–MHC class II complexes in the thymus, as described in section 7-10. Thus, entry into the T_{REG} effector type and negative selection are alternative fates of self-reactive developing T cells. The choice between these fates is driven by the forkhead family transcription factor FoxP3, and appears to depend on the nature of peptide–MHC recognition, but the details are not well understood. Alternatively, naïve CD4 T cells in the periphery can be induced by the cytokine TGF-β to express FoxP3 and become T_{REG} cells with the same properties as T_{REG} cells that develop in the thymus (see section 5-13). These induced T_{REG} cells, in contrast to central T_{REG} cells, are often not self-reactive, but rather are generated in response to foreign peptides that are present during a chronic infection. Their role, therefore, is thought to be the prevention of damaging pathology resulting from chronic and/or excessively active immune responses, as for example occurs in hepatitis B virus infection and in inflammatory bowel disease, which is described in section 13-6. Once generated, T_{REG} cells express CD25 (IL-2R α chain) and are long-lived in the periphery. They constitute approximately 10% of recirculating CD4 T cells.

Properties of Regulatory T Cells

	Naïve CD4 T cells	Regulatory T cells
Localization	recirculating	recirculating
Reactivity to self-peptide–MHC class II	low	moderate or strong
Requirement for costimulation	yes	yes
Function	cell-mediated immunity or promotion of antibody production	suppression of T cell responses
Expansion/survival	TCR + autocrine or paracrine IL-2 (or other cytokines)	TCR + paracrine IL-2
Response to cytokines of dendritic cells	effector differentiation IL-6 prevents suppression	?
Cytokines made	IL-2	IL-10, TGF-β (?)
Change in localization after expansion	to sites of inflammation	to sites of inflammation

Figure 12-7 Table of properties of regulatory T cells Comparison with the properties of naïve CD4 T cells that can develop into other types of effector T cells.

References

Belkaid, Y. and Rouse, B.T.: **Natural regulatory T cells in infectious disease.** *Nat. Immunol.* 2005, **6**:353–360.

Li, M.O. *et al.*: **Transforming growth factor-β regulation of immune responses.** *Annu. Rev. Immunol.* 2006, **24**:99–146.

Sakaguchi, S.: **Naturally arising Foxp3-expressing CD25⁺CD4⁺ regulatory T cells in immunological tolerance to self and non-self.** *Nat. Immunol.* 2005, **6**:345–352.

von Boehmer, H.: **Mechanisms of suppression by suppressor T cells.** *Nat. Immunol.* 2005, **6**:338–344.

Weaver, C.T. *et al.*: **Th17: an effector CD4 T cell lineage with regulatory T cell ties.** *Immunity* 2006, **24**:677–688.

In addition to T$_{REG}$ cells, other negative regulatory populations of T cells have been described and referred to as T$_R$1 and T$_H$3 cells; these appear to be distinct from T$_{REG}$ cells because they lack expression of FoxP3. These inhibitory cell types are not well understood and are not discussed further in this chapter.

Regulatory T cells are essential for avoiding pathological immune responses

Defects in FoxP3, the transcription factor responsible for programming the functional properties of T$_{REG}$ cells, result in both human and mouse in severe disease characterized by extensive inflammation in many organs. In humans, FoxP3 is encoded on the X chromosome, and deficiency in this disease is called immunodysregulation, polyendocrinopathy, enteropathy, X-linked syndrome (IPEX). In the mouse, FoxP3 deficiency causes an essentially similar severe autoimmune and inflammatory disease, and this disease can be treated simply by transferring a functional FoxP3 gene into some of the CD4 cells, demonstrating that it is the absence of T$_{REG}$ cells that is responsible for this severe loss of immune tolerance.

T$_{REG}$ cells differ from other CD4 T cells in the cytokines they produce and in how their activation and function are controlled (Figure 12-7). One key difference is that T$_{REG}$ cells do not produce IL-2 after TCR stimulation, and thus cannot promote their own expansion. In many other ways T$_{REG}$ cells do behave similarly to other CD4 T cells. For example, they respond to peptide–MHC class II complexes on antigen-presenting cells via their TCR, and this signaling is regulated positively by costimulators and CD28. Also, in the absence of infection, they express L-selectin and recirculate between the blood and secondary lymphoid organs. Although they are continually stimulated by self-peptide–MHC class II complexes presented by dendritic cells that have matured homeostatically and are present in the T cell zones of the secondary lymphoid organs, this recognition occurs in the presence of low levels of costimulators and in the absence of substantial IL-2. This stimulation is thought to support the survival of these cells (unlike the response of naïve CD4 T cells to such stimulation, as described in the previous section), and to maintain them in a recirculating mode. Like naïve CD4 T cells responding to antigen, T$_{REG}$ cells proliferate vigorously *in vitro* if provided with IL-2. This may reflect what occurs in a lymph node that is supporting the initiation of a T cell immune response. T$_{REG}$ cells that expand in such circumstances can subsequently alter their adhesion and homing molecules and go to sites of inflammation, in the same way that other types of effector T cells do.

Thus, like other CD4 T cells, T$_{REG}$ cells exist in recirculating or inflammation-seeking modes. Moreover, it appears that they can suppress responses of CD4 or CD8 T cells in either location and at multiple steps in the process of an immune response. Similarly, they appear to use several different mechanisms to inhibit T cells and/or antigen-presenting cells, including secretion of the inhibitory cytokines IL-10 and TGF-β and also by a cell-contact-dependent mechanism that is not defined but is independent of either of these two cytokines (Figure 12-8). T$_{REG}$ cells require TCR and costimulatory signaling to induce their suppressive action, but this action must be controlled to allow immune responses to proceed in the face of infection. It is thought that responding T cells become resistant to the actions of T$_{REG}$ cells upon receiving signals made by dendritic cells that have been directly stimulated via their TLRs, including secreted IL-6 and a cell-bound TNF family member called GITR ligand, which is induced by TLR stimulation of the dendritic cell. Thus, the suppressive function of T$_{REG}$ cells is blocked by innate immune recognition of infection, allowing antigen-recognizing CD4 and CD8 T cells to make immune responses to counter an active infection.

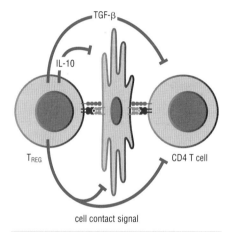

Figure 12-8 Proposed mechanisms of suppression by regulatory T cells Regulatory T cells inhibit T cell immune responses at multiple stages, including initial expansion in the lymph node, and effector actions at sites of inflammation. It is therefore likely that they have multiple modes of suppression. Depending on the circumstance, IL-10 and TGF-β have been implicated in the actions of regulatory T cells, but in some circumstances their suppressive actions have been shown to depend on cell–cell contact and not on IL-10 or TGF-β. Some of their effects are likely to be on dendritic cells or other antigen-presenting cells (for example, through IL-10), whereas other effects, including the effects of TGF-β, are likely to be directly on T cells.

12-4 Autoimmune Diseases: General Principles

Many diseases result from immune attack on uninfected tissues

Although immunological tolerance mechanisms are largely successful, they are not perfect and the failure of tolerance to a particular self component can lead to damage due to an immune attack directed against that component, which is then called an **autoantigen**. Such diseases are called **autoimmune diseases**. Altogether, over 40 autoimmune diseases are recognized that afflict approximately 3% of the population in developed countries (Figure 12-9).

Autoimmune diseases are often subdivided into those that are organ-specific and those that are systemic. For example, autoimmune diabetes affects only the insulin-secreting cells of the pancreas, while in systemic lupus erythematosus, antibodies form complexes with ubiquitous self antigens and these immune complexes activate effector mechanisms to cause widespread damage to tissues and organs. Although most of the effector mechanisms of adaptive immunity are activated in most autoimmune diseases, typically the main cause of the pathology in a particular disease can be antibody-mediated, cell-mediated, or both, as is described in the next two sections.

Autoimmunity is typically the result of genetic susceptibility and an environmental trigger

Autoimmune diseases are thought to result from the combination of a genetic susceptibility and an environmental trigger, but specific understanding is often not far advanced from that general statement. Studies with identical twins typically show that if one twin develops a particular autoimmune disease, the other twin is much more likely to develop the same autoimmune disease than is a member of the general population. This likelihood is between 10% and 50% for most of the major autoimmune diseases, indicating a strong genetic predisposition that is not sufficient in itself to cause disease.

Several rare single-gene defects have been identified that give rise to autoimmunity in a high proportion of individuals carrying the defective gene. These include mutations in the gene encoding Fas or its signaling components, and mutations in the transcriptional regulators Aire and FoxP3. Fas is important for deletion of self-reactive B cells (see section 12-2) and has also been implicated in the deletion of T cells after repetitive antigenic stimulation, Aire promotes the expression in the thymus of some proteins otherwise expressed only in particular endocrine organs, and thereby contributes to central tolerance of T cells (see section 12-1), and FoxP3 is required for the development of regulatory T cells (see section 12-3). Thus, single-gene defects leading to autoimmune disease are highly informative about the mechanisms of immunological tolerance.

In contrast, the common genetic contributions to autoimmune disease are multigenic and cause disease in a low proportion of individuals carrying them. The best-characterized of these genetic contributors are genes of the MHC. Typically, certain MHC alleles predispose to certain autoimmune diseases whereas others are protective. Since MHC molecules control T cell reactivity to antigen by binding peptides and presenting them to T cells, these associations strongly imply important roles for CD4 and/or CD8 T cells in the development of these autoimmune diseases; beyond that, however, the role of disease susceptibility or resistance alleles is not understood. In genetic studies of human and murine autoimmune diseases, over 100 genetic loci have been implicated in disease susceptibility, suggesting that many genes affect immune tolerance in more subtle ways than do total loss of function mutations of *FAS*, *AIRE* and *FOXP3*. Other than those encoding MHC molecules, few genes have been definitively

Frequency of Selected Autoimmune Diseases in the USA

Disease	Incidence (per 100,000 per yr)	Prevalence (per 100,000)
Systemic autoimmune diseases		
rheumatoid arthritis	23.7	860
systemic lupus erythematosus	7.3	23.8
Sjögren's syndrome	?	14.4
polymyositis/ dermatomyositis	1.8	5.1
scleroderma	0.8	4.4
Organ-specific autoimmune diseases		
thyroiditis	21.8	1,324
Graves' disease	13.9	1,152
vitiligo	18.9	400
autoimmune diabetes	12.2	192
pernicious anemia	?	151
multiple sclerosis	3.2	58
Addison's disease	?	5
myasthenia gravis	0.4	5
primary biliary cirrhosis	0.9	3.5
uveitis	?	1.7

Figure 12-9 Table showing the frequency of selected autoimmune diseases in the USA Note that most but not all autoimmune diseases are chronic, but with relatively low mortality, such that the prevalence represents the accumulation of newly arising cases over many years. Data from Jacobson, D.L. *et al.*: *Clin. Immunol. Immunopathol.* 1997, **84**: 223–243.

Definitions

autoantigen: a self component that is the target of the immune response in an **autoimmune disease**.

autoimmune disease: a disease in which the primary mechanism of pathology is an immune attack directed against components of the self (**autoantigens**).

epitope spreading: a phenomenon in which an autoimmune response is directed initially against one or a few self antigens, but later becomes broader and is directed at more **autoantigens**.

molecular mimicry: structural similarity between a self component and a component of an infectious agent, such that the immune response to the infectious agent cross-reacts with the self component, initiating the autoimmune attack.

References

Jacobson, D.L. *et al.*: **Epidemiology and estimated population burden of selected autoimmune diseases in the United States.** *Clin. Immunol. Immunopathol.* 1997, **84**:223–243.

Kohm, A.P. *et al.*: **Mimicking the way to autoimmunity: an evolving theory of sequence and structural homology.** *Trends Microbiol.* 2003, **11**:101–105.

Raman, K. and Mohan, C.: **Genetic underpinnings of autoimmunity — lessons from studies in arthritis, diabetes, lupus and multiple sclerosis.** *Curr. Opin. Immunol.* 2003, **15**:651–659.

Vanderlugt, C.L. and Miller, S.D.: **Epitope spreading in immune-mediated diseases: implications for immunotherapy.** *Nat. Rev. Immunol.* 2002, **2**:85–95.

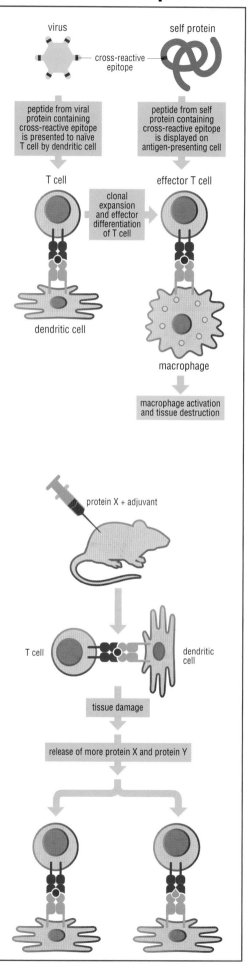

Figure 12-10 Proposed mechanism of autoimmunity due to molecular mimicry If a viral protein (top left) and a self protein (top right) contain peptide epitopes (red) recognized by the same T cell, then T cells activated against the viral epitope may also kill uninfected cells displaying the same epitope from the self protein. This would require that T cell tolerance to the self epitope be incomplete, perhaps because the self peptide is presented only at a low level in normal conditions. Shown is the example of cross-reactivity of a CD4 T cell; if a CD8 T cell is cross-reactive, then it could directly kill normal cells expressing the cross-reactive self protein.

identified, although intriguing evidence has recently been obtained that polymorphisms in the inhibitory costimulatory receptors CTLA4 and PD-1 may contribute to autoimmune diabetes, Graves' disease, autoimmune thyroiditis (all CTLA4) and systemic lupus erythematosus (PD-1).

Molecular mimicry and epitope spreading may be important for autoimmunity

Immune pathology of autoimmune diseases can generally be classified according to a scheme similar to that for hypersensitivity diseases, which are described in Chapter 13, with damage being due to cytotoxic IgG antibodies (type II), immune complex deposition (type III) or T cell-induced chronic inflammation (type IV). Many of the antigens recognized by autoantibodies have been identified, since antibodies can be readily used to isolate and characterize the antigens they recognize. In contrast, most autoantigens recognized by T cells are not defined and this is currently a major limitation in trying to understand how autoimmune diseases are triggered in genetically susceptible individuals. Moreover, often it is not established whether CD4 or CD8 T cells are mainly responsible for the autoimmune pathology.

The environmental trigger for most autoimmune diseases is likely to be an infection. This could cause the activation of an autoantigen-directed immune response in either of two ways: **molecular mimicry** or a tissue-damage-induced breakdown of tolerance. The molecular mimicry hypothesis postulates that an infectious agent has an antigen that is similar to a self component so that the immune response to the agent includes the activation of lymphocytes that react against both an antigen of the microbe or virus and against the similar autoantigen (Figure 12-10). In other words, there is a cross-reaction between the antigen of the infectious agent and a self component, and the cross-reacting antibody or T cells then initiate the autoimmune disease. Molecular mimicry is thought to account for the induction of rheumatic fever by group A *Streptococcus pyogenes* infection, in which antibodies are produced that react with the bacterial M protein and heart muscle.

The alternative hypothesis is that an infection leads to sufficient damage to a particular organ to break the tolerance of some self-reactive lymphocytes while the infection is being fought, and the resulting activated lymphocytes propagate the response even after the infection is over. In support of this mechanism is a phenomenon common to autoimmune diseases known as **epitope spreading** (Figure 12-11). An epitope is the part of a particular antigen that is recognized by a single antibody or T cell receptor. In animals, autoimmune disease can be induced in susceptible strains by immunization with a particular autoantigen so that the immune response to it and to other antigens of the target tissue can be followed. Typically, the immune response is initially restricted to the epitopes present in the immunogen. Subsequently, however, tissue damage releases tissue-specific self antigens that are not normally exposed in significant quantities, and which now can promote the activation of lymphocytes reactive to other epitopes of the protein or tissue. By analogy, if viral infection of a particular organ causes substantial damage to the tissue, it could cause epitope spreading from the viral epitopes to the self epitopes and thereby convert an antiviral immune response into an autoimmune response. Thus, epitope spreading is likely to be essential to the propagation of chronic autoimmune destruction.

Figure 12-11 Epitope spreading In animals, models of human autoimmune disease can often be generated by immunizing with a corresponding autoantigen (shown here as protein X) in a strong adjuvant. At first, the response is dominated by T cells responding to one peptide derived from protein X, but later, after tissue damage has released both protein X and other self proteins from damaged cells during an ongoing immune response, T cells responsive to other peptides of protein X, and to peptides from other proteins, become activated. Presumably, an analogous phenomenon occurs when an infection leads to tissue damage. In most individuals, the reaction to self components is presumably limited, ceasing as the original infecting agent is cleared and the tissue damage is repaired, but in some individuals the anti-self reaction may be excessive and continue, resulting in chronic disease.

Antibodies against self components can cause disease

The pathology of an autoimmune disease may be mediated by autoantibodies, by cell-mediated immune reactions, or by a combination of the two. Those diseases caused by autoantibodies are the best understood in terms of the nature of the self antigen and how the immune response causes the disease manifestations. In this section some of the organ-specific humoral autoimmune diseases are described, and in the next two sections the cell-mediated diseases are considered, followed by a discussion of the systemic autoimmune diseases.

Autoantibodies may cause disease by three general mechanisms: (1) binding to cell surface or basement membrane components and activation of effector mechanisms through complement and Fc receptors of phagocytes leading to cell and tissue damage (corresponding to type II hypersensitivity, as described in Chapter 13), (2) immune complex formation and deposition, again activating phagocytes and causing damage (corresponding to type III hypersensitivity), and (3) direct effects of autoantibodies on the autoantigen affecting its important functions (Figure 12-12). Of these, the first and third lead to organ-specific autoimmunity, whereas the second leads to systemic autoimmune diseases, especially systemic lupus erythematosus.

Autoantibodies can recognize cells or extracellular matrix and induce pathology

Organ-specific autoimmunity proceeding by the type II hypersensitivity-type mechanisms include **autoimmune hemolytic anemia** (antibodies against red blood cells cause their destruction), **autoimmune thrombocytic purpura** (antibodies against platelets cause their destruction), and **Goodpasture's syndrome** (antibodies against collagen type IV in the basement membranes of kidneys and lungs lead to damage to these organs). The lung component of the latter disease is seen primarily in smokers, reflecting one aspect of environmental contributions to autoimmune disease. Autoimmune hemolytic anemia can be induced by infection with the bacterium *Mycoplasma pulmonis*. In this case, the immune response to infection produces an IgM antibody that recognizes both the I antigen of red blood cells and a related self oligosaccharide on cells that is used for adhesion by the bacteria in the lung. A structurally related polysaccharide on the bacterium has not been identified. The binding of bacterial molecules to the I antigen-related oligosaccharide may permit epitope spreading of the antibody response from

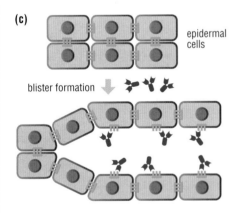

(a) red blood cell — macrophage — FcγR

(b) immune complex — neutrophil — proteases ROS

(c) epidermal cells — blister formation

Figure 12-12 **Three different mechanisms of autoantibody-mediated pathology** Pathological mechanisms analogous to type II and type III hypersensitivity diseases are illustrated in panels **(a)** and **(b)**, respectively. In panel (a), autoantibodies against red blood cells give rise to autoimmune hemolytic anemia. The antibodies bind to red blood cells and lead to their destruction primarily by macrophages through FcγR-mediated phagocytosis. Complement lysis of red blood cells can also occur. In panel (b), autoantibodies in systemic lupus erythematosus recognize soluble antigens, such as chromatin released from dying cells, and form immune complexes that can be deposited in capillaries and joints, leading to damage induced by degranulation by neutrophils. FcγR engagement induces the release of secretory granules containing proteases and also the release of reactive oxygen species (ROS). These components damage surrounding tissue. The third mechanism by which autoantibodies cause pathology is illustrated in panel **(c)**. In the example shown, autoantibodies against desmosomes (green) are found in the autoimmune disease pemphigus vulgaris, and these autoantibodies inhibit the adhesive ability of the desmosomes to hold together layers in the epidermis.

Definitions

autoimmune hemolytic anemia: an autoimmune disease in which antibodies reacting with red blood cells are produced, leading to premature clearance of the red blood cells by macrophages or complement-mediated lysis.

autoimmune thrombocytic purpura: an autoimmune disease in which autoantibodies against platelets promote their rapid clearance by macrophages, leading to a bleeding disorder.

Goodpasture's syndrome: an autoimmune disease often occurring with rapid onset in which antibodies are made against collagen type IV. These antibodies bind to basement membranes in the kidneys and lungs and induce damage through neutrophil secretion of proteases and reactive oxygen species.

Graves' disease: an autoimmune disease in which autoantibodies react with the receptor for thyroid-stimulating hormone (TSH) and activate it. The result is hyperthyroidism, as the thyroid cells continually secrete thyroid hormones in an unregulated fashion. Graves'

disease is typically treated by ablating the thyroid gland and giving thyroid hormone replacement therapy.

myasthenia gravis: an autoimmune disease in which autoantibodies react with the nicotinic acetylcholine receptor, which is the receptor in the neuromuscular synapse that receives the neurotransmitter from motor neurons to induce muscle contraction. The autoantibodies induce the internalization of receptors and can also block the binding of acetylcholine, and hence block receptor function and prevent muscle contraction. The name is Latin for severe muscle weakness.

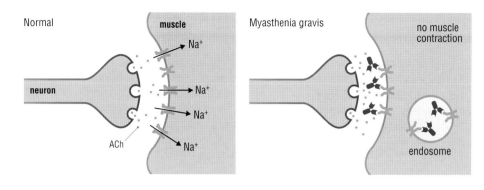

Figure 12-13 Mechanism of muscle weakness in myasthenia gravis In the normal neuro-muscular junction (left), acetylcholine receptor ion channels are clustered at the neuro-muscular synapse and respond to acetylcholine (ACh) by allowing the entry of sodium ions. This depolarizes the cell membrane, leading to activation of voltage-dependent calcium channels, influx of calcium, and muscle contraction. Auto-antibodies against the acetylcholine receptor induce its internalization and degradation and can also directly inhibit acetylcholine binding to the receptor (right). Antibodies against the acetylcholine receptor are observed in about 85% of myasthenia gravis patients; the remainder usually have autoantibodies against other neuromuscular synapse proteins.

the bacterial component to the associated self component. This form of hemolytic anemia generally resolves when the infection is cleared, but other, less well understood triggers lead to long-lasting IgG responses inducing anemia.

Autoantibodies can cause disease by affecting the function of the target autoantigen

Antibodies against cellular receptors mediate the disease processes evident in **myasthenia gravis** and **Graves' disease**. In myasthenia gravis (a Latin name meaning severe muscle weakness), autoantibodies are made against the nicotinic acetylcholine receptor. This is the ion channel that functions as a receptor in muscle, receiving input from motor neurons at the neuromuscular synapse and inducing muscle contraction. The autoantibodies can inhibit the binding of acetylcholine to the receptor and in addition induce the internalization and degradation of receptors, decreasing their number on the surface of the muscle cell (Figure 12-13). Thus, in myasthenia gravis, the autoantibodies inhibit receptor function but do not cause damage to the muscle. In Graves' disease, the autoantibodies are directed against the receptor for thyroid-stimulating hormone (TSH). Instead of blocking the TSH receptor function, these antibodies mimic ligand, causing continual stimulation of the thyroid cells, which then pump out thyroid hormones in excessive and unregulated amounts, causing hyperthyroidism (Figure 12-14).

Pemphigus vulgaris, which means ordinary blisters, is an antibody-mediated disease causing blisters in the skin and mucosal surfaces. In this disease, an autoantibody is made that recognizes desmoglein-3, a component of the desmosomes that mediates intercellular junctions in the epidermis and mucosal epithelium. For reasons that are obscure, this autoantibody is predominantly IgG4, which does not fix complement well, and hence it appears to act primarily by interfering with proper desmosome function, leading to weak adhesions of cells and the resulting blisters when those adhesions fail. Pregnant women with Graves' disease, pemphigus vulgaris or myasthenia gravis transmit disease symptoms to their babies in a transient fashion, providing strong evidence for the role of IgG autoantibodies in mediating these diseases.

Another example of an autoimmune disease with function-blocking autoantibodies is **pernicious anemia**, in which anemia results from vitamin B_{12} deficiency. This is an outcome of autoimmune gastritis, in which an autoimmune attack, possibly cell-mediated in common with the autoimmune diseases discussed in the next section, is made upon the stomach. In 10–20% of patients with gastritis, autoantibodies are made against a secreted protein called intrinsic factor, which is necessary for the uptake of vitamin B_{12} from the diet.

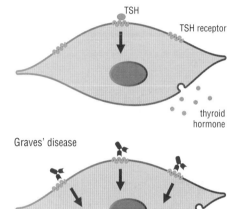

Figure 12-14 Mechanism of hyperthyroidism in Graves' disease In the normal thyroid gland, thyroid-stimulating hormone (TSH) coming from the pituitary gland acts on thyroid cells through the TSH receptor to stimulate the release of thyroid hormones, which regulate basal metabolism and body temperature. Thyroid hormones in turn negatively regulate TSH production, providing a tightly regulated system. In Graves' disease, autoantibodies against the TSH receptor bind to it and act like TSH to stimulate the cells to secrete thyroid hormones. This secretion is unregulated since the antibodies are always present, and stimulate the thyroid cells fully, so that thyroid hormones are overproduced.

pemphigus vulgaris: an autoimmune disease in which autoantibodies react with a component of the epidermal intercellular adhesion complexes, called desmosomes. This component, desmoglein-3, is a member of the cadherin family of cell adhesion molecules. These antibodies interfere with the adhesion of cell layers within the epidermis, so individuals with this disease easily form blisters of the skin. There are also manifestations in the oral tract.

pernicious anemia: an autoimmune disease in which autoantibodies are made that recognize intrinsic factor, a protein that is secreted by gastric parietal cells and is involved in the uptake of vitamin B_{12} from the diet.

References

Hertl, M. and Veldman, C.: **Pemphigus – paradigm of autoantibody-mediated autoimmunity.** *Skin Pharmacol. Appl. Skin Physiol.* 2001, **14**:408–418.

Hudson, B.G. *et al.*: **Alport's syndrome, Goodpasture's syndrome, and type IV collagen.** *N. Engl. J. Med.* 2003, **348**:2543–2556.

Lang, B. and Vincent, A.: **Autoantibodies to ion channels at the neuromuscular junction.** *Autoimmun. Rev.* 2003, **2**:94–100.

Prabhakar, B.S. *et al.*: **Current perspective on the pathogenesis of Graves' disease and ophthalmopathy.** *Endocr. Rev.* 2003, **24**:802–835.

Cell-mediated autoimmune diseases attack a variety of organs

The understanding of cell-mediated autoimmunity is in general less advanced than that of the autoantibody-mediated diseases mentioned above. Typically, the antigens that are the target of this type of autoimmunity are less well defined and more diverse, because of epitope spreading. Progress in understanding these diseases is coming from three different directions: genetic studies to define the genes that give susceptibility to a particular autoimmune disease, studies of animal models to define the mechanism of autoimmunity, and effects of novel therapeutic agents that target defined molecules or cells of the immune system. We shall first briefly survey the diseases in this category and then consider insights from these three lines of investigation.

In general, there may be autoantibodies in these diseases and they may contribute to the pathology that is seen, but they are thought to be at most a portion of the disease and the cell-mediated component is likely to be critical for understanding and treating the disease. For example, autoimmune gastritis is a disease that is thought to be a cell-mediated attack on the parietal cells of the stomach, which, as mentioned in the previous section, proceeds in a minority of cases to production of autoantibodies to intrinsic factor, leading to vitamin B_{12} deficiency and anemia.

Type 1 diabetes mellitus, also called insulin-dependent diabetes, is a disease in which a cell-mediated attack on the islets of Langerhans in the endocrine pancreas results in the death of the insulin-producing beta cells. The other hormone-secreting cells of the islets are spared, illustrating the specificity of the autoimmune process. The pancreas has a considerable regenerative capacity, but diabetes symptoms are unfortunately not manifest until nearly all of the beta cells have been killed. Autoantigens that have been implicated in the disease include glutamic acid decarboxylase (GAD) and insulin itself, but how the disease is initiated is largely mysterious. Viral infections of the pancreas may trigger the autoimmune response, but no single virus has been clearly implicated.

Autoimmune thyroiditis, also called Hashimoto's thyroiditis, involves a cell-mediated attack on the thyroid gland, characterized by inflammation and the formation of ectopic lymphoid tissue, as is also seen in rheumatoid arthritis (see section 12-9), with destruction of the gland leading to thyroid deficiency. This disease is accompanied by the production of antibodies against a number of thyroid-specific proteins, but these may be reflections of a continuing T cell-driven immune response, rather than being important for gland destruction. This disease has the opposite effect on the thyroid gland from that seen in Graves' disease, which results in hyperthyroidism.

Finally, **multiple sclerosis** is caused by immune attack on the myelin sheath that surrounds nerves in the central nervous system and supports the transmission of neural signals, so that damage leads to neurological deficits. Decline in neurological function can be gradual or it can be punctuated, with short periods of exacerbation spaced between periods of stable function (referred to as relapsing-remitting). Multiple sclerosis lesions are often focal, suggestive of some local event, such as virus replication. Interestingly, the risk of developing multiple sclerosis is strongly associated with latitude, with people living in temperate climates being much more likely to develop this disease. This could reflect a viral trigger for initiating the autoimmune response. Although there have been many efforts to implicate particular viruses in multiple sclerosis, at present none have been clearly established as the cause of the disease. One treatment that is effective at halting disease progression in some multiple sclerosis patients is injection of interferon-β, although the mechanism is not known.

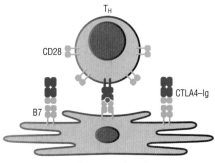

costimulation blocked by CTLA4–Ig

T_H

CD28

CTLA4–Ig

B7

antigen-presenting cell

Figure 12-15 Inhibition of autoimmune disease by blockade of B7 costimulation CTLA4–Ig is a fusion protein containing the extracellular domain of CTLA4, which binds to B7-1 and B7-2 with high affinity, fused to the Fc portion of IgG. The resulting fusion protein binds to B7-1 and B7-2 on mature dendritic cells, activated B cells, and macrophages to prevent the receipt of costimulatory signals by T cells, attenuating their activation. Note that by blocking B7-1 and B7-2, CTLA4–Ig prevents costimulatory signaling by CD28, and also inhibitory costimulatory signaling by CTLA4. The net effect is to decrease T cell activation, because of the importance of CD28 for the initial activation of T cells.

Definitions

autoimmune thyroiditis: an autoimmune disease in which the thyroid gland is destroyed. Also called Hashimoto's thyroiditis.

CTLA4–Ig: a chimeric or fusion protein containing at its N-terminus the extracellular domain of CTLA4, which binds with high affinity to the costimulatory molecules B7-1 and B7-2, and at its C-terminus the Fc portion of IgG, which gives the molecule a long half-life in blood.

multiple sclerosis: a presumed autoimmune disease in which there is an immune attack on the myelin sheaths around neuronal axons in the central nervous system, leading to neurological disease.

psoriasis: a cell-mediated inflammatory disease of the skin.

type 1 diabetes mellitus: an autoimmune disease in which there is inflammation of the endocrine pancreas leading to the destruction of beta cells of the islets of Langerhans and insulin insufficiency.

T cell costimulation is likely to be important for progression of cell-mediated autoimmune diseases

The success of anti-TNF therapies in treating rheumatoid arthritis (see section 12-9) and Crohn's disease (see section 13-6) encouraged analogous trials in the organ-specific autoimmune diseases, but in these diseases blocking TNF has not been useful (type 1 diabetes) or has even been harmful (multiple sclerosis). Although the exact meaning of these results is not known, at the least this indicates that the pathologic mechanisms in type 1 diabetes and rheumatoid arthritis have important differences.

More encouraging have been recent efforts with **CTLA4–Ig**, a chimeric protein containing the extracellular domain of the inhibitory costimulatory receptor CTLA4 (see section 5-7) fused to the Fc portion of IgG. The Ig portion of the molecule greatly prolongs its half-life in blood through interaction with FcRn (see section 6-2), which protects IgG from clearance or catabolism in the blood. By binding with high avidity to B7-1 and B7-2 on mature dendritic cells, and activated B cells and macrophages, CTLA4–Ig prevents these cells from activating T cells through the costimulatory receptor CD28 (see sections 4-1, 5-5 and 5-7), thus suppressing T cell-mediated immune responses (Figure 12-15). CTLA4–Ig also blocks the binding of B7-1 and B7-2 to CTLA4 itself on the T cell surface, an interaction that is normally important in limiting the proliferation of activated T cells (see section 5-7); but as CTLA4–Ig prevents T cell activation, proliferation is suppressed without the need for this braking mechanism. CTLA4–Ig has been found to be efficacious in clinical trials for several cell-mediated autoimmune diseases, including **psoriasis** (an inflammatory autoimmune disease of the skin) and rheumatoid arthritis, and also for blocking rejection of kidney transplants. Other novel therapeutics based on blocking T cell activation or homing to sites of inflammation are in various stages of testing in these diseases.

Interestingly, analyses of genes leading to susceptibility to type 1 diabetes, autoimmune thyroiditis and Graves' disease all have recently identified a particular polymorphism in the CTLA4 gene as predisposing to these diseases. This variant of CTLA4 leads to the production of less of a secreted splice variant of CTLA4 that may function naturally in an analogous fashion to CTLA4–Ig, described above. Many other genetic loci have been implicated in autoimmune diseases, but the exact genetic variations are not yet identified. An example of some genetic loci identified in multiple genetic screens is illustrated in Figure 12-16.

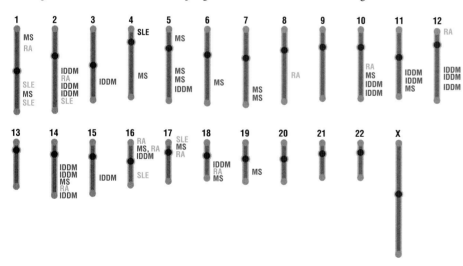

Figure 12-16 Genetic loci implicated as risk factors for the development of systemic lupus erythematosus (SLE), rheumatoid arthritis (RA), multiple sclerosis (MS) or insulin-dependent diabetes mellitus (IDDM) in humans Locations of some of the better-established loci contributing to the genetic susceptibility of these four common autoimmune diseases are shown schematically on the p or the q arms of the chromosomes (p arms are shown above the centromere (dark grey circle) and q arms below). Relative positions of genetic loci are not drawn to scale. CTLA4 is located at 2q33. Adapted from Raman, K. and Mohan, C.: *Curr. Opin. Immunol.* 2003, **15**:651–659.

References

Abrams, J.R. *et al.*: **CTLA4Ig-mediated blockade of T-cell costimulation in patients with psoriasis vulgaris.** *J. Clin. Invest.* 1999, **103**:1243–1252.

Raman, K. and Mohan, C.: **Genetic underpinnings of autoimmunity — lessons from studies in arthritis, diabetes, lupus and multiple sclerosis.** *Curr. Opin. Immunol.* 2003, **15**:651–659.

Ueda, H. *et al.*: **Association of the T-cell regulatory gene *CTLA4* with susceptibility to autoimmune disease.** *Nature* 2003, **423**:506–511.

12-7 Organ-Specific Autoimmunity: Animal Models

Animal models of cell-mediated autoimmune disease have been informative about the mechanisms of autoimmunity

A wide variety of animal models of autoimmune diseases have been developed and studied. Much of our current understanding of autoimmune disease mechanisms, particularly in cell-mediated autoimmunity, comes from such studies, and such models have also been used to test potential therapeutic agents, although they are not perfect predictors of the response in patients.

NOD mice spontaneously develop type 1 diabetes

One popular model of autoimmunity is the **non-obese diabetic (NOD) mouse** strain, which spontaneously develops diabetes with a high probability. Genetic studies in this strain have identified at least 20 genetic loci that contribute to this spontaneous disease, including the MHC class II allele $I\text{-}A^{g7}$. In humans, *HLA-DQ8* is a susceptibility gene, and there are structural similarities between the two molecules that may explain why these alleles predispose to autoimmunity; in both cases the molecules are relatively unstable, and their peptide-binding grooves have similar unusual preferences. The NOD MHC molecule appears to be atypical in that it allows the export of $I\text{-}A^{g7}$/self-peptide-reactive T cells to the periphery. Additional NOD genes also appear to contribute to poor thymic negative selection, suggesting that defects in central T cell tolerance contribute to spontaneous autoimmune diabetes development in NOD mice. A decrease in the expression of one CTLA4 isoform may also be a genetic contributor. Interestingly, a variant allele of CTLA4 has also been implicated as a contributor to genetic susceptibility to autoimmune diabetes in humans.

Experimental allergic encephalitis has many similarities to multiple sclerosis

Experimentally induced models of autoimmunity have also been highly informative. Immunization with a number of different tissue-restricted proteins in strong adjuvants can induce disease in susceptible strains of mice or other animals. Thus, these models reproduce the interaction between genetic susceptibility and environmental trigger, which must be strongly inflammatory. Particularly noteworthy is **experimental allergic encephalitis (EAE)**, which closely mimics the demyelination of the central nervous system (CNS) seen in multiple sclerosis. EAE can be induced by immunizing susceptible strains of mice with various myelin-derived proteins, or with peptides of those proteins that represent T cell epitopes. In the latter circumstance, the initial T cell response is restricted to the immunizing peptide, but epitope spreading occurs so that, as time goes on, T cells specific for other epitopes in the same protein

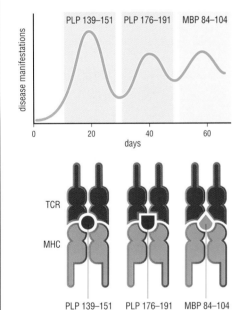

Figure 12-17 Epitope spreading in experimental allergic encephalitis Injection of a peptide corresponding to amino-acid residues 139–151 of the myelin-localized proteolipid protein (PLP) induces a remitting-relapsing form of encephalitis in SJL inbred mice. Injection of the PLP 139–151 peptide on day 0 leads to the appearance of neurological symptoms by day 20. At this point the T cell response is dominated by reactivity to the immunizing peptide. The disease then improves, until it is followed by a spontaneous relapse at about day 40. At this time, many T cells are responding to another epitope within PLP, corresponding to residues 176–191. The disease again wanes and a second relapse is observed around day 60, now containing many T cells responding to an epitope from another myelin protein, myelin basic protein (MBP) 84–104. Data are adapted from the work of S.D. Miller and colleagues, summarized in Fuller, K.G. *et al.*: *Methods Mol. Med.* 2004, **102**:339–362.

Definitions

EAE: see **experimental allergic encephalitis**.

experimental allergic encephalitis (EAE): an autoimmune disease induced in animals by immunization with a myelin protein in adjuvant. The pathologic mechanisms in EAE are quite similar to those in the human disease multiple sclerosis.

NOD mouse: see **non-obese diabetic mouse**.

non-obese diabetic (NOD) mouse: mouse of an inbred strain that spontaneously develops autoimmune diabetes at high frequency. NOD mice have at least 20 genetic loci that contribute to the spontaneous development of diabetes because of T cell-mediated pancreatic islet beta-cell destruction. MHC class II (I-A) is one of these, and a polymorphic variant of CTLA4 is also likely to be a contributor.

and for epitopes in other proteins of the myelin sheath can be detected (Figure 12-17). Recent studies have demonstrated that T_H17 effector cells are responsible for much of the pathology in mouse models of EAE.

EAE in mice can also be induced by infection with a neurotropic virus, Theiler's murine encephalomyelitis virus, which is a natural mouse pathogen. The damage caused by the virus in these mice could in principle be caused either by epitope spreading after infection-induced damage to neuronal tissue, or by molecular mimicry, but the T cell clones activated are the same ones that are activated by epitope spreading after immunization with myelin proteins or peptides, making epitope spreading the more likely mechanism. The first T cells activated in the autoimmune response are those with specificities that are present at the highest frequencies in the naïve T cell population. These cells recognize a peptide epitope, known as PLP 139–151, that is found in a proteolipid protein expressed in the myelin sheath. Because of alternative splicing, this peptide epitope is not present in the same protein in the thymus. Thus, the prevalence of T cells of this specificity, and perhaps their ability to promote autoimmune disease, may arise from a failure of negative selection during thymic development, leading to a later deficiency in T cell tolerance.

Molecular mimicry is responsible for a virus-induced autoimmune eye disease

Transgenic expression of a viral protein from LCMV in the beta cells of the pancreas, hepatocytes of the liver, or oligodendrocytes of the CNS has been used for another set of autoimmune models. In these mice, infection with LCMV (see Chapter 10) triggers an autoimmune attack on the tissue expressing the viral proteins as a transgene. In some situations tolerance is strong and the precursor frequency of responding T cells must first be elevated. These studies demonstrate the potential for molecular mimicry and suggest that particular aspects of the cross-reacting self antigen (for example, poor induction of tolerance) or genetic compromise of tolerance mechanisms may be needed in addition to infection.

In an animal model for autoimmune eye disease, infection of mouse eyes with murine herpesvirus 1 induces eye disease by a mechanism involving molecular mimicry. The initial T cell response to this virus is directed to a peptide from the virus UL6 protein, and these T cells cross-react with a corneal protein. The use of a virus with a point mutation in which this particular epitope of the UL6 protein is altered greatly decreases eye disease upon infection with a low dose of virus, but eye disease still occurs when a higher dose of virus is used (Figure 12-18). Thus, there are two mechanisms by which this virus can induce eye damage, one at low levels of virus infection via molecular mimicry, and a non-mimicry mechanism after infection with higher virus levels, which may be autoimmune or simply a pathologically strong reaction to the infection.

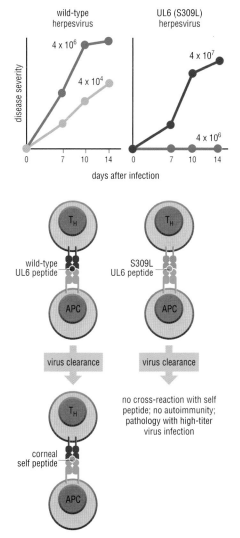

Figure 12-18 Molecular mimicry in a virally induced autoimmune eye disease Eye infection of mice with a wild-type murine herpesvirus 1 (KOS strain) induces autoimmune eye disease in 7–14 days at doses as low as 4×10^4 infectious virus particles, despite the fact that the virus infection has been cleared by day 5, indicating that an autoimmune process is occurring. T cells reactive to an epitope in the viral UL6 protein are observed. Analogous infection with a virus containing a point mutation in the UL6 epitope (S309L) does not lead to eye damage at low virus dose. Only after infection with 1,000-fold more virus is eye pathology seen. Data from Panoutsakopoulou, V. *et al.*: *Immunity* 2001, **15**:137–147.

References

Ercolini, A.M. and Miller, S.D.: **Mechanisms of immunopathology in murine models of central nervous system demyelinating disease.** *J. Immunol.* 2006, **176**:3293–3298.

Fuller, K.G. *et al.*: **Mouse models of multiple sclerosis: experimental autoimmune encephalomyelitis and Theiler's virus-induced demyelinating disease.** *Methods Mol. Med.* 2004, **102**:339–362.

Panoutsakopoulou, V. *et al.*: **Analysis of the relationship between viral infection and autoimmune disease.** *Immunity* 2001, **15**:137–147.

Systemic autoimmune diseases include SLE and rheumatoid arthritis

Systemic autoimmunity comprises a heterogeneous group of diseases in which autoimmunity is not restricted to a single organ or organ system. Examples include **systemic lupus erythematosus** (SLE or lupus), various forms of arthritis including rheumatoid arthritis and reactive arthritis, Sjögren's syndrome, in which there is inflammation of multiple exocrine glands, and perhaps scleroderma (systemic sclerosis).

As with the organ-specific autoimmune diseases, these diseases probably involve a multigenic predisposition and an environmental trigger. For example, an individual who has an identical twin with lupus has a much higher frequency of developing lupus than a member of the general population (28–57% versus about 0.1%). Reactive arthritis and ankylosing spondylitis (an arthritis mainly of the spinal column joints) both show a strikingly higher prevalence in individuals expressing at least one allele of the MHC class I molecule HLA-B27. The mechanism of this susceptibility is not known, but in reactive arthritis it is clear that a bout of bacterial food poisoning can be the trigger that induces this systemic autoimmune arthritis.

Rheumatoid arthritis is considered separately in the next section, and we shall focus the discussion here mainly on SLE, which is the best-understood systemic autoimmune disease.

SLE is characterized by the production of high levels of antibodies against nuclear components

SLE patients exhibit a variety of symptoms, which can include the characteristic butterfly skin rash on the face over the cheeks and nose (Figure 12-19), vasculitis, kidney disease, arthritis and central nervous system abnormalities. Most or all symptoms are thought to be due to autoantibodies, especially those against nuclear components, including double-stranded (ds) DNA, histones and various nuclear RNA–protein complexes. Kidney disease, which is one of the most common severe consequences of lupus, has been strongly associated with the deposition of anti-dsDNA–DNA immune complexes in the glomeruli of the kidneys. The hydrodynamics of blood flow may lead to the trapping of immune complexes in the glomeruli. Similarly, immune complexes may become trapped in joints for physical reasons, or they may be concentrated there by the release of nuclear components from apoptotic cells in that location. Local release of nuclear components from apoptotic cells may explain the skin rashes, which are dependent on sun exposure, and perhaps ultraviolet-induced apoptosis of skin cells.

A striking feature of the autoantigens in lupus is the presence of nucleic acids and proteins that associate with nucleic acids. Recently it has been suggested that the DNA in chromatin fragments released from apoptotic cells may induce specific antibody production by the simultaneous engagement of two important receptors on B cells: TLR9, which recognizes CpG-containing motifs in DNA (section 3-10), as well as the antigen receptor on DNA-specific B cells (Figure 12-20). Most people do not develop high titers of anti-DNA antibodies, however, so SLE must require additional elements, such as genetic compromise of B cell tolerance mechanisms.

Another striking feature of the autoantigens recognized by lupus antibodies is that they can be found displayed on the surface of apoptotic cells or the blebs coming from them. Indeed, defects in the clearance of apoptotic cells (see section 3-8) promote the development of a lupus-like disease in mice (Figure 12-21). This may explain the well documented tendency of individuals with genetic defects in early components of the classical pathway of complement (see section 3-3) to develop SLE. For example, nearly all humans and mice defective in C1q

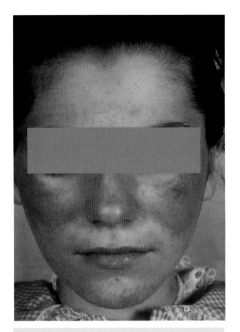

Figure 12-19 Characteristic butterfly skin rash of a female SLE patient Kindly provided by David Wofsy. ©1972-2004 American College of Rheumatology Clinical Slide Collection. Used with permission.

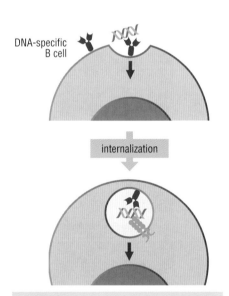

Figure 12-20 Possible role of TLR9 in the production of anti-dsDNA antibodies
A B cell with specificity for DNA binds to DNA-containing chromatin fragments released by apoptotic cells. Antigen-induced clustering of the B cell antigen receptor triggers antigen receptor signaling, which can promote B cell activation in conjunction with other signals coming from helper T cells or TLRs. Antigen binding to the antigen receptor also promotes highly efficient uptake of the antigen-containing chromatin fragment into endosomes, where it is recognized by TLR9. Although TLR9 is largely specific for unmethylated CpG-containing DNA, mammalian DNA appears capable of stimulating TLR9 in this context.

Definitions

rituximab: an anti-CD20 monoclonal antibody used to treat B cell malignancies and experimentally to treat certain autoimmune diseases, including lupus.

systemic lupus erythematosus: a systemic autoimmune disease characterized by anti-nuclear antibodies, often including antibodies reactive against double-stranded DNA. Lupus is Latin for wolf, although the exact reason for connecting SLE to the wolf is a matter of debate.

References

Cocca, B.A. *et al.*: **Blebs and apoptotic bodies are B cell autoantigens.** *J. Immunol.* 2002, **169**:159–166.

Cohen, P.L.: **Systemic autoimmunity** in *Fundamental Immunology*, 5th ed. Paul, W.E. ed. (Lippincott Williams & Wilkins, Philadelphia, 2003), 1371–1399.

Gorman, C. *et al.*: **Does B cell depletion have a role to play in the treatment of systemic lupus erythematosus?** *Lupus* 2004, **13**:312–316.

Figure 12-21 Table of mouse mutations leading to SLE-like disease The complement component deficiencies may result in poorer clearance of apoptotic cells because of the ability of C1q to bind to apoptotic cells and promote their clearance (see section 3-8). The CD45 wedge mutation leads to a hyperactive CD45, preventing negative regulation of Src-family kinases (see section 5-4). The mutations shown in the final category show a large variation in the extent of the disease and are listed in approximate order of severity. For example, the SHP-1 deficiency causes very severe disease (motheaten mice), causing death at an early age, whereas the coreceptor defects exhibit some autoantibody production but mild disease. Not included in the table, Fas deficiency gives rise to autoimmune lymphoproliferative syndrome in humans and a lupus-like disease coupled with lymphoproliferation in mice.

Mouse Loss of Function Mutations Leading to SLE-Like Disease
Defects in clearance of apoptotic cells
serum amyloid P protein
DNase I
Mer truncation
Defects in complement components
C1q
C4
complement receptor 2
Defects in regulation of BCR signaling
SHP-1
Lyn
CD45 wedge mutation
CD22
FcγRIIb
SHIP

develop SLE. C1q binds to apoptotic cells and aids their clearance. It is possible that among the other genetic predispositions to lupus are included partial defects in clearance of apoptotic cells.

SLE patients often have elevated levels of interferon-α in their serum, or exhibit evidence of *in vivo* responses to interferon-α. This appears to reflect ongoing production of cytokines by plasmacytoid dendritic cells, but the stimulus is unknown. Also, these patients often have elevated levels of acute phase reactants in the blood, reflecting a systemic response of the liver to IL-1 and IL-6 in the blood (see section 3-15). Elevated levels of cytokines may contribute to the autoantibody production by their effects on adaptive immune responses.

Mouse models provide insight into the genetic susceptibility of SLE

Mouse models of SLE and the development of SLE in various genetically altered mice have been quite informative about the mechanisms of development of this disease. Because of its similarities to the human disease, the NZM2410 mouse is a particularly interesting model of spontaneously developing SLE. Genetic breeding studies (outlined in Figure 12-22) have identified three genetic loci that combine to give a high probability of developing the disease. Each of these three loci, called *Sle1*, *Sle2* and *Sle3*, seems to contribute a different element to the development of SLE: *Sle1* leads to impaired tolerance to nuclear antigens, *Sle2* causes enhanced B cell activation, and *Sle3* seems to manifest in T cells, reflecting the fact that autoimmune antibody production is likely to be T cell dependent, as it exhibits somatic hypermutation which increases affinity for autoantigens. In addition, gene knockouts that interfere with processes that negatively regulate B cell antigen receptor signaling are strongly associated with the development of SLE in mice (see Figure 12-21). These include the B cell antigen receptor negative coreceptor pathways described in section 6-8 (for example, the coreceptors FcγRIIB and CD22; Lyn, which phosphorylates them; and the phosphatases that mediate inhibition, SHIP and SHP-1).

With current standard of care, SLE patients have an approximately 80% chance of survival 10 years after diagnosis. Amelioration of the disease is unfortunately often accompanied by severe side effects of the drugs used, namely immunosuppressive steroids and the DNA-synthesis inhibitor cyclophosphamide. Novel therapies currently being tested include **rituximab**, an anti-CD20 monoclonal antibody that depletes B cells and is used to treat B cell lymphomas. B cell depletion leads to about a 50% reduction in serum IgG levels, perhaps reflecting continued production of IgG by long-lived plasma cells, which do not express CD20. Also being tested currently in the clinic are antibodies that block the cytokine BAFF, which is required for the survival of mature follicular and marginal zone B cells (see section 7-6), and CTLA–Ig (see section 12-6), which blocks the costimulation of T cells through CD28 and is therefore expected to interfere with the helper T cell activation of B cells.

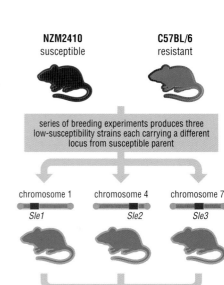

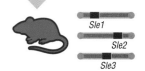

Figure 12-22 Mouse model of multigenic SLE A mouse strain, NZM2410, with a high incidence of SLE-like disease was crossed with disease-resistant C57BL/6 mice to produce animals that each had a different single region inherited from the susceptible parent (*Sle1* on chromosome 1, *Sle2* on chromosome 4 and *Sle3* on chromosome 7), that conferred low susceptibility to SLE-like symptoms. Further breeding experiments combining all three loci with the genetic background of C57BL/6 resulted in mice with a high spontaneous rate of SLE, similar to that of NZM2410, demonstrating directly the importance of these genetic loci and the ability of multiple genetic variations to combine to create a strong genetic propensity for autoimmunity.

Jego, G. *et al.*: **Plasmacytoid dendritic cells induce plasma cell differentiation through type I interferon and interleukin 6.** *Immunity* 2003, **19**:225–234.

Liu, C.C. *et al.*: **Apoptosis, complement and systemic lupus erythematosus: a mechanistic view.** *Curr. Dir. Autoimmun.* 2004, **7**:49–86.

Pascual, V. *et al.*: **The central role of dendritic cells and interferon-α in SLE.** *Curr. Opin. Rheumatol.* 2003, **15**:548–556.

Radic, M.Z. and Weigert, M.: **Genetic and structural evidence for antigen selection of anti-DNA antibodies.** *Annu. Rev. Immunol.* 1994, **12**:487–520.

Viglianti, G.A. *et al.*: **Activation of autoreactive B cells by CpG dsDNA.** *Immunity* 2003, **19**:837–847.

Wakeland, E.K. *et al.*: **Delineating the genetic basis of systemic lupus erythematosus.** *Immunity* 2001, **15**:397–408.

Yu, C.C. *et al.*: **Signaling mutations and autoimmunity.** *Curr. Dir. Autoimmun.* 2003, **6**:61–88.

12-9 Systemic Autoimmunity: Rheumatoid Arthritis

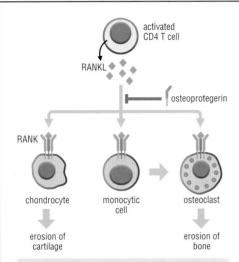

Figure 12-23 Postulated role of RANK ligand in the joint destruction of rheumatoid arthritis Activated autoimmune CD4 T cells produce RANK ligand (RANKL), along with other cytokines. RANK ligand stimulates activity of chondrocytes. Together with TNF, RANK ligand acts on monocytic precursors of osteoclasts to promote differentiation to osteoclasts, and together with IL-1 it stimulates the activity of mature osteoclasts. RANK ligand acts via a TNF receptor family member, RANK, and also binds another TNF receptor family member, osteoprotegerin, which is a soluble receptor-like protein produced by osteoblasts, the cells that make bone. Osteoprotegerin acts to block RANK ligand function by preventing its binding to RANK and is currently being tested as a therapeutic agent.

Rheumatoid arthritis is a common systemic autoimmune disease

The most common systemic autoimmune disease is **rheumatoid arthritis**, which affects between 0.3% and 1.5% of the population in developed countries. This disease afflicts females more commonly than males (3:1) and increases in incidence with age. The hallmark of rheumatoid arthritis is a symmetrical inflammation of multiple joints, as evidenced by pain, swelling, and morning stiffness. Related autoimmune diseases of joints include reactive arthritis, which can follow bouts of food poisoning in some individuals, and ankylosing spondylitis, in which the joints of the spine are especially affected. Rheumatoid arthritis and SLE can share many important features, including vasculitis, arthritis, and elevated levels of acute-phase reactants. Despite these similarities, characteristic autoantibodies are different. The underlying mechanisms of these two diseases are also likely to be quite different. Whereas in SLE the key autoimmune response appears to be the production of anti-nuclear antibodies, in rheumatoid arthritis the seminal disease process is probably a CD4 T cell response to one or more autoantigens found in the lining of the joint, the **synovium**.

In rheumatoid arthritis, there is an inflammatory accumulation of fluid in the joint, containing a large number of neutrophils, along with a characteristic hyperplasia of the synovium containing an infiltration primarily of CD4 T cells, macrophages and B cells. The macrophage inflammatory cytokines TNF, IL-1 and IL-6 are especially abundant, often accompanied by the T_H17 cell cytokine IL-17A. Commonly, the lymphocytes and macrophages organize themselves into lymph-node-like structures with T cell zones and germinal centers containing B cells undergoing a T cell-dependent immune response. Such tertiary lymphoid tissue reflects an ongoing chronic immune response, presumed to be against one or more self antigens contained in the synovium, but the T cell autoantigens of synovium are not well defined. Some of the B cells are producing **rheumatoid factor**, which is antibody, often IgM, directed against the Fc portion of IgG. Whether or not rheumatoid factor contributes to disease manifestations is unclear, and it is frequently produced in other conditions, including viral infections. Its function may be primarily to aid in the clearance of immune complexes.

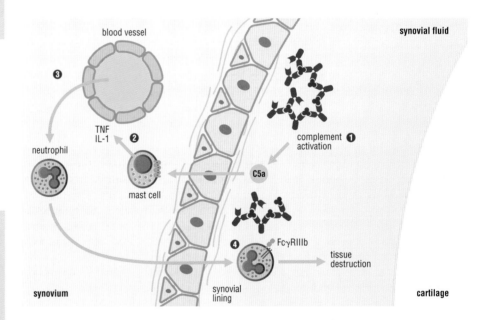

Figure 12-24 Proposed pathogenic mechanism of autoantibodies to glucose-6-phosphate isomerase Immune complexes deposit on the surface of the joint and activate the complement cascade (1). Perhaps because of their amplification role, the alternative pathway components are especially important. C5a is produced, which acts on synovial mast cells to induce inflammation. Mast cells have stored TNF and IL-1 for rapid release (2). These cytokines act on the neighboring endothelial cells to induce inflammation, including an influx of neutrophils (3). Neutrophils degranulate in response to stimulation of their FcγRIII by the immune complexes, releasing destructive proteases such as elastase (4). A small fraction of rheumatoid arthritis patients exhibit antibodies against glucose-6-phosphate isomerase in their blood, but whether this antibody contributes to the pathology of human rheumatoid arthritis is not established.

Definitions

chondrocyte: a cell type that degrades cartilage.

osteoclast: a differentiated cell of the monocytic lineage responsible for the resorption of bone. A distinct type of cell, the osteoblast, is responsible for building bone mass.

RANK ligand: a cytokine of the TNF family secreted by activated CD4 T cells and other cells that acts together with other cytokines to promote osteoclast differentiation and activation. RANK ligand also pro-motes the activity of chondrocytes in eroding cartilage and also acts on monocytes and dendritic cells to promote immune responses. Its receptor, RANK (receptor activator of NF-κB), is a member of the TNF receptor family.

rheumatoid arthritis: an autoimmune disease of joints, typically manifested in multiple peripheral joints in a symmetrical fashion. Rheumatoid arthritis also can include autoimmune disease in other tissues, such as blood vessels. Arthritis means joint inflammation.

The chronic T cell immune response and inflammation of the synovium leads to erosion of bone and cartilage and eventually ligaments, all leading to disability. Studies in animal models of arthritis suggest that joint destruction may be caused by a CD4 T cell cytokine of the TNF-family **RANK ligand**; this promotes the activity of **osteoclasts**, which resorb bone, and **chrondrocytes**, which degrade cartilage (Figure 12-23).

CD4 T cells are strongly implicated as a central player in causing rheumatoid arthritis; indeed, some of the genetic susceptibility of this disease is due to MHC class II molecules (in particular some alleles of HLA-DR4). In contrast, reactive arthritis and ankylosing spondylitis are almost completely restricted to individuals with HLA-B27 or very closely related MHC class I alleles, arguing for an important role for CD8 T cells.

The relative contributions of cell-mediated immunity and autoantibodies to the pathogenesis of rheumatoid arthritis are not well established, and it may well be that patients are heterogeneous. Four models of rheumatoid arthritis in mice are possibly informative in this regard. In one, mice are injected with type II collagen in adjuvant and this induces arthritis. This arthritis is dependent on IL-17A production by T_H17 effector cells and antibody does not seem to be a major factor. Genetic manipulation of mice to overexpress TNF also results in arthritis, along with inflammatory bowel disease, again indicating the importance of the inflammatory process. In contrast, a TCR transgene recognizing a peptide from glucose-6-phosphate isomerase bound to the I-A^{g7} class II MHC molecule results in a severe arthritis that is transferred by antibodies. T cells expressing the transgenic TCR induce the production of disease-causing autoantibodies against the ubiquitously expressed intracellular enzyme glucose-6-phosphate isomerase. This enzyme is released in small amounts from dying cells. Immune complexes of this enzyme and the specific antibody deposit on the surface of the joint cavity and induce an inflammation that is dependent on activation of the alternative pathway of complement, C5a, FcγRIII, mast cells, neutrophils, TNF and IL-1. A proposed model for this disease process is shown in Figure 12-24. This inflammation then leads to joint destruction that is very similar to that seen in rheumatoid arthritis. A small percentage of patients with rheumatoid arthritis have specific antibody against glucose-6-phosphate isomerase in their sera, but whether this is important for pathogenesis is not known. Finally, in the fourth model, rheumatoid arthritis results from a point mutation in the tyrosine kinase ZAP-70, which is a central player in TCR signaling (see section 5-3). In this model, there is thought to be a defect in negative selection of T cells in the thymus due to the decreased function of ZAP-70, illustrating again the important role of negative selection in protecting against autoimmunity.

Therapies for rheumatoid arthritis have focused on generally antiinflammatory and immunosuppressive drugs. In recent years, agents blocking TNF or IL-1 (Figure 12-25) have become approved therapies, and the former is highly efficacious in many but not all patients. In contrast, blocking IL-1 appears to be effective primarily in those patients who develop rheumatoid arthritis as juveniles. Thus, the inflammatory effects of these cytokines are probably important for maintaining the chronic autoimmune response in the synovia. Recently, CTLA4–Ig (see section 12-6) has also been found to be effective in treating rheumatoid arthritis and has been approved for this purpose in the USA. This fusion protein binds with high affinity to B7-1 and B7-2 and in this way blocks CD28-mediated costimulation for activation of T cells. In addition, depleting B cells with an anti-CD20 monoclonal antibody (rituximab) has been shown to improve arthritis in some patients, suggesting a pathogenic role for autoantibodies in these pateints.

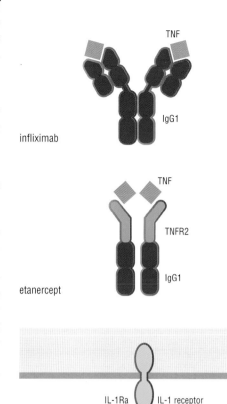

Figure 12-25 Recently introduced therapeutic drugs for rheumatoid arthritis based on blocking inflammatory cytokines Two types of therapeutic drugs blocking TNF have been introduced, a humanized monoclonal IgG1 antibody (infliximab) and a chimeric molecule between a TNF receptor (TNFR2) and the Fc portion of human IgG1 (etanercept). The IgG portion gives the molecule a much longer half-life in serum. Both of these, by binding to TNF, prevent it from reaching its receptor on the surface of cells. Finally, anakinra is a recombinant slightly modified form of a natural inhibitor of the interleukin 1 receptor I, called IL-1 receptor antagonist (IL-1Ra). It competes with IL-1 for binding to IL-1RI, but fails to trigger signaling reactions. For rheumatoid arthritis patients in general, blocking TNF is more likely to be efficacious than blocking IL-1, but recent studies indicate that blocking IL-1 is more efficacious in patients who developed rheumatoid arthritis as juveniles.

---

rheumatoid factor: autoantibody (either IgM or IgG) recognizing the Fc portion of self IgG molecules. Rheumatoid factor is commonly found in the sera of rheumatoid arthritis patients, but can also be found in some normal individuals, for example after a viral infection.

synovium: a specialized layer of tissue lining some joints.

References

Feldmann, M. and Maini, R.N.: **Anti-TNFα therapy of rheumatoid arthritis: what have we learned?** *Annu. Rev. Immunol.* 2001, **19**:163–196.

Firestein, G.S.: **The T cell cometh: interplay between adaptive immunity and cytokine networks in rheumatoid arthritis.** *J. Clin. Invest.* 2004, **114**:471–474.

Kontoyiannis, D. *et al.*: **Impaired on/off regulation of TNF biosynthesis in mice lacking TNF AU-rich elements: implications for joint and gut-associated immunopathologies.** *Immunity* 1999, **10**:387–398.

Monach, P.A. *et al.*: **The role of antibodies in mouse models of rheumatoid arthritis, and relevance to human disease.** *Adv. Immunol.* 2004, **82**:217–248.

Sakaguchi, N. *et al.*: **Altered thymic T-cell selection due to a mutation of the ZAP-70 gene causes autoimmune arthritis in mice.** *Nature* 2003, **426**:454–460.

Theill, L.E. *et al.*: **RANK-L and RANK: T cells, bone loss, and mammalian evolution.** *Annu. Rev. Immunol.* 2002, **20**:795–823.

13

Allergy and Hypersensitivity

Although the central function of the immune system is to defend us against infectious agents, in many individuals, an adaptive immune response is made against one or more non-infectious environmental agents, ranging from pollens to particular foods to reactive chemicals. These responses are collectively called hypersensitivities and can if they are vigorous enough cause tissue damage and can even be life-threatening. Of particular importance are the allergies, which are hypersensitivities resulting from IgE and activation of mast cells, basophils, and T_H2 cells.

Hypersensitivity diseases result from excessive responses to non-infectious environmental antigens

The recognition molecules of adaptive immunity, as we have seen, are generated by genetic recombination in developing lymphocytes and can recognize a very large diversity of molecules. One consequence of this is that immune reactions can be directed against harmless environmental or even self antigens, and not only to antigens of infectious agents. Innate immune recognition elements help to focus adaptive immunity on infectious agents, but in some individuals otherwise harmless environmental antigens elicit immune responses that can be uncomfortable, damaging and even sometimes fatal. These responses are known as **hypersensitivity reactions** and are traditionally divided into four types, according to the immune mechanism(s) involved.

Type I hypersensitivity, more commonly known as **allergy**, is characterized by immune responses driven by T_H2 cells and resulting in the production of IgE. IgE becomes bound to tissue mast cells, which on reappearance of the antigen are triggered to degranulate, with release of mediators that cause many of the manifestations of allergies. Allergic reactions include most cases of asthma, as well as hay fever and many skin rashes and hypersensitive responses to specific foods. Individuals allergic to one type of antigen often have allergic responses to others as well: this tendency to allergic responses is known as **atopy**.

Type II hypersensitivity and **type III hypersensitivity** are also due to antibody responses driven by helper T cells, but these reactions are due to IgG rather than IgE. Type II hypersensitivity reactions involve antibodies recognizing cell-surface components and disrupting cell function (most of these are autoimmune reactions rather than hypersensitivity to environmental antigens); type III hypersensitivity reactions involve the recognition of soluble antigen and the formation of damaging immune complexes.

Type IV hypersensitivity, rather than being antibody-mediated, is predominantly cell-mediated, usually by T_H1 cells, less often by cytotoxic T cells. The most familiar example of a type IV hypersensitivity reaction in the USA is the hypersensitive response to poison ivy, in which a blistering skin rash develops in response to skin contact with a substance contained in the leaves of *Toxicodendron radicans*; but foods, drugs and metals in jewellery also cause type IV hypersensitivity reactions.

We describe examples of all of these types of hypersensitivity reactions in this chapter. Although hypersensitive reactions are generally caused predominantly by one or other of the four mechanisms described above, they may have some elements of other types as well, so while this categorization is convenient it should not be regarded as absolute.

Hypersensitive responses occur only after prior sensitization

As with other adaptive immune responses, development of a hypersensitive response requires the activation and clonal expansion of naïve lymphocytes, which occurs on first encounter with the antigen and is known as **sensitization**. First exposure to the antigen often does not result in an overt response, because the antigen is usually no longer present by the time effector cells and antibodies are generated, and the hypersensitive reaction is noticed only on subsequent exposure to the same antigen. Sensitivity can then increase as the memory pool for the inducing antigen, or the numbers of antigen-specific antibodies, or both, are boosted with each encounter. The environmental antigens that elicit hypersensitive responses generally do not

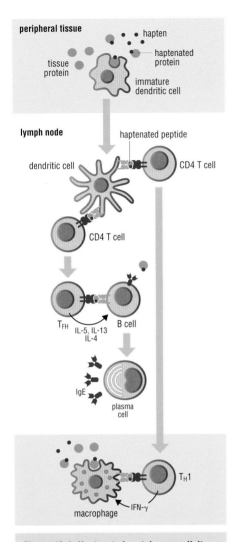

Figure 13-1 Haptenated proteins can elicit both cell-mediated and antibody-mediated hypersensitive responses A chemically reactive small organic molecule (the hapten) enters the tissues and becomes covalently bound to a host protein (which is then said to be haptenated). The haptenated protein is internalized by a tissue dendritic cell, which migrates to the lymph node, where it presents haptenated peptides derived from the ingested protein on its surface MHC molecules. CD4 T cells binding to the haptenated-peptide–MHC complex are activated and differentiate into effector cells. These may be T_{FH} cells that migrate to the follicles, where they encounter B cells that have internalized haptenated protein carried in on the lymph; the B cells present the haptenated protein to the T cells, which activate them to secrete antibodies specific for the hapten. Alternatively, the CD4 T cells may be T_H1 cells that migrate to the periphery and activate macrophages that have internalized haptenated proteins.

Definitions

allergen: antigen from an environmental non-infectious agent that can induce the production of IgE and thus an allergic condition.

allergy: immune responsiveness to any non-infectious environmental agent resulting in the production of IgE and a T_H2-driven inflammatory response.

atopy: the propensity to develop allergic disease, operationally defined by elevated levels of serum IgE or by positivity in the allergic skin test.

delayed-type hypersensitivity (DTH): immune response to a non-infectious environmental antigen resulting in the activation of T_H1 or cytotoxic T cells and inflammation and/or tissue destruction.

DTH: see **delayed-type hypersensitivity**.

hapten: small molecule that can become covalently bound to a protein and serve as a determinant that can be recognized by antibodies, or by T cells when bound to a peptide presented on an MHC molecule.

haptenation: the process whereby a chemically

contain elements recognized by TLRs, which promote the initiation of immune responses by dendritic cells, so it is unclear why these antigens activate T cells in some individuals. Predisposing genetic factors are likely to play a part: these are discussed later in this chapter.

Antigens that elicit hypersensitive responses can be haptens that modify host proteins

In many cases—for example allergy to penicillin, which often is a type 1 hypersensitivity, or poison ivy hypersensitivity, which is a T cell-mediated reaction—the sensitizing antigen is a small organic molecule that becomes covalently attached to proteins in the tissues. This process is known as **haptenation**, the small organic molecule being the **hapten**. Haptens are directly recognized by B cells, and haptenated peptides derived from the protein can be recognized as foreign by T cells, so that both antibody-mediated and cell-mediated hypersensitive responses can be activated by haptenated proteins (Figure 13-1).

The time course of a hypersensitive reaction depends on the mechanism

Type 1 hypersensitivity diseases, or allergies, are also known as **immediate-type hypersensitivity**, because once an individual is sensitized, subsequent exposure to the antigen, in this case known as the **allergen**, results in immediate release of mediators from tissue mast cells already armed with IgE, and the response develops within minutes (Figure 13-2, left-hand side).

By contrast, type IV hypersensitivity reactions, which depend on T_H1 cells or cytotoxic T cells rather than antibodies, are known as **delayed-type hypersensitivity (DTH)**, because local exposure to antigen must be followed either by migration of antigen-specific effector memory T cells into the site, or by activation of central memory T cells, and typically this takes days. Most delayed-type hypersensitivity reactions are thought to be elicited by haptenated proteins, although some antigens may bind directly to MHC molecules. The course of a delayed-type hypersensitivity response is schematically illustrated in Figure 13-2 (right-hand side).

In this chapter we describe some common hypersensitivity reactions, beginning with the allergies. Because mast cells are central to the pathology of allergic responses, we start by describing the activation of these cells and how they contribute to the pathology of allergic responses.

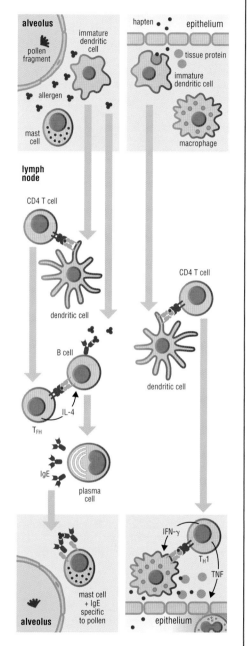

Figure 13-2 Initiation of hypersensitive immune responses Left-hand side: immediate-type hypersensitivity. An inhaled allergen, in this case a pollen particle, is inhaled into the lungs. Components of the pollen cross the epithelium and are taken up by immature dendritic cells. The dendritic cells are induced to mature and migrate to the draining lymph node, but without strong TLR signaling and hence with little production of IL-12 by the dendritic cells. In the lymph node, the dendritic cells present the allergen to naïve CD4 T cells, which are activated and differentiate in the absence of IL-12 or IFN-γ to T_H2 cells. Allergen-specific B cells are activated by allergen entering the lymph node in the lymph and by the specific CD4 cells which produce high levels of IL-4, promoting IgE production by the B cells. IgE in serum reaches inflammatory sites where it binds to FcεRI on mast cells. Subsequent allergen contact will induce an immediate-type hypersensitivity response. Right-hand panel: delayed-type hypersensitivity to a hapten. A small organic molecule (the hapten) becomes covalently bound to a protein in the epithelium and the haptenated protein is internalized by tissue-resident dendritic cells, which degrade the protein and present haptenated peptides to T cells in the lymph nodes. The T cells are activated to proliferate and differentiate into T_H1 cells, which migrate to the epithelium and activate macrophages displaying the same haptenated protein, and to secrete inflammatory cytokines.

reactive small molecule (the **hapten**) becomes covalently bound to a protein molecule, where it acts as a determinant that can be recognized by antibodies, or by T cells when bound to a peptide presented on an MHC molecule.

hypersensitivity reaction: immune response to a non-infectious environmental antigen resulting in inflammation or organ dysfunction.

immediate-type hypersensitivity: see **allergy**.

sensitization: initial exposure to an antigen leading to

the development of a hypersensitive response. For **allergy**, this involves the production of IgE and its binding to mast cells.

type I hypersensitivity: see **allergy**.

type II hypersensitivity: immune response to a cell-surface or matrix protein resulting in the production of IgG and consequent blocking of the function of the protein or immune damage to the cell or tissue.

type III hypersensitivity: immune response to a non-infectious soluble antigen resulting in the production

of IgG and the formation of damaging immune complexes.

type IV hypersensitivity: see **delayed-type hypersensitivity**.

References

Wills-Karp, M. and Hershey, G.K.K.: **Immunological mechanisms of allergic disorders** in *Fundamental Immunology* 5th ed. Paul, W.E. ed. (Lippincott Williams & Wilkins, Philadelphia, 2003), 1439–1479.

13-1 Mast Cells and Allergic Reactions

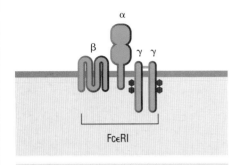

Figure 13-3 Structure of FcεRI on mast cells The α chain binds IgE with high affinity and the γ chain homodimer contains ITAMs, which mediate receptor signaling. The β chain amplifies receptor signaling.

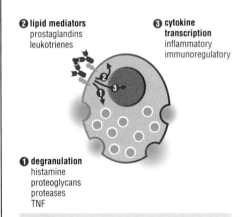

❷ lipid mediators
prostaglandins
leukotrienes

❸ cytokine transcription
inflammatory
immunoregulatory

❶ degranulation
histamine
proteoglycans
proteases
TNF

Figure 13-4 Activation of mast cells triggers an allergic reaction Mast cells can be stimulated by allergen binding to specific IgE on the surface of mast cells, which brings together FcεRI molecules and triggers signaling. This results in three types of effects: (1) release of stored mediators by exocytosis (degranulation), (2) generation of lipid mediators by release of arachidonic acid from phospholipids and conversion to prostaglandins and leukotrienes, and (3) *de novo* synthesis of additional cytokines.

Immediate-type hypersensitivity reactions are initiated by binding of allergen to IgE on sensitized mast cells

The first step in the development of allergies is the production of IgE specific for an allergen and the binding of this IgE onto tissue mast cells. This is referred to as sensitization. When the allergen is next encountered, it binds to the IgE that is attached to mast cells and induces them to release histamine, which has immediate effects, and leukotrienes and cytokines, both of which have slower and more prolonged effects.

IgE is produced in a T cell-dependent antibody response by follicular B cells in which T cell help is provided by T_H2 cells (see section 5-13). Secreted IgE is carried by the blood to inflammatory sites where the vasculature is leaky and allows plasma to enter the tissue. There it binds to the high-affinity FcεRI on tissue mast cells with a dissociation constant of about 10^{-10} M. Effectively, IgE can stay on mast cells for months. As a result, IgE levels in the blood of normal individuals are quite low, roughly 150 ng/ml. The other cell type that expresses FcεRI and binds IgE strongly is the basophil, which is attracted to sites of inflammation along with eosinophils by mast cells and T_H2 cells. Basophils bind IgE and can be triggered by allergens to release mediators similar to those secreted by mast cells and in this way amplify the response.

FcεRI is composed of the IgE-binding α chain, a β chain and a homodimer of the γ chain (Figure 13-3). The β and γ chains participate in signal transduction. The γ chain contains the ITAM sequence that is essential for signaling, as in signaling from antigen receptors of lymphocytes (see sections 2-2 and 6-8), whereas the β chain functions to amplify FcεRI signaling. The resulting ITAM signaling reactions induce the rapid exocytosis of mast cell granules (**degranulation**), releasing large amounts of histamine, the proteoglycans heparin and chondroitin sulfate, proteases called tryptase and chymase, and the inflammatory cytokine TNF (Figure 13-4). Histamine is a key mediator of the immediate effects of allergic reactions, acting directly on blood vessels to induce an influx of fluid, causing swelling, or edema (see section 3-13), and indirectly to induce the relaxation of the vascular smooth muscle cells, causing redness, or **erythema**, of the affected region. Together, these actions give a characteristic **wheal and flare** appearance which is the basis of the *allergen skin test*, in which small amounts of allergen are injected into the skin to test for an immediate mast cell response (Figure 13-5). This and elevated levels of serum IgE are the two tests for atopy. The allergen skin test is also used to test for reactivity to specific allergens. Histamine also acts on sensory nerves to induce the sensation of itch and has regulatory effects on several types of immune cells including dendritic cells and T cells.

The actions of histamine are both rapid and short-lived, as histamine is taken up and metabolized with a half-time of about 1 minute. Histamine action is blocked by **histamine receptor** antagonists, referred to as **antihistamines**. The functions of the proteoglycans and proteases are less well understood, but proteases provide some protection against snake and insect venoms.

Cytokines and lipid mediators are responsible for prolonged effects of mast cell triggering

Triggering FcεRI on mast cells also results in the rapid production of eicosanoid mediators derived from arachidonic acid, including prostaglandin D_2 (PGD_2) and leukotrienes (Figure 13-6). PGD_2 and leukotrienes, especially leukotrienes, have more prolonged actions than histamine. These include smooth muscle contraction, which leads, for example, to bronchoconstriction in asthma, as we shall see shortly. Recently, antagonists of leukotriene receptors have come into

---

Definitions

antihistamines: antagonists of H_1 **histamine receptors.** Antihistamines block some of the symptoms of allergic reactions, particularly the most rapid ones, but do not block inflammation or the actions of leukotrienes.

degranulation: stimulation-dependent exocytosis of mast cell granules. Mast cell degranulation is stimulated by cross-linking by allergen of specific IgE bound to FcεRI, but can also be induced by other stimuli which may account for non-atopic forms of asthma and edema caused by some insect bites.

erythema: redness of tissues due to increased blood flow.

histamine receptors: seven-transmembrane G-protein-coupled receptors that respond to histamine. Four types, called H_1, H_2, H_3 and H_4, are known. H_1 is primarily important for allergic reactions.

wheal and flare: small circular area of soft swelling in skin surrounded by a red rim: the result of a positive allergen skin test.

References

Gould, H.J. *et al.*: **The biology of IgE and the basis of**

allergic disease. *Annu. Rev. Immunol.* 2003, **21**:579–628.

Jutel, M. *et al.*: **Immune regulation by histamine.** *Curr. Opin. Immunol.* 2002, **14**:735–740.

Robbie-Ryan, M. and Brown, M.A.: **The role of mast cells in allergy and autoimmunity.** *Curr. Opin. Immunol.* 2002, **14**:728–733.

Saini, S.S. and MacGlashan, D.: **How IgE upregulates the allergic response.** *Curr. Opin. Immunol.* 2002, **14**:694–697.

use for asthma. Finally, stimulation of sensitized mast cells induces the synthesis and secretion of inflammatory cytokines, which influence the nature of the inflammatory response, as described below.

As with the other tissue-resident immune cells (macrophages and dendritic cells), the properties of mast cells vary somewhat depending on their location, with different populations making different amounts of proteases and other granule components and different relative amounts of the eicosanoid lipid mediators.

Mast cell responses can be stimulated in other ways as well, for example by complement C5a fragments, TLR ligands, some neuropeptides and tissue damage. In addition, the magnitude of mast cell responses is controlled by inhibitory receptors such as FcγRIIB (see section 6-3) and others not yet well understood. As with other inflammatory reactions, those induced by mast cells have the potential to cause considerable tissue injury, so inhibitory receptor restraints on mast cells are likely to be important for limiting acute or chronic injury.

Mast cells induce an inflammation rich in basophils, eosinophils and T_H2 cells

A major downstream effect of mast cell activation is the recruitment of immune cells to the sites of inflammation. TNF, present as a preformed molecule in granules, is released immediately and acts on the endothelium to attract leukocytes from the blood (see section 3-14). Subsequent production of the inducible cytokine IL-4 has effects on the mucosal epithelium as described in section 5-12. IgE and mast cell-mediated inflammation is characterized by the influx of eosinophils, basophils and T_H2 cells. Together, these inflammatory cells are responsible for the symptoms that occur between 6 and 48 hours after allergen contact.

Basophils are very similar to mast cells and can be viewed as continuing the response initiated by mast cells. Basophils do not make PGD_2, but make large amounts of leukotriene C_4, in part accounting for the importance of this mediator at later times. Basophils also produce large amounts of IL-4 and IL-13, which act on mucosal epithelium and smooth muscle cells. IL-13 in particular is important for stimulating mucus production by goblet cells of the mucosa.

Eosinophils are specialized for killing antibody-coated target cells, a process called antibody-dependent cellular cytotoxicity (ADCC) (see section 6-3), which is triggered in these cells by Fc receptors for IgG and IgA. The granules of eosinophils contain several highly basic proteins that are toxic for worms, microbes and surrounding tissue. Eosinophils also produce cytokines at sites of allergic reactions. In addition to the immune cytokines mentioned above, eosinophils produce cytokines involved in wound healing, including TGF-β, TGF-α, VEGF and PDGF, which act on tissue fibroblasts and endothelial cells.

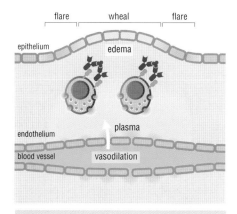

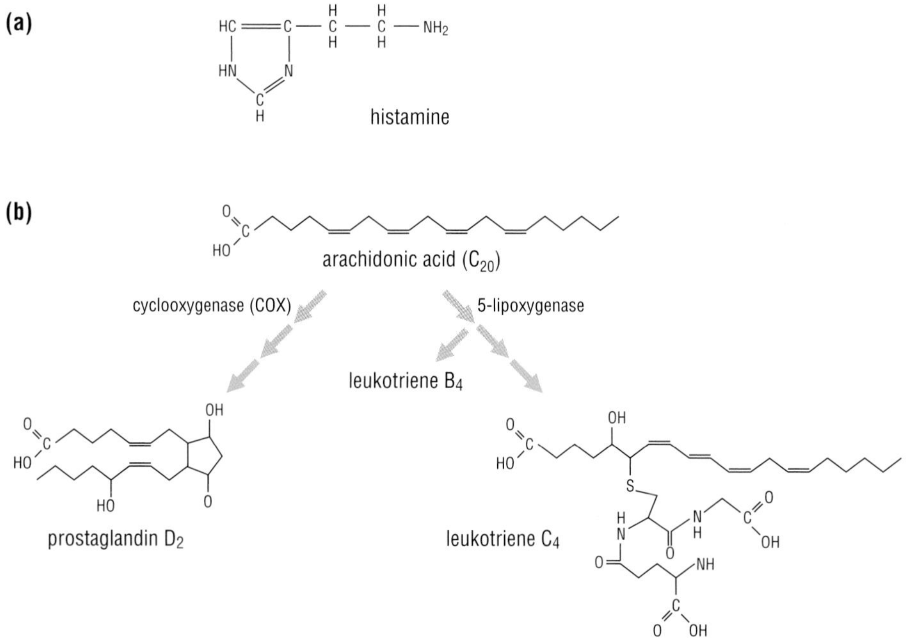

Figure 13-5 The allergen skin test A small amount of allergen is introduced into the skin by a pin prick and is recognized by specific IgE bound to mast cells via FcεRI. Cross-linking of mast cell FcεRI by IgE and allergen induces degranulation and the release of histamine, which acts on the vasculature to induce fluid to enter the tissue and the blood vessel to dilate, giving edema and erythema, respectively. The central area is the wheal and the surrounding area is the flare.

(a)

histamine

(b)

arachidonic acid (C_{20})

cyclooxygenase (COX)

5-lipoxygenase

leukotriene B_4

prostaglandin D_2

leukotriene C_4

Figure 13-6 Small molecule mediators made by mast cells (a) Structure of histamine, which is present at very high concentrations in mast cell granules (100 mM). Histamine is vasoactive, inducing swelling and greater blood flow. In rodents, mast cells also contain large amounts of serotonin, which is also vasoactive. **(b)** Lipid mediators made by stimulated mast cells. FcεRI clustering induced by allergen activates phospholipase A_2, which releases arachidonic acid from the second position of phospholipids. Arachidonic acid is then converted to prostaglandins by the action of cyclooxygenase (COX) and subsequent enzymes or converted to leukotrienes by the action of 5-lipoxygenase and subsequent enzymes. Cyclooxygenase is also called prostaglandin synthase. The major products of mast cells are PGD_2 and leukotriene C_4, which is also called a cysteinyl leukotriene. Leukotriene C_4 is converted to leukotriene D_4 and leukotriene E_4 (not shown). Collectively, the cysteinyl leukotrienes attract eosinophils and basophils, whereas leukotriene B_4 attracts neutrophils.

Allergic diseases have both environmental and genetic causes

Allergic diseases are quite prevalent, particularly in developed countries, where their incidence approaches 30%. Why some individuals make vigorous T_H2/IgE immune responses to common environmental allergens whereas others do not is unclear, but there is evidence for both genetic and environmental factors. Evidence that atopy is largely inherited comes from studies on identical twins: if one of a pair has an allergic disease, in about 75% of cases the other will also develop an allergic disease. Genetic studies on atopic people have established that multiple genetic loci contribute to atopy. Five loci are most often identified in such studies: (1) the MHC class II region, (2) the T_H2 cytokine locus encoding IL-4, IL-5 and IL-13, (3) the locus including the β chain of FcεRI, (4) the locus including STAT6 and interferon-γ, and (5) the locus encoding TCR α and δ chains. Although some of these genes are clearly important for T_H2 cell development and function, the identity of genetic variations that predispose toward atopy and how these variations may affect functions of the genes in question remain to be determined.

Environmental factors also contribute to the development of specific allergies. For example, in the Scandinavian countries, allergies to birch pollen, which is prevalent in the spring, are much more frequent in babies born in late winter or early spring than in babies born at other times of the year, suggesting a critical window early in life when sensitization is most likely to lead to allergies to this pollen. Interestingly, young humans and mice both exhibit a stronger T_H2 bias than do older individuals. Other environmental factors are less well understood. The prevalence of allergic disease is higher in developed countries than in less-developed countries and in developed countries it has been increasing for at least the past 40 years. These correlations have led to the *hygiene hypothesis*, which is described in the next section in the context of asthma.

Manifestations of allergic diseases are site-dependent

Allergic immune responses can occur at most sites where epithelial surfaces meet the environment. Responses to airborne allergens result in **allergic rhinitis** in the nasal passages and **asthma** in the lungs. **Atopic dermatitis** and **urticaria** result from allergic responses in the skin, and food allergies generally manifest as gastrointestinal discomfort, although they can become systemic, in which case the response is referred to as **anaphylaxis**, which can be life threatening. Some drugs and allergens delivered by insect bites can also cause anaphylaxis (discussed further below). Some common allergens are listed in Figure 13-7.

When a particular allergen can be identified as the cause of the symptoms, the ideal strategy for prevention is avoidance; but this is not always practicable (pollen and insects are hard to avoid). **Immunotherapy** in which the allergen is injected repeatedly can be beneficial where avoidance is difficult, although the mechanism is not established. Recently an anti-IgE monoclonal antibody has been introduced as a therapy for severe asthma. This antibody binds to IgE in a position that blocks its binding to FcεRI and in this way interferes with allergic reactions.

Anaphylaxis is a systemic response to an allergen

Most allergic responses compromise quality of life rather than threaten survival, but when such responses become systemic, in anaphylaxis, they have simultaneous effects in multiple locations, including the skin, lungs and cardiovascular and gastrointestinal systems and can be rapidly fatal as a consequence of cardiovascular collapse or pulmonary obstruction, or both. Anaphylaxis can result from allergies to drugs that are chemically reactive and haptenate

Some Common Allergens
Asthma and allergic rhinitis
dust mite feces
cockroach feces
animal dander
pollens (trees, grasses, ragweed, etc.)
mold spores
Food allergies
milk*
eggs*
soy*
wheat*
peanuts
tree nuts
fish
shellfish
Drugs
penicillins
cephalosporins
sulfonamides
imaging contrast media
Other
latex
insect bites
snake venom
*in young children

Figure 13-7 Table of examples of allergens

Definitions

allergic rhinitis: an allergic response affecting the nasal passages and often also the eyes. Symptoms include sneezing, runny nose, watery eyes and itch.

anaphylaxis: a potentially life-threatening systemic allergic reaction in which there is rapid and simultaneous onset of symptoms in several locations, often including skin, lungs and gastrointestinal tract.

asthma: an allergic disease affecting primarily the lungs.

atopic dermatitis: an inflammatory condition in the skin associated with atopy, also called eczema. In this condition, the inflammatory features of immediate-type hypersensitivity dominate over the immediate, histamine-mediated effects on the vasculature.

immunotherapy: (of allergies) a treatment for atopic diseases in which the allergen responsible for the condition is injected repeatedly. Immunotherapy is often successful in alleviating symptoms, but the mechanism is controversial.

urticaria: an allergic skin reaction analogous to the allergen skin test but occurring in multiple locations at once; also called hives.

endogenous proteins (see Figure 13-1): penicillins for example have highly reactive four-membered ring structures (Figure 13-8). Another common allergen that can cause anaphylaxis is present in latex gloves, which are increasingly used in medicine and dentistry to avoid spreading infection. Venoms, certain foods and radiocontrast media used in medical imaging can also cause anaphylaxis. Although serious, anaphylaxis is relatively rare, with an incidence of less than one per million per year. Once anaphylaxis is suspected, immediate injection of epinephrine (also known as adrenaline) is required to prevent shock. This acts via adrenergic receptors to relax bronchospasm and reverse vasodilation. Once an anaphylactic response has occurred, avoidance of the inciting agent is important, since subsequent exposure can result in a stronger response.

Allergic rhinitis occurs in response to allergens in the nasal passages

Allergic rhinitis, or hay fever, is the commonest of the allergic diseases and represents an immediate-type hypersensitivity reaction in the nasal passages. Allergen binding to IgE, FcɛRI cross-linking and mast cell degranulation result in sneezing, fluid discharge, nasal congestion and redness. Eye involvement (watery discharge, itch) is frequently associated with this allergic response. Often allergic rhinitis is seasonal, caused by pollens. Animal dander, dust mites and mold spores are also common causes. Therapies include allergen avoidance, antihistamines, α-adrenergic agonists to help open the breathing passages, and intranasal steroids.

Food allergy is commonest in young children

Food allergy can give rise to anaphalaxis but more commonly causes rapidly developing nausea, abdominal pain, colic, vomiting and diarrhea. Food allergies, particularly for fruit and vegetable allergens, can also manifest as symptoms in the lips, mouth and throat. Most foods do not induce immune responses and in part this reflects that route of exposure: introduction of foreign antigens by mouth often induces immune tolerance rather than an immune response. Thus, food allergies represent exceptions to the normal situation. Food allergies are more commonly seen in young children and less commonly in adults. Relatively few foods account for most food allergies and most are characteristic of one age group or another (see Figure 13-7). Most food allergens are water-soluble, heat- and acid-stable glycoproteins, and are also resistant to proteolytic degradation, properties that decrease degradation in the stomach and small intestines, which would preclude the induction of an IgE antibody response. Food allergy is addressed primarily by omission of the food from the diet.

Allergies often manifest in skin

Atopic dermatitis or eczema is a common skin disorder, particularly in children. It represents an inflammatory reaction, and typically patients have high IgE levels specific to inhaled or food allergens, but the cause of the dermatitis is not clear in most cases and it behaves more like a chronic (or *relapsing*) inflammatory disease than like immediate-type hypersensitivity. Treatment usually involves relief of symptoms with topical steroids and immunosuppressive drugs.

Urticaria or hives is another skin manifestation of allergic disease, typically resulting from food or drug allergy. The symptoms are characteristic of immediate-type hypersensitivity and resemble the wheal and flare reaction (Figure 13-9). The symptoms can last for 1–48 hours. Urticaria is generally thought to be caused primarily by histamine, although leukotrienes are also likely to contribute. Avoidance of the inciting antigen is generally sufficient to prevent disease, where the inciting antigen can be identified.

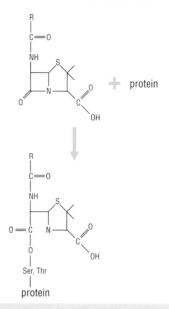

Figure 13-8 Haptenation by penicillin
Reaction of penicillin with an endogenous protein to generate an immunogenic hapten–protein complex. Penicillin is a highly reactive compound containing a four-membered ring. In its antibiotic function, this ring reacts with the active site of enzymes that synthesize the peptidoglycan layer of bacterial cell walls, inactivating them and thereby making growing bacteria subject to osmotic lysis. However, penicillin can also react with endogenous proteins to haptenate them. The resulting penicillin–protein conjugates are immunogenic in some individuals, leading to penicillin allergy, which can either be IgE-mediated or IgG-mediated, depending on the individual. R refers to a substituent that varies between different penicillins.

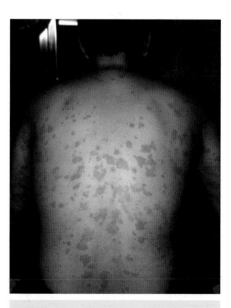

Figure 13-9 Photograph of the back of an individual showing a hives reaction Courtesy of Tim Berger and Kari Connolly.

References

Flicker, S. and Valenta, R.: **Renaissance of the blocking antibody concept in type I allergy.** *Int. Arch. Allergy Immunol.* 2003, **132**:13–24.

Holgate, S.T. and Broide, D.: **New targets for allergic rhinitis—a disease of civilization.** *Nat. Rev. Drug Discov.* 2003, **2**:903–915.

Terr, A.I.: **The atopic diseases** and **Anaphylaxis and urticaria** in *Medical Immunology*, 9th ed. Stites, D.P. *et al.* eds (Appleton and Lange, Stamford, 1997),

389–418.

Valenta, R. and Kraft, D.: **From allergen structure to new forms of allergen-specific immunotherapy.** *Curr. Opin. Immunol.* 2002, **14**:718–727.

Vercelli, D.: **Genetic polymorphism in allergy and asthma.** *Curr. Opin. Immunol.* 2003, **15**:609–613.

Asthma is a respiratory disease of increasing prevalence

Asthma is a disease characterized by persistent and chronic airway inflammation, which leads to episodic but reversible periods of **airway obstruction**, through the excessive production of airway mucus, and **bronchial hyperreactivity**, in which exaggerated narrowing of the airways is induced by excessive smooth muscle contraction in response to airway antigens or irritants. The familiar symptoms of the disease are cough, wheezing and shortness of breath.

Asthma is estimated to affect over 300 million people worldwide. Although particularly prevalent in children, it frequently persists into adulthood. In the USA, one in 6–10 children is believed to have asthma, and health costs from this disease have reached $6 billion per year. Family studies have revealed genetic predisposition to asthma, with concordance rates of 75% in monozygotic and 15% in dizygotic twins, but the increasing incidence in developed countries over the past three decades indicates major environmental contributions. These observations have led to the suggestion, termed the **hygiene hypothesis**, that improvements in hygiene in developed countries and the consequent decrease in childhood respiratory infections may leave the lung susceptible to dysregulated immune responses to common environmental antigens in later life.

Asthma reflects dysregulated T_H2-associated immunity

Asthma is strongly associated with common allergic diseases, including atopic dermatitis (eczema) and allergic rhinitis (hayfever), and shares with these diseases evidence for dysregulated T_H2 immune responses to common environmental antigens, or allergens. Allergens associated with asthma share properties of size, stability and solubility that enable them to become airborne and reach the lower airways. Major airborne allergens—typically proteins from dust mites, animal dander, cockroaches and pollens—are widely disseminated in the environment, leading to repeated exposures and chronic immune-mediated inflammation.

Examination of lung tissues, even in patients with mild disease, reveals infiltration with activated T_H2 cells and eosinophils, and cytokines and chemokines typical of allergic immunity are present. Elevated levels of serum IgE are common. The response to inhaled allergens is initiated when the allergen cross-links IgE molecules bound to Fc receptors on mast cells. The first stage

Figure 13-10 The three stages of asthma
Stage 1: Allergen cross-linking of IgE receptors on mast cells induces the release of histamine, prostaglandin and leukotrienes, leading to vascular engorgement and bronchoconstriction. The vascular effects are amplified by neurogenic peptides (calcitonin gene related peptide, substance P) released from sensory nerves (not shown). Stage 2: Cytokines and chemokines released from activated tissue cells recruit cells from the blood, including T_H2 cells, eosinophils and basophils, which amplify the type 2 immune response. Stage 3: Myeloid cells release contents of granules and reactive oxygen and nitrogen products that induce reactive trophic changes in tissues, including the appearance of mucus-secreting goblet cells (green) and the accumulation of collagen (cross-hatching).

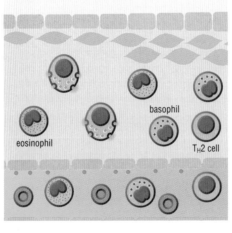
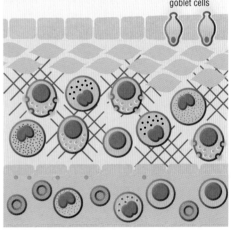

Stage 1

airway

bronchial epithelium
smooth muscle
allergen
mast cell
blood vessel
chemokine

Stage 2

eosinophil
basophil
T_H2 cell

Stage 3

goblet cells

Definitions

airway obstruction: (in asthma) a pathologic narrowing of the lower airways due to subepithelial edema, hypersecretion of mucus, and smooth muscle contraction.

bronchial hyperreactivity: exaggerated airway narrowing in response to airborne allergens or to nonspecific bronchoconstrictor agents, such as acetylcholine or methylcholine.

hygiene hypothesis: an epidemiologically-driven hypothesis that the increased prevalence of allergic diseases in developing countries is due to a decreased number of infections and/or exposure to microbial products early in life.

References

Braun-Fahrländer, C. et al.: **Environmental exposure to endotoxin and its relation to asthma in school-age children.** N. Engl. J. Med. 2002, **347**:869–877.

Eder, W. et al.: **The asthma epidemic.** N. Engl. J. Med. 2006, **355**:2226–2235.

Hysi, P. et al.: **NOD1 variation, immunoglobulin E and asthma.** Hum. Mol. Genet. 2005, **14**:935–941.

Sears, M.R. et al.: **A longitudinal, population-based, cohort study of childhood asthma followed to adulthood.** N. Engl. J. Med. 2003, **349**:1414–1422.

Wills-Karp, M. and Ewart, S.L.: **Time to draw breath: asthma-susceptibility genes are identified.** Nat. Rev. Genet. 2004, **5**:376–387.

Yazdanbakhsh, M. et al.: **Allergy, parasites and the hygiene hypothesis.** Science 2002, **296**:490–494.

(Figure 13-10, stage 1) involves immediate mast cell degranulation, with the release of histamine, prostaglandins, leukotrienes and vasoactive neuropeptides. This early-phase airway obstruction resolves after an hour, but is followed after 4–6 hours by a second phase mediated by newly infiltrating cells responding to cytokine and chemokine signals from cells in the inflamed tissues (Figure 13-10, stage 2). With repeated episodes over time, chronic changes occur. These include loss of epithelial cells, goblet cell hyperplasia, submucosal collagen accumulation with thickening of the airway walls, mucosal mast cell accumulation, and myofibroblast proliferation with increases in smooth muscle, collectively referred to as *airway remodeling* (Figure 13-10, stage 3). Slow, progressive, loss of lung function ensues if inflammation is not controlled. Viral and bacterial infections frequently trigger asthma attacks, with cytokines from T_H1 and T_H17 cells serving to augment inflammation, thus exacerbating the airway disease. Although rare, fatal cases do occur and are caused by severe hypoxia due to diffuse mucous plugging of the inflamed airways.

Asthma represents an interaction of allergens and genetics

Although allergic asthma develops in response to specific allergens, moving to avoid environmental allergens is rarely effective over time, suggesting that atopic individuals are predisposed to develop hyperreactivity to prevalent allergens wherever they are. Defects in lung T_{REG} cells, which may regulate tolerance to airborne allergens, have been postulated.

A number of genome-wide screens have been carried out among asthma populations around the world. Altogether, some 20 chromosomal linkages have been identified, although most associations are modest. Of much interest are those implicated in multiple studies, which include sites that encode type 2 cytokines and their receptor-signaling modules that have been associated with other allergic diseases (see section 13-2). More recent genome-wide screens have implicated specific gene polymorphisms in asthma, although these have yet to be replicated using diverse populations or supported in mouse models. The products of these genes include NOD1, which is a sensor for bacterial peptidoglycan fragments similar to NOD2 (section 3-12), a protein implicated in tissue remodeling, and several less well understood molecules. Further studies will be needed to determine whether and how these molecules may contribute to genetic susceptibility to asthma.

Therapy directed at the inflammation in the lung usually controls symptoms of the disease. Most commonly, inhaled beta adrenergic agents and corticosteroids are used to combat bronchial constriction and inflammation, respectively. Anti-IgE monoclonal antibodies and leukotriene inhibitors, including 5-lipoxygenase and cysteinyl leukotriene receptor I antagonists, have also been effective in some patients.

Mouse models mimic aspects of acute airway inflammation in human asthma

The most widely used animal model of airway sensitization involves challenge with ovalbumin (Figure 13-11), a commonly used experimental antigen. After intraperitoneal immunization with ovalbumin in the adjuvant alum twice over 14 days, mice receive aerosolized ovalbumin in saline at 2–7-day intervals. One day after the final aerosol, mice are analyzed using histology for inflammatory lung infiltrates, by staining for inflammatory cells in brocholaveolar lavage fluid washed from the lungs, and for evidence of T_H2 cell activation by serum IgE levels and cytokine analysis after ovalbumin-specific T cell activation. Treated animals demonstrate increased airway hyperreactivity to aerosolized or intravenously injected bronchoconstrictor agents, such as acetylcholine. T_H2-mediated inflammatory responses and airway hyperreactivity are markedly attenuated in T cell-deficient mice, as well as in mice deficient in IL-13, IL-4Rα or STAT6, and in animals expressing mutant GATA-3 proteins. Conversely, mice deficient in the T-bet transcription factor develop spontaneous allergic lung inflammation. Animals expressing IL-4, IL-5, IL-9 or IL-13 from lung-specific transgenes develop overlapping features of allergic inflammation, with IL-13 implicated specifically in lung fibrosis and IL-4 in inflammatory cell recruitment. Although many important differences exist between the animal model and human asthma, so that caution is needed in extrapolating from mouse to human, the model has been instrumental in directing attention to dysregulated immune pathways associated with asthma. Genetic polymorphisms implicated in the murine asthma model include Tim1, a member of the T cell immunoglobulin and mucin-domain containing molecules, and C5, complement factor 5.

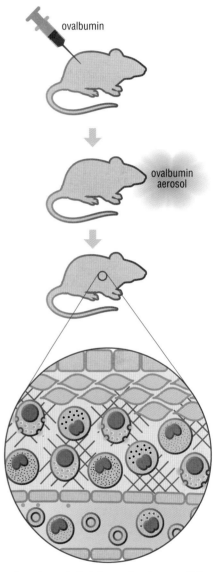

ovalbumin

ovalbumin aerosol

airway hyperplasia, inflammation, hyperreactivity

Figure 13-11 Mouse model of airway inflammation After intraperitoneal immunization with ovalbumin, mice are repeatedly exposed to aerosols containing ovalbumin. Examination of cells washed from the airways (bronchial alveolar lavage cells) reveals eosinophils and T_H2 cells, histologic examination of the lung shows inflammation, and physiological measurements after challenge with bronchoconstrictor agents reveal airway hyperreactivity.

A variety of hypersensitivities and autoimmune diseases exhibit immune pathology caused by IgG

Whereas allergies and asthma are mediated by IgE directed against otherwise innocuous environmental antigens, IgG can also mediate a variety of immune pathologies caused by immune responses to antibiotics, protein therapeutic agents and autoantigens. Antibodies can be directed against antigens that are present on the surface of cells or are part of tissue structures, such as basement membranes, and cause damage to those cells or tissues. Alternatively, pathology can result from the formation of immune complexes between IgG molecules and soluble antigens and their deposition in tissues and local damage. In general, immune pathology caused by IgG is due to activation of phagocytic cells through FcγRs, and to complement activation. In the standard classification of hypersensitivity diseases, the pathology caused by binding of IgG antibodies to cell surface molecules or tissue structures is referred to as type II hypersensitivity, whereas immune complex-mediated disease is called type III hypersensitivity. Autoantibodies are also important causes of immune pathology by both these mechanisms in subsets of autoimmune diseases: these have already been discussed in Chapter 12. Additionally, in some autoimmune diseases, IgG specific for cell-surface receptors may induce disease by mimicking ligands or by blocking their function simply by virtue of binding.

Hemolytic anemias can be caused by Ig produced in response to infection or materno-fetal Rhesus incompatibility

Antibodies to red blood cells and platelets lead to hemolytic anemia and thrombocytopenia, respectively. Red blood cells are very sensitive to complement-mediated lysis and are also cleared from blood via FcγR-mediated phagocytosis by macrophage populations of the spleen and liver (see section 1-3). For reasons that are not well understood, individuals whose lungs are infected with the bacterium *Mycoplasma pneumoniae* often make a transient IgM response to the I antigen of red blood cells, which is a carbohydrate epitope. In a small percentage of these individuals, IgM of sufficient quantity and/or affinity is made to induce hemolytic anemia. Hemolytic anemia is also sometimes seen in individuals with infectious mononucleosis (Epstein–Barr virus infection).

Hemolytic anemia can also occur with Rhesus (Rh) incompatibility, leading to *erythroblastosis fetalis*, or **hemolytic disease of the newborn**. This problem can occur when a Rhesus-negative mother is pregnant with a second Rhesus-positive child. The primary exposure of the mother to the Rh antigen occurs at the time of birth of the first Rhesus-positive child, when enough fetal blood is introduced to the mother's circulation to induce a primary antibody response and

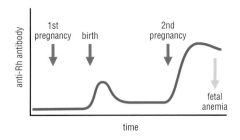

Figure 13-12 Induced humoral immune response to Rhesus antigen in a Rhesus-negative mother During the first pregnancy, antigen exposure induces a primary anti-Rh response that is asymptomatic. Much of this response occurs as a result of exposure of the mother to substantial numbers of the baby's red blood cells at the time of birth. Antibody titers drop substantially by the time of subsequent pregnancy. However, small amounts of antigen are presented to the maternal immune system during pregnancy and this may induce a secondary IgG antibody response. These IgG antibodies are transported across the placenta like other maternal IgG, and can then bind to fetal red blood cells and induce their clearance by phagocytic cells. This sequence can largely be prevented by the injection of human anti-Rh antibodies once during pregnancy and again around the time of birth. This is thought to remove the antigen before the mother becomes sensitized. The amount of antibody injected during pregnancy is insufficient to induce anemia on its own.

Definitions

cryoglobulinemia: a condition in which small immune complexes are present in the blood can be seen to precipitate when serum samples are chilled to refrigerator temperature. Cryoglobulins are antibodies that precipitate in the cold.

hemolytic disease of the newborn: hemolytic anemia in which IgG antibodies against red blood cell antigens, typically the Rh antigen, are made by the mother during pregnancy, cross the placenta, bind to fetal red blood cells and induce their clearance by phagocytes.

post-streptococcal acute glomerulonephritis: an immune complex disease seen after *Streptococcus pyogenes* infections in some individuals.

serum sickness: a systemic immune complex disease originally observed upon immunotherapy of various infectious diseases with serum from horses immunized against the infectious agent or its toxin. The large amounts of IgG antibodies produced by the patient in response to the horse serum proteins result in formation of small immune complexes, leading to immune complex deposition and resulting pathology.

References

Nordstrand, A. *et al.*: **Pathogenic mechanism of acute post-streptococcal glomerulonephritis.** *Scand. J. Infect. Dis.* 1999 **31**:523–537.

Sansonno, D. and Dammacco, F.: **Hepatitis C virus, cryoglobulinaemia, and vasculitis: immune complex relations.** *Lancet Infect. Dis.* 2005 **5**:227–236.

Urbaniak, S.J.: **Alloimmunity to RhD in humans.** *Transfus. Clin. Biol.* 2006 **13**:19–22.

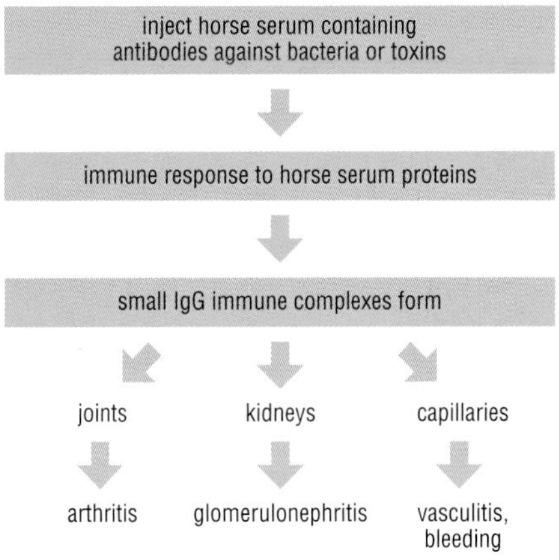

Figure 13-13 Serum sickness induced by early immunotherapy approaches In the late 19th century, infectious diseases such as scarlet fever and diphtheria were treated by injection of serum from horses that had been immunized with particular bacteria or toxins from them. This therapy induced an immune response to horse serum proteins. IgG antibodies produced, either from the initial treatment or a subsequent injection of horse serum, could combine with the horse proteins under conditions of antigen excess, leading to the formation of large amounts of soluble immune complexes. These soluble immune complexes then deposit in various locations, especially including the synovial lining of joints, glomeruli of kidneys, small capillaries in skin and other locations. The immune complexes activate phagocytes via FcγRs, activate complement and induce inflammation and tissue damage. In a murine experimental model of this disease, called the Arthus reaction, phagocytic responses via FcγRs are generally more important than complement. Modern therapies, such as monoclonal antibodies, can run into similar problems, particularly if murine monoclonal antibodies are used. Often the antibody responses against these therapeutics ends their efficacy, by promoting their clearance as well as possibly inducing some pathology. This problem has led to several approaches to create either fully human monoclonal antibodies or to take murine monoclonal antibodies and replace as much of their constant and framework regions as possible with their human counterparts, while retaining binding affinity for the desired antigen. Some individuals still make IgG responses to these largely human therapeutic agents but the problems are greatly decreased by these strategies.

inject horse serum containing antibodies against bacteria or toxins

↓

immune response to horse serum proteins

↓

small IgG immune complexes form

↓ joints → arthritis

↓ kidneys → glomerulonephritis

↓ capillaries → vasculitis, bleeding

generation of B cell memory (Figure 13-12). This can be prevented by injecting the mother once during pregnancy and also at the time of birth with human IgG anti-Rh antibodies. This antibody is thought to promote clearance of the antigen so that the woman does not make an anti-Rh antibody response. If sensitization occurs, then during a subsequent pregnancy, small amounts of the antigen are seen by the mother's immune system, leading to an IgG secondary response. Like other maternal IgG, these IgG anti-Rh antibodies are transported into the fetal circulation and can induce hemolytic anemia in the fetus. The Rh antigen is present at low density on red blood cells, so complement is not effectively activated, but clearance of the antibody-coated cells by phagocytes occurs.

Immune complexes can cause damage to joints, blood vessels and kidneys

Immune complex-mediated diseases are usually systemic because they typically result from the production of large amounts of soluble immune complexes, which circulate through the blood before being precipitated in the joints, kidneys and capillaries at many sites. One of the more common causes of such pathology is excessive IgG responses to antibiotics, particularly penicillin and related antibiotics. Penicillins are chemically reactive, resulting in covalent coupling to endogenous proteins (see Figure 13-9). Penicillin–protein conjugates can serve as immunogens and cause the production of IgE or IgG antibodies, depending on the individual. If the patient has been receiving a large dose of an antibiotic, for example intravenously, and develops an IgG response (probably a secondary response), then immune complexes form, are widely deposited, and cause severe systemic symptoms over the next several days. A related manifestation of systemic immune complex disease was initially recognized over a century ago, after the introduction of immunized horse serum to treat a variety of infectious diseases, including diphtheria (to protect against diphtheria toxin) and scarlet fever (caused by *Streptococcus pyogenes*). Some individuals, particularly those receiving a second injection of horse serum, developed severe symptoms referred to as **serum sickness** (Figure 13-13). A hallmark of these responses is that antigen–antibody complexes are formed at antigen excess (see section 6-5), resulting in small immune complexes. These complexes circulate until they deposit in the synovium of joints, the glomeruli of kidneys and the capillaries of many locations, including the skin, and can induce arthritis, glomerulonephritis and vasculitis, respectively. Horse serum is no longer widely used, but animal antibody preparations are still used as antivenoms to treat poisonous snake bites. Certain protein therapeutic agents, such as monoclonal antibodies, may also induce an antibody response, although the most common result is loss of efficacy rather than severe immune complex disease. Immune complex disease can also be associated with infections, for example infection with hepatitis C virus (serum sickness like) or *Streptococcus pyogenes* (**post-streptococcal acute glomerulonephritis**). In these latter conditions, the presence of small immune complexes in the blood can frequently be demonstrated by the precipitation of these immune complexes that occurs when serum samples are chilled; antibodies that precipitate in this way are called cryoglobulins, so this condition is called **cryoglobulinemia**. Finally, in systemic autoimmune diseases there are typically immune complexes leading to arthritis, vasculitis, and/or glomerulonephritis (see sections 12-8 and 12-9).

13-5 Delayed-Type Hypersensitivity Reactions

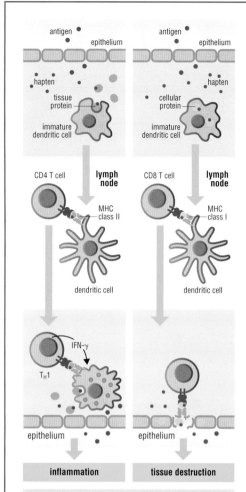

CD4 T cell — lymph node — MHC class II — dendritic cell — IFN-γ — T_H1 — epithelium — **inflammation**

CD8 T cell — lymph node — MHC class I — dendritic cell — epithelium — **tissue destruction**

Figure 13-14 Activation of CD4 and CD8 T cells by haptenated peptides Left-hand side: An antigen penetrates the skin and haptenates extracellular matrix proteins. They are internalized by immature dendritic cells, which mature and migrate to the lymph nodes where they present haptenated peptides to CD4 T cells, which differentiate into T_H1 cells and migrate to the epithelium where they activate macrophages. Right-hand side: A lipid-soluble antigen penetrates the skin, crosses the plasma membrane of immature dendritic cells and haptenates cytoplasmic proteins, which are degraded and presented on MHC class I molecules. When the dendritic cell matures and migrates to the lymph node, the haptenated-peptide–MHC complex is recognized by CD8 T cells, which differentiate into cytotoxic effector cells that migrate to the skin and destroy epithelial cells displaying the same haptenated peptide.

Delayed-type hypersensitivity reactions reflect the actions of T_H1 and cytotoxic T cells

Delayed-type hypersensitivity (DTH), or type IV hypersensitive reactions, occur when peptides from a haptenated or otherwise modified protein, or, more rarely, intact small molecules, are presented on MHC molecules and elicit T_H1 or cytotoxic T cell responses. Many of these hypersensitive reactions take the form of *contact dermatitis*, a rash that develops when the antigen comes into contact with the skin. Delayed-type hypersensitivity can however also occur in response to food antigens and to systemically administered drugs. Because the antigen-specific T cells that mediate the response are memory cells that must first be activated and migrate to the site of antigen exposure, these responses can take a day or more to develop, and this can make it difficult to identify the antigen responsible. Moreover, in many cases the hapten inducing the response is a metabolite of the original compound, and this means that the tissue in which the response occurs may be the one in which the metabolite is generated or accumulates, rather than the original site of exposure: this may explain for example why systemically administered drugs often give rise to skin rashes.

Delayed-type hypersensitivity reactions are commonly recorded in response to skin contact with specific plants, or to contact with metals such as nickel (for example in jewellery), where nickel sulfate is formed and can become bound to tissue proteins; they also occur in response to drugs and cosmetics. In most cases the proteins that become haptenated to produce delayed-type hypersensitivity reactions are extracellular, so that peptides derived from them are generated after internalization into the endosomal–lysosomal compartment of dendritic cells or macrophages and bind to MHC class II molecules to elicit a CD4 T cell response (Figure 13-14, left-hand side). Lipid-soluble antigens however can cross the cell membrane and haptenate cytoplasmic proteins: in this case the haptenated peptides bind to MHC class I molecules, and cytotoxic CD8 T cell responses are elicited (Figure 13-14, right-hand side). These responses are more destructive but are also less common. In any given delayed-type hypersensitivity response both types of cells may participate.

The two delayed-type hypersensitivity diseases we describe here are contact dermatitis in response to components of *urushiol*, an oily mixture contained in plants belonging to the Anacardiaceae, which include poison ivy and mango; and **celiac disease**, a gastrointestinal hypersensitive response to wheat in the diet.

Contact sensitivity to urushiol is an inflammatory response to haptenated proteins

Contact sensitivity to urushiol from the leaves of *Toxicodendron radicans*, or poison ivy, is the single commonest cause of contact dermatitis in the USA, where the plant grows wild in woodland. Its close relatives, poison oak (*Toxicodendron diversilobium*) and poison sumac (*Toxicodendron vernix*), and the bark and peel (but not the flesh) of the mango also contain urushiol, although mango contains much less and is a problem chiefly for fruit pickers. The antigenic compound in urushiol is a small hydrophobic molecule, pentadecacatechol, which easily penetrates the skin on contact and becomes covalently bound to extracellular proteins. It can also cross cell membranes and haptenate cytoplasmic proteins. Both CD4 and CD8 T cells, therefore, are activated in response to urushiol. Most haptenated proteins are relatively weak antigens and it may take several exposures for sensitivity to develop. Sensitivity to poison ivy is an exception and can develop on a single exposure, partly because of the large quantity of antigen present in the plant. Subsequent contact with the plant then

Definitions

celiac disease: a gastrointestinal delayed-type hypersensitivity reaction to gluten proteins present in wheat and other grains that is characterized by inflammatory destruction of the upper intestinal mucosa, resulting in malabsorption and diarrhea.

References

Di Sabatino, A.: **Epithelium derived interleukin 15 regulates intraepithelial lymphocyte Th1 cytokine production, cytotoxicity, and survival in coeliac** disease. *Gut* 2006, **55**:469–477.

Galli, S.J. et al.: **Mast cells as "tunable" effector and immunoregulatory cells: recent advances.** *Annu. Rev. Immunol.* 2005, **23**:749–786.

Jabri, B. and Sollid, L.M.: **Mechanisms of disease: immunopathogenesis of celiac disease.** *Nat. Clin. Pract. Gastroenterol. Hepatol.* 2006, **3**:516–525.

MacDonald, T.T. and Monteleone, G.: **Immunity, inflammation, and allergy in the gut.** *Science* 2005, **307**:1920–1925.

Qiao, S.-W. et al.: **Refining the rules of gliadin T cell epitope binding to the disease-associated DQ2 molecule in celiac disease: importance of proline spacing and glutamine deamidation.** *J. Immunol.* 2005, **175**:254–261.

Rosen, F. and Geha, R.: **Contact sensitivity to poison ivy** in *Case Studies in Immunology* 4th ed. (Garland Science, New York, 2004) 63–69.

Roychowdhury, S. and Svensson, C.K.: **Mechanisms of drug-induced delayed-type hypersensitivity reactions in skin.** *Am. Assoc. Pharmaceut. Sci. J.* 2005, **7**:E834–E846.

activates a delayed-type hypersensitive reaction that takes one or two days to develop. At least 50% of people are sensitive to urushiol, and possibly most, given sufficient exposure.

Activation of T$_H$1 cells at the site of exposure leads to macrophage activation and the release of inflammatory mediators, causing an itching rash that builds as more inflammatory cells are recruited by chemokines released by the activated macrophages. Destruction of skin cells by activated CD8 T cells recognizing haptenated peptides bound to MHC class I molecules on their surface causes tissue damage and blistering.

Pentadecacatechol is easily spread from the original site of contact through touching or scratching of the affected area (thorough washing helps prevent this), and haptenated proteins in the skin can persist for weeks, so the rash can be widespread and persistent. (Moreover, because the different Anacardiaceae that elicit hypersensitive responses all contain the same antigenic compounds, sensitization by one can lead to a hypersensitive response to another.) Poison ivy contact dermatitis is usually treated with corticosteroids applied to the skin or, in severe cases, taken by mouth, although antihistamines may also be prescribed to counter the effects of histamine released by mast cells activated by mediators other than IgE.

Celiac disease is a delayed-type hypersensitive response to gliadin peptides from dietary wheat

Celiac disease is a gastrointestinal hypersensitive reaction to gluten proteins in wheat, barley, oats and rye that occurs in about 0.5% of the population in wheat-eating regions of the world. It is characterized by a T$_H$1-driven response that causes destruction of the mucosal surface of the small intestine (Figure 13-15a) that is often asymptomatic but can also result in severe diarrhea and failure to absorb nutrients, including iron, with consequent debility and fatigue. Its cause was first recognized during the food shortages in Europe in World War II when a perspicacious physician in the Netherlands noticed the marked improvement in celiac patients when bread was scarce. It often develops in infancy, on introduction of wheat into the diet, but can also appear in young adulthood or later.

Gluten is a mixture of two types of proteins, *glutenins* and *gliadins*, that are rich in proline and glutamine. Celiac disease develops when peptides from these proteins—particularly gliadins—bind to either of two HLA-D alleles—HLA-DQ2, which is present in 90% of celiac patients, or HLA-DQ8, which is present in most of the others. Celiac disease does not develop in the absence of these HLA-D alleles. (The converse, however, is not true—only about one in 50 of those individuals carrying either allele develops the disease.) HLA-DQ2, which is present in the majority of celiac patients, has a strong preference for negatively charged peptides that is met by the deamidation of the glutamines in gliadin by the ubiquitous gut enzyme tissue transglutaminase to give the negatively charged glutamic acid. Presentation of these modified peptides by HLA-DQ2 or DQ8 molecules activates T$_H$1 cells, which launch an inflammatory response driven by interferon-γ in the lamina propria (the tissue underlying the epithelium) in the small intestine.

As well as the inflammatory response driven by T$_H$1 cells in the lamina propria, celiac disease is characterized by greatly increased numbers of intraepithelial lymphocytes (see section 8-5). In active disease, these cells express NKG2D, the receptor for the stress-induced molecules MICA and MICB, which are induced on the intestinal epithelial cells in celiac disease, and it is thought that cytotoxic activation of intraepithelial lymphocytes contributes to the destruction of the gut epithelium. Immune mechanisms in celiac disease are illustrated in Figure 13-15b. The only known effective treatment for celiac disease is removal of gluten from the diet.

Delayed-type hypersensitive reactions are used to test for T cell function or for the presence of specific infections or hypersensitivities

Since cellular immunity to any given antigen generally results in a delayed local inflammatory response to the injection of the antigen just under the skin, DTH reactions can be used to test for cellular immune responses, or for T cell function; and with antigen applied as a patch on the skin they are a standard test for contact hypersensitivity. The DTH reaction is also the basis for the PPD skin test for infection with the tuberculosis bacillus, *Mycobacterium tuberculosis* (see section 9-7), in which a purified protein from the bacterium is injected into the skin. Mycobacterial infections elicit T$_H$1 cell responses, and in infected individuals redness and swelling appear at the site of the injection within a few days.

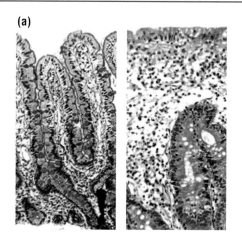

(a)

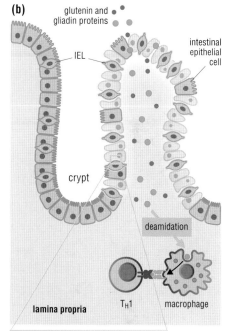

(b)

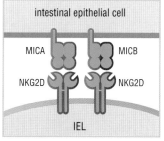

Figure 13-15 Celiac disease (a) Normal (left) and celiac (right) small intestine mucosa, showing flattening of the villi consequent on destruction of the epithelium in celiac disease. Courtesy of Coeliac UK. **(b)** Mechanism of inflammation and tissue destruction in celiac disease. Gluten proteins (gliadins and glutenins) are absorbed from the gut lumen, deamidated and internalized by lamina propria macrophages, which present them to peptide–MHC-specific T$_H$1 cells which drive local inflammation. Intestinal epithelial cells are induced by inflammatory stress to express MICA and MICB, which are recognized by NKG2D that appears on intraepithelial lymphocytes (see inset) in celiac disease and induces cytolytic destruction by the lymphocytes of the neighboring epithelial cells.

Crohn's disease and ulcerative colitis are chronic intestinal inflammatory diseases

Inflammatory bowel disease (IBD) is the term used for chronic, relapsing episodes of inflammation of the gastrointestinal tract of unknown cause. There are two major types of IBD: **Crohn's disease** and **ulcerative colitis**. IBD is most prevalent in developed countries of the Northern Hemisphere, including the USA, the United Kingdom and Scandinavia. Both diseases are generally sporadic, but they can be familial. When the diseases occur in monozygotic twins, both members of a pair are affected in 45–60% of cases in Crohn's disease and in 5–20% of cases in ulcerative colitis. Human and animal studies suggest that IBD is a polygenic disorder that reflects aberrant immune reactivity to normal intestinal microbial flora, perhaps related to diminished epithelial barrier function. Almost 20% of Caucasians with Crohn's disease are homozygous for variants of NOD2, a macrophage and Paneth cell protein that mediates NF-κB activation in response to muramyl dipeptides in bacterial cell wall peptidoglycans (see section 3-12), although family members with the same mutations can be healthy. Additional genomic regions broadly associated with other human autoimmune diseases have been linked with both forms of IBD. Both diseases typically present in young adults with diarrhea, abdominal pain, fever and weight loss. Crohn's disease most commonly involves the terminal ileum, where the small intestine enters the large bowel, or colon, and the adjacent region of ascending colon, as well as discontinuous (or *skip*) regions throughout the gastrointestinal tract. In contrast, ulcerative colitis is restricted to the colon, where it affects continuous regions of bowel, most commonly extending back from the rectum. Models for IBD in mice indicate that development of these diseases depends not only on predisposing genes but also on the presence of commensal intestinal bacteria. Functional intestinal regulatory T cells (T_{REG}) are also necessary to prevent unrestrained injury by immune cells. The episodic nature of IBD may reflect variations in underlying immunity, intestinal barrier function or the microbial flora, although the precise mechanism remains unknown.

Crohn's disease represents a dysregulated mucosal immune response

Pathologically, the lesions of Crohn's disease consist of regions of inflammation extending from the lumen through the entire bowel wall. Lymphocytes and activated macrophages predominate, and granulomas are common. Although the terminal ileum is most commonly involved, affected areas can occur throughout the gastrointestinal tract. Common complications are narrowing of the bowel, or *strictures*, where the bowel wall is thickened by inflammation and accompanying submucosal fibrosis, and the development of channels, or *fistulas*, that reflect extension of inflammation into adjoining tissues. Systemic inflammatory conditions, including fever, arthritis and lesions of the eyes, mouth and skin, can also accompany Crohn's disease.

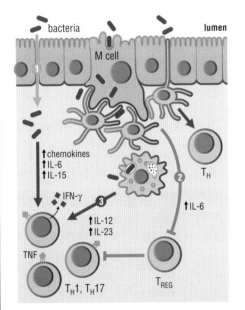

Figure 13-16 Proposed model of inflammation in Crohn's disease Loss of intestinal integrity may lead to increased bacterial translocation and increased recruitment and activation of inflammatory cells (1). Alternatively, dendritic cells that sample bacteria from the lumen by means of intraepithelial projections, or at the base of M cells, may promote activation of effector T cells or may inhibit activity of T_{REG} cells, perhaps by production of IL-6 (2). Finally, activation of macrophages in response to the bacterial products may drive the overproduction of IL-12 and IL-23, leading to amplification of T_H1 and T_H17 effector functions, with increased release of IFN-γ and the generation of membrane TNF (3). Variants of NOD2 associated with susceptibility to Crohn's disease may compromise the epithelial barrier by reducing secretion of antimicrobial peptides by Paneth cells and/or may alter the amount or types of inflammatory cytokines produced by dendritic cells and/or macrophages. Variants of one chain of the IL-23R have also been associated with risk for both Crohn's disease and ulcerative colitis, presumably reflecting the important role of IL-23 in T_H17 cell responses.

Definitions

Crohn's disease: an inflammatory disease of unknown cause typically affecting the terminal ileum and adjacent regions of the colon and characterized by episodes of diarrhea, weight loss, abdominal pain and fever. Crohn's disease is one of two kinds of inflammatory bowel disease, the other being **ulcerative colitis**.

crypt abscess: collection of neutrophils and inflammatory debris encased by collagenous reactive tissues that forms at the base (the crypt) of epithelial villi in the colon in **ulcerative colitis**. Crypts represent sites near

the epithelial stem cell compartment and where Paneth cells secrete defensins and other effector molecules that interact with the intestinal microbial flora.

ulcerative colitis: an inflammatory disease of unknown cause typically affecting the lower colon and rectum, and characterized by episodes of diarrhea, weight loss, abdominal pain and fever. Ulcerative colitis is one of two kinds of inflammatory bowel disease, the other being **Crohn's disease**.

Lesions from Crohn's disease patients suggest that activated T_H1 and T_H17 cells may both contribute to the pathologic process. Macrophages that express IL-12 and IL-23 are found, and CD4 T cells that express IFN-γ, IL-12Rβ2 and the transcriptional regulator T-bet, but also IL-17A and IL-6, are present (Figure 13-16). Mutations in NOD2 and a variant subunit of the IL-23 receptor have been associated with risk for IBD.

Typically, flares of Crohn's disease are treated with steroids and antimicrobial agents to reduce the intestinal bacterial load while suppressing inflammation. Maintenance therapy is usually with aspirin-based antiinflammatory compounds or stronger immunosuppressants, such as steroids and azathioprine. More recently, success has been achieved using recombinant reagents that target the immune system more selectively. Anti-TNF antibody has become widely used, and is believed not only to target TNF-mediated inflammation but also to induce apoptosis of T_H1 and T_H17 cells that express membrane TNF. Anti-IL-12 antibody has proven successful in limited early trials and may function to block both IL-12 and IL-23 through recognition of their shared p40 subunit. Antibodies against α4 integrin, which is required for lymphocyte migration into intestinal mucosa, can also prevent attacks of Crohn's disease.

Ulcerative colitis represents a dysregulated T_H2-associated mucosal immune response

Lesions of ulcerative colitis involve the rectum and extend variably but contiguously into the colon, where the disease is confined. Inflammation is present in the superficial mucosal layers, which become eroded by lymphocyte and neutrophil infiltration. The mucosae become denuded, with diarrhea, pain and bleeding common. Neutrophil accumulation at the bases of the epithelial villi results in the formation of **crypt abscesses**, with destruction of sites important for epithelial stem cell renewal. Chronic inflammation leaves patients at risk for colon cancer, which occurs more commonly among patients with ulcerative colitis than in those with Crohn's disease.

Lesions from patients with ulcerative colitis suggest that dysregulated T_H2 immune responses may contribute to the underlying pathogenesis. Although IL-4 is not typically present, IL-5 and IL-13 are increased. Elevated serum IgG4 is consistent with immunoglobulin isotype switching mediated by T_H2-associated cytokines. Like Crohn's disease, ulcerative colitis is treated during inflammatory episodes with steroids or immunosuppressants such as cyclosporin. Maintenance therapy with aspirin derivatives, steroids and general immunosuppressants is used to prevent attacks. Surgical removal of the colon may be required to control disease and remove the risk of colon cancer.

Animal models for inflammatory bowel disease are diverse

Mouse models for Crohn's disease and ulcerative colitis fall into three broad categories, although these frequently overlap: models resulting from disruption of epithelial integrity, models resulting from dysregulated immune cell activation, and models resulting from loss of T_{REG} cells (Figure 13-17). In all of the models, disease develops only in the presence of the intestinal microbial flora, which supports the hypothesis that these diseases reflect a loss of immune regulation at the bacterial/intestinal interface. Mice deficient in TLR-mediated bacterial sensing demonstrate abnormal colonic epithelial turnover, showing the importance of the bacteria/host interface in gut homeostasis. Murine models of ulcerative colitis induced by epithelial injury using oxazolone or dextran sodium sulfate can be abrogated by neutralizing IL-13 or blocking eosinophil function, respectively.

Representative Mouse Models of Inflammatory Bowel Disease
Impaired epithelial barrier function
N-cadherin dominant negative mice, intestinal trefoil factor-deficient mice, mdr1a-deficient mice
exogenous epithelial toxins: oxazolone, dextran sodium sulfate, trinitrobenzene sulfonic acid
Dysregulated polarized immune responses
TNF_{ARE} mutant mice, STAT4 transgenic mice, myeloid cell STAT3-deficient mice
Loss of T_{REG} cells or function
IL-2 or IL-2Rα deficiency, IL-10 or IL-10R deficiency, TGF-β deficiency or TGF-βRII dominant-negative, Smad3 deficiency
transfer of isolated CD4, CD45RB^{hi} T cells into Rag-deficient or SCID mice

Figure 13-17 Table of representative mouse models of inflammatory bowel disease

References

Bouma, G. and Strober, W.: **The immunological and genetic basis of inflammatory bowel disease.** *Nat. Rev. Immunol.* 2003, **3**:521–533.

Duerr, R.H. *et al.*: **A genome-wide association study identifies *IL23R* as an inflammatory bowel disease gene.** *Science* 2006, **314**:1461–1463.

Ghosh, S. *et al.*: **Natalizumab for active Crohn's disease.** *N. Engl. J. Med.* 2003, **348**:24–32.

Mannon, P.J. *et al.*: **Anti-interleukin-12 antibody for active Crohn's disease.** *N. Engl. J. Med.* 2004, **351**:2069–2079.

Podolsky, D.K.: **Inflammatory bowel disease.** *N. Engl. J. Med.* 2002, **347**:417–429.

Rakoff-Nahoum, S. *et al.*: **Recognition of commensal microflora by Toll-like receptors is required for intestinal homeostasis.** *Cell* 2004, **118**:229–241.

14

Transplantation Immunology, Tumor Immunity and Vaccination

There are many circumstances in which it is desirable to manipulate the immune response, and in the course of this book we have encountered some of the main therapeutic agents used to suppress inflammation, as well as those targeted more specifically to components of particular immune responses, for example in hypersensitivity reactions. In this chapter, we discuss the specific problems posed by organ and tissue transplantation, where the adaptive immune response must be suppressed to prevent rejection; and vaccination, in which natural immune responses are prophylactically elicited to protect individuals and populations from infectious disease. We also discuss the special issues to be confronted in the attempt to exploit immune mechanisms against cancer, and describe the main types of inherited immune deficiency, which have helped to identify key components of immune responses.

Modulating the immune response is often important for improving human health

In this book, we have seen how the immune system provides defense against infections by bacteria, viruses, fungi and parasites and how sometimes these immune defenses are mistakenly directed at the organism's own components or at harmless environmental antigens. In this chapter, we examine other situations where medical needs require modulation of the immune system, either to make it work better or to block it. First, we consider the special case of inherited immunodeficiency diseases, in which there is a defect in one or more components of the immune system. These diseases are interesting for what they tell us about how the immune system protects us against particular infections, a point that has been made throughout the book but is summarized here. Then we move on to discuss circumstances in which it is advantageous to boost immune responses of normal individuals to vaccinate against disease and combat cancer, or to inhibit them to prevent rejection of organ or tissue grafts.

Vaccination is an important tool in the fight against infectious diseases

Vaccination was the first major contribution of immunology to the improvement of human health, being developed by Jenner in the 18th century and Pasteur in the 19th (Figure 14-1a and b). Although the immune system is remarkably effective in providing protection against infectious agents, protection is not perfect and some pathogens have evolved mechanisms for outflanking the immune response against them to the point where they can cause severe disease and even death, particularly to those at the ends of the age spectrum. Vaccination, by pre-arming the adaptive immune system, prevents many of these diseases and saves millions of lives every year. But there remain a number of devastating infectious diseases for which we do not have effective vaccines. Among these are tuberculosis, AIDS and malaria, which cause chronic infections in large numbers of people, and many deaths. In this chapter we describe the requirements for a successful vaccine and discuss some of the reasons for obstacles to success where an effective vaccine has yet to be developed.

A further goal that has yet to be met is the vaccination of cancer patients with antigens from their cancers so as to enlist the immune system in the removal of residual cancer cells remaining after surgery, irradiation, and/or chemotherapy. It has been established that many cancers express antigens that can be recognized by the immune system, and increasing evidence supports the concept of immune surveillance, the idea that the immune system eliminates many nascent cancerous cells before they grow into tumors. This is especially true for those cancers, currently estimated to be approximately 17%, that result from infection with viruses that subvert cellular growth control or induce a chronic inflammation that promotes the development of cancer, but it is true for other types of cancers as well.

Immunosuppression is often required for the success of cell and organ therapies

Many therapeutic procedures depend on cell and tissue grafts. Blood transfusion is commonplace, and organ transplantation is increasingly successful. More recently, much effort has been focused on the possibility of harnessing stem cells to regenerate lost or defective tissues. All of these therapeutic strategies are limited by immunological recognition and rejection of the transferred cells or organs.

Blood transfusion, a mainstay of current medicine, became practicable when Landsteiner (Figure 14-1c), in the early 20th century, discovered the ABO antigen system. Subsequently other, less important but still significant, blood group antigens were discovered. Transplantation of solid organs has been a greater challenge, yet many organs are now successfully transplanted, including kidney, liver, heart, lung, and cornea, to name just a few. Whereas for blood transfusion antibodies represent the major rejection mechanism, for solid organs, cell-mediated immune rejection is often even more important, and it is in this context that the genetic locus encoding the major histocompatibility complex proteins was originally discovered by George Snell in the mid-20th century (Figure 14-1d). Moreover, the extremely high degree of polymorphism of MHC alleles within human populations means that most organ transplantation must be done in cases where the donor and recipient are not completely matched at these loci. Transplantation is therefore highly dependent upon effective immunosuppressive regimens to prevent immunological rejection.

Such regimens are now remarkably successful, but there remain life-threatening complications. Severe infections can occur in immunosuppressed patients after an organ transplant, and in the case of bone marrow transplantation, immune cells in the transplanted tissue (recall that some effector and memory cells home to bone marrow) may attack cells of the transplant recipient, a phenomenon known as *graft-versus-host disease*. Thus, the development of successful strategies for establishing immunological tolerance, as described in Chapter 12, remains an important goal of efforts to improve cell and tissue therapies.

(a)

(b)

(c)

(d)

Figure 14-1 Key historical figures
(a) Edward Jenner (1749–1823) discovered the principle of vaccination in the late 18th century when he used cowpox virus, which is harmless to humans, to protect against infection with the related smallpox virus. **(b)** Louis Pasteur (1822–1895) discovered that pathological agents grown in culture sometimes lose their ability to cause disease in their natural hosts, and that such attenuated pathogens can therefore be used for safe vaccines. This led to the further realization that dead or inactivated organisms could be used as even safer alternatives. **(c)** Karl Landsteiner (1868–1943) discovered the major blood group antigens early in the 20th century, opening the way for successful blood trans-fusions. **(d)** George Snell (1903–1996) discovered the human major histocompatability complex genes, a crucial step in the success of tissue transplantation from one individual to another, as matching major histocompatability complex alleles between donor and recipient can greatly reduce rejection of the grafted tissue. Panels (a) and (b) courtesy of the Wellcome Library, London. Panels (c) and (d) courtesy of the Nobel Foundation.

14-1 Immunodeficiencies

Genetic deficiencies in components of immunity often increase susceptibility to particular infections

Individuals with severe and recurrent infections starting early in life often are found to have genetic deficiencies in key components of the adaptive immune system or, less frequently, in components of innate immunity. Over 200 distinct immunodeficiencies are now recognized, and the genetic cause has been determined for at least 120 of them. The more common and some of the rare but informative immunodeficiencies are included in Figure 14-2. The types of infections that occur in individuals with particular immunodeficiencies have illuminated the biologically important role of the affected component of the immune system, since control of particular infectious agents is often highly dependent on one specific immune mechanism.

Mutations that strongly compromise both cell-mediated immunity and humoral immunity give rise to diseases categorized as severe combined immunodeficiencies (SCID). In the most severe forms of SCID, adaptive immunity is essentially absent, whereas in other cases, some

Figure 14-2 Table of selected inherited immunodeficiency diseases CMI: cell-mediated immunity; Orai1: a component of the calcium channel in the plasma membrane of T cells that contributes to elevation of intracellular calcium after TCR engagement; WASP: Wiskott–Aldrich syndrome protein.

Inherited Immunodeficiency Diseases

Type of immunodeficiency	Defective component	Process affected	Types of infections
Severe combined immunodeficiency			
lymphocyte development	γ_c cytokine receptor or Jak3	cytokine responses	bacteria, fungi, viruses
	IL-7Rα chain	T cell development	bacteria, fungi, viruses
	adenosine deaminase (ADA)	survival	bacteria, fungi, viruses
	RAG-1, RAG-2, Artemis	V(D)J recombination	bacteria, fungi, viruses
T cell development/activation	ZAP-70, Orai 1	TCR signaling	bacteria, fungi, viruses
Combined immunodeficiency (usually some residual T cell function)			
lymphocyte development	purine nucleoside phosphorylase	survival	bacteria, fungi, viruses
	DiGeorge syndrome	thymic development	bacteria, fungi, viruses
lymphocyte activation	WASP	antibody responses, CMI	encapsulated bacteria, some fungi, viruses
CD4 T cell development	MHC class II transcription	antigen presentation to CD4 T cells	bacteria, fungi, some viruses
Humoral immunity			
B cell development	Btk	pre-BCR, BCR signaling	extracellular bacteria
B cell activation	CD40L, CD40	T cell help for germinal center reaction	extracellular bacteria; some CMI defects
	AID	Ig class switch and hypermutation	extracellular bacteria
	ICOS	T cell help	extracellular bacteria
	BAFF R	mature B cell survival	extracellular bacteria
	(unknown)	class switch to IgA	bacteria infections of respiratory, gastrointestinal and urogenital tracts
complement	C3, factor H, factor I	major functions of complement	extracellular bacteria
phagocyte killing	phagocyte oxidase	reactive oxygen species production	subset of extracellular bacteria
Cell-mediated immunity			
CD8 T cells	CD8α, TAP1/TAP2, tapasin	cytotoxic T cell antigen recognition	respiratory tract
T_H1 immunity	IFN-γR, IL-12, IL-12R receptor	macrophage activation by T_H1	mycobacteria (other intracellular bacteria)
NK + T cell immunity	SAP	signaling by some CD2 family members	Epstein–Barr virus
Innate immunity			
inflammatory response	IRAK4	TLR signaling	bacteria
	CD18 ($\beta2$ integrins)	leukocyte extravasation, T cell adhesion	bacteria, fungi
	NEMO (IKK γ subunit)	NF-κB activation, inflammatory cytokines, antibody production	bacteria, some viruses
NK cell defect	(unknown)	absence of NK cells	herpesviruses

References

Aiuti, A. *et al.*: **Gene therapy for adenosine deaminase deficiency.** *Curr. Opin. Allergy Clin. Immunol.* 2003, **3**:461–466.

Buckley, R.H.: **Primary immunodeficiency diseases** in *Fundamental Immunology* 5th ed. Paul, W.E. ed. (Lippincott Williams & Wilkins, Philadelphia, 2003), 1593–1620.

Casanova, J.-L. and Abel, L.: **Genetic dissection of immunity to mycobacteria: the human model.** *Annu.* *Rev. Immunol.* 2002, **20**:581–620.

Casanova, J.-L. and Abel, L.: **Inborn errors of immunity to infection: the rule rather than the exception.** *J. Exp. Med.* 2005, **202**:197–201.

Cavazzana-Calvo, M. *et al.*: **Gene therapy for severe combined immunodeficiency.** *Annu. Rev. Med.* 2005, **56**:585–602.

Hacein-Bey-Abina, S. *et al.*: **Sustained correction of X-linked severe combined immunodeficiency by ex vivo gene therapy.** *N. Engl. J. Med.* 2002,

346:1185–1193.

Ku, C.-L. *et al.*: **Inherited disorders of human Toll-like receptor signaling: immunological implications.** *Immunol. Rev.* 2005, **203**:10–20.

Leonard, W.J.: **Cytokines and immunodeficiency diseases.** *Nat. Rev. Immunol.* 2001, **1**:200–208.

Ochs, H.D. *et al.*: *Primary Immunodeficiency Diseases: A Molecular and Genetic Approach* 2nd ed. (Oxford University Press, Oxford, 2006).

T cell-independent antibody responses remain but T cell immunity is completely absent. Diseases in which a small amount of residual T cell immunity remains are sometimes referred to as combined immunodeficiency, to indicate that the susceptibility to infection is somewhat less than in SCID. The majority of SCID patients have defects in one of seven genes, altogether adding up to between 1/50,000 and 1/100,000 live births. The most common of these is called X-SCID, which is caused by mutation of the γ_c cytokine receptor subunit (see section 2-8). This subunit combines with other polypeptides to form receptors for IL-2, IL-4, IL-7, IL-9, IL-15 and IL-21. Of these, the defective ability to respond to IL-7 is responsible for the complete block in T cell development. These individuals also lack NK cells, as a result of non-responsiveness to IL-15, and have B cells with poor functionality, because of defects in responses to IL-2, IL-4 and/or IL-21. An essentially identical form of SCID results from mutations in the gene encoding Jak3, which is a protein kinase that binds selectively to the cytoplasmic tail of γ_c and is required for its signaling function. Another form of SCID results from defects in the IL-7Rα chain, which combines with γ_c to confer responsiveness to IL-7.

Although the genes encoding Jak3 and the IL-7Rα chain are on autosomal chromosomes, these forms of SCID account for approximately 6% and 9% of SCID, respectively, from one American center. X-SCID accounts for about 45% of SCID, because of the ease of transmission of an X-linked disease to males. Also autosomally encoded is adenosine deaminase, which accounts for 16% of SCID. In this defect in purine metabolism, toxic intermediates build up that interfere with the survival of developing lymphocytes. Mutations in components of V(D)J recombination (RAG-1, RAG-2 or Artemis) account for another 3–10% of SCID, although Artemis defects are the predominant form of inherited SCID in several groups of native North Americans (Athabascans, Navajo and Apache). In general, SCID can be treated very effectively by bone marrow transplantation. In adenosine deaminase (ADA) deficiency, injection of the enzyme into the blood is also effective, although less successful than bone marrow transplantation. In addition, gene therapy has been successful for treating the ADA-deficiency form of SCID. In X-SCID there has also been considerable success with gene therapy, but also a high incidence of leukemia as a side effect, due to integration of the vector sequences next to a cancer-promoting gene (an oncogene) (Figure 14-3), so this therapeutic effort is currently reserved for patients who are not transplant candidates.

Defects that compromise some or all of T cell function can result from defects in TCR signaling, from deficient MHC class II gene expression, from defects in MHC class I antigen presentation, from defects in induction of interferon-γ production, and also from defects in formation of the thymus (DiGeorge syndrome). Deficient humoral immunity arises from defects in B cell development, in B cell activation through CD40 and CD40L, in Ig gene affinity maturation and class switch, in complement components, and in phagocyte killing of internalized bacteria. For defects in the production of antibodies, therapy is to provide injections at roughly monthly intervals of intravenous immune globulin (IVIG). IgG produced from many pooled donors contains specific antibodies against many pathogens and is broadly protective.

Some immunodeficiencies are acquired

In addition to the inherited immunodeficiencies, there are also acquired immunodeficiencies, of which the most important is AIDS, caused by HIV-1 (see section 10-4). Also of interest in this category is that loss of the spleen as a result of a traumatic injury leads to an increased risk of bacteremia, reflecting the role of the macrophages and/or marginal zone B cells of the spleen in protecting against microbial infections that reach the blood.

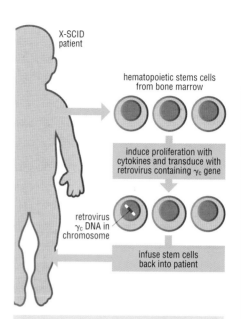

Figure 14-3 Gene therapy for X-SCID or ADA-SCID X-linked severe combined immunodeficiency (X-SCID) results from a defect in the gene encoding the γ_c cytokine receptor. Most such individuals can be successfully treated by bone marrow transplantation using a donor who is identical or partly mismatched at the MHC locus. However, a significant fraction of patients are not successfully treated. Thus, an effort was made to treat newly diagnosed infants with X-SCID by gene therapy. Hematopoietic stem cells were obtained from the patient's bone marrow and infected with a defective retrovirus encoding the γ_c gene in place of normal virus genes. Upon infection, the retrovirus made a DNA copy of its genome and inserted this copy into a cellular chromosome at a random location. These cells were then infused back into the patient. This procedure was successful in most of the infants, but three of the first ten developed a T cell leukemia. Two of these leukemias could be treated by chemotherapy, but this therapeutic modality is currently limited to infants failing bone marrow transplant therapy. For details, see Hacein-Bey-Abina, S. *et al.: N. Engl. J. Med.* 2002, **346**:1185–1193. ADA-SCID has also been treated by gene therapy, and so far has not exhibited leukemia as a side effect.

Reith, W. and Mach, B.: **The bare lymphocyte syndrome and the regulation of MHC expression.** *Annu. Rev. Immunol.* 2001, **19**:331–373.

Successful vaccines depend on the production of antibody

Vaccination is by far the most important contribution of immunology so far to public health. Smallpox has been eradicated worldwide, and other once-common diseases of childhood—polio, measles, mumps, rubella, diphtheria and pertussis—have largely disappeared in the industrialized West. The success of vaccines against these diseases is due to their ability to induce high-affinity neutralizing antibodies that rapidly bind to the pathogen and prevent it from adhering to or entering host cells (Figure 14-4), or to neutralize toxins. These antibodies are produced by long-lived plasma cells that are generated by interactions of B cells with T helper cells during the initial exposure to antigen, or by the progeny of memory B cells generated in the same way (see section 6-12), and thus depend on the induction of effective T cell help. Some vaccines also elicit CD8 T cells, and some elicit T cell-independent antibody production by B cells; but the most important property of the existing successful vaccines is the ability to induce T cell-dependent neutralizing antibodies.

The discovery of vaccination, usually attributed to Jenner in 1796, considerably predates the discovery of antibodies (in the late 19th century) or the elucidation of the role of helper T cells (in the mid-20th century), and until recently the development of vaccines has remained largely empirical. In this section we discuss the requirements for a successful vaccine; and in the next sections, we describe two recent successes in the rational design of new pediatric vaccines, and examine the reasons for the failure so far of efforts to develop effective vaccines against some important infections, including AIDS and malaria.

Adjuvants are required to potentiate and amplify immune responses to vaccines

A successful vaccine stimulates an effective adaptive immune response directed at appropriate target antigens without itself causing disease. Edward Jenner achieved this through the use of the animal virus vaccinia to immunize against the related human smallpox virus. Because the two viruses are closely related, they can be recognized by the same immune cells; but unlike the smallpox virus vaccinia does not cause disease in humans. Thus the vaccine is both antigenic and safe. The third crucial property of vaccinia virus is that as a natural pathogen, it contains components that are recognized by the cells of the innate immune system and is thus able to induce the production of cytokines and activation of dendritic cells, which are required to launch an effective adaptive immune response: that is, it is **immunogenic**.

Almost 100 years later, Pasteur discovered the principle of **attenuation**, whereby pathogenic agents that become adapted to growth in culture can sometimes, in adverse conditions or in cells from species they do not normally infect, lose the ability to cause disease in their natural host. This opened up the possibility of developing safe vaccines directly from the target pathogen, and soon led to the realization that killed or inactivated organisms could offer an even safer alternative. The advantage of a live vaccine is that the attenuated agent is still capable of replication and is thus able to persist for longer and thereby induce better lymphocyte proliferation, and consequently a larger pool of memory cells and long-lived plasma cells, than does a killed or inactivated vaccine. (Note that the term live vaccine commonly includes viral vaccines where the virus has not been inactivated, although strictly speaking viruses cannot be described as live.) The disadvantage is that, for the same reason, the live vaccine is more likely to give rise to inflammatory reactions or mild symptoms of disease. In the early vaccines, these could be damaging or dangerous; and while in modern attenuated vaccines the danger of serious adverse effects in an individual with an intact immune system is extremely small, any adverse

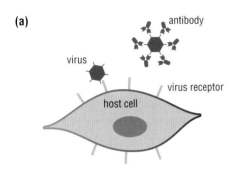

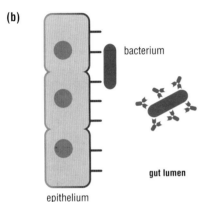

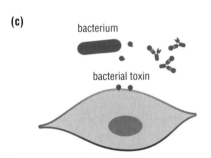

Figure 14-4 Actions of neutralizing antibodies (a) Viruses are prevented from entering cells by neutralizing antibodies that bind to proteins on their surface that are required for recognition of or entry into the host cell. (b) Often bacteria must adhere to mucosal epithelial cells in order to establish an infection: this is prevented by neutralizing antibodies that block the sites through which they adhere. (c) Neutralizing antibodies can also bind to bacterial toxins and inactivate them.

Definitions

adjuvant: component or property of a vaccine or inoculum that makes it **immunogenic**. The term was originally used for a component added to a vaccine or inoculum in order to produce an adaptive immune response, but it is now recognized that vaccines or inocula may have inherent adjuvant or immunogenic properties.

attenuation: (of a pathogen) loss of pathogenicity, usually through adaptation to growth in culture in adverse conditions or in cells from a species other than

that of the normal host. Attentuated pathogens are the basis of many vaccines.

immunogenic: able to induce an adaptive immune response. Many antigens recognized by lymphocytes of the adaptive immune system fail to elicit immune responses in the absence of **adjuvants** that promote adaptive immune responses, often by activating cells of innate immunity.

subunit vaccine: vaccine composed of defined components of a bacterium, virus or parasite rather than the whole infectious agent.

virus-like particle (VLP): particle that spontaneously assembles from viral coat proteins in the absence of other viral components. Virus-like particles generated from recombinant coat proteins are used in vaccines against hepatitis B virus and papillomavirus.

VLP: see **virus-like particle**.

effects will decrease the acceptability of a vaccine, especially one administered to infants in the absence, in an existing vaccinated population, of any immediate threat from the disease. In individuals with compromised immunity, live vaccines may cause disease and are dangerous. This has led to an increasing focus on the development of vaccines based on defined components of the pathogen, purified from the whole organism or produced by recombinant DNA technology: these are known as **subunit vaccines**. Figure 14-5 shows all the vaccines at present licensed for human use, grouped according to vaccine type.

On the whole, the safer a vaccine is the less effective it is likely to be in inducing an immune response. With whole-organism vaccines, live vaccines are generally more effective than killed or inactivated ones. Subunit vaccines are particularly safe, in part because they are free of components that can induce damaging inflammatory responses but may also be necessary to activate the innate immune cells required to initiate the adaptive response. These deficiencies are overcome in two ways. The first is to administer the vaccine in several doses over a period of weeks or months, thereby extending the period of stimulation and expanding the pool of memory cells or long-lived plasma cells. The second is by means of an **adjuvant**. An adjuvant is any component or property of a vaccine or inoculum that promotes or amplifies an adaptive immune response. The best-understood adjuvants are known ligands for the Toll-like receptors (TLRs) that recognize conserved components of bacteria and viruses and induce immune responses by activating macrophages and dendritic cells. LPS, for example, is an extremely potent adjuvant that is used in the laboratory to induce immune responses to non-immunogenic antigens, but it cannot be used clinically because of its powerful inflammatory effects, though we shall see later that safer derivatives have been developed and are on trial for human use. CpG DNA, also a TLR ligand (see section 3-10), is in clinical trials as an adjuvant in human vaccines. New adjuvants are the focus of much research interest, not only because they are important to the efficacy of vaccines that are weakly immunogenic, but also because they can influence the type of immune response induced: CpG DNA, for example, preferentially elicits T_H1 responses, and this may be important for the development of vaccines against infections that are controlled by T cells rather than by antibodies. The addition of cytokines to vaccines is also under investigation as an adjuvant strategy for both amplifying and directing immune responses. At present, however, only two adjuvants are licensed for human use: alum (a general term for precipitate of aluminum hydroxide and aluminum phosphate), which is by far the most commonly used; and an oil-in-water emulsion known as MF59. Both of these promote the production of neutralizing antibodies but they are not known to activate dendritic cells and this has led to a search for a possible distinct type of innate immune cell through which B cell activation is specifically promoted.

In practice, most vaccines have inherent adjuvant properties without alum, and these determine the number of doses and the amount of additional adjuvant required. Thus for example, the vaccines against yellow fever and anthrax are both live attenuated vaccines, but the yellow fever vaccine protects for 35 years after only one dose, whereas anthrax requires six doses and protects for about a year. The exceptional effectiveness of the yellow fever vaccine is thought to be due to the large number of TLR ligands contained in the yellow fever virus. Adjuvant properties can also be due to structural features of the vaccine. Thus subunit vaccines can be made from viral coat proteins that spontaneously assemble into **virus-like particles (VLPs)** and are immunogenic even without added adjuvant, presumably because of the avidity with which the repeated arrays of viral proteins bind to the antigen receptors of B cells. The recombinant vaccines against hepatitis B and papillomavirus, which we discuss later, are of this type. Clearly, an adjuvant is no single defined entity, and can work by any or all of the mechanisms discussed in earlier chapters of this book for promoting adaptive immune responses.

References

Ada, G.: **Vaccines and vaccination.** *N. Engl. J. Med.* 2001, **345**:1042–1053.

Jordan, M.B. *et al.*: **Promotion of B cell immune responses via an alum-induced myeloid cell population.** *Science* 2004, **304**:1808-1810.

O'Hagan, D.T. and Rappuoli, R.: **Novel approaches to pediatric vaccine delivery.** *Adv. Drug Deliv. Rev.* 2006, **58**:29–51.

Plotkin, S.A.: **Vaccines: past, present and future.** *Nat. Med.* 2005, **11**:S5–S11.

Pulendran, B. and Ahmed, R.: **Translating innate immunity into immunological memory: implications for vaccine development.** *Cell* 2006, **124**:849–863.

Approved Human Vaccines

Live vaccines

Vaccine	Type
smallpox	related animal virus
BCG	related animal bacterium attenuated
rabies	attenuated
anthrax	
yellow fever	
oral polio vaccine	
measles	
mumps	
adenovirus	
varicella	
rotavirus 89–12	
rubella	
live influenza	
Ty21a typhoid	
influenza seed	reassortants
live influenza	
rotavirus (bovine-human)	

Inactivated vaccines

Vaccine	Type
typhoid	inactivated whole organism
cholera	
plague	
whole-cell pertussis	
influenza	
inactivated polio vaccine	
hepatitis A	

Subunit vaccines

Vaccine	Type
Japanese encephalitis	extracts or subunits
influenza	
anthrax	
cell-culture rabies	
diphtheria	toxoids
tetanus	
anthrax	
pneumococcal	capsular polysaccharides
typhoid (Ty Vi)	
H. influenzae type b	protein-conjugated
pneumococcal	capsular polysaccharides
meningococcal	
staphylococcal	
hepatitis B	purified or recombinant
human papillomavirus	proteins
acellular pertussis	
Lyme disease	
hepatitis A/B	combined vaccines
tetanus/diphtheria/pertussis	

Figure 14-5 Table of licensed human vaccines Not all the vaccines in the table are licensed in all countries. For example, oral polio vaccine is not used in developed countries because of the extremely rare occurrence of reversion of the attenuated virus to the neurovirulent form in which it causes paralysis in unvaccinated individuals; but because it is inexpensive and easy to administer it is used in less developed countries where the benefits overwhelmingly outweigh the risk. Reassortants are vaccines against highly variable viruses made by combining the currently circulating variants to produce a vaccine that is broadly protective. The seed virus is the reassortant that is selected for growth into the vaccine. Toxoid, capsular polysaccharide and conjugated vaccines are discussed in the next section.

Declining incidence of infectious disease may encourage a decline in vaccination

A major incentive to the development of safer and more effective vaccines is the need to maintain the level of vaccination in the populations of developed countries at the 95% thought to be required for *herd immunity*—that is, the level of immunity in a population at which the disease cannot spread. Parents of young children in the industrialized West today have no experience of diseases such as diphtheria, pertussis, polio and measles that until the second half of the 20th century affected hundreds of thousands of children every year in the USA and killed or disabled thousands. Minor ill effects from vaccines can thus seem today more threatening than the possibility of disease, so that vaccination with the fewest possible injections and adverse reactions becomes important to encourage compliance by parents in infant vaccination schedules. The damaging effects of failure to maintain population immunity can be illustrated by the experience of the UK in the 1970s, when press reports attributing brain damage to the pertussis vaccine against whooping cough led to a drop in immunization levels to 30%. Subsequent analysis showed no evidence linking brain damage to the vaccine, but the decline in levels of immunity led to two outbreaks of whooping cough and more than 30 deaths. Pertussis, which was at that time administered as a whole killed organism, has since been developed as a subunit vaccine that is less prone to adverse reactions. We describe below two vaccines, one containing the pertussis subunit vaccine, designed for safety and efficacy in infants.

The combined diphtheria/tetanus/pertussis vaccine is based on toxoids with added adjuvants

Children are now routinely immunized against at least ten diseases in the first year or two of life, and since each vaccine requires two to four doses, this would mean a large number of injections for a small child if they were all administered separately. For this reason, measles, mumps and rubella are administered together as the so-called MMR vaccine, and diphtheria, tetanus and pertussis are administered together as DTP, or more recently DTaP. DTaP stands for diphtheria, tetanus and acellular pertussis, and is a subunit vaccine based on chemically inactivated toxins, or **toxoids**, of each bacterium. The toxoids have lost their toxic properties but still have the same antigenic properties as the active toxin, and antibodies that block their toxic action are a highly effective way of preventing disease.

Diphtheria and tetanus toxoids have been in use since the early 20th century as vaccines, and these were initially combined with whole killed pertussis to make the DTP vaccine. This vaccine has the advantage that pertussis has strong adjuvant properties, but for that reason it also has a relatively high frequency of adverse effects such as pain at the site of injection, fever and malaise. This (along with the publicity given to the possibility of more serious adverse effects) has led to the development of an acellular pertussis vaccine consisting of pertussis toxoid with other pertussis components implicated in bacterial adherence to host cells and therefore good targets for neutralizing antibodies. This acellular preparation is now used with the diphtheria and tetanus toxoids in the DTaP vaccine, and with added adjuvant to compensate for the loss of immunogenicity conferred by the whole persussis organism, the vaccine is as effective as DTP.

The success of the childhood vaccines has raised the question of how long immunity can last in the absence of natural boosters in the form of circulating pathogens (see section 5-16). It is thought, for example, that persistent cough in adults is often due to pertussis infections, and

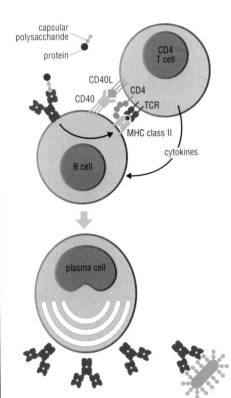

Figure 14-6 Conjugate vaccine against *Haemophilus influenzae* type b The capsular polysaccharide of *Haemophilus influenzae* type b is conjugated with a protein, usually diphtheria or tetanus toxoid, so that B cells recognizing the polysaccharide also internalize the protein, which is degraded and presented to T helper cells as peptide fragments bound to MHC class II molecules. The helper T cells in turn activate the B cells to produce antibody against the polysaccharide.

Definitions

conjugate vaccine: vaccine consisting of a polysaccharide from a capsulated bacterium chemically linked to a protein, usually a bacterial toxoid. Conjugate vaccines are used to elicit T cell-dependent antibody responses to capsular polysaccharides that on their own elicit largely T cell-independent antibodies: helper T cells generated in response to the protein activate B cells that recognize the polysaccharide and internalize the conjugate, and present peptide fragments to the helper T cell. Conjugate vaccines are particularly important for

protecting infants in whom the T cell-independent response to polysaccharides is absent for the first two years of life.

toxoid: chemically inactivated bacterial toxin that is no longer harmful. Toxoids that retain the antigenic features of the original toxin are used in vaccines to induce neutralizing antibodies against the original toxin.

References

Holmgren, J. and Czerkinsky, C.: **Mucosal immunity and vaccines.** *Nat. Med.* 2005, **11**:S45–S53.

Neutra, M.R. and Kozlowski, P.A.: **Mucosal vaccines: the promise and the challenge.** *Nat. Rev. Immunol.* 2006, **6**:148–158.

Nossal, G.J.V.: **Vaccines** in *Fundamental Immunology* 5th ed. Paul, W.E. ed. (Lippincott Williams & Wilkins, Philadelphia, 2003), 1319–1369.

booster injections are recommended for adolescents in the USA. Systematic booster programs may be required for other vaccines as natural disease declines.

Conjugate vaccines activate T cell help for B cell responses to polysaccharides

The pediatric vaccine against *Haemophilus influenzae* type b (Hib) is the best example so far of the application of basic immunology to vaccine design. Hib is an encapsulated bacterium (see section 3-2) that is the principal cause of bacterial meningitis in small children and until recently caused the death of 1,000 a year in the USA. The capsular polysaccharides of the encapsulated bacteria are attractive as subunit vaccines because they are potently immunogenic and are an effective target of protective antibodies. However, antibody responses elicited by purified polysaccharides are T cell independent (see section 6-9) which has several disadvantages: affinity maturation to produce high-affinity IgG, and immune memory, are limited; and most important, it means that polysaccharide vaccines cannot be used in young infants because T cell-independent responses to polysaccharides do not develop until the age of about two years. This deficiency was overcome in the early 1990s by the development of **conjugate vaccines**, in which the capsular polysaccharide is chemically coupled, or *conjugated*, to a protein—usually diphtheria or tetanus toxoid, which are already approved for safety. Peptides derived from the protein component of the conjugate are presented to T cells and a helper response is thereby generated. B cells recognizing the polysaccharide component internalize the conjugate and in turn present peptides from the protein to the helper T cells, which activate the B cell to differentiate into a plasma cell secreting antibodies that can recognize the polysaccharide on the encapsulated bacterium (Figure 14-6).

The route of immunization is important in determining the type of protection elicited

Most vaccines in current use are injected. They induce the production of high-affinity IgG, which is highly effective in providing systemic protection—that is, in protecting the blood and tissues—and in preventing severe disease from invasive pathogens. Most pathogens, however, enter through the mucosa, where infection elicits large quantities of dimeric IgA adapted to protect mucosal surfaces (see section 6-2); and although IgG produced in response to injected vaccines can be transferred across the mucosa of the respiratory and uro-genital tracts, it is not transferred across the mucosa of the gut, which is by far the largest mucosal surface and portal of entry for pathogens. Because effector and memory lymphocytes home to the tissues in which the adaptive immune response was initiated (see sections 5-8 and 5-15), many infections may be most effectively prevented by vaccines administered to the mucosa. These infections include those caused by *Vibrio cholerae* (the cause of cholera) and enteropathogenic *Escherichia coli*, which colonize the epithelial surface; viruses such as rotavirus (a major cause of diarrhea and infant death in the developing world) and influenza virus, which infect cells of the mucosal epithelium; and *Shigella flexneri* (another major cause of diarrhea) and *Salmonella typhi* (the cause of typhoid), which are invasive and infect the lamina propria of the gut.

Oral vaccines exist for polio (see Figure 14-5), *V. cholerae*, rotavirus and *S. typhi*. A nasal vaccine exists for influenza virus, which enters through the respiratory mucosa. (Oral administration does not elicit immune responses in the lung, and nasal administration does not elicit immune responses in the gut.)

Mucosal protection from *V. cholerae* and enteropathogenic *E. coli* is essential because they do not penetrate the tissues and their pathogenic effects are due to their actions in the gut; for invasive pathogens, which enter through the mucosa but whose principal pathogenic actions are in the tissues, an injected vaccine prevents systemic disease when the pathogen has entered the tissues, and the advantage of mucosal immunity is to abort the infection at the entry point so that it cannot spread to other individuals.

The defense of mucosal barriers by vaccines is an area of particular current interest because of the possibility that it could be used in the conquest of HIV, whose principal site of replication early in infection is the mucosal lymphoid system of the small intestine (see section 10-5), and whose evasive devices pose exceptional problems for the design of vaccines, as we see in the next section.

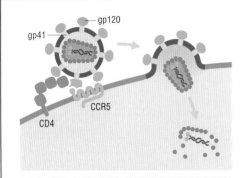

Figure 14-7 CD4 and CCR5 binding sites on HIV gp120 Highly schematic diagram of gp120/gp41 binding to CD4 and CCR5 at the surface of a cell: when the protein binds to CD4 it undergoes a conformational change that exposes the binding site for CCR5 and allows membrane fusion mediated by gp41 so that the viral contents can enter the cell.

Persistent pathogens present a special challenge for vaccine design

The successful vaccines we have discussed so far are all against acutely infecting pathogens—that is, pathogens that replicate rapidly and persist (when they are not lethal) by moving on to another host before they are eliminated by the adaptive immune response. Recovery from these infections is accompanied by lasting immunity. However many pathogens cause persistent or chronic infections by adapting to or evading host immune responses; they may either eventually kill the host or establish lifelong infection without disease, but are never cleared. Epstein–Barr virus and cytomegalovirus are two such pathogens that we have already discussed (see section 5-16). *Mycobacterium tuberculosis* is another (see section 9-7): although it can kill, especially if the infected individual is immunosuppressed, in a healthy population only about 10% of those infected develop active tuberculosis: in the rest, it persists for life without causing overt disease. Parasitic infections are generally of this type, as is AIDS. The widely used BCG vaccine for tuberculosis at best provides some protection for infants, but does not protect from pulmonary tuberculosis in adolescents and adults, which is the major cause of death; and there is no vaccine for AIDS or any of the parasitic infections. The principle of vaccination is to exploit the natural immune response. The challenge of these diseases is to establish whether it is possible to design a vaccine that can do better than the natural immune response. The issues are particularly well understood in the case of the human immunodeficiency virus (HIV) that causes AIDS, and are discussed here for AIDS and malaria.

HIV escapes both neutralizing antibodies and cytotoxic T cells

The natural immune response to HIV includes neutralizing antibodies and cytotoxic T cells, but because of the mutability of the virus, new viral variants arise faster than the adaptive immune system, increasingly crippled by viral destruction of CD4 T cells, can generate new lymphocytes that recognize them (see section 10-6). Ideally, a vaccine induces sterile immunity—that is, complete clearance of the infectious organism. For a virus that, like HIV, can establish a latent state inside cells, this is likely to be possible only if the virus can be blocked by neutralizing antibody before it enters cells. This has been shown to be possible in principle for HIV by experiments with SIV, the closely related virus that infects monkeys, in which passive injections with neutralizing antibodies can prevent infection. In practice however the virus has made it extraordinarily difficult.

Neutralizing antibodies act, as we have seen, by binding to the surface molecules whereby the virus enters cells. HIV enters cells by binding to CD4 and CCR5 through the envelope protein gp120 (see section 10-4), which undergoes a conformational change on binding CD4 that leads to a sequence of events culminating in the fusion of viral and cell membranes and the entry of the viral contents into the cell (see Figure 10-13). The CD4 and CCR5 binding sites on gp120 are well protected. The binding site for CD4 is in a deep pocket masked by highly glycosylated hypervariable loops that can (and do) undergo rapid mutation without affecting the function of the protein. The CCR5 binding site is buried in the interior of gp120 until it binds CD4, when it undergoes a conformational change that exposes the CCR5 binding site at the surface of the host cell, where it immediately binds to CCR5 (Figure 14-7). The hypervariable loops are not a good target for neutralizing antibodies because of the extreme mutability of the virus, so effective neutralizing antibodies would have to recognize the masked CD4 binding site or the buried CCR5 binding site. An extensive analysis of antibodies produced against gp120 has yielded only five that are neutralizing, and these are either very weakly neutralizing because of the difficulty of binding to these inaccessible sites, or of such unusual structure that the likelihood of eliciting such antibodies with a vaccine is negligible.

References

Blumberg, B.S.: **Hepatitis B virus, the vaccine, and the control of primary cancer of the liver.** *Proc. Natl Acad. Sci. USA* 1997, **94**:7121–7125.

Chen, B. *et al.*: **Structure of an unliganded simian immunodeficiency virus gp120 core.** *Nature* 2005, **433**:834–841.

Frazer, I.: **God's gift to women: the human papillomavirus vaccine.** *Immunity* 2006, **25**:179–184.

Hill, A.V.S.: **Pre-erythrocytic malaria vaccines: towards greater efficacy.** *Nat. Rev. Immunol.* 2006, **6**:21–32.

McMichael, A.J.: **HIV vaccines.** *Annu. Rev. Immunol.* 2006, **24**:227–255.

Moore, A.C. and Hill, A.V.S.: **Progress in DNA-based heterologous prime-boost immunization strategies for malaria.** *Immunol. Rev.* 2004, **199**:126–143.

Waters, A.: **Malaria: new vaccines for old?** *Cell* 2006, **124**:689–693.

Most HIV vaccines currently under trial are aimed not at eliciting antibodies but at stimulating mucosal cytotoxic CD8 T cell responses targeted at the infected cells, in particular CD4 T cells, in which early viral replication occurs in the gut-associated lymphoid tissue. Most viral vaccines are thought to elicit CD8 T cell responses as well as neutralizing antibodies, and CD8 T cells are known to be important in the control of HIV. Although CD8 T cells too are handicapped by the mutability of the virus, peptides derived from viral proteins offer a wider range of targets that the virus cannot mutate without loss of essential functions, and vaccination to generate CD8 T cells could give the immune system a head start on the pathogen. CD8 T cells however could not be expected to confer sterile immunity without effective neutralizing antibodies: whereas antibodies generated by vaccination are already circulating at the time of infection, effector CD8 T cells would almost certainly have first to be generated from a memory pool and then migrate to the mucosa, so that at best they might be able to control the virus at a level consistent with the survival of the host. Indeed, it is widely believed that the most realistic aim, both for HIV and for other pathogens adapted to long-term survival in human hosts, would be a vaccine that would allow the immune system to suppress disease without eliminating the pathogen, as naturally occurs in the case of, for example, EBV.

Malaria is a complex target for recombinant vaccines

Long-term infection with the malaria parasite can, as in infection with EBV, lead to nonsterile immunity in which the infection is controlled but not eliminated. However, many are killed by the disease before immunity can develop: it is estimated that 1–2 million people die of malaria worldwide every year, most of them children under 5 years old in Africa. The economic toll in those debilitated but not killed is enormous.

Malaria is caused by four species of the protozoon *Plasmodium*, the most important being *P. falciparum*, whose life cycle is summarized in Figure 14-8. It is introduced into the bloodstream by mosquitoes as a *sporozoite*, rapidly entering liver cells where it develops into *merozoites*. Overt disease begins after a few days when the merozoites are released into the bloodstream and invade red blood cells. The principal cause of mortality is cerebral malaria in which infected erythrocytes adhere to blood vessels in the brain through an adhesive molecule, *PfEMP1* (for *P. falciparum* erythrocyte membrane protein 1), produced by the parasite and causing inflammation and blocking the blood supply. Infected erythrocytes cannot be attacked by CD8 T cells because erythrocytes do not express MHC class I molecules, but PfEMP1 is in principle an attractive target for neutralizing antibodies. However, it is highly polymorphic, and the parasite genome contains about 50 copies of the gene encoding it; these recombine in the course of an infection and make the molecule so variable that this approach has been largely abandoned. The merozoites themselves are only exposed so fleetingly as they transfer between cells that most current trial vaccines are aimed at the pre-erythrocyte stages.

Vaccines currently in phase II trials are based on recombinant sporozoite antigens either fused to the hepatitis B coat protein (see section 14-2 and below) to form particles in which the sporozoite antigen is exposed; or expressed from genes inserted in attenuated bird or animal viruses. An important component of the hepatitis B-sporozoite vaccine is an adjuvant called AS02A, containing a derivative of the lipid A component of LPS (see Figure 9-6) that has immunostimulatory properties with less toxicity; and a saponin from tree bark that stimulates both T cell-dependent and T cell-independent antibodies as well as cellular immunity. The broadly immunogenic properties of this adjuvant also make it attractive for the assault on tuberculosis and AIDS. So far, however, only short-term immunity has been achieved with any malaria vaccine.

Cancer can be prevented by vaccines against tumorigenic pathogens

The principal success of vaccines against cancer has been due to vaccination against the hepatitis B virus, one of the commonest causes of human disease and a major cause of liver cancer worldwide. Recently, a vaccine against papillomavirus, the cause of cancer of the uterine cervix, has been approved for human use. Cervical cancer is the second-commonest cancer of females, causing 300,000 deaths a year. Both of these vaccines are recombinant vaccines composed of viral coat proteins assembled into particles (VLPs; see section 14-2). Tumor cells themselves can be thought of as an extreme case of a vaccine target that has adapted to live with its host, and are more difficult to tackle. We discuss in the next section the properties of tumor cells that might enable them to become the targets of immune responses, including immune responses provoked by vaccines.

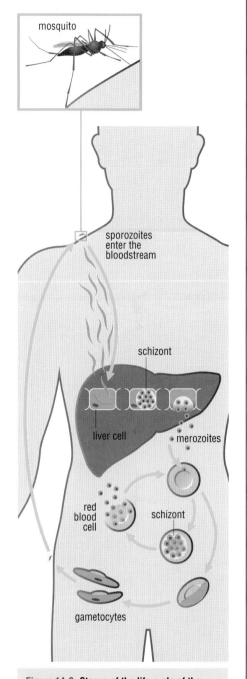

Figure 14-8 Stages of the life cycle of the malaria parasite that could be vaccine targets Malaria parasites enter humans in the bloodstream from mosquito bites, as sporozoites that invade liver cells and there develop into schizonts each containing 10,000–30,000 merozoites before rupturing and releasing the merozoites into the circulation. There they infect red blood cells inside which they multiply to produce more merozoites that infect other red blood cells, and also differentiate into gametocytes, the sexual form in which they infect mosquitoes. The subsequent fusion of the gametes and their development into sporozoites in the mosquito gut is not shown.

The immune system can eliminate many tumors before they become evident

It has been known for many years that the immune system can react against tumors. It has been postulated, therefore, that many tumors are eliminated by the immune system at an early stage, before they become clinically evident. This is the **immune surveillance** hypothesis. For cancers caused by viruses, such as cervical cancer caused by human papilloma viruses, it is easy to see that immune surveillance occurs, but in recent years strong evidence has accumulated that immune surveillance also occurs for common human tumors that are not virally induced. For example, mice lacking lymphocytes ($rag^{-/-}$ mice) or with defects in certain parts of the immune system have a higher incidence of chemically induced and spontaneously arising tumors. In humans, there is likewise strong, although less direct, evidence for a widespread role of the immune system in reducing cancers. For example, immunosuppressed recipients of organ transplants have substantially higher rates of cancers.

Both innate and adaptive immune cells can detect cancer cells

To make a response against tumor cells, the immune system must be alerted to features of tumor cells that distinguish them from normal cells. For the innate immune system, the distinctive feature of tumor cells is likely to be their high rate of apoptosis and tissue remodeling. If there are too many apoptotic cells to be phagocytosed in a timely fashion by the resident tissue macrophages, then some of these apoptotic cells release cellular components into the extracellular space. Some intracellular heat shock proteins released by apoptotic cells and proteolytic fragments of extracellular matrix components generated by active tissue remodeling are thought to be ligands for TLRs and to trigger an inflammatory reaction in this way. In addition, the oxidative stress associated with neoplastic growth of cells typically induces the expression of several stress-induced cell surface proteins recognized by activating receptors of NK cells, NK T cells and certain γδ T cells. These include the MHC class I-like MICA and MICB proteins that activate NK cells and some γδ T cells through NKG2D and the Vδ1 γδ TCR on a subset of γδ T cells (see Chapter 8). Thus, innate immune recognition probably initiates inflammation and cytokine production at the site of neoplastic cell growth. At the same time, the inflammatory cytokines mobilize immature dendritic cells to present antigens associated with tumor cells (see below) to naïve αβ T cells, and thereby initiate the adaptive immune response.

The interferon-γ produced by NK cells, γδ T cells or activated CD8 T cells is a key molecule for immune surveillance. Like RAG-deficient mice, mice in which interferon-γ, its receptor or the downstream signaling component STAT1 has been genetically ablated exhibit higher rates of cancers than normal mice. Interferon-γ is likely to play multiple roles in defense against cancers, but one of the more important is induced upregulation of class I MHC molecules and components of the peptide processing pathway. This response makes tumor cells more readily

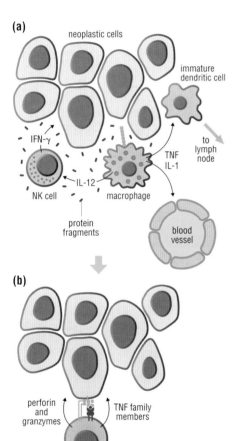

(a)

neoplastic cells

immature dendritic cell

IFN-γ

to lymph node

TNF IL-1

IL-12

NK cell

macrophage

protein fragments

blood vessel

(b)

perforin and granzymes

TNF family members

CD8 T cell

Figure 14-9 Immune response to a tumor (a) Protein fragments released from dying tumor cells can stimulate TLRs on tissue macrophages, leading to the secretion of TNF and IL-1, causing inflammation of neighboring endothelial cells and activation of immature dendritic cells that have taken up debris from dying tumor cells, and of IL-12, which activates NK cells that are attracted to the site by inflammatory signals. NK cells recognize stress-induced ligands on tumor cells and produce interferon-γ (IFN-γ) in response to IL-12. **(b)** IFN-γ promotes the expression of MHC class I molecules and their peptide-loading machinery in the tumor cells, increasing their effectiveness as targets for anti-tumor CD8 T cells, which kill them via perforin and granzyme mechanisms or via production of TNF family members, which act via death domain receptors to induce apoptosis.

Definitions

graft-versus-leukemia effect: the hypothesis that alloreactive T cells from a bone marrow transplant can kill residual leukemia cells remaining after chemotherapy or radiation treatment. This hypothesis is based on enhanced patient survival if mature T cells are not removed from allogeneic bone marrow before transplant, a situation that greatly increases the chance of graft-versus-host disease.

immune surveillance: the hypothesis that the immune system recognizes and eliminates many nascent tumors

before they become clinically evident.

References

Blattman, J.N. and Greenberg, P.D.: **Cancer immunotherapy: a treatment for the masses.** Science 2004, **305**:200–205.

Coussens, L.M. and Werb, Z.: **Inflammation and cancer.** Nature 2002, **420**:860–867.

Dunn, G.P. et al.: **The three Es of cancer immunoediting.**

Annu. Rev. Immunol. 2004, **22**:329–360.

Schreiber, H.: **Tumor immunology** in Fundamental Immunology 5th ed. Paul, W.E. ed. (Lippincott Williams & Wilkins, Philadelphia, 2003), 1557–1592.

recognized by CD8 T cells, which then kill neoplastic cells via perforin and granzymes and possibly also via TNF family members that act through receptors with death domains (see section 2-9). A scenario for the innate and adaptive immune responses to a typical tumor is illustrated in Figure 14-9, but it should be recognized that the details may differ somewhat depending on the type of tumor involved.

There is evidence for the importance of cytotoxic lymphocytes (CD8 T cells and/or NK cells), both from experiments on mice and from human patients. In mice, deficiencies in perforin lead to increased cancer incidence. In human patients, the number of CD8 T cells infiltrating a tumor is positively correlated with the chance that the patient will stay in remission after treatment by surgical removal and chemotherapy or radiation treatment to remove as much of the tumor as possible, suggesting that tumor-specific CD8 T cells function to kill residual tumor cells. The adaptive immune response also frequently leads to the production of antibodies that react to the tumor as well, but their role in immune surveillance is unclear.

Normal, mutant and foreign proteins can serve as tumor antigens

Many tumor-associated antigens recognized by T cells and antibodies have been identified by using tumor-reactive CD8 T cells from patients or tumor-specific antibodies from the patients' sera coupled with recombinant DNA cloning methods. These antigens fall into five categories: (1) viral proteins related to tumor induction (for example, human papilloma virus E6 or E7 oncogene products), (2) mutated cell proteins that may participate in unregulated growth or survival (for example, mutated p53 or Ras, frequently seen in many cancers), (3) overexpressed proteins that may promote tumor growth (for example, Her2/neu on breast cancer cells), (4) differentiation antigens usually but not always characteristic of the cell type from which the cancer arose (for example, tyrosinase expression by melanoma cells) and (5) the so-called cancer-testis antigens, which are ordinarily expressed only in the testis. In some cases, the immune response will react also with antigen expressed in normal tissue, leading to autoimmune phenomena, such as vitiligo (destruction of melanocytes in the skin) in individuals with melanoma.

Just as cancer cells can become resistant to chemotherapy drugs through genetic changes, they can also become resistant to immune recognition or attack, and such resistance mechanisms are likely to explain how many cancers that do arise manage to grow in the face of immune recognition and attack. Indeed, cancers that arise in immunodeficient mice are much more readily rejected when transplanted into normal mice than when transplanted into immunodeficient mice, whereas cancers arising in normal mice grow equally well in both types of mice, indicating that they have already been selected to be resistant to immune surveillance. Neoplastic cells may lose the expression of antigens that are the target of CD8 T cells, or the MHC class I allele that presents the main peptide epitopes; or they may become resistant to interferon-γ. For example, loss of at least one MHC class I allele has been reported in between 40% and 90% of cancers. In one study of lung adenocarcinomas, 25% were found to have lost responsiveness to interferon-γ. Production of the antiinflammatory and immunosuppressive cytokines TGF-β and IL-10 by tumors is frequently observed, and some cancer cells shed their MICA and MICB molecules, which may make them resistant to killing by NK cells.

Cytokines, antibodies and T cells are all potentially useful for cancer immunotherapy

A number of humanized or fully human monoclonal antibodies (see section 6-4) are currently used successfully in cancer therapy (Figure 14-10). Less advanced are efforts to boost T cell immunity to complement conventional therapies for cancer, although as mentioned above, those patients with relatively high numbers of CD8 T cells infiltrating their tumors appear to be doing this without aid. Likewise, treatment of many leukemias involves chemotherapy plus donor bone marrow grafts to replace hematopoietic stem cells, which are sensitive to chemotherapy. We shall see in the next section that mature donor T cells in bone marrow grafts can attack recipient tissue, causing *graft-versus-host disease*, which is potentially lethal. In leukemia patients however such mature donor T cells help remove residual leukemic cells, a phenomenon known as the **graft-versus-leukemia effect**. Currently many efforts are directed to boosting T cell immunity. These include using dendritic cells loaded with tumor material, and/or enhanced with cytokines, blocking of the negative costimulator CTLA4, and *in vitro* expansion and reintroduction of anti-tumor T cells from the patient's blood.

Monoclonal Antibody Immune Therapies

Monoclonal antibody	Molecule recognized	Cancer
Rituximab	CD20	B cell lymphomas
Trastuzumab	Her2/Neu	breast cancer
Gemtuzumab ozogamicin	CD33	AML
Alemtuzumab	CD52	B-CLL
Bevacizumab	VEGF	colon cancer
Cetuximab	EGFR	colon cancer

Figure 14-10 Monoclonal antibody therapeutics currently used to treat cancer Shown are monoclonal antibodies approved in the USA for treatment of cancer patients. Gemtruzumab ozogamicin is a conjugate in which a DNA-damaging antibiotic, calcheamicin, has been chemically coupled to the antibody. The other antibodies are all unmodified antibodies that are thought to work either by blocking the function of the molecule recognized, or by inducing cytotoxicity of the cells expressing the target antigen. Note that in the case of Bevacizumab the target is a protein that participates in angiogenesis, rather than a molecule on the surface of the tumor cells. Blocking angiogenesis inhibits tumor growth by limiting oxygen and possibly nutrients available to the tumor cells. AML: acute myelogenous leukemia; B-CLL: B cell chronic lymphocytic leukemia; Her2 (also called Neu): a close relative of the EGF receptor (EGFR); VEGF: vascular endothelial growth factor.

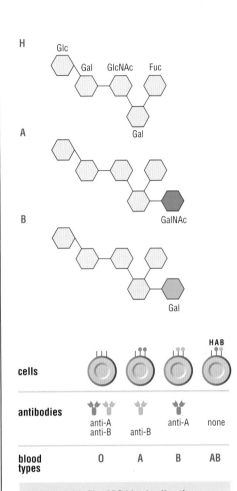

cells

antibodies
anti-A anti-B anti-A none
anti-B

blood
types   O   A   B   AB

Figure 14-11 The ABO blood cell antigen system Top: schematic representation of the carbohydrate structures corresponding to the A, B and H antigens. The H antigen is modified by the addition of a single sugar residue by the allelic glycosyltransferase enzymes that create the A and B antigens. The H antigen is present on the surface of A, B, AB and O individuals, as indicated in the lower panel, because not all H antigens are modified by the A or B glycosyltransferase. The H antigen is only recognized by antibodies made in rare individuals who lack enzymes that make this oligosaccharide structure, which is present on glycolipids and glycoproteins of red blood cells and some other cell types including endothelial cells. Also shown in the lower panel is the antibody to the A and B antigens present in individuals of different blood types.

Immunological rejection is a major barrier to transplantation

In medicine, there are many conditions that can be treated by providing cells, tissues or organs from a donor individual to the patient. Loss of blood in an accident or surgery can be treated by blood transfusion, a patient with failing kidneys can receive a good kidney from a relative, and so on. The major limitation in these situations is immunological rejection, which is due to genetic differences between the patient and donor. Transplants between genetically identical individuals such as identical twins, or *syngrafts*, are not rejected; transplants between genetically distinct members of the same species, or **allografts**, are rejected; and transplants between different species, or **xenografts**, are generally rejected much faster. Immunosuppressive drugs allow many allografts to succeed, but xenografts, which would alleviate the problem of shortage of donor organs, are currently unsuccessful with a few minor exceptions. Immunosuppression comes at a cost: patients run a high risk of severe infection; but nonetheless, organ transplants save many lives.

Transplantation antigens include the blood group antigens, MHC and minor histocompatibility antigens

Blood transfusions only became widely successful after Landsteiner in the early 20th century discovered the major blood group antigens, and methods were developed to test for them. The most important blood group antigens, the **ABO antigens**, are carbohydrate antigens that differ between individuals, because of allelic differences in expression of two **glycosyltransferase** enzymes (Figure 14-11) that modify a cell surface carbohydrate molecule, the *H antigen*, in unique ways. The A and B alleles encode very similar proteins, exhibiting only four amino-acid differences, which are nonetheless sufficient to change the sugar added by the enzyme. The O allele produces an enzymatically inactive truncated product. Consequently, A and B individuals have different patterns of glycosylation on their H oligosaccharides, giving rise to the A and B antigens; AB individuals have both A and B antigens; and O individuals have neither, and hence do not have an antigen that is recognized by individuals of any of the other three types. ABO antigens are expressed on many cell types and must be matched not only for blood transfusion but also for most organ and cell transplants.

A unique feature of the ABO antigens that makes them especially important for transfusion and transplantation is that gut bacteria induce the production of IgM antibodies to the A and B antigens in individuals who do not express these antigens and hence are not tolerant to them. That is, a person of blood type A has circulating antibodies that are directed to the B antigen and so on. As a result, a blood transfusion that is mismatched for ABO is usually an immediate failure, as the antibody rapidly binds to the surface of the transfused cells and leads to their elimination. The exceptions to this rule are very young recipients who have not yet made antibodies to A or B antigens, and recipients for whom extensive efforts are made to remove such antibodies from the blood before transplantation.

Also of medical importance are the **Rh antigens**. Individuals are either Rh+ or Rh−. Rh− individuals are like type O, in that their cells do not have an antigen that is recognized by an Rh+ individual. However, in contrast to the ABO antigens, most Rh− individuals do not have circulating anti-Rh antibodies, the main exception being Rh− women who have borne Rh+ babies. These women typically become immunized to the Rh antigen through exposure to fetal blood during birth of the child. This serves essentially as a primary immunization, which is often prevented by injection of anti-Rh antibodies within 72 hours of birth as well

Definitions

ABO antigens: the most important of the currently recognized 23 blood group systems. The A and B antigens are carbohydrate antigens resulting from **glycosyltransferase** enzymes, so individuals can be A, B, AB or O; O is the absence of A and B.

allogeneic: produced from an allele different from that of the host: allogeneic MHC molecules or **minor histocompatibility antigens** induce immune responses that cause the rejection of grafts bearing them.

allograft: a graft in which the donor is from the same species as the recipient, but not genetically identical. Most transplantation is of this type.

glycosyltransferase: enzyme that adds a particular sugar to an oligosaccharide in a particular place.

graft-versus-host disease (GVH): a disease in which mature T cells are transferred as a component of a bone marrow transplant and react strongly to the transplantation antigens of the recipient.

GVH: see **graft-versus-host disease.**

histocompatibility: the match between the tissue of a graft and that of a graft recipient, which determines whether or not a graft will be rejected.

minor histocompatibility antigens: peptides that are derived from polymorphic non-MHC proteins in transplanted tissue and can give rise to a slowly developing immune response that can lead to transplant rejection and requires immunosuppression.

Rh antigens: also known as the Rhesus blood group system: the second most important blood group after ABO. The major Rh antigen is also called D and there are

as once at 28 weeks of gestation. The specific IgG serves to sequester the antigen and also may induce B cell unresponsiveness due to the inhibitory properties of the negative coreceptor FcγRIIB on the B cells (see section 6-8). If this is not done, miscarriage can result during pregnancy with a second child that is Rh+. This occurs because small amounts of fetal blood enter the mother and induce a secondary IgG response. The IgG anti-Rh antibody crosses the placenta and can destroy the red blood cells of the fetus, a condition called hemolytic disease of the newborn (see section 13-4). This complication does not usually occur with ABO antigens because the preexisting antibodies are primarily IgM, which does not cross the placenta.

Organ transplants can also be recognized as foreign by both CD4 and CD8 T cells. Especially important in this regard are allelic differences of MHC class II or MHC class I molecules, respectively. Indeed, the genetic locus encoding the MHC molecules was discovered by its importance in skin graft rejection, hence the name major histocompatibility complex, **histocompatibility** meaning tissue compatibility (see section 4-2). MHC genes have many alleles, a property that has been driven evolutionarily by the need to protect a population against the spread of an infectious agent that can mutate to escape recognition by a single MHC allele (see section 4-2). For reasons that are discussed in the next section, a large fraction of T cells is responsive to cells expressing **allogeneic** MHC molecules—that is, MHC allelic variants genetically distinct from those of the host. This can be a few percent of T cells instead of the naïve precursor frequency for a typical foreign antigen (1 in 10,000 or lower). This high fraction of antigen-recognizing T cells means that the immune response is very vigorous and rapidly proceeds to graft rejection. A slower graft rejection can occur when the MHC alleles are matched between allogenic donor and recipient, due to **minor histocompatibility antigens**, which are proteins with allelic differences that provide distinctive peptides that can bind to the MHC molecules for presentation to T cells.

Bone marrow transplants are increasingly used to treat genetic diseases of bone marrow cells, such as severe combined immunodeficiency, and to avert failure of bone marrow resulting from aggressive chemotherapy to treat various types of cancers, or to replace cancerous bone marrow cells in leukemia. But bone marrow transplants are subject to an additional complication. Mature T cells contained in the donor bone marrow can react against major and minor histocompatibility antigens of the recipient and cause severe disease, a situation called **graft-versus-host disease** (Figure 14-12). It is possible to remove mature T cells from bone marrow before transplantation and this is done for treatment of severe combined immunodeficiency patients. For treatment of leukemias in conjunction with chemotherapy, however, mature T cells are left in the bone marrow because the mature T cells appear to have some ability to kill residual leukemia cells.

Bone marrow transplants are an example of treating disease with stem cell transplants, a modality that may become very important to medical treatment in the future, since stem cells have a strong regenerative capacity. Preventing immunological rejection is one of the challenges in developing such therapies. It is also a challenge for the future development of *gene therapy*, in which a defective gene is replaced with a functional one in the patient's own cells: in this case DNA of viral origin may be required as a *vector* to introduce the replacement gene into the cell, and immune responses may develop to the vector or even to the therapeutic gene product.

In the next two sections, we explain in more detail how grafted tissue is recognized and rejected, and describe the strategies currently used to prevent graft rejection.

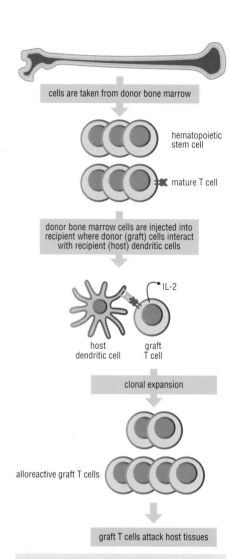

Figure 14-12 Graft-versus-host disease In bone marrow transplantation, hematopoietic stem cells are taken from the donor bone marrow but the donor marrow cells also include mature T cells. After injection into the recipient, the mature T cells from the graft are activated by MHC molecules on host dendritic cells, leading to clonal expansion followed by attack of host tissues.

also two minor Rh antigens called C and E. Clinically, Rh+ refers to the D antigen. 15% of Caucasians are Rh–, and the frequency is lower for other races.

xenograft: a graft coming from a different species.

References

Sykes, M. *et al.*: **Transplantation immunology** in *Fundamental Immunology*, 5th ed. Paul, W.E. ed. (Lippincott Williams & Wilkins, Philadelphia, 2003), 1481–1555.

Walsh, P.T. *et al.*: **Routes to transplant tolerance versus rejection: the role of cytokines.** *Immunity* 2004, **20**:121–131.

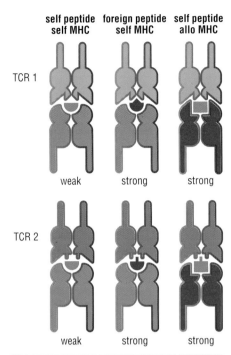

Figure 14-13 Recognition of allogeneic MHC molecules The upper and lower panels show two TCRs that recognize allogeneic MHC differently. For comparison, also shown (left and center interactions) are the same TCRs weakly recognizing complexes of self peptide bound to self MHC and strongly recognizing complexes of foreign peptide bound to self MHC. The stronger interaction results from one or more amino-acid side chains of the foreign peptide that contribute higher affinity to the interaction. On the right, recognition of allogeneic MHC in the upper panel is mediated primarily by a closer fit between the TCR and the MHC residues in the contact area; the peptide itself does not bind to the TCR better than does the self peptide in the left panel. In the lower panel, enhanced binding comes primarily from a closer fit of the peptide to the TCR, and hence is more analogous to the case of the foreign peptide + self MHC. Note that many different self peptides will bind to the allogeneic MHC molecule in distinct ways and hence there will be many different TCRs that can find activating ligands on the surface of allogenic dendritic cells. Both types of allorecognition have been demonstrated, and a hybrid involving elements of both is possible. The example in the upper panel has been observed by X-ray crystallography.

Organ transplant rejection occurs by three different mechanisms

All effector arms of adaptive immunity can participate in transplant rejection, and rejection of organ transplants appears to take place by three distinct routes. Preexisting antibody reactive with the organ leads to **hyperacute rejection**, in which antibody binding to small blood vessels induces clotting, hemorrhage, loss of blood flow and death of the tissue. Hyperacute rejection can occur when the recipient and donor are mismatched for ABO blood types, which are present on endothelial cells, or when the recipient has antibodies recognizing MHC class I proteins, which can be generated in a mother at the time of child birth (as with Rh), or in a recipient of a blood transfusion containing leukocytes. Hyperacute rejection is unstoppable, so great care is taken to test for preexisting antibodies before a transplant is done, a process called **cross-match**.

Although hyperacute rejection is generally avoided, all organ transplants other than those between identical twins may undergo **acute rejection** or **chronic rejection**. In both types of rejection, the role of CD4 and CD8 T cells is paramount, although antibodies are made to MHC differences and probably participate in chronic rejection. Acute rejection is associated with an intense inflammatory reaction in the graft tissue beginning as early as 7 days after transplantation, but immunosuppressive regimens (described in the next section) can now quite often successfully bring the reaction under control and prevent rejection. The risk of acute rejection is generally greatest within the first 3 months after the transplant, but can occur up to 1 year later. Currently approximately 90% of kidney transplants are successful for at least 1 year, but after that chronic rejection claims 3–5% a year. Most other solid organ transplants have similar failure rates. The rate of transplant loss due to chronic rejection has not really changed in 30 years, despite great improvements in managing acute rejection episodes. In animal models, this process can be caused by antibodies to MHC class I molecules of the graft or by CD4 T cells without antibody. In transplant recipients, the mechanism is not established, although there is a good correlation between levels of antibody recognizing donor MHC molecules and chronic rejection.

T cells can be stimulated by dendritic cells of the graft or of the host

T cells are activated by allogeneic MHC molecules in two distinct ways, referred to as **direct presentation** and **indirect presentation**. Direct presentation depends on recognition by host T cells of non-self MHC molecules on graft dendritic and tissue cells, a phenomenon known as **allorecognition**. Indirect presentation depends on presentation by host dendritic cells of peptides derived from MHC molecules or other polymorphic molecules that may differ between graft and host and are known as minor histocompatibility antigens (see section 14-6). In the period soon after transplantation, the immune response resulting from direct presentation is more vigorous than that resulting from indirect presentation, because of the remarkably high proportion of T cells that recognize non-self MHC. Nonetheless, both direct and indirect presentation are believed to generate T cells reactive against graft tissue, or *alloreactive T cells*, and thus to contribute to graft rejection. Before describing the processes whereby T cells are activated in each case, we will examine the basis for allorecognition.

MHC molecules produced from different alleles differ in two ways: first, they may have differences in the amino acids that lie on the surface that contacts the TCR; second, they may differ in the amino acids lining their peptide-binding grooves, and thus in the peptides that bind to them. Thus T cells bearing receptors that have been selected in the thymus to bind self MHC molecules with low affinity may, because of differences in the surface amino acids, bind non-self

Definitions

acute rejection: rapid inflammatory rejection of organ grafts starting as early as 7 days after transplantation and often decreasing in severity after 3 months, but can occur later. Generally due to CD4 and CD8 T cells, the rejection can be controlled by aggressive immunosuppressive therapy.

alloreactive T cells: T cells that react to antigenic differences in major or minor histocompatibility antigens from an allogenic transplant.

allorecognition: recognition by T cells of allelically variant MHC molecules or minor histocompatibility antigens presented by MHC molecules.

chronic rejection: a slower rejection not responsive to immunosuppressive therapy that is believed to be caused by antibody to graft MHC I molecules and/or CD4 T cells.

cross-match: test of serum from the patient against blood cells of the donor to rule out the presence of pre-existing antibodies that could cause **hyperacute rejection**.

direct presentation: allogeneic graft antigen-presenting cells activating a host T cell directly.

hyperacute rejection: very rapid rejection of an organ graft due to preexisting antibody that reacts to a cell surface molecule on the endothelium of the transplanted organ. This is typically due to ABO blood type differences or to antibodies to MHC I molecules of the graft.

indirect presentation: antigen presentation to **alloreactive T cells** by host antigen-presenting cells that have taken up antigens from a tissue graft.

MHC with higher affinity so that binding leads to T cell activation (TCR 1 in Figure 14-13). Alternatively, if the non-self MHC molecules bind peptides that are not bound by self MHC, these may be recognized as foreign, leading to activation of the T cell (TCR 2 in Figure 14-13).

In direct presentation, dendritic cells of the graft are mobilized, for example by inflammation of the graft tissue caused by the stress of the transplantation procedure, and migrate to the lymph nodes of the host. Here many T cells recognize the non-self MHC molecules of the graft and are activated to differentiate into alloreactive effector cells. The number of **alloreactive T cells** is very high, and so the response is very vigorous. Moreover, among the alloreactive T cells, probably some are memory T cells that recognize allogenic MHC and peptide by chance; and because they have a lower requirement for costimulation (see section 5-16), this adds further to the speed and vigor of the response.

Indirect presentation (Figure 14-14, right-hand panel) occurs when monocytes or dendritic cells of the host go to inflamed sites in the graft, take up material from the graft tissue, process the graft proteins into peptides, load those peptides onto self MHC molecules and present them to host T cells. Because of differences between polymorphic host and graft proteins, some of these peptides will be distinct from those of the host and will be recognized by T cells as foreign (that is, as the minor histocompatibility antigens mentioned above). The most polymorphic proteins made by cells are the MHC molecules themselves, and so they make a major contribution to indirect presentation as peptides loaded onto self MHC molecules. This route of stimulation of T cell immunity can mediate graft rejection but is likely to activate a lower fraction of recipient T cells than direct presentation.

Only direct presentation generates cytotoxic T cells that can directly damage the graft, and hence it is more destructive. In direct presentation, allogenic dendritic cells migrate from the graft and present peptides with allogenic MHC class I to host CD8 T cells in the lymph node, activating them to proliferate and differentiate into cytotoxic effectors that migrate into the graft and attack it. This route is illustrated in the left-hand panel of Figure 14-14. Over time, the number of allogenic dendritic cells within the graft declines, as bone marrow cells derived from the donor are replaced with those of the recipient. This leads to the indirect route of presentation, because the recipient dendritic cells can only present graft peptides from graft tissue debris on self MHC class II molecules. Thus, as time goes on, the immune response is thought to shift from direct presentation to indirect, in which alloreactive CD4 T cells are stimulated to activate macrophages and B cells, which are destructive in different ways (Figure 14-14, right-hand panel). It is therefore likely that T cells activated by direct presentation and those activated by indirect presentation have distinct and complementary roles in graft rejection.

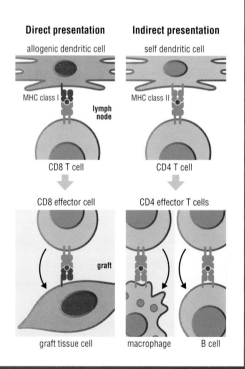

Figure 14-14 Modes of antigen presentation in transplant rejection In direct presentation, graft dendritic cells are mobilized to migrate from the organ to the draining lymph node and the spleen, where they present graft-derived peptides on allogenic MHC and can activate as many as several percent of the T cells present, resulting in a very vigorous immune response. After several days of stimulation, effector CD4 and CD8 T cells migrate to sites of inflammation, where the CD8 cells can kill parenchymal or endothelial cells of the graft and the CD4 cells can secrete cytokines and induce inflammation (not shown). CD4 T cells may also kill some graft cells via FasL (not shown). Indirect presentation occurs when host monocytes or dendritic cell precursors enter inflamed sites in the graft, ingest graft components and migrate to the draining lymph node, where they present graft-derived peptides on host MHC class II molecules. CD4 T cells become activated and can activate B cells specific for allogenic MHC class I molecules or can migrate out to sites of inflammation, where they may be activated by host macrophages to secrete cytokines and chemokines that contribute to tissue damage of the graft.

References

Braun, M.Y. *et al.*: **Acute rejection in the absence of cognate recognition of allograft by T cells.** *J. Immunol.* 2001, **166**:4879–4883.

Game, D.S. and Lechler, R.I.: **Pathways of allorecognition: implications for transplantation tolerance.** *Transpl. Immunol.* 2002, **10**:101–108.

Rosenberg, A.S. and Singer, A.: **Cellular basis of skin allograft rejection: an in vivo model of immune-mediated tissue destruction.** *Annu. Rev. Immunol.* 1992, **10**:333–360.

Valujskikh, A. and Heeger, P.S.: **Emerging roles of endothelial cells in transplant rejection.** *Curr. Opin. Immunol.* 2003, **15**:493–498.

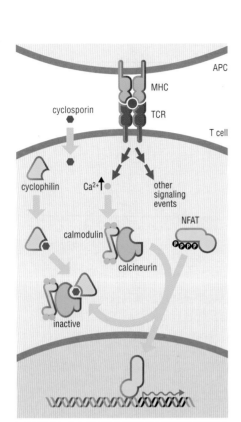

Figure 14-15 Mechanism of action of cyclosporin The 11 amino-acid cyclic peptide cyclosporin A binds to an intracellular protein, cyclophilin. The cyclosporin–cyclophilin complex binds to and inhibits the enzymatic activity of calcineurin, a calcium/calmodulin-dependent protein phosphatase consisting of a catalytic A subunit and a regulatory B subunit. A key target of calcineurin is NFAT, the nuclear factor of activated T cells. NFAT is present in the cytoplasm of unstimulated T cells in a phosphorylated form. Upon TCR engagement, calcium levels rise and this activates calcineurin to remove the inhibitory phosphates from NFAT, revealing a nuclear localization signal that takes dephosphorylated NFAT to the nucleus, where it binds to promoters of cytokine genes such as IL-2 and stimulates their transcription. This is blocked in cells treated with cyclosporin by inhibition of calcineurin. The macrolide FK506 acts in an analogous way, although instead of binding cyclophilin it binds to a separate partner, called FK binding protein 12 (FKBP-12), named in part for its molecular weight of 12 kD.

Transplantation is successful because of immunosuppressive drugs and antibodies

Given the vigorous nature of the T cell immune response to allogeneic cells, it is remarkable that allografts can succeed. Nonetheless, transplantation of many organs, including kidney, liver, heart, lung, cornea and small intestine, is remarkably successful. The 1-year graft survival for an allogeneic kidney from a live donor is now approximately 94%, despite a ferocious immune response. This success is due to the availability of drugs and monoclonal antibodies that can suppress the immune system. These treatments have side effects, the most important of which is decreased resistance to infections. Most of the immunosuppressive drugs also have effects on other tissues, and hence non-immunological side effects, but often their effects are greatest on T cells, as described below. Although with time the immune system becomes less aggressive in its response to transplanted tissue, life-long immunosuppression is currently required for nearly all organ transplants. In contrast, bone marrow transplantation is much more likely to induce long-lasting immune tolerance, and hence not to require continued immunosuppression. This outcome remains a key goal of transplantation research and there is currently considerable hope that contemporary knowledge of the immune system can be harnessed to induce specific immune tolerance to organ grafts.

Steroids, antiproliferative agents and calcineurin inhibitors are the mainstays of immunosuppression for transplantation

The strength of the immune response to an allograft has led to the common approach of combining at least three different types of immunosuppressants to decrease the chance of acute rejection. One type of drug typically used is a **corticosteroid** such as prednisone. These agents are antiinflammatory and at higher doses also inhibit T cell activation by inhibiting the production of some cytokines, including IL-2. Corticosteroids or glucocorticoids bind to the glucocorticoid receptor, which is a nuclear hormone receptor that regulates transcription upon ligand binding. Among the genes induced by the glucocorticoid receptor is I-κBα (see section 2-10). By increasing the production of I-κBα, glucocorticoids inhibit NF-κB activity, which is required for the transcription of many inflammation-related genes and for the production of IL-2 by T cells (see section 5-5). In addition, the glucocorticoid receptor has many effects on gene regulation independently of NF-κB.

Since immune responses require tremendous expansion of lymphocytes, immunosuppression for transplantation also typically includes one of several antiproliferative drugs borrowed from cancer chemotherapy. Most commonly used is azathioprine, which is converted in the body to 6-mercaptopurine, a compound that blocks the biosynthesis of the purine nucleotides required for DNA synthesis, and hence DNA replication. Lymphocytes are especially sensitive to blockage of purine biosynthesis because they are poor at utilizing the alternative, the salvage pathway whereby purines obtained from the diet are taken up by cells. A newer alternative to azathioprine is mycophenolate mofetil, which is metabolized to mycophenolic acid, which inhibits another enzyme in the synthesis of the purine guanosine.

The third type of immunosuppressant drug typically used is the calcineurin inhibitors **cyclosporin** or **FK506** . These natural products have an unusual mode of action, in that they do not directly inhibit calcineurin, but rather first form a complex with an intracellular protein and it is the complex that binds to and inhibits calcineurin (Figure 14-15). Calcineurin is a protein phosphatase that is activated by **calmodulin (CaM)** when calcium levels rise within the cell. Of course, one of the key early signaling events of TCR engagement is calcium elevation. One of the primary targets of calcineurin within T cells is the nuclear factor of activated T cells (NFAT), which has a key role in the transcription of cytokine genes, including IL-2 and IL-4 and also CD40L and FasL (see section 5-5). In resting T cells, NFAT exists in the cytoplasm in a highly phosphorylated and inactive form. Active calcineurin dephosphorylates NFAT, leading to its translocation to the nucleus and stimulation of gene transcription (see Figure 14-15). NFAT activity follows calcium elevation within the cell closely, because it is rapidly rephosphorylated within the nucleus and returned to the inactive state if calcineurin does not oppose this phosphorylation.

The name calcineurin reflects the high level of calcineurin expression in the brain and control of the protein by calcium signaling. Although calcineurin is most highly expressed in the brain, it is expressed in nearly all cells. Thus, the discovery that cyclosporin and FK506 selectively inhibit T cell activation by blocking calcineurin was quite surprising. The answer to this paradox is that T cells are very sensitive to inhibition by cyclosporin precisely because they express low levels of calcineurin. The mechanism of action of these drugs is via the formation of a very stable inhibitory complex, which means that at low levels these drugs inhibit a limited number of calcineurin molecules, which in T cells may be nearly all calcineurin molecules present, whereas in the brain it is a small proportion. Indeed, overexpression of calcineurin in T cells makes them resistant to the typical inhibitory concentrations of cyclosporin. Nonetheless, there are some side effects of cyclosporin, including kidney toxicity. These side effects are likely to be due to inhibition of calcineurin because they are very similar in treatment with FK506, which is chemically dissimilar and even binds to a different binding protein within cells.

A recently introduced immunosuppressant drug **rapamycin** has a mechanism of action that is similar to that of FK506, but distinct. Both compounds bind to the same protein within cells, a protein called **FK506 binding protein (FKBP)** (or sometimes FKBP-12, for its molecular weight of 12 kD), but whereas the FK506–FKBP complex binds to and inhibits calcineurin, the rapamycin–FKBP complex has a different target, called mTOR (mammalian target of rapamycin), a protein kinase that regulates the increase in the rate of protein synthesis during cellular proliferation. Thus, rapamycin is also antiproliferative, somewhat analogous to the purine metabolism inhibitors. The mechanisms of action of FK506 and rapamycin are schematically illustrated in Figure 14-16.

Anti-CD3 antibodies are used to combat rejection episodes

While the three types of drugs mentioned above are the mainstays of immunosuppression for transplant recipients, even with this aggressive approach acute rejection episodes occur with decreased organ function, and swelling and pain in the organ from inflammation containing many T cells. A common approach to blocking rejection in such flareups is to treat with a monoclonal antibody recognizing the CD3 component of the TCR, called OKT3. This treatment induces a rapid cytokine storm similar to that seen in bacterial sepsis with bacteria that produce superantigens (see section 9-3). It also leads to internalization of the TCR complex, making the T cells unresponsive to further antigen stimulation for some days. T cells remain, but the individual is temporarily without T cell immunity. Monoclonal antibodies such as OKT3 thus are highly effective at blocking acute rejection, although they do not induce long-lasting tolerance to the graft, so immunosuppressive drugs must continue to be taken after OKT3 treatment.

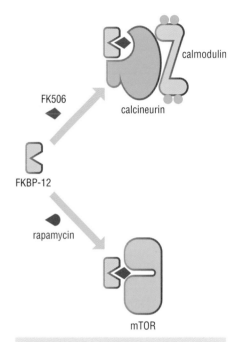

Figure 14-16 Distinct actions of FK506 and rapamycin FK506 and the chemically related molecule rapamycin both bind to FKBP and both inhibit T cell proliferation. However, the FK506–FKBP complex binds and inhibits calcineurin, whereas the rapamycin–FKBP complex binds and inhibits another enzyme, the protein kinase mTOR (mammalian target of rapamycin). The main function of calcineurin in stimulated T cells is in the activation of NFAT, although it also has a role in the activation of NF-κB. The function of mTOR is to regulate the increase in cell protein synthesis needed to support cell-cycle progression and multiplication. Levels of FKBP-12 in cells are high, so at pharmacological concentrations FK506 and rapamycin do not compete with one another to a significant degree.

References

Aw, M.M.: **Transplant immunology.** *J. Pediatr. Surg.* 2003, **38**:1275–1280.

Chatenoud, L.: **CD3-specific antibody-induced active tolerance: from bench to bedside.** *Nat. Rev. Immunol.* 2003, **3**:123–132.

Choi, J. *et al.*: **Structure of the FKBP12-rapamycin complex interacting with the binding domain of human FRAP.** *Science* 1996, **273**:239–242.

Crabtree, G.R.: **Generic signals and specific outcomes: signaling through Ca²⁺, calcineurin, and NF-AT.** *Cell* 1999, **96**:611–614.

Kirken, R.A. and Wang, Y.L.: **Molecular actions of sirolimus: sirolimus and mTor.** *Transplant. Proc.* 2003, **35**:S227–S230.

Schreiber, S.L. and Crabtree, G.R.: **The mechanism of action of cyclosporin A and FK506.** *Immunol. Today* 1992, **13**:136–142.

Winkelstein, A.: **Immunosuppressive therapy** in

Medical Immunology 9th ed. Stites, D.P. *et al.* eds (Appleton and Lange, Stamford, 1997), 827–845.

Wullschleger, S. *et al.*: **TOR signaling in growth and metabolism.** *Cell* 2006, **124**:471–484.

Glossary

12/23 rule: a property of V(D)J recombination whereby gene segments that are recombined are limited to combinations of one **RSS** with a 12-bp spacer and one **RSS** with a 23-bp spacer. For example, gene segments with 12-bp spacer **RSS**s never combine with each other. This is one of the mechanisms that limits V(D)J recombination to avoid undesirable combinations. (7-1)

α-dystroglycan: a heavily glycosylated protein associated noncovalently on the cell surface with the integral membrane protein β-dystroglycan; the two dystroglycans are derived by post-transcriptional processing of a single gene product and contribute to interactions of extracellular matrix ligands, including laminin and agrin, with the cell actin cytoskeleton. α-Dystroglycan is implicated in the binding of arenaviruses, including LCMV and Lassa fever virus, but also *Mycobacterium leprae*, the causative agent of leprosy, to cell surfaces. (10-9)

α-proteobacteria: large, metabolically diverse and widely dispersed, evolutionarily ancient group of Gram-negative bacteria that express an altered lipid A and fatty-acid side-chain pattern, resulting in an unusual LPS that can be several thousand-fold less stimulatory for TLRs than LPS derived from typical Gram-negative bacteria; some members of the family, such as *Sphingomonas*, totally lack LPS and utilize instead a family of **glycosphingolipids** in the outer membrane; α-proteobacteria include a large number of organisms that establish chronic infections in plants (*Agrobacterium*, *Rhizobia*), insects (*Wolbachia*) and vertebrates, including humans (*Bartonella*, *Brucella*, *Rickettsia*), and probably contains the mitochondrial ancestor. (8-3)

ABO antigens: the most important of the currently recognized 23 blood group systems. The A and B antigens are carbohydrate antigens resulting from **glycosyltransferase** enzymes, so individuals can be A, B, AB or O; O is the absence of A and B. (14-6)

activation-induced cell death: (of T lymphocytes) programmed death of activated T lymphocytes through proapoptotic signals delivered by activating stimuli. (5-15)

activation-induced cytidine deaminase (AID): a cytidine deaminase necessary for both **class switch recombination** and **somatic hypermutation**. AID is expressed in germinal center B cells and is related to APOBEC3G. (6-11)

acute-phase response: systemic response to infection mediated primarily by IL-1 and IL-6 acting on hepatocytes to increase the synthesis of blood proteins that include soluble recognition molecules for microbes or apoptotic cells, complement components, coagulation proteins, protease inhibitors, lipid-binding proteins and metal-binding proteins. These proteins are all likely to have value in helping fight infections. (3-15)

acute rejection: rapid inflammatory rejection of organ grafts starting as early as 7 days after transplantation and often decreasing in severity after 3 months, but can occur later. Generally due to CD4 and CD8 T cells, the rejection can be controlled by aggressive immunosuppressive therapy. (14-7)

adaptive immunity: immune responses mediated by **lymphocytes** and their products and requiring activation by **innate immune mechanisms** on first encounter with **antigen** but acting immediately on subsequent encounters. (1-0)

adaptor: component of a signaling pathway lacking enzymatic function but participating in signaling reactions by binding to other signaling molecules and thereby bringing them together. (2-0)

ADCC: see **antibody-dependent cellular cytotoxicity**. (6-3)

adjuvant: component or property of a vaccine or inoculum that makes it **immunogenic**. The term was originally used for a component added to a vaccine or inoculum in order to produce an adaptive immune response, but it is now recognized that vaccines or inocula may have inherent adjuvant or immunogenic properties. (14-2)

afferent lymphatics: lymphatic vessels entering **lymph nodes** from tissue spaces. (1-6)

affinity: strength of a nonconvalent binding interaction; the higher the affinity, the more likely two partners will exist in a complex. (2-4)

affinity maturation: the property of an immune response in which the average affinity of antibodies produced against an antigen increases as the response continues. This occurs over a period of several weeks. (1-8, 6-0)

AID: see **activation-induced cytidine deaminase**. (6-11)

Aire: see **autoimmune regulator**. (7-10)

airway obstruction: (in asthma) a pathologic narrowing of the lower airways due to subepithelial edema, hypersecretion of mucus, and smooth muscle contraction. (13-3)

allele: genetic variant (of gene). (4-3)

allelic exclusion: the property that a single lymphocyte expresses only one functionally rearranged allele of an antigen receptor gene. Both IgH and IgL loci show allelic exclusion, as does the TCR β locus, whereas TCR α and δ loci do not. (7-3)

allergen: antigen from an environmental non-infectious agent that can induce the production of IgE and thus an allergic condition. (13-0)

allergic rhinitis: an allergic response affecting the nasal passages and often also the eyes. Symptoms include sneezing, runny nose, watery eyes and itch. (13-2)

allergy: immune responsiveness to any non-infectious environmental agent resulting in the production of IgE and a T_H2-driven inflammatory response. (13-0)

allogeneic: produced from an allele different from that of the host: allogeneic MHC molecules or **minor histocompatibility antigens** induce immune responses that cause the rejection of grafts bearing them. (14-6)

allograft: a graft in which the donor is from the same species as the recipient, but not genetically identical. Most transplantation is of this type. (14-6)

alloreactive T cells: T cells that react to antigenic differences in major or minor histocompatibility antigens from an allogenic transplant. (14-7)

allorecognition: recognition by T cells of allelically variant MHC molecules or minor histocompatibility antigens presented by MHC molecules. (14-7)

ALPS: see **autoimmune lymphoproliferative syndrome**. (12-2)

altered peptide ligands: peptides that when bound to MHC interact with the TCR of a particular T cell with low to intermediate affinity and/or fast dissociation rate, activate the corresponding T cell only partly, or inhibit activation by a peptide–MHC complex that interacts more strongly with the TCR. (5-2)

alternatively activated macrophages: macrophages activated by T_H2 cells and which promote the recruitment of eosinophils and basophils and more T_H2 cells, and secrete antiinflammatory cytokines, as distinct from macrophages activated by T_H1 cells and which have inflammatory effects. (5-12)

alternative pathway: complement activation pathway initiated at membrane surfaces and that both leads to production of later effector components of the pathway and provides an amplification loop for cleavage of C3. (3-3)

alveolar macrophages: macrophages in the lung. (1-3)

anaphylaxis: a potentially life-threatening systemic allergic reaction in which there is rapid and simultaneous onset of symptoms in several locations, often including skin, lungs and gastrointestinal tract. (13-2)

anchor residue: an amino acid in an antigenic peptide whose side chain binds in a pocket formed by specific amino-acid side chains in the peptide-binding groove of an MHC molecule. The composition of the pocket can vary in different polymorphic variants of the MHC molecule. (4-5)

anergy: (of lymphocytes) an unresponsive state resulting from stimulation by antigen in the absence of additional signals, such as those provided by helper T cells (for B cells) or by costimulation (for T cells) that are generally necessary for activation. (12-2)

antibodies: highly variable proteins produced by the **B lymphocytes** of the immune system and that recognize **antigen** and target it for destruction. (1-0)

antibody-dependent cellular cytotoxicity (ADCC): a process whereby FcR-bearing cells encounter an antibody-coated target cell and degranulate, releasing contents that kill the antibody-coated cell. (6-3)

antigen: any molecule or part of a molecule recognized by the variable antigen receptors of **lymphocytes**. (1-0)

antigenic drift: (of influenza virus) point mutations, predominantly in **hemagglutinin** and **neuraminidase**, that affect recognition by neutralizing human antibodies. Antigenic drift gives rise to epidemic infections. (10-7)

antigenic shift: (of influenza virus) reassortment of independent RNA segments from two different influenza genomes to generate recombinant virus with new antigenic subtypes. Antigenic shift gives rise to pandemic outbreaks. (10-7)

antigenic variation: (in parasites) clonal expression of members of families of proteins among parasite progeny; examples include the major surface glycoproteins of trypanosomes and the red cell adhesins encoded by malaria parasites; variant expression allows parasites to evade immune recognition. (11-0)

antigen presentation: the binding of fragments of intracellular molecules, usually peptides derived from proteins of pathogens, by **major histocompatibility complex (MHC) molecules** and their presentation on the cell surface for recognition by T cells. (4-2)

antigen-presenting cells: cells capable of displaying antigen for recognition by T cells and of activating naïve T cells. (1-3)

antihistamines: antagonists of H_1 **histamine receptors**. Antihistamines block some of the symptoms of allergic reactions, particularly the most rapid ones, but do not block inflammation or the actions of leukotrienes. (13-1)

antimicrobial peptides: peptide antibiotics that provide defense against microbes and viruses by interacting with membranes of infectious agents and increasing their permeability. Human antimicrobial peptides are members of either the α-**defensin**, β-defensin or **cathelicidin** families. (3-1)

Apaf-1: see **apoptotic protease activating factor 1.** (2-11)

APOBEC3G: one of a family of **cytidine deaminases** that includes APOBEC1, an mRNA editing enzyme, and AID, a DNA-editing enzyme that is responsible for immunoglobulin gene somatic hypermutation and class switch recombination. APOBEC3G targets retrovirus nucleocapsids for catastrophic mutation of retroviral DNA. (3-19)

APOBEC family: family of cytidine deaminases including both RNA- and DNA-editing enzymes, members of which are responsible for the somatic hypermutation of Ig genes as well as viral defense through the introduction of catastrophic mutations into retroviral DNA. APOBEC3G, 3F and 3B, and perhaps others, have antiviral activity and are likely to have evolved to protect against viruses, including retroviruses and hepatitis B virus, that replicate through reverse transcription, and to suppress the transposition of endogenous retroelements in the genome. (10-5)

apoptosis: regulated cell death in which activation of specific proteases and nucleases leads to death characterized by chromatin condensation, protein and DNA degradation, loss of plasma membrane lipid asymmetry and disintegration of the cell into membrane-bounded fragments. (2-11)

apoptosome: molecular complex of procaspase 9, **Apaf-1** and cytochrome c that assembles in response to mitochondrial release of cytochrome c and results in the activation of caspase 9 and triggering of the **intrinsic pathway** of **apoptosis.** (2-11)

apoptotic protease activating factor 1 (Apaf-1): an adaptor molecule that oligomerizes on interaction with cytochrome c, leading to the binding and activation of caspase 9. (2-11)

arginase: arginase I and arginase II enzymes metabolize L-arginine to urea and L-ornithine, a polyamine precursor; in macrophages, arginase I is induced by IL-4, IL-13, IL-10 or TGF-β, and serves to remove arginine available for oxidation by NOS2 in the formation of L-citrulline and reactive nitric oxide. (11-3)

Arp2/3 complex: a complex including two actin-related proteins, Arp2 and Arp3 and that initiates the polymerization of branched-chain actin filaments. It is normally activated by proteins of the Wiskott–Aldrich syndrome (WASP) and Scar/WAVE families, thus linking cytoskeletal remodeling with various signaling pathways. *Listeria* ActA mimics these proteins, establishing bacterial control of the actin activating pathway. (9-6)

asthma: an allergic disease affecting primarily the lungs. (13-2)

atopic dermatitis: an inflammatory condition in the skin associated with atopy, also called eczema. In this condition, the inflammatory features of immediate-type hypersensitivity dominate over the immediate, histamine-mediated effects on the vasculature. (13-2)

atopy: the propensity to develop allergic disease, operationally defined by elevated levels of serum IgE or by positivity in the allergic skin test. (13-0)

attenuation: (of a pathogen) loss of pathogenicity, usually through adaptation to growth in culture in adverse conditions or in cells from a species other than that of the normal host. Attenuated pathogens are the basis of many vaccines. (14-2)

autoantigen: a self component that is the target of the immune response in an **autoimmune disease.** (12-4)

autocrine: (of an extracellular signaling molecule) acting on the cell that secreted it. (2-6)

autoimmune disease: a disease in which the primary mechanism of pathology is an immune attack directed against components of the self (**autoantigens**). (12-4)

autoimmune hemolytic anemia: an autoimmune disease in which antibodies reacting with red blood cells are produced, leading to premature clearance of the red blood cells by macrophages or complement-mediated lysis. (12-5)

autoimmune lymphoproliferative syndrome (ALPS): a genetic disease characterized by splenomegaly, lymphadenopathy, and autoantibody production, and resulting from missense mutations in genes encoding Fas, FasL or caspase 10. These mutations result in defective Fas signaling. The increased cellularity of the secondary lymphoid organs in these individuals is due primarily to the accumulation of an unusual type of T cell lacking the expression of CD4 or CD8, and an accumulation of activated B cells. (12-2)

autoimmune polyendocrine syndrome type 1: a rare inherited disease due to mutation of the gene encoding **Aire** and characterized by multiple-organ-specific autoimmune disease. (7-10)

autoimmune regulator (Aire): transcriptional regulator expressed in thymic medullary epithelial cells and responsible for expression in these cells of a number of proteins otherwise expressed only in one particular organ. Originally discovered as the mutant protein responsible for autoimmune polyendocrine syndrome (APS) type 1 in humans; mice in which the gene encoding Aire is disabled have a deficiency in central tolerance and develop diverse solid-organ autoimmune diseases. (7-10)

autoimmune thrombocytic purpura: an autoimmune disease in which autoantibodies against platelets promote their rapid clearance by macrophages, leading to a bleeding disorder. (12-5)

autoimmune thyroiditis: an autoimmune disease in which the thyroid gland is destroyed. Also called Hashimoto's thyroiditis. (12-6)

autophagosome: double-membrane-enclosed compartment generated in the cytoplasm as a response to either nutrient deprivation or a bacterium in the cytoplasm, a process known as **autophagy.** Autophagosomes fuse with endosomes and then lysosomes and so mature to a structure similar to the **phagolysosome.** (3-8)

autophagy: a process activated during cellular starvation or in response to a damaged organelle that involves the sequestration and degradation of cellular organelles and cytoplasm for recycling of macromolecules. (3-8)

avidity: increased apparent **affinity** of a molecule for its ligand due to the presence of multiple binding sites on both partners. (2-4)

β₂-microglobulin: the smaller of the two chains of the **MHC class I molecule** heterodimer. (4-2)

B1 cells: B cells generated early in life and responsible for most **natural antibody** and some T cell-independent antibody production. (8-6)

bacille Calmette–Guerin (BCG): attenuated *M. bovis* used to vaccinate against tuberculosis. The attenuated bacterium was derived by passage of a virulent strain 230 times between 1908 and 1921; seed lots from original isolates have been separately propagated under varying conditions for use in childhood BCG vaccines. (9-7)

BAFF: see **B cell-activating factor of the TNF family.** (7-6)

basophils: circulating myeloid lineage cells that are characterized by cytoplasmic granules that stain with basic dyes and contain inflammatory mediators and are believed to be important in defense against parasites as well as in inflammatory and allergic reactions. (1-2)

B cell-activating factor of the TNF family (BAFF): a TNF family member that has a key role in B cell survival, maturation and activation, and which is made by lymphoid tissue stromal cells and by myeloid cells. Also called BLyS, TALL-1, THANK, zTNF4 and TNFSF13B. (7-6)

B cell antigen receptor: the complex of membrane immunoglobulin that recognizes antigen and a heterodimer of Igα and Igβ that signals antigen recognition in B lymphocytes. (1-4, 6-7)

BCG: see **bacille Calmette–Guerin.** (9-7)

Bcl-2: antiapoptotic protein of the **Bcl-2 family** of regulators of the intrinsic pathway of apoptosis, of which it was the first member to be discovered. (2-12)

Bcl-2 family: a family of regulators of apoptosis with homology to **Bcl-2**, falling into three main structural and functional groups: a directly death inducing group of three members, an antiapoptotic group of six members including Bcl-2 itself, and the **BH3-only family**, which regulate members of the other groups to induce apoptosis in various conditions. (2-12)

BCL-6: a zinc finger family transcriptional regulator highly expressed in germinal center B cells that prevents differentiation of mature B cells to plasma cells, in part by repressing Blimp-1 transcription while maintaining expression of the machinery required for hypermutation and follicular homing. (5-13)

BCR: see **B cell antigen receptor.** (6-7)

beta sandwich: domain structure in which the flat sides of two beta sheets are stacked against one another, like two pieces of bread in a sandwich. **Ig domains** fold into this type of structure. (2-1)

BH3-only family: family of stress-sensing proteins that induce apoptosis by inhibiting antiapoptotic **Bcl-2 family** members or, in one case, by activating apoptotic Bcl-2 family members. (2-12)

B lymphocytes: lymphocytes that when activated differentiate into antibody-secreting cells. (1-0)

bronchial hyperreactivity: exaggerated airway narrowing in response to airborne allergens or to nonspecific bronchoconstrictor agents, such as acetylcholine or methylcholine. (13-3)

bursa of Fabricius: a specialized primary lymphoid organ of chickens that does not have a strict counterpart in mammals. In the bursa, newly formed B cells proliferate and undergo gene conversion events at their IgH and IgL loci before antigenic stimulation. (7-4)

C1 inhibitor: a soluble protein that dissociates active complement components C1r–C1s from C1q and thereby limits the classical pathway of complement. (3-6)

C1q: complement component that binds to antibodies in immune complexes and activates the classical pathway of complement activation. In addition to activating the complement cascade, C1q is recognized by a phagocytic receptor of macrophages (C1qRp), and so can mediate phagocytosis directly. (3-4)

CIITA: see **class II transactivator.** (4-8)

C3 convertase: either of two proteolytic enzymes of the **complement system** that cleave C3 to generate C3a and C3b. The C3 convertase of the classical and lectin pathways is a complex of C4b and C2b, whereas the C3 convertase of the **alternative pathway** is a complex of C3b and Bb. (3-3)

C4-binding protein (C4bp): a soluble protein that binds complement component C4b and recruits factor I, which targets C4b for degradation, thereby limiting complement activation. (3-6)

C4bp: see **C4-binding protein**. (3-6)

C5 convertase: any of several proteolytic enzymes that cleave complement component C5 to generate C5a and C5b. The enzyme is formed by the membrane attachment of a C3b molecule adjacent to either of the two types of C3 convertases. The C3b binds C5 which is then efficiently cleaved by the C3 convertase. (3-5)

calcineurin: calcium/calmodulin-activated serine/threonine protein phosphatase, an essential component of many calcium-activated signaling pathways. Calcineurin action is essential to activate the **NFAT** family of transcription factors and is the target of the immunosuppressive drugs cyclosporin and FK506. (5-5)

calmodulin (CaM): a small calcium-binding protein that confers calcium regulation on cellular signaling and effector molecules, including protein kinases and phosphatases. (14-8)

calnexin: a membrane-bound general cellular chaperone that associates with MHC class I alpha chains immediately on synthesis. (4-6)

calreticulin: a soluble general cellular chaperone homologous to **calnexin** that associates with newly synthesized MHC class I alpha chains when calnexin dissociates. (4-6)

CaM: see **calmodulin**. (14-8)

capsular polysaccharide: cell-surface polymers of repeating oligosaccharide units, usually linked through phosphodiester bonds, that form a capsule on the surface of many pathogenic bacteria and protect bacterial cells from recognition by phagocytes. For this reason the presence of a capsule is often associated with virulence. (3-2)

cartilaginous fish: primitive jawed fish with cartilage instead of true bone. Includes sharks and skates. (7-4)

caspase: any of a family of cysteine proteases that cleave after aspartic acid residues in proteins and have an essential role in either **apoptosis** or processing of inflammatory cytokines to active forms. (2-11)

caspase activation and recruitment domain (CARD): domain found in several proteins that activate **apoptosis** or regulate NF-κB, with strong structural similarity to death domains and death effector domains. CARD domains generally associate with CARD domains of other proteins. (2-11)

cathelicidins: family of cationic **antimicrobial peptides** generated in pre-pro forms that require processing to generate the active peptide; cathelicidins contain an amino-terminal cathelin-like domain and a carboxy-terminal antimicrobial domain. (3-1)

cathepsin: any of a family of aspartic and cysteine proteases resident in endocytic vesicles and lysosomes and some of which are implicated in antigen processing for presentation by MHC class II molecules. (4-7)

Cbl: any of several E3 ubiquitin ligases that bind signaling TCRs or other receptors and target them for ubiquitination and degradation. (5-4)

CD: see **cluster of differentiation**. (1-5)

CD2 subfamily: family of related Ig superfamily

cell–cell adhesion molecules that bind either to themselves or to other family members. Most members contain two Ig domains and the outer domains are involved in binding. (2-3)

CD35: see **complement receptor 1**. (3-6)

CD40: a receptor of the TNF receptor family that is expressed on mature dendritic cells, B cells and activated macrophages and plays an essential part in their interactions with T cells. (5-7)

CD40L: a TNF-family molecule, also known as CD154, that binds to receptor **CD40** and is induced on activated T cells and plays an essential part in the initiation of T cell responses and in many of the effector actions of CD4 T cells. (5-7)

CD45: a transmembrane protein tyrosine phosphatase that is expressed by hematopoietic lineage cells and that dephosphorylates the negative regulatory site on Src-family tyrosine kinases. (5-4)

CD46: see **membrane cofactor of proteolysis**. (3-6)

CD55: see **decay-accelerating factor**. (3-6)

CD59: see **protectin**. (3-6)

CD161: see **NK1.1**. (8-1)

CDR: see **complementarity-determining region**. (5-1)

celiac disease: a gastrointestinal delayed-type hypersensitivity reaction to gluten proteins present in wheat and other grains that is characterized by inflammatory destruction of the upper intestinal mucosa, resulting in malabsorption and diarrhea. (13-5)

central memory T cells (T$_{CM}$): long-lived circulating T lymphocytes that are generated from antigen-specific T cells during a primary immune response and can be activated to proliferate and differentiate into effector cells on reencounter with antigen. Central memory cells provide a reservoir of antigen-specific cells that can be rapidly activated on recurrence on an infection. (5-16)

central supramolecular activation complex (c-SMAC): central area of some **immunological synapses** in which TCRs and integrins have segregated into distinct areas. Included in the c-SMAC are most molecules involved in TCR signaling. (5-4)

central tolerance: immune non-responsiveness due to mechanisms operating during lymphocyte development to eliminate self-reactive cells. (12-0)

centroblast: rapidly proliferating B cell in the **dark zone** of the germinal center. (6-11)

centrocyte: post-mitotic germinal center B cell that is programmed to die by apoptosis unless it receives survival signals from the BCR and helper T cells. (6-11)

CGD: see **chronic granulomatous disease**. (3-9)

chemoattractant receptor homologous molecule expressed on T$_H$2 cells (CRT$_H$2): a G-protein-coupled seven-transmembrane receptor that is expressed on T$_H$2 cells, eosinophils and basophils and that mediates chemotaxis to prostaglandin D$_2$. It is homologous to the more broadly expressed prostaglandin D$_2$ receptor. (5-10)

chemokine: any of a family of closely related small, basic cytokines whose main function is as chemoattractants. The name is a contraction of chemotactic **cytokine**. (1-0, 2-13)

chondrocyte: a cell type that degrades cartilage. (12-9)

chromosomal translocation: alteration in chromo-

some structure in which one or more pieces of one chromosome become exchanged with or attached to another chromosome. (7-11)

chronic granulomatous disease (CGD): genetic immunodeficiency disease caused by inactivating mutations of one of the subunits of **NADPH oxidase**. In this disease, certain types of bacteria cannot be killed adequately. (3-9)

chronic rejection: a slower rejection not responsive to immunosuppressive therapy that is believed to be caused by antibody to graft MHC I molecules and/or CD4 T cells. (14-7)

cirrhosis: irreversible chronic injury to the liver manifested pathologically by extensive fibrosis. (11-5)

class II-associated invariant chain peptide (CLIP): a proteolytic fragment of the **invariant chain** that remains in the peptide-binding groove of newly synthesized MHC class II molecules until they associate with the chaperone **DM**. (4-7)

class II transactivator (CIITA): a non-DNA-binding transcriptional activator that regulates the expression of MHC class II molecules, both directly by activating the expression of the genes encoding the alpha and beta chains and indirectly by activating transcription of the genes encoding DO, DM and the invariant chain, which are essential for peptide loading. (4-8)

classical MHC molecules: highly polymorphic cell-surface glycoproteins encoded in the **major histocompatibility complex** and whose function is to present peptide antigens to T cells. (4-3)

classical pathway: complement activation pathway through which antibody–antigen complexes trigger the complement cascade. This pathway is also activated by the **pentraxins**. (3-3)

class switch recombination: genetic recombination between switch regions that are found 5' of the constant exons of each Ig heavy-chain gene. Usually this involves the switch region upstream of the C$_\mu$ exons (S$_\mu$) and a downstream S region. (6-11)

CLIP: see **class II-associated invariant chain peptide**. (4-7)

clonal deletion: elimination of potentially self-reactive lymphocytes. Immature lymphocytes undergo programmed cell death after binding to antigen; in this way, cells bearing receptors that recognize self are deleted before they are capable of participating in immune responses. This is a major mechanism of **immune tolerance**. (1-4)

clonal expansion: the selective proliferation of mature naïve lymphocytes that encounter antigen. Only those lymphocytes bearing receptors specifically recognizing antigen are activated to proliferate and differentiate into effector cells. (1-4)

clonal selection: the process whereby potentially self-reactive lymphocytes are eliminated during ontogeny whereas mature lymphocytes recognizing non-self antigens are selectively expanded. (1-4)

cluster of differentiation (CD): the basis of a system for identifying cell surface molecules of immune cells by the use of antibodies and in which each molecule is given a specific number prefixed by CD to form the basis of a systematic nomenclature. The term cluster reflects the fact that each molecule is usually recognized by a group, or cluster of antibodies; and the appearance of the molecules usually reflects different differentiated states of the cell, hence differentiation. Surface marker molecules of immune cells of different types and at different stages of differentiation or activation have been identified in this way and can be used to classify cells, or to follow their progress through development or their activation status. (1-5)

c-Maf: a transcriptional regulator implicated in IL-4 and IL-10 expression in T$_H$2 cells and in lens fiber cell differentiation, and the cellular homolog of a proto-oncogene from the avian musculoaponeurotic fibrosarcoma virus. (5-10)

coagulation cascade: proteolytic cascade triggered by damage to endothelium or by plasma entering a tissue site due to increased vascular permeability and that results in the generation of plasmin, which induces blood clotting. (3-13)

coding joint: region of the DNA where V and D, D and J or V and J segments of lymphocyte antigen receptor genes are joined. Often this occurs imprecisely with the inclusion of **P regions** or **N regions**, and/or the removal of a few bases. This property of coding joints generates junctional diversity. (7-2)

collectin: any of a family of structurally related, carbohydrate-recognizing proteins of innate immunity, including **mannose-binding lectin** and **surfactant** proteins A and D. The name refers to the presence of a collagen-like domain and a C-type lectin domain. (3-2)

colony-stimulating factor (CSF): growth factor or cytokine that induces the differentiation from a multipotent precursor of one or a few specific cell types. (1-1)

combinatorial diversity: the diversity of the antigen receptors of lymphocytes resulting from different combinations of V, D and J gene segments. (7-1)

commensal bacteria: bacteria that colonize a niche of the host. From the Latin for one who eats at the same table. (9-0)

common γ (γ$_c$) chain: common component of the receptors for IL-2, IL-4, IL-7, IL-9, IL-15 and IL-21 and the component that is defective in patients with **X-linked severe combined immunodeficiency** (X-SCID). (2-8)

complement: serum proteins activated directly or indirectly by conserved surface features of microorganisms, or by antibody, to destroy microorganisms or induce their destruction through a coordinated immune response including induction of inflammation, attraction of leukocytes, stimulation of phagocytosis and stimulation of antibody production. (1-0, 3-3)

complementarity-determining region (CDR): region of a lymphocyte receptor for antigen that participates in the antigen-binding site and determines its structural complementarity to the antigen. (5-1)

complement receptor 1 (CR1); CD35: a membrane protein of phagocytes that targets complement component C3b for proteolytic inactivation, to limit activation of the alternative pathway. (3-6)

complement system: see **complement**. (1-0, 3-3)

concomitant immunity: immunity to reinfection that coexists with a pathogen that remains viable in an otherwise asymptomatic host; common to many chronic infections, including latent tuberculosis, leishmaniasis and intestinal worms; regulatory T cells may contribute to concomitant immunity. (11-0)

conjugate vaccine: vaccine consisting of a polysaccharide from a capsulated bacterium chemically linked to a protein, usually a bacterial toxoid. Conjugate vaccines are used to elicit T cell-dependent antibody responses to capsular polysaccharides that on their own elicit largely T cell-independent antibodies: helper T cells

generated in response to the protein activate B cells that recognize the polysaccharide and internalize the conjugate, and present peptide fragments to the helper T cell. Conjugate vaccines are particularly important for protecting infants in whom the T cell-independent response to polysaccharides is absent for the first two years of life. (14-3)

constant (C) domain: Ig-like domain of the type found in Ig constant regions. (2-1)

constant region (C region): region of a lymphocyte receptor for antigen that does not participate in antigen binding and does not vary between cells of different antigen specificities. (1-4, 5-1)

coreceptor: (of T lymphocytes) receptor on a T cell that recognizes invariant parts of MHC molecules and forms a recognition complex with the antigen receptor and contributes to intracellular signaling. The coreceptor CD8 binds to MHC class I molecules and is generally expressed on cytotoxic cells; the coreceptor CD4 binds to MHC class II molecules and is generally found on helper cells. (1-5, 5-2)

corticosteroids: natural and synthetic steroid hormones that bind to and activate the glucocorticoid receptor. Corticosteroids are very effective antiinflammatory drugs and also are immunosuppressive. (14-8)

costimulatory molecule: any of the immunoglobulin-superfamily molecules B7-1 and B7-2, which are upregulated on dendritic cells and some other immune system cells in the presence of infection, or their receptor, CD28, which is constitutively expressed on naïve T cells. Signaling through CD28 is essential for activation of naïve T cells through the TCR. Other related molecules of the Ig superfamily, as well as some molecules of the TNF/TNFR family, are also sometimes called costimulators because they modulate signaling from the TCR. (4-1)

CpG-containing DNA: DNA containing unmethylated C followed by G. Note that the sequence of bases adjacent to the CG motif also affects the stimulatory activity. The presence of unmethylated CpG motifs at a high frequency is characteristic of microbial DNA as opposed to mammalian DNA. CpG-containing DNA is the ligand for the **Toll-like receptor** TLR9. (3-10)

CR1: see **complement receptor 1**. (3-6)

C region: see **constant region**. (5-1)

CRIg: complement receptor of the immunoglobulin superfamily that is highly expressed on liver macrophages (Kupffer cells) and that mediates the clearance of pathogens bound to complement components C3b and iC3b from the blood. (9-2)

Crohn's disease: an inflammatory disease of unknown cause typically affecting the terminal ileum and adjacent regions of the colon and characterized by episodes of diarrhea, weight loss, abdominal pain and fever. Crohn's disease is one of two kinds of inflammatory bowel disease, the other being **ulcerative colitis**. (13-6)

cross-match: test of serum from the patient against blood cells of the donor to rule out the presence of pre-existing antibodies that could cause **hyperacute rejection**. (14-7)

cross-presentation: the presentation on MHC class I molecules of exogenous peptides derived from proteins internalized by endocytosis, phagocytosis or macropinocytosis. (4-9)

CRT$_H$2: see **chemoattractant receptor homologous molecule expressed on T$_H$2 cells**. (5-10)

cryoglobulinemia: a condition in which small immune complexes are present in the blood can be seen to precipitate when serum samples are chilled to refrigerator temperature. Cryoglobulins are antibodies that precipitate in the cold. (13-4)

cryopyrin: also known as **NALP3**, an intracellular sensor of peptidoglycan that acts as a scaffold to promote the activation of caspase 1, which is required for the processing of precursor forms of IL-1β and IL-18. (3-12)

crypt abscess: collection of neutrophils and inflammatory debris encased by collagenous reactive tissues that forms at the base (the crypt) of epithelial villi in the colon in **ulcerative colitis**. Crypts represent sites near the epithelial stem cell compartment and where Paneth cells secrete defensins and other effector molecules that interact with the intestinal microbial flora. (13-6)

CSF: see **colony-stimulating factor**. (1-1)

Csk: see **C-terminal c-Src kinase**. (5-4)

c-SMAC: see **central supramolecular activation complex**. (5-4)

C-terminal c-Src kinase (Csk): protein tyrosine kinase that phosphorylates the C-terminal negative regulatory site on Lck and other Src-family tyrosine kinases, thereby inhibiting their activity. (5-4)

CTLA4: see **cytotoxic T lymphocyte antigen 4**. (5-7)

CTLA4–Ig: a chimeric or fusion protein containing at its N-terminus the extracellular domain of CTLA4, which binds with high affinity to the costimulatory molecules B7-1 and B7-2, and at its C-terminus the Fc portion of IgG, which gives the molecule a long half-life in blood. (12-6)

C-type lectin: cell-surface or secreted protein that is characterized by a conserved three-dimensional fold and a somewhat conserved amino-acid sequence and that binds to carbohydrate ligands, typically in a calcium-dependent manner. Some immune receptors with this conserved structure have been shown to bind to protein ligands in a carbohydrate-independent manner and are called **C-type lectin-like receptors**. (2-5)

C-type lectin-like receptors: cell-surface proteins containing **C-type lectin** domains that do not bind carbohydrate ligands and are thus distinct from **C-type lectin receptors**; C-type lectin-like receptors commonly mediate phagocytic or endocytic uptake and/or signaling into the cell. (2-5)

C-type lectin receptors: cell-surface proteins that contain a **C-type lectin** domain that binds to carbohydrate ligands and that mediate phagocytic or endocytic uptake and/or signaling to the interior of the cell. (2-5)

cyclosporin: an immunosuppressive drug that when bound to cyclophilin blocks calcineurin activity and ultimately the transcriptional regulator NFAT. (14-8)

cytidine deaminase: enzyme that converts cytidine to uridine. The APOBEC1 family of cytidine deaminases act only on cytidines in nucleic acids, not on free cytosine or CTP. (3-19)

cytokine: polypeptide signaling molecule that participates in immune responses. Cytokines often act locally (in an **autocrine** or a **paracrine** manner) but can act systemically. Most cytokines are secreted molecules, but membrane-bound versions also occur in most cytokine types. (1-0, 2-6)

cytolysin-mediated translocation: mechanism by which Gram-positive pathogens introduce host-modifying proteins into target cells through cytolysin-mediated pores. (9-2)

cytotoxic T cells: T lymphocytes specialized to kill cells infected with pathogens that replicate in the cytoplasm. (1-2)

cytotoxic T lymphocyte antigen 4 (CTLA4): an inhibitory member of the CD28 family of cell surface molecules that is induced after a delay on activated T cells and binds to the B7 costimulator molecules in competition with CD28, reducing or terminating the activating signal to T cells. (5-7)

DAF: see **decay-accelerating factor**. (3-6)

dark zone: (of lymphoid follicles) region of the germinal center full of rapidly dividing **centroblasts**. (6-11)

death by neglect: the fate of double-positive thymocytes that fail to receive a positive selection signal. This occurs after 3–4 days of residence at the DP stage. (7-8)

death domain: protein domain that is found in the cytoplasmic regions of members of a subfamily of the TNF receptor family that activate apoptosis, and in adaptor molecules, and that mediates the interactions between these receptors and their adaptors. (2-9)

decay-accelerating factor (DAF); CD55: a membrane protein that protects cells against attack by the alternative pathway of complement by binding C3b and causing dissociation of factor B. (3-6)

dectin-1: a phagocytic C-type lectin-like receptor recognizing microbial β-glucan polysaccharides. (3-7)

defective ribosomal products (DRiPs): newly synthesized proteins that fail to fold properly, usually because they are defective, and are quickly degraded by the **proteasome**. (4-6)

defensins: any of a family of cationic **antimicrobial peptides** of vertebrates with three disulfide bonds and a largely beta-sheet structure. Humans have α- and β-defensins; some other mammals also have circular θ defensins. Cationic, cysteine-rich antimicrobial peptide molecules also called defensins are found in insects and plants, although it is not known whether they are derived from a common ancestral gene. (3-1)

degranulation: stimulation-dependent exocytosis of mast cell granules. Mast cell degranulation is stimulated by cross-linking by allergen of specific IgE bound to FcεRI, but can also be induced by other stimuli which may account for non-atopic forms of asthma and edema caused by some insect bites. (13-1)

delayed-type hypersensitivity (DTH): immune response to a non-infectious environmental antigen resulting in the activation of T_H1 or cytotoxic T cells and inflammation and/or tissue destruction. The response can take days to develop because of the need to recruit T cells to the site of the antigen. (13-0)

dendritic cells: specialized cells that ingest debris and infectious agents in the peripheral tissues and migrate to lymphoid tissues where they present fragments of the ingested particles for recognition by T lymphocytes in the activation of adaptive immune responses. Some dendritic cells are resident in lymphoid tissues and function as phagocytes there before differentiating into antigen-presenting cells. (1-0)

dermal dendritic cells: immature dendritic cells resident in the dermal layer of the skin: they are the **interstitial dendritic cells** of skin. (1-3)

diapedesis: process by which cells migrate between endothelial cells out of the blood into tissues. (3-14)

Dicer: endonuclease that recognizes double-stranded RNA and cleaves it into short fragments 21–25 nucleotides in length, called **siRNAs**. (3-19)

DiGeorge syndrome: a genetic disease caused by a deletion in chromosome 22q11 and in which development of several organs is compromised and the thymus fails to develop, leading to T cell immunodeficiency. (5-0)

direct presentation: allogeneic graft antigen-presenting cells activating a host T cell directly. (14-7)

DM: a non-classical class II MHC molecule of humans that associates with newly synthesized MHC class II molecules after the degradation of the **invariant chain**, resulting in the release of **CLIP** from the peptide-binding groove and chaperoning the loading of peptide. In mouse it is called H-2M. (4-7)

DO: a non-classical MHC class II molecule of humans that associates with **DM** in B cells and inhibits the chaperone **DM** until the B cell is activated. In mouse it is called H-2O. (4-7)

double-negative (DN) thymocytes: immature T cells at the earliest stage of T cell development in the thymus, characterized by the absence of CD4 or CD8. (7-7)

double-positive (DP) thymocyte: immature T cells at an intermediate stage in the development of αβ T cells, during which the cells express CD4 and CD8. (7-7)

DRiPs: see **defective ribosomal products**. (4-6)

DTH: see **delayed-type hypersensitivity**. (13-0)

EAE: see **experimental allergic encephalitis**. (12-7)

ECM: see **extracellular matrix**. (2-4)

edema: swelling of tissues due to increased vascular permeability. (3-13)

effector caspases: caspases that are activated by **initiator caspases**, amplifying the processes that result in the stereotypical features of **apoptosis**, including loss of plasma membrane asymmetry, nuclear condensation and endonuclease attack on nuclear DNA. (2-11)

effector cells: cells that are equipped to activate or destroy other cells directly. (1-2)

effector memory T cells (T_EM): long-lived effector T lymphocytes that are generated from antigen-specific T cells activated during a primary immune response and that disperse to the tissues or the bone marrow and can provide immediate effector action on recurrence of infection. (5-16)

effector T cells: T cells that secrete cytokines (in the case of helper T cells) or deliver cytotoxic signals or effector molecules (in the case of cytotoxic T cells) immediately on activation through recognition of peptide–MHC by their antigen receptors. (1-0, 5-0)

efferent lymphatics: lymphatic vessels leaving **lymph nodes** and returning **lymph** to the bloodstream. (1-6)

ELISA: see **enzyme-linked immunosorbent assay**. (6-6)

ELISPOT: an assay similar to a capture **ELISA** in which cells are incubated in a plastic dish coated with the capture antibody, followed by incubation of the dish with an enzyme-conjugated detection antibody and finally incubation with a substrate that is converted to an insoluble colored product, so that colored spots appear where cells secreting the molecule of interest were present. (6-6)

encapsulated bacteria: bacteria with a thick polysaccharide capsule covering up other cell wall structures, such as LPS or peptidoglycan. (3-2)

endoplasmic reticulum aminopeptidase1 (ERAP1): an aminopeptidase that is resident in the endoplasmic

reticulum and can trim peptides to 8–10 amino-acid residues long, to fit in the peptide-binding groove of MHC class I molecules. (4-6)

endosomal sorting complex required for transport (ESCRT): cellular machinery for sorting ubiquitinated cargo proteins to endosomal membranes for degradation; mediates intra-luminal budding and extrusion to generate a multivesicular body, or MVB. (10-4)

endotoxin: (of bacteria) a non-secreted toxin inherent in the cell membrane and that induces strong innate inflammatory responses that can be fatal when systemic. (9-3)

enterotoxins: secreted toxins that target intestinal mucosal cells and induce diarrhea by interfering with normal water resorption and/or secretion. Some enterotoxins are **superantigens** and cause diarrhea through secondary effects on mucosal cells of the local release of high levels of host cytokines. (9-4)

enzyme-linked immunosorbent assay (ELISA): a technique for measuring the amount of a molecule in a sample in which the antigen is put or trapped on a solid support and then an antibody with an attached enzyme is used to bind to the trapped antigen. Alternatively, an ELISA can be used to measure the amount of a specific antibody in a sample. The amount of enzymatic activity is directly related to the amount of the antigen or antibody on the plate. (6-6)

eomesodermin: a T-box family transcriptional regulator (Tbr2) that directs trophoblast and mesoderm development in vertebrates and is implicated in the differentiation of effector programs in CD8 T cells and NK cells. (5-10)

eosinophils: cells containing cytoplasmic granules that stain with the dye eosin and contain inflammatory mediators that are released on activation by T cells or antibody-coated parasites. (1-2)

epigenetic: inherited through mechanisms that are not dependent on DNA sequence. Known epigenetic mechanisms often concern gene regulation and are dependent on modifications of the DNA and associated proteins that affect local chromatin structure. (5-9)

epitope: molecular feature of an antigen that is specifically recognized by a lymphocyte or an antibody. Epitopes recognized by T cell receptors are peptides. Epitopes for antibodies on proteins can be linear, as for T cells, or can be made up of different regions of a polypeptide that come together in the three-dimensional structure of the antigen. The latter kind of epitope is called a discontinuous or conformational epitope. (1-0, 6-5)

epitope spreading: a phenomenon in which an autoimmune response is directed initially against one or a few self antigens, but later becomes broader and is directed at more **autoantigens**. (12-4)

ERAP1: see **endoplasmic reticulum aminopeptidase1**. (4-6)

ERp57: a cellular thiol oxidoreductase that is thought to chaperone the formation of the intrachain disulfide bonds in newly synthesized MHC class I alpha chains. (4-6)

erythema: redness of tissues due to increased blood flow. (13-1)

erythrocytes: red blood cells containing the oxygen-carrying protein hemoglobin. (1-1)

escape variants: viral progeny that differ from the parental virus through the acquisition of mutations that prevent recognition by antibody or binding to MHC class I so that the virus is no longer bound by antibody or presented to cytotoxic CD8 T cells in a given host. (10-6)

ESCRT: see **endosomal sorting complex required for transport**. (10-4)

exotoxin: molecule secreted from pathogens that has toxic effects on host cells. (9-4)

experimental allergic encephalitis (EAE): an auto-immune disease induced in animals by immunization with a myelin protein in adjuvant. The pathologic mechanisms in EAE are quite similar to those in the human disease multiple sclerosis. (12-7)

extracellular matrix (ECM): matrix of proteins that forms in tissues between cells; it can include fibronectin, collagen, vitronectin, and so on. (2-4)

extravasation: process of moving out of the blood through an intact endothelial layer and into the tissue, as occurs during inflammation or lymphocyte recirculation. (3-14)

extrinsic pathway: (of **apoptosis**) pathway inducing apoptosis downstream of TNF-receptor family members with death domains in their cytoplasmic tails. (2-11)

Fab fragment: fragment of an antibody generated by limited digestion with the protease papain. The Fab fragment contains the light chain plus the V and C_H1 domain of the heavy chain, and represents a monomeric antigen-binding fragment without effector functions other than neutralization. (6-1)

FADD: see **Fas-associated death domain**. (2-9)

Fas: TNF receptor superfamily member that induces apoptosis in response to Fas ligand (FasL) made by T cells. Fas is one of the mechanisms by which virus-infected cells are killed and is also important for immune tolerance to self. (2-9)

Fas-associated death domain (FADD): adaptor molecule that links death-domain-containing receptors of the TNF receptor superfamily to apoptosis. (2-9)

Fc fragment: fragment of an antibody generated by limited digestion with the protease papain and containing the C_H2 and C_H3 domains that connect to complement and Fc receptors. (6-1)

Fc receptors: immunoreceptors that mediate many of the effects of antibodies. These receptors are so named because the part of the antibody molecule they recognize is called the Fc region. Fc receptors mediate cellular functions such as phagocytosis, cell killing, and degranulation (secretion of stored mediators). (2-2)

FcRn: neonatal FcR, involved in transport of **IgG** across endothelial cells. (6-2)

FDC: see **follicular dendritic cell**. (6-11)

ficolin: any of a family of a structurally related, carbohydrate-recognizing proteins of innate immunity, containing a collagen-like domain and a fibrinogen domain. The name is derived from these domains. (3-2)

FK506: an immunosuppressive drug that when bound to **FKBP** blocks calcineurin activity and ultimately the transcriptional regulator NFAT. (14-8)

FK506 binding protein (FKBP): an intracellular peptidyl-prolyl *cis–trans* isomerase that helps catalyze protein folding and also binds to the immunosuppressive drugs **FK506** and **rapamycin** and participates in their different immunosuppressive actions. (14-8)

FKBP: see **FK506 binding protein**. (14-8)

flow cytometry: a technique used to enumerate and analyze a sample of cells by incubating them with one or more fluorescently labeled antibodies and/or other molecules that can bind to cellular components and measuring the fluorescence intensity of each fluor for each cell. (6-6)

follicle: (of lymphoid tissue) B cell area of **secondary lymphoid tissue**. (1-6)

follicular B cell: a B lymphocyte that recirculates through the blood and follicular regions of the secondary lymphoid tissues and participates in **T cell-dependent antibody responses** leading to the production of high-affinity antibodies from **long-lived plasma cells** and memory B cells. (6-9)

follicular dendritic cell (FDC): specialized non-hematopoietic cell found in B cell areas of secondary lymphoid tissue and specialized for collecting antigen bound to antibody or complement components and presenting it to **centrocytes**, which can thereby be selected for high-affinity binding. (1-7, 6-11)

formyl peptide receptor (FPR): seven-transmembrane chemotactic receptor expressed on neutrophils and monocytes and mediating recruitment of these cells to sites of infection. (3-14)

FoxP3: a member of the winged helix-forkhead DNA-binding domain protein family highly conserved in vertebrates that acts as a transcriptional repressor. In immunology, FoxP3 is a lineage-specific marker for regulatory T cells (T_{REG} cells). Genetic deficiencies of FoxP3 in the scurfy mouse and human IPEX syndrome (immunodysregulation, polyendocrinopathy, enteropathy, X-linked syndrome) lead to fatal inflammatory autoimmune diseases. (5-13)

FPR: see **formyl peptide receptor**. (3-14)

framework residues: conserved amino-acid residues observed upon comparing Ig or TCR V regions with one another. These residues are involved in maintaining the structure of the **Ig domain**. (2-1)

γ_c chain: see **common γ chain**. (2-8)

$\gamma\delta$ T cells: T lymphocytes expressing receptors containing variable γ and δ chains rather than α and β chains, and thought to be specialized to respond to a limited range of damage-induced and environmental antigens. (8-4)

GAP: see **GTPase-activating protein**. (2-0)

GATA-3: a member of a family of six zinc-finger transcription factors that share a conserved DNA-binding domain, which interacts with a core GATA nucleotide sequence that gives the family its name. All members of the family are highly conserved in vertebrate evolution and play an important part in cell-fate specification and differentiation in many tissues. In immunology, GATA-3 is important for directing the expression of the cytokine set characteristic of T_H2 cells and thereby contributing centrally to the polarization of this cell type. (5-10)

GEF: see **guanine nucleotide exchange factor**. (2-0)

gene conversion: the nonreciprocal transfer of information between homologous sequences. (7-4)

germinal center: site of vigorous proliferation of B cells in the B cell follicles of secondary lymphoid organs, and where B cells activated by helper T cells are selected for antigen receptors with high affinity for antigen. (1-6)

germinal center reaction: the T cell-dependent antibody response of follicular B cells in which there is rapid proliferation combined with **somatic hypermutation**

and **class switch recombination**, resulting in the production of high-affinity IgG, IgA and/or IgE antibodies, long-lived plasma cells, and memory B cells. (6-11)

GITR: see **glucocorticoid-induced TNF-family receptor**. (5-13)

glucocorticoid-induced TNF-family receptor (GITR): a TNF-family receptor expressed on activated effector T cells and constitutively on regulatory T cells. GITR signals render activated effector T cells refractory to T_{REG}-mediated suppression. (5-13)

glycosphingolipids: amphiphilic plasma membrane molecules consisting of one or more sugar residues linked with sphingosine to long-chain fatty-acid side chains. (8-3)

glycosyltransferase: enzyme that adds a particular sugar to an oligosaccharide in a particular place. (14-6)

glyoxylate cycle: the reactions catalyzed by the enzymes isocitrate lyase and malate synthase, which provide a by-pass in the tricarboxylic acid (TCA) cycle to allow the assimilation of two-carbon compounds, allowing organisms to synthesize glucose in glucose-deficient environments. (11-1)

goblet cells: mucus-producing epithelial cells; induced *in vitro* and *in vivo* by the differentiation of epithelial cells with IL-4 or IL-13. (11-4)

Gomori methenamine silver stain: stain used to detect reactive aldehydes in fungal cell walls by the precipitation of black silver salts at aldehyde groups exposed by treatment with chromic acid. (11-2)

Goodpasture's syndrome: an autoimmune disease often occurring with rapid onset in which antibodies are made against collagen type IV. These antibodies bind to basement membranes in the kidneys and lungs and induce damage through neutrophil secretion of proteases and reactive oxygen species. (12-5)

gp130: type I cytokine-receptor family member that mediates signaling by the IL-6 receptor and several other growth and differentiation factors of diverse function. (2-8)

G protein: any of a large class of GTPases that act as molecular switches that are active when bound to GTP and inactive when bound to GDP. They may be heterotrimeric G proteins with α, β and γ subunits, which typically signal from seven-transmembrane receptors, or small G proteins of the Ras superfamily. G proteins are also known as **GTP-binding proteins** or guanine-nucleotide-binding proteins. (2-0)

graft-versus-host disease (GVH): a disease in which mature T cells are transferred as a component of a bone marrow transplant and react strongly to the transplantation antigens of the recipient. (14-6)

graft-versus-leukemia effect: the hypothesis that alloreactive T cells from a bone marrow transplant can kill residual leukemia cells remaining after chemotherapy or radiation treatment. This hypothesis is based on enhanced patient survival if mature T cells are not removed from allogeneic bone marrow before transplant, a situation that greatly increases the chance of graft-versus-host disease. (14-5)

Gram-negative bacteria: major subgroup of bacteria characterized by the presence of an outer membrane bilayer made up of phospholipids and **LPS**. (3-0)

Gram-positive bacteria: major subgroup of bacteria lacking an outer membrane but typically containing a thick **peptidoglycan** layer. (3-0)

granulocytes: collective term for **neutrophils**,

basophils, **eosinophils** and **mast cells**, all of which contain vesicles loaded with inflammatory mediators that stain to give a granular appearance under the light microscope. (1-2)

granuloma: tightly structured spherical inflammatory lesion characterized by central necrosis and containing microorganisms and infected macrophages in various stages of maturation and activation. Interspersed with these are activated CD4 T cells, and surrounding them are epithelioid fibroblasts and activated CD8 T cells. Granulomas ultimately undergo fibrosis and calcification of the outer mantle. (3-9)

granzymes: proteases that are contained in membrane-bounded granules of cytotoxic lymphocytes and are released on activation of the cytotoxic cell. Many and perhaps all induce programmed cell death on entry into target cells. (5-14)

Graves' disease: an autoimmune disease in which autoantibodies react with the receptor for thyroid-stimulating hormone (TSH) and activate it. The result is hyperthyroidism, as the thyroid cells continually secrete thyroid hormones in an unregulated fashion. Graves' disease is typically treated by ablating the thyroid gland and giving thyroid hormone replacement therapy. (12-5)

GTPase-activating protein (GAP): protein that accelerates the intrinsic GTPase activity of **G proteins**, thereby inactivating them. (2-0)

GTP-binding protein: see **G protein**. (2-0)

guanine nucleotide exchange factor (GEF): protein that facilitates the exchange of GDP for GTP in **G proteins**, thereby activating them. (2-0)

GVH: see **graft-versus-host disease**. (14-6)

H2-M3 restricted T cells: T lymphocytes recognizing the non-classical MHC class I molecule H2-M3. (8-5)

HA: see **hemagglutinin**. (10-7)

HAART: see **highly active antiretroviral therapy**. (10-6)

HAMA: see **human anti-mouse Ig antibody**. (6-4)

hapten: small molecule that can become covalently bound to a protein and serve as a determinant that can be recognized by antibodies, or by T cells when bound to a peptide presented on an MHC molecule. (13-0)

haptenation: the process whereby a chemically reactive small molecule (the **hapten**) becomes covalently bound to a protein molecule, where it acts as a determinant that can be recognized by antibodies, or by T cells when bound to a peptide presented on an MHC molecule. Production of antibodies recognizing the hapten can be induced by helper T cells recognizing peptides derived from the protein. The protein component is referred to as the carrier. T cells can respond to a haptenated peptide from the hapten–carrier conjugate or to an unhaptenated peptide if they are not tolerant to it. (13-0)

heavy chain: the immunoglobulin heavy chains are the larger of the two kinds of polypeptide chains in the immunoglobulin molecule. Each light chain has a **variable region** contributing to the antigen-binding site, and a **constant region** that mediates the effector function of the molecule. (1-4)

hemagglutination: the property by which antibodies binding to red blood cells cause them to agglutinate. (6-5)

hemagglutinin (HA): influenza homotrimeric envelope protein that binds to widely distributed host-cell-surface

glycoproteins and glycolipids that are modified by sialic acid additions and is required for viral entry into host cells. Binding by HA induces internalization and transport of the sialylated receptors to endosomal compartments, where low pH induces a marked conformational change in HA that mediates fusion of the viral and host membranes, allowing translocation of the influenza nucleocapsid to the cytosol. (10-7)

hematopoietic stem cell (HSC): self-renewing cell that gives rise to all the red and white blood cells. (1-1)

hematopoietins: cytokines that promote proliferation and/or lineage-specific differentiation of hematopoietic cells. Hematopoietins also can have functional effects on mature members of specific lineages. (1-1, 2-7)

hemoglobinopathies: genetic mutations affecting the structure of hemoglobin and, hence, its capacity to bind and release oxygen; examples include sickle cell disease and thalassemia. They are evolutionarily selected in areas of high malaria transmission as a result of a restricted ability of parasites to grow in red cells heterozygous for the abnormal hemoglobin. (11-0)

hemolytic disease of the newborn: hemolytic anemia in which IgG antibodies against red blood cell antigens, typically the Rh antigen, are made by the mother during pregnancy, cross the placenta, bind to fetal red blood cells and induce their clearance by phagocytes. (13-4)

herpesvirus: any member of a family of more than 100 large, enveloped, double-stranded DNA viruses that cause widespread infections that are commonly asymptomatic in healthy individuals. Members include (with occasional symptomatic infections) herpes simplex virus (HSV) 1 and 2 (causing cold sores and genital sores, respectively), cytomegalovirus (CMV) (causing systemic disease in immunosuppressed individuals), Epstein–Barr virus (EBV) (causing mononucleosis, lymphomas, and nasopharyngeal carcinomas), human herpesvirus 8 (HHV8) (causing Kaposi's sarcoma), and varicella zoster virus (causing chickenpox and shingles). (8-1)

heterotrimeric G protein: signaling G protein composed of three different subunits, an α subunit with GTPase activity, and associated regulatory β and γ subunits. These GTPases act to relay signals from seven-trans-membrane receptors to downstream targets. Exchange of bound GDP for GTP on the α subunit causes dissociation of the heterotrimer into a free α subunit and a βγ heterodimer; hydrolysis of the bound GTP causes reassociation of the subunits. (2-13)

heterozygosity: the possession by an individual of two different **alleles** of a gene. (4-3)

HEV: see **high endothelial venules**. (1-7)

high endothelial venules (HEV): specialized blood vessels supplying lymph nodes and expressing adhesive molecules and chemokines specifically recognized by naïve circulating lymphocytes, which are thereby enabled to enter the secondary lymphoid tissue. (1-7)

highly active antiretroviral therapy (HAART): a combination of an HIV aspartyl protease inhibitor with at least two reverse transcriptase inhibitors used to treat AIDS. Combination therapy targeting multiple enzymes is necessary to avoid the rapid selection of resistance mutants. (10-6)

highly pathogenic avian influenza (HPAI): influenza viruses with high virulence for chickens and other birds. These viruses commonly express H5, H7 or H9 **hemagglutinin** proteins with an expanded basic amino-acid repeat that increases the ease of proteolytic digestion required to activate the fusion domain. (10-7)

hinge region: flexible stretch of amino acids between two domains, C_H1 and C_H2, of an IgG molecule, that allows the antigen-binding sites mobility relative to one another. (6-1)

histamine receptors: seven-transmembrane G-protein-coupled receptors that respond to histamine. Four types, called H_1, H_2, H_3 and H_4, are known. H_1 is primarily important for allergic reactions. (13-1)

histocompatibility: the match between the tissue of a graft and that of a graft recipient, which determines whether or not a graft will be rejected. (14-6)

Hlx: a member of the homeodomain family of transcriptional regulators expressed in discrete cell types at specific times during development and in adult hematopoietic tissues and cells, as well as in diverse mesodermal tissues. In immunology, it is important for cooperating with T-bet to stabilize IFN-γ expression in T_H1 cells. (5-10)

homeostatic chemokine: constitutively expressed chemoattractant molecule that directs the migration of lymphocytes and dendritic cells into specialized regions of the secondary lymphoid tissues. Also known as **lymphoid chemokine**. (2-13)

HPAI: see **highly pathogenic avian influenza**. (10-7)

HSC: see **hematopoietic stem cell**. (1-1)

human anti-mouse Ig antibody (HAMA): antibody produced by an immune response by humans to determinants on mouse Ig, which limits the therapeutic usefulness of mouse **monoclonal antibodies**. (6-4)

humanized antibodies: monoclonal antibodies made by using mouse cells and subsequently manipulated by recombinant DNA technology to produce a molecule most of which is derived from human sequences, with only the antigen-binding loops of the mouse molecule. (6-4)

hybridoma: cell produced by fusing an activated B cell to a **plasmacytoma** cell line. (6-4)

hygiene hypothesis: an epidemiologically-driven hypothesis that the increased prevalence of allergic diseases in developing countries is due to a decreased number of infections and/or exposure to microbial products early in life. It is speculated that this leads either to greater T_H2 dominance of immune responses to allergens or to more vigorous immune responses to allergens, or both. (13-3)

hyperacute rejection: very rapid rejection of an organ graft due to preexisting antibody that reacts to a cell surface molecule on the endothelium of the transplanted organ. This is typically due to ABO blood type differences or to antibodies to MHC I molecules of the graft. (14-7)

hypersensitivity reaction: immune response to a non-infectious environmental antigen resulting in inflammation or organ dysfunction. Hypersensitivity reactions may result in inflammation of skin or mucosal epithelia due to cell- or antibody-mediated immune responses, or in organ dysfunction due to immune complexes. (13-0)

hypervariable region: region of a lymphocyte receptor for antigen, or the gene encoding it, that contains a large number of amino acids or nucleotides that vary between cells of different antigen specificities and determine the antigen specificity of the cell. These regions occur at the sites that in the folded protein participate in the antigen-binding site. (5-1)

IAP: see **inhibitor of apoptosis**. (2-12)

iC3b: a large proteolytic fragment of complement component C3b that has lost enzymatic activity in the complement pathway but is recognized by complement receptors 2, 3 and 4 (CR2, CR3, CR4) and activates **phagocytosis**. (3-7)

iccosomes: cell fragments released by follicular dendritic cells and containing immune complexes bound to their surface. Also called immune-complex-coated bodies. (6-12)

I chain: see **invariant chain**. (4-7)

ICOS: see **inducible costimulator**. (5-8)

ICOSL: the ligand for **ICOS**, belonging to the B7 subfamily of the immunoglobulin superfamily and expressed on the surface of activated dendritic cells, B cells, macrophages and some non-immune cells. (5-8)

IEL: see **intraepithelial lymphocytes**. (8-5)

Ig: see **immunoglobulin**. (6-0)

Igα/Igβ: the disulfide-linked heterodimer that serves as the signaling component of the antigen receptor of B lymphocytes. Both proteins have an amino-terminal single Ig-like domain as an extracellular domain, a transmembrane domain and a moderate length (~50 amino acids) cytoplasmic domain containing an immunoreceptor tyrosine-based activation motif (ITAM). (6-7)

Ig domain: stably folded region of an immunoglobulin, as defined originally by protease resistance. (2-1)

Ig-like domain: domain of protein structure with amino-acid sequence and structural homology to **Ig domains**. Ig-like domains can be identified by amino-acid sequence homology to those found in antibody molecules. Ig-like domains from diverse proteins seem to all fold into beta-sandwich structures, referred to as the immunoglobulin fold. (2-1)

Ii: see **invariant chain**. (4-7)

I-κB: see **inhibitor of NF-κB**. (2-10)

I-κB kinase (IKK): complex consisting of two related kinase subunits, called IKKα and IKKβ, and a scaffolding subunit, IKKγ or NEMO, that activates the transcriptional activator **NF-κB** by marking its inhibitor, **I-κB**, for degradation. There are two pathways of activation, the classical and the alternative pathway. IKKβ and IKKγ are necessary for responses via the classical pathway but not via the alternative pathway, which only requires IKKα. Conversely, IKKα is dispensable for the classical pathway. IKKγ (NEMO) is encoded on the X-chromosome and is partly defective in the disease X-linked anhidrotic ectodermal dysplasia with immunodeficiency. (2-10)

IKK: see **I-κB kinase**. (2-10)

IL-1 receptor associated kinase (IRAK): any of four protein kinases, IRAK-1, -2, -M and -4. IRAK-1 and IRAK-4 contribute to TLR and IL-1 receptor signaling. (3-11)

IL-17 family: a six-member family of homodimeric cytokines designated IL-17A to IL-17F; IL-17E is also called IL-25; IL-17A and IL-17F are most homologous and reside next to each as duplicated genes on human chromosome 6 (mouse chromosome 1), and these two members are secreted by effector T cells. T-tropic Herpesvirus saimiri encodes an IL-17A-like molecule (viral IL-17). (5-13)

IL-21: a type I four-helix-bundle-type cytokine that binds to IL-21R/common γ chain receptors on activated B cells, T cells and NK cells. (5-13)

immature dendritic cells: dendritic cells that have differentiated from their lymphoid or myeloid precursors

and actively internalize microbial components and cellular debris by phagocytosis as well as receptor-mediated endocytosis and macropinocytosis but do not present internalized antigen on the cell surface. (4-1)

immediate-type hypersensitivity: see **allergy**. (13-0)

immune complex: noncovalent molecular complex formed when an antibody binds to the antigen it specifically recognizes. (6-5)

immune exhaustion: a state defined by loss of virus-specific T cell responses despite persistent viral infection and more commonly seen using large viral inocula with retroviruses or RNA viruses with high rates of mutation, which allow escape variants to arise through evasion of immune detection. (10-9)

immune memory: rapid response of the **adaptive immune system** to exposure to **antigens** previously encountered. (1-0)

immune precipitate: precipitate formed by the complex of a polyclonal antibody with a complex antigen, when the ratio of the components is favorable for forming a large lattice. (6-5)

immune surveillance: the hypothesis that the immune system recognizes and eliminates many nascent tumors before they become clinically evident. (14-5)

immune tolerance: non-responsiveness of the adaptive immune system to an antigen. Tolerance is usually to self components but can also occur to transplanted tissue, infectious agents, therapeutic proteins, and so on. (1-4, 12-0)

immunoblotting: also called western blotting. A technique for analyzing the proteins in a sample in which the sample is fractionated by SDS-polyacrylamide gel electrophoresis, which resolves denatured polypeptides by their approximate molecular mass; the resolved proteins are transferred to a solid support such as a flexible membrane made out of nitrocellulose, which is then reacted with labeled antibody specific to the protein of interest. (6-6)

immunodysregulation, polyendocrinopathy, enteropathy, X-linked (IPEX) syndrome: an X-linked multiorgan autoimmune disease caused by mutations in FoxP3, a transcription factor required for the development of regulatory T cells. (7-10)

immunofluorescence: a technique in which fluorescently labeled antibodies are used to determine the location of the corresponding antigen in a tissue section or in cells. (6-6)

immunogenic: able to induce an adaptive immune response. Many antigens recognized by lymphocytes of the adaptive immune system fail to elicit immune responses in the absence of **adjuvants** that promote adaptive immune responses, often by activating cells of innate immunity. (14-2)

immunoglobulin (Ig): a class of proteins produced by B lymphocytes of the immune system and that recognizes and binds to foreign antigens. Also called an antibody. The most common form of an immunoglobulin in the blood has a dimeric structure with two antigen-binding sites. (6-0)

immunoglobulin A (IgA): a class of antibody that is secreted across epithelial layers. (6-2)

immunoglobulin classes: different **isotypes** of antibodies characterized by the type of heavy chain. (6-2)

immunoglobulin D (IgD): a class of antibody that is primarily an antigen receptor found on naïve B cells. (6-2)

immunoglobulin E (IgE): a class of antibody that binds with high affinity to mast cells and basophils and participates in immune defense against helminthic worms and also in the symptoms of allergies and asthma. (6-2)

immunoglobulin G (IgG): the class of antibody that is the major antibody in the blood and extravascular fluids. (6-2)

immunoglobulin M (IgM): the class of antibody that is made initially by newly generated B cells as a membrane protein and serves as a component of the B cell antigen receptor. Later, after antigen stimulation, IgM is the first class of antibody to be secreted, in which form it is a pentamer. (6-2)

immunoglobulin (Ig) superfamily: family of proteins containing at least one immunoglobulin-like domain and for which the **Ig-like domain** is a major structural element. MHC proteins have an Ig-like domain, but their main structural element is a unique peptide-binding domain and therefore, it is best to consider MHC molecules as a separate family of proteins. (2-1)

immunohistochemistry: a technique in which enzyme-linked antibodies and a substrate that is converted to a colored insoluble product are used to determine the location of the corresponding antigen in a tissue section. (6-6)

immunological synapse: area of contact between a T cell and an antigen-presenting cell occurring during antigen recognition by the T cell. This term is sometimes also used to refer to contact between two immune cells when neither is a T cell. (5-4)

immunoprecipitation: isolation of a protein or other macromolecule through the use of specific antibodies that bind to the molecule of interest and are usually collected by beads covalently coupled to an antibody that reacts to the antibody being used or to protein A from *Staphylococcus aureus*, a protein that binds to antibodies. (6-5)

immunoproteasome: a specialized proteasome that is present in specialized immune cells and induced by interferon-γ in other cells and that processes peptides more effectively than the constitutive proteasome for presentation by MHC class I molecules. (4-8)

immunoreceptor tyrosine-based activation motif (ITAM): a sequence in the cytoplasmic domains of activating immunoreceptors that when phosphorylated becomes a binding site for the tandem SH2 domains of Syk and ZAP-70. The ITAM consensus sequence is D/EXYXXL/IX₇YXXL/I (one-letter amino-acid code; X = any amino acid). (2-2)

immunoreceptor tyrosine-based inhibitory motif (ITIM): a sequence in the cytoplasmic domains of inhibitory immunoreceptors that, upon phosphorylation on tyrosine recruits signaling inhibitors (the consensus sequence is V/IXYXXL/V). (2-2)

immunotherapy: (of allergies) a treatment for atopic diseases in which the allergen responsible for the condition is injected repeatedly. Immunotherapy is often successful in alleviating symptoms, but the mechanism is controversial. (13-2)

indirect presentation: antigen presentation to **alloreactive T cells** by host antigen-presenting cells that have taken up antigens from a tissue graft. (14-7)

inducible costimulator (ICOS): a receptor belonging to the CD28 subfamily of the immunoglobulin superfamily that is induced on activated and effector T lymphocytes and is important in the maintenance of effector and memory CD4 T cells and especially those interacting with B cells. (5-8)

inducible nitric oxide synthase (iNOS): one of three isoforms of NOS, the enzyme that makes nitric oxide, and that is synthesized by phagocytic cells as part of the microbicidal response. The other forms are found in neurons and endothelial cells and are constitutively expressed and regulated by intracellular calcium levels. (3-9)

inflammasome: multiprotein complex that promotes inflammatory responses by processing precursor forms of inflammatory cytokines. (3-12)

inflammation: coordinated response to infection or tissue injury recognized since ancient Roman times as characterized by heat, pain, redness and swelling (*calor, dolor, rubor, tumor* in Latin). (3-13)

inflammatory chemokine: chemokine that is produced at an inflammatory site and mediates the attraction of immune cells from the blood to that location. Different chemokines attract different types of cells and thereby dictate the nature of the inflammation. (2-13)

inflammatory cytokines: cytokines that are released by phagocytes of the **innate immune system** in the presence of microorganisms, or by activated lymphoid cells, and that act on blood vessels and cells of the immune system to induce or amplify immune responses. (1-0)

inflammatory response: release of **cytokines** by leukocytes at a site of infection causing dilatation and increased permeability of blood vessels and the recruitment of immune cells. (1-0)

inhibitor of apoptosis (IAP): any of a family of structurally related molecules that bind to and inhibit effector caspases and active caspase 9 as well as promoting their degradation by the proteasome. These proteins are thought to set a threshold for the activation of caspases to induce apoptosis. (2-12)

inhibitor of NF-κB (I-κB): any of a family of proteins that inhibit the transcriptional activator **NF-κB** by binding to it and preventing it from translocating to the nucleus. It exists in multiple isoforms, all of which contain ankyrin repeat structures that mediate the interaction with NF-κB. (2-10)

inhibitory receptors: receptors that block cell activation, generally when they bind their ligands. In immune cells the inhibitory function of these receptors is usually mediated by a consensus sequence in the cytoplasmic domain called the **ITIM**. (2-2)

initiator caspases: caspases that can be activated by death-domain-containing receptors or by perturbations to internal compartments such as the mitochondria or the endoplasmic reticulum. (2-11)

innate immunity: immune responses mediated by cells and molecules recognizing conserved features of microorganisms and activated immediately on encounter with them. (1-0)

iNOS: see **inducible nitric oxide synthase**. (3-9)

inside-out signaling: process whereby stimulation of cells via receptors such as antigen receptors or chemokine receptors leads to intracellular signaling events that alter the **affinity** and/or lateral mobility of **integrin** molecules in the plasma membrane so as to increase adhesion of those integrins with ligands on other cells or of the **extracellular matrix**. The flow of information through the integrin is from its cytoplasmic tails to the extracellular portion. (2-4)

instruction model: hypothesis to explain the mechanism of positive selection of developing T lymphocytes in which a DP thymocyte can recognize the type of MHC

molecule that it has interacted with and downregulates the coreceptor that is not needed to interact with that class of MHC. (7-9)

integrins: large family of αβ heterodimeric molecules that participate in cell–cell and cell–matrix adhesion. Adhesion by integrins of cells of the immune system is generally regulated by activation signals or chemokine receptor signals. (2-4)

interferon: any of several cytokines that act in immunity to inhibit viral replication or promote killing activity of macrophages, natural killer cells and cytotoxic T cells. Interferons are divided into two types: **type 1 interferons (interferon-α** and **interferon-β)** made by any virus-infected cell and by macrophages, dendritic cells and **plasmacytoid dendritic cells**, and acting to inhibit viral replication; and **type 2 interferon (interferon-γ)**, which is made by natural killer cells and some T cells and whose main function is the activation of macrophages. (3-16)

interferon-α: any of a large number of closely related **type 1 interferons**. (3-16)

interferon-β: type 1 interferon that is made before **interferons-α** by virus-infected cells. (3-16)

interferon-γ: cytokine made by T cells and natural killer cells that promotes killing of internalized microbes by macrophages and also has important antiviral defense roles (also called **type 2 interferon**). (3-16)

interferon regulatory factor (IRF): any of a family of nine structurally related transcription factors, many of which play key roles in the induction of **interferon** and also in responses to interferon. IRFs also participate in TLR signaling. (3-16)

interleukin: cytokine participating in immune responses and originally thought always to be produced by leukocytes and to act on other leukocytes. Many cytokines have been given a systematic name of interleukin *x*, where *x* is a number from 1 to at least 29. It is now clear that often the range of action of interleukins extends to non-hematopoietic lineage cells. (2-6)

internalins: protein products of a *Listeria* gene family containing leucine-rich repeats; most are anchored covalently in the peptidoglycan cell wall; inlA and inlB are invasins sufficient to trigger uptake by cells expressing the appropriate receptors. (9-6)

interstitial dendritic cells: immature dendritic cells in the peripheral tissues, mucosa or dermal layers of the skin. Also known as **tissue-resident dendritic cells**. (1-3, 4-1)

intraepithelial lymphocytes (IELs): T lymphocytes localized between the epithelial cells in gut and lung mucosa. Some are conventional memory T cells; others are specialized cells some of which express γδ receptors instead of αβ receptors for antigen and recognize non-classical MHC molecules. (8-5)

intrinsic pathway: (of **apoptosis**) pathway of inducing apoptosis resulting from cell stress or loss of survival signals from outside the cell (typically adhesion and/or cytokines). (2-11)

invariant chain (I chain or **Ii):** a membrane-bound protein that associates with MHC class II molecules immediately on synthesis, stabilizing them and filling the peptide-binding groove so that unfolded proteins present in the lumen of the endoplasmic reticulum cannot bind in it. (4-7)

IPEX: see **immunodysregulation, polyendocrinopathy, enteropathy, X-linked syndrome**. (7-10)

IRAK: see **IL-1 receptor associated kinase**. (3-11)

isoglobotrihexosylceramide (iGb3): lysosomal degradation product of **glycosphingolipids**. Thought to be the CD1d ligand on which NKT cells are selected in the thymus. (8-3)

isotype: any of several highly related forms of a molecule. (6-2)

ITAM: see **immunoreceptor tyrosine-based activation motif**. (2-2)

ITIM: see **immunoreceptor tyrosine-based inhibitory motif**. (2-2)

Itk (IL-2-inducible tyrosine kinase): a TEC-family tyrosine kinase playing a central part in T cell activation through activation of calcium and diacylglycerol signaling. (5-3)

Jak (Janus kinase) tyrosine kinase: an intracellular tyrosine kinase belonging to a small family with two kinase domains, one of which is the active tyrosine kinase and the second of which does not have catalytic activity. For this reason they are named for the two-headed Roman god of gates and doorways, Janus. Jak kinases associate with cytokine type I and type II receptors and are essential for signaling by these receptors. (2-7)

junctional diversity: lymphocyte receptor chain diversity generated by partial digestion of sequences at the junctions of the V, D and J gene segments encoding the **variable region** during their assembly, with insertion of random nucleotide sequences of variable length at the same sites. (5-1)

Kaposi's sarcoma: multiple, highly vascular, nodules consisting of proliferating spindle-shaped cells, endothelial cells and infiltrating lymphocytes in the skin, mucous membranes and internal organs that occur most frequently in immunocompromised individuals as a consequence of human herpesvirus 8 infection. (10-3)

Katayama fever: febrile illness marked by fatigue and tissue inflammatory responses to schistosome egg antigens; most commonly occurs in nonimmune adult travelers who encounter heavy schistosome infection; typically due to *S. japonicum* or *S. haematobium* with onset at the time of egg-laying, perhaps reflecting a serum sickness reaction. (11-5)

kinetic signaling model: a recent variation of the **instruction model** for positive selection of developing T lymphocytes in which the fate of the cell is determined by whether or not TCR signaling declines when CD8 expression is downregulated. If it does, then the cell turns off CD4, turns CD8 back on and adopts the CD8 lineage; if signaling is maintained, the cell chooses the CD4 lineage. This model incorporates features of the **stochastic/selection model**, but there is no stochastic element. (7-9)

kinetoplastid protozoa: flagellated protozoa containing a single mitochondrion with mitochondrial DNA massed into a region called the kinetoplast; medically important kinetoplast parasites belong to the genera *Leishmania* or *Trypanosoma*, while *Phytomonas* species cause disease in plants. (11-3)

kinin cascade: proteolytic cascade that is initiated by injury to tissue and ends with the production of bradykinin, which increases the permeability of blood vessels. (3-13)

KIR family: family of receptors belonging to the immunoglobulin superfamily that are expressed on human NK cells. Most are inhibitory and recognize classical MHC class I molecules, thereby preventing activation of NK cytotoxicity by normal cells. The genes encoding KIR family receptors lie on human chromosome 19q13 and include 15 highly polymorphic genes and two pseudogenes. Inhibitory KIR family receptors have long cytoplasmic tails that contain ITIMs; the others are activating through short cytoplasmic tails that associate through their transmembrane domains with DAP12. (8-1)

Kupffer cells: macrophages in the liver. (1-3)

LACK: see *Leishmania* **homolog of receptor for activated C kinase.** (11-3)

Langerhans cells: immature dendritic cells in the epidermal layers of the skin. (1-3)

LAT: see **linker for activation of T cells.** (5-3)

LBP: see **LPS-binding protein.** (3-10)

Lck: a Src-family tyrosine kinase that associates with the coreceptors CD4 and CD8 and plays a central role in TCR signaling. (5-3)

LD$_{50}$: the dose at which 50% of treated individuals die. (9-3)

lectin: protein that binds specifically to particular polysaccharides or other carbohydrate structures. There are four major structural groups: **C-type lectins**, p-type lectins, I-type lectins and galectin-like lectins. (2-5)

lectin pathway: complement activation pathway through which mannose-binding lectin and ficolins stimulate the complement cascade. (3-3)

***Leishmania* homolog of receptor for activated C kinase (LACK):** a WD-repeat protein containing an immunodominant peptide for the CD4 T cell response. (11-3)

leucine-rich repeat (LRR): unit of protein structure in which there are many repeats of a basic unit of approximately 25 amino acids. The repeats stack against one another. They are present in **Toll-like receptors**, innate immune molecules of plants, NOD1 and NOD2, lamprey adaptive immune molecules and specialized proteins of *Listeria* that promote invasion of cells. (3-10)

leukocyte adhesion deficiency: inherited immunodeficiency disease characterized by a defect in neutrophil adhesion and **extravasation**. Can result from genetic deficiency in the β2 integrin chain, in ICAM-1 or in fucosylation enzymes that contribute to the synthesis of ligands for selectins. (3-14)

light chain: the immunoglobulin light chains are the smaller of the two kinds of chains in the immunoglobulin molecule. Each has a **variable region** contributing to the antigen-binding site, and a **constant region** containing a cysteine by which it makes a disulfide bond with the constant region of the **heavy chain.** (1-4)

light zone: (of lymphoid follicles) region of the germinal center containing **centrocytes, follicular dendritic cells** and antigen-specific helper T cells. (6-11)

linker for activation of T cells (LAT): a lipid-raft-localized transmembrane scaffold molecule that plays an essential early part in TCR signaling. (5-3)

lipid raft: specialized membrane region rich in glycosphingolipids, cholesterol and some membrane proteins. Lipid rafts are dynamic structures in which components are continually exchanged between raft and non-raft regions of the membrane. (5-3)

lipoarabinomannan: a major immunostimulatory component of the lipid-rich mycobacterial cell wall,

containing phosphatidylinositol linked to the carbohydrates mannose and arabinose. (3-10)

lipopolysaccharide (LPS): major component of the outer membrane of **Gram-negative bacteria** and an important recognition element for innate immunity. (3-0)

lipoteichoic acid (LTA): cell wall constituent of **Gram-positive bacteria** in which the lipid portion anchors the molecule to the plasma membrane and is attached to repeating polymers of glycerol or ribitol joined in phosphodiester linkages, often with additional sugars and attached amino acids such as D-alanine. (3-0)

listeriolysin O (LLO): member of the cholesterol-dependent cytolysin family that includes streptolysin O from *Streptococcus pyogenes.* (9-6)

LLO: see **listeriolysin O.** (9-6)

long-lived plasma cells: plasma cells generated during a germinal center reaction that typically migrate to the bone marrow, where they can live for years and continue to produce protective antibody. (6-9)

long terminal repeat (LTR): repeated sequences at the 5′ and 3′ ends of a proviral or transposon genome that, when integrated into the host genome, act as promoters or polyadenylation and termination sites, respectively. (10-4)

LPS: see **lipopolysaccharide.** (3-0)

LPS-binding protein (LBP): a lipid transfer protein of serum that can extract monomers of LPS from bacterial membranes and deliver them to the innate immune receptor CD14. (3-10)

LRR: see **leucine-rich repeat.** (3-10)

LT: see **lymphotoxin.** (2-9)

LTA: see **lipoteichoic acid.** (3-0)

LTR: see **long terminal repeat.** (10-4)

Ly49 family: family of C-type lectin receptors expressed on mouse NK cells. Most are inhibitory and recognize classical MHC class I molecules, thereby preventing activation of NK cytotoxicity by normal cells. The genes encoding the Ly49 family lie on mouse distal chromosome 6 and include up to 16 genes and pseudogenes. Inhibitory family members have long cytoplasmic tails that contain ITIMs; the others are activating through short cytoplasmic tails that associate through their transmembrane domains with DAP12 and DAP10. A single, nonfunctional Ly49 member is present in the human genome. (8-1)

lymph: fluid drained from the tissues and flowing through **lymphatic vessels.** (1-6)

lymphadenopathy: enlarged lymph nodes. (5-7)

lymphatic vessels: system of vessels draining fluid (**lymph**) from the tissues and in which dendritic cells and antigens are delivered to **lymph nodes.** (1-6)

lymph-node-resident dendritic cells: dendritic cells that migrate to the lymph node as immature cells and take up residence there, or that have migrated there from the periphery in the absence of infection. Lymph-node-resident dendritic cells are thought to be composed of both cells that have matured homeostatically in the periphery and immature cells that have migrated to the lymph nodes directly from the blood. (5-6)

lymph nodes: secondary lymphoid organs distributed widely in the body but especially in the groin, the axilla and the neck, and along the small intestine. (1-6)

lymphocytes: white blood cells bearing highly variable receptors for **antigen** that circulate in the blood and lymph and are the mediators of **adaptive immunity.** (1-0)

lymphoid chemokine: constitutively expressed chemoattractant molecule that directs the migration of lymphocytes and dendritic cells into specialized regions of the **secondary lymphoid tissues.** Also known as *homeostatic chemokine.* (1-6)

lymphoid lineage: hematopoietic cell lineage containing the lymphocytes of the immune system. (1-1)

lymphoid tissues: specialized tissues in which **lymphocytes** mature and immune responses are initiated. (1-0)

lymphopenia: a decrease in the numbers of circulating lymphocytes. (9-3)

lymphotoxin (LT): TNF family member that can have two forms, a trimer of the secreted LTα subunit, which binds to the TNF receptors, and a heterotrimer of one α subunit and two membrane-bound β subunits (LTα$_1$β$_2$), which binds to the LTβR and is important for the development of lymphoid structures. (2-9)

lysozyme: enzyme that degrades the bacterial cell wall component peptidoglycan. (3-9)

μm mRNA: mRNA encoding the membrane form of the μ heavy chain of immunoglobulin. (6-7)

μs mRNA: mRNA encoding the secreted form of the μ heavy chain of immunoglobulin. (6-7)

macrophages: phagocytic cells resident in tissues that detect microorganisms by means of receptors recognizing conserved components and ingest and destroy them, and function in tissue repair and maintenance. (1-2)

macropinocytosis: process whereby cells take up large amounts of extracellular fluid independently of receptor binding to ligands. (3-7)

MAIT cells: see **mucosal-associated invariant T (MAIT) cells.** (8-5)

major histocompatibility complex (MHC): cluster of genes encoding the **classical** and many **non-classical MHC molecules** and other structurally unrelated molecules, many with important functions in immunity. (1-5, 4-3)

major histocompatibility complex (MHC) molecules: cell-surface glycoproteins encoded in the **major histocompatibility complex** and which bind degraded fragments derived from intracellular proteins and display them on the cell surface. (1-5, 4-2)

MALT: see **mucosa-associated lymphoid tissues.** (1-6)

mannose-binding lectin (MBL): a **collectin** family member that recognizes terminal sugars with equatorial hydroxyls in the C3 and C4 positions, such as mannose and fucose. Also called mannan-binding protein (MBP). (3-2)

mannose receptor: endocytic receptor of macrophages and dendritic cells that binds to branched alpha-linked oligomannose-containing molecules through its C-type lectin domains. Also called CD206. (3-7)

marginal sinus: (of spleen) blood vessels running between the **marginal zone** and the white pulp of the spleen. (1-7)

marginal zone: narrow region at the outer boundary of the splenic lymphoid tissue, bounded by the **marginal sinus** and red pulp. (1-7)

marginal zone B cells: B cells resident in the marginal zone of the spleen and providing rapid T cell-independent antibody responses to capsular polysaccharides of bacteria, as well as early T cell-dependent responses to bloodborne antigens. (8-7)

marginal zone macrophage: specialized macrophage in the **marginal zone** of the spleen thought to be important in resistance to bloodborne pathogens. (1-7)

mast cells: tissue cells that release inflammatory mediators including histamine in response to antibody bound to receptors on their surface. (1-2)

mature dendritic cells: terminally differentiated dendritic cells that have downregulated the receptors that enable them to internalize microbial and cellular components in tissues, and are positioned in T cell zones in secondary lymphoid organs where antigen internalized in the tissues is displayed on their surface for recognition by T cells. Dendritic cells that mature in the presence of infection also express high levels of **costimulatory molecules** and secrete inflammatory cytokines, and are the only cell type to be capable of efficiently activating naïve T cells. (4-1)

MAVS: mitochondrial activating signaling protein, also known as IPS-1 (IFN-β promoter stimulator-1), VISA (virus-induced signaling adaptor), and Cardif (CARD adaptor inducing IFN-β), is a CARD domain-containing adaptor protein in the mitochondrial outer membrane, which links upstream activation of the RNA helicases RIG-I and Mda5 with downstream kinase complexes responsible for activation of the transcriptional regulators NF-κB, IRF3 and IRF7. (10-1)

MBL: see **mannose-binding lectin**. (3-2)

M cells: specialized epithelial cells in the small intestine that collect antigen at Peyer's patches. (1-7)

MCP: see **membrane cofactor of proteolysis**. (3-6)

MD-2: a polypeptide that associates with the extra-cellular domain of the **Toll-like receptor** TLR4 and is required for LPS responsiveness. (3-10)

Mda5: a cytoplasmic protein composed of a CARD and an RNA helicase domain that recognizes dsRNA in virus-infected cells and induces **type 1 interferon** production. (3-16)

MDP: see **muramyl dipeptide**. (3-12)

membrane-attack complex: membrane pore structure of 7–10 nm in internal diameter formed from complement proteins C6, C7, C8 and C9 recruited to a membrane by C5b, and that is important in immune defense against some classes of bacteria. (3-5)

membrane cofactor of proteolysis (MCP); CD46: a molecule that protects cells against attack by the alternative pathway of complement activation by binding to C3b and C4b and recruiting factor I to inactivate them by proteolysis. (3-6)

memory cells: long-lived lymphocytes that differentiate during the clonal expansion of antigen-specific lymphocytes during a primary immune response and provide a rapidly activated effector pool on subsequent challenge with the same antigen. (1-8)

metacyclic promastigotes: infectious forms of *Leishmania* that represent terminally differentiated promastigotes released from the intestinal epithelia and that migrate to the sandfly proboscis; metacyclic forms are covered with a lengthened form of surface lipophosphoglycan molecules that renders parasites more resistant to complement and macrophage oxidant killing. (11-3)

metallophilic macrophage: specialized macrophage in the **marginal zone** of the spleen thought to be important in scavenging debris. (1-7)

MHC: see **major histocompatibility complex**. (1-5, 4-3)

MHC class I molecules: cell surface glycoproteins most of which are encoded in the **major histocompatibility complex** and most of which bind peptide fragments of cytoplasmic and secreted proteins and display them on the cell surface. An important function of these molecules in immunity is to signal the presence of viral infection to CD8 T cells. (4-2)

MHC class II molecules: cell surface glycoproteins encoded in the **major histocompatibility complex** and most of which bind peptide fragments of proteins derived from internalized molecules, including those of extracellular pathogens, and display them for recognition by CD4 T cells. (4-2)

MHC tetramers: recombinant soluble MHC molecules either physically linked or subsequently loaded with a peptide epitope of interest and chemically tetramerized to generate a four-valent binding molecule with sufficient avidity to bind to TCRs on the T cell membrane in an antigen-specific manner. The method has been widely used to study the frequency of antigen-specific T cells to epitopes from viruses and other pathogenic organisms. (10-2)

microglia: prevalent type of macrophages in the brain, characterized by long processes. (1-3)

minor histocompatibility antigens: peptides that are derived from polymorphic non-MHC proteins in transplanted tissue and can give rise to a slowly developing immune response that can lead to transplant rejection and requires immunosuppression. Nonetheless, a complete match of MHC molecules between donor and recipient increases the chance of transplant success compared with a transplant that is mismatched at one or more MHC allele. (14-6)

mitogen-activated protein (MAP) kinase: any of a small family of closely related protein kinases activated in receptor signaling pathways downstream of Ras or Rho-family GTPases through related upstream kinases (MAP kinase kinases) that phosphorylate a characteristic site in the kinase domain, resulting in the activation of the enzymatic activity of the MAP kinase. MAP kinases include Erk (extracellular signal regulated kinase), JNK (c-Jun N-terminal kinase) and p38 MAP kinase subfamilies. They regulate AP-1 type transcription factors and other events in the cell, including stability of the mRNA encoding TNF. (3-11)

molecular mimicry: structural similarity between a self component and a component of an infectious agent, such that the immune response to the infectious agent cross-reacts with the self component, initiating the autoimmune attack. (12-4)

monoclonal antibodies: antibodies produced by a single clone of cells and therefore with a single specificity. (6-4)

monocytes: circulating precursors of **macrophages** and some dendritic cells. (1-2)

monocytopenia: a decrease in the numbers of circulating monocytes. (9-3)

mucin: glycoprotein with many *O*-linked glycans on serine or threonine residues. (2-14)

mucosa-associated lymphoid tissues (MALT): secondary lymphoid tissue in the walls of the gastrointestinal, respiratory and urogenital tracts. (1-6)

mucosal-associated invariant T (MAIT) cells: T lymphocytes of unknown function that are found in mucosa and express invariant receptors recognizing the non-classical MHC class I molecule MR1. (8-5)

multiple sclerosis: a presumed autoimmune disease in which there is an immune attack on the myelin sheaths around neuronal axons in the central nervous system, leading to neurological disease. (12-6)

multivesicular body (MVB): vesicles of the endosomal pathway created when sections of endosome membrane are pinched off into vesicles inside the endosome. The internalized material is targeted for degradation and also becomes sequestered away from the cytoplasm, which therefore terminates signaling reactions by ligand-bound receptors in that membrane. MVBs facilitate antigen degradation within a structured membrane-bounded compartment, and in antigen-presenting cells are a site of peptide–MHC complex assembly. (10-4)

muramyl dipeptide (MDP): a substructure of bacterial peptidoglycan containing the sugar muramic acid linked to ʟ-alanine and ᴅ-glutamic acid or ᴅ-glutamine. MDP is an effective adjuvant, but does not stimulate TLRs. (3-12)

MVB: see **multivesicular body**. (10-4)

Mx proteins: any of a series of antiviral proteins synthesized in response to interferon and structurally related to the GTPase dynamin. (3-17)

myasthenia gravis: an autoimmune disease in which autoantibodies react with the nicotinic acetylcholine receptor, which is the receptor in the neuromuscular synapse that receives the neurotransmitter from motor neurons to induce muscle contraction. The autoantibodies induce the internalization of receptors and can also block the binding of acetylcholine, and hence block receptor function and prevent muscle contraction. The name is Latin for severe muscle weakness. (12-5)

MyD88: see **myeloid differentiation factor 88**. (3-11)

myeloid differentiation factor 88 (MyD88): an adaptor molecule that contains a TIR domain and mediates the major proinflammatory signaling pathway of TLRs. (3-11)

myeloid lineage: hematopoietic cell lineage containing the phagocytic and inflammatory cells of the immune system. (1-1)

NA: see **neuraminidase**. (10-7)

NADPH oxidase: phagocyte enzyme that upon induced assembly creates **reactive oxygen intermediates**, also called phagocyte oxidase. (3-9)

naïve lymphocytes: mature circulating lymphocytes that have not yet encountered antigen. (1-2)

NALP3: see **cryopyrin**. (3-12)

natural antibody: antibody, primarily IgM and IgA, from individuals previously unexposed to antigen, that is produced mainly by **B1 cells** and binds to microbes or virus particles and participates in host defense. (8-6)

natural cytotoxicity: the process whereby NK cells kill virus-infected cells and some tumors without requiring prior immunization of the host. (8-1)

natural killer (NK) cells: cytotoxic lymphocytes lacking antigen-specific receptors but with invariant receptors that detect infected cells and some tumor cells and activate their destruction. (1-2, 8-1)

natural killer T (NKT) cells: T lymphocytes that express both αβ T cell receptors and NK cell receptors and recognize glycolipids presented on CD1d non-classical MHC class I molecules. (8-3)

necrosis: mode of cell-injury-induced cell death that often involves rupture of the plasma membrane and release of cytoplasmic contents into the extracellular space. (2-11)

negative selection: a process during the development of lymphocytes in which cells whose antigen receptors bind strongly to self antigens die. (7-0)

nematode: unsegmented roundworm. Most nematodes are free living, but together they are the largest cause of helminth infections of humans; hookworms are an example of a nematode parasite of humans. (11-4)

neuraminidase (NA): influenza homotetrameric sialidase expressed on the virion that cleaves terminal sialic acid residues from host glycoconjugates to allow the release of infectious progeny virus. (10-7)

neutralizing antibodies: antibodies that can directly block infection by viruses or attachment by bacteria or toxins, without need of complement or Fc receptors. (6-0)

neutropenia: low neutrophil numbers in blood; in humans, susceptibility to infectious diseases increases when neutrophil counts fall below 1,000 cells/μl and control of endogenous microflora is impaired when counts fall below 500 cells/μl. (11-1)

neutrophilia: a rise in the number of circulating neutrophils. (9-3)

neutrophils: phagocytic cells that circulate in the blood and detect microorganisms by means of receptors that recognize conserved components. (1-2)

NFAT: see **nuclear factor of activated T cells.** (5-5)

NF-κB: see **nuclear factor κB.** (2-10)

NK1.1: a member of the polymorphic NKR-P1 family of C-type lectins encoded in the NK complex on mouse chromosome 6. In the mouse, both activating and inhibitory members are present, and bind to C-type lectin ligands also encoded in the NK complex. In humans, a single NKR-P1 inhibitory receptor binds a homologous C-type lectin. (8-1)

NK cells: see **natural killer cells.** (1-2)

NKG2D: monogenic activating receptor of the C-type lectin receptor family that is expressed on NK cells as well as on some γδ and effector CD8 T cells. NKG2D signals through noncovalent association with DAP10 (and, in the mouse, with DAP12), a transmembrane signaling molecule lacking an ITAM motif but containing a cytoplasmic phosphorylatable YXXM motif shared by costimulatory molecules, such as CD28 and ICOS. (8-2)

NOD: see **nucleotide oligomerization domain.** (3-12)

NOD1 (nucleotide oligomerization domain 1): an intracellular innate immune recognition molecule for dipeptides, which contain diaminopimelic acid, a modified amino acid found in peptidoglycan. Also called CARD4. (3-12)

NOD2 (nucleotide oligomerization domain 2): an intracellular innate immune recognition molecule for muramyl dipeptide, a substructure of peptidoglycan. Also called CARD15. (3-12)

NOD mouse: see **non-obese diabetic mouse.** (12-7)

non-classical MHC molecules: MHC molecules that are relatively nonpolymorphic and may have functions other than the presentation of antigen to T cells. (4-3)

non-homologous end joining: mechanism for repairing double-strand breaks in DNA in which the broken ends are rejoined directly, usually with the loss of nucleotides at the join. (7-1)

non-obese diabetic (NOD) mouse: mouse of an inbred strain that spontaneously develops autoimmune diabetes at high frequency. NOD mice have at least 20 genetic loci that contribute to the spontaneous development of diabetes because of T cell-mediated pancreatic islet beta cell destruction. MHC class II (I-A) is one of these, and a polymorphic variant of CTLA4 is also likely to be a contributor. (12-7)

non-steroidal antiinflammatory drugs (NSAIDs): any of a class of drugs that inhibit inflammatory responses by inhibiting cyclooxygenases 1 and/or 2. Aspirin and most other NSAIDs block COX-1 and COX-2. COX-2-specific NSAIDs have been developed more recently; they have fewer gastrointestinal side effects but may carry an increased risk of cardiovascular side effects. (3-13)

non-structural protein-1 (NS1): a protein produced by influenza virus that downregulates host cell antiviral innate immune responses by binding to viral RNAs and sequestering them from cytosolic nucleic-acid-sensing proteins that trigger type 1 interferon production. (10-8)

Notch: any of a family of cell-surface receptors that often participate in cell fate determination during development. (7-0)

N region: nucleotides introduced into the coding joints of antigen receptor genes of lymphocytes by **TdT.** (7-2)

NS1: see **non-structural protein-1.** (10-8)

NSAIDs: see **non-steroidal antiinflammatory drugs.** (3-13)

nuclear factor κB (NF-κB): any of a small family of dimeric DNA-binding proteins that mostly function as transcriptional activators and have a central role in both innate and adaptive immune responses; originally described for their binding to the B site in the Ig κ intronic enhancer. (2-10)

nuclear factor of activated T cells (NFAT): any of a family of four closely related transcription factors that are responsive to calcium elevation. (5-5)

nucleotide oligomerization domain (NOD): nucleotide-binding domain that regulates protein–protein interactions in a variety of proteins that participate in immunity and apoptosis. (3-12)

opsonin: any soluble molecule that recognizes and coats microorganisms and thereby stimulates internalization of the microorganism by phagocytes. A particle coated with opsonins is said to be opsonized. (3-2)

opsonized: bound by soluble recognition elements of the innate or adaptive immune system, such as antibody, **iC3b** and collectins, that are recognized by **phagocytic receptors.** (3-7)

osteoclast: a differentiated cell of the monocytic lineage responsible for the resorption of bone. A distinct type of cell, the osteoblast, is responsible for building bone mass. (12-9)

outside-in signaling: intracellular signaling events that result from **integrin** binding to its extracellular ligands. The flow of information through the integrin flows from outside the cell where it binds ligand into the cell where the cytoplasmic domains of the integrins participate in generating signaling events that regulate cell activation, proliferation, and/or survival. Note that different integrins can mediate different effects on cell

behavior, indicating that outside-in signaling is not identical for all integrins. (2-4)

OX40: a TNFR-family molecule that is induced on activated and effector T lymphocytes and is important in the maintenance of effector and memory CD4 T cells, and especially those interacting with B cells. It is called OX40 because it was first identified in an Oxford laboratory as one of a number of molecules induced on activated T cells. (5-8)

OX40L: the TNF-family ligand for **OX40,** expressed on activated dendritic cells and B cells and some non-immune cells. (5-8)

p53: gene regulatory protein that orchestrates the long-term DNA damage response in multicellular organisms by activating a cell-cycle checkpoint and/or apoptosis in response to DNA damage or to DNA replication in an incorrect phase of the cell cycle. (3-18)

PAF receptor: see **platelet-activating factor receptor.** (9-5)

PAG1: see **phosphoprotein associated with glycophospholipid-enriched microdomains.** (5-4)

PALS: see **periarteriolar lymphoid sheath.** (1-6)

Paneth cell: specialized cell type of the epithelial layer of the small intestines. Paneth cells are localized to the base of the crypts of the intestines and secrete **antimicrobial peptides.** (3-1)

paracortical area: (of **lymph node**) area beneath the outer regions of the lymph node and where T lymphocytes accumulate. (1-6)

paracrine: produced by one cell and acting on a nearby cell. This is thought to be the main mode of action of **cytokines,** although they can act in an **autocrine** fashion or systemically in an endocrine fashion. (2-6)

pathogen: any microorganism that causes disease. (1-0)

pathogenicity islands: DNA encoding contiguous virulence genes and found in pathogenic bacteria. (9-2)

PD-1: see **programmed death 1.** (5-8)

PD-L1: a B7-family ligand for **PD-1** that is expressed widely on tissue cells and is thought to be important in limiting damage by activated T cells in the periphery. (5-8)

PD-L2: a B7-family ligand for **PD-1** that is expressed chiefly on dendritic cells and is thought to contribute to T cell tolerance. (5-8)

pemphigus vulgaris: an autoimmune disease in which autoantibodies react with a component of the epidermal intercellular adhesion complexes, called desmosomes. This component, desmoglein-3, is a member of the cadherin family of cell adhesion molecules. These antibodies interfere with the adhesion of cell layers within the epidermis, so individuals with this disease easily form blisters of the skin. There are also manifestations in the oral tract. (12-5)

pentraxin: any of a small family of pentameric serum proteins that participate in innate immunity by binding to membranes of microbes and apoptotic cells and activating the **classical pathway** of **complement.** C-reactive protein and serum amyloid P protein are pentraxins that participate in opsonization of apoptotic cells and chromatin. (3-3)

peptide-binding groove: cleft on the membrane-distal surface of **major histocompatibility complex (MHC) molecules** that binds to fragments of intracellular or internalized extracellular molecules, usually proteins, and in which these fragments are displayed on the cell surface. (4-2)

peptidoglycan: rigid polymer of repeating disaccharides cross-linked by short peptides that is a major structural element of most types of bacterial cell walls. (3-0)

perforin: a protein contained in membrane-bounded granules of cytotoxic lymphocytes that is released on activation of the cytotoxic cell and forms pores in the membrane of target cells. (5-14)

periarteriolar lymphoid sheath (PALS): T cell area in the **spleen**, formed around arterioles. (1-6)

peripheral node addressin (PNAd): oligosaccharide ligand for L-selectin attached to any of several different protein backbones. (2-14)

peripheral supramolecular activation complex (p-SMAC): peripheral area of an **immunological synapse** in which TCRs and integrins have segregated into distinct areas. The p-SMAC contains the integrin LFA-1. (5-4)

peripheral tolerance: immune non-responsiveness due to mechanisms operating on mature self-reactive lymphocytes not eliminated during development. (12-0)

pernicious anemia: an autoimmune disease in which autoantibodies are made that recognize intrinsic factor, a protein that is secreted by gastric parietal cells and is involved in the uptake of vitamin B_{12} from the diet. (12-5)

PEST sequence: amino-acid sequence (P, proline; E, glutamate; S, serine; T, threonine) exposed on cytosolic proteins and constituting one mechanism for targeting proteins for degradation by proteosomes. (9-6)

Peyer's patches: organized regions of **secondary lymphoid tissue** in the wall of the small intestine. (1-6)

phage display: a technique for isolating proteins with specific binding characteristics whereby the desired protein is expressed on the surface of a bacteriophage. This allows phage to be selected for binding to a tissue culture dish coated with the ligand. (6-4)

phagocytic cells: cells that recognize and ingest molecules and particles including microorganisms and destroy them. (1-2)

phagocytic receptor: cell surface molecule of phagocytes that binds microbes, viruses or apoptotic cells, either directly or through opsonins, and induces **phagocytosis**. (3-7)

phagocytosis: receptor-mediated internalization of cells or other particles larger than 1 μm in diameter. (3-7)

phagolysosome: phagosome that has fused with lysosomes. (3-8)

phagosome: vesicle generated by invagination and fusion of the plasma membrane of a phagocyte around a particle bound to phagocytic receptors. (3-8)

phase variation: phenotypic variation within a species of bacteria. (9-2)

phosphatidylserine: a phospholipid that is a major component of the internal half of the plasma membrane. In apoptotic cells, the asymmetry of phospholipids is lost as an early event, with phosphatidylserine appearing on the outside of the plasma membrane, where it is recognized by opsonins and phagocytic receptors of macrophages or dendritic cells, leading to phagocytosis of the apoptotic cell. (3-8)

phosphoantigens: small, phosphorus-containing non-peptides that collectively represent antigens capable of activating human Vγ9Vδ2 T cells. Several isoprenoid pyrophospate derivatives represent natural phosphoantigens produced by bacteria, including mycobacteria. (8-4)

phosphoinositide-3-kinase (PI3 kinase): a signaling enzyme that adds a phosphate to the 3 position of PIP_2 to generate the membrane-bound second messenger PIP_3. (5-5)

phospholipase C (PLC)-γ1: a signaling enzyme that hydrolyzes phosphatidylinositols, generating diacylglycerol and inositol trisphosphate, the latter of which induces elevation of intracellular free calcium. (5-3)

phosphoprotein associated with glycophospholipid-enriched microdomains (PAG1): a transmembrane scaffold protein that when phosphorylated on tyrosine recruits the protein kinase Csk to lipid rafts, which are also called glycophospholipid-enriched membrane microdomains. Also known as Cbp (Csk-binding protein). (5-4)

PI3 kinase: see **phosphoinositide-3-kinase**. (5-5)

PKCθ: see **protein kinase Cθ**. (5-5)

PKR: double-stranded RNA-dependent protein kinase important in intracellular defense against viruses. PKR is also activated by other stimuli, including Toll-like receptors and stress, but its main function is blocking virus replication. (3-17)

plasma cells: terminally differentiated B lineage cells secreting large quantities of antibody. (1-2, 6-7)

plasmacytoid dendritic cell: cell type of the dendritic cell family that produces very large amounts of **type 1 interferons** upon contact with viruses. These cells are found principally in the blood and home to secondary lymphoid organs after detecting viral nucleic acids through TLR7 or TLR9. Also called interferon-producing cells. (1-3, 3-16)

plasmacytoma: cancerous plasma cell. Human cancers of this type are often called multiple myelomas, from the propensity of these cells to grow in the bone marrow. (6-4)

platelet-activating factor receptor (PAF receptor): receptor for platelet-activating factor that is expressed on neutrophils and can enhance neutrophil–endothelial adhesion in response to endothelial activation and production of platelet-activating factor (PAF), a substituted glycerol phosphate derivative. (9-5)

platelets: cell fragments that are shed from megakaryocytes in the erythroid lineage and induce blood clotting. (1-1)

pleiotropy: property whereby one agent may have diverse effects on many different cell types. (2-6)

PNAd: see **peripheral node addressin**. (2-14)

Pneumocystis jirovecii: the human-adapted species of *Pneumocystis*, and preferred over the designation *Pneumocystis carinii* f. sp. *hominis*; honors Otto Jirovec, the Czech parasitologist who published the first description of human *Pneumocystis* infection; *Pneumocystis carinii* refers to the rat-adapted species. (11-2)

polyclonal antibodies: heterogeneous antibodies against an antigen that are obtained by immunizing an individual. (6-4)

polygenic: encoded by more than one gene. (4-3)

polymeric Ig receptor: receptor mediating transcytosis of **immunoglobulin A** across epithelial cells to mucosal surfaces. (6-2)

polymorphism: difference between individuals in a DNA or protein sequence in which the different sequences are present at a frequency greater than 1% in a population. (4-3)

polymorphonuclear leukocytes: another name for **neutrophils**. (1-2)

portal hypertension: elevated blood pressure in the portal vein draining from the intestines to the liver; normally low because of the distensible hepatic sinusoids, portal blood pressure is raised by any obstruction to flow, including the fibrosis induced by chronic egg deposition of schistosomiasis; the high blood pressures result in retrograde flow, leading to splenomegaly, ascites (accumulation of peritoneal fluid) and **varices**. (11-5)

positive selection: a process during the development of T cells in which cells that have a low but positive response to a self antigen are selected to continue development. Positive selection is generally coupled to differentiation into either the CD8 cytotoxic T cell lineage or the CD4 helper T cell lineage. B cells are thought to undergo antigen receptor-dependent positive selection as well. (7-0)

post-streptococcal acute glomerulonephritis: an immune complex disease seen after *Streptococcus pyogenes* infections in some individuals. (13-4)

poxvirus: any member of a family of large, enveloped, double-stranded DNA viruses distinguished by replication in the cytoplasm, rather than in the nucleus as occurs for other DNA viruses. Variola (smallpox) and vaccinia (cowpox) viruses are poxviruses. (10-0)

PPD skin test: a test for infection with *M. tuberculosis* using purified protein derivative from the bacterium. A standardized preparation, PPD-S, is injected intracutaneously on the arm using 5 tuberculin units (0.0001 mg of protein); delayed-type hypersensitivity is quantified after 48–72 h by measuring the width of the indurated (hardened infiltrate) response. (9-7)

pre-B cell: the stage of B cell development at which a functional Ig heavy chain protein is expressed, but a functional Ig light chain is not expressed. (7-0)

pre-BCR: receptor complex formed in pre-B cells between the μ chain and the **surrogate light-chain** subunits λ5 and VpreB, as well as Igα and Igβ. The pre-BCR signals analogously to engaged BCRs to promote B cell developmental progression to the pre-B cell stage. (7-3)

precursor cell: cell that is committed to a single specific differentiated state but has not yet terminally differentiated. (1-1)

P region: palindromic nucleotides introduced into the coding joints of antigen receptor genes of lymphocytes by asymmetric cleavage of the hairpin intermediate in V(D)J recombination and subsequent copying of the resulting overhanging bases. (7-2)

pre-TCR: receptor complex found in double-negative thymocytes between the TCR β chain, pre-TCRα and CD3 chains. The pre-TCR signals to promote maturation to the double-positive thymocyte stage. (7-3)

pre-TCRα: a transmembrane Ig superfamily member polypeptide that is expressed during the development of T lymphocytes and pairs with TCR β chain to form the pre-TCR. The gene encoding pre-TCRα does not undergo rearrangement. (7-7)

primary granule: specialized intracellular vesicle in neutrophils derived from lysosomes and which contains antimicrobial peptides, proteases and other components; also called azurophilic granule because of its dye-staining properties. (3-9)

primary lymphoid tissues: bone marrow and thymus, in which lymphocytes differentiate and mature. (1-6)

priming: activation of naïve T or B lymphocytes on first encounter with antigen. (5-7)

pro-B cell: the stage of B cell development at which no functional Ig chains are produced but the cell is committed to the B cell lineage and DNA at the IgH locus is undergoing rearrangement. (7-0)

procaspase: precursor form of **caspase** with low or no activity. As with many other proteases, proteolytic cleavage removes an amino-terminal pro-domain. In the **initiator caspases**, this pro-domain includes the **CARD**. (2-11)

progenitor cell: committed cell giving rise to a lineage of related but distinct cells. (1-1)

programmed death 1 (PD-1): an inhibitory receptor belonging to the CD28 subfamily of the immunoglobulin superfamily, with an important part in peripheral T cell tolerance and in limiting immune damage at the effector cell stage. PD-1 has two ligands: **PD-L1**, which is expressed by many stromal cells of tissues, and **PD-L2**, which is expressed by dendritic cells and macrophages. (5-8)

proinflammatory cytokines: cytokines that act on endothelial cells to induce **inflammation**. Also known as inflammatory cytokines. (3-13)

properdin: complement pathway component that binds to and stabilizes the alternative pathway C3 convertase, C3bBb. (3-4)

proteasome: large, multisubunit enzyme complex that degrades cytosolic proteins into short peptides. (4-6)

protectin (CD59): a membrane-bound complement regulatory protein that prevents formation of the membrane-attack complex. (3-6)

protein C: a vitamin K-dependent plasma serine protease that is synthesized in liver and activated by a complex of thrombin bound to thrombomodulin on membranes, including endothelial cells, which express an additional stabilizing receptor, the endothelial protein C receptor; activated protein C then cleaves coagulant factors VIIIa and Va, inactivating them. (9-3)

protein kinase Cθ (PKCθ): a member of the protein kinase C family. It is important in signaling from the antigen receptor of T cells, where it becomes localized adjacent to the T cell receptor and CD28 and activates NF-κB. (5-5)

protein tyrosine kinase: enzyme that phosphorylates tyrosine residues on other proteins and/or on itself. (2-0)

protein tyrosine phosphatase: enzyme that removes phosphate groups from phosphorylated tyrosine residues on proteins. Protein tyrosine phosphatases oppose the action of **protein tyrosine kinases**. (2-0)

proto-oncogene: gene whose protein product when dysregulated or mutated can promote cancer. Proto-oncogenes generally encode proteins that regulate cell growth, cell survival or cell differentiation. (7-11)

P-selectin glycoprotein ligand 1 (PSGL1): mucin of leukocytes that is the principal scaffold for the oligosaccharide ligand for P-selectin but can also serve as the scaffold for the oligosaccharides recognized by E-selectin and L-selectin. (2-14)

pseudogene: a non-functional gene in the genome. Pseudogenes provide genetic information for **gene conversion** mechanisms that provide much of the diversity of Ig genes in chickens. (7-4)

PSGL1: see **P-selectin glycoprotein ligand 1**. (2-14)

p-SMAC: see **peripheral supramolecular activation complex**. (5-4)

psoriasis: a cell-mediated inflammatory disease of the skin. (12-6)

R5 virus: HIV-1 isolates with gp120-binding affinity for CD4 and CCR5; most primary mucosal HIV-1 infections are due to R5 viruses. (10-5)

radioimmunoassay (RIA): a technique to measure the amount of an antigen of interest in a solution by its ability to inhibit the incorporation of radiolabeled antigen into an **immunoprecipitation**. (6-5)

RAG-1: see **recombination activating gene-1**. (7-0)

RAG-2: see **recombination activating gene-2**. (7-0)

RANK ligand: a cytokine of the TNF family secreted by activated CD4 T cells and other cells that acts together with other cytokines to promote osteoclast differentiation and activation. RANK ligand also promotes the activity of chondrocytes in eroding cartilage and also acts on monocytes and dendritic cells to promote immune responses. Its receptor, RANK (receptor activator of NF-κB), is a member of the TNF receptor family. (12-9)

rapamycin: an immunosuppressive drug that when bound to **FKBP** blocks mammalian target of rapamycin (mTOR), which increases protein synthesis to support cellular proliferation. (14-8)

reactive nitrogen intermediates: highly reactive and toxic compounds generated from L-arginine by **inducible nitric oxide synthase (iNOS)**. (3-9)

reactive oxygen intermediates: highly reactive and toxic compounds generated from molecular oxygen by **NADPH oxidase**, superoxide dismutase and/or myeloperoxidase. (3-9)

receptor editing: a process whereby immature B cells in the bone marrow that express antigen receptors recognizing self arrest maturation and undergo additional DNA rearrangements at the Ig light-chain loci resulting in a new antigen receptor that may not recognize self antigens. (7-5)

recombination activating gene-1 (RAG-1): a gene encoding one of two proteins that mediate initial DNA cleavages involved in V(D)J recombination. (7-0)

recombination activating gene-2 (RAG-2): a gene encoding one of two proteins that mediate initial DNA cleavages involved in V(D)J recombination. (7-0)

recombination signal sequence (RSS): the sequence elements adjacent to lymphocyte antigen receptor gene segments that direct V(D)J recombination. RSSs are composed of a palindromic heptamer adjacent to the coding region, a non-conserved spacer of 12 or 23 base pairs in length, and an AT-rich nonamer. (7-1)

red pulp: site of destruction of senescent red blood cells in the **spleen**. (1-6)

redundancy: (of properties of molecules) property whereby more than one molecule seems to have the same function. (2-6)

regulatory T cells (T_REG): T cells that act to inhibit responses of other T cells. Regulatory T cells are dependent upon a differentiation program driven by the transcriptional regulator Foxp and usually express CD4 and CD25. (7-10)

Rel homology domain: 300-amino-acid-long homology region in all five subunits of the transcriptional activator **NF-κB**. This homology forms two immunoglobulin-like beta-sheet sandwich structures. (2-10)

retinoic acid receptor-related orphan receptor-gamma thymus (RORγt): a largely T cell-specific isoform of the nuclear hormone receptor superfamily expressed predominantly in CD4 CD8 double-positive thymocytes and also in CD4-positive lymph node inducer cells. (5-13)

Rh antigens: also known as the Rhesus blood group system; the second most important blood group after ABO. The major Rh antigen is also called D and there are also two minor Rh antigens called C and E. Clinically, Rh+ refers to the D antigen. 15% of Caucasians are Rh−, and the frequency is lower for other races. (14-6)

rheumatoid arthritis: an autoimmune disease of joints, typically manifested in multiple peripheral joints in a symmetrical fashion. Rheumatoid arthritis also can include autoimmune disease in other tissues, such as blood vessels. Arthritis means joint inflammation. (12-9)

rheumatoid factor: autoantibody (either IgM or IgG) recognizing the Fc portion of self IgG molecules. Rheumatoid factor is commonly found in the sera of rheumatoid arthritis patients, but can also be found in some normal individuals, for example after a viral infection. (12-9)

Rho-family GTPases: family of small G proteins that includes Rac, Rho and Cdc42, all of which have primary functions involving regulation of actin polymerization or myosin-based contraction of actin filaments. (3-8)

RIA: see **radioimmunoassay**. (6-5)

RIG-I: a cytoplasmic protein composed of a CARD and an RNA helicase domain that recognizes dsRNA in virus-infected cells and induces **type 1 interferon** production. (3-16)

RISC: see **RNA-induced silencing complex**. (3-19)

rituximab: an anti-CD20 monoclonal antibody used to treat B cell malignancies and experimentally to treat certain autoimmune diseases, including lupus. (12-8)

RNAi: see **RNA interference**. (3-19)

RNA-induced silencing complex (RISC): multisubunit complex that binds **siRNA** and targets homologous mRNAs for nuclease digestion. (3-19)

RNA interference (RNAi): mechanism by which short fragments of double-stranded RNA lead to the degradation of homologous mRNAs. Also called RNA silencing. RNAi is widely used experimentally to decrease the expression of a gene of interest to study its function. Many plant and some insect viruses have been shown to encode proteins that block RNAi. (3-19)

RNase L: RNase that is activated by 2′,5′-oligoadenylate, which is synthesized in response to the presence in a cell of viral dsRNA. RNase L degrades both mRNA and ribosomal RNA and thus interferes with virus and host protein synthesis. (3-17)

RORγt: see **retinoic acid receptor-related orphan receptor-gamma thymus**. (5-13)

RSS: see **recombination signal sequence**. (7-1)

Sabouraud dextrose agar: a selective medium developed for the cultivation of fungi; antibiotics are frequently added to suppress growth of contaminating bacteria. (11-1)

SAP: see **SLAM-associated protein**. (2-3)

scaffold: protein that binds two or more other proteins and increases the efficiency or specificity with which they act upon each other. (2-0)

scavenger receptor: any of a structurally diverse group of receptors defined by their ability to bind polyanionic ligands such as oxidized low-density lipoprotein (LDL), as occurs in atherosclerotic plaques. Most scavenger receptors additionally bind either microbial ligands or apoptotic cells. (3-7)

SCID: see **severe combined immunodeficiency**. (7-11)

SCR protein: protein with short consensus repeats, which are seen on many complement regulatory proteins. (3-5)

secondary granule: specialized intracellular vesicle in neutrophils that contains the membrane-bound components of the **NADPH oxidase**; also called specific granule. (3-9)

secondary lymphoid tissues: tissues in which lymphocytes are brought together with antigen and adaptive immune responses are initiated. (1-6)

secretory component: a chain of secreted **IgA** that is derived from the **polymeric Ig receptor** and remains bound to polymeric IgA after proteolytic cleavage of the transport receptor. (6-2)

selectin: any of a family of three structurally related lectins that mediate interactions between leukocytes or lymphocytes and endothelial cells in the tissues into which they migrate. (2-14)

sensitization: initial exposure to an antigen leading to the development of a hypersensitive response. For **allergy**, this involves the production of IgE and its binding to mast cells. (13-0)

serotype: variant of a pathogenic microorganism or virus in which the major antigen recognized by protective antibodies has been changed to the extent that it is not recognized by the antibodies directed to the other serotypes. (6-0)

serum sickness: a systemic immune complex disease originally observed upon immunotherapy of various infectious diseases with serum from horses immunized against the infectious agent or its toxin. The large amounts of IgG antibodies produced by the patient in response to the horse serum proteins result in formation of small immune complexes, leading to immune complex deposition and resulting pathology. (13-4)

seven-transmembrane G-protein-coupled receptor: receptor protein that crosses the cell membrane seven times and relays signals to the interior of the cell through **heterotrimeric G proteins**. (2-13)

severe combined immunodeficiency (SCID): disease characterized by early onset of severe and frequent infections of many types and by severe defects in T cells and a partial or full defect in antibody production. (7-11)

SH2 (Src-homology 2) domain: globular domains of slightly less than 100 amino-acid residues that have binding pockets for tyrosine-phosphorylated regions of proteins. Binding is generally dependent upon phosphorylation and the particular amino-acid residues found immediately downstream of the phosphorylation site. Specificity for the adjacent sequences can vary between different SH2 domains. (2-0)

signal joint: DNA in which two recombination signal sequences (RSSs) in gene segments of lymphocyte antigen receptor genes have been joined, usually in a precise manner. Most signal joints are found in circles of excised DNA, but some V(D)J recombination events result in inversion of the DNA between two RSSs, in which case the signal joint remains in the chromosomal DNA of the developing lymphocyte. (7-2)

signal transducer and activator of transcription (STAT): any of a family of rapidly activated transcriptional regulators that are directly activated by cytokine and growth factor receptors and represent the major mode of signaling by the cytokine/hematopoietin receptor superfamily. (2-7)

single-positive (SP) thymocytes: thymocytes that have undergone positive selection successfully and express either CD4 or CD8 but not both. (7-7)

siRNA: see **small interfering RNA**. (3-19)

SLAM-associated protein (SAP): a signaling adaptor molecule that associates with the intracellular domains of several **CD2 subfamily** members, including SLAM (signaling lymphocytic activation molecule). SAP is also called SH2D1A. Defects in this molecule lead to increased susceptibility to Epstein–Barr virus infection. (2-3)

SMAD proteins: any of a related family of signaling proteins activated by TGF-β family receptors. Phosphorylation of SMAD proteins induces them to enter the nucleus and activate or inhibit the transcription of target genes. (2-6)

small interfering RNA (siRNA): double-stranded RNAs 21–25 nucleotides in length that participate in **RNA interference**. Base pairing between siRNAs and their targets leads to the degradation of the targets. (3-19)

SOCS protein: see **suppressor of cytokine signaling protein**. (2-7)

somatic hypermutation: a process whereby mutations are introduced into and near the V-domain-encoding exons of the IgH and IgL genes. (6-11)

spleen: secondary lymphoid organ in the abdomen collecting antigen from the bloodstream. (1-6)

Src-family tyrosine kinases: a family of eight membrane-linked intracellular protein tyrosine kinases first discovered as the product of a viral oncogene causing sarcoma in chickens. Src kinases participate in immune cell function primarily via **ITAM** and **ITIM** receptor signaling. (2-2)

S regions: highly repetitive sequences upstream of each immunoglobulin heavy-chain constant gene locus. **Class switch recombination** occurs between two S regions. (6-11)

STAT: see **signal transducer and activator of transcription**. (2-7)

sterile immunity: a state of immunity in which a pathogen is completely eradicated from the host. (11-0)

stochastic/selection model: hypothesis to explain the mechanism of positive selection of developing T lymphocytes in which one of the two T cell coreceptors is randomly turned off at the start of positive selection; additional signaling must be received once the cell only has one coreceptor, or the cell dies by neglect. (7-9)

strength of signal model: a form of the **instruction model** for positive selection of developing T lymphocytes in which TCR + CD4 recognition sends a stronger signal whereas TCR + CD8 recognition sends a weaker signal. This model generally incorporates a checking mechanism after downregulation of one coreceptor that is analogous to the selection step in the stochastic model. This is a way of ensuring that the instructional step has provided accurate information. (7-9)

subunit vaccine: vaccine composed of defined components of a bacterium, virus or parasite rather than the whole infectious agent. (14-2)

superantigen: protein produced by bacteria or viruses that bind both to the antigen receptor chains of T lymphocytes and to the MHC molecules that they recognize and thereby strongly activate T cells. Most known superantigens are bacterial **exotoxins** that cause septic shock or diarrhea. (9-4)

suppressor of cytokine signaling (SOCS) protein: any of a family of inhibitors of **Jak–STAT** signaling, also called CIS (cytokine-inducible SH2-containing) proteins, that have a central SH2 domain and a conserved carboxy-terminal motif called the SOCS box that mediates ubiquitination of bound proteins. (2-7)

surface immunoglobulin: see **B cell antigen receptor**. (1-4)

surfactant: any of a number of lung proteins of diverse functions. The **collectins** surfactant protein A (SP-A) and surfactant protein D (SP-D) are important molecules of innate immunity. (3-2)

surrogate light chain: an immunoglobulin light chain-like complex of the VpreB and λ5 chains and which bind to μ heavy chains analogously to the light chain to form the **pre-BCR**. Surrogate light-chain subunits are expressed only in developing B cells. (7-3)

Syk: an intracellular protein tyrosine kinase of most hematopoietic cell types that is recruited to phosphorylated ITAMs and has an essential role in ITAM-based signaling. ZAP-70 of T cells is a close relative. (6-8)

symbiosis: a mutually beneficial relationship that sustains a stable interaction between different organisms. (9-0)

synovium: a specialized layer of tissue lining some joints. (12-9)

systemic lupus erythematosus: a systemic autoimmune disease characterized by anti-nuclear antibodies, often including antibodies reactive against double-stranded DNA. Lupus is Latin for wolf, although the exact reason for connecting SLE to the wolf is a matter of debate. (12-8)

TAP: see **transporter associated with antigen processing**. (4-6)

tapasin: a membrane-bound chaperone encoded in the MHC and that binds both to the MHC class I alpha chain and to the **TAP** transporter and facilitates peptide loading onto MHC class I molecules in the lumen of the endoplasmic reticulum. (4-6)

T-bet: a member of the large T-box family of transcriptional regulators implicated in developmental cell-fate decisions important for vertebrate body plan and organogenesis and named for the mouse *Brachyury* (also known as T) and *Drosophila optomotor-blind* genes, which share a common DNA-binding domain designated the T-box. In immunology it is important for directing expression of the cytokine set characteristic of T$_H$1 cells, and thereby contributing centrally to the polarization of these cells. (5-10)

T cell-dependent (TD) antibody responses: antibody responses that do not occur in animals or people lacking T cells. (6-9)

T cell-independent (TI) antibody responses: antibody responses that can occur in the absence of helper T cells, usually stimulated by repetitive components of microorganisms. Most intact infectious agents have epitopes for CD4 T cells and therefore can induce T cell-dependent responses, whether or not they can induce TI responses. However, some vaccines, for example those that are solely polysaccharides, do not have epitopes for helper T cells and can therefore stimulate only TI responses. (6-9)

T cell receptor (TCR): The complex of variable chains whereby T cells recognize antigen and signaling chains whereby antigen recognition is signaled to the cell interior. (1-4, 5-1)

T$_{CM}$: see **central memory T cells**. (5-16)

TCR: see **T cell receptor**. (1-4, 5-1)

TD: see **T cell-dependent antibody responses**. (6-9)

TdT: see **terminal deoxynucleotidyl transferase**. (7-2)

T$_{EM}$: see **effector memory T cells**. (5-16)

terminal deoxynucleotidyl transferase (TdT): an enzyme expressed in developing lymphocytes that inserts random nucleotides into the sites of recombination between V and D segments, D and J segments, or V and J segments of antigen receptor genes in an untemplated fashion to generate **N regions** and much junctional diversity at the CDR3 loop, particularly of the Ig heavy chain and TCR β chain. (7-2)

T helper cells: T lymphocytes that activate other cells of the immune system, including phagocytes, **mast cells**, **basophils**, **eosinophils** and B cells, which when activated differentiate into antibody-producing cells. (1-2)

thoracic duct: main vessel collecting **lymph** for delivery to the heart. (1-6)

thymocytes: developing T cells in the thymus. (7-0)

thymus: primary lymphoid organ in which T lymphocytes mature. (1-2)

TI: see **T cell-independent antibody responses**. (6-9)

TI-1 antigens: antigens defined operationally as those that induce antibody responses in T cell-deficient or Btk-deficient mice (**XID**). (6-9)

TI-2 antigens: antigens defined operationally as those that do not induce antibody responses in Btk-deficient mice but do in T cell-deficient mice. In general, these antigens also do not induce robust antibody responses in young individuals. (6-9)

TipDCs: see **TNF/iNOS-producing dendritic cells**. (5-11)

TIR domain: see **Toll/interleukin 1 receptor (TIR) domain**. (3-10)

TIR domain-containing adaptor inducing interferon-β (TRIF): an adaptor that mediates signaling by some TLRs to IRF-3 and transcription of type 1 interferons. (3-11)

tissue-resident dendritic cells: see **interstitial dendritic cells**. (4-1)

tissue tropism: (of a virus) restriction of viral growth to a specific tissue, often reflecting tissue-restricted expression of the viral entry receptor, although numerous intracellular processes can further restrict viral infection. An example is hepatitis B virus, which grows only in liver cells. (10-0)

TLR: see **Toll-like receptor**. (3-10)

T lymphocytes: lymphocytes that mature in the thymus and different classes of which mediate cytotoxic responses against cells infected with viruses, activate **B lymphocytes** to produce antibodies, and activate phagocytes to ingest and destroy microorganisms. (1-0)

TNF: see **tumor necrosis factor**. (2-9)

TNF/iNOS-producing dendritic cells (TipDCs): monocytes with a specialized function in the early destruction of intracellular bacteria. These cells express the chemokine receptor CCR2 and enter sites of bacterial infection and mature in the presence of IFN-γ and

bacterial products into MHC-class-II- and CD11c-positive cells that express high levels of TNF and iNOS. (5-11)

TNF receptor-associated death domain (TRADD): adaptor molecule that links **TNF** receptor 1 to transcriptional activators or to caspases. (2-9)

TNF receptor-associated factor (TRAF): one of a family of signaling components most of which associate with **TNF** superfamily receptors. There are six TRAFs and all except TRAF4 are known to interact with the cytoplasmic domains of TNF receptor family members. In addition, TRAF6 is a key signaling component of IL-1 receptors and of Toll-like receptors. (2-9)

Toll/interleukin 1 receptor (TIR) domain: domain responsible for transmitting signals downstream of IL-1 receptors and **Toll-like receptors**. TIR domains are found in the cytoplasmic domain of TLRs, IL-1R and IL-18 receptor, and also in signaling adaptor molecules that associate with these receptors and mediate their signaling. TIR domains are also found in many plant disease resistance proteins. The name comes from the first letters of Toll, IL-1 and resistance. (3-10)

Toll-like receptors (TLRs): family of receptors that have leucine-rich repeats in their extracellular domains and the **TIR domain** in their cytoplasmic domains. (3-10)

toxoid: chemically inactivated bacterial toxin that is no longer harmful. Toxoids that retain the antigenic features of the original toxin are used in vaccines to induce neutralizing antibodies against the original toxin. (14-3)

TRADD: see **TNF receptor-associated death domain**. (2-9)

TRAF: see **TNF receptor-associated factor**. (2-9)

transitional B cells: immature B cells that have left the bone marrow and entered the periphery but have not yet matured to their fully differentiated state. Transitional B cells can be phenotypically divided into two subtypes, called T1 and T2. T1 cells have the property of being very sensitive to apoptosis induced through the antigen receptor and T2 cells seem to be further advanced toward the mature state. (7-6)

transporter associated with antigen processing (TAP): a heterodimeric transporter in the endoplasmic reticulum that is encoded in the MHC and transports peptides from the cytosol into the endoplasmic reticulum in an ATP-dependent manner for peptide loading onto MHC class I molecules in the endoplasmic reticulum. (4-6)

T$_{REG}$: see **regulatory T cells**. (7-10)

trematode: flatworms or flukes; parasitic trematodes of humans typically have oral and ventral suckers to assist anchoring while feeding, and alternate between an adult sexual cycle in humans and a larval asexual multiplication in an intermediate host. (11-5)

TRIF: see **TIR domain-containing adaptor inducing interferon-β**. (3-11)

TRIM5α: see **tripartite motif 5α**. (10-5)

tripartite motif 5α (TRIM5α): tripartite motif 5α, a protein that contributes to species-specific blocking of retroviral infection through interactions with capsid proteins, interfering with uncoating and reverse transcription. (10-5)

tumor necrosis factor (TNF): prototype of a family of signaling molecules and a key initiator of inflammatory reactions. (2-9)

type 1 alveolar epithelial cells: cells covering approximately 90% of the alveolar surface of the lung and participating in gas exchange between the air space

and the capillaries; type 2 alveolar epithelial cells secrete hydrophobic surfactant proteins (surfactants B and C) involved in moderating alveolar surface tension. (11-2)

type 1 diabetes mellitus: an autoimmune disease in which there is inflammation of the endocrine pancreas leading to the destruction of beta cells of the islets of Langerhans and insulin insufficiency. (12-6)

type 1 interferon: any of **interferons-α** or **interferon-β**, which are important antiviral cytokines. (3-16)

type I cytokines: cytokines with substantial amino-acid homology that form related four-helix bundles. (2-7)

type I cytokine receptors: receptors for **type I cytokines** that contain two or more beta-sandwich domains, one of which has a WSXWS motif, and that signal via **Jak tyrosine kinases** and **STAT** transcriptional regulators. (2-7)

type I hypersensitivity: see **allergy**. (13-0)

type 2 granuloma: granulomatous response orchestrated by T$_H$2 cells, and including T$_H$2 cells, eosinophils, alternatively activated macrophages and collagen-secreting mesenchymal cells organized in a dense spherical structure around a central inciting agent, such as a schistosome egg. (11-5)

type 2 interferon: interferon-γ, a cytokine important in adaptive immunity and especially in macrophage activation. (3-16)

type II cytokines: cytokines including interferons, IL-10 and IL-10-like cytokines, that form an evolutionarily conserved group whose members are structurally similar to those of the **type I cytokines**. (2-7)

type II cytokine receptors: receptors for **type II cytokines** that signal via **Jaks** and **STATs** and have structural similarity to **type I cytokine receptors**, but lack the WSXWS motif characteristic of the latter receptors. (2-7)

type II hypersensitivity: immune response to a cell-surface or matrix protein resulting in the production of IgG and consequent blocking of the function of the protein or immune damage to the cell or tissue. (13-0)

type II IL-4 receptor: cytokine receptor composed of the IL-4Rα chain which recognizes IL-4 and the IL-13Rα1 chain which recognizes IL-13 and exhibits high-affinity for both IL-13 and IL-4. The type II IL-4 receptor is expressed widely on non-hematopoietic cells. (2-8)

type III hypersensitivity: immune response to a non-infectious soluble antigen resulting in the production of IgG and the formation of damaging immune complexes. (13-0)

type III secretion system: mechanism by which Gram-negative bacteria translocate host-modifying proteins into the cytoplasm of target cells through a syringe-like structure penetrating the host cell membrane. The type III secretion system is encoded in **pathogenicity islands**. (9-2)

type IV hypersensitivity: see **delayed-type hypersensitivity**. (13-0)

ulcerative colitis: an inflammatory disease of unknown cause typically affecting the lower colon and rectum, and characterized by episodes of diarrhea, weight loss, abdominal pain and fever. Ulcerative colitis is one of two kinds of inflammatory bowel disease, the other being **Crohn's disease**. (13-6)

urticaria: an allergic skin reaction analogous to the allergen skin test but occurring in multiple locations at once; also called hives. (13-2)

variable (V) domain: Ig-like domain of the type found in immunoglobulin variable regions. V domains have two more beta strands in the **beta sandwich** than do **C domains**. In antigen receptors, the V domain contains the variable binding site for antigens. (2-1)

variable region (V region): region of a lymphocyte receptor for antigen that participates in antigen binding and varies between cells of different antigen specificities. (1-4, 5-1)

varices: collaterals that develop in the portal circulation to by-pass liver obstruction: the absence of valves in the portal blood system allows retrograde flow, which leads to the formation of varices in the lower esophagus and rectum. (11-5)

V(D)J recombination: process whereby separate gene segments encoding parts of the chains forming the antigen receptor of a lymphocyte assemble in different combinations in different cells to provide specificity for different antigens. (5-1)

viral set-point: the relatively steady-state level of plasma HIV reached after acute infection becomes controlled by CD8 T cell responses. Although viral and CD4 T cell levels are relatively stable for prolonged periods, these levels reflect dynamic turnover in both compartments. (10-6)

virus-like particle (VLP): particle that spontaneously assembles from viral coat proteins in the absence of other viral components. Virus-like particles generated from recombinant coat proteins are used in vaccines against hepatitis B virus and papillomavirus. (14-2)

VLP: see **virus-like particle**. (14-2)

V region: see **variable region**. (5-1)

WASP: see **Wiskott–Aldrich syndrome protein**. (5-3)

wheal and flare: small circular area of soft swelling in skin surrounded by a red rim: the result of a positive allergen skin test. (13-1)

white pulp: lymphoid area of the **spleen**. (1-6)

Wiskott–Aldrich syndrome: X-linked inherited disease in which there are bleeding problems due to platelet defects and also immunodeficiency. (5-3)

Wiskott–Aldrich syndrome protein (WASP): one of a family of five proteins that promote actin polymerization. (5-3)

X4 virus: HIV-1 isolates with gp120-binding affinity for CD4 and CXCR4. X4 viruses can emerge late in HIV infection and be associated with rapid immune deterioration. (10-5)

xenograft: a graft coming from a different species. (14-6)

XID: see **X-linked immunodeficiency**. (6-9)

X-linked agammaglobulinemia: an inherited immunodeficiency disease in which B cell development is severely defective because of a defect in the intracellular protein tyrosine kinase Btk, which participates in BCR and pre-BCR signaling and is encoded on the X chromosome. (7-11)

X-linked hyper-IgM syndrome: a human immunodeficiency characterized by elevated levels of serum IgM, but very low levels of serum IgG, IgA or IgE. This disease is caused by loss-of-function mutations in the gene encoding CD40L (CD154). Much less common are mutations of the autosomal genes encoding CD40 or activation-induced cytidine deaminase (AID), an enzyme required for class switch recombination and somatic hypermutation, which give rise to similar diseases. (6-10)

X-linked immunodeficiency (XID): a genetic immunodeficiency disease of mice caused by a mutation in the Btk gene leading to a partly immunodeficient phenotype characterized by failure to respond to a subset of T cell-independent antigens (**TI-2 antigens**). (6-9)

X-linked lymphoproliferative disease: genetic disease caused by mutations in the gene encoding **SAP**. (2-3)

X-linked severe combined immunodeficiency: severe immunodeficiency disease caused by defects in the gene encoding the cytokine-receptor γ_c chain. Affected individuals lack T cells and NK cells and have defective B cells. An indistinguishable autosomal recessive form of SCID is caused by genetic defects in the gene encoding Jak3, a downstream signaling component of the pathway activated by γ_c receptors. (2-8)

ZAP-70: see **zeta chain associated protein of 70 kDa**. (5-3)

zeta chain associated protein of 70 kDa (ZAP-70): intracellular tyrosine kinase that associates with phosphotyrosines in ITAMs on the ζ chain and CD3 chains of the T cell receptor and is essential in downstream signaling. (5-3)

zoonotic infection: an infection that crosses species, from an animal, to infect humans. (10-4)

zymosan: cell wall preparation of *Saccharomyces cerevisiae* composed of β-glucans, mannans, mannoproteins and chitin. (11-0)

References

Abrams, J.R., Lebwohl, M.G., Guzzo, C.A., Jegasothy, B.V., Goldfarb, M.T., Goffe, B.S., Menter, A., Lowe, N.J., Krueger, G., Brown, M.J., Weiner, R.S., Birkhofer, M.J., Warner, G.L., Berry, K.K., Linsley, P.S., Krueger, J.G., Ochs, H.D., Kelley, S.L. and Kang, S.: **CTLA4Ig-mediated blockade of T-cell costimulation in patients with psoriasis vulgaris.** *J. Clin. Invest.* 1999, **103**:1243–1252. (12-6)

Acuto, O. and Cantrell, D.: **T cell activation and the cytoskeleton.** *Annu. Rev. Immunol.* 2000. **18**:165–184. (5-3)

Acuto, O. and Michel, F.: **CD28-mediated co-stimulation: a quantitative support for TCR signaling.** *Nat. Rev. Immunol.* 2003, **3**:939–951. (5-5)

Ada, G.: **Vaccines and vaccination.** *N. Engl. J. Med.* 2001, **345**:1042–1053. (14-2)

Adams, E.J., Chien, Y.-H. and Garcia, K.C.: **Structure of a γδ T cell receptor in complex with the non-classical MHC T22.** *Science* 2005, **308**:227–231. (5-2, 8-4)

Adams, J.M.: **Ways of dying: multiple pathways to apoptosis.** *Genes Dev.* 2003, **17**:2481–2495. (2-11)

Aderem, A. and Underhill, D.M.: **Mechanisms of phago-cytosis in macrophages.** *Annu. Rev. Immunol.* 1999, **17**:593–623. (3-8)

Agace, W.W.: **Tissue-tropic effector T cells: generation and targeting opportunities.** *Nat. Rev. Immunol.* 2006, **6**:682–692. (5-9)

Aggarwal, B.B.: **Signaling pathways of the TNF Superfamily: a double-edged sword.** *Nat. Rev. Immunol.* 2003, **3**:745–756. (2-9)

Ahlquist, P.: **RNA-dependent RNA polymerases, viruses, and RNA silencing.** *Science* 2002, **296**:1270–1273. (3-19)

Ahmed, R., Lanier, J.G. and Pamer, E.: **Immunological memory and infection** in *Immunology of Infectious Diseases* Kaufman, S.H.E., Sher, A. and Ahmed, R. eds (ASM Press, Washington, D.C., 2002), 175–189. (5-16)

Ahonen, C.L., Manning, E.M., Erickson, L.D., O'Connor, B.P., Lind, E.F., Pullen, S.S., Kehry, M.R. and Noelle, R.J.: **The CD40-TRAF6 axis controls affinity maturation and the generation of long-lived plasma cells.** *Nat. Immunol.* 2002, **3**:451–456. (6-11)

Aiuti, A., Ficara, F., Cattaneo, F., Bordignon, C. and Roncarolo, M.G.: **Gene therapy for adenosine deami-nase deficiency.** *Curr. Opin. Allergy Clin. Immunol.* 2003, **3**:461–466. (14-1)

Akira, S. and Takeda, K.: **Toll-like receptor signalling.** *Nat. Rev. Immunol.* 2004, **4**:499–511. (3-11)

Alberts, B., Johnson, A., Lewis, J., Raff, M., Roberts, K. and Walter, P.: **Development of Multicellular Organisms** in *Molecular Biology of the Cell* 4th ed. (Garland Science, New York, 2002), 1283–1296. (1-1)

Alexander, W.S. and Hilton, D.J.: **The role of suppressors of cytokine signaling (SOCS) proteins in regulation of the immune response.** *Annu. Rev. Immunol.* 2004, **22**:503–529. (2-7)

Allison, T.J., Winter, C.C., Fournié, J.-J., Bonneville, M. and Garboczi, D.N.: **Structure of a human γδ T-cell antigen receptor.** *Nature* 2001, **411**:820–824. (8-4)

Amara, A. and Littman, D.R.: **After Hrs with HIV.** *J. Cell Biol.* 2003, **162**:371–375. (10-4)

Anderson, M.S., Venanzi, E.S., Klein, L., Chen, Z., Berzins, S.P., Turley, S.J., von Boehmer, H., Bronson, R., Dierich, A., Benoist, C. and Mathis, D.: **Projection of an immuno-logical self shadow within the thymus by the Aire protein.** *Science* 2002, **298**:1395–1401. (7-10)

Ansel, K.M., Djuretic, I., Tanasa, B. and Rao, A.: **Regulation of TH2 differentiation and Il4 locus accessibility.** *Annu. Rev. Immunol.* 2006, **24**:607–656. (5-10)

Ansel, K.M., Harris, R.B.S. and Cyster, J.G.: **CXCL13 is required for B1 cell homing, natural antibody pro-duction, and body cavity immunity.** *Immunity* 2002, **16**:67–76. (8-6)

Ansel, K.M., Lee, D.U. and Rao, A.: **An epigenetic view of helper T cell differentiation.** *Nat. Immunol.* 2003, **7**:616–623. (5-9)

Antoine, C., Muller, S., Cant, A., Cavazzana-Calvo, M., Veys, P., Vossen, J., Fasth, A., Heilmann, C., Wulffraat, N., Seger, R., Blanche, S., Friedrich, W., Abinun, M., Davies, G., Bredius, R., Schulz, A., Landais, P. and Fischer, A.: **Long-term survival and transplantation of haemopoietic stem cells for immunodeficiencies: report of the European experience 1968–99.** *Lancet* 2003, **361**:553–560. (7-11)

Arase, H., Mocarski, E.S., Campbell, A.E., Hill, A.B. and Lanier, L.L.: **Direct recognition of cytomegalovirus by activating and inhibitory NK cell receptors.** *Science* 2002, **296**:1323–1326. (10-1)

Arbour, N.C., Lorenz, E., Schutte, B.C., Zabner, J., Kline, J.N., Jones, M., Frees, K., Watt, J.L. and Schwartz, D.A.: **TLR4 mutations are associated with endotoxin hypo-responsiveness in humans.** *Nat. Genet.* 2000, **25**:187–191. (9-3)

Ardavín, C.: **Origin, precursors and differentiation of mouse dendritic cells.** *Nat. Rev. Immunol.* 2003, **3**:582–590. (1-3)

Arnheiter, H., Frese, M., Kambadur, R., Meier, E. and Haller, O.: **Mx transgenic mice—animal models of health.** *Curr. Top. Microbiol. Immunol.* 1996, **206**:119–147. (10-8)

Arstila, T.P., Casrouge, A., Baron, V., Even, J., Kanellopoulos, J. and Kourilsky, P.: **A direct estimate of the human αβ T cell receptor diversity.** *Science* 1999, **286**:958–961. (5-2)

Arulanandam, B.P., Raeder, R.H., Nedrud, J.G., Bucher, D.J., Le, J. and Metzger, D.W.: **IgA immunodeficiency leads to inadequate Th cell priming and increased suscep-tibility to influenza virus infection.** *J. Immunol.* 2001, **166**:226–231. (10-8)

Asselin-Paturel, C., Boonstra, A., Dalod, M., Durand, I., Yessaad, N., Dezutter-Dambuyant, C., Vicari, A., O'Garra, A., Biron, C., Brière, F. and Trinchieri, G.: **Mouse type I IFN-producing cells are immature APCs with plasmacytoid morphology.** *Nat. Immunol.* 2001, **2**:1144–1150. (1-3)

Aw, M.M.: **Transplant immunology.** *J. Pediatr. Surg.* 2003, **38**:1275–1280. (14-8)

Ayabe, T., Satchell, D.P., Wilson, C.L., Parks, W.C., Selsted, M.E. and Ouellette, A.J.: **Secretion of microbicidal α-defensins by intestinal Paneth cells in response to bacteria.** *Nat. Immunol.* 2000, **1**:113–118. (3-1)

Babior, B.M.: **NADPH oxidase.** *Curr. Opin. Immunol.* 2004, **16**:42–47. (3-9)

Bachmann, M.F. and Kopf, M.: **The role of B cells in acute and chronic infections.** *Curr. Opin. Immunol.* 1999, **11**:332–339. (6-0)

Bachmann, M.F. and Zinkernagel, R.M.: **The influence of virus structure on antibody responses and virus serotype formation.** *Immunol. Today* 1996, **17**:553–558. (6-0)

Badovinac, V.P., Porter, B.B. and Harty, J.T.: **Programmed contraction of CD8⁺ T cells after infection.** *Nat. Immunol.* 2002, **3**:619–626. (5-15)

Baigent, S.J. and McCauley, J.W.: **Influenza type A in humans, mammals and birds: determinants of virus virulence, host-range and interspecies transmission.** *Bioessays* 2003, **25**:657–671. (10-7)

Balaji, K.N., Schaschke, N., Machleidt, W., Catalfamo, M. and Henkart, P.A.: **Surface cathepsin B protects cytotoxic lymphocytes from self-destruction after degranulation.** *J. Exp. Med.* 2002, **196**:493–503. (5-14)

Balázs, M., Martin, F., Zhou, T. and Kearney, J.F.: **Blood dendritic cells interact with splenic marginal zone B cells to initiate T-independent immune responses.** *Immunity* 2002, **17**:341–352. (8-7)

Banchereau, J. and Steinman, R.M.: **Dendritic cells and the control of immunity** *Nature* 1998, **392**:245–252. (4-1)

Barber, D.L., Wherry, E.J., Masopust, D., Zhu, B., Allison, J.P., Sharpe, A.H., Freeman, G.J. and Ahmed, R.: **Restoring function in exhausted CD8 T cells during chronic viral infection.** *Nature* 2006, **439**:682–687. (10-9)

Bassing, C.H., Swat, W. and Alt, F.W.: **The mechanism and regulation of chromosomal V(D)J recombination.** *Cell* 2002, **109**:S45–S55. (7-1)

Baumgarth, N., Herman, O.C., Jager, G.C., Brown, L.E., Herzenberg, L.A. and Chen, J.: **B-1 and B-2 cell–derived immunoglobulin M antibodies are nonredundant components of the protective response to influenza virus infection.** *J. Exp. Med.* 2000, **192**:271–280. (6-9, 8-6)

Baumgarth, N., Tung, J.W. and Herzenberg, L.A.: **Inherent specificities in natural antibodies: a key to immune defense against pathogen invasion.** *Springer Semin. Immunopathol.* 2005, **26**:347–362. (8-6)

Belkaid, Y. and Rouse, B.T.: **Natural regulatory T cells in infectious disease.** *Nat. Immunol.* 2005, **6**:353–360. (12-3)

Belkaid, Y., Piccirillo, C.A., Mendez, S., Shevach, E.M. and Sacks, D.L.: **CD4⁺CD25⁺ regulatory T cells control *Leishmania major* persistence and immunity.** *Nature* 2002, **420**:502–507. (11-3)

Belyakov, I.M. and Berzofsky, J.A.: **Immunobiology of mucosal HIV infection and the basis for develop-ment of a new generation of mucosal AIDS vaccines.** *Immunity* 2004, **20**:247–253. (10-6)

Bendelac, A., Bonneville, M. and Kearney, J.F.: **Autoreactivity by design: innate B and T lymphocytes.** *Nat. Rev. Immunol.* 2001, **1**:177–186. (8-0)

Benedict, C.A., Banks, T.A. and Ware, C.F.: **Death and survival: viral regulation of TNF signaling pathways.** *Curr. Opin. Immunol.* 2003, **15**:59–65. (2-9, 3-18)

Benedict, C.A., Norris, P.S. and Ware, C.F.: **To kill or be killed: viral evasion of apoptosis.** *Nat. Immunol.* 2002, **3**:1013–1018. (10-3)

Berger, M., Shankar, V. and Vafai, A.: **Therapeutic applications of monoclonal antibodies.** *Am. J. Med. Sci.* 2002, **324**:14–30. (6-4)

Bergman, Y. and Cedar, H.: **A stepwise epigenetic process controls immunoglobulin allelic exclusion.** *Nat. Rev. Immunol.* 2004, **4**:753–761. (7-3)

Berland, R. and Wortis, H.H.: **Origins and functions of B-1 cells with notes on the role of CD5.** *Annu. Rev. Immunol.* 2002, **20**:253–300. (8-6)

Berman, J. and Sudbery, P.E.: ***Candida albicans*: a mole-cular revolution built on lessons from budding yeast.** *Nat. Rev. Genet.* 2002, **3**:918–930. (11-1)

Bernard, G.R., Vincent, J.-L., Laterre, P.-F., LaRosa, S.P., Dhainaut, J.-F., Lopez-Rodriguez, A., Steingrub, J.S., Garber, G.E., Helterbrand, J.D., Ely, E.W. and Fisher, C.J.:

Efficacy and safety of recombinant human activated protein C for severe sepsis. *N. Engl. J. Med.* 2001, **344**:699–709. (9-3)

Berzofsky, J.A., Berkower, I.J. and Epstein, S.L.: **Antigen–antibody interactions and monoclonal antibodies** in *Fundamental Immunology* 5th ed. Paul, W.E. ed. (Lippincott Williams & Wilkins, Philadelphia, 2003), 69–105. (6-5)

Bettelli, E., Carrier, Y., Gao, W., Korn, T., Strom, T.B., Oukka, M., Weiner, H.L. and Kuchroo, V.K.: **Reciprocal developmental pathways for the generation of pathogenic effector T$_H$17 and regulatory T cells.** *Nature* 2006, **441**:235–238. (5-13)

Beutler, B.: **Inferences, questions and possibilities in Toll-like receptor signalling.** *Nature* 2004, **430**:257–263. (3-11)

Bevan, M.J.: **Helping the CD8$^+$ T-cell response.** *Nat. Rev. Immunol.* 2004, **4**:595–602. (5-8)

Bidère, N., Su, H.C. and Lenardo, M.J.: **Genetic disorders of programmed cell death in the immune system.** *Annu. Rev. Immunol.* 2006, **24**:321–352. (12-2)

Bieniasz, P.D.: **Intrinsic immunity: a front line defense against viral attack.** *Nat. Immunol.* 2004, **5**:1109–1115. (3-19)

Bieniasz, P.D.: **Restriction factors: a defense against retroviral infection.** *Trends Microbiol.* 2003, **11**:286–291. (10-5)

Binder, D., van den Broek, M.F., Kägi, D., Bluethmann, H., Fehr, J., Hengartner, H. and Zinkernagel, R.M.: **Aplastic anemia rescued by exhaustion of cytokine-secreting CD8$^+$ T cells in persistent infection with lymphocytic choriomeningitis virus.** *J. Exp. Med.* 1998, **187**:1903–1920. (10-9)

Biron, C.A., Nguyen, K.B., Pien, G.C., Cousens, L.P. and Salazar-Mather, T.P.: **Natural killer cells in antiviral defense: function and regulation by innate cytokines.** *Annu. Rev. Immunol.* 1999, **17**:189–220. (8-1)

Bishop, G.A.: **The multifaceted roles of TRAFs in the regulation of B-cell function.** *Nat. Rev. Immunol.* 2004, **4**:775–786. (6-11)

Bishop, G.A. and Hostager, B.S.: **B lymphocyte activation by contact-mediated interactions with T lymphocytes.** *Curr. Opin. Immunol.* 2001, **13**:278–285. (6-10)

Bishop, L.R. and Kovacs, J.A.: **Quantitation of anti-*Pneumocystis jiroveci* antibodies in healthy persons and immunocompromised patients.** *J. Infect. Dis.* 2003, **187**:1844–1848. (11-2)

Bizebard, T., Gigant, B., Rigolet, P., Rasmussen, B., Diat, O., Bosecke, P., Wharton, S.A., Skehel, J.J. and Knossow, M.: **Structure of influenza virus haemagglutinin complexed with a neutralizing antibody.** *Nature* 1995, **376**:92–94. (10-8)

Bjorkman, P.J.: **MHC restriction in three dimensions: a view of T cell receptor/ligand interactions.** *Cell* 1997, **89**:167–170. (5-1)

Blattman, J.N. and Greenberg, P.D.: **Cancer immunotherapy: a treatment for the masses.** *Science* 2004, **305**:200–205. (14-5)

Bley, K.R., Hunter, J.C., Eglen, R.M. and Smith, J.A.: **The role of IP prostanoid receptors in inflammatory pain.** *Trends Pharmacol. Sci.* 1998, **19**:141–147. (3-13)

Bluestone, J.A., St. Clair, E.W. and Turka, L.A.: **CTLA4Ig: bridging the basic immunology with clinical application.** *Immunity* 2006, **24**:233–238. (12-0, 12-2)

Blumberg, B.S.: **Hepatitis B virus, the vaccine, and the control of primary cancer of the liver.** *Proc. Natl Acad. Sci. USA* 1997, **94**:7121–7125. (14-4)

Bogdan, C.: **The function of type I interferons in antimicrobial immunity.** *Curr. Opin. Immunol.* 2000, **12**:419–424. (3-16)

Bogdan, C., Röllinghoff, M. and Diefenbach, A.: **Reactive oxygen and reactive nitrogen intermediates in innate and specific immunity.** *Curr. Opin. Immunol.* 2000, **12**:64–76. (3-9)

Boise, L.H., Minn, A.J., Noel, P.J., June, C.H., Accavitti, M.A., Lindsten, T. and Thompson, C.B.: **CD28 costimulation can promote T cell survival by enhancing the expression of Bcl-x$_L$.** *Immunity* 1995, **3**:87–98. (5-15)

Boles, K.S., Stepp, S.E., Bennett, M., Kumar, V. and Mathew, P.A.: **2B4 (CD244) and CS1: novel members of the CD2 subset of the immunoglobulin superfamily molecules expressed on natural killer cells and other leukocytes.** *Immunol. Rev.* 2001, **181**:234–249. (2-3)

Bomsel, M. and Alfsen, A.: **Entry of viruses through the epithelial barrier: pathogenic trickery.** *Nat. Rev. Mol. Cell Biol.* 2003, **4**:57–68. (10-0)

Bonfoco, E., Stuart, P.M., Brunner, T., Lin, T., Griffith, T.S., Gao, Y., Nakajima, H., Henkart, P.A., Ferguson, T.A. and Green, D.R.: **Inducible nonlymphoid expression of Fas ligand is responsible for superantigen-induced peripheral deletion of T cells.** *Immunity* 1998, **9**:711–720. (9-4)

Bonizzi, G. and Karin, M.: **The two NF-κB activation pathways and their role in innate and adaptive immunity.** *Trends Immunol.* 2004, **25**:280–288. (2-10)

Bork, P., Holm, L. and Sander, C.: **The immunoglobulin fold. Structural classification, sequence patterns and common core.** *J. Mol. Biol.* 1994, **242**:309–320. (2-1)

Borowski, C., Martin, C., Gounari, F., Haughn, L., Aifantis, I., Grassi, F. and von Boehmer, H.: **On the brink of becoming a T cell.** *Curr. Opin. Immunol.* 2002, **14**:200–206. (7-7)

Boulay, J.-L., O'Shea, J.J. and Paul, W.E.: **Molecular phylogeny within type I cytokines and their cognate receptors.** *Immunity* 2003, **19**:159–163. (2-8)

Bouma, G. and Strober, W.: **The immunological and genetic basis of inflammatory bowel disease.** *Nat. Rev. Immunol.* 2003, **3**:521–533. (13-6)

Bourne, H.R., Sanders, D.A. and McCormick, F.: **The GTPase superfamily: conserved structure and molecular mechanism.** *Nature* 1991, **349**:117–127. (2-0)

Boursalian, T.E. and Bottomly, K.: **Survival of naïve T cells: roles of restricting versus selecting MHC class II and cytokine milieu.** *J. Immunol.* 1999, **162**:3795–3801. (5-15)

Bouvier, M. and Wiley, D.C.: **Importance of peptide amino and carboxyl termini to the stability of MHC class I molecules.** *Science* 1994, **265**:398–402. (4-5)

Bradley, J.R. and Pober, J.S.: **Tumor necrosis factor receptor-associated factors (TRAFs).** *Oncogene* 2001, **20**:6482–6491. (2-9)

Brander, C. and Walker, B.D.: **Gradual adaptation of HIV to human host populations: good or bad news?** *Nat. Med.* 2003, **9**:1359–1362. (10-6)

Brandt, V.L. and Roth, D.B.: **A recombinase diversified: new functions of the RAG proteins.** *Curr. Opin. Immunol.* 2002, **14**:224–229. (7-2)

Braud, V.M., Allan, D.S. and McMichael, A.J.: **Functions of nonclassical MHC and non-MHC-encoded class I molecules.** *Curr. Opin. Immunol.* 1999, **11**:100–108. (4-4)

Braun, M.Y., Grandjean, I., Feunou, P., Duban, L., Kiss, R., Goldman, M. and Lantz, O.: **Acute rejection in the absence of cognate recognition of allograft by T cells.** *J. Immunol.* 2001, **166**:4879–4883. (14-7)

Braun-Fahrländer, C., Riedler, J., Herz, U., Eder, W., Waser, M., Grize, L., Maisch, S., Carr, D., Gerlach, F., Bufe, A., Lauener, R.P., Schierl, R., Renz, H., Nowak, D. and von Mutius, E.: **Environmental exposure to endotoxin and its relation to asthma in school-age children.** *N. Engl. J. Med.* 2002, **347**:869–877. (13-3)

Brigl, M., Bry, L., Kent, S.C., Gumperz, J.E. and Brenner, M.B.: **Mechanism of CD1d-restricted natural killer T cells activation during microbial infection.** *Nat. Immunol.* 2003, **4**:1230–1237. (8-3)

Brosch, R., Gordon, S.V., Marmiesse, M., Brodin, P., Buchrieser, C., Eiglmeier, K., Garnier, T., Gutierrez, C., Hewinson, G., Kremer, K., Parsons, L.M., Pym, A.S., Samper, S., van Soolingen, D. and Cole, S.T.: **A new evolutionary scenario for the *Mycobacterium tuberculosis* complex.** *Proc. Natl Acad. Sci. USA* 2002, **99**:3684–3689. (9-7)

Brown, E.J. and Gresham, H.D.: **Phagocytosis** in *Fundamental Immunology* 5th ed. Paul, W.E. ed. (Lippincott Williams & Wilkins, Philadelphia, 2003), 1105–1126. (1-3)

Brown, G.D. and Gordon, S.: **Fungal β-glucans and mammalian immunity.** *Immunity* 2003, **19**:311–315. (11-0)

Buckley, R.H.: **Primary immunodeficiency diseases** in *Fundamental Immunology* 5th ed. Paul, W.E. ed. (Lippincott Williams & Wilkins, Philadelphia, 2003), 1593–1620. (14-1)

Bunting, M., Harris, E.S., McIntyre, T.M., Prescott, S.M. and Zimmerman, G.A.: **Leukocyte adhesion deficiency syndromes: adhesion and tethering defects involving β$_2$ integrins and selectin ligands.** *Curr. Opin. Hematol.* 2002, **9**:30–35. (3-14)

Burnet, F.M.: *The Clonal Selection Theory of Acquired Immunity* (Cambridge University Press, London, 1959). (1-4)

Busch, D.H., Kerksiek, K. and Pamer, E.G.: **Processing of *Listeria monocytogenes* antigens and the *in vivo* T cell response to bacterial infection.** *Immunol. Rev.* 1999, **172**:163–169. (8-5)

Butz, E.A. and Bevan, M.J.: **Massive expansion of antigen-specific CD8$^+$ T cells during an acute virus infection.** *Immunity* 1998, **8**:167–175. (10-9)

Cambi, A. and Figdor, C.G.: **Dual function of C-type lectin-like receptors in the immune system.** *Curr. Opin. Cell Biol.* 2003, **15**:539–546. (2-5)

Cao, W., Henry, M.D., Borrow, P., Yamada, H., Elder, J.H., Ravkov, E.V., Nichol, S.T., Compans, R.W., Campbell, K.P. and Oldstone, M.B.A.: **Identification of α-dystroglycan as a receptor for lymphocytic choriomeningitis virus and Lassa fever virus.** *Science* 1998, **282**:2079–2081. (10-9)

Carding, S.R. and Egan, P.J.: **γδ T cells: functional plasticity and heterogeneity.** *Nat. Rev. Immunol.* 2002, **2**:336–345. (8-4)

Cariappa, A. and Pillai, S.: **Antigen-dependent B-cell development.** *Curr. Opin. Immunol.* 2002, **14**:241–249. (7-6)

Carman, C.V. and Springer, T.A.: **Integrin avidity regulation: are changes in affinity and conformation underemphasized?** *Curr. Opin. Cell Biol.* 2003, **15**:547–556. (2-4)

Cartmell, T., Poole, S., Turnbull, A.V., Rothwell, N.J. and Luheshi, G.N.: **Circulating interleukin-6 mediates the febrile response to localised inflammation in rats.** *J. Physiol.* 2000, **526**:653–661. (3-15)

Casadevall, A.: **Antibody-mediated protection against intracellular pathogens.** *Trends Microbiol.* 1998, **6**:102–107. (6-0)

Casanova, J.-L. and Abel, L.: **Genetic dissection of immunity to mycobacteria: the human model.** *Annu. Rev. Immunol.* 2002, **20**:581–620. (9-7, 14-1)

Casanova, J.-L. and Abel, L.: **Inborn errors of immunity to infection: the rule rather than the exception.** *J. Exp. Med.* 2005, **202**:197–201. (14-1)

Casey, L.C., Balk, R.A. and Bone, R.C.: **Plasma cytokine and endotoxin levels correlate with survival in patients with the sepsis syndrome.** *Ann. Intern. Med.* 1993, **119**:771–778. (9-3)

Castellino, F. and Germain, R.N.: **Cooperation between CD4+ and CD8+ T cells: where, when, and how.** *Annu. Rev. Immunol.* 2006, **24**:519–540. (5-8)

Cavazzana-Calvo, M., Lagresle, C., Hacein-Bey-Abina, S. and Fischer, A.: **Gene therapy for severe combined immunodeficiency.** *Annu. Rev. Med.* 2005, **56**:585–602. (14-1)

Cella, M., Jarrossay, D., Facchetti, F., Alebardi, O., Nakajima, H., Lanzavecchia, A. and Colonna, M.: **Plasmacytoid monocytes migrate to inflamed lymph nodes and produce large amounts of type I interferon.** *Nat. Med.* 1999, **5**:919–923. (10-1)

Centers for Disease Control and Prevention (CDC): **Update: Influenza activity—United States and worldwide, 2003–04 season, and composition of the 2004–05 influenza vaccine.** *MMWR Morb. Mortal. Wkly Rep.* 2004, **53**:547–552. (10-7)

Cerwenka, A. and Lanier, L.L.: **NK cells, viruses and cancer.** *Nat. Rev. Immunol.* 2001, **1**:41–49. (8-1)

Chaplin, D.D.: **Lymphoid tissues and organs** in *Fundamental Immunology* 5th ed. Paul, W.E. ed. (Lippincott Williams & Wilkins, Philadelphia, 2003), 419–453. (1-7)

Chatenoud, L.: **CD3-specific antibody-induced active tolerance: from bench to bedside.** *Nat. Rev. Immunol.* 2003, **3**:123–132. (14-8)

Chatterjee, M., Osborne, J., Bestetti, G., Chang, Y. and Moore, P.S.: **Viral IL-6-induced cell proliferation and immune evasion of interferon activity.** *Science* 2002, **298**:1432–1435. (10-3)

Chen, W., Havell, E.A., Moldawer, L.L., McIntyre, K.W., Chizzonite, R.A. and Harmsen, A.G.: **Interleukin 1: an important mediator of host resistance against *Pneumocystis carinii*.** *J. Exp. Med.* 1992, **176**:713–718. (11-2)

Chico-Calero, I., Suárez, M., González-Zorn, B., Scortti, M., Slaghuis, J., Goebel, W., The European *Listeria* Consortium and Vázquez-Boland, J.A.: **Hpt, a bacterial homolog of the microsomal glucose-6-phosphate translocase, mediates rapid intracellular proliferation in *Listeria*.** *Proc. Natl Acad. Sci. USA* 2002, **99**:431–436. (9-6)

Chiu, Y.-L., Soros, V.B., Kreisberg, J.F., Stopak, K., Yonemoto, W. and Greene, W.C.: **Cellular APOBEC3G restricts HIV-1 infection in resting CD4+ T cells.** *Nature* 2005, **435**:108–114. (10-5)

Choi, J., Chen, J., Schreiber, S.L. and Clardy, J.: **Structure of the FKBP12-rapamycin complex interacting with the binding domain of human FRAP.** *Science* 1996, **273**:239–242. (14-8)

Choi, Y., Kappler, J.W. and Marrack, P.: **A superantigen encoded in the open reading frame of the 3′ long terminal repeat of mouse mammary tumour virus.** *Nature* 1991, **350**:203–207. (9-4)

Chow, A., Toomre, D., Garrett, W. and Mellman, I.: **Dendritic cell maturation triggers retrograde MHC class II transport from lysosomes to the plasma membrane.** *Nature* 2002, **418**:988–994. (4-9)

Clark, M.R., Massenburg, D., Siemasko, K., Hou, P. and Zhang, M.: **B-cell antigen receptor signaling requirements for targeting antigen to the MHC class II presentation pathway.** *Curr. Opin. Immunol.* 2004, **16**:382–387. (6-10)

Cloutier, J.-F. and Veillette, A.: **Cooperative inhibition of T-cell antigen receptor signaling by a complex between a kinase and a phosphatase.** *J. Exp. Med.* 1999, **189**:111–121. (5-4)

Cocca, B.A., Cline, A.M. and Radic, M.Z.: **Blebs and apoptotic bodies are B cell autoantigens.** *J. Immunol.* 2002, **169**:159–166. (12-8)

Cohen, P.L.: **Systemic autoimmunity** in *Fundamental Immunology*, 5th ed. Paul, W.E. ed. (Lippincott Williams & Wilkins, Philadelphia, 2003), 1371–1399. (12-8)

Cole, S.T., Brosch, R., Parkhill, J., Garnier, T., Churcher, C., Harris, D., Gordon, S.V., Eiglmeier, K., Gas, S., Barry, C.E. 3rd, Tekaia, F., Badcock, K., Basham, D., Brown, D., Chillingworth, T., Connor, R., Davies, R., Devlin, K., Feltwell, T., Gentles, S., Hamlin, N., Holroyd, S., Hornsby, T., Jagels, K., Krogh, A., McLean, J., Moule, S., Murphy, L., Oliver, K., Osborne, J., Quail, M.A., Rajandream, M.-A., Rogers, J., Rutter, S., Seeger, K., Skelton, J., Squares, R., Squares, S., Sulston, J.E., Taylor, K., Whitehead S. and Barrell, B.G.: **Deciphering the biology of Mycobacterium tuberculosis from the complete genome sequence.** *Nature* 1998, **393**:537–544. (9-7)

Collot-Teixeira, S., Bass, J., Denis, F. and Ranger-Rogez, S.: **Human tumor suppressor p53 and DNA viruses.** *Rev. Med. Virol.* 2004, **14**:301–319. (10-3)

Cone, L.A., Woodard, D.R., Schlievert, P.M. and Tomory, G.S.: **Clinical and bacteriologic observations of a toxic shock-like syndrome due to *Streptococcus pyogenes*.** *N. Eng. J. Med.* 1987, **317**:146–149. (9-4)

Cook, D.N., Pisetsky, D.S. and Schwartz, D.A.: **Toll-like receptors in pathogenesis of human disease.** *Nat. Immunol.* 2004, **5**:975–979. (3-10)

Coombes, B.K., Valdez, Y. and Finlay, B.B.: **Evasive maneuvers by secreted bacterial proteins to avoid innate immune responses.** *Curr. Biol.* 2004, **14**:R856–R867. (3-8, 9-2)

Cory, S. and Adams, J.M.: **The Bcl2 family: regulators of the cellular life-or-death switch.** *Nat. Rev. Cancer* 2002, **2**: 647–656. (2-12)

Cosman, D., Mullberg, J., Sutherland, C.L., Chin, W., Armitage, R., Fanslow, W., Kubin, M. and Chalupny, N.J.: **ULBPs, novel MHC class I-related molecules, bind to CMV glycoprotein UL16 and stimulate NK cytotoxicity through the NKG2D receptor.** *Immunity* 2001, **14**:123–133. (4-4)

Coussens, L.M. and Werb, Z.: **Inflammation and cancer.** *Nature* 2002, **420**:860–867. (14-5)

Crabtree, G.R.: **Generic signals and specific outcomes: signaling through Ca^{2+}, calcineurin, and NF-AT.** *Cell* 1999, **96**:611–614. (14-8)

Cresswell, P.: **The biochemistry and cell biology of antigen processing** in *Fundamental Immunology* 5th ed. Paul, W.E. ed. (Lippincott Williams & Wilkins, Philadelphia, 2003), 613–629. (4-6, 4-7)

Cyster, J.G.: **Chemokines and cell migration in secondary lymphoid organs.** *Science* 1999, **286**: 2098–2102. (2-13)

Cyster, J.G.: **Chemokines, sphingosine-1-phosphate, and cell migration in secondary lymphoid organs.** *Annu. Rev. Immunol.* 2005, **23**:127–159. (1-6)

Danial, N.N. and Korsmeyer, S.J.: **Cell death: critical control points.** *Cell* 2004, **116**:205–219. (2-11)

Davies, D.R., Padlan, E.A. and Sheriff, S.: **Antibody-antigen complexes.** *Annu. Rev. Biochem.* 1990, **59**:439–473. (6-1)

Davies, S.J., Grogan, J.L., Blank, R.B., Lim, K.C., Locksley, R.M. and McKerrow, J.H.: **Modulation of blood fluke development in the liver by hepatic CD4+ lymphocytes.** *Science* 2001, **294**:1358–1361. (11-5)

Davis, M.M. and Bjorkman, P.J.: **T-cell antigen receptor genes and T-cell recognition.** *Nature* 1988, **334**:395–402. (5-1)

Davis, M.M. and Chien, Y.-H.: **T-cell antigen receptors** in *Fundamental Immunology* 5th ed. Paul, W.E. ed. (Lippincott Williams & Wilkins, Philadelphia, 2003), 227–258. (5-1)

Decatur, A.L. and Portnoy, D.A.: **A PEST-like sequence in listeriolysin O essential for *Listeria monocytogenes* pathogenicity.** *Science* 2000, **290**:992–995. (9-6)

Degano, M., Garcia, K.C., Apostolopoulos, V., Rudolph, M.G., Teyton, L. and Wilson, I.A.: **A functional hot spot for antigen recognition in a superagonist TCR/MHC complex.** *Immunity* 2000, **12**:251–261. (5-2)

DeMali, K.A., Wennerberg, K. and Burridge, K.: **Integrin signaling to the actin cytoskeleton.** *Curr. Opin. Cell Biol.* 2003, **15**:572–582. (2-4)

Diefenbach, A. and Raulet, D.H.: **Strategies for target cell recognition by natural killer cells.** *Immunol. Rev.* 2001, **181**:170–184. (4-4)

Diehn, M., Alizadeh, A.A., Rando, O.J., Liu, C.L., Stankunas, K., Botstein, D., Crabtree, G.R. and Brown, P.O.: **Genomic expression programs and the integration of the CD28 costimulatory signal in T cell activation.** *Proc. Natl Acad. Sci. USA* 2002, **99**:11796–11801. (5-5)

Dimitrov, D.S.: **Virus entry: molecular mechanisms and biomedical applications.** *Nat. Rev. Microbiol.* 2004, **2**:109–122. (10-0)

Di Sabatino, A., Ciccocioppo, R., Cupelli, F., Cinque, B., Millimaggi, D., Clarkson, M.M., Paulli, M., Cifone M.G. and Corazza G.R.: **Epithelium derived interleukin 15 regulates intraepithelial lymphocyte Th1 cytokine production, cytotoxicity, and survival in celiac disease.** *Gut* 2006, **55**:469–477. (13-5)

Di Santo, J.P.: **Natural killer cell developmental pathways: a question of balance.** *Annu. Rev. Immunol.* 2006, **24**:257–286. (8-1)

Doherty, P.C. and Christensen, J.P.: **Accessing complexity: the dynamics of virus-specific T cell responses.** *Annu. Rev. Immunol.* 2000, **18**:561–592. (10-8)

Doherty, P.C., Turner, S.J., Webby, R.G. and Thomas, P.G.: **Influenza and the challenge for immunology.** *Nat. Immunol.* 2006, **7**:449–455. (5-16)

Dong, C., Davis, R.J. and Flavell, R.A.: **MAP kinases in the immune response.** *Annu. Rev. Immunol.* 2002, **20**:55–72. (5-10)

d'Ostiani, C.F., Del Sero, G., Bacci, A., Montagnoli, C., Spreca, A., Mencacci, A., Ricciardi-Castagnoli, P. and Romani, L.: **Dendritic cells discriminate between yeasts and hyphae of the fungus *Candida albicans*: implications for initiation of T helper cell immunity in vitro and in vivo.** *J. Exp. Med.* 2000, **191**:1661–1674. (11-1)

Duan, L., Reddi, A.L., Ghosh, A., Dimri, M. and Band, H.: **The Cbl family and other ubiquitin ligases: destructive forces in control of antigen receptor signaling.** *Immunity* 2004, **21**:7–17. (5-4)

Dubois, R.N., Abramson, S.B., Crofford, L., Gupta, R.A., Simon, L.S., Van De Putte, L.B.A. and Lipsky, P.E.: **Cyclooxygenases in biology and disease.** *FASEB J.* 1998, **12**:1063–1073. (3-13)

Duclos, S. and Desjardins, M.: **Subversion of a young phagosome: the survival strategies of intracellular pathogens.** *Cell. Microbiol.* 2000, **2**:365–377. (3-8)

Duerr, R.H., Taylor, K.D., Brant, S.R., Rioux, J.D., Silverberg, M.S., Daly, M.J., Steinhart, A.H., Abraham, C., Regueiro, M., Griffiths, A., Dassopoulos, T., Bitton, A., Yang, H., Targan S., Datta, L.W., Kistner, E.O., Schumm, L.P., Lee, A.T., Gregersen, P.K., Barmada, M.M., Rotter, J.I., Nicolae, D.L. and Cho, J.H.: **A genome-wide association study identifies *IL23R* as an inflammatory bowel disease gene.** *Science* 2006, **314**:1461–1463. (13-6)

Dunn, E., Sims, J.E., Nicklin, M.J.H. and O'Neill, A.J.: **Annotating genes with potential roles in the immune system: six new members of the IL-1 family.** *Trends Immunol.* 2001, **22**:533–536. (2-6)

Dunn, G.P., Old, L.J. and Schreiber, R.D.: **The three E's of cancer immunoediting.** *Annu. Rev. Immunol.* 2004, **22**:329–360. (14-5)

Durandy, A., Revy, P., Imai, K. and Fischer, A.: **Hyperimmunoglobulin M syndromes caused by intrinsic B-lymphocyte defects.** *Immunol. Rev.* 2005, **203**:67–79. (6-11)

Eberl, G. and Littman, D.R.: **Thymic origin of intestinal αβ T cells revealed by fate mapping of RORγt⁺ cells.** *Science* 2004, **305**:248–251. (8-5)

Eckmann, L. and Karin, M.: **NOD2 and Crohn's disease: loss or gain of function?** *Immunity* 2005, **22**:661–667. (3-12)

Edelman, G.M.: **Antibody structure and molecular immunology.** *Science* 1973, **180**:830–840. (6-1)

Eder, W., Ege, M.J. and von Mutius, E.: **The asthma epidemic.** *N. Engl. J. Med.* 2006, **355**:2226–2235. (13-3)

Edman, J.C., Kovacs, J.A., Masur, H., Santi, D.V., Elwood, H.J. and Sogin, M.L.: **Ribosomal RNA sequence shows *Pneumocystis carinii* to be a member of the Fungi.** *Nature* 1988, **334**:519–522. (11-2)

Egen, J.G., Kuhns, M.S. and Allison, J.P.: **CTLA-4: new insights into its biological function and use in tumor immunotherapy.** *Nat. Immunol.* 2002, **3**:611–618. (5-7)

Ehlers, M.R.: **CR3: a general purpose adhesion-recognition receptor essential for innate immunity.** *Microbes Infect.* 2000, **2**:289–294. (3-7)

Engel, P., Eck, M.J. and Terhorst, C.: **The SAP and SLAM families in immune responses and X-linked lymphoproliferative disease.** *Nat. Rev. Immunol.* 2003, **3**:813–821. (2-3)

Engelhard, V.H.: **Structure of peptides associated with class I and class II MHC molecules.** *Curr. Opin. Immunol.* 1994, **6**:13–23. (4-5)

Ercolini, A.M. and Miller, S.D.: **Mechanisms of immunopathology in murine models of central nervous system demyelinating disease.** *J. Immunol.* 2006, **176**:3293–3298. (12-7)

Ernst, J.D.: **Bacterial inhibition of phagocytosis.** *Cell. Microbiol.* 2000, **2**:379–386. (3-8)

Eskola, J., Kilpi, T., Palmu, A., Jokinen, J., Haapakoski, J., Herva, E., Takala, A., Kayhty, H., Karma, P., Kohberger, R., Siber, G. and Makelä, P.H.: **Efficacy of a pneumococcal conjugate vaccine against acute otitis media.** *N. Engl. J. Med.* 2001, **344**:403–409. (9-5)

Etienne-Manneville, S. and Hall, A.: **Rho GTPases in cell biology.** *Nature* 2002, **420**:629–635. (2-0)

Fallon, P.G., Jolin, H.E., Smith, P., Emson, C.L., Townsend, M.J., Fallon, R., Smith, P. and McKenzie, A.N.J.: **IL-4 induces characteristic Th2 responses even in the combined absence of IL-5, IL-9, and IL-13.** *Immunity* 2002, **17**:7–17. (5-12, 11-4)

Fallon, P.G., Richardson, E.J., McKenzie, G.J. and McKenzie, A.N.: **Schistosome infection of transgenic mice defines distinct and contrasting pathogenic roles for IL-4 and IL-13: IL-13 is a profibrotic agent.** *J. Immunol.* 2000, **164**:2585–2591. (11-5)

Farah, C.S., Elahi, S., Pang, G., Gotjamanos, T., Seymour, G.J., Clancy, R.L. and Ashman, R.B.: **T cells augment monocyte and neutrophil function in host resistance against oropharyngeal candidiasis.** *Infect. Immun.* 2001, **69**:6110–6118. (11-1)

Fauci, A.S.: **Emerging and re-emerging infectious diseases: influenza as a prototype of the host-pathogen balancing act.** *Cell* 2006, **124**:665–670. (10-7)

Fauci, A.S.: **Twenty-five years of HIV/AIDS.** *Science* 2006, **313**:409. (10-4)

Faurschou, M. and Borregaard, N.: **Neutrophil granules and secretory vesicles in inflammation.** *Microbes Infect.* 2003, **5**:1317–1327. (3-9)

Faust, S.N., Levin, M., Harrison, O.B., Goldin, R.D., Lockhart, M.S., Kondaveeti, S., Laszik, Z., Esmon, C.T. and Heyderman, R.S.: **Dysfunction of endothelial protein C activation in severe meningococcal sepsis.** *N. Engl. J. Med.* 2001, **345**:408–416. (9-3)

Fearon, D.T. and Carroll, M.C.: **Regulation of B lymphocyte responses to foreign and self-antigens by the CD19/CD21 complex.** *Annu. Rev. Immunol.* 2000, **18**:393–422. (6-8)

Fearon, D.T. and Locksley, R.M.: **The instructive role of innate immunity in the acquired immune response.** *Science* 1996, **272**:50–53. (3-0)

Fearon, D.T., Carr, J.M., Telaranta, A., Carrasco, M.J. and Thaventhiran, J.E.D.: **The rationale for the IL-2-independent generation of the self-renewing central memory CD8⁺ T cells.** *Immunol. Rev.* 2006, **211**:104–118. (5-16)

Feldmann, M. and Maini, R.N.: **Anti-TNFα therapy of rheumatoid arthritis: what have we learned?** *Annu. Rev. Immunol.* 2001, **19**:163–196. (3-13, 12-9)

Feldmann, M. and Steinman, L.: **Design of effective immunotherapy for human autoimmunity.** *Nature* 2005, **435**:612–619. (12-0)

Fesik, S.W.: **Insights into programmed cell death through structural biology.** *Cell* 2000, **103**:273–282. (2-11)

Firestein, G.S.: **The T cell cometh: interplay between adaptive immunity and cytokine networks in rheumatoid arthritis.** *J. Clin. Invest.* 2004, **114**:471–474. (12-9)

Fischer, A., Hacein-Bey, S. and Cavazzana-Calvo, M.: **Gene therapy of severe combined immunodeficiencies.** *Nat. Rev. Immunol.* 2002, **2**:615–621. (7-11)

Fitzgerald, J.R., Sturdevant, D.E., Mackie, S.M., Gill, S.R. and Musser, J.M.: **Evolutionary genomics of *Staphylococcus aureus*: insights into the origin of methicillin-resistant strains and the toxic shock syndrome epidemic.** *Proc. Natl Acad. Sci. USA* 2001, **98**:8821–8826. (9-4)

Fitzgerald, K.A., McWhirter, S.M., Faia, K.L., Rowe, D.C., Latz, E., Golenbock, D.T., Coyle, A.J., Liao, S.-M. and Maniatis, T.: **IKKε and TBK1 are essential components of the IRF3 signaling pathway.** *Nat. Immunol.* 2003, **4**:491–496. (3-11)

Flicker, S. and Valenta, R.: **Renaissance of the blocking antibody concept in type I allergy.** *Int. Arch. Allergy Immunol.* 2003, **132**:13–24. (13-2)

Flynn, J.L. and Chan, J.: **Immunology of tuberculosis.** *Annu. Rev. Immunol.* 2001, **19**:93–129. (9-7)

Flynn, K.J., Belz, G.T., Altman, J.D., Ahmed, R., Woodland, D.L. and Doherty, P.C.: **Virus-specific CD8⁺ T cells in primary and secondary influenza pneumonia.** *Immunity* 1998, **8**:683–691. (10-8)

Fontenot, J.D. and Rudensky, A.Y.: **A well adapted regulatory contrivance: regulatory T cell development and the forkhead transcription factor Foxp3.** *Nat. Immunol.* 2005, **6**:331–337. (5-13)

Foxman, E.F., Kunkel, E.J. and Butcher, E.C.: **Integrating conflicting chemotactic signals. The role of memory in leukocyte navigation.** *J. Cell Biol.* 1999, **147**:577–588. (3-14)

Fraser, C.C., Howie, D., Morra, M., Qiu, Y., Murphy, C., Shen, Q., Gutierrez-Ramos, J.-C., Coyle, A., Kingsbury, G.A. and Terhorst, C.: **Identification and characterization of SF2000 and SF2001, two new members of the immune receptor SLAM/CD2 family.** *Immunogenetics* 2002, **53**:843–850. (2-3)

Frazer, I.: **God's gift to women: the human papillomavirus vaccine.** *Immunity* 2006, **25**:179–184. (14-4)

Frazer, J.K. and Capra, J.D.: **Structure and function of immunoglobulins** in *Fundamental Immunology* 4th ed. Paul, W.E. ed. (Lippincott-Raven, New York, 1999), 64–70. (1-4)

Freed, E.O.: **HIV-1 and the host cell: an intimate association.** *Trends Microbiol.* 2004, **12**:170–177. (10-4)

Frenette, P.S., Mayadas, T.N., Rayburn, H., Hynes, R.O. and Wagner, D.D.: **Susceptibility to infection and altered hematopoiesis in mice deficient in both P- and E-selectins.** *Cell* 1996, **84**:563–574. (9-1)

Fugmann, S.D., Lee, A.I., Shockett, P.E., Villey, I.J. and Schatz, D.G.: **The RAG proteins and V(D)J recombination: complexes, ends, and transposition.** *Annu. Rev. Immunol.* 2000, **18**:495–527. (7-2)

Fujita, T., Matsushita, M. and Endo, Y.: **The lectin-complement pathway – its role in innate immunity and evolution.** *Immunol. Rev.* 2004, **198**:185–202. (3-2, 3-3)

Fuller, K.G., Olson, J.K., Howard, L.M., Croxford, J.L. and Miller, S.D.: **Mouse models of multiple sclerosis: experimental autoimmune encephalomyelitis and Theiler's virus-induced demyelinating disease.** *Methods Mol. Med.* 2004, **102**:339–362. (12-7)

Gabay, C. and Kushner, I.: **Acute-phase proteins and other systemic responses to inflammation.** *N. Engl. J. Med.* 1999, **340**:448–454. (3-15)

Gadina, M., Hilton, D., Johnston, J.A., Morinobu, A., Lighvani, A., Zhou, Y.J., Visconti, R. and O'Shea, J.J.: **Signaling by type I and II cytokine receptors: ten years after.** *Curr. Opin. Immunol.* 2001, **13**:363–373. (2-7)

Gadjeva, M., Thiel, S. and Jensenius, J.C.: **The mannan-binding-lectin pathway of the innate immune response.** *Curr. Opin. Immunol.* 2001, **13**:74–78. (3-4)

Galli, S.J., Kalesnikoff, J., Grimbaldeston, M.A., Piliponsky, A.M., Williams, C.M.M. and Tsai, M.: **Mast cells as "tunable" effector and immunoregulatory cells: recent advances.** *Annu. Rev. Immunol.* 2005, **23**:749–786. (13-5)

Game, D.S. and Lechler, R.I.: **Pathways of allorecognition: implications for transplantation tolerance.** *Transpl. Immunol.* 2002, **10**:101–108. (14-7)

Gantner, B.N., Simmons, R.M., Canavera, S.J., Akira, S. and Underhill, D.M.: **Collaborative induction of inflammatory responses by dectin-1 and Toll-like receptor 2.** *J. Exp. Med.* 2003, **197**:1107–1117. (11-1)

Ganz, T.: **Fatal attraction evaded: how pathogenic bacteria resist cationic peptides.** *J. Exp. Med.* 2001, **193**:F31–F33. (9-2)

Gao, X., Nelson, G.W., Karacki, P., Martin, M.P., Phair, J., Kaslow, R., Goedert, J.J., Buchbinder, S., Hoots, K., Vlahov, D., O'Brien, S.J. and Carrington, M.: **Effect of a single amino acid change in MHC class I molecules on the rate of progression to AIDS.** *N. Engl. J. Med.* 2001, **344**:1668–1675. (10-6)

Garcia, K.C. and Adams, E.J.: **How the T cell receptor sees antigens—a structural view.** *Cell* 2005, **122**:333–336. (5-2)

Garcia, K.C., Teyton, L. and Wilson, I.A.: **Structural basis of T cell recognition.** *Annu. Rev. Immunol.* 1999, **17**:369–397. (4-2)

García-Sastre, A., Egorov, A., Matassov, D., Brandt, S., Levy, D.E., Durbin, J.E., Palese, P. and Muster, T.: **Influenza A virus lacking the NS1 gene replicates in interferon-deficient systems.** *Virology* 1998, **252**:324–330. (10-8)

Gasque, P.: **Complement: a unique innate immune sensor for danger signals.** *Mol. Immunol.* 2004, **41**:1089–1098. (3-5)

Gazit, R., Gruda, R., Elboim, M., Arnon, T.I., Katz, G., Achdout, H., Hanna, J., Qimron, U., Landau, G., Greenbaum, E., Zakay-Rones, Z., Porgador, A. and Mandelboim, O.: **Lethal influenza infection in the absence of the natural killer receptor gene Ncr1.** *Nat. Immunol.* 2006, **7**:517–523. (8-2)

Geijtenbeek, T.B.H., Torensma, R., van Vliet, S.J., van Duijnhoven, G.C.F., Adema, G.J., van Kooyk, Y. and Figdor, C.G.: **Identification of DC-SIGN, a novel dendritic cell-specific ICAM-3 receptor that supports primary immune responses.** *Cell* 2000, **100**:575–585. (5-6)

Geijtenbeek, T.B.H., Van Vliet, S.J., Engering, A., 't Hart, B.A. and Van Kooyk, Y.: **Self- and nonself-recognition by C-type lectins on dendritic cells.** *Annu. Rev. Immunol.* 2004, **22**:33–54. (2-5)

Geiss, G.K., Salvatore, M., Tumpey, T.M., Carter, V.S., Wang, X., Basler, C.F., Taubenberger, J.K., Bumgarner, R.E., Palese, P., Katze, M.G. and García-Sastre, A.: **Cellular transcriptional profiling in influenza A virus-infected lung epithelial cells: the role of nonstructural NS1 protein in the evasion of the host innate defense and its potential contribution to pandemic influenza.** *Proc. Natl Acad. Sci. USA* 2002, **99**:10736–10741. (10-8)

Gellert, M.: **V(D)J recombination: RAG proteins, repair factors, and regulation.** *Annu. Rev. Biochem.* 2002, **71**:101–132. (7-2)

Germain, R.N.: **T-cell development and the CD4–CD8 lineage decision.** *Nat Rev. Immunol.* 2002, **2**:309–322. (7-9)

Germain, R.N.: **The T cell receptor for antigen: signaling and ligand discrimination.** *J. Biol. Chem.* 2001, **276**:35223–35226. (5-4)

Germain, R.N. and Jenkins, M.K.: *In vivo* **antigen presentation.** *Curr. Opin. Immunol.* 2004, **16**:120–123. (5-6)

Gewurz, B.E., Gaudet, R., Tortorella, D., Wang, E.W. and Ploegh, H.L.: **Virus subversion of immunity: a structural perspective.** *Curr. Opin. Immunol.* 2001, **13**:442–450. (10-3)

Ghosh, S. and Karin, M.: **Missing pieces in the NF-κB puzzle.** *Cell* 2002, **109**:S81–S96. (2-10)

Ghosh, S., Goldin, E., Gordon, F.H., Malchow, H.A., Rask-Madsen, J., Rutgeerts, P., Vyhnálek, P., Zádorová, Z., Palmer, T. and Donoghue, S.: **Natalizumab for active Crohn's disease.** *N. Engl. J. Med.* 2003, **348**:24–32. (13-6)

Girardin, S.E., Hugot, J.-P. and Sansonetti, P.J.: **Lessons from Nod2 studies: towards a link between Crohn's disease and bacterial sensing.** *Trends Immunol.* 2003, **24**:652–658. (3-12)

Glazier, K.S., Hake, S.B., Tobin, H.M., Chadburn, A., Schattner, E.J. and Denzin, L.K.: **Germinal center B cells regulate their capability to present antigen by modulation of HLA-DO.** *J. Exp. Med.* 2002, **195**:1063–1069. Erratum in: *J. Exp. Med.* 2003, **198**:1765. (4-7)

Gleimer, M. and Parham, P.: **Stress management, MHC class I and class I-like molecules as reporters of cellular stress.** *Immunity* 2003, **19**:469–477. (8-0)

Glusman, G., Rowen, L., Lee, I., Boysen, C., Roach, J.C., Smit, A.F., Wang, K., Koop, B.F. and Hood, L.: **Comparative genomics of the human and mouse T cell receptor loci.** *Immunity* 2001, **15**:337–349. (5-1)

Goldberg, A.L., Cascio, P., Saric, T. and Rock, K.L.: **The importance of the proteasome and subsequent proteolytic steps in the generation of antigenic peptides.** *Mol. Immunol.* 2002, **39**:147–164. (4-8)

Goodnow, C.C., Sprent, J., Fazekas de St Groth, B. and Vinuesa, C.G.: **Cellular and genetic mechanisms of self tolerance and autoimmunity.** *Nature* 2005, **435**:590–597. (12-1)

Gordon, S.: **Alternative activation of macrophages.** *Nat. Rev. Immunol.* 2003, **3**:23–35. (5-12)

Gordon, S. and Taylor, P.R.: **Monocyte and macrophage heterogeneity.** *Nat. Rev. Immunol.* 2005, **5**:953–964. (1-3)

Gorman, C., Leandro, M. and Isenberg, D.: **Does B cell depletion have a role to play in the treatment of systemic lupus erythematosus?** *Lupus* 2004, **13**:312–316. (12-8)

Gould, H.J., Sutton, B.J., Beavil, A.J., Beavil, R.L., McCloskey, N., Coker, H.A., Fear, D. and Smurthwaite, L.: **The biology of IGE and the basis of allergic disease.** *Annu. Rev. Immunol.* 2003, **21**:579–628. (13-1)

Goulder, P.J.R., Brander, C., Tang, Y., Tremblay, C., Colbert, R.A., Addo, M.M., Rosenberg, E.S., Nguyen, T., Allen, R., Trocha, A., Altfeld, M., He, S., Bunce, M., Funkhouser, R., Pelton, S.I., Burchett, S.K., McIntosh, K., Korber, B.T.M. and Walker, B.D.: **Evolution and transmission of stable CTL escape mutations in HIV infection.** *Nature* 2001, **412**:334–338. (10-0)

Grashoff, C., Thievessen, I., Lorenz, K., Ussar, S. and Fässler, R.: **Integrin-linked kinase: integrin's mysterious partner.** *Curr. Opin. Cell Biol.* 2004, **16**:565–571. (2-4)

Grawunder, U. and Harfst, E.: **How to make ends meet in V(D)J recombination.** *Curr. Opin. Immunol.* 2001, **13**:186–194. (7-2)

Green, D.R. and Kroemer, G.: **The pathophysiology of mitochondrial cell death.** *Science* 2004, **305**:626–629. (2-12)

Groh, V., Rhinehart, R., Randolph-Habecker, J., Topp, M.S., Riddell, S.R. and Spies, T.: **Costimulation of CD8αβ T cells by NKG2D *via* engagement by MIC induced on virus-infected cells.** *Nat. Immunol.* 2001, **2**:255–260. (10-1)

Grossman, Z. and Paul, W.E.: **Autoreactivity, dynamic tuning and selectivity.** *Curr. Opin. Immunol.* 2001, **13**:687–698. (5-4)

Gumperz, J.E. and Brenner, M.B.: **CD1-specific T cells in microbial immunity.** *Curr. Opin. Immun.* 2001, **13**:471–478. (4-4)

Guo, R.-F. and Ward, P.A.: **Role of C5a in inflammatory responses.** *Annu. Rev. Immunol.* 2005, **23**:821–852. (3-5)

Gupta, A., Gartner, J.J., Sethupathy, P., Hatzigeorgiou, A.G. and Fraser, N.W.: **Anti-apoptotic function of a microRNA encoded by the HSV-1 latency-associated transcript.** *Nature* 2006, **442**:82–85. (10-0)

Gutierrez, M.G., Master, S.S., Singh, S.B., Taylor, G.A., Colombo, M.I. and Deretic, V.: **Autophagy is a defense mechanism inhibiting BCG and *Mycobacterium tuberculosis* survival in infected macrophages.** *Cell* 2004, **119**:753–766. (9-7)

Hacein-Bey-Abina, S., Le Deist, F., Carlier, F., Bouneaud, C., Hue, C., De Villartay, J.-P., Thrasher, A.J., Wulffraat, N., Sorensen, R., Dupuis-Girod, S., Fischer, A., Davies, E.G., Kuis, W., Leiva, L. and Cavazzana-Calvo, M.: **Sustained correction of X-linked severe combined immunodeficiency by ex vivo gene therapy.** *N. Engl. J. Med.* 2002, **346**:1185–1193. (14-1)

Hahn, B.H., Shaw, G.M., De Cock, K.M. and Sharp, P.M.: **AIDS as a zoonosis: scientific and public health implications.** *Science* 2000, **287**:607–614. (10-4)

Halaby, D.M. and Mornon, J.P.E.: **The immunoglobulin superfamily: an insight on its tissular, species, and functional diversity.** *J. Mol. Evol.* 1998, **46**:389–400. (2-1)

Halaby, D.M., Poupon, A. and Mornon, J.P.: **The immunoglobulin fold family: sequence analysis and 3D structure comparisons.** *Prot. Eng.* 1999, **12**:563–571. (2-1)

Haldane, J.B.S.: **Disease and Evolution.** *Ric. Sci.* 1949. **19** suppl.:68–76. (4-3)

Hangartner, L., Zinkernagel, R.M. and Hengartner, H.: **Antiviral antibody responses: the two extremes of a wide spectrum.** *Nat. Rev. Immunol.* 2006, **6**:231–243. (6-0, 6-9)

Hardy, R.R.: **B-cell commitment: deciding on the players.** *Curr. Opin. Immunol.* 2003, **15**:158–165. (7-5)

Hardy, R.R. and Hayakawa, K.: **B cell development pathways.** *Annu. Rev. Immunol.* 2001, **19**:595–621. (7-5, 8-6)

Harris, R.S. and Liddament, M.T.: **Retroviral restriction by APOBEC proteins.** *Nat. Rev. Immunol.* 2004, **4**:868–877. (10-5)

Harris, R.S., Bishop, K.N., Sheehy, A.M., Craig, H.M., Petersen-Mahrt, S.K., Watt, I.N., Neuberger, M.S. and Malim, M.H.: **DNA deamination mediates innate immunity to retroviral infection.** *Cell* 2003, **113**:803–809. (3-19)

Hataye, J., Moon, J.J., Khoruts, A., Reilly, C. and Jenkins, M.K.: **Naïve and memory CD4+ T cell survival controlled by clonal abundance.** *Science* 2006, **312**:114–116. (5-16)

Hayday, A., Theodoridis, E., Ramsburg, E. and Shires, J.: **Intraepithelial lymphocytes: exploring the Third Way in immunology.** *Nat. Immunol.* 2001, **2**:997–1003. (8-5)

Hayden, M.S. and Ghosh, S.: **Signaling to NF-κB.** *Genes Dev.* 2004, **18**:2195–2224. (2-10)

He, Y.-W.: **Orphan nuclear receptors in T lymphocyte development.** *J. Leukoc. Biol.* 2002, **72**:440–446. (7-10)

Heeney, J.L., Dalgleish, A.G. and Weiss, R.A.: **Origins of HIV and the evolution of resistance to AIDS.** *Science* 2006, **313**:462–466. (10-4)

Held, W., Waanders, G.A., Shakhov, A.N., Scarpellino, L., Acha-Orbea, H. and MacDonald, H.R.: **Superantigen-induced immune stimulation amplifies mouse mammary tumor virus infection and allows virus transmission.** *Cell* 1993, **74**:529–540. (9-4)

Helmy, K.Y., Katschke, K.J. Jr, Gorgani, N.N., Kljavin, N.M., Elliott, J.M., Diehl, L., Scales, S.J., Ghilardi, N. and van Lookeren Campagne, M.: **CRIg: a macrophage complement receptor required for phagocytosis of circulating pathogens.** *Cell* 2006, **124**:915–927. (3-5, 9-2)

Henkart, P.A. and Sitkovsky, M.V.: **Cytotoxic T lymphocytes** in *Fundamental Immunology* 5th ed. Paul, W.E. ed. (Lippincott Williams & Wilkins, Philadelphia, 2003), 1127–1150. (5-14)

Hennecke, J. and Wiley, D.C.: **T cell receptor–MHC interactions up close.** *Cell* 2001, **104**:1–4. (5-2)

Herbert, D.R., Hölscher, C., Mohrs, M., Arendse, B., Schwegmann, A., Radwanska, M., Leeto, M., Kirsch, R., Hall, P., Mossmann, H., Claussen, B., Förster, I. and Brombacher, F.: **Alternative macrophage activation is essential for survival during schistosomiasis and downmodulates T helper 1 responses and immunopathology.** *Immunity* 2004, **20**:623–635. (11-5)

Hertl, M. and Veldman, C.: **Pemphigus – paradigm of autoantibody-mediated autoimmunity.** *Skin Pharmacol. Appl. Skin Physiol.* 2001, **14**:408–418. (12-5)

Hill, A.V.S.: **Pre-erythrocytic malaria vaccines: towards greater efficacy.** *Nat. Rev. Immunol.* 2006, **6**:21–32. (14-4)

Hla, T.: **Dietary factors and immunological consequences.** *Science* 2005, **309**:1682–1683. (2-14)

Hoebe, K., Du, X., Georgel, P., Janssen, E., Tabeta, K., Kim, S.O., Goode, J., Lin, P., Mann, N., Mudd, S., Crozat, K., Sovath, S., Han, J. and Beutler, B.: **Identification of *Lps2* as a key transducer of MyD88-independent TIR signalling.** *Nature* 2003, **424**:743–748. (3-11)

Hoffmann, A., Leung, T.H. and Baltimore, D.: **Genetic analysis of NF-κB/Rel transcription factors defines functional specificities.** *EMBO J.* 2003, **22**:5530–5539. (2-10)

Hoffmann, A., Levchenko, A., Scott, M.L. and Baltimore, D.: **The IκB–NF-κB signaling module: temporal control and selective gene activation.** *Science* 2002, **298**:1241–1245. (2-10)

Hoffmann, J.A., Kafatos, F.C., Janeway, C.A. Jr and Ezekowitz, R.A.B.: **Phylogenetic perspectives in innate immunity.** *Science* 1999, **284**:1313–1318. (3-0)

Hoffmann, K.F., Cheever, A.W. and Wynn, T.A.: **IL-10 and the dangers of immune polarization: excessive type 1 and type 2 cytokine responses induce distinct forms of lethal immunopathology in murine schisto-**somiasis. *J. Immunol.* 2000, **164**:6406–6416. (11-5)

Hogquist, K.A.: **Signal strength in thymic selection and lineage commitment.** *Curr. Opin. Immunol.* 2001, **13**:225–231. (7-9)

Hogquist, K.A., Baldwin, T.A. and Jameson, S.C.: **Central tolerance: learning self-control in the thymus.** *Nat. Rev. Immunol.* 2005, **5**:772–782. (7-10)

Holgate, S.T. and Broide, D.: **New targets for allergic rhinitis—a disease of civilization.** *Nat. Rev. Drug Discov.* 2003, **2**:903–915. (13-2)

Holmgren, J. and Czerkinsky, C.: **Mucosal immunity and vaccines.** *Nat. Med.* 2005, **11**:S45–S53. (14-3)

Holt, B.F. 3rd, Hubert, D.A. and Dangl, J.L.: **Resistance gene signaling in plants—complex similarities to animal innate immunity.** *Curr. Opin. Immunol.* 2003, **15**:20–25. (3-12)

Honda, K., Ohba, Y., Yanai, H., Negishi, H., Mizutani, T., Tanaoka, A., Taya, C. and Taniguchi, T.: **Spatiotemporal regulation of MyD88–IRF-7 signaling for robust type-I interferon induction.** *Nature* 2005, **434**:1035–1040. (10-1)

Honda, K., Sakaguchi, S., Nakajima, C., Watanabe, A., Yanai, H., Matsumoto, M., Ohteki, T., Kaisho, T., Takaoka, A., Akira, S., Seya, T. and Taniguchi, T.: **Selective contributions of IFN-α/β signaling to the maturation of dendritic cells induced by double-stranded RNA or viral infection.** *Proc. Natl Acad. Sci. USA* 2003, **100**:10872–10877. (4-1)

Hooper, L.V. and Gordon, J.I.: **Commensal host-bacterial relationships in the gut.** *Science* 2001, **292**:1115–1118. (9-0)

Hooper, L.V., Midtvedt, T. and Gordon, J.I.: **How host-microbial interactions shape the nutrient environment of the mammalian intestine.** *Annu. Rev. Nutr.* 2002, **22**:283–307. (9-0)

Hori, S., Nomura, T. and Sakaguchi, S.: **Control of regulatory T cell development by the transcription factor Foxp3.** *Science* 2003, **299**:1057–1061. (7-10)

Horton, R., Wilming, L., Rand, V., Lovering, R.C., Bruford, E.A., Khodiyar, V.K., Lush, M.J., Povey, S., Talbot, C.C. Jr, Wright, M.W., Wain, H.M., Trowsdale, J., Ziegler, A. and Beck, S.: **Gene map of the extended MHC.** *Nat. Rev. Genet.* 2004, **5**:889–899. (4-3)

Hořejší, V., Zhang, W. and Schraven, B.: **Transmembrane adaptor proteins: organizers of immunoreceptor signaling.** *Nat. Rev. Immunol.* 2004, **4**:603–608. (5-3)

Hsieh, C.-S., Macatonia, S.E., Tripp, C.S., Wolf, S.F., O'Garra, A. and Murphy, K.M.: **Development of TH1 CD4+ T cells through IL-12 produced by *Listeria*-induced macrophages.** *Science* 1993, **260**:547–549. (9-6)

Hsu, E., Pulham, N., Rumfelt, L.L. and Flajnik, M.F.: **The plasticity of immunoglobulin gene systems in evolution.** *Immunol. Rev.* 2006, **210**:8–26. (7-4)

Huang, W., Na, L., Fidel, P.L. and Schwartzenberger, P.: **Requirement of interleukin-17A for systemic anti–*Candida albicans* host defense in mice.** *J. Infect. Dis.* 2004, **190**:624–631. (11-1)

Huang, Y. and Wange, R.L.: **T cell receptor signaling: beyond complex complexes.** *J. Biol. Chem.* 2004, **279**:28827–28830. (5-3)

Hudson, B.G., Tryggvason, K., Sundaramoorthy, M. and Neilson, E.G.: **Alport's syndrome, Goodpasture's syndrome, and type IV collagen.** *N. Engl. J. Med.* 2003, **348**:2543–2556. (12-5)

Hurst, S.M., Wilkinson, T.S., McLoughlin, R.M., Jones, S., Horiuchi, S., Yamamoto, N., Rose-John, S., Fuller, G.M., Topley, N. and Jones, S.A.: **IL-6 and its soluble receptor orchestrate a temporal switch in the pattern of leukocyte recruitment seen during acute inflammation.** *Immunity* 2001, **14**:705–714. (3-15)

Huseby, E.S., White, J., Crawford, F., Vass, T., Becker, D., Pinilla, C., Marrack, P. and Kappler, J.W.: **How the T cell repertoire becomes peptide and MHC specific.** *Cell* 2005, **122**:247–260. (5-0)

Hynes, R.O.: **Integrins: bidirectional allosteric signaling machines.** *Cell* 2002, **110**:7673–7687. (2-4)

Hynes, R.O. and Zhao, Q.: **The evolution of cell adhesion.** *J. Cell Biol.* 2000, **150**:F89–F96. (2-4)

Hysi, P., Kabesch, M., Moffatt, M.F., Schedel, M., Carr, D., Zhang, Y., Boardman, B., von Mutius, E., Weiland, S.K., Leupold, W., Fritzsch, C., Klopp, N., Musk, A.W., James, A., Nunez, G., Inohara, N. and Cookson, W.O.C.: **NOD1 variation, immunoglobulin E and asthma.** *Hum. Mol. Genet.* 2005, **14**:935–941. (13-3)

Inohara, N. and Nunez, G.: **NODs: intracellular proteins involved in inflammation and apoptosis.** *Nat. Rev. Immunol.* 2003, **3**:371–382. (3-12)

Inohara, N., Ogura, Y., Fontalba, A., Gutierrez, O., Pons, F., Crespo, J., Fukase, K., Inamura, S., Kusumoto, S., Hashimoto, M., Foster, S.J., Moran, A.P., Fernandez-Luna, J.L. and Nuñez, G.: **Host recognition of bacterial muramyl dipeptide mediated through NOD2. Implications for Crohn's disease.** *J. Biol. Chem.* 2003, **278**:5509–5512. (3-12)

Intlekofer, A.M., Takemoto, N., Wherry, E.J., Longworth, S.A., Northrup, J.T., Palanivel, V.R., Mullen, A.C., Gasink, C.R., Kaech, S.M., Miller, J.D., Gapin, L., Ryan, K., Russ, A.P., Lindsten, T., Orange, J.S., Goldrath, A.W., Ahmed, R. and Reiner, S.L.: **Effector and memory CD8+ T cell fate coupled by T-bet and eomesodermin.** *Nat. Immunol.* 2005, **6**:1236–1244. (5-10)

Itano, A.A., McSorley, S.J., Reinhardt, R.L., Ehst, B.D., Ingulli, E., Rudensky, A.Y. and Jenkins, M.K.: **Distinct dendritic cell populations sequentially present antigen to CD4 T cells and stimulate different aspects of cell-mediated immunity.** *Immunity* 2003, **19**:47–57. (5-6)

Ito, T., Couceiro, J.N.S.S., Kelm, S., Baum, L.G., Krauss, S., Castrucci, M.R., Donatelli, I., Kida, H., Paulson, J.C., Webster, R.G. and Kawaoka, Y.: **Molecular basis for the generation in pigs of influenza A viruses with pandemic potential.** *J. Virol.* 1998, **72**:7367–7373. (10-7)

Ivanov, I.I., McKenzie, B.S., Zhou, L., Tadokoro, C.E., Lepelley, A., Lafaille, J.J., Cua, D.J. and Littman, D.R.: **The orphan nuclear receptor RORγt directs the differentiation program of proinflammatory IL-17+ T helper cells.** *Cell* 2006, **126**:1121–1133. (5-13)

Iwasaki, A. and Medzhitov, R.: **Toll-like receptor control of the adaptive immune responses** *Nat. Immunol.* 2004, **5**:987–995. (4-1)

Jabri, B. and Sollid, L.M.: **Mechanisms of disease: immunopathogenesis of celiac disease.** *Nat. Clin. Pract. Gastroenterol. Hepatol.* 2006, **3**:516–525. (13-5)

Jacobs, M.D. and Harrison, S.D.: **Structure of an IκBα/NF-κB complex.** *Cell* 1998, **95**:749–758. (2-10)

Jacobson, D.L., Gange, S.J., Rose, N.R. and Graham, N.M.H.: **Epidemiology and estimated population burden of selected autoimmune diseases in the United States.** *Clin. Immunol. Immunopathol.* 1997, **84**:223–243. (12-4)

Jameson, J., Ugarte, K., Chen, N., Yachi, P., Fuchs, E., Boismenu, R. and Havran, W.L.: **A role for skin γδ T cells in wound repair.** *Science* 2002, **296**:747–749. (8-4)

Jameson, S.C.: **Maintaining the norm: T-cell homeostasis.** *Nat. Rev. Immunol.* 2002, **2**:547–556. (5-15)

Janeway, C.A. Jr: **The immune system evolved to discriminate infectious nonself from noninfectious self.** *Immunol. Today* 1992, **13**:11–16. (3-0)

Jansen, W.T.M., Breukels, M.A., Snippe, H., Sanders, L.A.M., Verheul, A.F.M. and Rijkers, G.T.: **Fcγ receptor polymorphisms determine the magnitude of in vitro phagocytosis of *Streptococcus pneumoniae* mediated by pneumococcal conjugate sera.** *J. Infect. Dis.* 1999, **180**:888–891. (9-5)

Jego, G., Palucka, A.K., Blanck, J.-P., Chalouni, C., Pascual, V. and Bancherau, J.: **Plasmacytoid dendritic cells induce plasma cell differentiation through type I interferon and interleukin 6.** *Immunity* 2003, **19**:225–234. (12-8)

Jenkins, M.K.: **Peripheral T-lymphocyte responses and function** in *Fundamental Immunology* 5th ed. Paul, W.E. ed. (Lippincott Williams & Wilkins, Philadelphia, 2003), 303–319. (5-7)

Jordan, M.B., Mills, D.M., Kappler, J., Marrack, P. and Cambier, J.C.: **Promotion of B cell immune responses via an alum-induced myeloid cell population.** *Science* 2004, **304**:1808–1810. (14-2)

Jordan, M.S., Singer, A.L. and Koretzky, G.A.: **Adaptors as central mediators of signal transduction in immune cells.** *Nat. Immunol.* 2003, **4**:110–116. (5-3)

Jost, S., Bernard, M.C., Kaiser, L., Yerly, S., Hirschel, B., Samri, A., Autran, B., Goh, L.E. and Perrin, L.: **A patient with HIV-1 superinfection.** *N. Engl. J. Med.* 2002, **347**:731–736. (10-6)

Julia, V., Rassoulzadegan, M. and Glaichenhaus, N.: **Resistance to *Leishmania major* induced by tolerance to a single antigen.** *Science* 1996, **274**:421–423. (11-3)

Jung, D., Giallourakis, C., Mostoslavsky, R. and Alt, F.W.: **Mechanism and control of V(D)J recombination at the immunoglobulin heavy chain locus.** *Annu. Rev. Immunol.* 2006, **24**:541–570. (7-3)

Jutel, M., Watanabe, T., Akdis, M., Blaser, K. and Akdis, C.A.: **Immune regulation by histamine.** *Curr. Opin. Immunol.* 2002, **14**:735–740. (13-1)

Kaech, S.M., Wherry, E.J. and Ahmed, R.: **Effector and memory T-cell differentiation: implications for vaccine development.** *Nat. Rev. Immunol.* 2002, **2**:251–262. (10-2)

Karst, S.M., Wobus, C.E., Lay, M., Davidson, J. and Virgin, H.W. 4th: **STAT1-dependent innate immunity to a Norwalk-like virus.** *Science* 2003, **299**:1575–1578. (10-1)

Kaushansky, K.: **Lineage-specific hematopoietic growth factors.** *N. Engl. J. Med.* 2006, **354**:2034–2045. (1-1)

Kawai, T. and Akira, S.: **Innate immune recognition of viral infection.** *Nat. Immunol.* 2006, **7**:131–137. (3-11, 10-1)

Kawasaki, T.: **Structure and biology of mannan-binding protein, MBP, an important component of innate immunity.** *Biochim. Biophys. Acta* 1999, **1473**:186–195. (3-2)

Keane, J., Gershon, S., Wise, R.P., Mirabile-Levens, E., Kasznica, J., Schwieterman, W.D., Siegel, J.N. and Braun, M.M.: **Tuberculosis associated with infliximab, a tumor necrosis factor α-neutralizing agent.** *N. Engl. J. Med.* 2001, **345**:1098–1104. (9-7)

Keefe, D., Shi, L., Feske, S., Massol, R., Navarro, F., Kirchhausen, T. and Lieberman, J.: **Perforin triggers a plasma membrane-repair response that facilitates CTL induction of apoptosis.** *Immunity* 2005, **23**:249–262. (5-14)

Keele, B.F., Van Heuverswyn, F., Li, Y., Bailes, E., Takehisa, J., Santiago, M.L., Bibollet-Ruche, F., Chen, Y., Wain, L.V., Liegeois, F., Loul, S., Ngole, E.M., Bienvenue, Y., Delaporte, E., Brookfield, J.F., Sharp, P.M., Shaw, G.M., Peeters, M. and Hahn, B.H.: **Chimpanzee reservoirs of pandemic and nonpandemic HIV-1.** *Science* 2006, **313**:523–526. (10-4)

Kelly, M.E. and Chan, A.C.: **Regulation of B cell function by linker proteins.** *Curr. Opin. Immunol.* 2000, **12**:267–275. (6-8)

Kerksiek, K.M., Busch, D.H., Pilip, I.M., Allen, S.E. and Pamer, E.G.: **H2-M3–restricted T cells in bacterial infection: rapid primary but diminished memory responses.** *J. Exp. Med.* 1999, **190**:195–204. (8-5)

Khor, B. and Sleckman, B.P.: **Allelic exclusion at the TCRβ locus.** *Curr. Opin. Immunol.* 2002, **14**:230–234. (7-3)

Kim, S., Poursine-Laurent, J., Truscott, S.M., Lybarger, L., Song, Y.-J., Yang, L., French, A.R., Sunwoo, J.B., Lemieux, S., Hansen, T.H. and Yokoyama, W.M.: **Licensing of natural killer cells by host major histocompatibility complex class I molecules.** *Nature* 2005, **436**:709–713. (8-1)

Kinjo, Y., Wu, D., Kim, G., Xing, G.-W., Poles, M.A., Ho, D.D., Tsuji, M., Kawahara, K., Wong, C.-H. and Kronenberg, M.: **Recognition of bacterial glycosphingolipids by natural killer T cells.** *Nature* 2005, **434**:520–525. (8-3)

Kirkegaard, K., Taylor, M.P. and Jackson, W.T.: **Cellular autophagy: surrender, avoidance and subversion of microorganisms.** *Nat. Rev. Microbiol.* 2004, **2**:301–314. (3-8)

Kirken, R.A. and Wang, Y.L.: **Molecular actions of sirolimus: sirolimus and mTor.** *Transplant. Proc.* 2003, **35**:S227–S230. (14-8)

Klausner, R.D., Fauci, A.S., Corey, L., Nabel, G.J., Gayle, H., Berkley, S., Haynes, B.F., Baltimore, D., Collins, C., Douglas, R.G., Esparza, J., Francis, D.P., Ganguly, N.K., Gerberding, J.L., Johnston, M.I., Kazatchkine, M.D., McMichael, A.J., Makgoba, M.W., Pantaleo, G., Piot, P., Shao, Y., Tramont, E., Varmus, H. and Wasserheit, J.N.: **The need for a global HIV vaccine enterprise.** *Science* 2003, **300**:2036–2039. (10-6)

Klenerman, P. and Hill, A.: **T cells and viral persistence: lessons from diverse infections.** *Nat. Immunol.* 2005, **6**:873–879. (5-16)

Kobayashi, K.S., Chamaillard, M., Ogura, Y., Henegariu, O., Inohara, N., Nuñez, G. and Flavell, R.A.: **Nod2-dependent regulation of innate and adaptive immunity in the intestinal tract.** *Science* 2005, **307**:731–734. (3-12)

Kogelberg, H. and Feizi, T.: **New structural insights into lectin-type proteins of the immune system.** *Curr. Opin. Struct. Biol.* 2001, **11**:635–643. (2-5)

Köhler, G. and Milstein, C.: **Continuous cultures of fused cells secreting antibody of predefined specificity.** *Nature* 1975, **256**:495–497. (6-4)

Kohm, A.P., Fuller, K.G. and Miller, S.D.: **Mimicking the way to autoimmunity: an evolving theory of sequence and structural homology.** *Trends Microbiol.* 2003, **11**:101–105. (12-4)

Kondo, M., Wagers, A.J., Manz, M.G., Prohaska, S.S., Scherer, D.C., Beilhack, G.F., Shizuru, J.A. and Weissman, I.L.: **Biology of hematopoietic stem cells and progenitors: Implications for clinical application.** *Annu. Rev. Immunol.* 2003, **21**:759–806. (1-1)

Kontoyiannis, D., Pasparakis M., Pizarro, T.T., Cominelli, F. and Kollias, G.: **Impaired on/off regulation of TNF biosynthesis in mice lacking TNF AU-rich elements: implications for joint and gut-associated immunopathologies.** *Immunity* 1999, **10**:387–398. (3-15, 12-9)

Korfhagen, T.R., LeVine, A.M. and Whitsett, J.A.: **Surfactant protein A (SP-A) gene targeted mice.** *Biochim. Biophys. Acta* 1998, **1408**:296–302. (3-2)

Kotwal, G.J.: **Poxviral mimicry of complement and chemokine system components: what's the end game?** *Immunol. Today* 2000, **21**: 242–248. (3-6)

Krajcsi, P. and Wold, W.S.M.: **Inhibition of tumor necrosis factor and interferon triggered responses by DNA viruses.** *Semin. Cell Dev. Biol.* 1998, **9**:351–358. (3-18)

Kroczek, R.A., Mages, H.W. and Hutloff, A.: **Emerging paradigms of T-cell costimulation.** *Curr. Opin. Immunol.* 2004, **16**:321–327. (5-8)

Kronenberg, M.: **Toward an understanding of NKT cell biology: progress and paradoxes.** *Annu. Rev. Immunol.* 2005, **23**:877–900. (8-3)

Ku, C.-L., Yang, K., Bustamante, J., Puel, A., von Bernuth, H., Santos, O.F., Lawrence, T., Chang, H.-H., Al-Mousa, H., Picard, C. and Casanova, J.-L.: **Inherited disorders of human Toll-like receptor signaling: immunological implications.** *Immunol. Rev.* 2005, **203**:10–20. (14-1)

Kühnel, F., Zender, L., Paul, Y., Tietze, M.K., Trautwein, C., Manns, M. and Kubicka, S.: **NFκB mediates apoptosis through transcriptional activation of Fas (CD95) in adenoviral hepatitis.** *J. Biol. Chem.* 2000, **275**: 6421–6427. (3-18)

Kuhns, D.B., Alvord, W.G. and Gallin, J.I.: **Increased circulating cytokines, cytokine antagonists, and E-selectin after intravenous administration of endotoxin in humans.** *J. Infect. Dis.* 1995, **171**:145–152. (9-3)

Kurosaki, T.: **Regulation of B cell fates by BCR signaling components.** *Curr. Opin. Immunol.* 2002, **14**:341–347. (6-8, 7-5)

Kyewski, B. and Klein, L.: **A central role for central tolerance.** *Annu. Rev. Immunol.* 2006, **24**:571–606. (12-1)

Kyewski, B., Derbinski, J., Gotter, J. and Klein, L.: **Promiscuous gene expression and central T-cell tolerance: more than meets the eye.** *Trends Immunol.* 2002, **23**:364–371. (7-10)

Lambrecht, B.N.: **Alveolar macrophage in the driver's seat.** *Immunity* 2006, **24**:366–368. (1-3)

Lang, B. and Vincent, A.: **Autoantibodies to ion channels at the neuromuscular junction.** *Autoimmun. Rev.* 2003, **2**:94–100. (12-5)

Lanier, L.L.: **NK cell recognition.** *Annu. Rev. Immunol.* 2005, **23**:225–274. (2-5, 8-2)

Lanning, D., Zhu, X., Zhai, S.-K. and Knight, K.L.: **Development of the antibody repertoire in rabbit: gut-associated lymphoid tissue, microbes, and selection.** *Immunol. Rev.* 2000, **175**:214–228. (7-4)

Lauber, K., Blumenthal, S.G., Waibel, M. and Wesselborg, S.: **Clearance of apoptotic cells: getting rid of the corpses.** *Mol. Cell* 2004, **14**:277–287. (3-8)

Lecuit, M., Vandormael-Pournin, S., Lefort, J., Huerre, M., Gounon, P., Dupuy, C., Babinet, C. and Cossart, P.: **A transgenic model for listeriosis: role of internalin in crossing the intestinal barrier.** *Science* 2001, **292**:1722–1725. (9-6)

Lee, G.R., Kim, S.T., Spilianakis, C.G., Fields, P.E. and Flavell, R.A.: **T helper cell differentiation: regulation by** *cis* **elements and epigenetics.** *Immunity* 2006, **24**:369–379. (5-9)

Lehrer, R.I.: **Primate defensins.** *Nat. Rev. Microbiol.* 2004, **2**:727–738. (3-1)

Lennon-Duménil, A.-M., Bakker, A.H., Wolf-Bryant, P., Ploegh, H.L. and Lagaudriere-Gesbert, C.: **A closer look at proteolysis and MHC-class-II-restricted antigen presentation.** *Curr. Opin. Immunol.* 2002, **14**:15–21. (4-7)

Leonard, W.J.: **Cytokines and immunodeficiency diseases.** *Nat. Rev. Immunol.* 2001, **1**:200–208. (2-8, 7-11, 14-1)

Leonard, W.J.: **Type 1 cytokines and interferons and their receptors** in *Fundamental Immunology* 5th ed. Paul, W.E. ed. (Lippincott Williams & Wilkins, Philadelphia, 2003), 701–747. (2-7)

Leonard, W.J.: **Type I cytokines and their receptors** in *Fundamental Immunology* 4th ed. Paul, W.E. ed. (Lippincott-Raven, New York, 1999), 741–774. (1-1)

Leonard, W.J. and Spolski, R.: **Interleukin-21: a modulator of lymphoid proliferation, apoptosis and differentiation.** *Nat. Rev. Immunol.* 2005 **5**:688–698. (6-10)

Levine, B.: **Eating oneself and uninvited guests: autophagy-related pathways in cellular defense.** *Cell* 2005, **120**:159–162. (3-8)

Levy, D.E. and García-Sastre, A.: **The virus battles: IFN induction of the antiviral state and mechanisms of viral evasion.** *Cytokine Growth Factor Rev.* 2001, **12**:143–156. (3-17, 10-3)

Ley, K. and Kansas, G.S.: **Selectins in T-cell recruitment to non-lymphoid tissues and sites of inflammation.** *Nat. Rev. Immunol.* 2004, **4**:325–335. (2-14)

Li, H., Llera, A., Tsuchiya, D., Leder, L., Ysern, X., Schlievert, P.M., Karjalainen, K. and Mariuzza, R.A.: **Three-dimensional structure of the complex between a T cell receptor β chain and the superantigen staphylococcal enterotoxin B.** *Immunity* 1998, **9**:807–816. (9-4)

Li, K.S., Guan, Y., Wang, J., Smith, G.J.D., Xu, K.M., Duan, L., Rahardjo, A.P., Puthavathana, P., Buranathai, C., Nguyen, T.D., Estoepangestie, A.T.S., Chaisingh, A., Auewarakul, P., Long, H.T., Hanh, N.T.H., Webby, R.J., Poon, L.L.M., Chen, H., Shortridge, K.F., Yuen, K.Y., Webster, R.G. and Peiris, J.S.M.: **Genesis of a highly pathogenic and potentially pandemic H5N1 influenza virus in eastern Asia.** *Nature* 2004, **430**:209–213. (10-7)

Li, M.O., Wan, Y.Y., Sanjabi, S., Robertson, A.-K.L. and Flavell, R.A.: **Transforming growth factor-β regulation of immune responses.** *Annu. Rev. Immunol.* 2006, **24**:99–146. (12-3)

Li, W., Shi, Z., Yu, M., Ren, W., Smith, C., Epstein, J.H., Wang, H., Crameri, G., Hu, Z., Zhang, H., Zhang, J., McEachern, J., Field, H., Daszak, P., Eaton. B.T., Zhang, S. and Wang. L.-F.: **Bats are natural reservoirs of SARS-like coronaviruses.** *Science* 2005, **310**:676–679. (10-0)

Li, W.-X., Li, H., Lu, R., Li, F., Dus, M., Atkinson, P., Brydon, E.W., Johnson, K.L., García-Sastre, A., Ball, L.A., Palese, P. and Ding, S.-W.: **Interferon antagonist proteins of influenza and vaccinia virus are suppressors of RNA silencing.** *Proc. Natl Acad. Sci. USA* 2004, **101**:1350–1355. (10-8)

Li, Y., Li, H., Dimasi, N., McCormick, J.K., Martin, R., Schuck, P., Schlievert, P.M. and Mariuzza, R.A.: **Crystal structure of a superantigen bound to the high-affinity, zinc-dependent site on MHC class II.** *Immunity* 2001, **14**:93–104. (9-4)

Liaw, P.C.Y., Mather, T., Oganesyan, N., Ferrell, G.L. and Esmon, C.T.: **Identification of the protein C/activated protein C binding sites on the endothelial cell protein C receptor.** *J. Biol. Chem.* 2001, **276**:8364–8370. (4-4)

Lin, J., Miller, M.J. and Shaw, A.S.: **The c-SMAC: sorting it all out (or in).** *J. Cell Biol.* 2005, **170**:177–182. (5-4)

Linehan, S.A., Martínez-Pomares, L. and Gordon, S.: **Macrophage lectins in host disease.** *Microbes Infect.* 2000, **2**:279–288. (3-7)

Liston, P., Fong W. G. and Korneluk, R.G.: **The inhibitors of apoptosis: there is more to life than Bcl2.** *Oncogene* 2003, **22**: 8568–8580. (2-12)

Litman, G.W., Anderson, M.K. and Rast, J.P.: **Evolution of antigen binding receptors.** *Annu. Rev. Immunol.* 1999, **17**:109–147. (7-4)

Liu, C.C., Navratil, J.S., Sabatine, J.M. and Ahearn, J.M.: **Apoptosis, complement and systemic lupus erythematosus: a mechanistic view.** *Curr. Dir. Autoimmun.* 2004, **7**:49–86. (12-8)

Liu, R., Paxton, W.A., Choe, S., Ceradini, D., Martin, S.R., Horuk, R., MacDonald, M.E., Stuhlmann, H., Koup, R.A. and Landau, N.R.: **Homozygous defect in HIV-1 coreceptor accounts for resistance of some multiply-exposed individuals to HIV-1 infection.** *Cell* 1996, **86**:367–377. (10-5)

Liu, Y.-J.: **IPC: professional type 1 interferon-producing cells and plasmacytoid dendritic cell precursors.** *Annu. Rev. Immunol.* 2005, **23**:275–306. (3-16)

Locksley, R.M., Killeen, N. and Lenardo, M.J.: **The TNF and TNF receptor superfamilies: integrating mammalian biology.** *Cell* 2001, **104**:487–501. (2-9)

Lodoen, M., Ogasawara, K., Hamerman, J.A., Arase, H., Houchins, J.P., Mocarski, E.S. and Lanier, L.L.: **NKG2D-mediated natural killer cell protection against cytomegalovirus is impaired by viral gp40 modulation of retinoic acid early inducible 1 gene molecules.** *J. Exp. Med.* 2003, **197**:1245–1253. (10-3)

Lohrum, M.A.E. and Vousden, K.H.: **Regulation and function of the p53-related proteins: same family, different rules.** *Trends Cell Biol.* 2000, **10**:197–202. (3-18)

Lopes-Carvalho, T., Foote, J. and Kearney, J.F.: **Marginal zone B cells in lymphocyte activation and regulation.** *Curr. Opin. Immunol.* 2005, **17**:244–250. (8-7)

Lord, G.M., Lechler, R.I. and George, A.J.T.: **A kinetic differentiation model for the action of altered TCR ligands.** *Immunol. Today* 1999, **20**:33–39. (5-2)

Lorenz, M.C. and Fink, G.R.: **The glyoxylate cycle is required for fungal virulence.** *Nature* 2001, **412**:83–86. (11-1)

Loukas, A. and Prociv, P.: **Immune responses in hookworm infections.** *Clin. Microbiol. Rev.* 2001, **14**:689–703. (11-0)

Lu, T.T. and Cyster, J.G.: **Integrin-mediated long-term B cell retention in the splenic marginal zone.** *Science* 2002, **297**:409–412. (8-7)

Luster, A.D.: **Chemokines—chemotactic cytokines that mediate inflammation.** *N. Engl. J. Med.* 1998, **338**:436–445. (2-13)

MacDonald, H.R., Radtke, F. and Wilson, A.: **T cell fate specification and αβ/γδ lineage commitment.** *Curr. Opin. Immunol.* 2001, **13**:219–224. (7-7)

MacDonald, T.T. and Monteleone, G.: **Immunity, inflammation, and allergy in the gut.** *Science* 2005, **307**:1920–1925. (13-5)

Macian, F.: **NFAT proteins: key regulators of T-cell development and function.** *Nat. Rev. Immunol.* 2005, **5**:472–484. (5-5)

MacLennan, I.C.M., Toellner, K.-M., Cunningham, A.F., Serre, K., Sze, D.M.-Y., Zúñiga, E., Cook, M.C. and Vinuesa, C.G.: **Extrafollicular antibody responses.** *Immunol. Rev.* 2003, **194**:8–18. (6-10)

Macpherson, A.J., Gatto, D., Sainsbury, E., Harriman, G.R., Hengartner, H. and Zinkernagel, R.M.: **A primitive T cell-independent mechanism of intestinal mucosal IgA responses to commensal bacteria.** *Science* 2000, **288**:2222–2226. (9-1)

Madakamutil, L.T., Christen, U., Lena, C.J., Wang-Zhu, Y., Attinger, A., Sundarrajan, M., Ellmeier, W., von Herrath, M.G., Jensen, P., Littman, D.R. and Cheroutre, H.: **CD8αα-mediated survival and differentiation of CD8 memory T cell precursors.** *Science.* 2004, **304**:590–593. (4-4)

Madden, J.C., Ruiz, N. and Caparon, M.: **Cytolysin-mediated translocation (CMT): a functional equivalent of type III secretion in gram-positive bacteria.** *Cell* 2001, **104**:143–152. (9-2)

Mahajan-Miklos, S., Tan, M.-W., Rahme, L.G. and Ausubel, F.M.: **Molecular mechanisms of bacterial virulence elucidated using a** *Pseudomonas aeruginosa-Caenorhabditis elegans* **pathogenesis model.** *Cell* 1999, **96**:47–56. (9-2)

Maizels, N.: **Immunoglobulin gene diversification.** *Annu. Rev. Genet.* 2005, **39**:23–46. (6-11)

Maizels, R.M. and Yazdanbakhsh, M.: **Immune regulation by helminth parasites: cellular and molecular mechanisms.** *Nat. Rev. Immunol.* 2003, **3**:733–744. (11-0)

Mangeat, B., Turelli, P., Caron, G., Friedli, M., Perrin, L. and Trono, D.: **Broad antiretroviral defence by human APOBEC3G through lethal editing of nascent reverse transcripts.** *Nature* 2003, **424**:99–103. (3-19)

Mannon, P.J., Fuss, I.J., Mayer, L., Elson, C.O., Sandborn, W.J., Present, D., Dolin, B., Goodman, N., Groden, C., Hornung, R.L., Quezado, M., Neurath, M.F., Salfeld, J., Veldman, G.M., Schwertschlag, U. and Strober, W.: **Anti-interleukin-12 antibody for active Crohn's disease.** *N. Engl. J. Med.* 2004, **351**:2069–2079. (13-6)

Manz, R.A., Hauser, A.E., Hiepe, F. and Radbruch, A.: **Maintenance of serum antibody levels.** *Annu. Rev. Immunol.* 2005, **23**:367–386. (6-12)

Marculescu, R., Le, T., Simon, P., Jaeger, U. and Nadel, B.: **V(D)J-mediated translocations in lymphoid neoplasms: a functional assessment of genomic instability by cryptic sites.** *J. Exp. Med.* 2002, **195**:85–98. (7-11)

Margulies, D.H.: **The major histocompatibility complex** in *Fundamental Immunology* 4th ed. Paul, W.E. ed. (Lippincott-Raven, New York, 1999), 263–285. (1-5)

Margulies, D.H. and McCluskey, J.: **The major histocompatibility complex and its encoded proteins** in *Fundamental Immunology* 5th ed. Paul, W.E. ed. (Lippincott Williams & Wilkins, Philadelphia, 2003), 571–612. (4-3)

Maric, M., Arunachalam, B., Phan, U.T., Dong, C., Garrett, W.S., Cannon, K.S., Alfonso, C., Karlsson, L., Flavell, R.A. and Cresswell, P.: **Defective antigen processing in GILT-free mice.** *Science* 2001, **294**:1361–1365. (4-7)

Marrack, P. and Kappler, J.: **Control of T cell viability.** *Annu. Rev. Immunol.* 2004, **22**:765–787. (5-15)

Marshall, J.S.: **Mast-cell responses to pathogens.** *Nat. Rev. Immunol.* 2004, **4**:787–799. (3-13)

Martin, F. and Kearney, J.F.: **B1 cells: similarities and differences with other B cell subsets.** *Curr. Opin. Immunol.* 2001, **13**:195–201. (7-6, 8-6)

Martin, F. and Kearney, J.F.: **B-cell subsets and the mature preimmune repertoire. Marginal zone and B1 B cells as part of a "natural immune memory".** *Immunol. Rev.* 2000, **175**:70–79. (8-6)

Martin, F. and Kearney, J.F.: **Marginal-zone B cells.** *Nat. Rev. Immunol.* 2002, **2**:323–335. (8-7)

Martin, F., Oliver, A.M. and Kearney, J.F.: **Marginal zone and B1 B cells unite in the early response against T-independent blood-borne particulate antigens.** *Immunity* 2001, **14**:617–629. (8-7, 9-5)

Martinon, F. and Tschopp, J.: **Inflammatory caspases: linking an intracellular innate immune system to autoinflammatory diseases.** *Cell* 2004, **117**:561–574. (2-11)

Martinon, F. and Tschopp, J.: **NLRs join TLRs as innate sensors of pathogens.** *Trends Immunol.* 2005, **26**:447–454. (3-12)

Martinvalet, D., Zhu, P. and Lieberman, J.: **Granzyme A induces caspase-Independent mitochondrial damage, a required first step for apoptosis.** *Immunity* 2005, **22**:355–370. (5-14)

Maruyama, M., Lam, K.-P. and Rajewsky. K.: **Memory B-cell persistence is independent of persisting immunizing antigen.** *Nature* 2000, **407**:636–642. (6-12)

Matsushita, M., Endo, Y., Hamasaki, N. and Fujita, T.: **Activation of the lectin complement pathway by ficolins.** *Int. Immunopharmacol.* 2001, **1**:359–363. (3-4)

Matsuuchi, L. and Gold, M.R.: **New views of BCR structure and organization**. *Curr. Opin. Immunol.* 2001, **13**:270–277. (6-7)

Matthias, P. and Rolink, A.G.: **Transcriptional networks in developing and mature B cells.** *Nat. Rev. Immunol.* 2005, **5**:497–508. (7-5)

Mattner, J., DeBord, K.L., Ismail, N., Goff, R.D., Cantu. C. 3rd, Zhou, D., Saint-Mezard, P., Wang, V., Gao, Y., Yin, N., Hoebe, K., Schneewind, O., Walker, D., Beutler, B., Teyton, L., Savage, P.B. and Bendelac, A.: **Exogenous and endogenous glycolipid antigens activate NKT cells during microbial infections.** *Nature* 2005, **434**:525–529. (8-3)

Mazumdar, P.M.H.: **History of Immunology** in *Fundamental Immunology* 5th ed. Paul, W.E. ed. (Lippincott Williams & Wilkins, Philadelphia, 2003), 23–46. (4-2)

McConville, M.J., Turco, S.J., Ferguson, M.A.J. and Sacks, D.L.: **Developmental modification of lipophosphoglycan during the differentiation of *Leishmania major* promastigotes to an infectious stage.** *EMBO J.* 1992, **11**:3593–3600. (11-3)

McCune, J.M.: **The dynamics of CD4+ T-cell depletion in HIV disease.** *Nature* 2001, **410**:974–979. (10-6)

McKenzie, B.S., Kastelein, R.A. and Cua, D.J.: **Understanding the IL-23–IL-17 immune pathway.** *Trends Immunol.* 2006, **27**:17–23. (5-13)

McMichael, A.J.: **HIV vaccines.** *Annu. Rev. Immunol.* 2006, **24**:227–255. (14-4)

McMichael, A.J. and Rowland-Jones, S.L.: **Cellular immune responses to HIV.** *Nature* 2001, **410**:980–987. (10-6)

Mebius, R.E. and Kraal, G.: **Structure and function of the spleen.** *Nature Rev. Immunol.* 2005, **5**:606–616. (1-7)

Meffre, E., Casellas, R. and Nussenzweig, M.C.: **Antibody regulation of B cell development.** *Nat. Immunol.* 2000, **1**:379–385. (7-5)

Melby, P.C.: **Recent developments in leishmaniasis.** *Curr. Opin. Infect. Dis.* 2002, **15**:485–490. (11-3)

Melchers, F.: **The pre-B-cell receptor: selector of fitting immunoglobulin heavy chains for the B-cell repertoire.** *Nat. Rev. Immunol.* 2005, **5**:578–584. (7-3)

Mellado, M., Rodriguez-Frade, J.M., Manes, S. and Martinez-A, C.: **Chemokine signaling and functional responses: the role of receptor dimerization and TK pathway activation.** *Annu. Rev. Immunol.* 2001, **19**:397–421. (2-13)

Mellman, I. and Steinman, R.M.: **Dendritic cells: specialized and regulated antigen processing machines.** *Cell* 2001, **106**:255–258. (4-1, 9-1)

Mempel, T.R., Henrickson, S.E. and von Andrian, U.H.: **T-cell priming by dendritic cells in lymph nodes occurs in three distinct phases.** *Nature* 2004, **427**:154–159. (5-6)

Mempel, T.R., Scimone, M.L., Mora, J.R. and von Andrian, U.H.: ***In vivo* imaging of leukocyte trafficking in blood vessels and tissues.** *Curr. Opin. Immunol.* 2004, **16**:406–417. (3-14)

Mendez, M.J., Green, L.L., Corvalan, J.R.F., Jia, X.-C., Maynard-Currie, C.E., Yang, X.-d., Gallo, M.L., Louie, D.M., Lee, D.V., Erickson, K.L., Luna, J., Roy, C.M.-N., Abderrahim, H., Kirschenbaum, F., Noguchi, M., Smith, D.H., Fukushima, A., Hales, J.F., Finer, M.H., Davis, C.G., Zsebo, K.M. and Jakobovits, A.: **Functional transplant of megabase human immunoglobulin loci recapitulates human antibody response in mice.** *Nat. Genet.* 1997, **15**:146–156. (6-4)

Meng, G., Wei, X., Wu, X., Sellers, M.T., Decker, J.M., Moldoveanu, Z., Orenstein, J.M., Graham, M.F., Kappes, J.C., Mestecky, J., Shaw, G.M. and Smith. P.D.: **Primary intestinal epithelial cells selectively transfer R5 HIV-1 to CCR5+ cells.** *Nat. Med.* 2002, **8**:150–156. (10-5)

Meylan, E., Curran, J., Hoffmann, K., Moradpour, D., Binder, M., Bartenschlager, R. and Tschapp, J.: **Cardif is an adaptor protein in the RIG-I antiviral pathway and is targeted by hepatitis C virus.** *Nature* 2005, **437**:1167–1172. (10-3)

Mi, Q.-S., Zhou, L., Schulze, D.H., Fischer, R.T., Lustig, A., Rezanka, L.J., Donovan, D.M., Longo, D.L. and Kenny, J.J.: **Highly reduced protection against *Streptococcus pneumoniae* after deletion of a single heavy chain gene in mouse.** *Proc. Natl Acad. Sci. USA* 2000, **97**:6031–6036. (9-5)

Miller, L.H., Baruch, D.I., Marsh, K. and Doumbo, O.K.: **The pathogenic basis of malaria.** *Nature* 2002, **415**:673–679. (11-0)

Mira, J.-P., Cariou, A., Grall, F., Delclaux, C., Losser, M.-R., Heshmati, F., Cheval, C., Monchi, M., Teboul, J.-L., Riché, F., Leleu, G., Arbibe, L., Mignon, A., Delpech, M. and Dhainaut, J.-F.: **Association of TNF2, a TNF-α promoter polymorphism, with septic shock susceptibility and mortality: a multicenter study.** *JAMA* 1999, **282**:561–568. (9-3)

Mitin, N., Rossman, K.L. and Der, C.J.: **Signaling interplay in Ras superfamily function.** *Curr. Biol.* 2005, **15**:R563–R574. (2-0)

Miwa, T. and Song, W.-C.: **Membrane complement regulatory proteins: insight from animal studies and relevance to human diseases.** *Int. Immunopharmacol.* 2001, **1**:445–459. (3-6)

Mold, C.: **Role of complement in host defense against bacterial infection.** *Microbes Infect.* 1999, **1**:633–638. (3-3)

Mombaerts, P., Arnoldi, J., Russ, F., Tonegawa, S. and Kaufmann, S.H.E: **Different roles of αβ and γδ T cells in immunity against an intracellular bacterial pathogen.** *Nature* 1993, **365**:53–56. (9-6)

Monach, P.A., Benoist, C. and Mathis, D.: **The role of antibodies in mouse models of rheumatoid arthritis, and relevance to human disease.** *Adv. Immunol.* 2004, **82**:217–248. (12-9)

Monroe, J.G.: **ITAM-mediated tonic signaling through pre-BCR and BCR complexes.** *Nat. Rev. Immunol.* 2006, **6**:283–294. (7-5)

Moore, A.C. and Hill, A.V.S.: **Progress in DNA-based heterologous prime-boost immunization strategies for malaria.** *Immunol. Rev.* 2004, **199**:126–143. (14-4)

Mullen, A.C., Hutchins, A.S., High, F.A., Lee, H.W., Sykes, K.J., Chodosh, L.A. and Reiner, S.L.: **Hlx is induced by and genetically interacts with T-bet to promote heritable T$_H$1 gene induction.** *Nat. Immunol.* 2002, **3**:652–658. (5-10)

Murali-Krishna, K., Altman, J.D., Suresh, M., Sourdive, D.J.D., Zajac, A.J., Miller, J.D., Slansky, J. and Ahmed, R.: **Counting antigen-specific CD8 T cells: a reevaluation of bystander activation during viral infection.** *Immunity* 1998, **8**:177–187. (10-9)

Murphy, K.M. and Reiner, S.L.: **The lineage decisions of helper T cells.** *Nat. Rev. Immunol.* 2002, **2**:933–944. (5-9)

Musher, D.M.: **Infections caused by *Streptococcus pneumoniae*: clinical spectrum, pathogenesis, immunity and treatment.** *Clin. Infect. Dis.* 1992, **14**:801–807. (9-5)

Nair, M.G., Guild, K.J. and Artis, D.: **Novel effector molecules in type 2 inflammation: lessons drawn from helminth infection and allergy.** *J. Immunol.* 2006, **177**:1393–1399. (5-12)

Nakagawa, T., Zhu, H., Morishima, N., Li, E., Xu, J., Yankner, B.A. and Yuan, J.: **Caspase-12 mediates endoplasmic-reticulum-specific apoptosis and cytotoxicity by amyloid-β.** *Nature* 2000, **403**:98–103. (3-18)

Nathan, C.: **Points of control in inflammation.** *Nature* 2002, **420**:846–852. (3-13)

Navarro, F. and Landau, N.R.: **Recent insights into HIV-1 Vif.** *Curr. Opin. Immunol.* 2004, **16**:477–482. (10-5)

Neuberger, M.S., Harris, R.S., Di Noia, J. and Petersen-Mahrt, S.K.: **Immunity through DNA deamination.** *Trends Biochem. Sci.* 2003, **28**:305–312. (6-11)

Neutra, M.R. and Kozlowski, P.A.: **Mucosal vaccines: the promise and the challenge.** *Nat. Rev. Immunol.* 2006, **6**:148–158. (14-3)

Neuzil, K.M., Mellen, B.G., Wright, P.F., Mitchel, E.F. and Griffin, M.R.: **The effect of influenza on hospitalizations, outpatient visits, and courses of antibiotics in children.** *N. Engl. J. Med.* 2000, **342**:225–231. (10-7)

Nicholson, K.G., Wood, J.M. and Zambon, M.: **Influenza.** *Lancet* 2003, **362**:1733–1745. (10-7)

Nikolich-Žugich, J., Slifka, M.K. and Messaoudi, I.: **The many important facets of T-cell repertoire diversity.** *Nat. Rev. Immunol.* 2004, **4**:123–132. (5-2)

Nimmerjahn, F. and Ravetch, J.V.: **Fcγ receptors: old friends and new family members.** *Immunity* 2006, **24**:19–28. (6-3)

Noben-Trauth, N., Lira, R., Nagase, H., Paul, W.E. and Sacks, D.L.: **The relative contribution of IL-4 receptor signaling and IL-10 to susceptibility to** *Leishmania major.* J. Immunol. 2003, **170**:5152–5158. (11-3)

Noelle, R.J. and Erickson, L.D.: **Determinations of B cell fate in immunity and autoimmunity.** Curr. Dir. Autoimmun. 2005, **8**:1–24. (12-2)

Nonaka, M.: **Evolution of the complement system.** Curr. Opin. Immunol. 2001, **13**:69–73. (3-4)

Nordstrand, A., Norgren, M. and Holm, S.E.: **Pathogenic mechanism of acute post-streptococcal glomerulonephritis.** Scand. J. Infect. Dis. 1999 **31**:523–537. (13-4)

Nossal, G.J.V.: **Vaccines** in *Fundamental Immunology* 5th ed. Paul, W.E. ed. (Lippincott Williams & Wilkins, Philadelphia, 2003), 1319–1369. (14-3)

Nyholm, S.V. and McFall-Ngai, M.J.: **The winnowing: establishing the squid–Vibrio symbiosis.** Nat. Rev. Microbiol. 2004, **2**:632–642. (9-0)

Ochs, H.D., Smith, C.I.E. and Puck, J.M. (eds): *Primary Immunodeficiency Diseases: A Molecular and Genetic Approach* 2nd ed. (Oxford University Press, Oxford, 2007). (14-1)

Ochsenbein, A.F. and Zinkernagel, R.M.: **Natural antibodies and complement link innate and acquired immunity.** Immunol. Today 2000, **21**:624–630. (6-9)

O'Hagan, D.T. and Rappuoli, R.: **Novel approaches to pediatric vaccine delivery.** Adv. Drug Deliv. Rev. 2006, **58**:29–51. (14-2)

Okada, T. and Cyster, J.G.: **B cell migration and interactions in the early phase of antibody responses.** Curr. Opin. Immunol. 2006, **18**:278–285. (6-10)

Oldenborg, P.-A., Zhelezznyak, A., Fang, Y.-F., Lagenaur, C.F., Gresham, H.D. and Lindberg, F.P.: **Role of CD47 as a marker of self on red blood cells.** Science 2000, **288**:2051–2054. (3-8)

Ouellette, A.J. and Selsted, M.E.: **Paneth cell defensins: endogenous peptide components of intestinal host defense.** FASEB J. 1996, **10**:1280–1289. (13-1)

Ozaki, K. and Leonard, W.J.: **Cytokine and cytokine receptor pleiotropy and redundancy.** J. Biol. Chem. 2002, **277**:29355–29358. (2-8)

Palmer, E.: **Negative selection — clearing out the bad apples from the T-cell repertoire.** Nat. Rev. Immunol. 2003, **3**:383–391. (7-10)

Pamer, E.G.: **Immune responses to** *Listeria monocytogenes.* Nat. Rev. Immunol. 2004, **4**:812–823. (9-6)

Pancer, Z., Amemiya, C.T., Ehrhardt, G.R.A., Ceitlin, J., Gartland, G.L. and Cooper, M.D.: **Somatic diversification of variable lymphocyte receptors in the agnathan sea lamprey.** Nature 2004, **430**:174–180. (7-4)

Panoutsakopoulou, V., Sanchirico, M.E., Huster, K.M., Jansson, M., Granucci, F., Shim, D.J., Wucherpfennig, K.W. and Cantor, H.: **Analysis of the relationship between viral infection and autoimmune disease.** Immunity. 2001, **15**:137–147. (12-7)

Pascual, V., Banchereau, J. and Palucka, A.K.: **The central role of dendritic cells and interferon-α in SLE.** Curr. Opin. Rheumatol. 2003, **15**:548–556. (12-8)

Paul, W.E.: **Pleiotropy and redundancy: T cell-derived lymphokines in the immune response.** Cell 1989, **57**:521–524. (2-6)

Pawson, T.: **Specificity in signal transduction: from phosphotyrosine-SH2 domain interactions to complex cellular systems.** Cell 2004, **116**:191–203. (2-0)

Pawson, T. and Scott, J.D.: **Protein phosphorylation in signaling – 50 years and counting.** Trends Biochem. Sci. 2005, **30**:286–290. (2-0)

Pearce, E.J. and MacDonald, A.S.: **The immunobiology of schistosomiasis.** Nat. Rev. Immunol. 2002, **2**:499–511. (11-5)

Peggs, K.S. and Allison, J.P.: **Co-stimulatory pathways in lymphocyte regulation: the immunoglobulin superfamily.** Br. J. Haematol. 2005, **130**:809–824. (5-8)

Peiser, L. and Gordon, S.: **The function of scavenger receptors expressed by macrophages and their role in the regulation of inflammation.** Microbes Infect. 2001, **3**:149–159. (3-7)

Pelanda, R. and Torres, R.M.: **Receptor editing for better or for worse.** Curr. Opin. Immunol. 2006, **18**:184–190. (12-1)

Pelayo, R., Welner, R., Perry, S.S., Huang, J., Baba, Y., Yokota, T. and Kincade, P.W.: **Lymphoid progenitors and primary routes to becoming cells of the immune system.** Curr. Opin. Immunol. 2005, **17**:100–107. (7-0)

Perelson, A.S., Neumann, A.U., Markowitz, M., Leonard, J.M. and Ho, D.D.: **HIV-1 dynamics in vivo: virion clearance rate, infected cell life-span, and viral generation time.** Science 1996, **271**:1582–1586. (10-6)

Peschel, A., Jack, R.W., Otto, M., Collins, L.V., Staubitz, P., Nicholson, G., Kalbacher, H., Nieuwenhuizen, W.F., Jung, G., Tarkowski, A., van Kessel, K.P.M. and van Strijp, J.A.G.: *Staphylococcus aureus* **resistance to human defensins and evasion of neutrophil killing via the novel virulence factor MprF is based on modification of membrane lipids with L-lysine.** J. Exp. Med. 2001, **193**:1067–1076. (13-1)

Pestka, S., Krause, C.D., Sarkar, D., Walter, M.R., Shi, Y. and Fisher, P.B.: **Interleukin-10 and related cytokines and receptors.** Annu. Rev. Immunol. 2004, **22**:929–979. (2-7)

Picard, C., Puel, A., Bonnet, M., Ku, C.-L., Bustamante, J., Yang, K., Soudais, C., Dupuis, S., Feinberg, J., Fieschi, C., Elbim, C., Hitchcock, R., Lammas, D., Davies, G., Al-Ghonaium, A., Al-Rayes, H., Al-Jumaah, S., Al-Hajjar, S., Al-Mohsen, I.Z., Frayha, H.H., Rucker, R., Hawn, T.R., Aderem, A., Tufenkeji, H., Haraguchi, S., Day, N.K., Good, R.A., Gougerot-Pocidalo, M.-A., Ozinsky, A. and Casanova, J.-L.: **Pyogenic bacterial infections in humans with IRAK-4 deficiency.** Science 2003, **299**:2076–2079. (3-11)

Picker, L.J. and Siegelbaum, M.H.: **Lymphoid tissues and organs** in *Fundamental Immunology* 4th ed. Paul, W.E. ed. (Lippincott-Raven, New York, 1999), 449–531. (1-6)

Pillai, S: *Lymphocyte Development: Cell Selection Events and Signals During Immune Ontogeny* (Springer-Verlag, New York, 1997). (7-0)

Pillai, S., Cariappa, A. and Moran, S.T.: **Marginal zone B cells.** Annu. Rev. Immunol. 2005, **23**:161–196. (8-7)

Pinschewer, D.D., Perez, M., Jeetendra, E., Bächi, T., Horvath, E., Hengartner, H., Whitt, M.A., de la Torre, J.C. and Zinkernagel, R.M.: **Kinetics of protective antibodies are determined by the viral surface antigen.** J. Clin. Invest. 2004, **114**:988–993. (10-2)

Plantanias, L.C.: **Mechanisms of type-I- and type-II-interferon-mediated signalling.** Nat. Rev. Immunol. 2005, **5**:375–386. (3-16)

Planz, O., Seiler, P., Hengartner, H. and Zinkernagel, R.M.: **Specific cytotoxic T cells eliminate cells producing neutralizing antibodies.** Nature 1996, **382**:726–729. (10-2)

Plasterk, R.H.A.: **RNA silencing: the genome's immune system.** Science 2002, **296**:1263–1265. (3-19)

Plotkin, S.A.: **Vaccines: past, present and future.** Nat. Med. 2005, **11**:S5–S11. (14-2)

Plow, E.F., Haas, T.A., Zhang, L., Loftus, J. and Smith J.W.: **Ligand binding to integrins.** J. Biol. Chem. 2000, **275**:21785–21788. (2-4)

Podolsky, D.K.: **Inflammatory bowel disease.** N. Engl. J. Med. 2002, **347**:417–429. (13-6)

Poltorak, A., He, X., Smirnova, I., Liu, M.-Y., Van, Huffel, C., Du, X., Birdwell, D., Alejos, E., Silva, M., Galanos, C., Freudenberg, M., Ricciardi-Castagnoli, P., Layton, B. and Beutler, B.: **Defective LPS signaling in C3H/HeJ and C57BL/10ScCr mice: mutations in** *Tlr4* **gene.** Science 1998, **282**:2085–2088. (3-10)

Porter, R.R.: **Structural studies of immunoglobulins.** Science 1973, **180**:713–716. (6-1)

Prabhakar, B.S., Bahn, R.S. and Smith, T.J.: **Current perspective on the pathogenesis of Graves' disease and ophthalmopathy.** Endocr. Rev. 2003, **24**:802–835. (12-5)

Pribila, J.T., Quale, A.C., Mueller, K.L., Shimizu, Y.: **Integrins and T cell-mediated immunity.** Annu. Rev. Immunol. 2004, **22**:157–180. (2-4)

Prinz, I., Sansoni, A., Kissenpfennig, A., Ardouin, L., Malissen, M. and Malissen, B.: **Visualization of the earliest steps of γδ T cell development in the adult thymus.** Nat. Immunol. 2006, **7**:995–1003. (8-4)

Puel, A., Picard, C., Ku, C.L., Smahi, A. and Casanova, J.L.: **Inherited disorders of NF-κB-mediated immunity in man.** Curr. Opin. Immunol. 2004, **16**:34–41. (2-10)

Pulendran, B. and Ahmed, R.: **Translating innate immunity into immunological memory: implications for vaccine development.** Cell 2006, **124**:849–863. (14-2)

Qiao, S.-W., Bergseng, E., Molberg, Ø., Jung, G., Fleckenstein, B. and Sollid, L.M.: **Refining the rules of gliadin T cell epitope binding to the disease-associated DQ2 molecule in celiac disease: importance of proline spacing and glutamine deamidation.** J. Immunol. 2005, **175**:254–261. (13-5)

Rabbitts, T.H.: **Chromosomal translocations in human cancer.** Nature 1994, **372**:143–149. (7-11)

Radic, M.Z. and Weigert, M.: **Genetic and structural evidence for antigen selection of anti-DNA antibodies.** Annu. Rev. Immunol. 1994, **12**:487–520. (12-8)

Rakoff-Nahoum, S., Paglino, J., Eslami-Varzaneh, F., Edberg, S. and Medzhitov, R.: **Recognition of commensal microflora by Toll-like receptors is required for intestinal homeostasis.** Cell 2004, **118**:229–241. (13-6)

Raman, K. and Mohan, C.: **Genetic underpinnings of autoimmunity — lessons from studies in arthritis, diabetes, lupus and multiple sclerosis.** Curr. Opin. Immunol. 2003, **15**:651–659. (12-4, 12-6)

Randall, T.D., Heath, A.W., Santos-Argumedo, L., Howard, M.C., Weissman, I.L. and Lund, F.E.: **Arrest of B lymphocyte terminal differentiation by CD40 signaling: mechanism for lack of antibody-secreting cells in germinal centers.** Immunity 1998, **8**:733–742. (6-12)

Raulet, D.H. and Vance, R.E.: **Self-tolerance in natural killer cells.** Nat. Rev. Immunol. 2006, **6**:520–531. (8-1)

Rautemaa, R. and Meri, S.: **Complement-resistance mechanisms of bacteria.** *Microbes Infect.* 1999, **1**:785–794. (3-6)

Ravetch, J.V. and Bolland, S.: **IgG Fc receptors.** *Annu. Rev. Immunol.* 2001, **19**:275–290. (6-3)

Ravetch, J.V. and Lanier, L.L.: **Immune inhibitory receptors.** *Science* 2000, **290**:84–89. (2-2)

Recher, M., Lang, K.S., Hunziker, L., Freigang, S., Eschli, B., Harris, N.L., Navarini, A., Senn, B.M., Fink, K., Lötscher, M., Hangartner, L., Zellweger, R., Hersberger, M., Theocharides, A., Hengartner, H. and Zinkernagel, R.M.: **Deliberate removal of T cell help improves virus-neutralizing antibody production.** *Nat. Immunol.* 2004, **5**:934–942. (10-2)

Reif, K., Ekland, E.H., Ohl, L., Nakano, H., Lipp, M., Förster, R. and Cyster, J.G.: **Balanced responsiveness to chemoattractants from adjacent zones determines B-cell position.** *Nature* 2002, **416**:94–99. (6-10)

Reis e Sousa, C., Hieny, S., Scharton-Kersten, T., Jankovic, D., Charest, H., Germain, R.N. and Sher, A.: *In vivo* **microbial stimulation induces rapid CD40 ligand-independent production of interleukin 12 by dendritic cells and their redistribution to T cell areas.** *J. Exp. Med.* 1997, **186**:1819–1829. (4-1)

Reith, W. and Mach, B.: **The bare lymphocyte syndrome and the regulation of MHC expression.** *Annu. Rev. Immunol.* 2001, **19**:331–373. (14-1)

Reith, W., LeibundGut-Landmann, S. and Waldburger, J.M.: **Regulation of MHC class II gene expression by the class II transactivator.** *Nat. Rev. Immunol.* 2005, **5**:793–806. (4-8)

Reits, E., Neijssen, J., Herberts, C., Benckhuijsen, W., Janssen, L., Drijfhout, J.W. and Neefjes, J.: **A major role for TPPII in trimming proteasomal degradation products for MHC class I antigen presentation.** *Immunity* 2004, **20**:495–506. (4-6)

Reits, E.A.J., Vos, J.C., Gromme, M. and Neefjes, J.: **The major substrates for TAP** *in vivo* **are derived from newly synthesized proteins.** *Nature* 2000, **404**:774–778. (4-6)

Ren, B., McCrory, M.A., Pass, C., Bullard, D.C., Ballantyne, C.M., Xu, Y., Briles, D.E. and Szalai, A.J.: **The virulence function of** *Streptococcus pneumoniae* **surface protein A involves inhibition of complement activation and impairment of complement receptor-mediated protection.** *J. Immunol.* 2004, **173**:7506–7512. (9-5)

Rescigno, M.: **CCR6+ dendritic cells: the gut tactical-response unit.** *Immunity* 2006, **24**:508–510. (1-3)

Reynaud, C.-A., Bertocci, B., Dahan, A. and Weill, J.-C.: **Formation of the chicken B-cell repertoire: ontogenesis, regulation of Ig gene rearrangement, and diversification by gene conversion.** *Adv. Immunol.* 1994, **57**:353–378. (7-4)

Reynaud, C.-A., Garcia, C., Hein, W.R. and Weill, J.-C.: **Hypermutation generating the sheep immunoglobulin repertoire is an antigen-independent process.** *Cell* 1995, **80**:115–125. (7-4)

Rioux, J.D. and Abbas, A.K.: **Paths to understanding the genetic basis of autoimmune disease.** *Nature* 2005, **435**:584–589. (12-0)

Robbie-Ryan, M. and Brown, M.A.: **The role of mast cells in allergy and autoimmunity.** *Curr. Opin. Immunol.* 2002, **14**:728–733. (13-1)

Robbins, J.B., Schneerson, R. and Szu, S.C.: **Perspective: hypothesis: serum IgG antibody is sufficient to confer protection against infectious diseases by inactivating the inoculum.** *J. Infect. Dis.* 1995, **171**:1387–1398. (10-2)

Robertson, J.M., MacLeod, M., Marsden, V.S., Kappler, J.W. and Marrack, P.: **Not all CD4+ memory T cells are long lived.** *Immunol. Rev.* 2006, **211**:49–57. (5-16)

Robey, E.A. and Bluestone, J.A.: **Notch signaling in lymphocyte development and function.** *Curr. Opin. Immunol.* 2004, **16**:360–366. (7-7)

Rock, K.L., York, I.A. and Goldberg, A.L.: **Post-proteasomal antigen processing for major histocompatibility class I presentation.** *Nat. Immunol.* 2004, **5**:670–677. (4-6)

Rolink, A.G. and Melchers, F.: **BAFFled B cells survive and thrive: roles of BAFF in B-cell development.** *Curr. Opin. Immunol.* 2002, **14**:266–275. (7-6)

Romani, L.: **Immunity to fungal infections.** *Nat. Rev. Immunol.* 2004, **4**:1–13. (11-0)

Romani, L., Bistoni, F. and Puccetti, P.: **Adaptation of** *Candida albicans* **to the host environment: the role of morphogenesis in virulence and survival in mammalian hosts.** *Curr. Opin. Microbiol.* 2003, **6**:338–343. (11-1)

Rosen, F. and Geha, R.: **Congenital asplenia** in *Case Studies in Immunology. A Clinical Companion* 4th ed. (Garland Science, New York, 2004), 1–6. (1-7)

Rosen, F. and Geha, R.: **Contact sensitivity to poison ivy** in *Case Studies in Immunology* 4th ed. (Garland Science, New York, 2004) 63–69. (13-5)

Rosen, S.D.: **Ligands for L-selectin: homing, inflammation, and beyond.** *Annu. Rev. Immunol.* 2004, **22**:129–156. (2-14, 3-14)

Rosenberg, A.S. and Singer, A.: **Cellular basis of skin allograft rejection: an in vivo model of immune-mediated tissue destruction.** *Annu. Rev. Immunol.* 1992, **10**:333–360. (14-7)

Ross, A.G.P., Bartley, P.B., Sleigh, A.C., Olds, G.R., Li, Y., Williams, G.M. and McManus, D.P.: **Schistosomiasis.** *N. Engl. J. Med.* 2002, **346**:1212–1220. (11-5)

Ross, J. and Tittensor, A.M.: **The establishment and spread of myxomatosis and its effect on rabbit populations.** *Philos. Trans. R. Soc. Lond., B, Biol. Sci.* 1986, **314**:599–606. (3-0)

Rot, A. and von Andrian, U.H.: **Chemokines in innate and adaptive host defense: basic chemokinese grammar for immune cells.** *Annu. Rev. Immunol.* 2004, **22**:891–928. (2-13, 3-14)

Roth, D.B. and Roth, S.Y.: **Unequal access: regulating V(D)J recombination through chromatin remodeling.** *Cell* 2000, **103**:699–702. (7-3)

Roulston, A., Marcellus, R.C. and Branton, P.E.: **Viruses and apoptosis.** *Annu. Rev. Microbiol.* 1999, **53**:577–628. (3-18)

Roychowdhury, S. and Svensson, C.K.: **Mechanisms of drug-induced delayed-type hypersensitivity reactions in skin.** *Am. Assoc. Pharmaceut. Sci. J.* 2005, **7**:E834–E846. (13-5)

Rudensky, A.Y., Preston-Hurlburt, P., Hong, S.-C., Barlow, A. and Janeway, C.A. Jr: **Sequence analysis of peptides bound to MHC class II molecules.** *Nature* 1991, **353**:622–627. (4-5)

Rudolph, M.G. and Wilson, I.A.: **The specificity of the TCR/pMHC interaction.** *Curr. Opin. Immunol.* 2002, **14**:52–65. (5-2)

Rudolph, M.G., Stanfield, R.L. and Wilson, I.A.: **How TCRs bind MHCs, peptides, and coreceptors.** *Annu. Rev. Immunol.* 2006, **24**:419–466. (5-2)

Russell, D.G.: *Mycobacterium tuberculosis:* **here today, and here tomorrow.** *Nat. Rev. Mol. Cell Biol.* 2001, **2**:569–586. (9-7)

Sacks, D. and Sher, A.: **Evasion of innate immunity by parasitic protozoa.** *Nat. Immunol.* 2002, **3**:1041–1047. (11-0)

Saini, S.S. and MacGlashan, D.: **How IgE upregulates the allergic response.** *Curr. Opin. Immunol.* 2002, **14**:694–697. (13-1)

Sakaguchi, N., Takahashi, T., Hata, H., Nomura, T., Tagami, T., Yamazaki, S., Sakihama, T., Matsutani, T., Negishi, I., Nakatsuru, S. and Sakaguchi, S.: **Altered thymic T-cell selection due to a mutation of the ZAP-70 gene causes autoimmune arthritis in mice.** *Nature* 2003, **426**:454–460. (12-9)

Sakaguchi, S.: **Naturally arising Foxp3-expressing CD25+CD4+ regulatory T cells in immunological tolerance to self and non-self.** *Nat. Immunol.* 2005, **6**:345–352. (12-3)

Sallusto, F., Geginat, J. and Lanzavecchia, A.: **Central memory and effector memory T cell subsets: Function, generation, and maintenance.** *Annu. Rev. Immunol.* 2004, **22**:745–763. (5-16)

Sallusto, F., Lenig, D., Forster, R., Lipp, M. and Lanzavecchia, A.: **Two subsets of memory T lymphocyts with distinct homing potentials and effector functions.** *Nature* 1999, **401**:708–712. (5-16)

Sallusto, F., Mackay, C.R. and Lanzavecchia, A.: **The role of chemokine receptors in primary, effector, and memory immune responses.** *Annu. Rev. Immunol.* 2000, **18**:593–620. (2-13)

Salzman, N.H., Ghosh, D., Huttner, K.M., Paterson, Y. and Bevins, C.L.: **Protection against enteric salmonellosis in transgenic mice expressing a human intestinal defensin.** *Nature* 2003, **422**:522–526. (9-1)

Sansonetti, P.J.: **War and peace at mucosal surfaces.** *Nat. Rev. Immunol.* 2004, **4**:953–964. (3-1)

Sansonno, D. and Dammacco, F.: **Hepatitis C virus, cryoglobulinaemia, and vasculitis: immune complex relations.** *Lancet Infect. Dis.* 2005 **5**:227–236. (13-4)

Santori, F.R., Kieper, W.C., Brown, S.M., Lu, Y., Neubert, T.A., Johnson, K.L., Naylor, S., Vukmanović, S., Hogquist, K.A. and Jameson, S.C.: **Rare, structurally homologous self-peptides promote thymocyte positive selection.** *Immunity* 2002, **17**:131–142. (7-8)

Schaible, U.E., Winau, F., Sieling, P.A., Fischer, K., Collins, H.L., Hagens, K., Modlin, R.L., Brinkmann, V. and Kaufmann, S.H.E.: **Apoptosis facilitates antigen presentation to T lymphocytes through MHC-I and CD1 in tuberculosis.** *Nat. Med.* 2003, **9**:1039–1046. (9-7)

Schatz, D.G. and Baltimore, D.: **Stable expression of immunoglobulin gene V(D)J recombinase activity by gene transfer into 3T3 fibroblasts.** *Cell* 1988, **53**:107–115. (7-2)

Schindler, M., Münch, J., Kutsch, O., Li, H., Santiago, M.L., Bibollet-Ruche, F., Müller-Trutwin, M.C., Novembre, F.J., Peeters, M., Courgnaud, V., Bailes, E., Roques, P., Sodora, D.L., Silvestri, G., Sharp, P.M., Hahn, B.H. and Kirchhoff, F.: **Nef-mediated suppression of T cell activation was lost in a lentiviral lineage that gave rise to HIV-1.** *Cell* 2006, **125**:1055–1067. (10-5)

Schlech, W.F. 3rd: **Foodborne listeriosis.** *Clin. Infect. Dis.* 2000, **31**:770–775. (9-6)

Schlessinger, J. and Ullrich, A.: **Growth factor signaling by receptor tyrosine kinases.** *Neuron* 1992, **9**:383–391. (2-6)

Schmitz, J.E., Kuroda, M.J., Santra, S., Sasseville, V.G., Simon, M.A., Lifton, M.A., Racz, P., Tenner-Racz, K., Dalesandro, M., Scallon, B.J., Ghrayeb, J., Forman, M.A., Montefiori, D.C., Rieber, E.P., Letvin, N.L. and Reimann, K.A.: **Control of viremia in simian immunodeficiency virus infection by CD8+ lymphocytes.** *Science* 1999, **283**:857–860. (10-6)

Schneider, P.: **The role of APRIL and BAFF in lymphocyte activation.** *Curr. Opin. Immunol.* 2005, **12**:282–289. (7-6)

Schreiber, H.: **Tumor immunology** in *Fundamental Immunology* 5th ed. Paul, W.E. ed. (Lippincott Williams & Wilkins, Philadelphia, 2003), 1557–1592. (14-5)

Schreiber, S.L. and Crabtree, G.R.: **The mechanism of action of cyclosporin A and FK506.** *Immunol. Today* 1992, **13**:136–142. (14-8)

Schroder, K., Hertzog, P.J., Ravasi, T. and Hume, D.A.: **Interferon-γ: an overview of signals, mechanisms and functions.** *J. Leukoc. Biol.* 2004, **75**:163–189. (5-11)

Schubert, U., Anton, L.C., Gibbs, J., Norbury, C.C., Yewdell, J.W. and Bennink, J.R.: **Rapid degradation of a large fraction of newly synthesized proteins by proteasomes.** *Nature* 2000, **404**:770–774. (4-6)

Schwab, S.R., Pereira, J.P., Matloubian, M., Xu, Y., Huang, Y. and Cyster, J.G.: **Lymphocyte sequestration through S1P lyase inhibition and disruption of S1P gradients.** *Science* 2005, **309**:1735–1739. (2-14)

Schwartz, M.A.: **Integrin signaling revisited.** *Trends Cell Biol.* 2001, **11**:466–470. (2-4)

Schwartz, R.H.: **T cell anergy.** *Annu. Rev. Immunol.* 2003, **21**:305–334. (5-6)

Scott, R.S., McMahon, E.J., Pop, S.M., Reap, E.A., Caricchio, R., Cohen, P.L., Earp, H.S. and Matsushima, G.K.: **Phagocytosis and clearance of apoptotic cells is mediated by MER.** *Nature* 2001, **411**:207–211. (3-8)

Seaman, M.S., Wang, C.R. and Forman, J.: **MHC class Ib-restricted CTL provide protection against primary and secondary *Listeria monocytogenes* infection.** *J. Immunol.* 2000, **165**:5192–5201. (4-4)

Sears, M.R., Greene, J.M., Willan, A.R., Wiecek, E.M., Taylor, D.R., Flannery, E.M., Cowan, J.O., Herbison, G.P., Silva, P.A. and Poulton, R.: **A longitudinal, population-based, cohort study of childhood asthma followed to adulthood.** *N. Engl. J. Med.* 2003, **349**:1414–1422. (13-3)

Seet, B.T., Johnston, J.B., Brunetti, C.R., Barrett, J.W., Everett, H., Cameron, C., Sypula, J., Nazarian, S.H., Lucas, A. and McFadden, G.: **Poxviruses and immune evasion.** *Annu. Rev. Immunol.* 2003, **21**:377–423. (10-0)

Segal, A.W.: **How neutrophils kill microbes.** *Annu. Rev. Immunol.* 2005, **23**:197–223. (3-9)

Seo, S.H., Hoffmann, E. and Webster, R.G.: **Lethal H5N1 influenza viruses escape host anti-viral cytokine responses.** *Nat. Med.* 2002, **8**:950–954. (10-8)

Serbina, N.V., Salazar-Mather, T.P., Biron, C.A., Kuziel, W.A. and Pamer, E.G.: **TNF/iNOS-producing dendritic cells mediate innate immune defense against bacterial infection.** *Immunity* 2003, **19**:59–70. (1-3)

Sevilla, N., Kunz, S., McGavern, D. and Oldstone, M.B.: **Infection of dendritic cells by lymphocytic choriomeningitis virus.** *Curr. Top. Microbiol. Immunol.* 2003, **276**:125–144. (10-9)

Shan, H., Shlomchik, M. and Weigert, M.: **Heavy-chain class switch does not terminate somatic mutation.** *J. Exp. Med.* 1990, **172**:531–536. (6-12)

Shands, K.N., Schmid, G.P., Dan, B.B., Blum, D., Guidotti, R.J., Hargrett, N.T., Anderson, R.L., Hill, D.L., Broome, C.V., Band, J.D. and Fraser, D.W.: **Toxic-shock syndrome in menstruating women: association with tampon use and *Staphylococcus aureus* and clinical features in 52 cases.** *N. Engl. J. Med.* 1980, **303**:1436–1442. (9-4)

Shapiro-Shelef, M. and Calame, K.: **Regulation of plasma-cell development.** *Nat. Rev. Immunol.* 2005, **5**:230–242. (6-12)

Sheehy, A.M., Gaddis, N.C., Choi, J.D. and Malim, M.H.: **Isolation of a human gene that inhibits HIV-1 infection and is suppressed by the viral Vif protein.** *Nature* 2002, **418**:646–650. (3-19)

Shellito, J., Suzara, V.V., Blumenfeld, W., Beck, J.M., Steger, H.J. and Ermak, T.H.: **A new model of *Pneumocystis carinii* infection in mice selectively depleted of helper T lymphocytes.** *J. Clin. Invest.* 1990, **85**:1686–1693. (11-2)

Shen, L., Sigal, L.J., Boes, M. and Rock, K.L.: **Important role of cathepsin S in generating peptides for TAP-independent MHC class I crosspresentation in vivo** *Immunity* 2004, **21**:155–165. (4-9)

Shepherd, V.L.: **Distinct roles for lung collectins in pulmonary host defense.** *Am. J. Respir. Cell Mol. Biol.* 2002, **26**:257–260. (3-2)

Shi, Y.: **Mechanisms of caspase activation and inhibition during apoptosis.** *Mol. Cell* 2002, **9**:459–470. (2-11)

Shibuya, A., Sakamoto, N., Shimizu, Y., Shibuya, K., Osawa, M., Hiroyama, T., Eyre, H.J., Sutherland, G.R., Endo, Y., Fujita, T., Miyabayashi, T., Sakano, S., Tsuji, T., Nakayama, E., Phillips, J.H., Lanier, L.L. and Nakauchi, H.: **Fcα/μ receptor mediates endocytosis of IgM-coated microbes.** *Nat. Immunol.* 2000, **1**:441–446. (9-1)

Shiina, T., Inoko, H. and Kulski, J.K.: **An update of the HLA genomic region, locus information and disease associations: 2004.** *Tissue Antigens* 2004, **64**:631–649. (4-3)

Shiloh, M.U., MacMicking, J.D., Nicholson, S., Brause, J.E., Potter, S., Marino, M., Fang, F., Dinauer, M. and Nathan, C.: **Phenotype of mice and macrophages deficient in both phagocyte oxidase and inducible nitric oxide synthase.** *Immunity* 1999, **10**:29–38. (9-1)

Shinkai, K., Mohrs, M. and Locksley, R.M.: **Helper T cells regulate type 2 innate immunity *in vivo.*** *Nature* 2002, **420**:825–829. (11-4)

Shiow, L.R., Rosen, D.B., Brdičková, N., Xu, Y., An, J., Lanier, L.L., Cyster, J.G. and Matloubian, M.: **CD69 acts downstream of interferon-α/β to inhibit S1P₁ and lymphocyte egress from lymphoid organs.** *Nature* 2006, **440**:540–544. (5-7)

Shortman, K. and Liu, Y.-J.: **Mouse and human dendritic cell subtypes.** *Nat. Rev. Immunol.* 2002, **2**:151–161. (1-1, 1-3)

Shuai, K.: **Modulation of STAT signaling by STAT-interacting proteins.** *Oncogene* 2000, **19**:2638–2644. (2-7)

Sidorenko, S.P. and Clark, E.A.: **The dual-function CD150 receptor subfamily: the viral attraction.** *Nat. Immunol.* 2003, **4**:19–24. (10-0)

Siegal, F.P., Kadowaki, N., Shodell, M., Fitzgerald-Bocarsly, P.A., Shah, K., Ho, S., Antonenko, S. and Liu, Y.-J.: **The nature of the principal type 1 interferon-producing cells in human blood.** *Science* 1999, **284**:1835–1837. (10-1)

Sigalov, A.: **Multi-chain immune recognition receptors: spatial organization and signal transduction.** *Semin. Immunol.* 2005, **17**:51–64. (2-2)

Siggs, O.M., Makaroff, L.E. and Liston, A.: **The why and how of thymocyte negative selection.** *Curr. Opin. Immunol.* 2006, **18**:175–183. (7-10)

Singer, A.: **New perspectives on a developmental dilemma: the kinetic signaling model and the importance of signal duration for the CD4/CD8 lineage decision.** *Curr. Opin. Immunol.* 2002, **14**:207–215. (7-9)

Sixt, M., Kanazawa, N., Selg, M., Samson, T., Roos, G., Reinhardt, D.P., Pabst, R., Lutz, M.B. and Sorokin, L.: **The conduit system transports soluble antigens from the afferent lymph to resident dendritic cells in the T cell area of the lymph node.** *Immunity* 2005, **22**:19–29. (5-6)

Snapper, C.M., Shen, Y., Khan, A.Q., Colino, J., Zelazowski, P., Mond, J.J., Gause, W.C. and Wu, Z.-Q.: **Distinct types of T-cell help for the induction of a humoral immune response to *Streptococcus pneumoniae*.** *Trends Immunol.* 2001, **22**:308–311. (6-10)

Song G., Yang, Y., Liu, J.-h., Casasnovas, J.M., Shimaoka, M., Springer, T.A. and Wang, J.-h.: **An atomic resolution view of ICAM recognition in a complex between the binding domains of ICAM-3 and integrin αLβ2.** *Proc. Natl Acad. Sci. USA* 2005, **102**:3366–3371. (2-3)

Song, J., So, T., Cheng, M., Tang, X. and Croft, M.: **Sustained survivin expression from OX40 costimulatory signals drives T cell clonal expansion.** *Immunity* 2005, **22**:621–631. (5-8)

Sonnenburg, J.L., Angenent, L.T. and Gordon, J.I.: **Getting a grip on things: how do communities of bacterial symbionts become established in our intestine?** *Nat. Immunol.* 2004, **5**:569–573. (9-0)

Sourdive, D.J.D., Murali-Krishna, K., Altman, J.D., Zajac, A.J., Whitmire, J.K., Pannetier, C., Kourilsky, P., Evavold, B., Sette, A. and Ahmed, R.: **Conserved T cell receptor repertoire in primary and memory CD8 T cell responses to an acute viral infection.** *J. Exp. Med.* 1998, **188**:71–82. (10-9)

Sprent, J. and Kishimoto, H.: **The thymus and negative selection.** *Immunol. Rev.* 2002, **185**:126–135. (7-10)

Springer, T.A.: **Traffic signals for lymphocyte recirculation and leukocyte emigration: the multistep paradigm.** *Cell* 1994, **76**:301–314. (3-14)

Springer, T.A. and Wang, J.-H.: **The three-dimensional structure of integrins and their ligands and conformational regulation of cell adhesion.** *Adv. Protein Chem.* 2004, **68**:29–63. (2-3)

Staib, P., Kretschmar, M., Nichterlein, T., Hof, H. and Morschhäuser, J.: **Differential activation of *Candida albicans* virulence gene family during infection.** *Proc. Natl Acad. Sci. USA* 2000, **97**:6102–6107. (11-1)

Starr, T.K., Jameson, S.C. and Hogquist, K.A.: **Positive and negative selection of T cells.** *Annu. Rev. Immunol.* 2003, **21**:139–176. (7-8)

Stark, G.R., Kerr, I.M., Williams, B.R.G., Silverman, R.H. and Schreiber, R.D.: **How cells respond to interferons.** *Annu. Rev. Biochem.* 1998, **67**:227–264. (3-16)

Stark, M.A., Huo, Y., Burcin, T.L., Morris, M.A., Olson, T.S. and Ley, K.: **Phagocytosis of apoptotic neutrophils regulates granulopoiesis via IL-23 and IL-17.** *Immunity* 2005, **22**:285–294. (5-13)

Stebbing, J., Gazzard, B. and Douek, D.C.: **Where does HIV live?** *N. Engl. J. Med.* 2004, **350**:1872–1880. (10-6)

Stebbins, C.E. and Galán, J.E.: **Structural mimicry in bacterial virulence.** *Nature* 2001, **412**:701–705. (9-2)

Steel, D.M. and Whitehead, A.S.: **The major acute phase reactants: C-reactive protein, serum amyloid P component and serum amyloid A protein.** *Immunol. Today* 1994, **15**:81–88. (3-15)

Stein, M.-P., Edberg, J.C., Kimberly, R.P., Managan, E.K., Bharadwaj, D., Mold, C. and Du Clos, T.W.: **C-reactive protein binding to FcγRIIa on human monocytes and neutrophils is allele-specific.** *J. Clin. Invest.* 2000, **105**:369–376. (9-5)

Steinman, R.: **Dendritic cells** in *Fundamental Immunology* 4th ed. Paul, W.E. ed. (Lippincott-Raven, New York, 1999), 547–573. (1-7)

Steinman, R.M., Hawiger, D. and Nussenzweig, M.C.: **Tolerogenic dendritic cells.** *Annu. Rev. Immunol.* 2003, **21**:685–711. (1-3)

Stenger, S., Donhauser, N., Thüring, H., Röllinghoff, M. and Bogdan, C.: **Reactivation of latent leishmaniasis by inhibition of inducible nitric oxide synthase.** *J. Exp. Med.* 1996, **183**:1501–1514. (11-3)

Stern, L.J., Brown, J.H., Jardetzky, T.S., Gorga, J.C., Urban, R.G., Strominger, J.L. and Wiley, D.C.: **Crystal structure of the human class II MHC protein HLA-DR1 complexed with an influenza virus peptide.** *Nature* 1994, **368**:215–221. (4-5)

Stetson, D.B., Mohrs, M., Mallet-Designe, V., Teyton, L. and Locksley, R.M.: **Rapid expansion and IL-4 expression by *Leishmania*-specific naive helper T cells in vivo.** *Immunity* 2002, **17**:191–200. (11-3)

Stevenson, M.: **HIV-1 pathogenesis.** *Nat. Med.* 2003, **9**:853–860. (10-5)

Stewart-Jones, G.B.E., McMichael, A.J., Bell, J.I., Stuart, D.I. and Jones, E.Y.: **A structural basis for immunodominant human T cell receptor recognition.** *Nat. Immunol.* 2003, **4**:657–663. (10-8)

Strasser, A.: **The role of BH3-only proteins in the immune system.** *Nat. Rev. Immunol.* 2005, **5**:189–200. (2-12)

Stuart, L.M. and Ezekowitz, R.A.: **Phagocytosis: elegant complexity.** *Immunity* 2005, **22**:539–550. (3-8)

Subbarao, K., Murphy, B.R. and Fauci, A.S.: **Development of effective vaccines against pandemic influenza.** *Immunity* 2006, **24**:5–9. (10-8)

Suffredini, A.F., Harpel, P.C. and Parillo, J.E.: **Promotion and subsequent inhibition of plasminogen activation after administration of intravenous endotoxin to normal subjects.** *N. Engl. J. Med.* 1989, **320**:1165–1172. (9-3)

Sykes, M., Auchincloss, H. Jr and Sachs, D.H.: **Transplantation immunology** in *Fundamental Immunology* 5th ed. Paul, W.E. ed. (Lippincott Williams & Wilkins, Philadelphia, 2003), 1481–1555. (14-6)

Szabo, S.J., Sullivan, B.M., Peng, S.L. and Glimcher, L.H.: **Molecular mechanisms regulating Th1 immune responses.** *Annu. Rev. Immunol.* 2003, **21**:713–758. (5-10)

Tabeta, K., Georgel, P., Janssen, E., Du, X., Hoebe, K., Crozat, K., Mudd, S., Shamel, L., Sovath, S., Goode, J., Alexopoulou, L., Flavell, R.A. and Beutler, B.: **Toll-like receptors 9 and 3 as essential components of innate immune defense against mouse cytomegalovirus infection.** *Proc. Natl Acad. Sci. USA* 2004, **101**:3516–3521. (3-11)

Takahama, Y.: **Journey through the thymus: stromal guides for T-cell development and selection.** *Nat. Rev. Immunol.* 2006, **6**:127–135. (7-7)

Takai, T., Ono, M., Hikida, M., Ohmori, H. and Ravetch, J.V.: **Augmented humoral and anaphylactic responses in FcγRII-deficient mice.** *Nature* 1996, **379**:346–349. (6-8)

Takeda, K., Kaisho, T. and Akira, S.: **Toll-like receptors.** *Annu. Rev. Immunol.* 2003, **21**:335–376. (3-10)

Takeuchi, O., Hoshino, K., Kawai, T., Sanjo, H., Takada, H., Ogawa, T., Takeda, K. and Akira, S.: **Differential roles of TLR2 and TLR4 in recognition of Gram-negative and Gram-positive bacterial cell wall components.** *Immunity* 1999, **11**:443–451. (3-10)

Tangye, S.G., Phillips, J.H., Lanier, L.L. and Nichols, K.E.: **Functional requirement for SAP in 2B4-mediated activation of human natural killer cells as revealed in the X-linked lymphoproliferative syndrome.** *J. Immunol.* 2000, **165**:2932–2936. (8-2)

Tanigaki, K., Han, H., Yamamoto, N., Tashiro, K., Ikegawa, M., Kuroda, K., Suzuki, A., Nakano, T. and Honjo, T.: **Notch–RBP-J signaling is involved in cell fate determination of marginal zone B cells.** *Nat. Immunol.* 2002, **3**:443–450. (8-7)

Taniguchi, T., Ogasawara, K., Takaoka, A. and Tanaka, N.: **IRF family of transcription factors as regulators of host defense.** *Annu. Rev. Immunol.* 2001, **19**:623–655. (3-16, 10-1)

Tarlinton, D.M. and Smith, K.G.C.: **Dissecting affinity maturation: a model explaining selection of antibody-forming cells and memory B cells in the germinal centre.** *Immunol. Today* 2000, **21**:436–441. (6-12)

Terr, A.I.: **The atopic diseases** and **Anaphylaxis and urticaria** in *Medical Immunology*, 9th ed. Stites, D.P., Terr, A.I. and Parslow, T.G. eds (Appleton and Lange, Stamford, 1997), 389–418. (13-2)

Tettelin, H., Nelson, K.E., Paulsen, I.T., Eisen, J.A., Read, T.D., Peterson, S., Heidelberg, J., DeBoy, R.T., Haft, D.H., Dodson, R.J., Durkin, A.S., Gwinn, M., Kolonay, J.F., Nelson, W.C., Peterson, J.D., Umayam, L.A., White, O., Salzberg, S.L., Lewis, M.R., Radune, D., Holtzapple, E., Khouri, H., Wolf, A.M., Utterback, T.R., Hansen, C.L., McDonald, L.A., Feldblyum, T.V., Angiuoli, S., Dickinson, T., Hickey, E.K., Holt, I.E., Loftus, B.J., Yang, F., Smith, H.O., Venter, J.C., Dougherty, B.A., Morrison, D.A., Hollingshead, S.K. and Fraser, C.M.: **Complete genome sequence of a virulent isolate of *Streptococcus pneumoniae*.** *Science* 2001, **293**:498–506. (9-5)

Tew, J.G., Wu, J., Fakher, M., Szakal, A.K. and Qin, D.: **Follicular dendritic cells: beyond the necessity of T-cell help.** *Trends Immunol.* 2001, **22**:361–367. (6-11)

Theill, L.E., Boyle, W.J. and Penninger J.M.: **RANK-L and RANK: T cells, bone loss, and mammalian evolution.** *Annu. Rev. Immunol.* 2002, **20**:795–823. (12-9)

The MHC Sequencing Consortium: **Complete sequence and gene map of a human major histocompatibility complex.** *Nature* 1999, **401**:921–923. (4-3)

Thomas, C.F. Jr and Limper, A.H.: **Pneumocystis pneumonia.** *N. Engl. J. Med.* 2004, **350**:2487–2498. (11-2)

Thome, M.: **CARMA1, BCL-10 and MALT1 in lymphocyte development and activation.** *Nat. Rev. Immunol.* 2004, **4**:348–359. (5-5)

Tilley, S.L., Coffman, T.M. and Koller, B.H.: **Mixed messages: modulation of inflammation and immune responses by prostaglandins and thromboxanes.** *J. Clin. Invest.* 2001, **108**:15–23. (3-13)

Tkalcevic, J., Novelli, M., Phylactides, M., Iredale, J.P., Segal, A.W. and Roes, J.: **Impaired immunity and enhanced resistance to endotoxin in the absence of neutrophil elastase and cathepsin G.** *Immunity* 2000, **12**:201–210. (3-9)

Tortorella, D., Gewurz, B.E., Furman, M.H., Schust, D.J. and Ploegh, H.L.: **Viral subversion of the immune system.** *Annu. Rev. Immunol.* 2000, **18**:861–926. (3-18)

Totet, A., Respaldiza, N., Pautard, J.-C., Raccurt, C. and Nevez, G.: ***Pneumocystis jiroveci* genotypes and primary infection.** *Clin. Infect. Dis.* 2003, **36**:1340–1342. (11-2)

Tough, D.F. and Sprent, J.: **Immunologic memory** in *Fundamental Immunology* 5th ed. Paul, W.E. ed. (Lippincott Williams & Wilkins, Philadelphia, 2003), 865–899. (5-16)

Tracey, K.J.: **The inflammatory reflex.** *Nature* 2002, **420**:853–859. (3-15)

Trambas, C.M. and Griffiths, G.M.: **Delivering the kiss of death.** *Nat. Immunol.* 2003, **4**:399–403. (5-14)

Tran, T.H., Nguyen, T.L., Nguyen, T.D., Luong, T.S., Pham, P.M., Nguyen, van V.C., Pham, T.S., Vo, C.D., Le, T.Q., Ngo, T.T., Dao, B.K., Le, P.P., Nguyen, T.T., Hoang, T.L., Cao, V.T., Le, T.G., Nguyen, D.T., Le, H.N., Nguyen, T.K.T., Le, H.S., Le, V.T., Christiane, D., Tran, T.T., Menno, de J., Schultsz, C., Cheng, P., Lim, W., Horby, P., Farrar, J.: **Avian influenza A (H5N1) in 10 patients in Vietnam.** *N. Engl. J. Med.* 2004, **350**:1179–1188. (10-7)

Traver, D., Akashi, K., Manz, M., Merad, M., Miyamoto, T., Engleman, E.G. and Weissman, I.L.: **Development of CD8α-positive dendritic cells from a common myeloid progenitor.** *Science* 2000, **290**:2152–2154. (1-1)

Travis, S.M., Singh, P.K. and Welsh, M.J.: **Antimicrobial peptides and proteins in the innate defense of the airway surface.** *Curr. Opin. Immunol.* 2001, **13**:89–95. (3-1)

Treanor, J.: **Influenza vaccine – outmaneuvering antigenic shift and drift.** *N. Engl. J. Med.* 2004, **350**:218–220. (10-8)

Treiner, E., Duban, L., Bahram, S., Radosavljevic, M., Wanner, V., Tilloy, F., Affaticati, P., Gilfillan, S. and Lantz, O.: **Selection of evolutionarily conserved mucosal-associated invariant T cells by MR1.** *Nature* 2003, **422**:164–169. (4-4, 8-5)

Trinchieri, G., Pflanz, S. and Kastelein, R.A.: **The IL-12 family of heterodimeric cytokines: new players in the regulation of T cell responses.** *Immunity* 2003, **19**:641–644. (2-8)

Trombetta, E.S. and Mellman, I.: **Cell biology of antigen processing in vitro and in vivo.** *Annu. Rev. Immunol.* 2005, **23**:975–1028. (4-9)

Trombetta, E.S., Ebersold, M., Garrett, W., Pypaert, M. and Mellman, I.: **Activation of lysosomal function during dendritic cell maturation.** *Science* 2003, **299**:1400–1403. (4-9)

Trowsdale, J. and Parham, P.: **Defense strategies and immunity-related genes.** *Eur. J. Immunol.* 2004, **34**:7–17. (4-3)

Tuomanen, E.I., Austrian, R. and Masure, H.R.: **Pathogenesis of pneumococcal infection.** *N. Engl. J. Med.* 1995, **332**:1280–1284. (9-5)

Turville, S.G., Cameron, P.U., Handley, A., Lin, G., Pöhlmann, S., Doms, R.W. and Cunningham, A.L.: **Diversity of receptors binding HIV on dendritic cell subsets.** *Nat. Immunol.* 2002, **3**:975–983. (10-5)

Ueda, H., Howson, J.M.M., Esposito, L., Heward, J., Snook, H., Chamberlain, G., Rainbow, D.B., Hunter, K.M.D., Smith, A.N., Di Genova, G., Herr, M.H., Dahlman, I., Payne, F., Smyth, D., Lowe, C., Twells, R.C.J., Howlett, S., Healy, B., Nutland, S., Rance, H.E., Everett, V., Smink, L.J., Lam, A.C.,

Cordell, H.J., Walker, N.M., Bordin, C., Hulme, J., Motzo, C., Cucca, F., Hess, J.F., Metzker, M.L., Rogers, J., Gregory, S., Allahabadia, A., Nithiyananthan, R., Tuomilehto-Wolf, E., Tuomilehto, J., Bingley, P., Gillespie, K.M., Undlien, D.E., Rønningen, K.S., Guja, C., Ionescu-Tîrgoviste, C., Savage, D.A., Maxwell, A.P., Carson, D.J., Patterson, C.C., Franklyn, J.A., Clayton, D.G., Peterson, L.B., Wicker, L.S., Todd, J.A. and Gough, S.C.L.: **Association of the T-cell regulatory gene *CTLA4* with susceptibility to autoimmune disease.** *Nature* 2003, **423**:506–511. (12-6)

Ugolini, S. and Vivier, E.: **Multifaceted roles of MHC class I and MHC class I-like molecules in T cell activation.** *Nat. Immunol.* 2001, **2**:198–200. (4-4)

Ulevitch, R.J. and Tobias, P.S.: **Receptor-dependent mechanisms of cell stimulation by bacterial endotoxin.** *Annu. Rev. Immunol.* 1995, **13**:437–457. (3-10)

Urban, J.F. Jr, Noben-Trauth, N., Donaldson, D.D., Madden, K.B., Morris, S.C., Collins, M. and Finkelman, F.D.: **IL-13, IL-4Rα, and Stat6 are required for the expulsion of the gastrointestinal nematode parasite *Nippostrongylus brasiliensis*.** *Immunity* 1998, **8**:255–264. (11-4)

Urbaniak, S.J.: **Alloimmunity to RhD in humans.** *Transfus. Clin. Biol.* 2006 **13**:19–22. (13-4)

Urdahl, K.B., Sun, J.C. and Bevan, M.J.: **Positive selection of MHC class Ib–restricted CD8⁺ T cells on hematopoietic cells.** *Nat. Immunol.* 2002, **3**:772–779. (8-5)

Valenta, R. and Kraft, D.: **From allergen structure to new forms of allergen-specific immunotherapy.** *Curr. Opin. Immunol.* 2002, **14**:718–727. (13-2)

Valujskikh, A. and Heeger, P.S.: **Emerging roles of endothelial cells in transplant rejection.** *Curr. Opin. Immunol.* 2003, **15**:493–498. (14-7)

van den Broek, M.F., Muller, U., Huang, S., Zinkernagel, R.M. and Aguet, M.: **Immune defence in mice lacking type I and/or type II interferon receptors.** *Immunol. Rev.* 1995, **148**:5–18. (3-16)

Vanderlugt, C.L. and Miller, S.D.: **Epitope spreading in immune-mediated diseases: implications for immunotherapy.** *Nat. Rev. Immunol.* 2002, **2**:85–95. (12-4)

van der Poll, T., Buller, H.R., ten Cate, H., Wortel, C.H., Bauer, K.A., van Deventer, S.J., Hack, C.E., Sauerwein, H.P., Rosenberg, R.D. and ten Cate, J.W.: **Activation of coagulation after administration of tumor necrosis factor to normal subjects.** *N. Engl. J. Med.* 1990, **322**:1622–1627. (9-3)

van der Woude, M.W. and Bäumler, A.J.: **Phase and antigenic variation in bacteria.** *Clin. Microbiol. Rev.* 2004, **17**:581–611. (9-2)

van Egmond, M., Damen, C.A., van Spriel, A.B., Vidarsson, G., van Garderen, E. and van de Winkel, J.G.J.: **IgA and the IgA Fc receptor.** *Trends Immunol.* 2001, **22**:205–211. (6-3, 9-1)

van Rie, A., Warren, R., Richardson, M., Victor, T.C., Gie, R.P., Enarson, D.A., Beyers, N. and van Helden, P.D.: **Exogenous reinfection as a cause of recurrent tuberculosis after curative treatment.** *N. Engl. J. Med.* 1999, **341**:1174–1179. (9-7)

Vázquez-Boland, J.A., Kuhn, M., Berche, P., Chakraborty, T., Domínguez-Bernal, G., Goebel, W., González-Zorn, B., Wehland, J. and Kreft, J.: ***Listeria* pathogenesis and molecular virulence determinants.** *Clin. Microbiol. Rev.* 2001, **14**:584–640. (9-6)

Veazey, R.S., DeMaria, M., Chalifoux, L.V., Shvetz, D.E., Pauley, D.R., Knight, H.L., Rosenzweig, M., Johnson, R.P., Desrosiers, R.C. and Lackner, A.A.: **Gastrointestinal tract as a major site of CD4⁺ T cell depletion and viral replication in SIV infection.** *Science* 1998, **280**:427–431. (10-5)

Veillette, A.: **SLAM family receptors regulate immunity with and without SAP-related adaptors.** *J. Exp. Med.* 2004, **199**:1175–1178. (2-3)

Veldhoen, M., Hocking, R.J., Atkins, C.J., Locksley, R.M. and Stockinger, B.: **TGFβ in the context of an inflammatory cytokine milieu supports de novo differentiation of IL-17-producing T cells.** *Immunity* 2006, **24**:179–189. (5-13)

Venkitaraman, A.R., Williams, G.T., Dariavach, P. and Neuberger, M.S.: **The B-cell antigen receptor of the five immunoglobulin classes.** *Nature* 1991, **352**:777–781. (6-7)

Vercelli, D.: **Genetic polymorphism in allergy and asthma.** *Curr. Opin. Immunol.* 2003, **15**:609–613. (13-2)

Viala, J., Chaput, C., Boneca, I.G., Cardona, A., Girardin, S.E., Moran, A.P., Athman, R., Mémet, S., Huerre, M.R., Coyle, A.J., DiStefano, P.S., Sansonetti, P.J., Labigne, A., Bertin, J., Philpott, D.J. and Ferrero, R.L.: **Nod1 responds to peptidoglycan delivered by the *Helicobacter pylori cag* pathogenicity island.** *Nat. Immunol.* 2004; **5**:1166–1174. (3-12)

Viglianti, G.A., Lau, C.M., Hanley, T.M., Miko, B.A., Shlomchik, M.J. and Marshak-Rothstein, A.: **Activation of autoreactive B cells by CpG dsDNA.** *Immunity* 2003, **19**:837–847. (12-8)

Vilches, C. and Parham, P.: **KIR: diverse, rapidly evolving receptors of innate and adaptive immunity.** *Annu. Rev. Immunol.* 2002, **20**:217–251. (8-2)

Villadangos, J.A. and Ploegh, H.L.: **Proteolysis in MHC class II antigen processing presentation: who's in charge?** *Immunity* 2000, **12**:233–239. (4-7)

Villadangos, J.A., Schnorrer, P. and Wilson, N.S.: **Control of MHC class II antigen presentation in dendritic cells: a balance between creative and destructive forces.** *Immunol. Rev.* 2005, **207**:191–205. (4-9)

Villaseñor, J., Benoist, C. and Mathis, D.: **AIRE and APECED: molecular insights into an autoimmune disease.** *Immunol. Rev.* 2005, **204**:156–164. (12-1)

Vinuesa, C.G., Tangye, S.G., Moser, B. and Mackay, C.R.: **Follicular B helper T cells in antibody responses and autoimmunity.** *Nat. Rev. Immunol.* 2005, **5**:853–865. (5-13, 6-11)

Voehringer, D., Shinkai, K. and Locksley, R.M.: **Type 2 immunity reflects orchestrated recruitment of cells committed to IL-4 production.** *Immunity* 2004, **20**:267–277. (5-12, 11-4)

Vogelmann, R., Amieva, M.R., Falkow, S. and Nelson, W.J.: **Breaking into the epithelial apical–junctional complex—news from pathogen hackers.** *Curr. Opin. Cell Biol.* 2004, **16**:86–93. (9-2)

Vogt, A.B. and Kropshofer, H.: **HLA-DM—an endosomal and lysosomal chaperone for the immune system.** *Trends Biochem. Sci.* 1999, **24**:150–154. (4-7)

Voinnet, O.: **RNA silencing as a plant immune system against viruses.** *Trends Genet.* 2001, **17**:449–459. (3-19)

von Andrian, U.H. and Mackay, C.R.: **T-cell function and migration — two sides of the same coin.** *N. Engl. J. Med.* 2000, **343**:1020–1034. (5-6)

von Andrian, U.H. and Mempel, T.R.: **Homing and cellular traffic in lymph nodes.** *Nat. Rev. Immunol.* 2003, **3**:867–878. (4-1, 5-6)

von Boehmer, H.: **Mechanisms of suppression by suppressor T cells.** *Nat. Immunol.* 2005, **6**:338–344. (12-3)

von Boehmer, H.: **Unique features of the pre-T-cell receptor α-chain: not just a surrogate.** *Nat. Rev. Immunol.* 2005, **5**:571–577. (7-7)

Vos, Q., Lees, A., Wu, Z.Q., Snapper, C.M. and Mond, J.J.: **B-cell activation by T-cell-independent type 2 antigens as an integral part of the humoral immune response to pathogenic microorganisms.** *Immunol. Rev.* 2000, **176**:154–170. (6-9)

Wada, M., Sunkin, S.M., Stringer, J.R. and Nakamura, Y.: **Antigenic variation by positional control of major surface glycoprotein gene expression in *Pneumocystis carinii*.** *J. Infect. Dis.* 1995, **171**:1563–1568. (11-2)

Wakefield, L.M. and Roberts, A.B.: **TGF-β signaling: positive and negative effects on tumorigenesis.** *Curr. Opin. Genet. Devel.* 2002, **12**:22–29. (2-6)

Wakeland, E.K., Liu, K., Graham, R.R. and Behrens, T.W.: **Delineating the genetic basis of systemic lupus erythematosus.** *Immunity* 2001, **15**:397–408. (12-8)

Wald, O., Weiss, I.D., Wald, H., Shoham, H., Bar-Shavit, Y., Beider, K., Galun, E., Weiss, L., Flaishon, L., Shachar, I., Nagler, A., Lu, B., Gerard, C., Gao, J.L., Mishani, E., Farber, J. and Peled, A.: **IFNγ acts on T cells to induce NK cell mobilization and accumulation in target organs.** *J. Immunol.* 2006, **176**:4716–4729. (5-11)

Walport, M.J.: **Complement.** *N. Engl. J. Med.* 2001, **344**:1058–1066 and 1140–1144. (3-3)

Walsh, P.T., Strom T.B. and Turka L.A.: **Routes to transplant tolerance versus rejection: the role of cytokines.** *Immunity* 2004, **20**:121–131. (14-6)

Wang, H. and Clarke, S.H.: **Regulation of B-cell development by antibody specificity.** *Curr. Opin. Immunol.* 2004, **16**:246–250. (8-6)

Wang, J.-h. and Springer, T.A.: **Structural specializations of immunoglobulin superfamily members for adhesion to integrins and viruses.** *Immunol. Rev.* 1998, **163**:197–215. (2-3)

Wang, J.-h., Smolyar, A., Tan, K., Liu, J.-h., Kim, M., Sun, Z.-y.J, Wagner, G. and Reinherz, E.L.: **Structure of a heterophilic adhesion complex between the CD2 and CD58 (LFA-3) counterreceptors.** *Cell* 1999, **97**:791–803. (2-3)

Waters, A.: **Malaria: new vaccines for old?** *Cell* 2006, **124**:689–693. (14-4)

Watts, C.: **Antigen processing in the endocytic compartment.** *Curr. Opin. Immunol.* 2001, **13**:26–31. (4-7)

Weaver, C.T., Harrington, L.E., Mangan, P.R., Gavrieli, M. and Murphy, K.M.: **Th17: an effector CD4 T cell lineage with regulatory T cell ties.** *Immunity* 2006, **24**:677–688. (12-3)

Weber, F., Kochs, G. and Haller, O.: **Inverse interference: how viruses fight the interferon system.** *Viral Immunol.* 2004, **17**:498–515. (3-17)

Weiss, A. and Littman, D.R.: **Signal transduction by lymphocyte antigen receptors.** *Cell* 1994, **76**:263–274. (5-3)

Weiss, A. and Schlessinger, J.: **Switching signals on or off by receptor dimerization.** *Cell* 1998, **94**:277–280. (2-6)

Wherry, E.J. and Ahmed, R.: **Memory CD8 T-cell differentiation during viral infection.** *J. Virol.* 2004, **78**:5535–5545. (10-2)

Whiteway, M. and Oberholzer, U.: **Candida** **morphogenesis and host–pathogen interactions.** Curr. Opin. Microbiol. 2004, **7**:350–357. (11-1)

Whitmire, J.K., Asano, M.S., Murali-Krishna, K., Suresh, M. and Ahmed, R.: **Long-term CD4 Th1 and Th2 memory following acute lymphocytic choriomeningitis virus infection.** J. Virol. 1998, **72**:8281–8288. (10-9)

Williams, A.F. and Barclay, A.N.: **The immunoglobulin superfamily—domains for cell surface recognition.** Annu. Rev. Immunol. 1988, **6**:381–405. (2-1)

Williams, B.R.G.: **PKR; a sentinel kinase for cellular stress.** Oncogene 1999, **18**:6112–6120. (3-17)

Williams, M.A. and Bevan, M.J.: **Effector and memory CTL differentiation.** Immunol. Rev. 2007, **25**:171–192. (5-8)

Williams, M.A., Tyznik, A.J. and Bevan, M.J.: **Interleukin-2 signals during priming are required for secondary expansion of CD8⁺ memory T cells.** Nature 2006, **441**:890–893. (5-8)

Williamson, P. and Schlegel, R.A.: **Hide and seek: the secret identity of the phosphatidylserine receptor.** J. Biol. 2004, **3**:14. (3-8)

Wills-Karp, M. and Ewart, S.L.: **Time to draw breath: asthma-susceptibility genes are identified.** Nat. Rev. Genet. 2004, **5**:376–387. (13-3)

Wills-Karp, M. and Hershey, G.K.K.: **Immunological Mechanisms of Allergic Disorders** in Fundamental Immunology 5th ed. Paul, W.E. ed. (Lippincott Williams & Wilkins, Philadelphia, 2003), 1439–1479. (13-0)

Wilson, C.L., Ouellette, A.J., Satchell, D.P., Ayabe, T., López-Boado, Y.S., Stratman, J.L., Hultgren, S.J., Matrisian, L.M. and Parks, W.C.: **Regulation of intestinal α-defensin activation by the metalloproteinase matrilysin in innate host defense.** Science 1999, **286**:113–117. (3-1, 9-1)

Winkelstein, A.: **Immunosuppresive therapy** in Medical Immunology 9th ed. Stites, D.P., Terr, A.I. and Parslow, T.G. eds (Appleton and Lange, Stamford, 1997), 827–845. (14-8)

Winter, G., Griffiths, A.D., Hawkins, R.E. and Hoogenboom, H.R.: **Making antibodies by phage display technology.** Annu. Rev. Immunol. 1994, **12**:433–455. (6-4)

Wong, P. and Pamer, E.G.: **CD8 T cell responses to infectious pathogens.** Annu. Rev. Immunol. 2003, **21**:29–70. (5-14)

Woof, J.M. and Burton, D.R.: **Human antibody–Fc receptor interactions illuminated by crystal structures.** Nat. Rev. Immunol. 2004, **4**:89–99. (6-3)

Wright, J.R.: **Immunoregulatory functions of surfactant proteins.** Nat. Rev. Immunol. 2005, **5**:58–68. (3-2)

Wullschleger, S., Loewith, R. and Hall, M.N.: **TOR signaling in growth and metabolism.** Cell 2006, **124**:471–484. (14-8)

Wynn, T.A., Thompson, R.W., Cheever, A.W. and Mentink-Kane, M.M.: **Immunopathogenesis of schistosomiasis.** Immunol. Rev. 2004, **201**:156–167. (11-5)

Xiong, J.-P., Stehle, T., Diefenbach, B., Zhang, R., Dunker, R., Scott, D.L., Joachimiak, A., Goodman, S.L. and Arnaout, M.A.: **Crystal structure of the extracellular segment of integrin αVβ3.** Science 2001, **294**:339–345. (2-4)

Xu, W. and Zhang, J.J.: **Stat1-dependent synergistic activation of T-bet for IgG2a production during early stage of B cell activation.** J. Immunol. 2005, **175**:7419–7424. (5-11)

Yamashita, M., Shinnakasu, R., Asou, H., Kimura, M., Hasegawa, A., Hashimoto, K., Hatano, N., Ogata, M. and Nakayama, T.: **Ras-ERK MAPK cascade regulates GATA3 stability and Th2 differentiation through ubiquitin-proteosome pathway.** J. Biol. Chem. 2005, **280**:29409–29419. (5-10)

Yang, D., Biragyn, A., Hoover, D.M., Lubkowski, J. and Oppenheim, J.J.: **Multiple roles of antimicrobial defensins, cathelicidins, and eosinophil-derived neurotoxin in host defense.** Annu. Rev. Immunol. 2004, **22**:181–215. (3-1)

Yang, D., Chertov, O., Bykovskaia, S.N., Chen, Q., Buffo, M.J., Shogan, J., Anderson, M., Schröder, J.M., Wang, J.M., Howard, O.M.Z. and Oppenheim, J.J.: **β-defensins: linking innate and adaptive immunity through dendritic and T cell CCR6.** Science 1999, **286**:525–528. (9-1)

Yazdanbakhsh, M., Kremsner, P.G. and van Ree, R.: **Allergy, parasites and the hygiene hypothesis.** Science 2002, **296**:490–494. (13-3)

Yoneyama, M., Kikuchi, M., Natsukawa, T., Shinobu, N., Imaizumi, T., Miyagishi, M., Taira, K., Akira, S. and Fujita, T.: **The RNA helicase RIG-I has an essential function in double-stranded RNA-induced antiviral responses.** Nat. Immunol. 2004, **5**:730–737. (3-16)

Young, A.C.M., Zhang, W., Sacchettini, J.C. and Nathenson, S.G.: **The three-dimensional structure of H-2Dᵇ at 2.4 Å resolution: implications for antigen-determinant selection.** Cell 1994, **76**:39–50. (4-5)

Yu, C.C., Mamchak, A.A. and DeFranco, A.L.: **Signaling mutations and autoimmunity.** Curr. Dir. Autoimmun. 2003, **6**:61–88. (12-8)

Zhang, J.-R., Mostov, K.E., Lamm, M.E., Nanno, M., Shimida, S.-i., Ohwaki, M. and Tuomanen, E.: **The polymeric immunoglobulin receptor translocates pneumococci across human nasopharyngeal epithelial cells.** Cell 2000, **102**:827–837. (9-5)

Zhao, A., McDermott, J., Urban, J.F. Jr, Gause, W., Madden, K.B., Yeung, K.A., Morris, S.C., Finkelman, F.D. and Shea-Donohue, T.: **Dependence of IL-4, IL-13 and nematode-induced alterations in murine small intestinal smooth muscle contractility on Stat6 and enteric nerves.** J. Immunol. 2003, **171**:948–954. (11-4)

Zheng, H., Fletcher, D., Kozak, W., Jiang, M., Hofmann, K.J., Conn, C.A., Soszynski, D., Grabiec, C., Trumbauer, M.E., Shaw, A., Kostura, M.J., Stevens, K., Rosen, H., North, R.J., Chen, H.Y., Tocci, M.J., Kluger, M.J. and Van der Ploeg, L.H.T.: **Resistance to fever induction and impaired acute-phase response in interleukin-1β-deficient mice.** Immunity 1995, **3**:9–19. (3-15)

Zhou, D., Mattner, J., Cantu, C. 3rd, Schrantz, N., Yin, N., Gao, Y., Sagiv, Y., Hudspeth, K., Wu, Y.-P., Yamashita, T., Teneberg, S., Wang, D., Proia, R.L., Levery, S.B., Savage, P.B., Teyton, L. and Bendelac, A.: **Lysosomal glycosphingolipid recognition by NKT cells.** Science 2004, **306**:1786–1789. (8-3)

Ziegler, S.F.: **FOXP3: of mice and men.** Annu. Rev. Immunol. 2006, **24**:209–226. (5-13)

Zinkernagel, R.M. and Hengartner, H.: **Protective 'immunity' by pre-existent neutralizing antibody titers and preactivated T cells but not by so-called 'immunological memory'.** Immunol. Rev. 2006, **211**:310–319. (5-16)

Zinkernagel, R.M. and Doherty, P.C.: **The discovery of MHC restriction.** Immunol. Today 1997, **18**:14–17. (1-5)

Zlotnik, A. and Yoshie, O.: **Chemokines: a new classification system and their role in immunity.** Immunity 2000, **12**:121–127. (2-13)

Zolla-Pazner, S.: **Identifying epitopes of HIV-1 that induce protective antibodies.** Nat. Rev. Immunol. 2004, **4**:199–210. (10-6)

Rho-family GTPases 70
ribonucleic acid *see* RNA
RIG-I
in defense against viruses 87, **F3-45**, 246
in influenza virus infections 260, **F10-18**
RIP-2 79
RISC *see* RNA-induced silencing complex
rituximab **F6-14,** 296, 297, **F14-10**
RNA
cellular sensors 246, **F10-4**
double stranded *see* double-stranded RNA
single-stranded, recognition by TLRs 75, 246
RNA-induced silencing complex (RISC) 92, 93, **F3-53**
RNA interference (RNAi) 93, **F3-53**
RNase L 88–89, **F3-46**
RNA viruses **F3-1,** 245, 252–253
RORγt 144–145
rotaviruses 325
roundworms *see* nematodes
RSS *see* recombination signal sequence

S1P *see* sphingosine 1-phosphate
Sabouraud dextrose agar 268, 269
Salmonella
evasion of host defenses 70, 230, 231
role of CD8 T cells 146
TraT protein 67
Salmonella typhi 325
Salmonella typhimurium 56, 229
SAP *see* serum amyloid P; SLAM-associated protein
scaffolds 22–23
scavenger receptors 68
A (SRA) family 68
AI (SRA I) 68, **F3-23**
schistosomiasis 276–277, **F11-11, F11-12**
SCID *see* severe combined immunodeficiency
scleroderma **F12-9,** 296
SCR proteins 64, 66
S domain 24, **F2-3**
secretory component 158, 159, **F6-9,** 229
secretory leukocyte protease inhibitor (SLPI) 228
selectins 32, 50–51
expression and ligands **F2-41**
in immune-cell homing 50–51, **F2-40**
in inflammatory response 82, **F3-41**
ligand binding 50–51, **F2-38**
self antigens/self peptides
expression in thymus 202, 283
loss of tolerance to 280–281
in maturation of B cells 194
in negative selection of T cells 202–203, **F7-30**
in positive selection of thymocytes 198–199, **F7-25**
in selection of B1 cells 221, **F8-16**
tolerance *see* tolerance
see also autoantigens
self peptide–MHC complexes
fate of T cells reactive to 280, **F12-1**
inactivation of T cells binding to 130
negative selection of T cells 202, **F7-30**
positive selection of T cells 198–199, 201
promoting survival of naïve T cells 148–149
sensitization 302–303, 304
sepsis syndrome (sepsis) 232–234
animal models 233
bacterial superantigens 234–235
septicemia **F9-2**
serotypes 154, 155, 236
serum amyloid P (SAP) 84
serum sickness 310, 311, **F13-13**
severe acute respiratory syndrome (SARS) 244
severe combined immunodeficiency (SCID) 187,
204–205, **F7-31,** 320–321, **F14-2**
Jak3 deficiency 38, 39, **F7-31,** 205, 321
treatment 321, **F14-3,** 331
X-linked (X-SCID) 38–39, 204–205, 321, **F14-3**

sexually transmitted diseases **F9-2**
SH2-containing inositol phosphatase (SHIP) 27
SH2-containing tyrosine phosphatase 1 (SHP-1) 27,
F2-8, 37
deficiency **F12-21**
SH2D1A *see* SLAM-associated protein
SH2 (Src-homology) domains **F2-1,** 23
sharks 190–191, **F7-13**
sheep 191
Shigella 230
Shigella flexneri 325
shock
induced by superantigens 235
septic 233
SHP-1 *see* SH2-containing tyrosine phosphatase 1
sialic acid 58, **F3-15,** 66
sialyl Lewis^x **F2-15, F2-38**
signaling lymphocytic activation molecule (SLAM) 29
signaling molecules 21–51
signal joint 186–187, **F7-6**
signal transducers and activators of transcription
(STATs) 36–37, **F2-21**
in antiviral response 246
in interferon signaling 86, **F3-44,** 87, **F3-45**
phosphorylation **F2-1**
see also STAT1; STAT2; STAT4; STAT6
simian immunodeficiency virus (SIV) 252, 255, 257, 326
siRNA *see* small interfering RNAs
SIRP-1α 71
SIV *see* simian immunodeficiency virus
Sjögren's syndrome **F12-9,** 296
skin
allergic disorders 307
barriers to infection 2, 228, **F9-3,** 229
commensal bacteria 226, **F9-1**
dendritic cells 9, **F1-8**
homing of effector T cells 137, **F5-24**
test, allergen 304, **F13-5**
SLAM 29
SLAM-associated protein (SAP; SH2D1A) 28, 29
mutations 213, **F14-2**
in NK cell killing 213
SLE *see* systemic lupus erythematosus
SLP-76 **F5-9**
SMAC *see* supramolecular activation complex
Smac 46, **F2-32**
SMAD3 deficiency **F9-4**
SMAD proteins 34, 35
small G proteins (GTPases) 23, 70
small interfering RNAs (siRNA) 93, **F3-53**
small intestine
antigen delivery to lymphoid tissue 17, **F1-22**
barriers to infection 56, 228, **F9-3**
celiac disease 313, **F13-15**
commensal bacteria 226, **F9-1**
defensins 56, **F3-6**
dendritic cells 9, **F1-9**
in HIV infection 254, **F10-14**
intraepithelial lymphocytes 218
M cells *see* M cells
Peyer's patches 14, 15, **F1-19**
T cells **F8-13**
smallpox vaccination 322
smooth muscle, T_H2 cell actions 142, 143
Snell, George **F14-1**
SOCS proteins *see* suppressor of cytokine signaling
proteins
somatic hypermutation 176–177
Spätzle dimers 266
sphingosine 1-phosphate (S1P) 51, 130
sphingosine 1-phosphate (S1P) receptor (S1P₁) 51,
130, 132
spleen 14, 15, **F1-19**
architecture 16–17, **F1-21**
B1 cells 221

B cell and T cell regions 15
B cell tolerance 282
filtration of blood 16–17, **F1-21,** 222, **F8-17**
macrophages 8–9
marginal sinus 16, **F1-21**
marginal zone *see* marginal zone
NKT cells 214–215
periarteriolar lymphoid sheath (PALS) 15, **F1-19**
red pulp 15
white pulp 15, 16
splenectomy 17, 237, 321
Src-family tyrosine kinases 27
in BCR signaling 170, **F6-20, F6-21**
ITAM phosphorylation **F2-6,** 27
see also Lck
Src-homology domains *see* SH2 domains
Staphylococcus aureus 226
antimicrobial peptide resistance 57
chronic infections 73
exotoxins 234–235, **F9-9**
protein A 161, 164
STAT1 37, **F3-44,** 87, **F3-45,** 246
deficiency 86, 141, 241, 321
in T cell subset differentiation **F5-25,** 139
viral evasion mechanisms 89, 251
STAT2 37, **F3-44,** 87, **F3-45,** 246
STAT4 **F5-25,** 139
STAT6
allergic disease and 306, 309
in *Nippostrongylus* infections 275
in T cell subset differentiation 139, **F5-26**
STATs *see* signal transducers and activators of transcription
stem cell factor (SCF) **F1-4**
stem cells, hematopoietic 4, **F1-3, F1-4**
stem cell transplants 331
sterile immunity 266, 267, 272
steroids *see* corticosteroids
stochastic/selection model 200–201, **F7-26**
strength of signal model 200, 201, **F7-27**
streptococci
exotoxins 234–235, **F9-9**
group A *see* *Streptococcus pyogenes*
IgA-binding proteins 161
Streptococcus pneumoniae (pneumococcus) 236–237
Hic protein **F3-21**
mouse models **F9-12,** 237, **F9-13**
surface protein A (PspA) 237
vaccine 237
Streptococcus pyogenes (group A streptococcus)
complement evasion mechanisms **F3-21**
exotoxins 234–235
-induced immune complex disease 311
induction of rheumatic fever 289
SIC protein 67
substance P 85
superantigens 234
activation of T cells 234–235, **F9-10**
bacterial 234–235, **F9-9**
endogenous retroviral 235, **F9-10**
superoxide anion 73
suppressor of cytokine signaling (SOCS) proteins 36, 37
supramolecular activation complex (SMAC)
central (c-SMAC) 126, **F5-11**
peripheral (p-SMAC) 127, **F5-11**
surfactant protein A (SP-A) 58–59, **F3-8**
surfactant protein D (SP-D) 58–59, **F3-8**
surfactants, lung 58–59
surrogate light chain 188–189, **F7-10, F7-16**
switched (S) domain 24, **F2-3**
Syk
in BCR signaling 170, **F6-20,** 171
in ITAM signaling 26–27, **F2-6, F2-7**
in NK cell signaling 212, **F8-6**
symbiosis 226

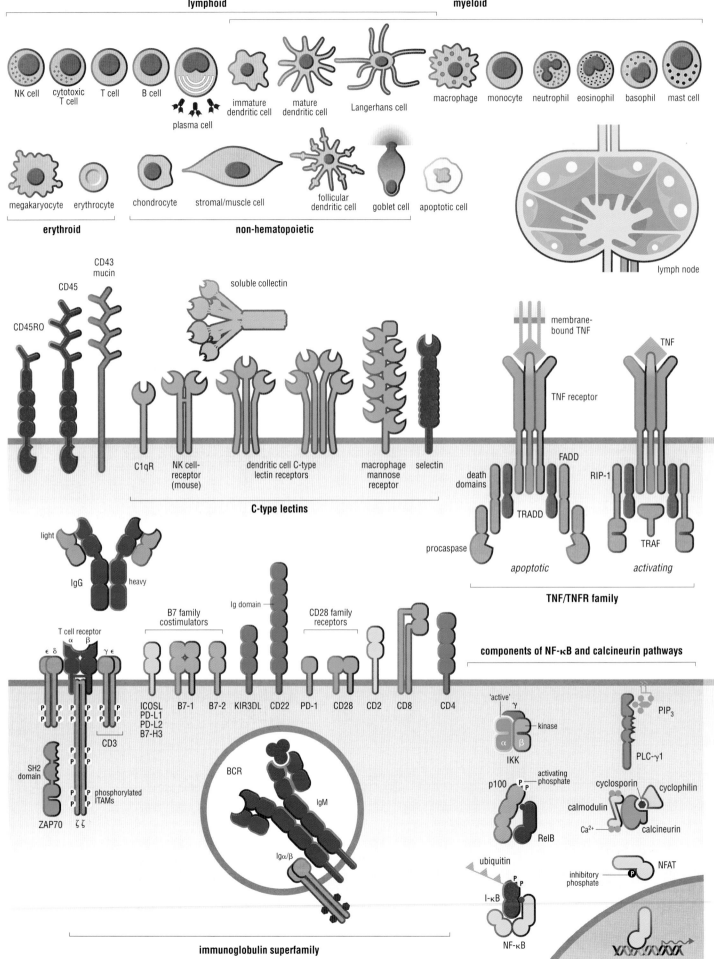

lymphoid

myeloid

NK cell
cytotoxic T cell
T cell
B cell
plasma cell
immature dendritic cell
mature dendritic cell
Langerhans cell
macrophage
monocyte
neutrophil
eosinophil
basophil
mast cell

megakaryocyte
erythrocyte
chondrocyte
stromal/muscle cell
follicular dendritic cell
goblet cell
apoptotic cell

erythroid

non-hematopoietic

lymph node

CD45RO
CD45
CD43 mucin
soluble collectin
membrane-bound TNF
TNF
TNF receptor
C1qR
NK cell-receptor (mouse)
dendritic cell C-type lectin receptors
macrophage mannose receptor
selectin
FADD
death domains
RIP-1
TRADD
procaspase
TRAF
apoptotic
activating

C-type lectins

TNF/TNFR family

light
IgG
heavy

Ig domain

B7 family costimulators
CD28 family receptors

components of NF-κB and calcineurin pathways

T cell receptor
ε δ
α β
γ ε
ICOSL PD-L1 PD-L2 B7-H3
B7-1
B7-2
KIR3DL
CD22
PD-1
CD28
CD2
CD8
CD4

'active'
γ
kinase
α β
IKK
PIP₃

CD3

SH2 domain
phosphorylated ITAMs
ZAP70
ζ ζ

BCR
IgM
Igα/β

PLC-γ1

p100
activating phosphate
RelB
cyclosporin
cyclophilin
calmodulin
Ca²⁺
calcineurin

ubiquitin
I-κB
NF-κB
inhibitory phosphate
NFAT

immunoglobulin superfamily

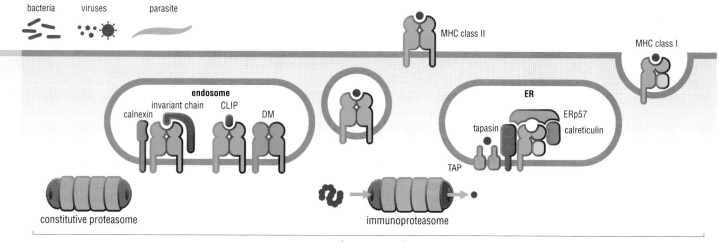

bacteria   viruses   parasite

MHC class II

MHC class I

endosome

calnexin   invariant chain   CLIP   DM

ER

ERp57

tapasin   calreticulin

TAP

constitutive proteasome

immunoproteasome

antigen presentation

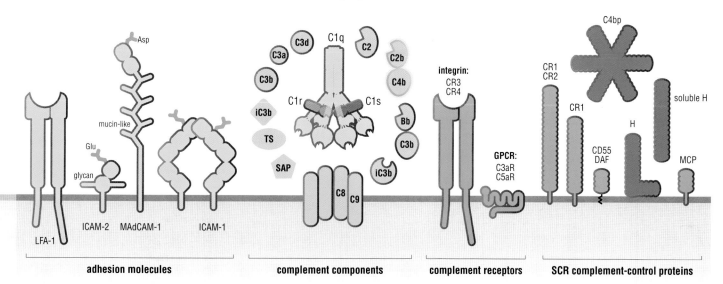

Asp

mucin-like

Glu

glycan

LFA-1   ICAM-2   MAdCAM-1   ICAM-1

C3d   C1q   C2

C3a   C2b

C3b   C4b

C1r   C1s

iC3b   Bb

TS   C3b

SAP   iC3b

C8   C9

integrin:
CR3
CR4

GPCR:
C3aR
C5aR

C4bp

CR1
CR2

CR1

soluble H

H

CD55
DAF

MCP

adhesion molecules   **complement components**   **complement receptors**   **SCR complement-control proteins**

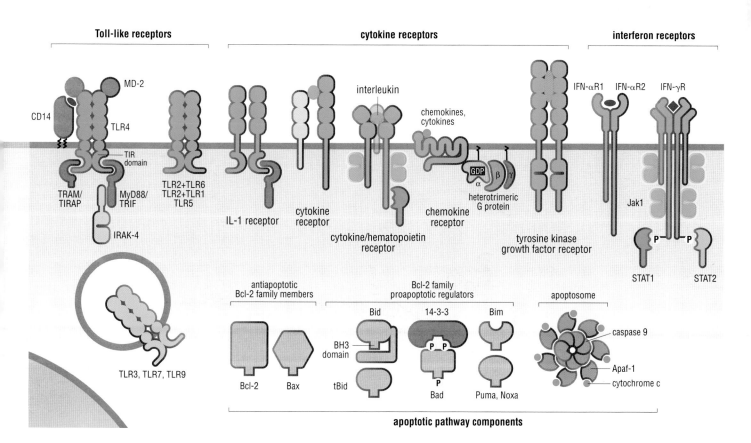

Toll-like receptors   **cytokine receptors**   **interferon receptors**

MD-2

CD14

TLR4

TIR domain

TRAM/ TIRAP

MyD88/ TRIF

IRAK-4

TLR2+TLR6
TLR2+TLR1
TLR5

IL-1 receptor

cytokine receptor

interleukin

cytokine/hematopoietin receptor

chemokines, cytokines

GDP
α   β   γ

heterotrimeric G protein

chemokine receptor

tyrosine kinase growth factor receptor

IFN-αR1   IFN-αR2   IFN-γR

Jak1

P   P

STAT1   STAT2

TLR3, TLR7, TLR9

antiapoptotic Bcl-2 family members

Bcl-2 family proapoptotic regulators

apoptosome

Bid   14-3-3   Bim

BH3 domain

P   P

caspase 9

Apaf-1

cytochrome c

Bcl-2   Bax   tBid   Bad   Puma, Noxa

apoptotic pathway components